Texts in Applied Mathematics 34

T0178501

Texts in Applied Mathematics

(continued after index)

Carmen Chicone

Ordinary Differential Equations with Applications

With 73 Illustrations

 Springer

Carmen Chicone
Department of Mathematics
University of Missouri
Columbia, MO 65211

Series Editors
J.E. Marsden
Control and Dynamical Systems, 107–81
California Institute of Technology
Pasadena, CA 91125
USA
marsden@cds.caltech.edu

S.S. Antman
Department of Mathematics
and
Institute for Physical Science
   and Technology
University of Maryland
College Park, MD 20742-4015
USA
ssa@math.umd.edu

L. Sirovich
Laboratory of Applied Mathematics
and
Department of Biomathematics
Mt. Sinai School of Medicine
New York, NY 10029-6574
chico@camelot.mssm.edu

Mathematics Subject Classification (2000): 34-01, 37-01

ISBN: 978-1-4419-2151-2
eISBN: 0-387-22623-0
eISBN: 978-0-387-35794-2

Printed on acid-free paper.

© 2006 Springer Science+Business Media, Inc.
Softcover reprint of the hardcover 2nd edition 2006

9 8 7 6 5 4 3 2 1

springer.com

Texts in Applied Mathematics

To Jenny, for giving me the gift of time.

Series Preface

Mathematics is playing an ever more important role in the physical and biological sciences, provoking a blurring of boundaries between scientific disciplines and a resurgence of interest in the modern as well as the classical techniques of applied mathematics. This renewal of interest, both in research and teaching, has led to the establishment of the series *Texts in Applied Mathematics* (TAM).

The development of new courses is a natural consequence of a high level of excitement on the research frontier as newer techniques, such as numerical and symbolic computer systems, dynamical systems, and chaos, mix with and reinforce the traditional methods of applied mathematics. Thus, the purpose of this textbook series is to meet the current and future needs of these advances and to encourage the teaching of new courses.

TAM will publish textbooks suitable for use in advanced undergraduate and beginning graduate courses, and will complement the *Applied Mathematical Sciences* (AMS) series, which will focus on advanced textbooks and research-level monographs.

Pasadena, California   J.E. Marsden
New York, New York   L. Sirovich
College Park, Maryland   S.S. Antman

Preface

This book is based on a two-semester course in ordinary differential equations that I have taught to graduate students for two decades at the University of Missouri. The scope of the narrative evolved over time from an embryonic collection of supplementary notes, through many classroom tested revisions, to a treatment of the subject that is suitable for a year (or more) of graduate study.

If it is true that students of differential equations give away their point of view by the way they denote the derivative with respect to the independent variable, then the initiated reader can turn to Chapter 1, note that I write $\dot{x}$, not x', and thus correctly deduce that this book is written with an eye toward dynamical systems. Indeed, this book contains a thorough introduction to the basic properties of differential equations that are needed to approach the modern theory of (nonlinear) dynamical systems. But this is not the whole story. The book is also a product of my desire to demonstrate to my students that differential equations is the least insular of mathematical subjects, that it is strongly connected to almost all areas of mathematics, and it is an essential element of applied mathematics.

When I teach this course, I use the first part of the first semester to provide a rapid, student-friendly survey of the standard topics encountered in an introductory course of ordinary differential equations (ODE): existence theory, flows, invariant manifolds, linearization, omega limit sets, phase plane analysis, and stability. These topics, covered in Sections 1.1–1.8 of Chapter 1 of this book, are introduced, together with some of their important and *interesting* applications, so that the power and beauty of the subject is immediately apparent. This is followed by a discussion of linear systems theory and the proofs of the basic theorems on linearized stability in Chapter 2. Then, I conclude the first semester by presenting one or two realistic applications from Chapter 3. These applications provide a capstone for the course as well as an excellent opportunity to teach the mathematics graduate students some physics, while giving the engineering and physics students some exposure to applications from a mathematical perspective.

In the second semester, I introduce some advanced concepts related to existence theory, invariant manifolds, continuation of periodic orbits, forced oscillators, separatrix splitting, averaging, and bifurcation theory. Since there is not enough time in one semester to cover all of this material in depth, I usually choose just one or two of these topics for presentation in class. The material in the remaining chapters is assigned for private study according to the interests of my students.

My course is designed to be accessible to students who have only studied differential equations during one undergraduate semester. While I do assume some knowledge of linear algebra, advanced calculus, and analysis, only the most basic material from these subjects is required: eigenvalues and eigenvectors, compact sets, uniform convergence, the derivative of a function of several variables, and the definition of metric and Banach spaces. With regard to the last prerequisite, I find that some students are afraid to take the course because they are not comfortable with Banach space theory. These students are put at ease by mentioning that no deep properties of infinite dimensional spaces are used, only the basic definitions.

Exercises are an integral part of this book. As such, many of them are placed strategically within the text, rather than at the end of a section. These interruptions of the flow of the narrative are meant to provide an opportunity for the reader to absorb the preceding material and as a guide to further study. Some of the exercises are routine, while others are sections of the text written in "exercise form." For example, there are extended exercises on structural stability, Hamiltonian and gradient systems on manifolds, singular perturbations, and Lie groups. My students are strongly encouraged to work through the exercises. How is it possible to gain an understanding of a mathematical subject without doing some mathematics? Perhaps a mathematics book is like a musical score: by sight reading you can pick out the notes, but practice is required to hear the melody.

The placement of exercises is just one indication that this book is not written in axiomatic style. Many results are used before their proofs are provided, some ideas are discussed without formal proofs, and some advanced topics are introduced without being fully developed. The pure axiomatic approach forbids the use of such devices in favor of logical order. The other extreme would be a treatment that is intended to convey the ideas of the subject with no attempt to provide detailed proofs of basic results. While the narrative of an axiomatic approach can be as dry as dust, the excitement of an idea-oriented approach must be weighed against the fact that it might leave most beginning students unable to grasp the subtlety of the arguments required to justify the mathematics. I have tried to steer a middle course in which careful formulations and complete proofs are given for the basic theorems, while the ideas of the subject are discussed in depth and the path from the pure mathematics to the physical universe is clearly marked. I am reminded of an esteemed colleague who mentioned that a certain textbook "has lots of fruit, but no juice." Above all, I have tried to avoid this criticism.

Application of the implicit function theorem is a recurring theme in the book. For example, the implicit function theorem is used to prove the rectification theorem and the fundamental existence and uniqueness theorems for solutions of differential equations in Banach spaces. Also, the basic results of perturbation and bifurcation theory, including the continuation of subharmonics, the existence of periodic solutions via the averaging method, as well as the saddle node and Hopf bifurcations, are presented as applications of the implicit function theorem. Because of its central role, the implicit function theorem and the terrain surrounding this important result are discussed in detail. In particular, I present a review of calculus in a Banach space setting and use this theory to prove the contraction mapping theorem, the uniform contraction mapping theorem, and the implicit function theorem.

This book contains some material that is not encountered in most treatments of the subject. In particular, there are several sections with the title "Origins of ODE," where I give my answer to the question "What is this good for?" by providing an explanation for the appearance of differential equations in mathematics and the physical sciences. For example, I show how ordinary differential equations arise in classical physics from the fundamental laws of motion and force. This discussion includes a derivation of the Euler–Lagrange equation, some exercises in electrodynamics, and an extended treatment of the perturbed Kepler problem. Also, I have included some discussion of the origins of ordinary differential equations in the theory of partial differential equations. For instance, I explain the idea that a parabolic partial differential equation can be viewed as an ordinary differential equation in an infinite dimensional space. In addition, traveling wave solutions and the Galërkin approximation technique are discussed. In a later "origins" section, the basic models for fluid dynamics are introduced. I show how ordinary differential equations arise in boundary layer theory. Also, the ABC flows are defined as an idealized fluid model, and I demonstrate that this model has chaotic regimes. There is also a section on coupled oscillators, a section on the Fermi–Ulam–Pasta experiments, and one on the stability of the inverted pendulum where a proof of linearized stability under rapid oscillation is obtained using Floquet's method and some ideas from bifurcation theory. Finally, in conjunction with a treatment of the multiple Hopf bifurcation for planar systems, I present a short introduction to an algorithm for the computation of the Lyapunov quantities as an illustration of computer algebra methods in bifurcation theory.

Another special feature of the book is an introduction to the fiber contraction principle as a powerful tool for proving the smoothness of functions that are obtained as fixed points of contractions. This basic method is used first in a proof of the smoothness of the flow of a differential equation where its application is transparent. Later, the fiber contraction principle appears in the nontrivial proof of the smoothness of invariant manifolds at a rest point. In this regard, the proof for the existence and smoothness of stable and center manifolds at a rest point is obtained as a corollary of a more general existence theorem for invariant manifolds in the presence of a "spectral gap." These proofs can be extended to infinite dimensions. In particular, the applications of the fiber contraction principle and the Lyapunov–Perron method in this book provide an introduction to some of the basic tools of invariant manifold theory.

The theory of averaging is treated from a fresh perspective that is intended to introduce the modern approach to this classical subject. A complete proof of the averaging theorem is presented, but the main theme of the chapter is partial averaging at a resonance. In particular, the "pendulum with torque" is shown to be a universal model for the motion of a nonlinear oscillator near a resonance. This approach to the subject leads naturally to the phenomenon of "capture into resonance," and it also provides the necessary background for students who wish to read the literature on multifrequency averaging, Hamiltonian chaos, and Arnold diffusion.

I prove the basic results of one-parameter bifurcation theory—the saddle node and Hopf bifurcations—using the Lyapunov–Schmidt reduction. The fact that degeneracies in a family of differential equations might be unavoidable is explained together with a brief introduction to transversality theory and jet spaces. Also, the multiple Hopf bifurcation for planar vector fields is discussed. In particular, and the Lyapunov quantities for polynomial vector fields at a weak focus are defined and this subject matter is used to provide a link to some of the algebraic techniques that appear in normal form theory.

Since almost all of the topics in this book are covered elsewhere, there is no claim of originality on my part. I have merely organized the material in a manner that I believe to be most beneficial to my students. By reading this book, I hope that you will appreciate and be well prepared to use the wonderful subject of differential equations.

Columbia, Missouri   Carmen Chicone
June 1999

Preface to the Second Edition

This edition contains new material, new exercises, rewritten sections, and corrections.

There are at least three nontrivial mathematical errors in the first edition: The proof of the Trotter product formula (Theorem 2.24) is valid only in case $e^{A+B} = e^A e^B$; the Floquet theorem (Theorem 2.47) on the existence of logarithms for matrices is valid only if the square of the real matrix in question has all positive eigenvalues; and the proof of the smoothness of invariant manifolds (Theorem 4.1) has a gap because the continuity of a certain fiber contraction with respect to its base space is assumed. The first two errors were pointed out by Mark Ashbaugh, the third by Mohamed ElBialy. These and many other less serious errors are corrected.

While much of the narrative has been revised, the most substantial additions and revisions not already mentioned are the following: the introductory Section 1.9.3 on contraction is rewritten to include a discussion of the continuity of fiber contractions and a more informative first application of the fiber contraction theorem, which is the proof of the smoothness of the solution of the functional equation $F \circ \phi - \phi = G$ (Theorem 1.234); Section 3.1 on the Euler-Lagrange equation is rewritten and expanded to include a more detailed discussion of Hamilton's theory, a presentation of Noether's Theorem, and several new exercises on the calculus of variations; Section 3.2 on classical mechanics has been revised by including more details; the application (in Section 3.5) of Floquet theory to the stability of the inverted pendulum is rewritten to incorporate a more elegant dimensionless model; a new Section 4.3.3 introduces the Lie derivative and applies it to prove the Hartman-Grobman theorem for flows; multidimensional continuation theory for periodic orbits in the presence of first integrals is discussed in the new Section 5.3.8, the basic result on the continuation of manifolds of periodic orbits in the presence of first integrals in involution is proved, and the Lie derivative is used again to characterize commuting flows; and the subject of dynamic bifurcation theory is introduced in a new Section 8.4 where the fundamental idea of delayed bifurcation is presented with applications to the pitchfork bifurcation and bursting.

Over 160 new exercises are included, most with multiple parts. While a few routine exercises are provided where I expect them to be helpful, most of the exercises are meant to challenge students on their understanding of the theory, stimulate interest, extend topics introduced in the narrative, and point the way to applications. Also, most exercises now have lettered parts for easy identification of portions of exercises for homework assignments.

As described in the Preface, the core first graduate course in ODE is contained in selections from the first three chapters. The instructor should budget class time so that all of the language and basic concepts of the subject (existence theory, flows, invariant manifolds, linearization, omega limit sets, phase plane analysis, and stability) are introduced and some applications are discussed in detail.

In my experience, sensitivity to the preparation of students is essential for a successful first graduate course in differential equations. Although the prerequisites are minimal, there are certainly some students who are unprepared for the challenges of a course based on this book if their exposure to differential equations is limited to no more than one undergraduate course where they studied only solution methods for linear second order equations. I have included some review (see Exercise 1.6) to serve as a bridge from their first course to this book. In addition, I often use some class time to review a few fundamental concepts (especially, the derivative as a linear transformation, compactness, connectedness, uniform convergence, linear spaces, eigenvalues, and Jordan canonical form) before they are encountered in context.

The second edition contains plenty of material for second semester courses, master's projects, and reading courses. Professionals might also find something of value.

I remain an enthusiastic teacher of the rich and important subject of differential equations. I hope that instructors will find this book a useful addition to their class design and preparation, and students will have a clear and faithful guide during their quest to learn the subject.

Columbia, Missouri   Carmen Chicone
August 2005

Acknowledgments

I thank all the people who have offered valuable suggestions for corrections and additions to this book, especially, Mark Ashbaugh, James Benson, Oksana Bihun, Tanya Christiansen, Timothy Davis, Jack Dockery, Michael Heitzman, Mohamed S. ElBialy, Sergei Kosakovsky, Marko Koselj, M. B. H. Rhouma, Stephen Schecter, Douglas Shafer, Richard Swanson, and Chongchun Zeng. Also, I thank Dixie Fingerson for much valuable assistance in the mathematics library.

Invitation

Please send your corrections or comments.
E-mail: carmen@chicone.math.missouri.edu
Mail: Department of Mathematics
     University of Missouri
     Columbia, MO 65211
Web: http://www.math.missouri.edu/~carmen/Book/book.html

Contents

1
Introduction to Ordinary Differential Equations

This chapter is about the most basic concepts of the theory of differential equations. We will answer some fundamental questions: What is a differential equation? Do differential equations always have solutions? Are solutions of differential equations unique? But, the most important goal of this chapter is to introduce a geometric interpretation for the space of solutions of a differential equation. Using this geometry, we will introduce some of the elements of the subject: rest points, periodic orbits, and invariant manifolds. Finally, we will review the calculus in a Banach space setting and use it to prove the classic theorems on the existence, uniqueness, and extension of solutions. References for this chapter include [10], [13], [57], [59], [101], [106], [123], [141], [183], [209], and [226].

1.1 Existence and Uniqueness

Let $J \subseteq \mathbb{R}$, $U \subseteq \mathbb{R}^n$, and $\Lambda \subseteq \mathbb{R}^k$ be open subsets, and suppose that $f : J \times U \times \Lambda \to \mathbb{R}^n$ is a smooth function. Here the term "smooth" means that the function f is continuously differentiable. An *ordinary differential equation* (ODE) is an equation of the form

$$\dot{x} = f(t, x, \lambda) \tag{1.1}$$

where the dot denotes differentiation with respect to the independent variable t (usually a measure of time), the dependent variable x is a vector of state variables, and λ is a vector of parameters. As convenient terminology, especially when we are concerned with the components of a vector differential equation, we will say that equation (1.1) is a *system of differential equations*. Also, if we are interested in changes with respect to parameters, then the differential equation is called a *family of differential equations*.

Example 1.1. The forced van der Pol oscillator

$$\dot{x}_1 = x_2,$$
$$\dot{x}_2 = b(1 - x_1^2)x_2 - \omega^2 x_1 + a \cos \Omega t$$

is a differential equation with $J = \mathbb{R}$, $x = (x_1, x_2) \in U = \mathbb{R}^2$,

$$\Lambda = \{(a, b, \omega, \Omega) : (a, b) \in \mathbb{R}^2, \omega > 0, \Omega > 0\},$$

and $f : \mathbb{R} \times \mathbb{R}^2 \times \Lambda \to \mathbb{R}^2$ defined in components by

$$(t, x_1, x_2, a, b, \omega, \Omega) \mapsto (x_2, b(1 - x_1^2)x_2 - \omega^2 x_1 + a \cos \Omega t).$$

If $\lambda \in \Lambda$ is fixed, then a *solution* of the differential equation (1.1) is a function $\phi : J_0 \to U$ given by $t \mapsto \phi(t)$, where J_0 is an open subset of J, such that

$$\frac{d\phi}{dt}(t) = f(t, \phi(t), \lambda) \tag{1.2}$$

for all $t \in J_0$.

Although, in this context, the words "trajectory," "phase curve," and "integral curve" are also used to refer to solutions of the differential equation (1.1), it is useful to have a term that refers to the image of the solution in $\mathbb{R}^n$. Thus, we define the *orbit* of the solution ϕ to be the set $\{\phi(t) \in U : t \in J_0\}$.

When a differential equation is used to model the evolution of a state variable for a physical process, a fundamental problem is to determine the future values of the state variable from its initial value. The mathematical model is then given by a pair of equations

$$\dot{x} = f(t, x, \lambda), \qquad x(t_0) = x_0$$

where the second equation is called an *initial condition*. If the differential equation is defined as equation (1.1) and $(t_0, x_0) \in J \times U$, then the pair of equations is called an *initial value problem*. Of course, a solution of this initial value problem is just a solution ϕ of the differential equation such that $\phi(t_0) = x_0$.

If we view the differential equation (1.1) as a family of differential equations depending on the parameter vector and perhaps also on the initial condition, then we can consider corresponding families of solutions—if they exist—by listing the variables under consideration as additional arguments. For example, we will write $t \mapsto \phi(t, t_0, x_0, \lambda)$ to specify the dependence of a solution on the initial condition $x(t_0) = x_0$ and on the parameter vector λ.

The fundamental issues of the general theory of differential equations are the existence, uniqueness, extension, and continuity with respect to parameters of solutions of initial value problems. Fortunately, all of these issues are resolved by the following foundational results of the subject: *Every initial value problem has a unique solution that is smooth with respect to initial conditions and parameters. Moreover, the solution of an initial value problem can be extended in time until it either reaches the boundary of the domain of definition of the differential equation or blows up to infinity.*

The next three theorems are the formal statements of the foundational results of the subject of differential equations. They are, of course, used extensively in all that follows.

Theorem 1.2 (Existence and Uniqueness). *If $J \subseteq \mathbb{R}$, $U \subseteq \mathbb{R}^n$, and $\Lambda \subseteq \mathbb{R}^k$ are open sets, $f : J \times U \times \Lambda \to \mathbb{R}^n$ is a smooth function, and $(t_0, x_0, \lambda_0) \in J \times U \times \Lambda$, then there exist open subsets $J_0 \subseteq J$, $U_0 \subseteq U$, $\Lambda_0 \subseteq \Lambda$ with $(t_0, x_0, \lambda_0) \in J_0 \times U_0 \times \Lambda_0$ and a function $\phi : J_0 \times J_0 \times U_0 \times \Lambda_0 \to \mathbb{R}^n$ given by $(t, s, x, \lambda) \mapsto \phi(t, s, x, \lambda)$ such that for each point $(t_1, x_1, \lambda_1) \in J_0 \times U_0 \times \Lambda_0$, the function $t \mapsto \phi(t, t_1, x_1, \lambda_1)$ is the unique solution defined on J_0 of the initial value problem given by the differential equation (1.1) and the initial condition $x(t_1) = x_1$.*

Recall that if $k = 1, 2, \ldots, \infty$, a function defined on an open set is called C^k if the function together with all of its partial derivatives up to and including those of order k are continuous on the open set. Similarly, a function is called *real analytic* if it has a convergent power series representation with a positive radius of convergence at each point of the open set.

Theorem 1.3 (Continuous Dependence). *If, for the system (1.1), the hypotheses of Theorem 1.2 are satisfied, then the solution $\phi : J_0 \times J_0 \times U_0 \times \Lambda_0 \to \mathbb{R}^n$ of the differential equation (1.1) is a smooth function. Moreover, if f is C^k for some $k = 1, 2, \ldots, \infty$ (respectively, f is real analytic), then ϕ is also C^k (respectively, real analytic).*

As a convenient notation, we will write $|x|$ for the usual Euclidean norm of $x \in \mathbb{R}^n$. But, because all norms on $\mathbb{R}^n$ are equivalent, the results of this section are valid for an arbitrary norm on $\mathbb{R}^n$.

Theorem 1.4 (Extension). *If, for the system (1.1), the hypotheses of Theorem 1.2 hold, and if the maximal open interval of existence of the solution $t \mapsto \phi(t)$ (with the last three of its arguments suppressed) is given by (α, β) with $-\infty \leq \alpha < \beta < \infty$, then $|\phi(t)|$ approaches ∞ or $\phi(t)$ approaches the boundary of U as $t \to \beta$.*

In case there is some finite T and $\lim_{t \to T} |\phi(t)|$ approaches ∞, we say the solution *blows up in finite time*.

The existence and uniqueness theorem is so fundamental in science that it is sometimes called the "principle of determinism." The idea is that if we know the initial conditions, then we can predict the future states of the system. Although the principle of determinism is validated by the proof of the existence and uniqueness theorem, the interpretation of this principle for physical systems is not as clear as it might seem. The problem is that solutions of differential equations can be very complicated. For example, the future state of the system might depend sensitively on the initial state of the system. Thus, if we do not know the initial state exactly, the final state may be very difficult (if not impossible) to predict.

The variables that we will specify as explicit arguments for the solution ϕ of a differential equation depend on the context, as we have mentioned above. We will write $t \mapsto \phi(t, x)$ to denote the solution such that $\phi(0, x) = x$. Similarly, when we wish to specify the parameter vector, we will use $t \mapsto \phi(t, x, \lambda)$ to denote the solution such that $\phi(0, x, \lambda) = x$.

Example 1.5. The solution of the differential equation $\dot{x} = x^2$, $x \in \mathbb{R}$, is given by the elementary function

$$\phi(t, x) = \frac{x}{1 - xt}.$$

For this example, $J = \mathbb{R}$ and $U = \mathbb{R}$. Note that $\phi(0, x) = x$. If $x > 0$, then the corresponding solution only exists on the interval $J_0 = (-\infty, x^{-1})$. Also, we have that $|\phi(t, x)| \to \infty$ as $t \to x^{-1}$. This illustrates one of the possibilities mentioned in the extension theorem, namely, blow up in finite time.

Exercise 1.6. [Review Problems] This exercise consists of a few key examples that can be solved using techniques from elementary courses in differential equations. (a) Solve the initial value problem

$$\dot{x} = -\frac{1}{1+t}x + 2, \qquad x(0) = 1.$$

(a) Recall the methods to solve forced second-order linear differential equations with constant coefficients; that is, differential equations of the form

$$m\ddot{x} + \lambda\dot{x} + \omega^2 x = A\cos\Omega t.$$

Determine the general solution of

$$\ddot{x} + \dot{x} + x = 2\cos t.$$

Also, determine the solution of the corresponding initial value problem for the initial conditions $x(0) = 1$ and $\dot{x}(0) = 0$. (b) Find the general solution of

$$\frac{dy}{dx} = -x/y.$$

(c) Find the general solution of $\dot{x} = x(1 - x)$. (d) Find the general solution of

$$(1 + \frac{a^2}{r^2}) \sin\theta \, \frac{dr}{d\theta} + (r - \frac{a^2}{r}) \cos\theta = 0.$$

(e) Find the general solution of the differential equation

$$\frac{dy}{dx} = \frac{y}{(y + 2)e^y - 2x}.$$

Show that the initial value problem with the initial condition $y(2) = 0$ has a unique solution and find this solution. Solve the initial value problem

$$\dot{x} = 2x - (2 + y)e^y, \quad \dot{y} = -y, \quad x(0) = 2, \quad y(0) = \ln 2$$

and show that $\lim_{t \to \infty}(x(t), y(t)) = (1, 0)$. (f) Find f so that

$$\phi(x, z, t) = f(z) \cos(kx - \omega t),$$

$\phi_{xx} + \phi_{zz} = 0$, $\phi_z(x, -h) = 0$, and $\phi_x(x, 0) = -a\omega \sin(kx - \omega t)$.

Exercise 1.7. Solve the following differential equation from fluid dynamics (see [137, p. 328]):

$$f'''(x) - (f'(x))^2 = -1, \quad f(0) = 0, \quad f'(0) = 0, \quad \lim_{x \to \infty} f'(x) = 1.$$

Hint: Let $g := f'$ and solve $g'' = g^2 - 1$ with $g(0) = 0$ and $\lim_{x \to \infty} g(x) = 1$. Guess that $\lim_{x \to \infty} g'(x) = 0$. Multiply both sides of the differential equation by g' and integrate once. Solve for the constant of integration using the guess and solve the resulting first-order differential equation by separation of variables. Answer:

$$g(x) = 3\left(\frac{\beta e^{\sqrt{2}x} - 1}{\beta e^{\sqrt{2}x} + 1}\right)^2 - 2, \quad \beta := \frac{\sqrt{3} + \sqrt{2}}{\sqrt{3} - \sqrt{2}}.$$

Note: It is not clear if the original problem has a unique solution.

Exercise 1.8. Consider the differential equation $\dot{x} = -\sqrt{x}$, $x \geq 0$. Find the general solution, and discuss the extension of solutions.

Exercise 1.9. (a) Determine the maximal open interval of existence of the solution of the initial value problem

$$\dot{x} = 1/x, \quad x(0) = 1.$$

(b) What is the maximal interval of existence? (c) Discuss your answer with respect to Theorem 1.4.

Exercise 1.10. Construct infinitely many different solutions of the initial value problem

$$\dot{x} = x^{1/3}, \quad x(0) = 0.$$

Why does Theorem 1.2 fail to apply in this case?

1.2   Types of Differential Equations

Differential equations may be classified in several different ways. In this section we note that the independent variable may be implicit or explicit, and that higher order derivatives may appear.

An *autonomous* differential equation is given by

$$\dot{x} = f(x, \lambda), \qquad x \in \mathbb{R}^n, \quad \lambda \in \mathbb{R}^k; \tag{1.3}$$

that is, the function f does not depend explicitly on the independent variable. If the function f does depend explicitly on t, then the corresponding differential equation is called *nonautonomous*.

In physical applications, we often encounter equations containing second, third, or higher order derivatives with respect to the independent variable. These are called second order differential equations, third order differential equations, and so on, where the *order* of the equation refers to the order of the highest order derivative with respect to the independent variable that appears explicitly in the equation.

Recall that Newton's second law—the rate of change of the linear momentum acting on a body is equal to the sum of the forces acting on the body—involves the second derivative of the position of the body with respect to time. Thus, in many physical applications the most common differential equations used as mathematical models are second order differential equations. For example, the natural physical derivation of van der Pol's equation leads to a second order differential equation of the form

$$\ddot{u} + b(u^2 - 1)\dot{u} + \omega^2 u = a \cos \Omega t. \tag{1.4}$$

An essential fact is that every differential equation is equivalent to a first order system. To illustrate, let us consider the conversion of van der Pol's equation to a first order system. For this, we simply define a new variable $v := \dot{u}$ so that we obtain the following system:

$$\begin{aligned}
\dot{u} &= v, \\
\dot{v} &= -\omega^2 u + b(1 - u^2)v + a \cos \Omega t.
\end{aligned} \tag{1.5}$$

Clearly, this system is equivalent to the second order equation in the sense that every solution of the system determines a solution of the second order van der Pol equation, and every solution of the van der Pol equation determines a solution of this first order system.

Let us note that there are many possibilities for the construction of equivalent first order systems—we are not required to define $v := \dot{u}$. For example, if we define $v = a\dot{u}$ where a is a nonzero constant, and follow the same procedure used to obtain system (1.5), then we will obtain a family of equivalent first order systems. Of course, a differential equation of order m can be converted to an equivalent first order system by defining $m - 1$ new variables in the obvious manner.

If our model differential equation is a nonautonomous differential equation of the form $\dot{x} = f(t, x)$, where we have suppressed the possible dependence on parameters, then there is an "equivalent" autonomous system obtained by defining a new variable as follows:

$$\dot{x} = f(\tau, x),$$
$$\dot{\tau} = 1. \tag{1.6}$$

For example, if $t \mapsto (\phi(t), \tau(t))$ is a solution of this system with $\phi(t_0) = x_0$ and $\tau(t_0) = t_0$, then $\tau(t) = t$ and

$$\dot{\phi}(t) = f(t, \phi(t)), \qquad \phi(t_0) = x_0.$$

Thus, the function $t \mapsto \phi(t)$ is a solution of the initial value problem

$$\dot{x} = f(t, x), \qquad x(t_0) = x_0.$$

In particular, every solution of the nonautonomous differential equation can be obtained from a solution of the autonomous system (1.6).

We have just seen that all ordinary differential equations correspond to first order autonomous systems. As a result, we will pay special attention to the properties of autonomous systems. In most, but not all cases, the conversion of a higher order differential equation to a first order system is useful. In particular, if a nonautonomous system is given by $\dot{x} = f(t, x)$ where f is a periodic function of t, then, as we will see, the conversion to an autonomous system is very often the best way to analyze the system.

Exercise 1.11. Find a first order system that is equivalent to the third order differential equation

$$\epsilon x''' + x x'' - (x')^2 + 1 = 0$$

where ϵ is a parameter and the $'$ denotes differentiation with respect to the independent variable.

Exercise 1.12. Find the solution of the initial value problem

$$\ddot{x} = -x^2, \quad x(0) = -1, \quad \dot{x}(0) = \sqrt{\frac{2}{3}}.$$

Hint: Write the second order equation as a first order system using $\dot{x} = y$, remove the time dependence using the chain rule to obtain a first order equation for y with independent variable x, solve the first order equation for y, substitute the solution into $\dot{x} = y$, and solve for x as a function of t.

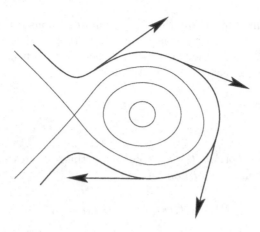

Figure 1.1: Tangent vector field and associated integral curve.

1.3   Geometric Interpretation of Autonomous Systems

In this section we will describe a very important geometric interpretation of the autonomous differential equation

$$\dot{x} = f(x), \qquad x \in \mathbb{R}^n. \tag{1.7}$$

The function given by $x \mapsto (x, f(x))$ defines a *vector field* on $\mathbb{R}^n$ associated with the differential equation (1.7). Here the first component of the function specifies the base point and the second component specifies the vector at this base point. A solution $t \mapsto \phi(t)$ of (1.7) has the property that its tangent vector at each time t is given by

$$(\phi(t), \dot{\phi}(t)) = (\phi(t), f(\phi(t))).$$

In other words, if $\xi \in \mathbb{R}^n$ is on the orbit of this solution, then the tangent line to the orbit at ξ is generated by the vector $(\xi, f(\xi))$, as depicted in Figure 1.1.

We have just mentioned two essential facts: (i) There is a one-to-one correspondence between vector fields and autonomous differential equations. (ii) Every tangent vector to a solution curve is given by a vector in the vector field. These facts suggest that the geometry of the associated vector field is closely related to the geometry of the solutions of the differential equation when the solutions are viewed as curves in a Euclidean space. This geometric interpretation of the solutions of autonomous differential equations provides a deep insight into the general nature of the solutions of differential equations, and at the same time suggests the "geometric method" for studying differential equations: qualitative features

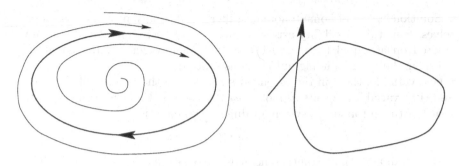

Figure 1.2: Closed trajectory (left) and fictitious trajectory (right) for an autonomous differential equation.

expressed geometrically are paramount; analytic formulas for solutions are of secondary importance. Finally, let us note that the vector field associated with a differential equation is given explicitly. Thus, one of the main goals of the geometric method is to derive qualitative properties of solutions directly from the vector field without "solving" the differential equation.

As an example, let us consider the possibility that the solution curve starting at $x_0 \in \mathbb{R}^n$ at time $t = 0$ returns to the point x_0 at $t = \tau > 0$. Clearly, the tangent vector of the solution curve at the point $\phi(0) = x_0$ is the same as the tangent vector at $\phi(\tau)$. The geometry suggests that the points on the solution curve defined for $t > \tau$ retraces the original orbit. Thus, it is possible that the orbit of an autonomous differential equation is a closed curve as depicted in the left panel of Figure 1.2. On the other hand, an orbit cannot cross itself as in the right panel of Figure 1.2. If there were such a crossing, then there would have to be two different tangent vectors of the same vector field at the crossing point.

The vector field corresponding to a nonautonomous differential equation changes with time. In particular, if a solution curve "returns" to its starting point, the direction specified by the vector field at this point generally depends on the time of arrival. Thus, the curve will generally "leave" the starting point in a different direction than it did originally. For example, suppose that $t \mapsto (g(t), h(t))$ is a curve in $\mathbb{R}^2$ that has a transverse crossing as in the right panel of Figure 1.2, and consider the following system of differential equations

$$\frac{dx}{dt} = g'(t), \qquad \frac{dy}{dt} = h'(t). \tag{1.8}$$

We have just defined a differential equation with the given curve as a solution. Thus, every smooth curve is a solution of a differential equation, but not every curve is a solution of an *autonomous* differential equation.

Solution curves of nonautonomous differential equations can cross themselves. But, this possibility arises because the explicit time variable is not treated on an equal footing with the dependent variables. Indeed, if we consider the corresponding autonomous system formed by adding time as a new variable, then, in the extended state space (the domain of the state and time variables), orbits cannot cross themselves. For example, the state space of the autonomous system of differential equations

$$\dot{x} = g'(\tau), \quad \dot{y} = h'(\tau), \quad \dot{\tau} = 1,$$

corresponding to the nonautonomous differential equation (1.8), is $\mathbb{R}^3$. The system's orbits in the extended state space cannot cross—the corresponding vector field in $\mathbb{R}^3$ is autonomous.

If the autonomous differential equation (1.7) has a closed orbit and $t \mapsto \phi(t)$ is a solution with its initial value on this orbit, then it is clear that there is some $T > 0$ such that $\phi(T) = \phi(0)$. In fact, as we will show in the next section, even more is true: The solution is T-periodic; that is, $\phi(t + T) = \phi(t)$ for all $t \in \mathbb{R}$. For this reason, closed orbits of autonomous systems are also called *periodic orbits*.

Another important special type of orbit is called a *rest point*. To define this concept, note that if $f(x_0) = 0$ for some $x_0 \in \mathbb{R}^n$, then the constant function $\phi : \mathbb{R} \to \mathbb{R}^n$ defined by $\phi(t) \equiv x_0$ is a solution of the differential equation (1.7). Geometrically, the corresponding orbit consists of exactly one point. Thus, if $f(x_0) = 0$, then x_0 is a rest point. Such a solution is also called a steady state, a critical point, an equilibrium point, or a zero (of the associated vector field).

What are all the possible orbit types for autonomous differential equations? The answer depends on what we mean by "types." But we have already given a partial answer: An orbit can be a point, a simple closed curve, or the homeomorphic image of an interval. A geometric picture of all the orbits of an autonomous differential equation is called its *phase portrait* or *phase diagram*. This terminology comes from the notion of *phase space* in physics, the space of positions and momenta. For the record, the *state space* in physics is the space of positions and velocities. But, in the present context, the terms state space and phase space are synonymous; they are used to refer to the domain of the vector field that defines the autonomous differential equation. At any rate, the fundamental problem of the geometric theory of differential equations is evident: Given a differential equation, determine its phase portrait.

Because there are essentially only the three types of orbits mentioned in the last paragraph, it might seem that phase portraits would not be too complicated. But, as we will see, even the portrait of a single orbit can be very complex. Indeed, the homeomorphic image of an interval can be a very complicated subset in a Euclidean space. As a simple but important example of a complex geometric feature of a phase portrait, let us note the curve that crosses itself in Figure 1.1. Such a curve cannot be an orbit

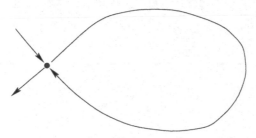

Figure 1.3: A curve in phase space consisting of four orbits of an autonomous differential equation.

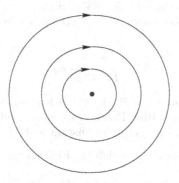

Figure 1.4: Phase portrait of the harmonic oscillator

of an autonomous differential equation. On the other hand, if the crossing point on the depicted curve is a rest point of the differential equation, then such a curve can exist in the phase portrait as a union of the four orbits indicated in Figure 1.3.

Exercise 1.13. Consider the harmonic oscillator (a model for an undamped spring) given by the second order differential equation $\ddot{u} + \omega^2 u = 0$ with the equivalent first order system

$$\dot{u} = \omega v, \qquad \dot{v} = -\omega u. \tag{1.9}$$

The phase portrait, in the phase plane, consists of one rest point at the origin of $\mathbb{R}^2$ with all other solutions being simple closed curves as in Figure 1.4. Solve the differential equation and verify these facts. Find the explicit time dependent solution that passes through the point $(u, v) = (1, 1)$ at time $t = 0$. Note that the system

$$\dot{u} = v, \qquad \dot{v} = -\omega^2 u$$

Figure 1.5: Phase portrait of $\dot{x} = \mu - x^2$ for $\mu = 0$.

is also equivalent to the harmonic oscillator. Is its phase portrait different from the phase portrait of the system (1.9)? Can you make precise the notion that two phase portraits are the same?

Exercise 1.14. Suppose that $F : \mathbb{R} \to \mathbb{R}$ is a smooth, positive, periodic function with period $p > 0$. (a) Prove: If $t \mapsto x(t)$ is a solution of the differential equation $\dot{x} = F(x)$ and

$$T := \int_0^p \frac{1}{F(y)}\, dy,$$

then $x(t+T) - x(t) = p$ for all $t \in \mathbb{R}$. (b) What happens for the case where F is periodic but not of fixed sign? Hint: Define G to be an antiderivative of $1/F$. Show that the function $y \to G(y+p) - G(y)$ is constant and $G(x(b)) - G(x(a)) = b - a$.

Exercise 1.15. [Predator-Prey Model] (a) Find all rest points of the system

$$\dot{x} = -x + xy, \qquad \dot{y} = ry(1 - \frac{y}{k}) - axy.$$

This is a model for a predator-prey interaction, where x is the quantity of predators and y is the quantity of prey. The positive parameter r represents the growth rate of the prey, the positive parameter k represents the carrying capacity of the environment for the prey, and the positive parameter a is a measure of how often a predator is successful in catching its prey. Note that the predators will become extinct if there are no prey. (b) What happens to the population of prey if there are no predators?

In case our system depends on parameters, the collection of the phase portraits corresponding to each choice of the parameter vector is called a *bifurcation diagram*.

As a simple but important example, consider the differential equation $\dot{x} = \mu - x^2$, $x \in \mathbb{R}$, that depends on the parameter $\mu \in \mathbb{R}$. If $\mu = 0$, then the phase portrait, on the phase line, is depicted in Figure 1.5. If we put together all the phase portrait "slices" in $\mathbb{R} \times \mathbb{R}$, where a slice corresponds to a fixed value of μ, then we produce the bifurcation diagram, Figure 1.6. Note that if $\mu < 0$, there is no rest point. When $\mu = 0$, a rest point is born in a "blue sky catastrophe." As μ increases from $\mu = 0$, there is a

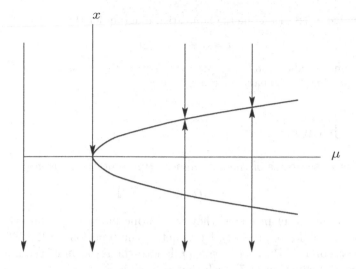

Figure 1.6: Bifurcation diagram $\dot{x} = \mu - x^2$.

"saddle-node" bifurcation; that is, two rest points appear. If $\mu < 0$, this picture also tells us the fate of each solution as $t \to \infty$. No matter which initial condition we choose, the solution goes to $-\infty$ in finite positive time. When $\mu = 0$ there is a steady state. If $x_0 > 0$, then the solution $t \mapsto \phi(t, x_0)$ with initial condition $\phi(0, x_0) = x_0$ approaches this steady state; that is, $\phi(t, x_0) \to 0$ as $t \mapsto \infty$. Whereas, if $x_0 < 0$, then $\phi(t, x_0) \to 0$ as $t \mapsto -\infty$. In this case, we say that x_0 is a *semistable* rest point. But if $\mu > 0$ and $x_0 > 0$, then the solution $\phi(t, x_0) \to \sqrt{\mu}$ as $t \mapsto \infty$. Thus, $x_0 = \sqrt{\mu}$ is a *stable* steady state. The point $x_0 = -\sqrt{\mu}$ is an *unstable* steady state.

Exercise 1.16. Fix $r > 0$ and $k > 0$, and consider the family of differential equations

$$\dot{x} = rx(1 - x/k) - \lambda$$

with parameter $\lambda \in \mathbb{R}$. (a) Draw the bifurcation diagram. (b) This differential equation is a phenomenological model of the number of individuals x in a population with per capita growth rate r and carrying capacity k such that individuals are removed from the population at the rate λ per unit time. For example, x could represent the number of fish in a population and λ the rate of their removal by fishing. Interpret your bifurcation diagram using this population model. Determine a critical value for λ such that if this value is exceeded, then extinction of the population is certain. (c) What would happen if the rate λ varies with time?

Exercise 1.17. (a) Draw the bifurcation diagram for the family $\dot{x} = \lambda x - x^3$. The bifurcation is called the supercritical pitchfork (see Chapter 8). (b) Draw the bifurcation diagram for the subcritical pitchfork $\dot{x} = \lambda x + x^3$.

Exercise 1.18. Describe the bifurcation diagram for the family

$$\dot{x} = \lambda - x^2, \qquad \dot{y} = -y.$$

The name "saddle-node bifurcation" comes from the name of the rest point type corresponding to the parameter value $\lambda = 0$.

1.4   Flows

The set of solutions of the autonomous differential equation (1.7)

$$\dot{x} = f(x), \qquad x \in \mathbb{R}^n$$

have an important property: they form a one-parameter group that defines a *phase flow*. More precisely, let us define the function $\phi : \mathbb{R} \times \mathbb{R}^n \to \mathbb{R}^n$ as follows: For $x \in \mathbb{R}^n$, let $t \mapsto \phi(t, x)$ denote the solution of the autonomous differential equation (1.7) such that $\phi(0, x) = x$.

We know that solutions of a differential equation may not exist for all $t \in \mathbb{R}$. But, for simplicity, let us assume that every solution does exist for all time. If this is the case, then each solution is called *complete*, and group property for ϕ is expressed concisely as follows:

$$\phi(t + s, x) = \phi(t, \phi(s, x)).$$

In view of this equation, if the solution starting at time zero at the point x is continued until time s, when it reaches the point $\phi(s, x)$, and if a new solution at this point with initial time zero is continued until time t, then this new solution will reach the same point that would have been reached if the original solution, which started at time zero at the point x, is continued until time $t + s$.

The prototypical example of a flow is provided by the general solution of the ordinary differential equation $\dot{x} = ax$, $x \in \mathbb{R}$, $a \in \mathbb{R}$. The solution is given by $\phi(t, x_0) = e^{at} x_0$, and it satisfies the group property

$$\phi(t + s, x_0) = e^{a(t+s)} x_0 = e^{at}(e^{as} x_0) = \phi(t, e^{as} x_0) = \phi(t, \phi(s, x_0)).$$

For the general case, let us suppose that for each $x \in \mathbb{R}^n$ $t \mapsto \phi(t, x)$ is the solution of the differential equation (1.7) such that $\phi(0, x) = 0$. Fix $s \in \mathbb{R}$, $x \in \mathbb{R}^n$, and define

$$\psi(t) := \phi(t + s, x), \qquad \gamma(t) := \phi(t, \phi(s, x)).$$

Note that $\phi(s, x)$ is a point in $\mathbb{R}^n$. Therefore, γ is a solution of the differential equation (1.7) with $\gamma(0) = \phi(s, x)$. The function ψ is also a solution of the differential equation because

$$\frac{d\psi}{dt} = \frac{d\phi}{dt}(t + s, x) = f(\phi(t + s, x)) = f(\psi(t)).$$

Finally, note that $\psi(0) = \phi(s,x) = \gamma(0)$. We have proved that both $t \mapsto \psi(t)$ and $t \mapsto \gamma(t)$ are solutions of the same initial value problem. Thus, by the uniqueness theorem, $\gamma(t) \equiv \psi(t)$. The idea of this proof—two functions that satisfy the same initial value problem are identical—is often used in the theory and the applications of differential equations.

By the theorem on continuous dependence, ϕ is a smooth function. In particular, for each fixed $t \in \mathbb{R}$, the function $x \mapsto \phi(t,x)$ is a smooth transformation of $\mathbb{R}^n$. In particular, if $t = 0$, then $x \mapsto \phi(0,x)$ is the identity transformation. Let us also note that

$$x = \phi(0,x) = \phi(t-t,x) = \phi(t,\phi(-t,x)) = \phi(-t,\phi(t,x)).$$

In other words, $x \mapsto \phi(-t,x)$ is the inverse of the function $x \mapsto \phi(t,x)$. Thus, in fact, $x \mapsto \phi(t,x)$ is a diffeomorphism for each fixed $t \in \mathbb{R}$.

In general, suppose that $J \times U$ is a product open subset of $\mathbb{R} \times \mathbb{R}^n$.

Definition 1.19. A function $\phi : J \times U \to \mathbb{R}^n$ given by $(t,x) \mapsto \phi(t,x)$ is called a flow if $\phi(0,x) \equiv x$ and $\phi(t+s,x) = \phi(t,\phi(s,x))$ whenever both sides of the equation are defined.

Of course, if $t \mapsto \phi(t,x)$ defines the family of solutions of the autonomous differential equation (1.7) such that $\phi(0,x) \equiv x$, then ϕ is a flow.

Suppose that $x_0 \in \mathbb{R}^n$, $T > 0$, and that $\phi(T,x_0) = x_0$; that is, the solution returns to its initial point after time T. Then $\phi(t+T,x_0) = \phi(t,\phi(T,x_0)) = \phi(t,x_0)$. In other words, $t \mapsto \phi(t,x_0)$ is a periodic function with period T. The smallest number $T > 0$ with this property is called the *period* of the periodic orbit through x_0.

In the mathematics literature, the notations $t \mapsto \phi_t(x)$ and $t \mapsto \phi^t(x)$ are often used in place of $t \mapsto \phi(t,x)$ for the solution of the differential equation

$$\dot{x} = f(x), \qquad x \in \mathbb{R}^n,$$

that starts at x at time $t = 0$. These notations emphasize that a flow is a one-parameter family of transformations. Indeed, for each t, ϕ_t maps an open subset of $\mathbb{R}^n$ (its domain) to $\mathbb{R}^n$. Moreover, ϕ_0 is the identity transformation and $\phi_t \circ \phi_s = \phi_{s+t}$ whenever both sides of this equation are defined. We will use all three notations. The only possible confusion arises when subscripts are used for partial derivatives. But the meaning of the notation will always be clear from the context in which it appears.

Exercise 1.20. For each integer p, construct the flow of the differential equation $\dot{x} = x^p$.

Exercise 1.21. Consider the differential equation $\dot{x} = t$. Construct the family of solutions $t \mapsto \phi(t,\xi)$ such that $\phi(0,\xi) = \xi$ for $\xi \in \mathbb{R}$. Does ϕ define a flow? Explain.

Exercise 1.22. Suppose that ϕ_t is a (smooth) flow. (a) Prove that there is a differential equation whose flow is ϕ. (b) Can two different differential equations have the same flow?

Exercise 1.23. (a) Show that the family of functions $\phi_t : \mathbb{R}^2 \to \mathbb{R}^2$ given by

$$\phi_t(x,y) = \begin{pmatrix} \cos t & -\sin t \\ \sin t & \cos t \end{pmatrix} \begin{pmatrix} x \\ y \end{pmatrix}$$

defines a flow on $\mathbb{R}^2$. (b) Find a differential equation whose flow is ϕ_t. (c) Repeat parts (a) and (b) for

$$\phi_t(x,y) = e^{-2t} \begin{pmatrix} \cos t & -\sin t \\ \sin t & \cos t \end{pmatrix} \begin{pmatrix} x \\ y \end{pmatrix}.$$

Exercise 1.24. Write $\ddot{u} + \alpha u = 0$, $u \in \mathbb{R}$, $\alpha \in \mathbb{R}$ as a first order system. (a) Determine the flow of the system, and verify the flow property directly. (b) Describe the bifurcation diagram of the system. (c) Show that the system has periodic orbits if $\alpha > 0$ and determine their period(s).

Exercise 1.25. (a) Determine the flow of the first order system

$$\dot{x} = y^2 - x^2, \qquad \dot{y} = -2xy.$$

Hint: Define $z := x + iy$. (b) Show that (almost) every orbit lies on an circle and the flow gives rational parameterizations for these orbits.

Exercise 1.26. Fluid moves through a round pipe with radius a. Suppose that the center of the pipe is on the z-axis, the radial and angular fluid velocities both vanish, and the axial velocity is given by $u_3(x,y,z) = x^2 + y^2 - a^2$. (a) What is the (three-dimensional) fluid velocity at the pipe. (b) Determine the flow of the fluid.

Exercise 1.27. [Evolution Families] Consider the nonautonomous differential equation $\dot{x} = f(t,x)$ and the first order system $\dot{\tau} = 1$, $\dot{x} = f(\tau, x)$. Show that if ψ is the solution of the nonautonomous equation with initial condition $x(t_0) = x_0$, then $t \mapsto (t, \psi(t))$ is the solution of the system with initial condition $\tau(t_0) = t_0$, $x(t_0) = x_0$. Prove the converse. Let $\phi(t, \tau, x)$ denote the flow of the autonomous system and define T and U by $(T(t,\tau,x), U(t,\tau,x)) = \phi(t - \tau, \tau, x)$. Show that $T(\tau, \tau, x) = \tau$ and $U(\tau, \tau, x) = x$. By using the definition of the flow as the solution of the system, show that $T(t, \tau, x) = t$. Finally, use the group property of the flow to show that $U(t, \tau, x) = U(t, s, U(s, \tau, x))$ whenever both sides of the identity are defined. A family of functions U such that $U(\tau, \tau, x) = x$ and $U(t, \tau, x) = U(t, s, U(s, \tau, x))$ for $t, s, \tau \in \mathbb{R}$ and $x \in \mathbb{R}^n$ is called an evolution family. This exercise shows that the solution $t \mapsto U(t, \tau, x)$ of the nonautonomous differential equation $\dot{x} = f(t,x)$ such that $U(\tau, \tau, x) = x$ is an evolution family.

1.5 Reparametrization of Time

Suppose that U is an open set in $\mathbb{R}^n$, $f : U \to \mathbb{R}^n$ is a smooth function, and $g : U \to \mathbb{R}$ is a *positive* smooth function. What is the relationship among the solutions of the differential equations

$$\dot{x} = f(x), \tag{1.10}$$
$$\dot{x} = g(x)f(x)? \tag{1.11}$$

The vector fields defined by f and gf have the same direction at each point in U, only their lengths are different. Thus, by our geometric interpretation of autonomous differential equations, it is intuitively clear that the differential equations (1.10) and (1.11) have the same phase portraits in U. This fact is a corollary of the next proposition.

Proposition 1.28. *If $J \subset \mathbb{R}$ is an open interval containing the origin and $\gamma : J \to \mathbb{R}^n$ is a solution of the differential equation (1.10) with $\gamma(0) = x_0 \in U$, then the function $B : J \to \mathbb{R}$ given by*

$$B(t) = \int_0^t \frac{1}{g(\gamma(s))}\, ds$$

is invertible on its range $K \subseteq \mathbb{R}$. If $\rho : K \to J$ is the inverse of B, then the identity

$$\rho'(t) = g(\gamma(\rho(t)))$$

holds for all $t \in K$, and the function $\sigma : K \to \mathbb{R}^n$ given by $\sigma(t) = \gamma(\rho(t))$ is the solution of the differential equation (1.11) with initial condition $\sigma(0) = x_0$.

Proof. The function $s \mapsto 1/g(\gamma(s))$ is continuous on J. So B is defined on J and its derivative is everywhere positive. Thus, B is invertible on its range. If ρ is its inverse, then

$$\rho'(t) = \frac{1}{B'(\rho(t))} = g(\gamma(\rho(t))),$$

and

$$\sigma'(t) = \rho'(t)\gamma'(\rho(t)) = g(\gamma(\rho(t)))f(\gamma(\rho(t))) = g(\sigma(t))f(\sigma(t)). \qquad \square$$

Exercise 1.29. Use Proposition 1.28 to prove that differential equations (1.10) and (1.11) have the same phase portrait in U.

Because the function ρ in Proposition 1.28 is the inverse of B, we have the formula

$$t = \int_0^\rho \frac{1}{g(\gamma(s))}\, ds.$$

Thus, if we view ρ as a new time-like variable (that is, a variable that increases with time), then we have

$$\frac{dt}{d\rho} = \frac{1}{g(\gamma(\rho))},$$

and therefore the differential equation (1.11), with the change of independent variable from t to ρ, is given by

$$\frac{dx}{d\rho} = \frac{dx}{dt}\frac{dt}{d\rho} = f(x).$$

In particular, this is just differential equation (1.10) with the independent variable renamed.

The next proposition expresses the same results in the language of flows.

Proposition 1.30. *Suppose that* $f : \mathbb{R}^n \to \mathbb{R}^n$, $g : \mathbb{R}^n \to \mathbb{R}$ *is a positive function, and* ϕ *is the flow of the differential equation* $\dot{x} = f(x)$. *If the family of solutions of the family of initial value problems*

$$\dot{y} = g(\phi(y,\xi)), \qquad y(0) = 0,$$

with parameter $\xi \in \mathbb{R}^n$, *is given by* $\rho : \mathbb{R} \times \mathbb{R}^n \to \mathbb{R}$, *then* ψ, *defined by* $\psi(t,\xi) = \phi(\rho(t,\xi),\xi)$ *is the flow of the differential equation* $\dot{x} = g(x)f(x)$.

Proof. By definition $\psi(0,\xi) \equiv \xi$, and by the chain rule

$$\frac{d}{dt}\psi(t,\xi) = g(\psi(t,\xi))f(\psi(t,\xi)). \qquad \square$$

As a convenient expression, we say that the differential equation (1.10) is obtained from the differential equation (1.11) by a *reparametrization of time*.

In the most important special cases the function g is constant. If its constant value is $c > 0$, then the reparametrization of the differential equation $\dot{x} = cf(x)$ by $\rho = ct$ results in the new differential equation

$$\frac{dx}{d\rho} = f(x).$$

Reparametrization in these cases is also called *rescaling*.

Note that rescaling, as in the last paragraph, of the differential equation $\dot{x} = cf(x)$ produces a differential equation in which the parameter c has been eliminated. This idea is often used to simplify differential equations. Also, the same rescaling is used in applied mathematics to render the independent variable dimensionless. For example, if the original time variable t is measured in seconds, and the scale factor c has the units of $1/\text{sec}$, then the new variable ρ is dimensionless.

The next proposition is a special case of the following claim: Every autonomous differential equation has a complete reparametrization (see Exercise 1.36).

Proposition 1.31. *If the differential equation* $\dot{x} = f(x)$ *is defined on* $\mathbb{R}^n$, *then the differential equation*

$$\dot{x} = \frac{1}{1 + |f(x)|^2} f(x) \tag{1.12}$$

is defined on $\mathbb{R}^n$ *and its flow is complete.*

Proof. The vector field corresponding to the differential equation (1.12) is smoothly defined on all of $\mathbb{R}^n$. If σ is one of its solutions with initial value $\sigma(0) = x_0$ and t is in the domain of σ, then, by integration with respect to the independent variable, we have that

$$\sigma(t) - \sigma(0) = \int_0^t \frac{1}{1 + |f(\sigma(s))|^2} f(\sigma(s)) \, ds.$$

Note that the integrand has norm less than one and use the triangle inequality (taking into account the possibility that t might be negative) to obtain the following estimate:

$$|\sigma(t)| \leq |x_0| + |t|.$$

In particular, the solution does not blow up in finite time. By the extension theorem, the solution is complete. $\qquad\square$

Exercise 1.32. Consider the scalar differential equations $\dot{x} = x$ and $\dot{x} = x^2$ for $x > 0$. Find explicit expressions for the corresponding flows ϕ and ψ and for the reparametrization function ρ, as in Proposition 1.30, so that $\psi(t, \xi) = \phi(\rho(t, \xi), \xi)$. Show that no such relation holds if the restriction $x > 0$ is removed. Does it matter that ψ is not complete? Consider instead the differential equations $\dot{x} = x$ and $\dot{x} = x^3$.

Exercise 1.33. Consider the function $g : (0, \infty) \to \mathbb{R}$ given by $g(x) = x^{-n}$ for a fixed positive integer n. Construct the flow ϕ_t of the differential equation $\dot{x} = -x$ and the flow ψ_t of $\dot{x} = -g(x)x$ on $(0, \infty)$, and find the explicit expression for the reparametrization function ρ such that $\psi_t(x) = \phi_{\rho(t)}(x)$ (see [53]).

Exercise 1.34. Suppose that n is an integer. Solve the initial value problem

$$\dot{x} = y(x + y)^n, \quad \dot{y} = x(x + y)^n, \quad x(0) = 1, \quad y(0) = 0.$$

Is the solution complete?

Exercise 1.35. Suppose that the solution γ of the differential equation $\dot{x} = f(x)$ is reparametrized by arc length; that is, in the new parametrization the velocity vector at each point of the solution curve has unit length. Find an implicit formula for the reparametrization ρ, and prove that if $t > 0$, then

$$|\gamma(\rho(t))| \leq |\gamma(0)| + t.$$

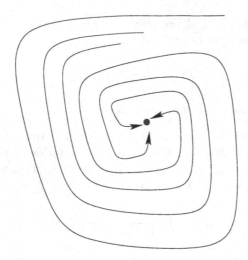

Figure 1.7: Phase portrait of an asymptotically stable (spiral) sink.

Exercise 1.36. Suppose that $\dot{x} = f(x)$ is a differential equation defined on an open subset U of $\mathbb{R}^n$. Show that the differential equation has a complete reparametrization.

Exercise 1.37. [Solute Transport] A model for transport of a solute (moles of salt) and solvent (volume of water) across a permeable membrane has the form

$$\dot{W} = A(k - \frac{M}{W}), \qquad \dot{M} = B(k - \frac{M}{W})$$

where k is a parameter representing the bulk solute concentration and A and B are parameters that represent the permeability of the membrane (see [40]). (a) The water volume W is a positive quantity. Show that the system can be made linear by a reparametrization. (b) Determine the transformation between solutions of the linear and nonlinear systems.

1.6   Stability and Linearization

Although rest points and periodic orbits correspond to very special solutions of autonomous differential equations, these are often the most important orbits in applications. In particular, common engineering practice is to run a process in "steady state." If the process does not stay near the steady state after a small disturbance, then the control engineer will have to face a difficult problem. We will not solve the control problem here, but we will introduce the mathematical definition of stability and the classic methods that can be used to determine the stability of rest points and periodic orbits.

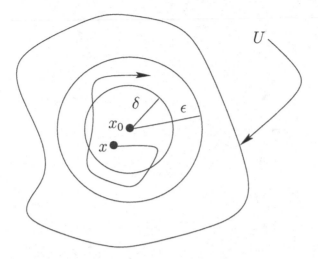

Figure 1.8: The open sets required in the definition of Lyapunov stability. The trajectory starting at x can leave the ball of radius δ but it must stay in the ball of radius ϵ.

The concept of *Lyapunov stability* is meant to capture the intuitive notion of stability—an orbit is stable if solutions that start nearby stay nearby. To give the formal definition, let us consider the autonomous differential equation

$$\dot{x} = f(x) \qquad (1.13)$$

defined on an open set $U \subset \mathbb{R}^n$ and its flow ϕ_t.

Definition 1.38. A rest point x_0 of the differential equation (1.13) is *stable* (in the sense of Lyapunov) if for each $\epsilon > 0$, there is a number $\delta > 0$ such that $|\phi_t(x) - x_0| < \epsilon$ for all $t \geq 0$ whenever $|x - x_0| < \delta$ (see Figure 1.8).

There is no reason to restrict the definition of stability to rest points. It can also refer to arbitrary solutions of the autonomous differential equation.

Definition 1.39. Suppose that x_0 is in the domain of definition of the differential equation (1.13). The solution $t \mapsto \phi_t(x_0)$ of this differential equation is *stable* (in the sense of Lyapunov) if for each $\epsilon > 0$, there is a $\delta > 0$ such that $|\phi_t(x) - \phi_t(x_0)| < \epsilon$ for all $t \geq 0$ whenever $|x - x_0| < \delta$.

Figure 1.7 shows a typical phase portrait of an autonomous system in the plane near a type of stable rest point called a *sink*. The stable rest point depicted in Figure 1.4 is called a *center*. More precisely, a rest point is a center if it is contained in an open set where every orbit (except the rest point) is periodic.

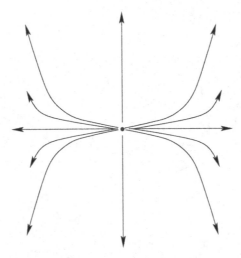

Figure 1.9: Phase portrait of an unstable rest point.

A solution that is not stable is called *unstable*. A typical phase portrait for an unstable rest point, a *source,* is depicted in Figure 1.9 (see also the saddle point in Figure 1.1).

Definition 1.40. A solution $t \to \phi_t(x_0)$ of the differential equation (1.13) is *asymptotically stable* if it is stable and there is a constant $a > 0$ such that $\lim_{t \to \infty} |\phi_t(x) - \phi_t(x_0)| = 0$ whenever $|x - x_0| < a$.

We have just defined the notion of stability for solutions in case a definite initial point is specified. The concept of stability for orbits is slightly more complicated. For example, we have the following definition of stability for periodic orbits (see also Section 2.4.4).

Definition 1.41. A periodic orbit Γ of the differential equation (1.13) is *stable* if for each open set $V \subseteq \mathbb{R}^n$ that contains Γ, there is an open set $W \subseteq V$ such that every solution, starting at a point in W at $t = 0$, stays in V for all $t \geq 0$. The periodic orbit is called *asymptotically stable* if, in addition, there is a subset $X \subseteq W$ such that every solution starting in X is asymptotic to Γ as $t \to \infty$.

The definitions just given capture the essence of the stability concept, but they do not give any indication of how to determine if a given solution or orbit is stable. We will study two general methods, called the indirect and the direct methods by Lyapunov, that can be used to determine the stability of rest points and periodic orbits. In more modern language, the indirect method is called the method of linearization and the direct method is called the method of Lyapunov. Before we discuss these methods in detail, let us note that for the case of the stability of special types of orbits, for example rest points and periodic orbits, there are two main problems: (i) Locating the special solutions. (ii) Determining their stability.

For the remainder of this section and the next, the discussion will be restricted to the analysis for rest points. Our introduction to the methods for locating and determining the stability of periodic orbits must be postponed until some additional concepts have been introduced.

Let us note that the problem of the location of rest points for the differential equation $\dot{x} = f(x)$ is exactly the problem of finding the roots of the equation $f(x) = 0$. Of course, finding roots may be a formidable task, especially if the function f depends on parameters and we wish to find its bifurcation diagram. In fact, in the search for rest points, sophisticated techniques of algebra, analysis, and numerical analysis are often required. This is not surprising when we stop to think that solving equations is one of the fundamental themes in mathematics. For example, it is probably not too strong to say that the most basic problem in linear algebra, abstract algebra, and algebraic geometry is the solution of systems of polynomial equations. The results of all of these subjects are sometimes needed to solve problems in differential equations.

Let us suppose that we have identified some point $x_0 \in \mathbb{R}^n$ such that $f(x_0) = 0$. What can we say about the stability of the corresponding rest point? One of the great ideas in the subject of differential equations—not to mention other areas of mathematics—is *linearization*. This idea, in perhaps its purest form, is used to obtain the premier method for the determination of the stability of rest points. The linearization method is based on two facts: (i) Stability analysis for linear systems is "easy." (ii) Nonlinear systems can be approximated by linear systems. These facts are just reflections of the fundamental idea of differential calculus: A nonlinear function is essentially linear if we consider its behavior in a sufficiently small neighborhood of a point in its domain. Indeed, it often suffices to approximate the graph of a function by its tangent lines.

To describe the linearization method for rest points, let us consider (homogeneous) linear systems of differential equations; that is, systems of the form $\dot{x} = Ax$ where $x \in \mathbb{R}^n$ and A is a linear transformation of $\mathbb{R}^n$. If the matrix A does not depend on t—so that the linear system is autonomous—then there is an effective method that can be used to determine the stability of its rest point at $x = 0$. In fact, we will show in Chapter 2 that if all of the eigenvalues of A have negative real parts, then $x = 0$ is an asymptotically stable rest point for the linear system. (The eigenvalues of a linear transformation are defined on page 154.)

If x_0 is a rest point for the nonlinear system $\dot{x} = f(x)$, then there is a natural way to produce a linear system that approximates the nonlinear system near x_0: Simply replace the function f in the differential equation with the linear function $x \mapsto Df(x_0)(x - x_0)$ given by the first nonzero term of the Taylor series of f at x_0. The linear differential equation

$$\dot{x} = Df(x_0)(x - x_0) \tag{1.14}$$

is called the *linearized system associated with* $\dot{x} = f(x)$ *at* x_0. By applying the change of variables $w := x - x_0$, the linearized system has the more convenient form $\dot{w} = Df(x_0)w$.

Alternatively, we may consider a family of solutions $t \mapsto \phi(t, \epsilon)$ of $\dot{x} = f(x)$ with parameter ϵ such that $\phi(t, 0) = x_0$. In other words, the family of solutions contains the constant solution corresponding to the rest point at the parameter value $\epsilon = 0$. By Taylor's theorem (with respect to ϵ at $\epsilon = 0$),

$$\phi(t, \epsilon) = x_0 + \epsilon\eta(t, 0) + \epsilon^2 R(t, \epsilon)$$

for some function R. Hence, the family of first-order approximations of the solutions in the family ϕ is given by

$$\phi(t, \epsilon) = x_0 + \epsilon\eta(t, 0).$$

To determine the function $t \mapsto \eta(t, 0)$, we use the differential equation and Taylor's theorem (with respect to ϵ at $\epsilon = 0$) to obtain

$$\begin{aligned}
\epsilon\dot{\eta}(t, \epsilon) &= \dot{\phi}(t, \epsilon) \\
&= f(\phi(t, \epsilon)) \\
&= f(x_0 + \epsilon\eta(t, \epsilon)) \\
&= \epsilon Df(x_0)\eta(t, 0) + \epsilon^2 \mathcal{R}(t, \epsilon)
\end{aligned}$$

where $\epsilon^2 \mathcal{R}(t, \epsilon)$ is the remainder in the Taylor expansion of $\epsilon \mapsto f(x_0 + \epsilon\eta(t, \epsilon))$. After dividing by ϵ, we obtain the equation

$$\dot{\eta}(t, \epsilon) = Df(x_0)\eta(t, \epsilon) + \epsilon R(t, \epsilon),$$

and, by taking the limit as $\epsilon \to 0$, it follows that

$$\dot{\eta}(t, 0) = Df(x_0)\eta(t, 0).$$

Of course, the initial condition for this equation is determined by the choice of the family ϕ; in fact,

$$\eta(0, 0) = \frac{\partial\phi}{\partial\epsilon}(0, \epsilon)\Big|_{\epsilon=0},$$

Thus, the linearized differential equation at x_0 (that is, $\dot{w} = Df(x_0)w$) is the same for every family ϕ, and every vector v in $\mathbb{R}^n$ is the initial vector ($w(0) = v$) for some such family.

The "principle of linearized stability" states that if the linearization of a differential equation at a steady state has a corresponding *stable* steady state, then the original steady state is stable. In the notation of this section, this principle states that if $w(t) = 0$ is a stable steady state for $\dot{w} = Df(x_0)w$, then x_0 is a stable steady state for $\dot{x} = f(x)$. *The principle of linearized stability is not a theorem*, but it is the motivation for several important results in the theory of stability of differential equations.

Exercise 1.42. Prove that the rest point at the origin for the differential equation $\dot{x} = ax$, $a < 0$, $x \in \mathbb{R}$ is asymptotically stable. Also, determine the stability of this rest point in case $a = 0$ and in case $a > 0$.

Exercise 1.43. Use the principle of linearized stability to determine the stability of the rest point at the origin for the system

$$\dot{x} = 2x + 2y - 3\sin x, \qquad \dot{y} = -2y + xy.$$

Let us suppose that x_0 is a rest point of the differential equation $\dot{x} = f(x)$. By the change of variables $u = x - x_0$, this differential equation is transformed to the equivalent differential equation $\dot{u} = f(u+x_0)$ where the rest point corresponding to x_0 is at the origin. For $g(u) := f(u + x_0)$, we have $\dot{u} = g(u)$ and $g(0) = 0$. Thus, it should be clear that there is no loss of generality if we assume that a rest point is at the origin. In this case, the linearized equation is $\dot{w} = Dg(0)w$.

If f is smooth at $x = 0$ and $f(0) = 0$, then

$$f(x) = f(0) + Df(0)x + R(x) = Df(0)x + R(x)$$

where $Df(0) : \mathbb{R}^n \to \mathbb{R}^n$ is the linear transformation given by the derivative of f at $x = 0$ and, for the remainder R, there is a constant $k > 0$ and an open neighborhood U of the origin such that

$$|R(x)| \leq k|x|^2$$

whenever $x \in U$. Because the stability of a rest point is a local property (that is, a property that is determined by the values of the restriction of the function f to an arbitrary open subset of the rest point) and in view of the estimate for the size of the remainder, it is reasonable to expect that the stability of the rest point at the origin of the linear system $\dot{x} = Df(0)x$ will be the same as the stability of the original rest point. This expectation is not always realized. But we do have the following fundamental stability theorem.

Theorem 1.44. *If x_0 is a rest point for the differential equation $\dot{x} = f(x)$ and if all eigenvalues of the linear transformation $Df(x_0)$ have negative real parts, then x_0 is asymptotically stable.*

Proof. See Theorem 2.78.   □

It turns out that if x_0 is a rest point and $Df(x_0)$ has at least one eigenvalue with positive real part, then x_0 is not stable. If some eigenvalues of $Df(x_0)$ lie on the imaginary axis, then the stability of the rest point may be very difficult to determine. Also, we can expect qualitative changes to occur in the phase portrait of a system near such a rest point as the parameters of the system are varied. These bifurcations are the subject of Chapter 8.

Exercise 1.45. Prove: If $\dot{x} = 0$, $x \in \mathbb{R}$, then $x = 0$ is Lyapunov stable. Consider the differential equations $\dot{x} = x^3$ and $\dot{x} = -x^3$. Prove that whereas the origin is not a Lyapunov stable rest point for the differential equation $\dot{x} = x^3$, it is Lyapunov stable for the differential equation $\dot{x} = -x^3$. Note that the linearized differential equation at $x = 0$ in both cases is the same; namely, $\dot{x} = 0$.

Exercise 1.46. Use Theorem 1.44 to show the asymptotic stability of the rest point at the origin of the system

$$\dot{x} = -2y - 4z + (x+y)x^2, \quad \dot{y} = x - 3y - z + (x+y)y^2, \quad \dot{z} = -4z + (x+y)z^2.$$

If x_0 is a rest point for the differential equation (1.13) and if the linear transformation $Df(x_0)$ has all its eigenvalues off the imaginary axis, then we say that x_0 is a *hyperbolic rest point*. Otherwise x_0 is called *nonhyperbolic*. In addition, if x_0 is hyperbolic and all eigenvalues have negative real parts, then the rest point is called a *hyperbolic sink*. If all eigenvalues have positive real parts, then the rest point is called a *hyperbolic source*. A hyperbolic rest point that is neither a source nor a sink is called a *hyperbolic saddle*. If the rest point is nonhyperbolic with all its eigenvalues on the punctured imaginary axis (that is, the imaginary axis with the origin removed), then the rest point is called a *linear center*. If zero is not an eigenvalue, then the corresponding rest point is called *nondegenerate*.

If every eigenvalue of a linear transformation A has nonzero real part, then A is called *infinitesimally hyperbolic*. If none of the eigenvalues of A have modulus one, then A is called *hyperbolic*. This terminology can be confusing: For example, if A is infinitesimally hyperbolic, then the rest point at the origin of the linear system $\dot{x} = Ax$ is hyperbolic. The reason for the terminology is made clear by consideration of the scalar linear differential equation $\dot{x} = ax$ with flow given by $\phi_t(x) = e^{at}x$. If $a \neq 0$, then the linear transformation $x \to ax$ is infinitesimally hyperbolic and the rest point at the origin is hyperbolic. In addition, if $a \neq 0$ and $t \neq 0$, then the linear transformation $x \mapsto e^{ta}x$ is hyperbolic. Moreover, the linear transformation $x \mapsto ax$ is obtained by differentiation with respect to t at $t = 0$ of the family of linear transformations $x \mapsto e^{ta}x$. Thus, in effect, differentiation— an infinitesimal operation on the family of hyperbolic transformations— produces an infinitesimally hyperbolic transformation.

The relationship between the dynamics of a nonlinear system and its linearization at a rest point is deeper than the relationship between the stability types of the corresponding rest points. The next theorem, called the Hartman–Grobman theorem, is an important result that describes this relationship in case the rest point is hyperbolic.

Theorem 1.47. *If x_0 is a hyperbolic rest point for the autonomous differential equation (1.13), then there is an open set U containing x_0 and a homeomorphism H with domain U such that the orbits of the differential equation (1.13) are mapped by H to orbits of the linearized system $\dot{x} = Df(x_0)(x - x_0)$ in the set U.*

Proof. See Section 4.3.   □

In other words, the linearized system has the same phase portrait as the original system in a sufficiently small neighborhood of the hyperbolic rest point. Moreover, the homeomorphism H in the theorem can be chosen to preserve not just the orbits as point sets, but their time parameterizations as well.

Exercise 1.48. In the definition of asymptotic stability for rest points, the first requirement is that the rest point be stable; the second requirement is that all solutions starting in some open set containing the rest point be asymptotic to the rest point. Does the first requirement follow from the second? Answer this question for flows on the line and the circle. Hint: Consider flows on the circle first. They are obtained by solving differential equations of the form $\dot{\theta} = f(\theta)$ where $f : \mathbb{R} \to \mathbb{R}$ is 2π-periodic. Find a differential equation on the circle with exactly one rest point that is semistable. See Figure 1.22 and Exercise 1.145 for an explicit example of a planar system with a rest point such that every orbit is attracted to the rest point, but the rest point is not asymptotically stable.

Exercise 1.49. Consider the mathematical pendulum given by the second order differential equation $\ddot{u} + \sin u = 0$. (a) Find the corresponding first order system. (b) Find all rest points of the first order system, and characterize these rest points according to their stability type. (c) Draw the phase portrait of the system in a neighborhood at each rest point. (d) Solve the same problems for the second order differential equation given by

$$\ddot{x} + (x^2 - 1)\dot{x} + \omega^2 x - \lambda x^3 = 0.$$

Exercise 1.50. (a) Linearize at each rest point of the predator-prey model in Exercise 1.15. What condition on the parameters (if any) implies the existence of an asymptotically stable rest point? (b) Interpret the result of part (a) as a statement about the fate of the predators and prey.

Exercise 1.51. Show that the origin is an asymptotically stable rest point for the system

$$\dot{x} = -11x - 48y - 16z + xyz,$$
$$\dot{y} = x + 3y + 2z + x^2 - yz,$$
$$\dot{z} = 2y + 2z + \sin x.$$

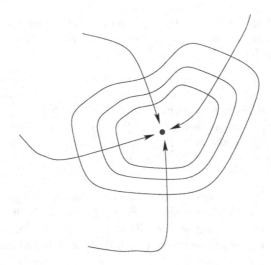

Figure 1.10: Level sets of a Lyapunov function.

Exercise 1.52. Use the Hartman-Grobman theorem to describe (geometrically) the behavior near the origin of the system

$$\dot{x} = 4y - x^2 z,$$
$$\dot{y} = -x + 4y + xy^2,$$
$$\dot{z} = -10z + yz^2.$$

Exercise 1.53. Prove that there are open intervals U and V containing the origin and a differentiable map $H : U \to V$ with a differentiable inverse such that the flow ϕ_t of $\dot{x} = -x$ is conjugate to the flow ψ_t of $\dot{x} = -x + x^2$; that is, $H(\phi_t(H^{-1}(x))) = \psi_t(x)$ whenever $x \in V$.

Exercise 1.54. [Reversible systems] A planar system $\dot{x} = f(x,y)$, $\dot{y} = g(x,y)$ is called reversible if it is invariant under the change of variables $t \mapsto -t$ and $y \mapsto -y$. For example, $\dot{x} = -y$, $\dot{y} = x$ is reversible. (a) Prove that a linear center of a reversible system is a center. (b) Construct a planar system with exactly three rest points all of which are centers. Hint: The system $\dot{x} = (x+2)x(x-2)y$, $\dot{y} = (x-1)(x+1)$ has exactly two rest points both of which are centers.

1.7 Stability and the Direct Method of Lyapunov

Let us consider a rest point x_0 for the autonomous differential equation

$$\dot{x} = f(x), \qquad x \in \mathbb{R}^n. \tag{1.15}$$

A continuous function $V : U \to \mathbb{R}$, where $U \subseteq \mathbb{R}^n$ is an open set with $x_0 \in U$, is called a *Lyapunov function* for the differential equation (1.15) at x_0 if

(i) $V(x_0) = 0$,

(ii) $V(x) > 0$ for $x \in U \setminus \{x_0\}$,

(iii) the function V is continuously differentiable on the set $U \setminus \{x_0\}$, and, on this set, $\dot{V}(x) := \operatorname{grad} V(x) \cdot f(x) \le 0$.

The function V is called a *strict Lyapunov function* if, in addition,

(iv) $\dot{V}(x) < 0$ for $x \in U \setminus \{x_0\}$.

Theorem 1.55 (Lyapunov's Stability Theorem). *If there is a Lyapunov function defined in an open neighborhood of a rest point of the differential equation (1.15), then the rest point is stable. If, in addition, the Lyapunov function is a strict Lyapunov function, then the rest point is asymptotically stable.*

The *idea* of Lyapunov's method is very simple. In many cases the level sets of V are "spheres" surrounding the rest point x_0 as in Figure 1.10. Suppose this is the case and let ϕ_t denote the flow of the differential equation (1.15). If y is in the level set $\mathcal{S}_c = \{x \in \mathbb{R}^n : V(x) = c\}$ of the function V, then, by the chain rule, we have that

$$\frac{d}{dt} V(\phi_t(y)) \Big|_{t=0} = \operatorname{grad} V(y) \cdot f(y) \le 0. \qquad (1.16)$$

The vector $\operatorname{grad} V$ is an outer normal for $\mathcal{S}_c$ at y. (Do you see why it must be the *outer* normal?) Thus, V is not increasing on the curve $t \mapsto \phi_t(y)$ at $t = 0$, and, as a result, the image of this curve either lies in the level set $\mathcal{S}_c$, or the set $\{\phi_t(y) : t > 0\}$ is a subset of the set in $\mathbb{R}^n$ with outer boundary $\mathcal{S}_c$. The same result is true for every point on $\mathcal{S}_c$. Therefore, a solution starting on $\mathcal{S}_c$ is trapped; it either stays in $\mathcal{S}_c$, or it stays in the set $\{x \in \mathbb{R}^n : V(x) < c\}$. The stability of the rest point follows easily from this result. If V is a strict Lyapunov function, then the solution curve definitely crosses the level set $\mathcal{S}_c$ and remains inside the set $\{x \in \mathbb{R}^n : V(x) < c\}$ for all $t > 0$. Because the same property holds at all level sets "inside" $\mathcal{S}_c$, the rest point x_0 is asymptotically stable.

If the level sets of our Lyapunov function are as depicted in Figure 1.10, then the argument just given proves the stability of the rest point. But the level sets of a Lyapunov function may not have this simple configuration. For example, some of the level sets may not be bounded.

The *proof* of Lyapunov's stability theorem requires a more delicate analysis. Let us use the following notation. For $\alpha > 0$ and $\zeta \in \mathbb{R}^n$, define

$$S_\alpha(\zeta) := \{x \in \mathbb{R}^n : |x - \zeta| = \alpha\},$$
$$B_\alpha(\zeta) := \{x \in \mathbb{R}^n : |x - \zeta| < \alpha\},$$
$$\bar{B}_\alpha(\zeta) := \{x \in \mathbb{R}^n : |x - \zeta| \le \alpha\}.$$

Proof. Suppose that $\epsilon > 0$ is given, and note that, in view of the definition of Lyapunov stability, it suffices to assume that $\bar{B}_\epsilon(x_0)$ is contained in the domain U of the Lyapunov function V. Because $S_\epsilon(x_0)$ is a compact set not containing x_0, there is a number $m > 0$ such that $V(x) \geq m$ for all $x \in S_\epsilon(x_0)$. Also, there is some $\delta > 0$ with $\delta < \epsilon$ such that $V(x) \leq m/2$ for x in the compact set $\bar{B}_\delta(x_0)$. If not, then for each $k \geq 2$ there is a point x_k in $\bar{B}_{\epsilon/k}(x_0)$ such that $V(x_k) > m/2$. The sequence $\{x_k\}_{k=2}^\infty$ converges to x_0. Using the continuity of the Lyapunov function V at x_0, we have $\lim_{k \to \infty} V(x_k) = V(x_0) = 0$, in contradiction.

Let ϕ_t denote the flow of (1.15). If $x \in B_\delta(x_0)$, then

$$\frac{d}{dt} V(\phi_t(x)) = \operatorname{grad} V(\phi_t(x)) \cdot f(\phi_t(x)) \leq 0.$$

Thus, the function $t \to V(\phi_t(x))$ is not increasing. Since $V(\phi_0(x)) \leq M < m$, we must have $V(\phi_t(x)) < m$ for all $t \geq 0$ for which the solution $t \mapsto \phi_t(x)$ is defined. But, for these values of t, we must also have $\phi_t(x) \in B_\epsilon(x_0)$. If not, there is some $T > 0$ such that $|\phi_T(x) - x_0| \geq \epsilon$. Since $t \mapsto |\phi_t(x) - x_0|$ is a continuous function, there must then be some τ with $0 < \tau \leq T$ such that $|\phi_\tau(x) - x_0| = \epsilon$. For this τ, we have $V(\phi_\tau(x)) \geq m$, in contradiction. Thus, $\phi_t(x) \in B_\epsilon(x_0)$ for all $t \geq 0$ for which the solution through x exists. By the extension theorem, if the solution does not exist for all $t \geq 0$, then $|\phi_t(x)| \to \infty$ as $t \to \infty$, or $\phi_t(x)$ approaches the boundary of the domain of definition of f. Since neither of these possibilities occur, the solution exists for all positive time with its corresponding image in the set $B_\epsilon(x_0)$. Thus, x_0 is stable.

If, in addition, the Lyapunov function is strict, we will show that x_0 is asymptotically stable.

Let $x \in B_\delta(x_0)$. By the compactness of $\bar{B}_\epsilon(x_0)$, either $\lim_{t \to \infty} \phi_t(x) = x_0$, or there is a sequence $\{t_k\}_{k=1}^\infty$ of real numbers $0 < t_1 < t_2 \cdots$ with $t_k \to \infty$ such that the sequence $\{\phi_{t_k}(x)\}_{k=1}^\infty$ converges to some point $x_* \in \bar{B}_\epsilon(x_0)$ with $x_* \neq x_0$. If x_0 is not asymptotically stable, then such a sequence exists for at least one point $x \in B_\delta(x_0)$.

Using the continuity of V, it follows that $\lim_{k \to \infty} V(\phi_{t_k}(x)) = V(x_*)$. Also, V decreases on orbits. Thus, for each natural number k, we have that $V(\phi_{t_k}(x)) > V(x_*)$. But, because the function $t \mapsto V(\phi_t(x_*))$ is strictly decreasing, we have

$$\lim_{k \to \infty} V(\phi_{1+t_k}(x)) = \lim_{k \to \infty} V(\phi_1(\phi_{t_k}(x))) = V(\phi_1(x_*)) < V(x_*).$$

Thus, there is some natural number ℓ such that $V(\phi_{1+t_\ell}(x)) < V(x_*)$. Clearly, there is also an integer $j > \ell$ such that $t_j > 1 + t_\ell$. For this integer, we have the inequalities $V(\phi_{t_j}(x)) < V(\phi_{1+t_\ell}(x)) < V(x_*)$, in contradiction. $\square$

The next result can be used to prove the instability of a rest point.

Theorem 1.56. *Suppose that V is a smooth function defined on an open neighborhood U of the rest point x_0 of the autonomous system $\dot{x} = f(x)$ such that $V(x_0) = 0$ and $\dot{V}(x) > 0$ on $U \setminus \{x_0\}$. If V has a positive value somewhere in each open set containing x_0, then x_0 is not stable.*

Proof. Suppose that x_0 is stable, and let ϕ_t denote the flow of the differential equation. Choose $\epsilon > 0$ such that $\bar{B}_\epsilon(x_0) \subset U$ and $\bar{B}_\epsilon(x_0)$ is also in the domain of f. There is some positive δ such that $\delta < \epsilon$ and $\phi_t(x)$ is in $B_\epsilon(x_0)$ whenever $x \in B_\delta(x_0)$ and $t \geq 0$. Also, by the continuity of V, there is some $\alpha > 0$ such that $V(x) \leq \alpha$ whenever $x \in \bar{B}_\epsilon(x_0)$.

By hypothesis, there is some $x \in B_\delta(x_0)$ such that $V(x) > 0$. Also, by the hypotheses, $\beta := \inf_{t \geq 0} \dot{V}(\phi_t(x)) \geq 0$. Suppose that $\beta = 0$. In this case, there is a sequence $t_1 < t_2 < t_3 < \cdots$ such that $\lim_{j \to \infty} t_j = \infty$ and $\lim_{j \to \infty} \dot{V}(\phi_{t_j}(x)) = 0$. By the stability and the compactness of $\bar{B}_\epsilon(x_0)$, $\{\phi_{t_j}(x)\}_{j=0}^\infty$ has a convergent subsequence. Without loss of generality, we can assume that the sequence itself converges to some $x_* \in \bar{B}_\epsilon(x_0)$. By the continuity of $\dot{V}$, we have $\lim_{j \to \infty} \dot{V}(\phi_{t_j}(x)) = \dot{V}(x_*) = 0$. Since $\dot{V}$ does not vanish on $U \setminus \{x_0\}$, it follows that $x_* = x_0$. Since V is continuous, $\lim_{j \to \infty} V(\phi_{t_j}(x)) = V(x_0) = 0$. But, $V(\phi_{t_j}(x)) > V(x) > 0$, in contradiction. Hence, $\beta > 0$.

Note that

$$V(\phi_t(x)) = V(x) + \int_0^t \dot{V}(\phi_s(x))\, ds \geq V(x) + \beta t$$

for all $t \geq 0$. If t is sufficiently large, then $V(\phi_t(x)) > \alpha$, in contradiction to the stability of x_0. $\qquad\square$

Example 1.57. The linearization of $\dot{x} = -x^3$ at $x = 0$ is $\dot{x} = 0$. It provides no information about stability. Define $V(x) = x^2$ and note that $\dot{V}(x) = 2x(-x^3) = -2x^4$. Thus, V is a strict Lyapunov function, and the rest point at $x = 0$ is asymptotically stable.

Example 1.58. Consider the harmonic oscillator $\ddot{x} + \omega^2 x = 0$ with $\omega > 0$. The equivalent first order system

$$\dot{x} = y, \qquad \dot{y} = -\omega^2 x$$

has a rest point at $(x, y) = (0, 0)$. Define the total energy (kinetic energy plus potential energy) of the harmonic oscillator to be

$$V = \frac{1}{2}\dot{x}^2 + \frac{\omega^2}{2}x^2 = \frac{1}{2}(y^2 + \omega^2 x^2).$$

A computation shows that $\dot{V} = 0$. Thus, the rest point is stable. The energy of a physical system is often a good choice for a Lyapunov function.

Exercise 1.59. As a continuation of example (1.58), consider the equivalent first order system

$$\dot{x} = wy, \qquad \dot{y} = -wx.$$

Study the stability of the rest point at the origin using Lyapunov's direct method.

Exercise 1.60. Consider a Newtonian particle of mass m moving under the influence of the potential U. The equation of motion $(F = ma)$ is given by

$$m\ddot{q} = -\operatorname{grad} U(q)$$

where the position coordinate is denoted by $q = (q_1, \ldots, q_n)$. If q_0 is a strict local minimum of the potential, show that the equilibrium $(q, \dot{q}) = (q_0, 0)$ is Lyapunov stable. Hint: Consider the total energy of the particle.

Exercise 1.61. Determine the stability of the rest points of the following systems. Formulate properties of the unspecified scalar function g so that the system has a rest point at the origin which is respectively stable, asymptotically stable, and unstable.

1. $\dot{x} = y - x^3$,
   $\dot{y} = -x - y^3$
2. $\dot{x} = y + \alpha x(x^2 + y^2)$,
   $\dot{y} = -x + \alpha y(x^2 + y^2)$
3. $\dot{x} = 2xy - x^3$,
   $\dot{y} = -x^2 - y^5$
4. $\dot{x} = y - xg(x, y)$,
   $\dot{y} = -x - yg(x, y)$
5. $\dot{x} = y + xy^2 - x^3 + 2xz^4$,
   $\dot{y} = -x - y^3 - 3x^2y + 3yz^4$,
   $\dot{z} = -\frac{5}{2}y^2z^3 - 2x^2z^3 - \frac{1}{2}z^7$

Exercise 1.62. (a) Determine the stability of all rest points for the following differential equations. For the unspecified scalar function g determine conditions so that the origin is a stable and/or asymptotically stable rest point.

1. $\ddot{x} + \epsilon\dot{x} + w^2x = 0$, $\epsilon > 0$, $w > 0$
2. $\ddot{x} + \sin x = 0$
3. $\ddot{x} + x - x^3 = 0$
4. $\ddot{x} + g(x) = 0$
5. $\ddot{x} + \epsilon\dot{x} + g(x) = 0$, $\epsilon > 0$
6. $\ddot{x} + \dot{x}^3 + x = 0$.

(b) The total energy is a good choice for the strict Lyapunov function required to study system 5. It almost works. Modify the total energy to obtain a strict Lyapunov function. Hint: See Exercise 2.80. (c) Prove the following refinement of Theorem 1.55: Suppose that x_0 is a rest point for the differential equation $\dot{x} = f(x)$ with flow ϕ_t and V is a Lyapunov function at x_0. If, in addition, there is a neighborhood W of the rest point x_0 such that for each point $p \in W \setminus \{x_0\}$, the function V is not constant on the set $\{\phi_t(p) : t \geq 0\}$, then x_0 is asymptotically stable (see Exercise 1.171). (d) Apply part (c) to system 5.

Exercise 1.63. Suppose that in addition to the hypotheses of Lyapunov's stability theorem 1.55, the strict Lyapunov function V is defined on all of $\mathbb{R}^n$ and $\lim_{|x|\to\infty} V(x) = \infty$. (a) Prove that the rest point is globally asymptotically stable; that is, the rest point is the ω-limit set of every point in $\mathbb{R}^n$. (b) Prove that if $\sigma > 0$, $b > 0$, and $0 < r < 1$, then the origin is globally asymptotically stable for the Lorenz system

$$\dot{x} = \sigma(y - x), \quad \dot{y} = rx - y - xz, \quad \dot{z} = xy - bz.$$

Exercise 1.64. Suppose that f is a function such that $f''(x) + f'(x) + f(x)^3 = 0$ for all $x \geq 0$. Show that $\lim_{x\to\infty} f(x) = 0$ and $\lim_{x\to\infty} f'(x) = 0$.

Exercise 1.65. [Basins of Attraction] Consider system 5 in the previous exercise, and note that if $g(0) = 0$ and $g'(0) > 0$, then there is a rest point at the origin that is asymptotically stable. Moreover, this fact can be proved by the principle of linearization. (a) Construct a strict Lyapunov function for this system. The construction of a strict Lyapunov function is not necessary to determine the stability of the rest point, but a Lyapunov function can be used to estimate the basin of attraction of the rest point; that is the set of all points in the space that are asymptotic to the rest point. Consider the (usual) first order system corresponding to the differential equation

$$\ddot{x} + \epsilon \dot{x} + x - x^3 = 0$$

for $\epsilon > 0$. (b) Describe the basin of attraction of the origin. (c) Define a subset of the basin of attraction, which you have described, and *prove* that it is contained in the basin of attraction. (d) Prove the following general theorem. Let x_0 be a rest point of the system $\dot{x} = f(x)$ in $\mathbb{R}^n$ with flow ϕ_t, and suppose that $V : U \to \mathbb{R}$ is a Lyapunov function at x_0. If B is a closed neighborhood of x_0 contained in U such that $\phi_t(b) \in B$ whenever $b \in B$ and $t \geq 0$, and there is no complete orbit in $B \setminus \{x_0\}$ on which V is constant, then x_0 is asymptotically stable and B is in the basin of attraction of x_0 (see [123]).

Note: In engineering practice, physical systems (for example a chemical plant or a power electronic system) are operated in steady state. When a disturbance occurs in the system, the control engineer wants to know if the system will return to the steady state. If not, she will have to take drastic action. Do you see why theorems of the type mentioned in this exercise (possible projects for the rest of your mathematical life) might have practical value?

1.8 Manifolds

In this section we will define the concept of a manifold as a generalization of a linear subspace of $\mathbb{R}^n$, and we will begin our discussion of the central role that manifolds play in the theory of differential equations.

Let us note that the fundamental definitions of calculus are local in nature. For example, the derivative of a function at a point is determined once we know the values of the function in some neighborhood of that point. This fact is the basis for the manifold concept: Informally, a manifold is a subset of $\mathbb{R}^n$ such that, for some fixed integer $k \geq 0$, each point in the subset has a neighborhood that is essentially the same as the Euclidean space $\mathbb{R}^k$. To make this definition precise we will have to define what is meant by a neighborhood in the subset, and we will also have to understand the meaning of the phrase "essentially the same as $\mathbb{R}^k$." But these notions should be intuitively clear: In effect, a neighborhood in the manifold is an open subset that is diffeomorphic to $\mathbb{R}^k$.

Points, lines, planes, arcs, spheres, and tori are examples of manifolds. Some of these manifolds have already been mentioned. Let us recall that a curve is a smooth function from an open interval of real numbers into $\mathbb{R}^n$. An arc is the image of a curve. Every solution of a differential equation is a curve; the corresponding orbit is an arc. Thus, every orbit of a differential equation is a manifold. As a special case, let us note that a periodic orbit is a one-dimensional torus. The statements in this paragraph are true as long as the objects are viewed without reference to how they are "embedded" in some larger space. For example, although an arc in $\mathbb{R}^n$ where $n > 1$ is a manifold, it may not be a submanifold of $\mathbb{R}^n$ because the arc may accumulate on itself (see Exercise 1.97). Thus, although the intuitive notion of a manifold is a useful way to begin our study of these objects, it should be clear from the last remarks that there are complications that can only be fully understood using the precise definitions. On the other hand, an intuitive understanding is sufficient to appreciate the importance of manifolds in the theory of differential equations.

We will discuss invariant manifolds, a precise approach to submanifolds of Euclidean space, tangent spaces, coordinate transformations, and polar coordinates as they relate to differential equations.

1.8.1 Introduction to Invariant Manifolds

Consider the differential equation

$$\dot{x} = f(x), \qquad x \in \mathbb{R}^n, \tag{1.17}$$

with flow ϕ_t, and let S be a subset of $\mathbb{R}^n$ that is a union of orbits of this flow. If a solution has its initial condition in S, then the corresponding orbit stays in S for all time, past and future. The concept of a set that is the union of orbits of a differential equation is formalized in the next definition.

Definition 1.66. A set $S \subseteq \mathbb{R}^n$ is called an *invariant set* for the differential equation (1.17) if, for each $x \in S$, the solution $t \mapsto \phi_t(x)$, defined on its maximal interval of existence, has its image in S. Alternatively, the orbit

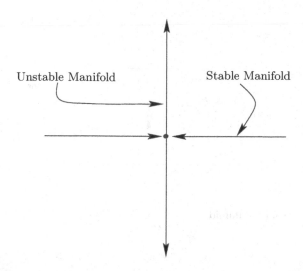

Figure 1.11: Stable and unstable manifolds for the linear saddle at the origin for the system $\dot{x} = -x$, $\dot{y} = y$.

passing through each $x \in S$ lies in S. In addition, S is called an *invariant manifold* if S is a manifold.

We will illustrate the notion of invariant manifolds for autonomous differential equations by describing two important examples: the stable, unstable, and center manifolds of a rest point, and the energy surfaces of Hamiltonian systems.

The stable manifold concept is perhaps best introduced by discussing a concrete example. Thus, let us consider the planar first order system

$$\dot{x} = -x, \qquad \dot{y} = y,$$

and note that the x-axis and the y-axis are invariant one-dimensional manifolds. The invariance of these sets follows immediately by inspection of the solution of the uncoupled linear system. Note that a solution with initial value on the x-axis approaches the rest point $(x, y) = (0, 0)$ as time increases to $+\infty$. On the other hand, a solution with initial value on the y-axis approaches the rest point as time decreases to $-\infty$. Solutions on the x-axis move toward the rest point; solutions on the y-axis move away from the rest point. For this example, the x-axis is called the stable manifold of the rest point, and the y-axis is called the unstable manifold (see Figure 1.11).

Similar invariant linear subspaces exist for all linear systems $\dot{x} = Ax$, $x \in \mathbb{R}^n$. In fact, the space $\mathbb{R}^n$ can always be decomposed as a direct sum of linear subspaces: the stable eigenspace (stable manifold) defined to be the A-invariant subspace of $\mathbb{R}^n$ such that the eigenvalues of the restriction of

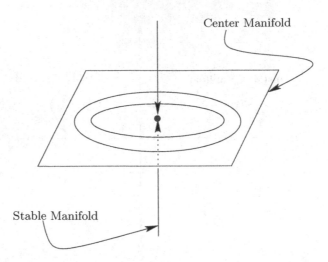

Figure 1.12: Phase portrait for a linear system with a one-dimensional stable and a two-dimensional center manifold.

A to this space are exactly the eigenvalues of A with negative real parts, the unstable eigenspace (unstable manifold) corresponding similarly to the eigenvalues of A with positive real parts, and the center eigenspace (center manifold) corresponding to the eigenvalues with zero real parts. It turns out that these linear subspaces are also invariant sets for the linear differential equation $\dot{x} = Ax$. Thus, they determine its phase portrait. For example, Figure 1.12 shows the phase portrait of a linear system on $\mathbb{R}^3$ with a one-dimensional stable manifold and a two-dimensional center manifold. Of course, some of these invariant sets might be empty. In particular, if A is infinitesimally hyperbolic (equivalently, if the rest point at the origin is hyperbolic), then the linear system has an empty center manifold at the origin.

Exercise 1.67. Discuss the existence of stable, unstable, and center manifolds for the linear systems with the following system matrices:

$$\begin{pmatrix} -1 & 1 & 0 \\ 0 & -1 & 0 \\ 0 & 0 & 2 \end{pmatrix}, \quad \begin{pmatrix} 1 & 2 & 3 \\ 4 & 5 & 6 \\ 7 & 8 & 9 \end{pmatrix}, \quad \begin{pmatrix} 0 & 1 & 0 \\ -1 & 0 & 0 \\ 0 & 0 & -2 \end{pmatrix}.$$

Two important theorems in the subject of differential equations, the stable manifold theorem and the center manifold theorem, will be proved in Chapter 4. We have the following formal definition.

Definition 1.68. The *stable manifold* of a rest point x_0 for an autonomous differential equation with (locally defined) flow ϕ_t is the set of all points x

in the domain of definition of ϕ_t such that $\lim_{t\to\infty} \phi_t(x) = x_0$. The *unstable manifold* of x_0 is the set of all points x in the domain of definition of ϕ_t such that $\lim_{t\to-\infty} \phi_t(x) = x_0$.

The stable manifold theorem states that a hyperbolic rest point has a unique stable manifold (respectively, unstable manifold) that is tangent to the corresponding stable (respectively, unstable) eigenspace of the corresponding linearized system at the rest point and that these invariant sets are indeed smooth manifolds.

The Hartman–Grobman theorem implies that a hyperbolic rest point has stable and unstable invariant sets that are homeomorphic images of the corresponding invariant manifolds for the corresponding linearized system, but it gives no indication that these invariant sets are smooth manifolds.

The existence of stable and unstable invariant manifolds is essential to our understanding of many features of the dynamics of differential equations. For example, their existence provides a theoretical basis for determining the analytic properties of the flow of a differential equation in the neighborhood of a hyperbolic rest point. They also serve to bound other invariant regions in the phase space. Thus, the network of all stable and unstable manifolds forms the "skeleton" for the phase portrait. Finally, the existence of the stable and unstable manifolds in the phase space, especially their intersection properties, lies at the heart of an explanation of the complex motions associated with many nonlinear ordinary differential equations. In particular, this phenomenon is fundamental in the study of deterministic chaos (see Chapter 6).

For rest points of a differential equation that are not hyperbolic, the center manifold theorem states the existence of an invariant manifold tangent to the corresponding center eigenspace. This center manifold is not necessarily unique, but the differential equation has the same (arbitrarily complicated) phase portrait when restricted to any one of the center manifolds at the same rest point. In particular, center manifolds cannot be characterized by a simple dynamical property as in Definition 1.68. Analysis using center manifolds is often required to understand many of the most delicate problems that arise in the theory and applications of differential equations. For example, the existence and smoothness properties of center manifolds are foundational results in bifurcation theory (see Chapter 8).

Other types of invariant sets, for example, periodic orbits can have stable manifolds, unstable manifolds and center manifolds. Indeed, the extension of Definition 1.68 to a general invariant set is clear.

Exercise 1.69. Determine the stable and unstable manifolds for the rest point of the system

$$\dot{x} = 2x - (2+y)e^y, \quad \dot{y} = -y.$$

Hint: See Exercise 1.6.

Exercise 1.70. (a) Determine a stable and a center manifold for the rest point of the system

$$\dot{x} = x^2, \quad \dot{y} = -y.$$

(b) Show that the system has infinitely many center manifolds.

Invariant manifolds, called energy surfaces, are useful in the study of Hamiltonian systems of differential equations. To define this important class of differential equations, let $H : \mathbb{R}^n \times \mathbb{R}^n \to \mathbb{R}$ be a smooth function given by

$$(q_1, \dots, q_n, p_1, \dots, p_n) \mapsto H(q_1, \dots, q_n, p_1, \dots, p_n),$$

and define the associated Hamiltonian system on $\mathbb{R}^{2n}$ with Hamiltonian H by

$$\dot{q}_i = \frac{\partial H}{\partial p_i}, \quad \dot{p}_i = -\frac{\partial H}{\partial q_i}, \quad i = 1, \dots, n.$$

Let us note that the dimension of the phase space of a Hamiltonian system is required to be even. The reason for this restriction will soon be made clear.

As a prototypical example of a Hamiltonian system, let $H : \mathbb{R}^2 \to \mathbb{R}$ be given by $H(x, y) := \frac{1}{2}(y^2 + \omega^2 x^2)$. The associated Hamiltonian system is the harmonic oscillator

$$\dot{x} = y, \quad \dot{y} = -\omega^2 x.$$

More generally, suppose that $U : \mathbb{R}^n \to \mathbb{R}$ and let $H : \mathbb{R}^n \times \mathbb{R}^n \to \mathbb{R}$ be given by

$$H(q, p) = \frac{p^2}{2m} + U(q)$$

where $p^2 := p_1^2 + \cdots + p_n^2$. A Hamiltonian in this form is called a *classical Hamiltonian*. The corresponding Hamiltonian system

$$\dot{q} = \frac{1}{m}p, \quad \dot{p} = -\operatorname{grad} U(q)$$

is equivalent to Newton's equation of motion for a particle influenced by a conservative force (see Exercise 1.60). The vector quantity $p := m\dot{q}$ is called the (generalized) momentum, the function U is called the potential energy, and the function $p \mapsto \frac{1}{2m}p^2 = \frac{m}{2}\dot{q}^2$ is called the kinetic energy.

The *configuration space* for a classical mechanical system is the space consisting of all possible positions of the system, and the corresponding Hamiltonian system is said to have n *degrees of freedom* if the configuration space is locally specified by n coordinates $(q_1, \ldots, q_n)$. For example, for the pendulum, the configuration space can be taken to be $\mathbb{R}$ with the coordinate q_1 specifying the angular position of the bob relative to the downward vertical. It is a system with one degree of freedom. Of course, for this example, the physical positions are specified by the angular coordinate q_1 modulo 2π. Thus, the configuration space can also be viewed as a nonlinear manifold—namely, the unit circle in the plane. This is yet another way in which manifolds arise in the study of mechanical systems.

The *phase space* of a Hamiltonian system is the subset of $\mathbb{R}^n \times \mathbb{R}^n$ of all positions and momenta specified by the coordinates $(q_1, \ldots, q_n, p_1, \ldots, p_n)$. The dimension of the phase space is therefore even; it is the space in which the Hamiltonian system evolves. The *state space* is also a subset of $\mathbb{R}^n \times \mathbb{R}^n$, but it is the space of positions and velocities with the coordinates $(q_1, \ldots, q_n, \dot{q}_1, \ldots, \dot{q}_n)$ (see Chapter 3).

For $c \in \mathbb{R}$ and the Hamiltonian $H : \mathbb{R}^n \times \mathbb{R}^n \to \mathbb{R}$, the corresponding *energy surface* with energy c is defined to be the set

$$S_c = \{(q, p) \in \mathbb{R}^n \times \mathbb{R}^n : H(q, p) = c\}.$$

If $\operatorname{grad} H(q, p) \neq 0$ for each $(q, p) \in S_c$, then the set S_c is called a *regular energy surface*.

Note that the vector field given by

$$\operatorname{grad} H = (\frac{\partial H}{\partial q}, \frac{\partial H}{\partial p})$$

is orthogonal to the Hamiltonian vector field given by

$$(\frac{\partial H}{\partial p}, -\frac{\partial H}{\partial q})$$

at each point in the phase space. Thus, the Hamiltonian vector field is everywhere tangent to each regular energy surface. As a consequence of this fact—a proof will be given later in this section—every energy surface S_c is an invariant set for the flow of the corresponding Hamiltonian system. Moreover, every regular energy surface is an invariant manifold.

The structure of energy surfaces and their invariance is important. Indeed, the phase space of a Hamiltonian system is the union of its energy surfaces. Or, as we say, the space is *foliated* by its energy surfaces. Moreover, each regular energy surface of a Hamiltonian system with n degrees of freedom has "dimension" $2n - 1$. Thus, we can reduce the dimension of the phase space by studying the flow of the original Hamiltonian system restricted to each of these invariant subspaces. For example, the analysis of a Hamiltonian system with one degree of freedom can be reduced to the consideration of just one space dimension where the solution of the Hamiltonian differential equation can be reduced to a *quadrature*. To see what this means, consider the classical Hamiltonian $H(q, p) = \frac{1}{2}p^2 + U(q)$

on $\mathbb{R} \times \mathbb{R}$ and a regular energy surface of H with energy h. Notice that, by using the Hamiltonian differential equations and the energy relation, we have the following scalar differential equations

$$\dot{q} = p = \frac{dq}{dt} = \pm(2(h - U(q)))^{1/2}$$

for solutions whose initial conditions are on this energy surface. By separation of variables and a specification of the initial condition, the ambiguous sign is determined and the solution of the corresponding scalar differential equation is given implicitly by the integral (=quadrature)

$$\int_{q(0)}^{q(t)} (2(h - U(q)))^{-1/2} \, dq = \pm t.$$

This result "solves" the original system of Hamiltonian differential equations. The same idea works for systems with several degrees of freedom, only the equations are more complicated.

Let us also note that the total energy of a Hamiltonian system might not be the only conserved quantity. In fact, if F is a function on the phase space with the property that grad $F(q, p)$ is orthogonal to the Hamiltonian vector field at every point (q, p) in an open subset of the phase space, then the level sets of F are also invariant sets. In this case F is called an *integral*, or *first integral*, of the Hamiltonian system. Thus, the intersection of an energy surface and a level set of F must also be invariant, and, as a consequence, the space is foliated with $(2n - 2)$-dimensional invariant sets. If there are enough first integrals, then the solution of the original system can be expressed in quadratures. In fact, for an n-degree-of-freedom Hamiltonian system, it suffices to determine n "independent" first integrals (see [12, §49]). For these reasons, it should be clear that energy surfaces, or more generally, level sets of first integrals, are important objects that are worthy of study. They are prime examples of smooth manifolds.

While the notion of an energy surface is naturally associated with Hamiltonian systems, the underlying idea for proving the invariance of energy surfaces easily extends to general autonomous systems. In fact, if $\dot{x} = f(x)$ is an autonomous system with $x \in \mathbb{R}^n$ and the function $G : \mathbb{R}^n \to \mathbb{R}$ is such that the vector grad $G(x)$ is orthogonal to $f(x)$ for all x in some open subset of $\mathbb{R}^n$, then every level set of G that is contained in this open set is invariant. Thus, just as for Hamiltonian systems, some of the dynamical properties of the differential equation $\dot{x} = f(x)$ can be studied by restricting attention to a level set of G, a set that has codimension one in the phase space (see Exercise 1.77).

Exercise 1.71. Find the Hamiltonian for a first order system equivalent to the model equation for the pendulum given by $\ddot{\theta} + k \sin \theta = 0$ where k is a parameter. Describe the energy surfaces.

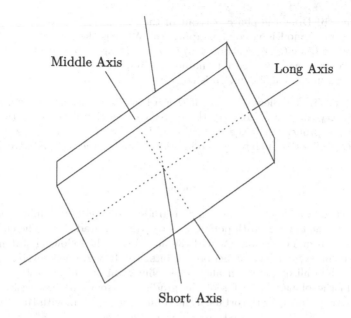

Figure 1.13: A rigid body and its three axes of symmetry

Exercise 1.72. Reduce the solution of the harmonic oscillator $H(q, p) = \frac{1}{2}(p^2 + \omega^2 q^2)$ where $\omega > 0$ to a quadrature on each of its regular energy surfaces and carry out the integration explicitly. (This is not the simplest way to solve the equations of motion, but you will learn a valuable method that is used, for example, in the construction of the solution of the equations of motion for the Hamiltonian system mentioned in the previous exercise.)

Exercise 1.73. (a) Show that

$$\dot{I}_1 = I_1 \cos(\theta_1 - \theta_2),$$
$$\dot{I}_2 = -I_1 \cos(\theta_1 - \theta_2),$$
$$\dot{\theta}_1 = -1 - \sin(\theta_1 - \theta_2),$$
$$\dot{\theta}_2 = 1$$

is a Hamiltonian system. (b) Find a first integral that is independent of the Hamiltonian.

Exercise 1.74. (a) The Hamiltonian system with Hamiltonian $H = q_2 p_1 - q_1 p_2 + q_4 p_3 - q_3 p_4 + q_2 q_3$ has a rest point at the origin (see [145, p. 212]). Linearize at the origin and determine the eigenvalues. What can you conclude about the stability of the rest point? (b) Prove that the rest point at the origin is unstable.

Exercise 1.75. [Basins of Attraction] Consider the differential equation

$$\ddot{x} + \epsilon \dot{x} - x + x^3 = 0$$

with parameter ϵ. (a) Show that the system with $\epsilon = 0$ corresponds to a classical Hamiltonian system with a double-well potential and draw the phase portrait of this system. (b) Draw the phase portrait of the system for $\epsilon > 0$. Note: In this case, the term $\epsilon \dot{x}$ models viscous damping. (c) What is the fate of the solution with initial condition $(x(0), \dot{x}(0)) = (4, 0)$ for $\epsilon = 0.1$? Note: To solve this problem you will probably have to resort to numerics. How do we know that the result obtained by a numerical simulation is correct?

Exercise 1.76. [Gradient Systems] If H is a Hamiltonian, then the vector field grad H is everywhere orthogonal to the corresponding Hamiltonian vector field. What are the properties of the flow of grad H? More generally, for a smooth function $G : \mathbb{R}^n \to \mathbb{R}$ (maybe n is odd), let us define the associated gradient system

$$\dot{x} = \operatorname{grad} G(x).$$

Because a conservative force is the negative gradient of a potential, many authors define the gradient system with potential G to be $\dot{x} = -\operatorname{grad} G(x)$. The choice of sign simply determines the direction of the flow. Prove the following statements: (a) A gradient system has no periodic orbits. (b) If a gradient system has a rest point, then all of the eigenvalues of its linearization at the rest point are real. (c) In the plane, the orbits of the gradient system with potential G are orthogonal trajectories for the orbits of the Hamiltonian system with Hamiltonian G. (d) If $x_0 \in \mathbb{R}^n$ is an isolated maximum of the function $G : \mathbb{R}^n \to \mathbb{R}$, then x_0 is an asymptotically stable rest point of the corresponding gradient system $\dot{x} = \operatorname{grad} G(x)$.

Exercise 1.77. [Rigid Body Motion] A system that is not Hamiltonian, but closely related to this class, is given by Euler's equations for rigid body motion. The angular momentum $M = (M_1, M_2, M_3)$ of a rigid body, relative to a coordinate frame rotating with the body with axes along the principal axes of the body and with origin at its center of mass, is related to the angular velocity vector Ω by $M = A\Omega$, where A is a symmetric matrix called the *inertia matrix*. Euler's equation is $\dot{M} = M \times \Omega$. Equivalently, the equation for the angular velocity is $A\dot{\Omega} = (A\Omega) \times \Omega$. If A is diagonal with diagonal components (moments of inertia) (I_1, I_2, I_3), show that Euler's equations for the components of the angular momentum are given by

$$\dot{M}_1 = -\left(\frac{1}{I_2} - \frac{1}{I_3}\right) M_2 M_3,$$

$$\dot{M}_2 = \left(\frac{1}{I_1} - \frac{1}{I_3}\right) M_1 M_3,$$

$$\dot{M}_3 = -\left(\frac{1}{I_1} - \frac{1}{I_2}\right) M_1 M_2.$$

Assume that $0 < I_1 \leq I_2 \leq I_3$. Find some invariant manifolds for this system. Can you use your results to find a qualitative description of the motion? As a physical example, take this book and hold its covers together with a rubber band. Then, toss the book vertically three times, imparting a rotation in turn about each of its axes of symmetry (see Figure 1.13). Are all three rotary motions Lyapunov stable? Do you observe any other interesting phenomena associated with the motion? For example, pay attention to the direction of the front cover of the book after each toss. Hint: Look for invariant quadric surfaces; that is, manifolds defined as level sets of quadratic polynomials (first integrals) in the variables (M_1, M_2, M_3). For example, show that the kinetic energy given by $\frac{1}{2}\langle A\Omega, \Omega \rangle$ is constant along orbits. The total angular momentum (length of the angular momentum) is also conserved. For a complete mathematical description of rigid body motion, see [12]. For a mathematical description of the observed "twist" in the rotation of the tossed book, see [20]. Note that Euler's equations do not describe the motion of the book in space. To do so would require a functional relationship between the coordinate system rotating with the body and the position coordinates relative to a fixed coordinate frame in space.

1.8.2 Smooth Manifolds

Because the modern definition of a smooth manifold can appear quite formidable at first sight, we will formulate a simpler equivalent definition for the class of manifolds called the *submanifolds* of $\mathbb{R}^n$. Fortunately, this class is rich enough to contain the manifolds that are met most often in the study of differential equations. In fact, every manifold can be "embedded" as a submanifold of some Euclidean space. Thus, the class that we will study can be considered to contain all manifolds.

Recall that a manifold is supposed to be a set that is locally the same as $\mathbb{R}^k$. Thus, whatever is meant by "locally the same," every open subset of $\mathbb{R}^k$ must be a manifold.

If $W \subseteq \mathbb{R}^k$ is an open set and $g : W \to \mathbb{R}^{n-k}$ is a smooth function, then the graph of g is the subset of $\mathbb{R}^n$ defined by

$$\mathrm{graph}(g) := \{(w, g(w)) \in \mathbb{R}^n : w \in W\}.$$

The set $\mathrm{graph}(g)$ is the same as $W \subseteq \mathbb{R}^k$ up to a nonlinear change of coordinates. By this we mean that there is a smooth map G with domain W and image $\mathrm{graph}(g)$ such that G has a smooth inverse. In fact, such a map $G : W \to \mathrm{graph}(g)$ is given by $G(w) = (w, g(w))$. Clearly, G is smooth. Its inverse is the linear projection on the first k coordinates of the point $(w, g(w)) \in \mathrm{graph}(g)$; that is, $G^{-1}(w, g(w)) = w$. Thus, G^{-1} is smooth as well.

Open subsets and graphs of smooth functions are the prototypical examples of what we will call submanifolds. But these classes are too restrictive; they include objects that are *globally* the same as some Euclidean space. The unit circle $\mathbb{T}$ in the plane, also called the one-dimensional torus, is an example of a submanifold that is not of this type. Indeed, $\mathbb{T} := \{(x, y) : x^2 + y^2 = 1\}$ is not the graph of a scalar function defined on an open subset of $\mathbb{R}$. On the other hand, every point of $\mathbb{T}$ is contained in a *neighborhood in $\mathbb{T}$ that is the graph of such a function*. In other words, $\mathbb{T}$ is *locally* the same as $\mathbb{R}$. In fact, each point in $\mathbb{T}$ is in one of the four sets

$$S_{\pm} := \{(x, y) \in \mathbb{R}^2 : y = \pm\sqrt{1 - x^2}, \quad |x| < 1\},$$
$$S^{\pm} := \{(x, y) \in \mathbb{R}^2 : x = \pm\sqrt{1 - y^2}, \quad |y| < 1\}.$$

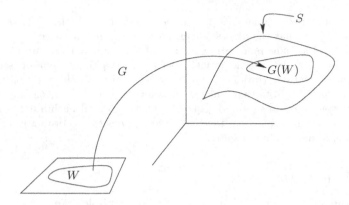

Figure 1.14: A chart for a two-dimensional submanifold in $\mathbb{R}^3$.

Submanifolds of $\mathbb{R}^n$ are subsets with the same basic property: Every point in the subset is in a neighborhood that is the graph of a smooth function.

To formalize the submanifold concept for subsets of $\mathbb{R}^n$, we must deal with the problem that, in the usual coordinates of $\mathbb{R}^n$, not all graphs are given by sets of the form

$$\{(x_1,\dots,x_k,g_{k+1}(x_1,\dots,x_k),\dots,g_n(x_1,\dots,x_k)):$$
$$(x_1,\dots,x_k) \in W \subseteq \mathbb{R}^k\}.$$

Rather, we must allow, as in the example provided by $\mathbb{T}$, for graphs of functions that are not functions of the first k coordinates of $\mathbb{R}^n$. To overcome this technical difficulty we will build permutations of the variables into our definition.

Definition 1.78. Suppose that $S \subseteq \mathbb{R}^n$ and $x \in S$. The pair (W,G) where W is an open subset of $\mathbb{R}^k$ for some $k \leq n$ and $G : W \to \mathbb{R}^n$ is a smooth function is called a *k-dimensional submanifold chart for S at x* (see Figure 1.14) if there is an open set $U \subseteq \mathbb{R}^n$ with $x \in U \cap S$ such that $U \cap S = G(W)$ and one of the following two properties is satisfied:
1) The integer k is equal to n and G is the identity map.
2) The integer k is less than n and G has the form

$$G(w) = A\binom{w}{g(w)}$$

where $g : W \to \mathbb{R}^{n-k}$ is a smooth function and A is a nonsingular $n \times n$ matrix.

Definition 1.79. The set $S \subseteq \mathbb{R}^n$ is called a *k-dimensional smooth submanifold* of $\mathbb{R}^n$ if there is a k-dimensional submanifold chart for S at every point x in S.

The map G in a submanifold chart (W, G) is called a *submanifold coordinate map*. If S is a submanifold of $\mathbb{R}^n$, then (even though we have not yet defined the concept), let us also call a submanifold S of $\mathbb{R}^n$ a smooth manifold.

As an example, let us show that $\mathbb{T}$ is a one-dimensional manifold. Consider a point in the subset $S^+ = \{(x, y) : x = \sqrt{1 - y^2}, |y| < 1\}$ of $\mathbb{T}$. Define the set $W := \{t \in \mathbb{R} : |t| < 1\}$, the function $g : W \to \mathbb{R}$ by $g(t) = \sqrt{1 - t^2}$, the set $U := \{(x, y) \in \mathbb{R}^2 : (x - 1)^2 + y^2 < 2\}$, and the matrix

$$A := \begin{pmatrix} 0 & 1 \\ 1 & 0 \end{pmatrix}.$$

Then we have

$$\mathbb{T} \cap U = \left\{ \begin{pmatrix} x \\ y \end{pmatrix} \in \mathbb{R}^2 : \begin{pmatrix} x \\ y \end{pmatrix} = \begin{pmatrix} 0 & 1 \\ 1 & 0 \end{pmatrix} \begin{pmatrix} t \\ g(t) \end{pmatrix}, \quad t \in W \right\}.$$

Similarly, $\mathbb{T}$ is locally the graph of a smooth function at points in the subsets S^- and $S_{\pm}$, as required.

A simple but important result about submanifold charts is the following proposition.

Proposition 1.80. *If (W, G) is a submanifold chart for a k-dimensional submanifold of $\mathbb{R}^n$, then the function $G : W \to G(W) \subseteq S$ is invertible. Moreover, the inverse of G is the restriction of a smooth function that is defined on all of $\mathbb{R}^n$.*

Proof. The result is obvious if $k = n$. If $k < n$, then define $\Pi : \mathbb{R}^n \to \mathbb{R}^k$ to be the linear projection on the first k-coordinates; that is, $\Pi(x_1, \ldots, x_n) = (x_1, \ldots, x_k)$, and define

$$F : G(W) \to W$$

by

$$F(s) = \Pi A^{-1} s.$$

Clearly, F is smooth as a function defined on all of $\mathbb{R}^n$. Also, if $w \in W$, then

$$F \circ G(w) = F\left(A\begin{pmatrix} w \\ g(w) \end{pmatrix}\right) = \Pi A^{-1} A \begin{pmatrix} w \\ g(w) \end{pmatrix} = w.$$

If $s \in G(W)$, then $s = A\begin{pmatrix} w \\ g(w) \end{pmatrix}$ for some $w \in W$. Hence, we also have

$$G(F(s)) = G(w) = s.$$

This proves that F is the inverse of G. $\square$

If S is a submanifold, then we can use the submanifold charts to define the open subsets of S.

Definition 1.81. Suppose that S is a submanifold. The *open* subsets of S are all possible unions of all sets of the form $G(W)$ where (W, G) is a submanifold chart for S.

The next proposition is an immediate consequence of the definitions.

Proposition 1.82. *If S is a submanifold of $\mathbb{R}^n$ and if V is an open subset of S, then there is an open set U of $\mathbb{R}^n$ such that $V = S \cap U$; that is, the topology defined on S using the submanifold charts agrees with the subspace topology on S.*

As mentioned above, one of the main reasons for defining the manifold concept is to distinguish those subsets of $\mathbb{R}^n$ on which we can use the calculus. To do so, let us first make precise the notion of a smooth function.

Definition 1.83. Suppose that S_1 is a submanifold of $\mathbb{R}^m$, S_2 is a submanifold of $\mathbb{R}^n$, and F is a function $F : S_1 \to S_2$. We say that F is *differentiable* at $x_1 \in S_1$ if there are submanifold charts (W_1, G_1) at x_1 and (W_2, G_2) at $F(x_1)$ such that the map $G_2^{-1} \circ F \circ G_1 : W_1 \to W_2$ is differentiable at $G_1^{-1}(x_1) \in W_1$. If F is differentiable at each point of an open subset V of S_1, then we say that F is differentiable on V.

Definition 1.84. Suppose that S_1 and S_2 are manifolds. A smooth function $F : S_1 \to S_2$ is called a *diffeomorphism* if there is a smooth function $H : S_2 \to S_1$ such that $H(F(s)) = s$ for every $s \in S_1$ and $F(H(s)) = s$ for every $s \in S_2$. The function H is called the inverse of F and is denoted by F^{-1}.

With respect to the notation in Definition 1.83, we have defined the concept of differentiability for the function $F : S_1 \to S_2$, but we have not yet defined what we mean by its derivative. We have, however, determined the derivative relative to the submanifold charts used in the definition. Indeed, the *local representative* of the function F is given by $G_2^{-1} \circ F \circ G_1$, a function defined on an open subset of a Euclidean space with range in another Euclidean space. By definition, the *local representative of the derivative* of F relative to the given submanifold charts is the usual derivative in Euclidean space of this local representative of F. In the next subsection, we will interpret the derivative of F without regard to the choice of a submanifold chart; that is, we will give a coordinate-free definition of the derivative of F (see also Exercise 1.85).

Exercise 1.85. Prove: The differentiability of a function defined on a manifold does not depend on the choice of submanifold chart.

Exercise 1.86. (a) Show that $\dot{\theta} = f(\theta)$ can be viewed as a (smooth) differential equation on the unit circle if and only if f is periodic. To be compatible with the usual (angular) coordinate on the circle it is convenient to consider only 2π-periodic functions (see Section 1.8.5). (b) Describe the bifurcations that occur for the family $\dot\theta = 1 - \lambda \sin\theta$ with $\lambda \geq 0$. (c) For each $\lambda < 1$, the corresponding differential equation has a periodic orbit. Determine the period of this periodic orbit and describe the behavior of the period as $\lambda \to 1$ (see [218, p. 98]).

We have used the phrase "smooth function" to refer to a function that is continuously differentiable. In view of Definition 1.83, the smoothness of a function defined on a manifold is determined by the smoothness of its local representatives—functions that are defined on open subsets of Euclidean spaces. It is clear that smoothness of all desired orders can be defined in the same manner by imposing the requirement on local representatives. More precisely, if F is a function defined on a manifold S, then we will say that F is an element of $C^r(S)$, for r a nonnegative integer, $r = \infty$, or $r = \omega$, provided that at each point of S there is a local representative of F all of whose partial derivatives up to and including those of order r are continuous. If $r = \infty$, then all partial derivatives are required to be continuous. If $r = \omega$, then all local representatives are all required to have convergent power series representations valid in a neighborhood of each point of their domains. A function in C^ω is called *real analytic*.

In the subject of differential equations, specifying the minimum number of derivatives of a function required to obtain a result often obscures the main ideas that are being illustrated. Thus, as a convenient informality, we will often use the phrase "smooth function" to mean that the function in question has as many continuous derivatives as needed. In cases where the exact requirement for the number of derivatives is essential, we will refer to the appropriate class of C^r functions.

The next definition formalizes the concept of a coordinate system.

Definition 1.87. Suppose that S is a k-dimensional submanifold. The pair (V, Ψ) is called a *coordinate system* or *coordinate chart* on S if V is an open subset of S, W is an open subset of $\mathbb{R}^k$, and $\Psi : V \to W$ is a diffeomorphism.

Exercise 1.88. Prove: If (W, G) is a submanifold chart for a manifold S, then $(G(W), G^{-1})$ is a coordinate chart on S.

The abstract definition of a manifold is based on the concept of coordinate charts. Informally, a set S together with a collection of subsets $\mathcal{S}$ is defined to be a k-dimensional manifold if every point of S is contained in at least one set in $\mathcal{S}$ and if, for each member V of $\mathcal{S}$, there is a corresponding open subset W of $\mathbb{R}^k$ and a function $\Psi : V \to W$ that is bijective. If two such subsets V_1 and V_2 overlap, then the domain of the map

$$\Psi_1 \circ \Psi_2^{-1} : \Psi_2(V_1 \cap V_2) \to W_1$$

is an open subset of $\mathbb{R}^k$ whose range is contained in an open subset of $\mathbb{R}^k$. The set S is called a manifold provided that all such "overlap maps" are smooth (see [120] for the formal definition). This abstract notion of a manifold has the advantage that it does not require a manifold to be a subset of a Euclidean space.

Exercise 1.89. Prove: If $F : \mathbb{R}^m \to \mathbb{R}^n$ is smooth and $F(S_1) \subseteq S_2$ for submanifolds S_1 and S_2, then the restriction of F to S_1 is differentiable.

Exercise 1.90. Prove: If $\alpha \in \mathbb{R}$, then the map $\mathbb{T} \to \mathbb{T}$ given by

$$(x, y) \mapsto (x \cos \alpha - y \sin \alpha, x \sin \alpha + y \cos \alpha)$$

is a diffeomorphism.

Now that we know the definition of a manifold, we are ready to prove that linear subspaces of $\mathbb{R}^n$ and regular level sets of smooth functions are manifolds.

Proposition 1.91. *A linear subspace of $\mathbb{R}^n$ is a submanifold.*

Proof. Let us suppose that S is the span of the k linearly independent vectors $v_1, \ldots, v_k$ in $\mathbb{R}^n$. We will show that S is a k-dimensional submanifold of $\mathbb{R}^n$.

Let $e_1, \ldots, e_n$ denote the standard basis of $\mathbb{R}^n$. By a basic result from linear algebra, there is a set consisting of $n - k$ standard basis vectors $f_{k+1}, \ldots, f_n$ such that the vectors

$$v_1, \ldots, v_k, f_{k+1}, \ldots, f_n$$

are a basis for $\mathbb{R}^n$. (Why?) Let us denote the remaining set of standard basis vectors by $f_1, \ldots, f_k$. For each $j = 1, \ldots, k$, there are scalars λ_i^j and μ_i^j such that

$$f_j = \sum_{i=1}^k \lambda_i^j v_i + \sum_{i=k+1}^n \mu_i^j f_i.$$

Hence, if $(t_1, \ldots, t_k) \in \mathbb{R}^k$, then the vector

$$\sum_{j=1}^k t_j f_j - \sum_{j=1}^k t_j \left(\sum_{i=k+1}^n \mu_i^j f_i \right) = \sum_{j=1}^k t_j \left(\sum_{i=1}^k \lambda_i^j v_i \right)$$

is in S; and, relative to the basis $f_1, \cdots, f_n$, the vector

$$\left(t_1, \ldots, t_k, -\sum_{j=1}^k t_j \mu_{k+1}^j, \ldots, -\sum_{j=1}^k t_j \mu_n^j \right)$$

is in S.

Define $g : \mathbb{R}^k \to \mathbb{R}^{n-k}$ by

$$g(t_1, \ldots, t_k) := \Big(-\sum_{j=1}^{k} t_j \mu_{k+1}^j, \ldots, -\sum_{j=1}^{k} t_j \mu_n^j \Big)$$

and let A denote the permutation matrix given by $Ae_j = f_j$. It follows that the pair $(\mathbb{R}^k, G)$, where $G : \mathbb{R}^k \to \mathbb{R}^n$ is defined by

$$G(w) = A\Big(\begin{matrix} w \\ g(w) \end{matrix} \Big),$$

is a k-dimensional submanifold chart such that $G(\mathbb{R}^k) = \mathbb{R}^n \cap S$. In fact, by the construction, it is clear that the image of G is a linear subspace of S. Moreover, because the image of G has dimension k as a vector space, the subspace $G(\mathbb{R}^k)$ is equal to S. □

As mentioned previously, linear subspaces often arise as invariant manifolds of differential equations. For example, consider the differential equation given by $\dot{x} = Ax$ where $x \in \mathbb{R}^n$ and A is an $n \times n$ matrix. If S is an invariant subspace for the matrix A, for example, one of its generalized eigenspaces, then, by Proposition 1.91, S is a submanifold of $\mathbb{R}^n$. Also, S is an invariant set for the corresponding linear system of differential equations. Although a complete proof of this proposition requires some results from linear systems theory that will be presented in Chapter 2, the essential features of the proof are simply illustrated in the special case where the linear transformation A restricted to S has a complete set of eigenvectors. In other words, S is a k-dimensional subspace of $\mathbb{R}^n$ spanned by k linearly independent eigenvectors $v_1, \ldots, v_k$ of A. Under this assumption, if $Av_i = \lambda_i v_i$, then $t \to e^{\lambda_i t} v_i$ is a solution of $\dot{x} = Ax$. Also, note that $e^{\lambda_i t} v_i$ is an eigenvector of A for each $t \in \mathbb{R}$. Therefore, if $x_0 \in S$, then there are scalars $(a_1, \ldots, a_k)$ such that $x_0 = \sum_{i=1}^{k} a_i v_i$ and

$$t \mapsto \sum_{i=1}^{k} e^{\lambda_i t} a_i v_i$$

is the solution of the ordinary differential equation with initial condition $x(0) = x_0$. Clearly, the corresponding orbit stays in S for all $t \in \mathbb{R}$.

Linear subspaces can be invariant sets for nonlinear differential equations. For example, consider the Volterra–Lotka system

$$\dot{x} = x(a - by), \qquad \dot{y} = y(cx - d).$$

In case a, b, c, and d are all positive, this system models the interaction of the population y of a predator and the population x of its prey. For this system, the x-axis and the y-axis are each invariant sets. Indeed, suppose that $(0, y_0)$ is a point on the y-axis corresponding to a population of predators with no prey, then $t \mapsto (0, e^{-dt}y_0)$ is the solution of the system starting at this point that models this population for all future time. This solution stays on the y-axis for all time, and, as there are is no spontaneous generation of prey, the predator population dies out in positive time.

Let us now discuss level sets of functions. Recall that the *level set with energy c* of a smooth function $H : \mathbb{R}^n \to \mathbb{R}$ is the set

$$S_c := \{x \in \mathbb{R}^n : H(x) = c\}.$$

Moreover, the level set S_c is called a *regular level set* if $\text{grad } H(x) \neq 0$ for each $x \in S_c$.

Proposition 1.92. *If $H : \mathbb{R}^n \to \mathbb{R}$ is a smooth function, then each of its regular level sets is an $(n-1)$-dimensional submanifold of $\mathbb{R}^n$.*

It is instructive to outline a proof of this result because it provides our first application of a nontrivial and very important theorem from advanced calculus, namely, the implicit function theorem.

Suppose that S_c is a regular level set of H, choose $a \in S_c$, and define $F : \mathbb{R}^n \to \mathbb{R}$ by

$$F(x) = H(x) - c.$$

Let us note that $F(a) = 0$. Also, because $\text{grad } H(a) \neq 0$, there is at least one integer $1 \leq i \leq n$ such that the corresponding partial derivative $\partial F / \partial x_i$ does not vanish when evaluated at a. For notational convenience let us suppose that $i = 1$. All other cases can be proved in a similar manner.

We are in a typical situation: We have a function $F : \mathbb{R} \times \mathbb{R}^{n-1} \to \mathbb{R}$ given by $(x_1, x_2, \ldots, x_n) \mapsto F(x_1, \ldots, x_n)$ such that

$$F(a_1, \ldots, a_n) = 0, \qquad \frac{\partial F}{\partial x_1}(a_1, a_2, \ldots, a_n) \neq 0.$$

This calls for an application of the implicit function theorem. A preliminary version of the theorem is stated here; a more general version will be proved later (see Theorem 1.259).

If $f : \mathbb{R}^\ell \times \mathbb{R}^m \to \mathbb{R}^n$ is given by $(p, q) \mapsto f(p, q)$, then, for fixed $b \in \mathbb{R}^m$, consider the function $\mathbb{R}^\ell \to \mathbb{R}^n$ defined by $p \mapsto f(p, b)$. Its derivative at $a \in \mathbb{R}^\ell$ will be denoted by $f_p(a, b)$. Of course, with respect to the usual bases of $\mathbb{R}^\ell$ and $\mathbb{R}^n$, this derivative is represented by an $n \times \ell$ matrix of partial derivatives.

Theorem 1.93 (Implicit Function Theorem). *Suppose that $F : \mathbb{R}^m \times \mathbb{R}^k \to \mathbb{R}^m$ is a smooth function given by $(p, q) \mapsto F(p, q)$. If (a, b) is in $\mathbb{R}^m \times \mathbb{R}^k$ such that $F(a, b) = 0$ and the linear transformation $F_p(a, b) : \mathbb{R}^m \to \mathbb{R}^m$ is invertible, then there exist two open metric balls $U \subseteq \mathbb{R}^m$ and $V \subseteq \mathbb{R}^k$ with $(a, b) \in U \times V$ together with a smooth function $g : V \to U$ such that $g(b) = a$ and $F(g(v), v) = 0$ for each $v \in V$. Moreover, if $(u, v) \in U \times V$ and $F(u, v) = 0$, then $u = g(v)$.*

Continuing with our outline of the proof of Proposition 1.92, let us observe that, by an application of the implicit function theorem to F, there is an open set $Z \subseteq \mathbb{R}$ with $a_1 \in Z$, an open set $W \subseteq \mathbb{R}^{n-1}$ containing the point $(a_2, \dots, a_n)$, and a smooth function $g : W \to Z$ such that $g(a_2, \dots, a_n) = a_1$ and

$$H(g(x_2, \dots, x_n), x_2, \dots, x_n) - c \equiv 0.$$

The set

$$U := \{(x_1, \dots, x_n) \in \mathbb{R}^n : x_1 \in Z \text{ and } (x_2, \dots, x_n) \in W\} = Z \times W$$

is open. Moreover, if $x = (x_1, \dots, x_n) \in S_c \cap U$, then $x_1 = g(x_2, \dots, x_n)$. Thus, we have that

$$S_c \cap U = \{(g(x_2, \dots, x_n), x_2, \dots, x_n) : (x_2, \dots, x_n) \in W\}$$

$$= \{u \in \mathbb{R}^n : u = A\binom{w}{g(w)} \text{ for some } w \in W\}$$

where A is the permutation of $\mathbb{R}^n$ given by

$$(y_1, \dots, y_n) \mapsto (y_n, y_1, \dots, y_{n-1}).$$

In particular, it follows that S_c is an $(n-1)$-dimensional manifold.

Exercise 1.94. Show that $\mathbb{S}^{n-1} := \{(x_1, \dots, x_n) \in \mathbb{R}^n : x_1{}^2 + \dots + x_n{}^2 = 1\}$ is an $(n-1)$-dimensional manifold.

Exercise 1.95. For $p \in \mathbb{S}^2$ and $p \neq \pm e_3$ (the north and south poles) define $f(p) = v$ where $\langle v, p \rangle = 0$, $\langle v, e_3 \rangle = 1 - z^2$, and $\langle p \times e_3, v \rangle = 0$. Define $f(\pm e_3) = 0$ Prove that f is a smooth function $f : \mathbb{S}^2 \to \mathbb{R}^3$.

Exercise 1.96. Show that the surface of revolution S obtained by rotating the circle given by $(x - 2)^2 + y^2 = 1$ around the y-axis is a two-dimensional manifold. This manifold is diffeomorphic to a (two-dimensional) torus $\mathbb{T}^2 := \mathbb{T} \times \mathbb{T}$. Construct a diffeomorphism.

Exercise 1.97. Suppose that J is an interval in $\mathbb{R}$ and $\gamma : J \to \mathbb{R}^n$ is a smooth function. The image $\mathcal{C}$ of γ is, by definition, a curve in $\mathbb{R}^n$. Is $\mathcal{C}$ a one-dimensional submanifold of $\mathbb{R}^n$? Formulate and prove a theorem that gives sufficient conditions for $\mathcal{C}$ to be a submanifold. Hint: Consider the function $t \mapsto (t^2, t^3)$ for $t \in \mathbb{R}$ and the function $t \mapsto (1 - t^2, t - t^3)$ for two different domains: $t \in \mathbb{R}$ and $t \in (-\infty, 1)$. Can you imagine a situation where the image of a smooth curve is a dense subset of a manifold with dimension $n > 1$? Hint: Consider curves mapping into the two-dimensional torus.

Exercise 1.98. Show that the closed unit disk in $\mathbb{R}^2$ is not a manifold. Actually, it is a *manifold with boundary*. How should this concept be formalized?

Exercise 1.99. Prove that for $\epsilon > 0$ there is a $\delta > 0$ and a root r of the polynomial $x^3 - ax + b$ such that $|r| < \epsilon$ whenever $|a - 1| + |b| < \delta$.

Exercise 1.100. Show that if $f : \mathbb{R}^n \to \mathbb{R}^n$, A is a nonsingular $(n \times n)$-matrix and $|\epsilon|$ is sufficiently small, then the differential equation $\dot{x} = Ax + \epsilon f(x)$ has a rest point.

1.8.3   Tangent Spaces

We have used, informally, the following proposition: *If S is a manifold in $\mathbb{R}^n$, and $(x, f(x))$ is tangent to S for each $x \in S$, then S is an invariant manifold for the differential equation $\dot{x} = f(x)$.* To make this proposition precise, we will give a definition of the concept of a tangent vector on a manifold. This definition is the main topic of this section.

Let us begin by considering some examples where the proposition on tangents and invariant manifolds can be applied.

The vector field on $\mathbb{R}^3$ associated with the system of differential equations given by

$$
\begin{aligned}
\dot{x} &= x(y + z), \\
\dot{y} &= -y^2 + x \cos z, \\
\dot{z} &= 2x + z - \sin y
\end{aligned}
\tag{1.18}
$$

is "tangent" to the linear two-dimensional submanifold $S := \{(x, y, z) : x = 0\}$ in the following sense: If $(a, b, c) \in S$, then the value of the vector function

$$
(x, y, z) \mapsto (x(y + z), y^2 + x \cos z, 2x + z - \sin y)
$$

at (a, b, c) is a vector in the linear space S. Note that the vector assigned by the vector field depends on the point in S. For this reason, we will view the vector field as the function

$$
(x, y, z) \mapsto (x, y, z, x(y + z), -y^2 + x \cos z, 2x + z - \sin y)
$$

where the first three component functions specify the *base point*, and the last three components, called the *principal part*, specify the vector that is assigned at the base point.

To see that S is an invariant set, choose $(0, b, c) \in S$ and consider the initial value problem

$$
\dot{y} = -y^2, \quad \dot{z} = z - \sin y, \qquad y(0) = b, \quad z(0) = c.
$$

Note that if its solution is given by $t \mapsto (y(t), z(t))$, then the function $t \mapsto (0, y(t), z(t))$ is the solution of system (1.18) starting at the point $(0, b, c)$. In particular, the orbit corresponding to this solution is contained in S. Hence, S is an invariant set. In this example, the solution is not defined for all $t \in (-\infty, \infty)$. (Why?) But, every solution that starts in S stays in S, as required by Definition 1.66.

The following system of differential equations,

$$\dot{x} = x^2 - (x^3 + y^3 + z^3)x,$$
$$\dot{y} = y^2 - (x^3 + y^3 + z^3)y,$$
$$\dot{z} = z^2 - (x^3 + y^3 + z^3)z \qquad (1.19)$$

has a nonlinear invariant submanifold; namely, the unit sphere

$$\mathbb{S}^2 := \{(x,y,z) \in \mathbb{R}^3 : x^2 + y^2 + z^2 = 1\}.$$

This fact follows from our proposition, provided that the vector field associated with the differential equation is everywhere tangent to the sphere. To prove this requirement, recall from Euclidean geometry that a vector in space is defined to be tangent to the sphere if it is orthogonal to the normal line passing through the base point of the vector. Moreover, the normal lines to the sphere are generated by the outer unit normal field given by the restriction of the vector field

$$\eta(x,y,z) := (x,y,z,x,y,z)$$

to $\mathbb{S}^2$. By a simple computation, it is easy to check that the vector field associated with the differential equation is everywhere orthogonal to η on $\mathbb{S}^2$; that is, at each base point on $\mathbb{S}^2$ the corresponding principal parts of the two vector fields are orthogonal, as required.

We will give a definition for tangent vectors on a manifold that generalizes the definition given in Euclidean geometry for linear subspaces and spheres. Let us suppose that S is a k-dimensional submanifold of $\mathbb{R}^n$ and (G, W) is a submanifold coordinate chart at $p \in S$. Our objective is to define the tangent space to S at p.

Definition 1.101. The *tangent space* to $\mathbb{R}^k$ with base point at $w \in \mathbb{R}^k$ is the set

$$T_w \mathbb{R}^k := \{w\} \times \mathbb{R}^k.$$

We have the following obvious proposition: If $w \in \mathbb{R}^k$, then the tangent space $T_w \mathbb{R}^k$, with addition defined by

$$(w, \xi) + (w, \zeta) := (w, \xi + \zeta)$$

and scalar multiplication defined by

$$a(w, \xi) := (w, a\xi),$$

is a vector space that is isomorphic to the vector space $\mathbb{R}^k$.

To define the *tangent space of the submanifold S at $p \in S$*, denoted $T_p S$, we simply move the space $T_w \mathbb{R}^k$, for an appropriate choice of w, to S

with a submanifold coordinate map. More precisely, suppose that (W, G) is a submanifold chart at p. By Proposition 1.80, the coordinate map G is invertible. If $q = G^{-1}(p)$, then define

$$T_pS := \{p\} \times \{v \in \mathbb{R}^n : v = DG(q)\xi, \ \xi \in \mathbb{R}^k\}. \qquad (1.20)$$

Note that the set

$$\mathcal{V} := \{v \in \mathbb{R}^n : v = DG(q)\xi, \ \xi \in \mathbb{R}^k\}$$

is a k-dimensional subspace of $\mathbb{R}^n$. If $k = n$, then $DG(q)$ is the identity map. If $k < n$, then $DG(q) = AB$ where A is a nonsingular matrix and the $n \times k$ block matrix

$$B := \begin{pmatrix} I_k \\ Dg(q) \end{pmatrix}$$

is partitioned by rows with I_k the $k \times k$ identity matrix and g a map from W to $\mathbb{R}^{n-k}$. Thus, we see that $\mathcal{V}$ is just the image of a linear map from $\mathbb{R}^k$ to $\mathbb{R}^n$ whose rank is k.

Proposition 1.102. *If S is a manifold and $p \in S$, then the vector space T_pS is well-defined.*

Proof. If K is a second submanifold coordinate map at p, say $K : Z \to S$ with $K(r) = p$, then we must show that the tangent space defined using K agrees with the tangent space defined using G. To prove this fact, let us suppose that $(p, v) \in T_pS$ is given by

$$v = DG(q)\xi.$$

Using the chain rule, it follows that

$$v = \frac{d}{dt}G(q + t\xi)\Big|_{t=0}.$$

In other words, v is the directional derivative of G at q in the direction ξ. To compute this derivative, we simply choose a curve, here $t \mapsto q + t\xi$, that passes through q with tangent vector ξ at time $t = 0$, move this curve to the manifold by composing it with the function G, and then compute the tangent to the image curve at time $t = 0$.

The curve $t \mapsto K^{-1}(G(q + t\xi))$ is in Z (at least this is true for $|t|$ sufficiently small). Thus, we have a vector $\alpha \in \mathbb{R}^k$ given by

$$\alpha := \frac{d}{dt}K^{-1}(G(q + t\xi))\Big|_{t=0}.$$

We claim that $DK(r)\alpha = v$. In fact, we have

$$K^{-1}(G(q)) = K^{-1}(p) = r,$$

and

$$DK(r)\alpha = \frac{d}{dt} K(K^{-1}(G(q + t\xi)))\Big|_{t=0}$$

$$= \frac{d}{dt} G(q + t\xi)\Big|_{t=0}$$

$$= v.$$

In particular, $T_p S$, as originally defined, is a subset of the "tangent space at p defined by K." But this means that this subset, which is itself a k-dimensional affine subspace (the translate of a subspace) of $\mathbb{R}^n$, must be equal to $T_p S$, as required.   □

Exercise 1.103. Prove: If $p \in \mathbb{S}^2$, then the tangent space $T_p \mathbb{S}^2$, as in Definition 1.20, is equal to

$$\{p\} \times \{v \in \mathbb{R}^3 : \langle p, v \rangle = 0\}.$$

Definition 1.104. The *tangent bundle* TS of a manifold S is the union of its tangent spaces; that is, $TS := \bigcup_{p \in S} T_p S$. Also, for each $p \in S$, the vector space $T_p S$ is called the *fiber* of the tangent bundle over the base point p.

Definition 1.105. Suppose that S_1 and S_2 are manifolds, and $F : S_1 \to S_2$ is a smooth function. The *derivative*, also called the *tangent map*, of F is the function $F_* : TS_1 \to TS_2$ defined as follows: For each $(p, v) \in T_p S_1$, let (W_1, G_1) be a submanifold chart at p in S_1, (W_2, G_2) a submanifold chart at $F(p)$ in S_2, $(G_1^{-1}(p), \xi)$ the vector in $T_{G_1^{-1}(p)} W_1$ such that $DG_1(G_1^{-1}(p))\xi = v$, and $(G_2^{-1}(F(p)), \zeta)$ the vector in $T_{G_2^{-1}(F(p))} W_2$ such that

$$\zeta = D(G_2^{-1} \circ F \circ G_1)(G_1^{-1}(p))\xi.$$

The tangent vector $F_*(p, v)$ in $T_{F(p)} S_2$ is defined by

$$F_*(p, v) = \big(F(p), DG_2(G_2^{-1}(F(p)))\zeta\big).$$

Although definition 1.105 seems to be rather complex, the idea is natural: we simply use the local representatives of the function F and the definition of the tangent bundle to define the derivative F_* as a map with two component functions. The first component is F (to ensure that base points map to base points) and the second component is defined by the derivative of a local representative of F at each base point.

The following proposition is obvious from the definitions.

Proposition 1.106. *The tangent map is well-defined and it is linear on each fiber of the tangent bundle.*

The derivative, or tangent map, of a function defined on a manifold has a geometric interpretation that is the key to understanding its applications in the study of differential equations. We have already discussed this interpretation several times for various special cases. But, because it is so important, let us consider the geometric interpretation of the derivative in the context of the notation introduced in Definition 1.105. If $t \mapsto \gamma(t)$ is a curve—a smooth function defined on an open set of $\mathbb{R}$—with image in the submanifold $S_1 \subseteq \mathbb{R}^m$ such that $\gamma(0) = p$, and if

$$v = \dot{\gamma}(0) = \frac{d}{dt}\gamma(t)\Big|_{t=0},$$

then $t \mapsto F(\gamma(t))$ is a curve in the submanifold $S_2 \subseteq \mathbb{R}^n$ such that $F(\gamma(0)) = F(p)$ and

$$F_*(p,v) = \Big(F(p), \frac{d}{dt}F(\gamma(t))\Big|_{t=0}\Big).$$

We simply find a curve that is tangent to the vector v at p and move the curve to the image of the function F to obtain a curve in the range. The tangent vector to the new curve at $F(p)$ is the image of the tangent map.

Proposition 1.107. *A submanifold S of $\mathbb{R}^n$ is an invariant manifold for the ordinary differential equation $\dot{x} = f(x)$, $x \in \mathbb{R}^n$ if and only if*

$$(x, f(x)) \in T_x S$$

for each $x \in S$. If, in addition, S is compact, then each orbit on S is defined for all $t \in \mathbb{R}$.

Proof. Suppose that S is k-dimensional, $p \in S$, and (W, G) is a submanifold chart for S at p. The idea of the proof is to change coordinates to obtain an ordinary differential equation on W.

Recall that the submanifold coordinate map G is invertible and G^{-1} is the restriction of a linear map defined on $\mathbb{R}^n$. In particular, we have that $w \equiv G^{-1}(G(w))$ for $w \in W$. If we differentiate both sides of this equation and use the chain rule, then we obtain the relation

$$I = DG^{-1}(G(w))DG(w) \tag{1.21}$$

where I denotes the identity transformation of $\mathbb{R}^n$. In particular, for each $w \in W$, we have that $DG^{-1}(G(w))$ is the inverse of the linear transformation $DG(w)$.

Under the hypothesis, we have that $(x, f(x)) \in T_x S$ for each $x \in S$. Hence, the vector $f(G(w))$ is in the image of $DG(w)$ for each $w \in W$. Thus, it follows that

$$(w, DG^{-1}(G(w))f(G(w))) \in T_w \mathbb{R}^k,$$

and, as a result, the map

$$w \mapsto (w, DG^{-1}(G(w))f(G(w)))$$

defines a vector field on $W \subseteq \mathbb{R}^n$. The associated differential equation on W is given by

$$\dot{w} = DG^{-1}(G(w))f(G(w)). \tag{1.22}$$

Suppose that $G(q) = p$, and consider the initial value problem on W given by the differential equation (1.22) together with the initial condition $w(0) = q$. By the existence theorem, this initial value problem has a unique solution $t \mapsto \omega(t)$ that is defined on an open interval containing $t = 0$.

Define $\phi(t) = G(\omega(t))$. We have that $\phi(0) = p$ and, using equation (1.21), that

$$\frac{d\phi}{dt}(t) = DG(\omega(t))\dot{\omega}(t)$$
$$= DG(\omega(t)) \cdot DG^{-1}(G(\omega(t)))f(G(\omega(t)))$$
$$= f(\phi(t)).$$

Thus, $t \mapsto \phi(t)$ is the solution of $\dot{x} = f(x)$ starting at p. Moreover, this solution is in S because $\phi(t) = G(\omega(t))$. The solution remains in S as long as it is defined within the submanifold chart. The same result is true for every submanifold chart. Thus, the solution remains in S as long as it is defined.

Suppose that S is compact and note that the solution just defined is a solution of the differential equation $\dot{x} = f(x)$ defined on $\mathbb{R}^n$. By the extension theorem, if a solution of $\dot{x} = f(x)$ does not exist for all time, for example, if it exists only for $0 \le t < \beta < \infty$, then it approaches the boundary of the domain of definition of f or it blows up to infinity as t approaches β. As long as the solution stays in S, both possibilities are excluded if S is compact. Since the manifold S is covered by coordinate charts, the solution stays in S and it is defined for all time.

If S is invariant, $p \in S$ and $t \mapsto \gamma(t)$ is the solution of $\dot{x} = f(x)$ with $\gamma(0) = p$, then the curve $t \to G^{-1}(\gamma(t))$ in $\mathbb{R}^k$ has a tangent vector ξ at $t = 0$ given by

$$\xi := \frac{d}{dt}G^{-1}(\gamma(t))\Big|_{t=0}.$$

As before, it is easy to see that $DG(q)\xi = f(p)$. Thus, $(p, f(p)) \in T_pS$, as required. $\square$

Exercise 1.108. Show that the function $f(\theta) = 1 - \lambda \sin \theta$ defines a (smooth) vector field on $\mathbb{T}^1$, but $f(\theta) = \theta - \lambda \sin \theta$ does not.

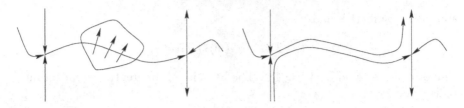

Figure 1.15: The left panel depicts a heteroclinic saddle connection and a locally supported perturbation. The right panel depicts the phase portrait of the perturbed vector field.

Exercise 1.109. State and prove a proposition that is analogous to Proposition 1.107 for the case where the submanifold S is not compact.

Exercise 1.110. We have mentioned several times the interpretation of the derivative of a function whereby a curve tangent to a given vector at a point is moved by the function to obtain a new curve whose tangent vector is the directional derivative of the function applied to the original vector. This interpretation can also be used to define the tangent space at a point on a manifold. In fact, let us say that two curves $t \mapsto \gamma(t)$ and $t \mapsto \nu(t)$, with image in the same manifold S, are equivalent if $\gamma(0) = \nu(0)$ and $\dot{\gamma}(0) = \dot{\nu}(0)$. Prove that this is an equivalence relation. A tangent vector at $p \in S$ is defined to an equivalence class of curves all with value p at $t = 0$. As a convenient notation, let us write $[\gamma]$ for the equivalence class containing the curve γ. The tangent space at p in S is defined to be the set of all equivalence classes of curves that have value p at $t = 0$. Prove that the tangent space at p defined in this manner can be given the structure of a vector space and this vector space has the same dimension as the manifold S. Also prove that this definition gives the same tangent space as defined in equation 1.20. Finally, for manifolds S_1 and S_2 and a function $F : S_1 \to S_2$, prove that the tangent map F_* is given by $F_*[\gamma] = [F \circ \gamma]$.

Exercise 1.111. Let A be an invertible symmetric $(n \times n)$-matrix. (a) Prove that the set $M := \{x \in \mathbb{R}^2 : \langle Ax, x \rangle = 1\}$ is a submanifold of $\mathbb{R}^n$. (b) Suppose that $x_0 \in M$. Describe the tangent space to M at x_0. Hint: Apply Exercise 1.110.

Exercise 1.112. [General Linear Group] The general linear group $GL(\mathbb{R}^n)$ is the set of all invertible real $n \times n$-matrices where the group structure is given by matrix multiplication (see also Exercise 2.55). (a) Prove that $GL(\mathbb{R}^n)$ is a submanifold of $\mathbb{R}^{n^2}$. Hint: Consider the determinant function. (b) Determine the tangent space of $GL(\mathbb{R}^n)$ at its identity. Hint: Apply Exercise 1.110. (c) Prove that the map $GL(\mathbb{R}^n) \times GL(\mathbb{R}^n) \to GL(\mathbb{R}^n)$ given by $(A, B) \mapsto AB$ is smooth. (d) Prove that the map $GL(\mathbb{R}^n) \to GL(\mathbb{R}^n)$ given by $A \mapsto A^{-1}$ is smooth. Note: A Lie group is a group that is also a smooth manifold such that the group operations are smooth. The vector space $T_I GL(\mathbb{R}^n)$ is called the Lie algebra of the Lie group when endowed with the multiplication $[A, B] = AB - BA$.

Exercise 1.113. (a) Prove that the tangent bundle of the torus $\mathbb{T}^2$ is trivial; that is, it can be viewed as $T\mathbb{T}^2 = \mathbb{T}^2 \times \mathbb{R}^2$. (b) (This exercise requires some knowledge of topology) Prove that the tangent bundle of $\mathbb{S}^2$ is not trivial.

Exercise 1.114. Suppose that $f : \mathbb{R}^n \to \mathbb{R}^n$ is smooth and the differential equation $\dot{x} = f(x)$ has a first integral all of whose level sets are compact. Prove that the corresponding flow is complete.

Exercise 1.115. Prove: The diagonal

$$\{(x,y) \in \mathbb{R}^n \times \mathbb{R}^n : x = y\}$$

in $\mathbb{R}^n \times \mathbb{R}^n$ is an invariant set for the system

$$\dot{x} = f(x) + h(y - x), \qquad \dot{y} = f(y) + g(x - y)$$

where $f, g, h : \mathbb{R}^n \to \mathbb{R}^n$ and $g(0) = h(0)$.

Exercise 1.116. [An Open Problem in Structural Stability] Let $H(x, y, z)$ be a homogeneous polynomial of degree n and η the outer unit normal on the unit sphere $\mathbb{S}^2 \subset \mathbb{R}^3$. Show that the vector field $X_H = \operatorname{grad} H - nH\eta$ is tangent to $\mathbb{S}^2$.

Call a rest point *isolated* if it is the unique rest point in some open set. Prove that if n is fixed, then the number of isolated rest points of X_H is uniformly bounded over all homogeneous polynomials H of degree n. Suppose that $n = 3$, the uniform bound for this case is B, and m is an integer such that $0 \le m \le B$. What is B? Is there some H such that X_H has exactly m rest points? If not, then for which m is there such an H? What if $n > 3$?

Note that the homogeneous polynomials of degree n form a finite dimensional vector space $\mathcal{H}_n$. What is its dimension? Is it true that for an open and dense subset of $\mathcal{H}_n$ the corresponding vector fields on $\mathbb{S}^2$ have only hyperbolic rest points?

In general, if X is a vector field in some class of vector fields $\mathcal{H}$, then X is called *structurally stable* with respect to $\mathcal{H}$ if X is contained in some open subset $U \subset \mathcal{H}$ such that the phase portrait of every vector field in U is the same; that is, if Y is a vector field in U, then there is a homeomorphism of the phase space that maps orbits of X to orbits of Y. Let us define $\mathcal{X}_n$ to be the set of all vector fields on $\mathbb{S}^2$ of the form X_H for some $H \in \mathcal{H}_n$. It is an interesting unsolved problem to determine the structurally stable vector fields in $\mathcal{X}_n$ with respect to $\mathcal{X}_n$.

One of the key issues that must be resolved to determine the structural stability of a vector field on a two-dimensional manifold is the existence of *heteroclinic orbits*. A heteroclinic orbit is an orbit that is contained in the stable manifold of a saddle point q and in the unstable manifold of a different saddle point p. If $p = q$, such an orbit is called *homoclinic*. A basic fact from the theory of structural stability is that if two saddle points are connected by a heteroclinic orbit, then the local phase portrait near this orbit can be changed by an arbitrarily small smooth perturbation. In effect, a perturbation can be chosen such that, in the phase portrait of the perturbed vector field, the saddle connection is broken (see Figure 1.15). Thus, in particular, a vector field with two saddle points connected by a heteroclinic orbit is not structurally stable with respect to the class of all smooth vector fields. Prove that a vector field X_H in $\mathcal{X}_n$ cannot have a homoclinic orbit. Also, prove that X_H cannot have a periodic orbit. Construct a homogeneous

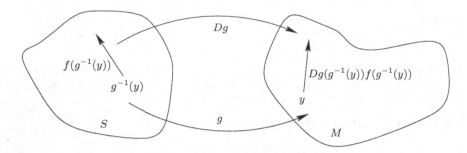

Figure 1.16: The "push forward" of a vector field f by a diffeomorphism $g : S \to M$.

polynomial $H \in \mathcal{H}_3$ such that X_H has hyperbolic saddle points p and q connected by a heteroclinic orbit.

Is every heteroclinic orbit of a vector field $X_H \in \mathcal{X}_3$ an arc of a great circle? The answer to this question is not known. But if it is true that all heteroclinic orbits are arcs of great circles, then the structurally stable vector fields, with respect to the class $\mathcal{X}_3$, are exactly those vector fields with all their rest points hyperbolic and with no heteroclinic orbits. Moreover, this set is open and dense in $\mathcal{X}_n$. A proof of these facts requires some work. But the main idea is clear: if X_H has a heteroclinic orbit that is an arc of a great circle, then there is a homogeneous polynomial K of degree $n = 3$ such that the perturbed vector field $X_{H+\epsilon K}$ has no heteroclinic orbits for $|\epsilon|$ sufficiently small. In fact, K can be chosen to be of the form

$$K(x, y, z) = (ax + by + cz)(x^2 + y^2 + z^2)$$

for suitable constants a, b, and c. (Why?) Of course, the conjecture that heteroclinic orbits of vector fields in $\mathcal{H}_3$ lie on great circles is just one approach to the structural stability question for $\mathcal{X}_3$. Can you find another approach?

There is an extensive and far-reaching literature on the subject of structural stability (see, for example, [192] and [204]).

1.8.4   Change of Coordinates

The proof of Proposition 1.107 contains an important computation that is useful in many other contexts; namely, the formula for changing coordinates in an autonomous differential equation. To reiterate this result, suppose that we have a differential equation $\dot{x} = f(x)$ where $x \in \mathbb{R}^n$, and $S \subseteq \mathbb{R}^n$ is an invariant k-dimensional submanifold. If g is a diffeomorphism from S to some k-dimensional submanifold $M \subseteq \mathbb{R}^m$, then the ordinary differential equation (or, more precisely, the vector field associated with the differential equation) can be "pushed forward" to M. In fact, if $g : S \to M$ is the diffeomorphism, then

$$\dot{y} = Dg(g^{-1}(y))f(g^{-1}(y)) \tag{1.23}$$

is a differential equation on M. Since g is a diffeomorphism, the new differential equation is the same as the original one up to a change of coordinates as schematically depicted in Figure 1.16.

Example 1.117. Consider $\dot{x} = x - x^2$, $x \in \mathbb{R}$. Let $S = \{x \in \mathbb{R} : x > 0\}$, $M = S$, and let $g : S \to M$ denote the diffeomorphism defined by $g(x) = 1/x$. Here, $g^{-1}(y) = 1/y$ and

$$\dot{y} = Dg(g^{-1}(y))f(g^{-1}(y))$$
$$= -\left(\frac{1}{y}\right)^{-2}\left(\frac{1}{y} - \frac{1}{y^2}\right)$$
$$= -y + 1.$$

The diffeomorphism g defines the change of coordinates $y = 1/x$ used to solve this special form of Bernoulli's equation; it is encountered in elementary courses on differential equations.

Exercise 1.118. According to the Hartman-Grobman theorem 1.47, there is a homeomorphism (defined on some open neighborhood of the origin) that maps orbits of $\dot{y} = y$ to orbits of $\dot{x} = x - x^2$. In this case, the result is trivial; the homeomorphism h given by $h(y) = y$ satisfies the requirement. For one and two-dimensional systems (which are at least twice continuously differentiable) a stronger result is true: There is a diffeomorphism h defined on a neighborhood of the origin with $h(0) = 0$ such that h transforms the linear system into the nonlinear system. Find an explicit formula for h and describe its domain.

Exercise 1.119. [Bernoulli's Equation] Show that the differential equation

$$\dot{x} = g(t)x - h(t)x^n$$

is transformed to a linear differential equation by the change of coordinates $y = 1/x^{n-1}$.

Coordinate transformations are very useful in the study of differential equations. New coordinates can reveal unexpected features. As a dramatic example of this phenomenon, we will show that all autonomous differential equations are the same, up to a smooth change of coordinates, near each of their regular points. Here, a *regular point* of $\dot{x} = f(x)$ is a point $p \in \mathbb{R}^n$, such that $f(p) \neq 0$. The following precise statement of this fact, which is depicted in Figure 1.17, is called the *rectification lemma*, the *straightening out theorem*, or the *flow box theorem*.

Lemma 1.120 (Rectification Lemma). *Suppose that* $\dot{x} = f(x)$, $x \in \mathbb{R}^n$. *If* $p \in \mathbb{R}^n$ *and* $f(p) \neq 0$, *then there are open sets* U, V *in* $\mathbb{R}^n$ *with* $p \in U$, *and a diffeomorphism* $g : U \to V$ *such that the differential equation in the new coordinates, that is, the differential equation*

$$\dot{y} = Dg(g^{-1}(y))f(g^{-1}(y)),$$

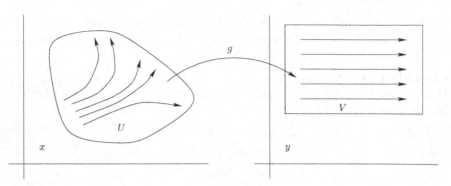

Figure 1.17: The flow of a differential equation is rectified by a change of coordinates $g : U \to V$.

is given by $(\dot{y}_1, \dots, \dot{y}_n) = (1, 0, 0, \dots, 0)$.

Proof. The idea of the proof is to "rectify" at one point, and then to extend the rectification to a neighborhood of this point.

Let $\mathbf{e}_1, \dots, \mathbf{e}_n$ denote the usual basis of $\mathbb{R}^n$. There is an invertible (affine) map $H_1 : \mathbb{R}^n \to \mathbb{R}^n$ such that $H_1(p) = 0$ and $DH_1(p)f(p) = \mathbf{e}_1$. (Why?) Here, an affine map is just the composition of a linear map and a translation. Let us also note that $\mathbf{e}_1$ is the transpose of the vector $(1, 0, 0, \dots, 0) \in \mathbb{R}^n$. If the formula (1.23) is used with $g = H_1$, then the differential equation $\dot{x} = f(x)$ is transformed to the differential equation denoted by $\dot{z} = f_1(z)$ where $f_1(0) = \mathbf{e}_1$. Thus, we have "rectified" the original differential equation at the single point p.

Let ϕ_t denote the flow of $\dot{z} = f_1(z)$, define $H_2 : \mathbb{R}^n \to \mathbb{R}^n$ by

$$(s, y_2, \dots, y_n) \mapsto \phi_s(0, y_2, \dots, y_n),$$

and note that $H_2(0) = 0$. The action of the derivative of H_2 at the origin on the standard basis vectors is

$$DH_2(0, \dots, 0)\mathbf{e}_1 = \frac{d}{dt}H_2(t, 0, \dots, 0)\Big|_{t=0} = \frac{d}{dt}\phi_t(0, \dots, 0)\Big|_{t=0} = \mathbf{e}_1,$$

and, for $j = 2, \dots, n$,

$$DH_2(0, \dots, 0)\mathbf{e}_j = \frac{d}{dt}H_2(t\mathbf{e}_j)\Big|_{t=0} = \frac{d}{dt}t\mathbf{e}_j\Big|_{t=0} = \mathbf{e}_j.$$

In particular, $DH_2(0)$ is the identity, an invertible linear transformation of $\mathbb{R}^n$.

To complete the proof we will use the inverse function theorem.

Theorem 1.121 (Inverse Function Theorem). *If* $F : \mathbb{R}^n \to \mathbb{R}^n$ *is a smooth function,* $F(p) = q$, *and* $DF(p)$ *is an invertible linear transformation of* $\mathbb{R}^n$, *then there exist two open sets* U *and* V *in* $\mathbb{R}^n$ *with* $(p, q) \in U \times V$, *together with a smooth function* $G : V \to U$, *such that* $G(q) = p$ *and* $G = F^{-1}$; *that is,* $F \circ G : V \to V$ *and* $G \circ F : U \to U$ *are identity functions.*

Proof. Consider the function $H : \mathbb{R}^n \times \mathbb{R}^n \to \mathbb{R}^n$ given by $H(x, y) = F(x) - y$. Note that $H(p, q) = 0$ and that $H_x(p, q) = DF(p)$ is invertible. By the implicit function theorem, there are open balls $\tilde{U}$ and V contained in $\mathbb{R}^n$, and a smooth function $G : V \to \tilde{U}$ such that $(p, q) \in \tilde{U} \times V$, $G(q) = p$, and $F(G(y)) = y$ for all $y \in V$. In particular, the function $F \circ G : V \to V$ is the identity.

Because F is continuous, the set $U := F^{-1}(V) \cap \tilde{U}$ is an open subset of $\tilde{U}$ with $p \in U$ and $F(U) \subset V$. If $x \in U$, then $(x, F(x)) \in \tilde{U} \times V$ and $H(x, F(x)) = 0$. Thus, by the uniqueness of the implicit solution (as stated in the implicit function theorem), $G(F(x)) = x$ for all $x \in U$. In other words $G \circ F : U \to U$ is the identity function. $\qquad\square$

By the inverse function theorem, there are two neighborhoods U and V of the origin such that $H_2 : U \to V$ is a diffeomorphism. The new coordinate, denoted y, on U is related to the old coordinate, denoted z, on V by the relation $y = H_2^{-1}(z)$. The differential equation in the new coordinates has the form

$$\dot{y} = (DH_2(y))^{-1} f_1(H_2(y)) := f_2(y).$$

Equivalently, at each point $y \in U$, we have $f_1(H_2(y)) = DH_2(y) f_2(y)$.

Suppose that $y = (s, y_2, \dots, y_n)$ and consider the tangent vector

$$(y, \mathbf{e}_1) \in T_y \mathbb{R}^n.$$

Also, note that $(y, \mathbf{e}_1)$ is tangent to the curve $\gamma(t) = (s + t, y_2, \dots, y_n)$ in $\mathbb{R}^n$ at $t = 0$ and

$$DH_2(y)\mathbf{e}_1 = \frac{d}{dt} H_2(\gamma(t))\Big|_{t=0} = \frac{d}{dt} \phi_t(\phi_s(0, y_2, \dots, y_n))\Big|_{t=0}$$

$$= f_1(H_2(s, y_2, \dots, y_n)) = f_1(H_2(y)).$$

Because $DH_2(y)$ is invertible, it follows that $f_2(y) = \mathbf{e}_1$.

The map $g := H_2^{-1} \circ H_1$ gives the required change of coordinates. $\qquad\square$

The idea that a change of coordinates may simplify a given problem is a far-reaching idea in many areas of mathematics; it certainly plays a central role in the study of differential equations.

Exercise 1.122. Show that the implicit function theorem is a corollary of the inverse function theorem.

Exercise 1.123. Suppose that $f : \mathbb{R}^n \to \mathbb{R}^n$ is smooth. Prove that if $|\epsilon|$ is sufficiently small, then the function $F : \mathbb{R}^n \to \mathbb{R}^n$ given by $F(x) := x + \epsilon f(x)$ is invertible in a neighborhood of the origin. Also, determine $DF^{-1}(0)$.

Exercise 1.124. [Newton's Method] Recall Newton's method: Suppose that $f : \mathbb{R}^n \to \mathbb{R}^n$ is twice continuously differentiable and $f(r) = 0$. We have $f(r) \approx f(x) + Df(x)(x - r)$ (see Theorem 1.237) and $0 \approx f(x) + Df(x)(x - r)$. Solve for r to obtain $r \approx x - Df(x)^{-1}f(x)$. Finally turn this into an iterative procedure to approximate r; that is, $x_{n+1} = x_n - Df(x_n)^{-1}f(x_n)$. Note: To implement this procedure (on a computer) it is usually better to solve for w in the equation $Df(x_n)w = -f(x_n)$ and then put $x_{n+1} = x_n + w$ (Why?). (a) Is the function $F(x) := x - Df(x)^{-1}f(x)$ invertible near $x = r$? (b) Prove that if $Df(r)$ is invertible and $|x_0 - r|$ is sufficiently small, then there is a constant $K > 0$ such that $|x_{n+1} - r| \leq K|x_n - r|^2$ and $\lim_{n\to\infty} x_n = r$. (c) A sequence $\{x_n\}_{n=0}^{\infty}$ converges linearly to r if there is a constant $\lambda > 0$ (the asymptotic error) such that $\lim_{n\to\infty} |x_{n+1}-r|/|x_n-r| = \lambda$; it converges quadratically if $\lim_{n\to\infty} |x_{n+1} - r|/|x_n - r|^2 = \lambda$. Show by discussing an explicit example, that quadratically convergent sequences converge much faster than linearly convergent sequences. (d) Compare the rates of convergence, to the positive zero of the function $f(x) = (x^2 - 2)/4$, of Newton's method and the iterative scheme $x_{n+1} = x_n - f(x_n)$. (e) The solution of the initial value problem

$$\ddot{\theta} + \sin\theta = 0, \quad \theta(0) = \pi/4, \quad \dot{\theta}(0) = 0$$

is periodic. Approximate the period (correct to three decimal places) using Newton's method.

Exercise 1.125. [Flow Box with Section] Prove the following modification of the rectification lemma. *Suppose that $\dot{x} = f(x)$, $x \in \mathbb{R}^2$. If $p \in \mathbb{R}^2$, the vector $f(p)$ is not zero, and there is a curve Σ in $\mathbb{R}^2$ such that $p \in \Sigma$ and $f(p)$ is not tangent to Σ, then there are open sets U, V in $\mathbb{R}^2$ with $p \in U$ and a diffeomorphism $g : U \to V$ such that the differential equation in the new coordinates, that is, the differential equation*

$$\dot{y} = Dg(g^{-1}(y))f(g^{-1}(y)),$$

is given by $(\dot{y}_1, \dot{y}_2) = (1, 0)$. Moreover, the image of $\Sigma \cap U$ under g is the line segment $\{(y_1, y_2) \in V : y_1 = 0\}$. Generalize the result to differential equations on $\mathbb{R}^n$.

Exercise 1.126. Prove that the function given by

$$(x, y) \mapsto \frac{x^2 + 2y + 1}{(x^2 + y + 1)^2}$$

is constant on the trajectories of the differential equation

$$\dot{x} = -y, \quad \dot{y} = x + 3xy + x^3.$$

Show that the function

$$(x, y) \mapsto \left(\frac{x}{x^2 + y + 1}, \frac{x^2 + y}{x^2 + y + 1} \right)$$

is birational—that is, the function and its inverse are both defined by rational functions. Finally, show that the change of coordinates given by this birational map linearizes the differential equation (see [198]).

Exercise 1.127. [Ważewski's Equation] Suppose that $\Omega \subseteq \mathbb{R}^n$ is open and $f : \Omega \to \mathbb{R}^n$ is a smooth map. (a) Prove that if $t \mapsto x(t)$ is a solution of Ważewski's equation $\dot{x} = Df(x)^{-1}v$, where $v \in \mathbb{R}^n$, then $f(x(t)) = f(x(0)) + tv$. (b) Prove a similar formula for the differential equation $\dot{x} = (Df(x))^{-1}(f(x) - f(v))$. (c) Suppose that $0 \in \Omega$, $f(0) = 0$, $Df(x)$ is invertible for every $x \in \Omega$, and the initial value problem $\dot{x} = (Df(x))^{-1}v$, $x(0) = 0$ has a solution, which exists at least for $|t| \le 1$, for every choice of $v \in \mathbb{R}^n$. Prove that f is invertible with a smooth inverse. Hint: Define $t \mapsto x(t, \xi)$ to be the solution of the initial value problem and $g(\xi) = x(1, \xi)$. Note: Ważewski's equation can be used to prove more general results on the invertibility of smooth maps (see, for example, [174] and [210]).

1.8.5 Polar Coordinates

There are several special "coordinate systems" that are important in the analysis of differential equations, especially, polar coordinates, cylindrical coordinates, and spherical coordinates. In this section we will consider the meaning of these coordinates in the language of differentiable manifolds, and we will also explore a few applications, especially blowup of a rest point and compactification at infinity. But, the main purpose of this section is to provide a deeper understanding and appreciation for the manifold concept in the context of the study of differential equations.

What are polar coordinates?

Perhaps the best way to understand the meaning of polar coordinates is to recall the "angular wrapping function" definition of angular measure from elementary trigonometry. We have proved that the unit circle $\mathbb{T}$ is a one-dimensional manifold. The wrapping function $P : \mathbb{R} \to \mathbb{T}$ is given by

$$P(\theta) = (\cos\theta, \sin\theta).$$

Clearly, P is smooth and surjective. But P is not injective. In particular, P is not a diffeomorphism (see Exercise 1.128).

The function P is a *covering map;* that is, each point of $\mathbb{T}$ is contained in an open set on which a local inverse of P is defined. Each such open set, together with its corresponding inverse function, is a coordinate system, as defined in Definition 1.87, that we will call an *angular coordinate system*. The image of a point of $\mathbb{T}$ under an angular coordinate map is called its angular coordinate, or simply its angle, relative to the angular coordinate system. For example, the pair (V, Ψ) where

$$V := \{(x, y) \in \mathbb{T} : x > 0\}$$

and $\Psi : V \to (-\frac{\pi}{2}, \frac{\pi}{2})$ is given by $\Psi(x, y) = \arctan(y/x)$ is an angular coordinate system. The number $\theta = \Psi(x, y)$ is the angle assigned to (x, y) in *this* angular coordinate system. Of course, there are infinitely many different angular coordinate systems defined on the same open set V. For example, the function given by $(x, y) \mapsto 4\pi + \arctan(y/x)$ on V also determines an angular coordinate system on $\mathbb{T}$ for which the corresponding angles belong to the interval $(\frac{7\pi}{2}, \frac{9\pi}{2})$.

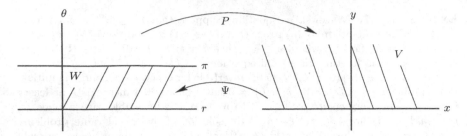

Figure 1.18: The polar wrapping function $P : \mathbb{R}^2 \to \mathbb{R}^2$ and a polar coordinate system $\Psi : V \to W$ on the upper half-plane.

As we have just seen, each point of $\mathbb{T}$ is assigned infinitely many angles. But all angular coordinate systems are compatible in the sense that they all determine local inverses of the wrapping function P. The totality of these charts might be called the angular coordinates on $\mathbb{T}$.

Exercise 1.128. Prove that $\mathbb{T}$ is not diffeomorphic to $\mathbb{R}$.

Exercise 1.129. Find a collection of angular coordinate systems that cover the unit circle.

Let us next consider coordinates on the plane compatible with the polar wrapping function $P : \mathbb{R}^2 \to \mathbb{R}^2$ given by

$$P(r, \theta) = (r \cos \theta, r \sin \theta).$$

The function P is a smooth surjective map that is not injective. Thus, P is not a diffeomorphism. Also, this function is *not* a covering map. For example, P has no local inverse at the origin of its range. On the other hand, P does have a local inverse at every point of the punctured plane (that is, the set $\mathbb{R}^2$ with the origin removed). Thus, in analogy with the definition of the angular coordinate on $\mathbb{T}$, we have the following definition of polar coordinates.

Definition 1.130. A *polar coordinate system* on the punctured plane is a coordinate system (V, Ψ) where $V \subset \mathbb{R}^2 \setminus \{(0,0)\}$, the range W of the coordinate map Ψ is contained in $\mathbb{R}^2$, and $\Psi : V \to W$ is the inverse of the polar wrapping function P restricted to the set W. The collection of all polar coordinate systems is called *polar coordinates*.

If

$$V := \{(x, y) \in \mathbb{R}^2 : y > 0\}, \quad W := \{(r, \theta) \in \mathbb{R}^2 : r > 0, 0 < \theta < \pi\},$$

and $\Psi : V \to W$ is given by

$$\Psi(x,y) = \left(\sqrt{x^2+y^2},\ \frac{\pi}{2} - \arctan\left(\frac{x}{y}\right)\right),$$

then (V, Ψ) is a polar coordinate system on the punctured plane (see Figure 1.18). By convention, the two slot functions defined by Ψ are named as follows

$$\Psi(x,y) = (r(x,y), \theta(x,y)),$$

and the point (x,y) is said to have polar coordinates $r = r(x,y)$ and $\theta = \theta(x,y)$.

The definition of cylindrical and spherical coordinates is similar to Definition 1.130 where the respective wrapping functions are given by

$$(r, \theta, z) \mapsto (r\cos\theta, r\sin\theta, z),$$
$$(\rho, \phi, \theta) \mapsto (\rho\sin\phi\cos\theta, \rho\sin\phi\sin\theta, \rho\cos\phi). \qquad (1.24)$$

To obtain covering maps, the z-axis must be removed in the target plane in both cases. Moreover, for spherical coordinates, the second variable must be restricted so that $0 \le \phi \le \pi$.

Let us now consider a differential equation $\dot{u} = f(u)$ defined on $\mathbb{R}^2$ with the usual Cartesian coordinates $u := (x,y)$. If (V, Ψ) is a polar coordinate system on the punctured plane such that $\Psi : V \to W$, then we can push forward the vector field f to the open set W by the general change of variables formula $\dot{y} = Dg(g^{-1}(y))f(g^{-1}(y))$ (see page 60). The new differential equation corresponding to the push forward of f is then said to be expressed in polar coordinates.

Specifically, the (principal part of the) new vector field is given by

$$F(r, \theta) = D\Psi(P(r,\theta))f(P(r,\theta)).$$

Of course, because the expressions for the components of the Jacobian matrix corresponding to the derivative $D\Psi$ are usually more complex than those for the matrix DP, the change to polar coordinates is usually easier to compute if we use the chain rule to obtain the identity

$$D\Psi(P(r,\theta)) = [DP(r,\theta)]^{-1} = \frac{1}{r}\begin{pmatrix} r\cos\theta & r\sin\theta \\ -\sin\theta & \cos\theta \end{pmatrix}$$

and recast the formula for F in the form

$$F(r,\theta) = [DP(r,\theta)]^{-1}f(P(r,\theta)).$$

In components, if $f(x,y) = (f_1(x,y), f_2(x,y))$, then

$$F(r,\theta) = \begin{pmatrix} \cos\theta f_1(r\cos\theta, r\sin\theta) + \sin\theta f_2(r\cos\theta, r\sin\theta) \\ -\frac{\sin\theta}{r}f_1(r\cos\theta, r\sin\theta) + \frac{\cos\theta}{r}f_2(r\cos\theta, r\sin\theta) \end{pmatrix}. \qquad (1.25)$$

Note that the vector field F obtained by the push forward of f in formula (1.25) does not depend on the choice of the polar coordinate system; that is, it does not depend on the choice of the local inverse Ψ. Thus, the vector field F is globally defined except on the line in the coordinate plane given by $\{(r,\theta) \in \mathbb{R}^2 : r = 0\}$. In general this is the best that we can do because the second component of the vector field F has a singularity at $r = 0$.

In practice, perhaps the simplest way to change to polar coordinates is to first differentiate in the formulas $r^2 = x^2 + y^2$ and $\theta = \arctan(y/x)$ to obtain the components of F in the form

$$r\dot{r} = x\dot{x} + y\dot{y} = xf_1(x,y) + yf_2(x,y),$$
$$r^2\dot{\theta} = x\dot{y} - y\dot{x} = xf_2(x,y) - yf_1(x,y),$$

and then substitute for x and y using the identities $x = r\cos\theta$ and $y = r\sin\theta$.

Exercise 1.131. Change the differential equations to polar coordinates:

1. $\dot{x} = -y + x(1 - x^2 - y^2), \quad \dot{y} = x + y(1 - x^2 - y^2).$
2. $\dot{x} = 1 - y^2, \quad \dot{y} = x.$
3. $\dot{x} = (x^2 + y^2)y, \quad \dot{y} = -(x^2 + y^2)x.$
4. $\dot{x} = y, \quad \dot{y} = -x - \epsilon(x^2 - 1)y.$

Exercise 1.132. Change the differential equations to Cartesian coordinates:

1. $\dot{r} = ar, \quad \dot{\theta} = b.$
2. $\dot{r} = r^3 \sin\theta \cos^3\theta, \quad \dot{\theta} = -1 + r^2 \cos^4\theta.$
3. $\dot{r} = -r^3, \quad \dot{\theta} = 1.$

Exercise 1.133. Show that the system

$$\dot{x} = y + xy, \quad \dot{y} = -x - x^2$$

has a center at the origin.

Exercise 1.134. Show that the systems

$$\dot{x} = y + 2xy, \quad \dot{y} = -x + xy$$

and

$$\dot{u} = v, \quad \dot{v} = -u - \frac{1}{\sqrt{5}}(2u^2 + 3uv - 2v^2)$$

are the same up to a change of variables. Hint: The transformation is a rotation.

A change to polar coordinates in a planar differential equation introduces a singularity on the line $\{(r,\theta) \in \mathbb{R}^2 : r = 0\}$. The next proposition states that if the differential equation has a rest point at the origin, then this singularity is removable (see [76]).

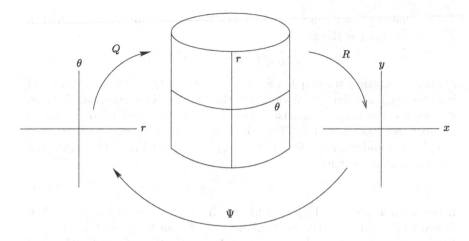

Figure 1.19: The polar wrapping function factored through the phase cylinder.

Proposition 1.135. *If $\dot{u} = f(u)$ is a differential equation on the plane and $f(0) = 0$, then the corresponding differential equation in polar coordinates has a removable singularity. Also, if f is class C^r, then the desingularized vector field in polar coordinates is in class C^{r-1}.*

Proof. Apply Taylor's theorem to the Taylor expansions of the components of the vector field f at the origin.   □

Even if Proposition 1.135 applies, and we do obtain a smooth vector field defined on the whole polar coordinate plane, the desingularized vector field is *not* the push forward of the original vector field; that is, the desingularized vector field is not obtained merely by a change of coordinates. Remember that there is no polar coordinate system at the origin of the Cartesian plane. In fact, the desingularized vector field in polar coordinates is an *extension* of the push forward of the original vector field to the singular line $\{(r, \theta) \in \mathbb{R}^2 : r = 0\}$.

It is evident from formula (1.25) that the desingularized vector field is 2π periodic in θ; that is, for all (r, θ) we have

$$F(r, \theta + 2\pi) = F(r, \theta).$$

In particular, the phase portrait of this vector field is periodic with period 2π. For this reason, let us change the point of view one last time and consider the vector field to be defined on the *phase cylinder;* that is, on $\mathbb{T} \times \mathbb{R}$ with θ the angular coordinate on $\mathbb{T}$ and r the Cartesian coordinate on $\mathbb{R}$.

The phase cylinder can be realized as a two-dimensional submanifold in $\mathbb{R}^3$, for example, as the set

$$\mathcal{C} := \{(x, y, z) \in \mathbb{R}^3 : x^2 + y^2 = 1\}.$$

For this realization, the map $Q : \mathbb{R}^2 \to \mathcal{C}$ defined by $Q(r, \theta) = (\cos\theta, \sin\theta, r)$ is a covering map. Here, $\mathbb{R}^2$ is viewed as the "polar coordinate plane." Thus, we can use the map Q to push forward the vector field F to the phase cylinder (see Exercise 1.138). There is also a natural covering map R, from the phase cylinder minus the set $\{(x, y, z) \in \mathcal{C} : z = 0\}$ onto the punctured Cartesian plane, defined by

$$R(x, y, z) = (xz, yz). \tag{1.26}$$

If the original vector field f vanishes at the origin, then it can be pushed forward by Ψ to F on the polar plane, and F can be pushed forward by Q to a vector field h on the phase cylinder. If finally, h is pushed forward by R to the punctured Cartesian plane, then we recover the original vector field f. In fact, by Exercise 1.138, the composition $R \circ Q \circ \Psi$, depicted in Figure 1.19, is the identity map.

Even though the phase cylinder can be realized as a manifold in $\mathbb{R}^3$, most often the best way to consider a vector field in polar coordinates is to view the polar coordinates abstractly as coordinates on the cylinder; that is, to view θ as the angular variable on $\mathbb{T}$ and r as the Cartesian coordinate on $\mathbb{R}$.

Exercise 1.136. Prove the following statements. If F is the push forward to the polar coordinate plane of a smooth vector field on the Cartesian plane, then F has the following symmetry:

$$F(-r, \theta + \pi) = -F(r, \theta).$$

If F can be desingularized, then its desingularization retains the symmetry.

Exercise 1.137. Prove that the cylinder $\{(x, y, z) \in \mathbb{R}^3 : x^2 + y^2 = 1\}$ is a two-dimensional submanifold of $\mathbb{R}^3$.

Exercise 1.138. Suppose that F is the push forward to the polar coordinate plane of a smooth vector field on the Cartesian plane that vanishes at the origin. Find the components of the push forward h of F to the phase cylinder realized as a submanifold in $\mathbb{R}^3$. Show that the push forward of h to the Cartesian plane via the natural map (1.26) is the original vector field f.

Exercise 1.139. [Hamiltonians and Gradients on Manifolds] Let

$$G : \mathbb{R}^3 \to \mathbb{R}$$

be a smooth map and consider its gradient. We have tacitly assumed that the *definition* of the gradient in $\mathbb{R}^3$ is

$$\operatorname{grad} G = \left(\frac{\partial G}{\partial x}, \frac{\partial G}{\partial y}, \frac{\partial G}{\partial z}\right). \tag{1.27}$$

But this expression for the gradient of a function is correct only on Euclidean space, that is, $\mathbb{R}^3$ together with the *usual inner product*. The definition of the gradient for a scalar function defined on a manifold, to be given below, is coordinate-free.

Recall that if $G : \mathbb{R}^n \to \mathbb{R}$, then its derivative can be viewed as a function from the tangent bundle $T\mathbb{R}^n$ to $T\mathbb{R}$. If $T\mathbb{R}$ is identified with $\mathbb{R}$, then on each tangent space of $\mathbb{R}^n$, the derivative of G is a linear functional. In fact, if we work locally at $p \in \mathbb{R}^n$, then $DG(p)$ is a map from the vector space $\mathbb{R}^n$ to $\mathbb{R}$. Moreover, the assignment of the linear functional corresponding to the derivative of G at each point of the manifold varies smoothly with the base point. From this point of view, the derivative of the scalar-valued function G is a differential 1-form on $\mathbb{R}^n$ that we will denote by dG. Finally, the derivative of G may be interpreted as the the *differential* of G. In this interpretation, if V is a tangent vector at $p \in \mathbb{R}^n$ and γ is a curve such that $\gamma(0) = p$ and $\dot{\gamma}(0) = V$, then

$$dG(V) = \frac{d}{ds}G(\gamma(s))\Big|_{s=0}.$$

If G is a scalar function defined on a manifold, then all of our interpretations for the derivative of G are still viable.

The definition of the gradient requires a new concept: A *Riemannian metric* on a manifold is a smooth assignment of an inner product in each tangent space of the manifold. Of course, the usual inner product assigned in each tangent space of $\mathbb{R}^n$ is a Riemannian metric for $\mathbb{R}^n$. Moreover, the manifold $\mathbb{R}^n$ together with this Riemannian metric is called *Euclidean space*. Note that the Riemannian metric can be used to define length. For example, the norm of a vector is the square root of the inner product of the vector with itself. It follows that the shortest distance between two points is a straight line. Thus, the geometry of Euclidean space is Euclidean geometry, as it should be. These notions can be generalized. For example, let γ be a curve in Euclidean space that connects the points p and q; that is, $\gamma : [a, b] \to \mathbb{R}^n$ such that $\gamma(a) = p$ and $\gamma(b) = q$. The length of γ is defined to be

$$\int_a^b \sqrt{\langle \dot{\gamma}(t), \dot{\gamma}(t) \rangle}\, dt,$$

where the angle brackets denote the usual inner product. The distance between p and q is the infimum of the set of all lengths of curves joining these points. A curve is called a *geodesic* joining p and q if its length equals the distance from p to q. For instance, the curve $\gamma(t) = tq + (1 - t)p$ is a geodesic. Of course, all geodesics lie on straight lines (see Exercise 3.9). Similarly, suppose that g is a Riemannian metric on a manifold M and $p, q \in M$. The length of a curve γ that connects p and q is defined to be

$$\int_a^b \sqrt{g_{\gamma(t)}(\dot{\gamma}(t), \dot{\gamma}(t))}\, dt$$

where $g_r(v, w)$ denotes the inner product of the vectors (r, v) and (r, w) in $T_r M$. Geodesics play the role of lines in the "Riemannian geometry" defined on a manifold by a Riemannian metric.

The gradient of $G : M \to \mathbb{R}$ with respect to the Riemannian metric g is the vector field, denoted by grad G, such that

$$dG_p(V) = g_p(V, \operatorname{grad} G) \qquad (1.28)$$

for each point $p \in M$ and every tangent vector $V \in T_p M$. The associated gradient system on the manifold is the differential equation $\dot{p} = \operatorname{grad} G(p)$.

(a) Prove that the gradient vector field is uniquely defined.

(b) Prove that if the Riemannian metric g on $\mathbb{R}^3$ is the usual inner product at each point of $\mathbb{R}^3$, then the invariant definition (1.28) of gradient agrees with the Euclidean gradient.

Consider the upper half-plane of $\mathbb{R}^2$ with the Riemannian metric

$$g_{(x,y)}(V, W) = y^{-2} \langle V, W \rangle \qquad (1.29)$$

where the angle brackets denote the usual inner product. The upper half-plane with the metric g is called the Poincaré or Lobachevsky plane; its geodesics are vertical lines and arcs of circles whose centers are on the x-axis. The geometry is non-Euclidean; for example, if p is a point not on such a circle, then there are infinitely many such circles passing through p that are parallel to (do not intersect) the given circle (see Exercise 3.11).

(c) Determine the gradient of the function $G(x, y) = x^2 + y^2$ with respect to the Riemannian metric (1.29) and draw the phase portrait of the corresponding gradient system on the upper half-plane. Also, compare this phase portrait with the phase portrait of the gradient system with respect to the usual metric on the plane.

If S is a submanifold of $\mathbb{R}^n$, then S inherits a Riemannian metric from the usual inner product on $\mathbb{R}^n$.

(d) Suppose that $F : \mathbb{R}^n \to \mathbb{R}$. What is the relationship between the gradient of F on $\mathbb{R}^n$ and the gradient of the function F restricted to S with respect to the inherited Riemannian metric (see Exercise 1.116)?

Hamiltonian systems on manifolds are defined in essentially the same way as gradient systems except that the Riemannian metric is replaced by a symplectic form. Although these objects are best described and analyzed using the calculus of differential forms (see [12], [89], and [213]), they are easy to define. Indeed, a *symplectic form* on a manifold is a smooth assignment of a bilinear, skew-symmetric, nondegenerate 2-form in each tangent space. A 2-form ω on a vector space X is nondegenerate provided that $y = 0$ is the only element of X such that $\omega(x, y) = 0$ for all $x \in X$. Prove: If a manifold has a symplectic form, then the dimension of the manifold is even.

Suppose that M is a manifold and ω is a symplectic form on M. The Hamiltonian vector field associated with a smooth scalar function H defined on M is the unique vector field X_H such that, for every point $p \in M$ and all tangent vectors V at p, the following identity holds:

$$dH_p(V) = \omega_p(X_H, V). \qquad (1.30)$$

(e) Let $M := \mathbb{R}^{2n}$, view $\mathbb{R}^{2n}$ as $\mathbb{R}^n \times \mathbb{R}^n$ so that each tangent vector V on M is decomposed as $V = (V_1, V_2)$ with $V_1, V_2 \in \mathbb{R}^n$, and define

$$\omega(V, W) := (V_1, V_2) \begin{pmatrix} 0 & I \\ -I & 0 \end{pmatrix} \begin{pmatrix} W_1 \\ W_2 \end{pmatrix}.$$

Show that ω is a symplectic form on M and Hamilton's equations are produced by the invariant definition (1.30) of the Hamiltonian vector field.

(f) Push forward the Euclidean gradient (1.27) of the function $G : \mathbb{R}^3 \to \mathbb{R}$ to the image of a cylindrical coordinate map, define

$$\mathcal{G}(r,\theta,z) = G(r\cos\theta, r\sin\theta, z),$$

and show that the push forward gives the result

$$\operatorname{grad}\mathcal{G} = \left(\frac{\partial\mathcal{G}}{\partial r}, \frac{1}{r^2}\frac{\partial\mathcal{G}}{\partial\theta}, \frac{\partial\mathcal{G}}{\partial z}\right). \tag{1.31}$$

(In practice, the function $\mathcal{G}$ is usually again called G. These two functions are local representations of the same function in two different coordinate systems.)

(g) Recall the formula for the gradient in cylindrical coordinates from vector analysis; namely,

$$\operatorname{grad}\mathcal{G} = \frac{\partial\mathcal{G}}{\partial r}\mathbf{e}_r + \frac{1}{r}\frac{\partial\mathcal{G}}{\partial\theta}\mathbf{e}_\theta + \frac{\partial\mathcal{G}}{\partial z}\mathbf{e}_z. \tag{1.32}$$

Show that the gradient vector fields (1.31) and (1.32) coincide.

(h) Express the usual inner product in cylindrical coordinates, and use the invariant definition of the gradient to determine the gradient in cylindrical coordinates.

(i) Repeat part (h) for spherical coordinates.

Exercise 1.140. [Electrostatic Potential] Suppose that two point charges with opposite signs, each with charge q, placed a units apart and located symmetrically with respect to the origin on the z-axis in space, produce the electrostatic potential

$$G_0(x,y,z) = kq\left[(x^2 + y^2 + (z - \tfrac{a}{2})^2)^{-1/2} - (x^2 + y^2 + (z + \tfrac{a}{2})^2)^{-1/2}\right]$$

where $k > 0$ is a constant and $q > 0$. If we are interested only in the field far from the charges, the "far field," then a is relatively small and therefore the first nonzero term of the Taylor series of the electrostatic potential with respect to a at $a = 0$ gives a useful approximation of G_0. This approximation, an example of a "far field approximation," is called the *dipole potential* in Physics (see [87, Vol. II, 6-1]). Show that the dipole potential is given by

$$G(x,y,z) = kqaz(x^2 + y^2 + z^2)^{-3/2}.$$

By definition, the electric field E produced by the dipole potential associated with the two charges is $E := -\operatorname{grad}G$. Draw the phase portrait of the differential equation $\dot{u} = E(u)$ whose orbits are the "dipole" lines of force. Discuss the stability of all rest points. Hint: Choose a useful coordinate system that reduces the problem to two dimensions.

Blow Up at a Rest Point

As an application of polar coordinates, let us determine the phase portrait of the differential equation in the Cartesian plane given by

$$\dot{x} = x^2 - 2xy, \qquad \dot{y} = y^2 - 2xy, \tag{1.33}$$

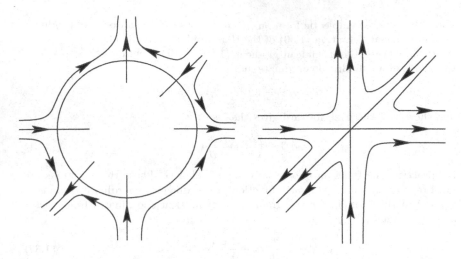

Figure 1.20: Phase portrait for the differential equation (1.34) on the upper half of the phase cylinder and its "blowdown" to the Cartesian plane.

(see [76]). This system has a unique rest point at the origin that is not hyperbolic. In fact, the system matrix for the linearization at the origin vanishes. Thus, linearization provides no information about the phase portrait of the system near the origin.

Because the polar coordinate representation of a plane vector field is always singular at the origin, we might expect that the polar coordinate representation of a planar vector field is not particularly useful to determine the phase portrait near the origin. But this is not the case. Often polar coordinates are the best way to analyze the vector field near the origin. The reason is that the desingularized vector field in polar coordinates is a smooth extension to the singular line represented as the equator of the phase cylinder. All points on the equator are collapsed to the single rest point at the origin in the Cartesian plane. Or, as we say, the equator is the *blowup* of the rest point. This extension is valuable because the phase portrait of the vector field near the original rest point corresponds to the phase portrait on the phase cylinder near the equatorial circle. Polar coordinates and desingularization provide a mathematical microscope for viewing the local behavior near the "Cartesian" rest point.

The desingularized polar coordinate representation of system (1.33) is

$$\dot{r} = r^2(\cos^3\theta - 2\cos^2\theta\sin\theta - 2\cos\theta\sin^2\theta + \sin^3\theta),$$
$$\dot{\theta} = 3r(\cos\theta\sin^2\theta - \cos^2\theta\sin\theta). \tag{1.34}$$

For this particular example, both components of the vector field have r as a common factor. From our discussion of reparametrization, we know that the system with this factor removed has the same phase portrait as the original differential equation in the portion of the phase cylinder where $r > 0$. Of course, when we "blow down" to the Cartesian plane, the push forward of the reparametrized vector field has the same phase portrait as the original vector field in the punctured plane; exactly the set where the original phase portrait is to be constructed.

Let us note that after division by r, the differential equation (1.34) has several *isolated* rest point on the equator of the phase cylinder. In fact, because this differential equation restricted to the equator is given by

$$\dot{\theta} = 3\cos\theta\sin\theta(\sin\theta - \cos\theta),$$

we see that it has six rest points with the following angular coordinates:

$$0, \quad \frac{\pi}{4}, \quad \frac{\pi}{2}, \quad \pi, \quad \frac{5\pi}{4}, \quad \frac{3\pi}{2}.$$

The corresponding rest points for the reparametrized system are all hyperbolic. For example, the system matrix at the rest point $(r, \theta) = (0, \frac{\pi}{4})$ is

$$\frac{1}{\sqrt{2}}\begin{pmatrix} -1 & 0 \\ 0 & 3 \end{pmatrix}.$$

It has the negative eigenvalue $-1/\sqrt{2}$ in the positive direction of the Cartesian variable r on the cylinder and the positive eigenvalue $3/\sqrt{2}$ in the positive direction of the angular variable. This rest point is a hyperbolic saddle. If each rest point on the equator is linearized in turn, the phase portrait on the cylinder and the corresponding blowdown of the phase portrait on the Cartesian plane are found to be as depicted in Figure 1.20. Hartman's theorem can be used to construct a proof of this fact.

The analysis of differential equation (1.33) is very instructive, but perhaps somewhat misleading. Often, unlike this example, the blowup procedure produces a vector field on the phase cylinder where some or all of the rest points are not hyperbolic. Of course, in these cases, we can treat the polar coordinates near one of the nonhyperbolic rest points as Cartesian coordinates; we can translate the rest point to the origin; and we can blow up again. If, after a finite number of such blowups, all rest points of the resulting vector field are hyperbolic, then the local phase portrait of the original vector field at the original nonhyperbolic rest point can be determined. For masterful treatments of this subject and much more, see [19], [75], [76], and [219].

The idea of blowup and desingularization are far-reaching ideas in mathematics. For example, these ideas seem to have originated in algebraic geometry, where they play a fundamental role in understanding the structure of algebraic varieties [29].

Compactification at Infinity

The orbits of a differential equation on $\mathbb{R}^n$ may be unbounded. One way to obtain some information about the behavior of such solutions is to (try to) *compactify* the Cartesian space, so that the vector field is extended to a new manifold that contains the "points at infinity." This idea, due to Henri Poincaré [185], has been most successful in the study of planar systems given by polynomial vector fields, also called polynomial systems (see [7, p. 219] and [99]). In this section we will give a brief description of the compactification process for such planar systems. We will again use the manifold concept and the idea of reparametrization.

Let us consider a plane vector field, which we will write in the form

$$\dot{x} = f(x,y), \qquad \dot{y} = g(x,y). \tag{1.35}$$

To study its phase portrait "near" infinity, let us consider the unit sphere $\mathbb{S}^2$; that is, the two-dimensional submanifold of $\mathbb{R}^3$ defined by

$$\mathbb{S}^2 := \{(x,y,z) : x^2 + y^2 + z^2 = 1\},$$

and the tangent plane Π at its north pole; that is, the point with coordinates $(0,0,1)$. The push forward of system (1.35) to Π by the natural map $(x,y) \mapsto (x,y,1)$ is

$$\dot{x} = f(x,y), \qquad \dot{y} = g(x,y), \qquad \dot{z} = 0. \tag{1.36}$$

The idea is to "project" differential equation (1.36) to the unit sphere by central projection; then the behavior of the system near infinity is the same as the behavior of the projected system near the equator of the sphere.

Central projection is defined as follows: A point $p \in \Pi$ is mapped to the sphere by assigning the unique point on the sphere that lies on the line segment from the origin in $\mathbb{R}^3$ to the point p. To avoid a vector field specified by *three* components, we will study the projected vector field restricted to a coordinate system on the sphere where the vector field is again planar. Also, to obtain the desired compactification, we will choose local coordinates defined in open sets that contain portions of the equator of the sphere.

The central projection map $Q : \Pi \to \mathbb{S}^2$ is given by

$$Q(x,y,1) = (x(x^2 + y^2 + 1)^{-1/2}, y(x^2 + y^2 + 1)^{-1/2}, (x^2 + y^2 + 1)^{-1/2}).$$

One possibility for an appropriate coordinate system on the Poincaré sphere is a spherical coordinate system; that is, one of the coordinate charts that is compatible with the map

$$(\rho, \phi, \theta) \mapsto (\rho \sin\phi \cos\theta, \rho \sin\phi \sin\theta, \rho \cos\phi) \tag{1.37}$$

(see display (1.24)). For example, if we restrict to the portion of the sphere where $x > 0$, then one such coordinate map is given by

$$\Psi(x, y, z) := (\arccos(z), \arctan\left(\frac{y}{x}\right)).$$

The transformed vector field on the sphere is the push forward of the vector field X that defines the differential equation on Π by the map $\Psi \circ Q$. In view of equation (1.37) and the restriction to the sphere, the inverse of this composition is the transformation P given by

$$P(\phi, \theta) = \left(\frac{\sin\phi}{\cos\phi}\cos\theta,\ \frac{\sin\phi}{\cos\phi}\sin\theta\right).$$

Thus, the push forward of the vector field X is given by

$$DP(\phi, \theta)^{-1}X(P(\phi, \theta)).$$

Of course, we can also find the transformed vector field simply by differentiating with respect to t in the formulas

$$\phi = \arccos((x^2 + y^2 + 1)^{-1/2}), \qquad \theta = \arctan\left(\frac{y}{x}\right).$$

If the vector field is polynomial with maximal degree k, then after we evaluate the polynomials f and g in system (1.36) at $P(\phi, \theta)$ and take into account multiplication by the Jacobian matrix, the denominator of the resulting expressions will contain $\cos^{k-1}\phi$ as a factor. Note that $\phi = \frac{\pi}{2}$ corresponds to the equator of the sphere and $\cos(\frac{\pi}{2}) = 0$. Thus, the vector field in spherical coordinates is desingularized by a reparametrization of time that corresponds to multiplication of the vector field defining the system by $\cos^{k-1}\phi$. This desingularized system ([53])

$$\dot\phi = (\cos^{k+1}\phi)(\cos\theta f + \sin\theta g), \qquad \dot\theta = \frac{\cos^k\phi}{\sin\phi}(\cos\theta g - \sin\theta f) \qquad (1.38)$$

is smooth at the equator of the sphere, and it has the same phase portrait as the original centrally projected system in the upper hemisphere. Therefore, we can often determine the phase portrait of the original vector field "at infinity" by determining the phase portrait of the desingularized vector field on the equator. Note that because the vector field corresponding to system (1.38) is everywhere tangent to the equator, the equator is an invariant set for the desingularized system.

Spherical coordinates are global in the sense that all the spherical coordinate systems have coordinate maps that are local inverses for the fixed spherical wrapping function (1.37). Thus, the push forward of the original vector field will produce system (1.38) in every spherical coordinate system.

There are other coordinate systems on the sphere that have also proved useful for the compactification of plane vector fields. For example, the right hemisphere of $\mathbb{S}^2$; that is, the subset $\{(x, y, z) : y > 0\}$ is mapped diffeo-morphically to the plane by the coordinate function defined by

$$\Psi_1(x, y, z) = \left(\frac{x}{y}, \frac{z}{y}\right).$$

Also, the map $\Psi_1 \circ Q$, giving the central projection in these coordinates, is given by

$$(x, y, 1) \mapsto \left(\frac{x}{y}, \frac{1}{y}\right).$$

Thus, the local representation of the central projection in this chart is obtained using the coordinate transformations

$$u = \frac{x}{y}, \qquad v = \frac{1}{y}.$$

Moreover, a polynomial vector field of degree k in these coordinates can again be desingularized at the equator by a reparametrization correspond-ing to multiplication of the vector field by v^{k-1}. In fact, the desingularized vector field has the form

$$\dot{u} = v^k \left(f\left(\frac{u}{v}, \frac{1}{v}\right) - ug\left(\frac{u}{v}, \frac{1}{v}\right)\right), \qquad \dot{v} = -v^{k+1} g\left(\frac{u}{v}, \frac{1}{v}\right).$$

The function Ψ_1 restricted to $y < 0$ produces the representation of the central projection in the left hemisphere. Similarly, the coordinate map

$$\Psi_2(x, y, z) = \left(\frac{y}{x}, \frac{z}{x}\right)$$

on the sphere can be used to cover the remaining points, near the equator in the upper hemisphere, with Cartesian coordinates (x, y, z) where $y = 0$ but $x \neq 0$.

The two pairs of charts just discussed produce two different local vector fields. Both of these are usually required to analyze the phase portrait near infinity. Also, it is very important to realize that if the degree k is even, then multiplication by v^{k-1} in the charts corresponding respectively to $x < 0$ and $y < 0$ *reverses* the original direction of time.

As an example of compactification, let us consider the phase portrait of the quadratic planar system given by

$$\dot{x} = 2 + x^2 + 4y^2, \qquad \dot{y} = 10xy. \tag{1.39}$$

This system has no rest points in the finite plane.

In the chart corresponding to $v > 0$ with the chart map Ψ_1, the desin-gularized system is given by

$$u' = 2v^2 - 9u^2 + 4, \qquad v' = -10uv \tag{1.40}$$

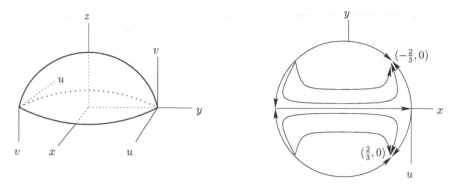

Figure 1.21: Phase portrait on the Poincaré sphere for the differential equation (1.39).

where the symbol " ′ " denotes differentiation with respect to the new independent variable after reparametrization. The first order system (1.40) has rest points with coordinates $(u, v) = (\pm\frac{2}{3}, 0)$. These rest points lie on the u-axis: the set in our chart that corresponds to the equator of the Poincaré sphere. Both rest points are hyperbolic. In fact, $(\frac{2}{3}, 0)$ is a hyperbolic sink and $(-\frac{2}{3}, 0)$ is a hyperbolic source.

In the chart with $v < 0$ and chart map Ψ_1, the reparametrized local system is given by the differential equation (1.40). But, because $k = 2$, the direction of "time" has been reversed. Thus, the sink at $(\frac{2}{3}, 0)$ in this chart corresponds to a source for the original vector field centrally projected to the Poincaré sphere. The rest point $(-\frac{2}{3}, 0)$ corresponds to a sink on the Poincaré sphere.

We have now considered all points on the Poincaré sphere except those on the great circle given by the equation $y = 0$. For these points, we must use the charts corresponding to the map Ψ_2. In fact, there is a hyperbolic saddle point at the origin of each of these coordinate charts, and these rest points correspond to points on the equator of the Poincaré sphere. Of course, the other two points already discussed are also rest points in these charts.

The phase portrait of the compactification of system (1.39) is shown in Figure 1.21. Because the x-axis is an invariant manifold for the original vector field, the two saddles at infinity are connected by a heteroclinic orbit.

Exercise 1.141. Prove that $\mathbb{S}^2$ is a two-dimensional submanifold of $\mathbb{R}^3$.

Exercise 1.142. Use spherical coordinates to determine the compactification of the differential equation (1.39) on the Poincaré sphere.

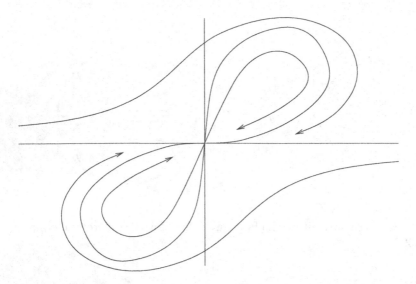

Figure 1.22: Phase portrait of Vinograd's system (1.41).

Exercise 1.143. Find the compactification of the differential equation

$$\dot{x} = x + y - y^3, \qquad \dot{y} = -x + y + x^3$$

on the Poincaré sphere using spherical coordinates. Show that the equator is a periodic orbit. See [53, p. 411] for a stability analysis of this periodic orbit, but note that there is a typographical error in the formula given for the desingularized projection of this vector field.

Exercise 1.144. Draw the phase portrait of the vector field

$$\dot{x} = x^2 + y^2 - 1, \qquad \dot{y} = 5(xy - 1).$$

This example is studied by Poincaré in his pioneering memoir on differential equations ([185, Oeuvre, p. 66]; see also [141, p. 204]).

Exercise 1.145. [Vinograd's System] Show that the phase portrait of Vinograd's system

$$\dot{x} = x^2(y - x) + y^5, \qquad \dot{y} = y^2(y - 2x) \tag{1.41}$$

agrees with Figure 1.22. In particular, show that while every orbit is attracted to the origin, the origin is not asymptotically stable (see [111, p. 191] and [227]). (a) Prove that the system has exactly one rest point. (b) Prove that the system is invariant with respect to the transformation $x \to -x$ and $y \to -y$. In particular, the phase portrait is symmetric relative to the origin. (c) Prove that the x-axis is invariant and all points on this set are attracted to the origin. (d) Prove that there is an open set in the plane containing the origin such that every solution starting in the set is attracted to the rest point at the origin. Hint: Consider the isocline $x^2(y - x) + y^5 = 0$. Prove that every solution which enters the region bounded by the isocline and the positive x-axis is attracted to the origin. Prove that every solution starting near the origin in the upper half-plane eventually enters this region. (e) Prove that the rest point at the origin has an elliptic sector; that is, there are two solutions (one approaching the rest point in forward time, one in backward time) such that, in one of the two regions subtended by the corresponding orbits, every solution starting sufficiently close to the rest point in this region is doubly asymptotic to the rest point (that is, such solutions are asymptotic to the rest point in both the forward and backward directions). Hint: Blow up the rest point. (f) Prove that every trajectory is attracted to the origin? Hint: Compactify the system. Show that the system, written in the coordinates given by $x = 1/v$ and $y = u/v$, takes the form

$$\dot{u} = -u(u^5 - v^2u^2 + 3v^2u - v^2), \qquad \dot{v} = -v(u^5 + v^2u - v^2) \tag{1.42}$$

where the equation of the line at infinity is $v = 0$. Since the rest points at infinity are highly degenerate, special weighted blowups compatible with the weighted polar blowup $(r, \theta) \mapsto (r^2 \cos\theta, r^5 \sin\theta)$ are the best choice (see [19]). In fact, it suffices to use the chart given by $u = pq^2$ and $v = q^5$ where system (1.42) has the form

$$\dot{p} = -\frac{1}{5}p(3p^5 - 5q^4p^2 + 13q^2p - 3), \quad \dot{q} = -\frac{1}{5}q(p^5 + q^2p - 1).$$

This system has a semi-hyperbolic rest point (that is, the linearization has exactly one zero eigenvalue). The local phase portrait at this rest point can be determined by an additional blowup. (See Exercise 4.5 for an alternative method.) To finish the proof, use the Poincaré-Bendixson theorem 1.174. (g) Draw the global phase portrait including the circle at infinity.

Exercise 1.146. [Singular Differential Equations] Consider the first order system

$$x' = y, \quad y' = z, \quad \epsilon z' = y^2 - xz - 1,$$

which is equivalent to the third order differential equation in Exercise 1.11. Suppose that the independent variable is $\tau \in \mathbb{R}$ and ϵ is a small parameter. (a) For the new independent variable $t = \tau/\epsilon$, show that the system is transformed to

$$\dot{x} = \epsilon y, \quad \dot{y} = \epsilon z, \quad \dot{z} = y^2 - xz - 1.$$

Note that a change in t of one unit is matched by a change in τ of ϵ units. For this reason, the variable τ is called *slow* and t is called *fast*. (b) Set $\epsilon = 0$ in the fast time system and prove that this system has an invariant manifold S, called the slow-manifold, that consists entirely of rest points. Identify this manifold as a quadric surface. Draw a picture. (c) Determine the stability type(s) of the rest points on the slow manifold. (d) For $\epsilon = 0$, the original slow-time system is "singular." In fact, if we set $\epsilon = 0$ in the slow-time system, then we obtain two differential equations coupled with an algebraic equation, namely,

$$x' = y, \quad y' = z, \quad y^2 - xz - 1 = 0.$$

Prove that the set $S := \{(x, y, z) : y^2 - xz - 1 = 0\}$ is a manifold in $\mathbb{R}^3$. For $W := \{(x, y) : x > 0\}$ and $G(x, y) := (x, y, (y^2 - 1)/x)$, show that (W, G) is a coordinate chart on S. The vector field

$$(x, y) \mapsto \left(x, y, y, \frac{y^2 - 1}{x} \right)$$

is derived from the unperturbed ($\epsilon = 0$) singular system by solving the algebraic equation for z. Draw the phase portrait of the corresponding differential equation. Show that the line $y = 1$ is invariant.

While the slow-time system is singular at $\epsilon = 0$, the fast-time system is regular. By understanding the fate of the slow-manifold for small $\epsilon > 0$ (it is not completely destroyed by the perturbation) the dynamics of the original slow-time system can be partially determined. For example, the original slow-time system has solutions that are attracted to the perturbed slow-manifold. Some of these are attracted to the line $y = 1$.

The subject of this exercise is called *singular perturbation theory*. A fundamental idea of the theory is to make appropriate changes of coordinates that are not defined when the small parameter ϵ is set to zero (for example, the change from slow to fast time $t = \tau/\epsilon$). After such a change of coordinates, the new system is equivalent to the old system only for nonzero values of ϵ. But, the dynamics of the "singular limit" system obtained by setting $\epsilon = 0$ in the transformed system might be easily understood; and, more importantly, some of this dynamical behavior might persist for small nonzero values of ϵ for which the transformed system is equivalent to the original system. Thus, dynamical information about the original system can be obtained by perturbing from (perhaps different choices of) singular limiting systems.

See Section 6.3 and equation (6.71) for the origin of this exercise, the book [179] for an introduction to singular perturbation theory, and the survey [132] for an introduction to geometric singular perturbation theory.

1.9   Periodic Solutions

We have seen that the stability of a rest point can often be determined by linearization or by an application of Lyapunov's direct method. In both cases, the stability can be determined by analysis in an arbitrary open ball that contains the rest point. For this reason, we say that the stability of a rest point is a *local* problem. On the other hand, to determine the stability of a periodic solution, we must consider the behavior of the corresponding vector field in a neighborhood of the entire periodic orbit. Because *global* methods must be employed, the analysis of periodic solutions is much more difficult (and more interesting) than the analysis of rest points. Some of the basic ideas that are used to study the existence and stability of periodic solutions will be introduced in this section.

1.9.1   The Poincaré Map

A very powerful concept in the study of periodic orbits is the Poincaré map. It is a corner stone of the "geometric theory" of Henri Poincaré [185], the

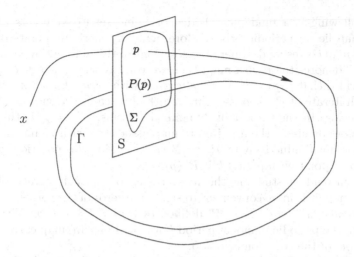

Figure 1.23: A Poincaré section Σ and the corresponding Poincaré return map. The trajectory starting at x is asymptotic to a periodic orbit Γ. The trajectory passes through the section Σ at the point p and first returns to the section at the point $P(p)$.

father of our subject. To define the Poincaré map, also called the return map, let ϕ_t denote the flow of the differential equation $\dot{x} = f(x)$, and suppose that $S \subseteq \mathbb{R}^n$ is an $(n-1)$-dimensional submanifold. If $p \in S$ and $(p, f(p)) \notin T_pS$, then we say that the vector $(p, f(p))$ is *transverse* to S at p. If $(p, f(p))$ is transverse to S at each $p \in S$, we say that S is a *section* for ϕ_t. If p is in S, then the curve $t \mapsto \phi_t(p)$ "passes through" S as t passes through $t = 0$. Perhaps there is some $T = T(p) > 0$ such that $\phi_T(p) \in S$. In this case, we say that the point p *returns* to S at time T. If there is an open subset $\Sigma \subseteq S$ such that each point of Σ returns to S, then Σ is called a *Poincaré section*. In this case, let us define $P : \Sigma \to S$ as follows: $P(p) := \phi_{T(p)}(p)$ where $T(p) > 0$ is the time of the first return to S. The map P is called the *Poincaré map*, or the *return map* on Σ and $T : \Sigma \to \mathbb{R}$ is called the *return time map* (see Figure 1.23). Because the solution of a differential equation is smoothly dependent on its initial value, the implicit function theorem can be used to prove that both P and T are smooth functions on Σ (see Exercise 1.147).

Exercise 1.147. Prove that the return-time map T is smooth. Hint: Find a function $F : \mathbb{R}^n \to \mathbb{R}$ so that $F(u) = 0$ if and only if $u \in \Sigma$ and define $G(t, u) = F(\phi_t(u))$. If $p \in \Sigma$ and T is the time of its first return, then apply the implicit function theorem to G at (T, p) to solve for T as a function of p.

The following is a fundamental idea of Poincaré: Fixed points of the return map lie on periodic orbits. More generally, periodic points of the Poincaré map correspond to periodic solutions of the differential equation. Here, if P denotes the return map, then we will say that p is a *fixed point* of P provided that $P(p) = p$. A *periodic point* with *period* k is a fixed point of the kth iterate of P—it passes through the Poincaré section $k - 1$ times before closing. In the subject of dynamical systems, $P^1 := P$ is the first iterate; more precisely, the first iterate map associated with P and the kth iterate is defined inductively by $P^k := P \circ P^{k-1}$. Using this notation, $p \in \Sigma$ is a periodic point with period k if $P^k(p) = p$.

Often, instead of studying the fixed points of the kth iterate of the Poincaré map, it is more convenient to study the zeros of the corresponding *displacement function* $\delta : \Sigma \to \mathbb{R}^n$ defined by $\delta(p) = P^k(p) - p$. With this definition, the periodic points of period k for the Poincaré map correspond to the roots of the equation $\delta(p) = 0$.

If $p \in \Sigma$ is a periodic point of the Poincaré map of period k, then the stability of the corresponding periodic orbit of the differential equation is determined by computing the eigenvalues of the linear map $DP^k(p)$. In fact, an important theorem, which we will prove in Section 2.4.4, states that if $P^k(p) = p$ and $DP^k(p)$ has all its eigenvalues inside the unit circle, then the periodic orbit with initial point p is asymptotically stable.

Exercise 1.148. Suppose that A is an 2×2 matrix and consider the linear transformation of $\mathbb{R}^2$ given by $x \mapsto Ax$ as a dynamical system. (a) Prove: If the spectrum of A lies inside the unit circle in the complex plane, then $A^k x \to 0$ as $k \to \infty$ for every $x \in \mathbb{R}^2$. (b) Prove: If at least one eigenvalue of A lies outside the unit circle, then there is a point $x \in \mathbb{R}^2$ such that $\|A^k x\| \to \infty$ as $k \to \infty$. (c) Define the notion of stability and asymptotic stability for discrete dynamical systems, and show that the origin is asymptotically stable for the linear dynamical system associated with A if and only if the spectrum of A lies inside the unit circle. (d) Suppose that the spectrum of A does not lie inside the unit circle. Give a condition that implies the origin is stable. See Section 2.4.4 for the $n \times n$ case.

Exercise 1.149. [One-dimensional Dynamics] A discrete dynamical system need not be invertible. For example, consider the quadratic family $f : [0,1] \to [0,1]$ defined by $f(x) = \lambda x(1-x)$, for λ in the interval $(0, 4]$. It defines a dynamical system via $x_{n+1} = f(x_n)$. (a) Prove: If $\lambda < 1$, then f has a globally attracting fixed point. (b) If $1 < \lambda < 3$, then f has a globally attracting nonzero fixed point. (c) Prove: A bifurcation occurs at $\lambda = 3$ such that for $\lambda > 3$ there is a periodic orbit with period two and this orbit is asymptotically stable for $3 < \lambda < 1 + \sqrt{6}$. (d) Prove: A bifurcation occurs at $\lambda = 1 + \sqrt{6}$ such that for $\lambda > 1 + \sqrt{6}$ there is a periodic orbit with period four. (e) In fact, a countable sequence of such bifurcations occur at $\lambda_1 = 3$, $\lambda_2 = 1 + \sqrt{6}$, ... so that a periodic orbit of period 2^n is born at λ_n. The sequence $\{\lambda_n\}_{n=1}^\infty$ converges to a number $\lambda_\infty \approx 3.57$. It turns out that $\lim_{n\to\infty}(\lambda_n - \lambda_{n-1})/(\lambda_{n+1} - \lambda_n) = 4.67 \cdots$. This is a universal constant (for families whose graphs have a unique nondegenerate maximum, e.g. $x \mapsto \lambda \sin(\pi x)$ for $x \in [0,1]$) called the Feigenbaum number. Verify these statements with numerical experiments. (f) For $\lambda = \lambda_\infty$, the dynamical system has periodic points of all periods. It is not difficult to prove that the system is chaotic at $\lambda = 4$. At least, it is not difficult to prove that the dynamics are as random as a coin toss. The ideas for a proof are in the next few exercises. (g) Prove that the tent map $h : [0, 1] \to [0, 1]$ given by $h(x) = 2x$, for $0 \le x \le 1/2$ and $h(x) = 2 - 2x$, for $1/2 \le x \le 1$ is semiconjugate to the quadratic map $f(x) = 4x(1 - x)$ via $g : [0, 1] \to [0, 1]$ given by $g(x) = \sin^2 \pi x$; that is $f(g(x)) = g(h(x))$ for $x \in [0, 1]$. (h) Prove: Every point in $[0, 1]$ can be represented by a binary decimal expansion $x = .x_1 x_2 x_3 \cdots$, where $x_n = 0$ or $x_n = 1$. (i) Prove: The map h acts on binary sequences by the rule $h(.x_1 x_2 x_3 \cdots) = .(x_1 \oplus x_2)(x_1 \oplus x_3)(x_1 \oplus x_4) \cdots$, where $\oplus$ is addition base two. (j) Use the binary representation of h to prove that for an arbitrary sequence of coin tosses, say $HHTTHT \cdots$, there is a point in $[0, 1]$ such that its iterates under h fall in the intervals $H = [0, 1/2)$ and $T = [1/2, 1]$ in the order specified by the coin tosses. Note: Much more can be said about one-dimensional dynamical systems (see, for example, [72],[96], [103], and [218]).

Although it is very difficult, in general, to find a suitable Poincaré section and to analyze the associated Poincaré map, there are many situations where these ideas can be used to great advantage. For example, suppose that there is a Poincaré section Σ and a closed ball $B \subseteq \Sigma$ such that $P : B \to B$. Recall Brouwer's fixed point theorem (see any book on algebraic topology, for example, [150] or [159]).

Theorem 1.150 (Brouwer's Fixed Point Theorem). *Every continuous map of a closed (Euclidean) ball into itself has at least one fixed point.*

By this theorem, the map P must have at least one fixed point. In other words, the associated differential equation has a periodic orbit passing through the set B. This idea is used in the following "toy" example. See Exercise 1.158 for an application.

Consider the nonautonomous differential equation

$$\dot{y} = (a \cos t + b)y - y^3, \qquad a > 0, \quad b > 0 \qquad (1.43)$$

and note that the associated vector field is time periodic with period 2π. To take advantage of this periodicity property, let us recast this differential equation—using the standard "trick"—as the first order system

$$\dot{y} = (a \cos \tau + b)y - y^3,$$
$$\dot{\tau} = 1. \qquad (1.44)$$

Also, for each $\xi \in \mathbb{R}$, let $t \mapsto (\tau(t, \xi), y(t, \xi))$ denote the solution of system (1.44) with the initial value

$$\tau(0, \xi) = 0, \quad y(0, \xi) = \xi$$

and note that $\tau(t, \xi) \equiv t$. Here, the order of the variables is reversed to conform with two conventions: The angular variable is written second in

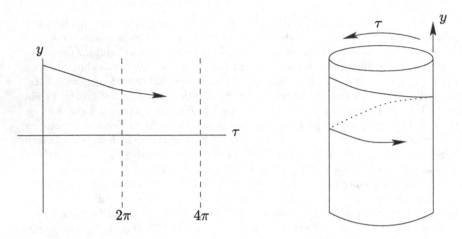

Figure 1.24: The phase cylinder for the differential equation (1.43).

a system of this type, but the phase portrait is depicted on a plane where the angular coordinate axis is horizontal.

The vector field corresponding to the system (1.44) is the same in every vertical strip of width 2π in the plane considered with coordinates (τ, y). Thus, from our geometric point of view, it is convenient to consider system (1.44) as a differential equation defined on the cylinder $\mathbb{T} \times \mathbb{R}$ obtained by identifying the line $\Sigma := \{(\tau, y) : \tau = 0\}$ with each line $\{(\tau, y) : \tau = 2\pi\ell\}$ where ℓ is an integer (see Figure 1.24). On this cylinder, Σ is a section for the flow. Moreover, if $\xi \in \mathbb{R}$ is the coordinate of a point on Σ, then the associated Poincaré map is given by

$$P(\xi) = y(2\pi, \xi)$$

whenever the solution $t \mapsto (\tau(t, \xi), y(t, \xi))$ is defined on the interval $[0, 2\pi]$.

By the definition of a Poincaré map, the fixed points of P correspond to periodic orbits of the differential equation defined on the phase cylinder. Let us prove that the fixed points of P correspond to periodic solutions of the original differential equation (1.43). In fact, it suffices to show that if $y(2\pi, \xi_0) = \xi_0$ for some $\xi_0 \in \mathbb{R}$, then $t \mapsto y(t, \xi_0)$ is a 2π-periodic solution of the differential equation (1.43).

By the extension theorem, there is some $t_* > 0$ such that the function $t \mapsto z(t)$ given by $z(t) := y(t + 2\pi, \xi_0)$ is defined on the interval $[0, t_*)$. Note

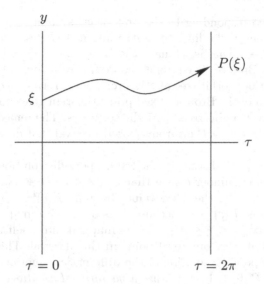

Figure 1.25: The Poincaré map for the system (1.44).

that $z(0) = y(2\pi, \xi_0) = \xi_0$ and

$$\begin{aligned}
\dot{z}(t) &= \dot{y}(t + 2\pi, \xi_0) \\
&= (a(\cos(t + 2\pi)) + b)y(t + 2\pi, \xi_0) - y^3(t + 2\pi, \xi_0) \\
&= (a\cos t + b)y(t + 2\pi, \xi_0) - y^3(t + 2\pi, \xi_0) \\
&= (a\cos t + b)z(t) - z^3(t).
\end{aligned}$$

Thus, $t \mapsto z(t)$ is a solution of the differential equation (1.43) with the same initial value as the solution $t \mapsto y(t, \xi_0)$. By the uniqueness theorem, it follows that $z(t) = y(t, \xi_0)$ for $0 \le t < t_*$. Hence, if $t \mapsto y(t + 2\pi, \xi_0)$ blows up on the interval $t_* \le t \le 2\pi$, then so does the function $t \mapsto y(t, \xi_0)$, contrary to the hypothesis. Thus, $t \mapsto y(t, \xi_0)$ is defined on the interval $[0, 4\pi]$ and $y(t + 2\pi, \xi_0) = y(t, \xi_0)$ for $0 \le t \le 2\pi$. By repeating the argument inductively with $z(t) = y(t + k2\pi, \xi_0)$ for the integers $k = 2, 3, \dots$, it follows that $t \mapsto y(t, \xi_0)$ is a 2π-periodic solution of the differential equation (1.43), as required.

Because $y(t, 0) \equiv 0$, it follows immediately that $P(0) = 0$; that is, the point $\xi = 0$ corresponds to a periodic orbit. To find a nontrivial periodic solution, note that $a\cos t + b \le a + b$, and consider the line given by $y = a + b + 1$ in the phase cylinder. The y-component of the vector field on this line is

$$(a + b + 1)(a\cos\tau + b - (a + b + 1)^2).$$

Since

$$a\cos\tau + b - (a + b + 1)^2 \le (a + b + 1) - (a + b + 1)^2 < 0,$$

the vector field corresponding to the first order system "points" into the region that lies below the line. In particular, if $0 \le \xi \le a + b + 1$, then $0 \le P(\xi) \le a + b + 1$; that is, P maps the closed interval $[0, a + b + 1]$ into itself. Hence, the Brouwer fixed point theorem can be applied to prove the existence of a periodic orbit (see also Exercise 1.151). But, because $P(0) = 0$, this application of the Brouwer fixed point theorem gives no information about the existence of nontrivial periodic solutions. The remedy, as we will soon see, is to construct a P invariant closed interval that does not contain $\xi = 0$.

Suppose that $P'(0) > 1$; that is, the trivial periodic solution is unstable. Then, there is some number c such that $0 < c < a + b + 1$ and $P'(\xi) > 1$ as long as $0 \le \xi \le c$. By the mean value theorem, $P(c) = P'(\xi)c$ for some ξ, $0 < \xi < c$. Thus, $P(c) > c$. Because P is a Poincaré map, it is easy to see that the interval $c \le \xi \le a + b + 1$ is mapped into itself by P and, as a result, there is at least one fixed point in this interval. This fixed point corresponds to a periodic solution of the differential equation (1.43).

To prove that $P'(0) > 1$ we will use a *variational equation*. This method is employed very often in the analysis of differential equations. The present elementary example is a good place to learn the basic technique. The idea is simple: The derivative of the solution of a differential equation with respect to its initial value is itself the solution of a differential equation.

Recall that $P(\xi) = y(2\pi, \xi)$. Since

$$\frac{d}{dt}y(t,\xi) = (a\cos t + b)y(t,\xi) - y^3(t,\xi)$$

we have that

$$\frac{d}{dt}y_\xi(t,\xi) = (a\cos t + b)y_\xi(t,\xi) - 3y^2(t,\xi)y_\xi(t,\xi).$$

Because $y(0,\xi) = \xi$, we also have the initial condition $y_\xi(0,\xi) = 1$. Moreover, at the point $\xi = 0$ the function $t \mapsto y(t,\xi)$ is identically zero. Thus, if $t \to w(t)$ is the solution of the variational initial value problem

$$\dot{w} = (a\cos t + b)w, \qquad w(0) = 1,$$

then $P'(0) = w(2\pi)$.

The variational differential equation is linear. Its solution is given by

$$w(t) = e^{\int_0^t (a\cos s + b)\,ds} = e^{a\sin t + bt}.$$

In particular, we have

$$P'(0) = w(2\pi) = e^{2\pi b} > 1,$$

as required. Moreover, this computation shows that the periodic solution given by $y(t) \equiv 0$ is unstable. (Why?)

Exercise 1.151. Prove Brouwer's fixed point theorem for a closed interval in $\mathbb{R}$. Hint: Use the intermediate value theorem.

Exercise 1.152. Find the initial point for the nontrivial periodic solution in the interval $0 < \xi < a+b+1$ for (1.43) as a function of a and b. Are there exactly two periodic solutions?

Exercise 1.153. Find conditions on $a(t)$ and on f that ensure the existence of at least one (nontrivial) periodic solution for a differential equation of the form

$$\dot{y} = a(t)y + f(y).$$

Exercise 1.154. Consider the differential equation (1.43) on the cylinder, and the transformation given by $u = (y+1)\cos\tau$, $v = (y+1)\sin\tau$ that maps the portion of the cylinder defined by the inequality $y > -1$ into the plane. What is the image of this transformation? Find the differential equation in the new coordinates, and draw its phase portrait.

We have proved that there is at least one 2π-periodic solution of the differential equation (1.43) with initial condition in the interval $0 < \xi < a + b + 1$. But even more is true: This periodic orbit is stable and unique. To prove this fact, let us suppose that $0 < \xi_0 < a+b+1$ and $P(\xi_0) = \xi_0$, so that the corresponding solution $t \mapsto y(t, \xi_0)$ is 2π-periodic.

To determine the stability type of the solution with initial value ξ_0, it suffices to compute $P'(\xi_0)$. As before, $P'(\xi_0) = w(2\pi)$ where $t \mapsto w(t)$ is the solution of the variational initial value problem

$$\dot{w} = [(a\cos t + b) - 3y^2(t, \xi_0)]w, \qquad w(0) = 1.$$

It follows that

$$P'(\xi_0) = w(2\pi)$$
$$= e^{\int_0^{2\pi} a\cos t + b - 3y^2(t, \xi_0)\, dt}$$
$$= e^{2\pi b - 3\int_0^{2\pi} y^2(t, \xi_0)\, dt}.$$

To compute $\int_0^{2\pi} y^2\, dt$, note that because $y(t, \xi_0) > 0$ for all t, we have the following equality

$$\frac{\dot{y}(t, \xi_0)}{y(t, \xi_0)} = a\cos t + b - y^2(t, \xi_0).$$

Using this formula and the periodicity of the solution $t \mapsto y(t, \xi_0)$, we have that

$$\int_0^{2\pi} y^2(t, \xi_0)\, dt = 2\pi b - \int_0^{2\pi} \frac{\dot{y}(t, \xi_0)}{y(t, \xi_0)}\, dt = 2\pi b,$$

and, as a result,

$$P'(\xi_0) = e^{2\pi b - 3(2\pi b)} = e^{-4\pi b} < 1. \qquad (1.45)$$

Hence, every periodic solution in the interval $[0, a + b + 1]$ is stable. The uniqueness of the periodic solution is a consequence of this result. In fact, the map P is real analytic. Thus, if P has infinitely many fixed points in a compact interval, then P is the identity. This is not true, so P has only a finite number of fixed points. If ξ_0 and ξ_1 are the coordinates of two consecutive fixed points, then the displacement function, that is, $\xi \mapsto P(\xi) - \xi$, has negative slope at two consecutive zeros, in contradiction.

Exercise 1.155. Find an explicit formula for the solution of the differential equation (1.43) and use it to give a direct proof for the existence of a nontrivial periodic solution.

Exercise 1.156. Prove that $P''(\xi) < 0$ for $\xi > 0$, where P is the Poincaré map defined for the differential equation (1.43). Use this result and the inequality (1.45) to prove the uniqueness of the nontrivial periodic solution of the differential equation.

Exercise 1.157. Show that the (stroboscopic) Poincaré map for the differential equation (1.43) has exactly one fixed point on the interval $(0, \infty)$. How many fixed points are there on $(-\infty, \infty)$?

Exercise 1.158. Suppose that $h : \mathbb{R} \to \mathbb{R}$ is a T-periodic function, and $0 < h(t) < 1/4$ for every $t \in \mathbb{R}$. Show that the differential equation $\dot{x} = x(1-x) - h(t)$ has exactly two T-periodic solutions. The differential equation can be interpreted as a model for the growth of a population in a limiting environment that is subjected to periodic harvesting (cf. [200]).

Exercise 1.159. Is it possible for the Poincaré map for a scalar differential equation not to be the identity map on a fixed compact interval and at the same time have infinitely many fixed points in the interval?

Exercise 1.160. [Boundary Value Problem] (a) Prove that the Dirichlet boundary value problem

$$x'' = 1 - x^2, \quad x(0) = 0, \quad x(2) = 0$$

has a solution. Hint: Use the phase plane. Show that the first positive time T such that the orbit with initial conditions $x(0) = 0$ and $x'(0) = 0$ reaches the x-axis is $T < 2$ and for the initial conditions $x(0) = 0$ and $x'(0) = 2/\sqrt{3}$, $T > 2$. To show this fact use the idea in the hint for Exercise 1.12 to construct an integral representation for T. (b) Find a solution of the boundary value problem by shooting and Newton's method (see Exercise 1.124). Hint: Use the phase plane with $x' = y$. Consider the solution $t \mapsto (x(t, \eta), y(t, \eta))$ with initial conditions $x(0) = 0$ and $y(0) = \eta$ and use Newton's method to solve the equation $y(2, \eta) = 0$. Note: The solutions with different choices for the velocity are viewed as shots. The velocity is adjusted until the target is hit.

Exercise 1.161. Consider the linear system

$$\dot{x} = ax, \quad \dot{y} = -by$$

where $a > 0$ and $b > 0$ in the open first quadrant of the phase plane and let ϕ_t denote its flow. (a) Show that $L := \{(\xi, 1) : \xi > 0\}$ and $M := \{(1, \eta) : \eta > 0\}$ are transverse sections for the system. (b) Find a formula for the section map h from L to M. (c) Find a formula for $T : L \to \mathbb{R}$, called the time-of-flight map, which is defined by $\phi_{T(\xi)}(\xi, 1) = (1, h(\xi))$.

Exercise 1.162. Compute the time required for the solution of the system

$$\dot{x} = x(1 - y), \qquad \dot{y} = y(x - 1)$$

with initial condition $(x, y) = (1, 0)$ to arrive at the point $(x, y) = (2, 0)$. Note that this system has a section map $y \mapsto h(y)$ defined from a neighborhood of $(x, y) = (1, 0)$ on the line given by $x = 1$ to the line given by $x = 2$. Compute $h'(0)$.

Exercise 1.163. Observe that the x-axis is invariant for the system

$$\dot{x} = 1 + xy, \qquad \dot{y} = 2xy^2 + y^3,$$

and the trajectory starting at the point $(1, 0)$ crosses the line $x = 3$ at $(3, 0)$. Thus, there is a section map h and a time-of-flight map T from the line $x = 1$ to the line $x = 3$ with both functions defined on some open interval about the point $(1, 0)$ on the line $x = 1$. Compute $T'(0)$ and $h'(0)$.

Exercise 1.164. Research Problem: Consider the second order differential equation

$$\ddot{x} + f(x)\dot{x} + g(x) = 0$$

where f and g are 2π-periodic functions. Determine conditions on f and g that ensure the existence of a periodic solution.

1.9.2   Limit Sets and Poincaré–Bendixson Theory

The general problem of finding periodic solutions for differential equations is still an active area of mathematical research. Perhaps the most well developed theory for periodic solutions is for differential equations defined on the plane. But, even in this case, the theory is far from complete. For example, consider the class of planar differential equations of the form

$$\dot{x} = f(x, y), \qquad \dot{y} = g(x, y)$$

where f and g are quadratic polynomials. There are examples of such "quadratic systems" that have four isolated periodic orbits—"isolated" means that each periodic orbit is contained in an open subset of the plane that contains no other periodic orbits (see Exercise 1.194). But, no one knows at present if there is a quadratic system with more than four isolated periodic orbits. The general question of the number of isolated periodic orbits for a polynomial system in the plane has been open since 1905; it is called Hilbert's 16th problem (see [55], [126], [187], and [197]).

Although there are certainly many difficult issues associated with periodic orbits of planar systems, an extensive theory has been developed that has been successfully applied to help determine the dynamics of many mathematical models. Some of the basic results of this theory will be explained later in this section after we discuss some important general properties of flows of autonomous, not necessarily planar, systems.

The properties that we will discuss enable us to begin to answer the question "What is the long term behavior of a dynamical system?" This is often the most important question about a mathematical model. Ask an engineer what he wants to know about a model ordinary differential equation. Often his response will be the question "What happens if we start the system running and then wait for a long time?" or, in engineering jargon, "What is the steady state behavior of the system?" We already know how to answer these questions in some special circumstances where the steady state behavior corresponds to a rest point or periodic orbit. The following definitions will be used to precisely describe the limiting behavior of an arbitrary orbit.

Definition 1.165. Suppose that ϕ_t is a flow on $\mathbb{R}^n$ and $p \in \mathbb{R}^n$. A point x in $\mathbb{R}^n$ is called an *omega limit point* (ω-limit point) of the orbit through p if there is a sequence of numbers $t_1 \le t_2 \le t_3 \le \cdots$ such that $\lim_{i \to \infty} t_i = \infty$ and $\lim_{i \to \infty} \phi_{t_i}(p) = x$. The collection of all such omega limit points is denoted $\omega(p)$ and is called the *omega limit set* (ω-limit set) of p. Similarly, the α-limit set $\alpha(p)$ is defined to be the set of all limits $\lim_{i \to \infty} \phi_{t_i}(p)$ where $t_1 \ge t_2 \ge t_3 \ge \cdots$ and $\lim_{i \to \infty} t_i = -\infty$.

Definition 1.166. The orbit of the point p with respect to the flow ϕ_t is called *forward complete* if $t \to \phi_t(p)$ is defined for all $t \ge 0$. Also, in this case, the set $\{\phi_t(p) : t \ge 0\}$ is called the *forward orbit* of the point p. The orbit is called *backward complete* if $t \to \phi_t(p)$ is defined for all $t \le 0$ and the backward orbit is $\{\phi_t(p) : t \le 0\}$.

Proposition 1.167. *The omega limit set of a point is closed and invariant.*

Proof. The empty set is closed and invariant.

Suppose that $\omega(p)$ is not empty for the flow ϕ_t and $x \in \omega(p)$. Consider $\phi_T(x)$ for some fixed $T \in \mathbb{R}$. There is a sequence $t_1 \le t_2 \le t_3 \le \cdots$ with $t_i \to \infty$ and $\phi_{t_i}(p) \to x$ as $i \to \infty$. Note that $t_1 + T \le t_2 + T \le t_3 + T \le \cdots$ and that $\phi_{t_i + T}(p) = \phi_T(\phi_{t_i}(p))$. By the continuity of the flow, we have that $\phi_T(\phi_{t_i}(p)) \to \phi_T(x)$ as $i \to \infty$. Thus, $\phi_T(x) \in \omega(p)$, and therefore $\omega(p)$ is an invariant set.

To show $\omega(p)$ is closed, it suffices to show that $\omega(p)$ is the intersection of closed sets. In fact, we have that

$$\omega(p) = \bigcap_{\tau \ge 0} \text{closure}\, \{\phi_t(p) : t \ge \tau\}. \qquad \square$$

Proposition 1.168. *Suppose that $p \in \mathbb{R}^n$ and the orbit of the flow ϕ_t through the point p is forward complete. If the forward orbit of p has compact closure, then $\omega(p)$ is nonempty, compact, and connected.*

Proof. The sequence $\{\phi_n(p)\}_{n=1}^{\infty}$ is contained in the compact closure of the orbit through p. Thus, it has at least one limit point x. In fact, there is an infinite sequence of integers $n_1 \leq n_2 \leq \cdots$ such that $\phi_{n_i}(p) \to x$ as $i \to \infty$. Hence, $x \in \omega(p)$, and therefore $\omega(p) \neq \emptyset$.

Since $\omega(p)$ is a closed subset of the compact closure of the orbit through p, the set $\omega(p)$ is compact.

To prove that $\omega(p)$ is connected, suppose to the contrary that there are two disjoint open sets U and V whose union contains $\omega(p)$ such that $\omega(p) \cap U \neq \emptyset$ and $\omega(p) \cap V \neq \emptyset$. There is some $t_1 > 0$ such that $\phi_{t_1}(p) \in U$ and some $t_2 > t_1$ such that $\phi_{t_2}(p) \in V$. But the set $K = \{\phi_t(p) : t_1 \leq t \leq t_2\}$ is the continuous image of an interval, hence a connected set. Thus K cannot be contained in $U \cup V$. In particular, there is at least one $\tau_1 > 0$ such that $\phi_{\tau_1}(p)$ is not in this union.

Similarly we can construct a sequence $\tau_1 \leq \tau_2 \leq \cdots$ such that

$$\lim_{i \to \infty} \tau_i = \infty$$

and for each i the point $\phi_{\tau_i}(p)$ is in the complement of $U \cup V$. By the compactness, the sequence $\{\phi_{\tau_i}(p)\}_{i=1}^{\infty}$ has a limit point x. Clearly, x is also in $\omega(p)$ and in the complement of $U \cup V$. This is a contradiction. $\square$

Exercise 1.169. Construct examples to show that the compactness hypothesis of Proposition 1.168 is necessary.

Exercise 1.170. Show that a reparametrization of a flow does not change its omega limit sets. Thus, an omega limit set is determined by an orbit and its direction, not the parametrization of the orbit.

Exercise 1.171. Suppose that x_0 is a rest point for the differential equation $\dot{x} = f(x)$ with flow ϕ_t, and V is a Lyapunov function at x_0. If, in addition, there is a neighborhood W of the rest point x_0 such that, for each point $p \in W \setminus \{x_0\}$, the function V is not constant on the forward orbit of p, then x_0 is asymptotically stable. Hint: The point x_0 is Lyapunov stable. If it is not asymptotically stable, then there is a point p in the domain of V whose omega limit set $\omega(p)$ is also in the domain of V such that $\omega(p) \neq \{x_0\}$. Show that V is constant on this omega limit set (the constant is the greatest lower bound of the range of V on the forward orbit through p).

Exercise 1.172. Suppose that the differential equation $\dot{x} = f(x)$ with flow ϕ_t has a compact invariant set K, and $V : K \to \mathbb{R}$ is a continuously differentiable function such that $\dot{V}(x) \leq 0$ for every $x \in K$. If Ω is the largest invariant set in $\{x \in K : \dot{V}(x) = 0\}$, then every solution in K approaches Ω as $t \to \infty$.

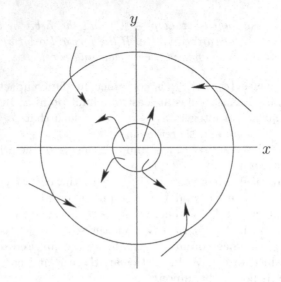

Figure 1.26: A positively invariant annular region for a flow in the plane.

The ω-limit set of a point for a flow in $\mathbb{R}^n$ with $n \geq 3$ can be very complicated; for example, it can be a fractal. But the situation in $\mathbb{R}^2$ is much simpler. The reason is the deep fact about the geometry of the plane stated in the next theorem.

Theorem 1.173 (Jordan Curve Theorem). *A simple closed (continuous) curve in the plane divides the plane into two connected components, one bounded and one unbounded, each with the curve as boundary.*

Proof. Modern proofs of this theorem use algebraic topology (see for example [212]). □

This result will play a central role in what follows.

The fundamental result about limit sets for flows of planar differential equations is the Poincaré–Bendixson theorem. There are several versions of this theorem; we will state two of them. The main ingredients of their proofs will be presented later in this section beginning with Lemma 1.187.

Theorem 1.174 (Poincaré–Bendixson). *If Ω is a nonempty compact ω-limit set of a flow in $\mathbb{R}^2$, and if Ω does not contain a rest point, then Ω is a periodic orbit.*

A set S that contains the forward orbit of each of its elements is called *positively invariant*. An orbit whose α-limit set is a rest point p and whose ω-limit is a rest point q is said to *connect* p and q. Note: the definition of a connecting orbit allows $p = q$.

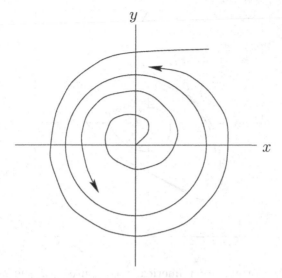

Figure 1.27: A limit cycle in the plane.

Theorem 1.175. *Suppose that ϕ_t is a flow on $\mathbb{R}^2$ and $S \subseteq \mathbb{R}^2$ is a positively invariant set with compact closure. If $p \in S$ and ϕ_t has at most a finite number of rest points in the closure of S, then $\omega(p)$ is either (i) a rest point, (ii) a periodic orbit, or (iii) a union of finitely many rest points and a nonempty finite or countable infinite set of connecting orbits.*

Exercise 1.176. Illustrate possibility (iii) of the last theorem with an example having an infinite set of connecting orbits.

Exercise 1.177. We have assumed that all flows are smooth. Is this hypothesis required for all the theorems in this section on ω-limit sets?

Definition 1.178. A *limit cycle* Γ is a periodic orbit that is either the ω-limit set or the α-limit set of some point that is in the phase space but not in Γ.

A "conceptual" limit cycle is illustrated in Figure 1.27. In this figure, the limit cycle is the ω-limit set of points in its interior (the bounded component of the plane with the limit cycle removed) and its exterior (the corresponding unbounded component of the plane). A limit cycle that is generated by numerically integrating a planar differential equation is depicted in Figure 1.28 (see [33]).

Sometimes the following alternative definition of a limit cycle is given. A "limit cycle" is an isolated periodic orbit; that is, the unique periodic orbit

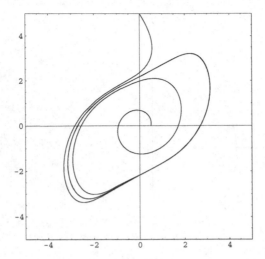

Figure 1.28: Two orbits are numerically computed for the system $\dot{x} = 0.5x - y + 0.1(x^2 - y^2)(x - y)$, $\dot{y} = x + 0.5y + 0.1(x^2 - y^2)(x + y)$: one with initial value $(x, y) = (0.5, 0)$, the other with initial value $(x, y) = (0, 5)$. Both orbits approach a stable limit cycle.

in some open subset of the phase space. This definition is *not* equivalent to Definition 1.178 in general. The two definitions, however, are equivalent for real analytic systems in the plane (see Exercise 1.182).

An annular region is a subset of the plane that is homeomorphic to the closed annulus bounded by the unit circle at the origin and the concentric circle whose radius is two units in length.

The following immediate corollary of the Poincaré–Bendixson theorem is often applied to prove the existence of limit cycles for planar systems.

Theorem 1.179. *If a flow in the plane has a positively invariant annular region S that contains no rest points of the flow, then S contains at least one periodic orbit. If in addition, some point in S is in the forward orbit of a point on the boundary of S, then S contains at least one limit cycle.*

We will discuss two applications of Theorem 1.179 where the main idea is to find a rest-point free annular region as depicted in Figure 1.26.

The first example is provided by the differential equation

$$\dot{x} = -y + x(1 - x^2 - y^2), \quad \dot{y} = x + y(1 - x^2 - y^2). \tag{1.46}$$

Note that the annulus S bounded by the circles with radii $\frac{1}{2}$ and 2, respectively, contains no rest points of the system. Let us show that S is positively invariant. To prove this fact, consider the outer normal vector N on ∂S that is the restriction of the vector field $N(x, y) = (x, y, x, y) \in \mathbb{R}^2 \times \mathbb{R}^2$ to ∂S and compute the dot product of N with the vector field corresponding to the differential equation. In fact, the dot product

$$x^2(1 - x^2 - y^2) + y^2(1 - x^2 - y^2) = (x^2 + y^2)(1 - x^2 - y^2)$$

is positive on the circle with radius $\frac{1}{2}$ and negative on the circle with radius 2. Therefore, S is positively invariant and, by Theorem 1.179, there is at least one limit cycle in S.

The differential equation (1.46) is so simple that we can find a formula for its flow. In fact, by changing to polar coordinates (r, θ), the transformed system

$$\dot{r} = r(1 - r^2), \qquad \dot{\theta} = 1$$

decouples, and its flow is given by

$$\phi_t(r, \theta) = \left(\left(\frac{r^2 e^{2t}}{1 - r^2 + r^2 e^{2t}} \right)^{\frac{1}{2}}, \theta + t \right). \qquad (1.47)$$

Note that $\phi_t(1, \theta) = (1, \theta + t)$ and, in particular, $\phi_{2\pi}(1, \theta) = (1, \theta + 2\pi)$. Thus, the unit circle in the plane is a periodic orbit with period 2π. Here, of course, we must view θ as being defined modulo 2π, or, better yet, we must view the polar coordinates as coordinates on the cylinder $\mathbb{T} \times \mathbb{R}$ (see Section 1.8.5).

If the formula for the flow (1.47) is rewritten in rectangular coordinates, then the periodicity of the unit circle is evident. In fact, the periodic solution starting at the point $(\cos \theta, \sin \theta) \in \mathbb{R}^2$ (in rectangular coordinates) at $t = 0$ is given by

$$t \mapsto (x(t), y(t)) = (\cos(\theta + t), \sin(\theta + t)).$$

It is easy to see that if $r \neq 0$, then the ω-limit set $\omega((r, \theta))$ is the entire unit circle. Thus, the unit circle is a limit cycle.

If we consider the positive x-axis as a Poincaré section, then we have

$$P(x) = \left(\frac{x^2 e^{4\pi}}{1 - x^2 + x^2 e^{4\pi}} \right)^{\frac{1}{2}}.$$

Here $P(1) = 1$ and $P'(1) = e^{-4\pi} < 1$. In other words, the intersection point of the limit cycle with the Poincaré section is a hyperbolic fixed point of the Poincaré map; that is, the linearized Poincaré map has no eigenvalue on the unit circle of the complex plane. In fact, here the single eigenvalue of the linear transformation of $\mathbb{R}$ given by $x \mapsto P'(1)x$ is inside the unit circle. It should be clear that in this case the limit cycle is an asymptotically stable periodic orbit. We will also call such an orbit a *hyperbolic stable limit cycle*. (The general problem of the stability of periodic orbits is discussed in Chapter 2.)

As a second example of the application of Theorem 1.179, let us consider the very important differential equation

$$\ddot{\theta} + \lambda\dot{\theta} + \sin\theta = \mu$$

where $\lambda > 0$ and μ are constants, and θ is an angular variable; that is, θ is defined modulo 2π. This differential equation is a model for an unbalanced rotor or pendulum with viscous damping $\lambda\dot{\theta}$ and external torque μ.

Consider the equivalent first order system

$$\dot{\theta} = v, \quad \dot{v} = -\sin\theta + \mu - \lambda v, \tag{1.48}$$

and note that, since θ is an angular variable, the natural phase space for this system is the cylinder $\mathbb{T} \times \mathbb{R}$. With this interpretation we will show the following result: *If $|\mu| > 1$, then system (1.48) has a globally attracting limit cycle.* The phrase "globally attracting limit cycle" means that there is a limit cycle Γ on the cylinder and Γ is the ω-limit set of every point on the cylinder. In other words, the steady state behavior of the unbalanced rotor, with viscous damping and sufficiently large torque, is stable periodic motion. (See [143] for the existence of limit cycles in case $|\mu| \leq 1$.)

The system (1.48) with $|\mu| > 1$ has no rest points. (Why?) Also the quantity $-\sin\theta + \mu - \lambda v$ is negative for sufficiently large positive values of v, and it is positive for negative values of v that are sufficiently large in absolute value. Therefore, there are numbers $v_- < 0$ and $v_+ > 0$ such that every forward orbit is contained in the compact subset of the cylinder $A := \{(r, \theta) : v_- \leq v \leq v_+\}$. In addition, A is diffeomorphic to an annular region in the plane. It follows that the Poincaré–Bendixson theorem is valid in A, and therefore the ω-limit set of every point on the cylinder is a limit cycle.

Although there are several ways to prove that the limit cycle is unique, let us consider a proof based on the following propositions: (i) *If the divergence of a vector field is everywhere negative, then the flow of the vector field contracts volume* (see Exercise 2.22). (ii) *Every periodic orbit in the plane surrounds a rest point* (see Exercise 1.189). (A replacement for the first proposition is given in Exercise 1.200; an alternate method of proof is suggested in Exercise 1.202.)

To apply the propositions, note that the divergence of the vector field for system (1.48) is the negative number $-\lambda$. Also, if $|\mu| > 1$, then this system has no rest points. By the second proposition, no periodic orbit of the system is contractable on the cylinder (see panel (a) of Figure 1.29). Thus, if there are two periodic orbits, they must bound an *invariant* annular region on the cylinder as in panel (b) of Figure 1.29. But this contradicts the fact that the area of the annular region is contracted by the flow. It follows that there is a unique periodic orbit on the cylinder that is a globally attracting limit cycle.

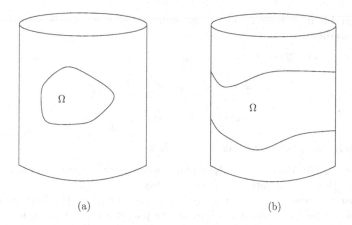

(a)   (b)

Figure 1.29: Panel (a) depicts a contractable periodic orbit on a cylinder. Note that the region Ω in panel (a) is simply connected. Panel (b) depicts two periodic orbits that are not contractable; they bound a multiply connected region Ω on the cylinder.

Exercise 1.180. Give a direct proof that the point $(1/\sqrt{2}, 1/\sqrt{2})$ on the unit circle is an ω-limit point of the point $(3, 8)$ for the flow of system (1.46).

Exercise 1.181. Discuss the phase portrait of system (1.48) for $|\mu| < 1$.

Exercise 1.182. (a) Show that the set containing "limit cycles" defined as isolated periodic orbits is a proper subset of the set of limit cycles. Also, if the differential equation is a real analytic planar autonomous system, then the two concepts are the same. Hint: Imagine an annular region consisting entirely of periodic orbits. The boundary of the annulus consists of two periodic orbits that might be limit cycles, but neither of them is isolated. To prove that an isolated periodic orbit Γ is a limit cycle, show that every section of the flow at a point $p \in \Gamma$ has a subset that is a Poincaré section at p. For an analytic system, again consider a Poincaré section and the associated Poincaré map P. Zeros of the *analytic* displacement function $\xi \mapsto P(\xi) - \xi$ correspond to periodic orbits. (b) Show that the polynomial (hence real analytic) system in $\mathbb{R}^3$ given by

$$\dot{x} = -y + x(1 - x^2 - y^2),$$
$$\dot{y} = x + y(1 - x^2 - y^2),$$
$$\dot{z} = 1 - x^2 - y^2 \qquad\qquad (1.49)$$

has limit cycles that are not isolated. (c) Determine the long-term behavior of the system (1.49). In particular, show that

$$\lim_{t \to \infty} z(t) = z(0) - \frac{1}{2}\ln(x^2(0) + y^2(0)).$$

Exercise 1.183. Show that the system

$$\dot{x} = ax - y + xy^2, \quad \dot{y} = x + ay + y^3$$

has an unstable limit cycle for $a < 0$ and no limit cycle for $a > 0$. Hint: Change to polar coordinates.

Exercise 1.184. Show that the system

$$\dot{x} = y + x(x^2 + y^2 - 1)\sin\frac{1}{x^2 + y^2 - 1},$$

$$\dot{y} = -x + y(x^2 + y^2 - 1)\sin\frac{1}{x^2 + y^2 - 1}$$

has infinitely many limit cycles in the unit disk.

Exercise 1.185. Prove: An analytic planar system cannot have infinitely many limit cycles that accumulate on a periodic orbit. Note: This (easy) exercise is a special case of a deep result: An analytic planar system cannot have infinitely many limit cycles in a compact subset of the plane; and, a polynomial system cannot have infinitely many limit cycles (see [79] and [126]).

Exercise 1.186. Consider the differential equation

$$\dot{x} = -ax(x^2 + y^2)^{-1/2}, \quad \dot{y} = -ay(x^2 + y^2)^{-1/2} + b$$

where a and b are positive parameters. The model represents the flight of a projectile, with speed a and heading toward the origin, that is moved off course by a constant force with strength b. Determine conditions on the parameters that ensure the solution starting at the point $(x, y) = (p, 0)$, for $p > 0$, reaches the origin. Hint: Change to polar coordinates and study the phase portrait of the differential equation on the cylinder. Explain your result geometrically. The differential equation is not defined at the origin. Is this a problem?

The next two lemmas are used in the proof of the Poincaré Bendixson theorem. The first lemma is a corollary of the Jordan curve theorem.

Lemma 1.187. *If Σ is a section for the flow ϕ_t and if $p \in \mathbb{R}^2$, then the orbit through the point p intersects Σ in a monotone sequence; that is, if $\phi_{t_1}(p)$, $\phi_{t_2}(p)$, and $\phi_{t_3}(p)$ are on Σ and if $t_1 < t_2 < t_3$, then $\phi_{t_2}(p)$ lies strictly between $\phi_{t_1}(p)$ and $\phi_{t_3}(p)$ on Σ or $\phi_{t_1}(p) = \phi_{t_2}(p) = \phi_{t_3}(p)$.*

Proof. The proof is left as an exercise. Hint: Reduce to the case where t_1, t_2, and t_3 correspond to consecutive crossing points. Then, consider the curve formed by the union of $\{\phi_t(p) : t_1 \leq t \leq t_2\}$ and the subset of Σ between $\phi_{t_1}(p)$ and $\phi_{t_2}(p)$. Draw a picture.   □

Lemma 1.188. *If Σ is a section for the flow ϕ_t and if $p \in \mathbb{R}^2$, then $\omega(p) \cap \Sigma$ contains at most one point.*

Proof. The proof is by contradiction. Suppose that $\omega(p) \cap \Sigma$ contains at least two points, x_1 and x_2. By rectification of the flow at x_1 and at x_2, that is, by the rectification lemma (Lemma 1.120), it is easy to see that there are sequences $\{\phi_{t_i}(p)\}_{i=1}^{\infty}$ and $\{\phi_{s_i}(p)\}_{i=1}^{\infty}$ in Σ such that $\lim_{i \to \infty} \phi_{t_i}(p) = x_1$ and $\lim_{i \to \infty} \phi_{s_i}(p) = x_2$. By the rectification lemma in Exercise 1.125, such sequences can be found in Σ. Indeed, we can choose the rectifying neighborhood so that the image of the Poincaré section is a line segment transverse to the rectified flow. In this case, it is clear that if an orbit has one of its points in the rectifying neighborhood, then this orbit passes through the Poincaré section.

By choosing a local coordinate on Σ, let us assume that Σ is an open interval. Working in this local chart, there are open subintervals J_1 at x_1 and J_2 at x_2 such that $J_1 \cap J_2 = \emptyset$. Moreover, by the definition of limit sets, there is an integer m such that $\phi_{t_m}(p) \in J_1$; an integer n such that $s_n > t_m$ and $\phi_{s_n}(p) \in J_2$; and an integer ℓ such that $t_\ell > s_n$ and $\phi_{t_\ell}(p) \in J_1$. By Lemma 1.187, the point $\phi_{s_n}(p)$ must be between the points $\phi_{t_m}(p)$ and $\phi_{t_\ell}(p)$ on Σ. But this is impossible because the points $\phi_{t_m}(p)$ and $\phi_{t_\ell}(p)$ are in J_1, whereas $\phi_{s_n}(p)$ is in J_2. $\square$

We are now ready to prove the Poincaré–Bendixson theorem (Theorem 1.174): *If Ω is a nonempty compact ω-limit set of a flow in $\mathbb{R}^2$, and if Ω does not contain a rest point, then Ω is a periodic orbit.*

Proof. Suppose that $\omega(p)$ is nonempty, compact, and contains no rest points. Choose a point $q \in \omega(p)$. We will show first that the orbit through q is closed.

Consider $\omega(q)$. Note that $\omega(q) \subseteq \omega(p)$ and $\omega(q)$ is not empty. (Why?) Let $x \in \omega(q)$. Since x is not a rest point, there is a section Σ at x and a sequence on Σ consisting of points on the orbit through q that converges to x. These points are in $\omega(p)$. But, by the last corollary, this is impossible unless every point in this sequence is the point x. Since q is not a rest point, this implies that q lies on a closed orbit Γ, as required. In particular, the limit set $\omega(p)$ contains the closed orbit Γ.

To complete the proof we must show $\omega(p) \subseteq \Gamma$. If $\omega(p) \neq \Gamma$, then we will use the connectedness of $\omega(p)$ to find a sequence $\{p_n\}_{n=1}^{\infty} \subset \omega(p) \setminus \Gamma$ that converges to a point z on Γ. To do this, consider the union A_1 of all open balls with unit radius centered at some point in Γ. The set $A_1 \setminus \Gamma$ must contain a point in $\omega(p)$. If not, consider the union $A_{1/2}$, (respectively $A_{1/4}$) of all open balls with radius $\frac{1}{2}$ (respectively $\frac{1}{4}$) centered at some point in Γ. Then the set $A_{1/4}$ together with the complement of the closure of $A_{1/2}$ "disconnects" $\omega(p)$, in contradiction. By repeating the argument with balls whose radii tend to zero, we can construct a sequence of points in $\omega(p) \setminus \Gamma$ whose distance from Γ tends to zero. Using the compactness of $\omega(p)$, there is a subsequence, again denoted by $\{p_n\}_{n=1}^{\infty}$, in $\omega(p) \setminus \Gamma$ that converges to a point $z \in \Gamma$.

Let U denote an open set at z such that the flow is rectified in a diffeomorphic image of U. There is some integer n such that $p_n \in U$. But, by using the rectification lemma, it is easy to see that the orbit through p_n has a point y of intersection with some Poincaré section Σ at z. Because

p_n is not in Γ, the points y and z are distinct elements of the set $\omega(p) \cap \Sigma$, in contradiction to Lemma 1.188. $\qquad\qquad\qquad\qquad\qquad\qquad\qquad\square$

Exercise 1.189. Suppose that γ is a periodic orbit of a smooth flow defined on $\mathbb{R}^2$. Use Zorn's lemma to prove that γ surrounds a rest point of the flow. That is, the bounded component of the plane with the periodic orbit removed contains a rest point. Note: See Exercise 1.217 for an alternative proof.

Exercise 1.190. Use Exercise 1.189 to prove Brouwer's fixed point theorem for the closed unit disk $\mathbb{D}$ in $\mathbb{R}^2$. Hint: First prove the result for a smooth function $f : \mathbb{D} \to \mathbb{D}$ by considering the vector field $f(x) - x$, and then use the following result: A continuous transformation of $\mathbb{D}$ is the uniform limit of smooth transformations [123, p. 253].

Exercise 1.191. Suppose that a closed ball in $\mathbb{R}^n$ is positively invariant under the flow of an autonomous differential equation on $\mathbb{R}^n$. Prove that the ball contains a rest point or a periodic orbit. Hint: Apply Brouwer's fixed point theorem to the time-one map of the flow. Explain the differences between this result and the Poincaré-Bendixson theorem.

Exercise 1.192. Construct an example of an (autonomous) differential equation defined on all of $\mathbb{R}^3$ that has an (isolated) limit cycle but no rest points.

Exercise 1.193. Prove: A nonempty ω-limit set of an orbit of a gradient system consists entirely of rest points.

Exercise 1.194. Is a limit cycle isolated from all other periodic orbits? Hint: Consider planar vector fields of class C^1 and those of class C^ω—real analytic vector fields. Study the Poincaré map on an associated transversal section.

The next theorem can often be used to show that no periodic orbits exist.

Proposition 1.195 (Dulac's Criterion). *Consider a smooth differential equation on the plane*

$$\dot{x} = g(x, y), \qquad \dot{y} = h(x, y).$$

If there is a smooth function $B(x, y)$ defined on a simply connected region $\Omega \subseteq \mathbb{R}^n$ such that the quantity $(Bg)_x + (Bh)_y$ is not identically zero and of fixed sign on Ω, then there are no periodic orbits in Ω.

Proof. We will prove Bendixson's criterion, which is the special case of the theorem where $B(x, y) \equiv 1$ (see Exercise 1.198 for the general case). In other words, we will prove that if the divergence of $f := (g(x, y), h(x, y))$ given by

$$\operatorname{div} f(x, y) := g_x(x, y) + h_y(x, y)$$

is not identically zero and of fixed sign in a simply connected region Ω, then there are no periodic orbits in Ω.

Suppose that Γ is a closed orbit in Ω and let G denote the bounded region of the plane bounded by Γ. Note that the line integral of the one form $g\,dy - h\,dx$ over Γ vanishes. (Why?) On the other hand, by Green's theorem, the integral can be computed by integrating the two-form (div f) $dxdy$ over G. Since, by the hypothesis, the divergence of f does not vanish, the integral of the two-form over G does not vanish, in contradiction. Thus, no such periodic orbit can exist. $\square$

The function B mentioned in the last proposition is called a Dulac function.

We end this section with a result about global asymptotic stability in the plane.

Theorem 1.196. *Consider a smooth differential equation on the plane*

$$\dot{x} = g(x,y), \qquad \dot{y} = h(x,y)$$

that has the origin as a rest point. Let J denote the Jacobian matrix for the transformation $(x,y) \mapsto (g(x,y), h(x,y))$, and let ϕ_t denote the flow of the differential equation. If the following three conditions are satisfied, then the origin is globally asymptotically stable.

Condition 1.   For each $(x,y) \in \mathbb{R}^2$, the trace of J given by $g_x(x,y) + h_y(x,y)$ is negative.

Condition 2.   For each $(x,y) \in \mathbb{R}^2$, the determinant of J given by $g_x(x,y)h_y(x,y) - g_y(x,y)h_x(x,y)$ is positive.

Condition 3.   For each $(x,y) \in \mathbb{R}^2$, the forward orbit $\{\phi_t(x,y) : 0 \leq t < \infty\}$ is bounded.

Proof. From the hypotheses on the Jacobian matrix, if there is a rest point, the eigenvalues of its associated linearization all have negative real parts. Therefore, each rest point is a hyperbolic attractor; that is, the basin of attraction of the rest point contains an open neighborhood of the rest point. This fact follows from Hartman's theorem (Theorem 1.47) or Theorem 2.61. In particular, the origin is a hyperbolic attractor.

By the hypotheses, the trace of the Jacobian (the divergence of the vector field) is negative over the entire plane. Thus, by Bendixson's criterion, there are no periodic solutions.

Let Ω denote the basin of attraction of the origin. Using the continuity of the flow, it is easy to prove that Ω is open. In addition, it is easy to prove that the boundary of Ω is closed and contains no rest points.

We will show that the boundary of Ω is positively invariant. If not, then there is a point p in the boundary and a time $T > 0$ such that either $\phi_T(p)$ is in Ω or such that $\phi_T(p)$ is in the complement of the closure of Ω in the plane. In the first case, since $\phi_T(p)$ is in Ω, it is clear that $p \in \Omega$, in contradiction. In the second case, there is an open set V in the complement of the closure of Ω that contains $\phi_T(p)$. The inverse image of V under the continuous map ϕ_T is an open set U containing the boundary point p. By the definition of boundary, U contains a point $q \in \Omega$. But then, q is mapped to a point in the complement of the closure of Ω, in contradiction to the fact that q is in the basin of attraction of the origin.

If the boundary of Ω is not empty, consider one of its points. The (bounded) forward orbit through the point is precompact and contained in the (closed) boundary of Ω. Thus, its ω-limit set is contained in the boundary of Ω. Since the boundary of Ω contains no rest points, an application of the Poincaré–Bendixson theorem shows this ω-limit set is a periodic orbit, in contradiction. Thus, the boundary is empty and Ω is the entire plane.   □

Theorem 1.196 is a (simple) special case of the "Markus-Yamabe problem." In fact, the conclusion of the theorem is true without assuming Condition 3 (see [104]).

Exercise 1.197. Prove: If $\delta > 0$, then the origin is a global attractor for the system

$$\dot{u} = (u - v)^3 - \delta u, \qquad \dot{v} = (u - v)^3 - \delta v.$$

Also, the origin is a global attractor of orbits in the first quadrant for the system

$$\dot{u} = uv(u - v)(u + 1) - \delta u, \qquad \dot{v} = vu(v - u)(v + 1) - \delta v.$$

(Both of these first order systems are mentioned in [229].)

Exercise 1.198. [Dulac's Criterion] (a) Prove Proposition 1.195. (b) Use Dulac's criterion to prove a result due to Nikolai N. Bautin: The system

$$\dot{x} = x(a + bx + cy), \quad \dot{y} = y(\alpha + \beta x + \gamma y)$$

has no limit cycles. Hint: Show that no periodic orbit crosses a coordinate axis. Reduce the problem to showing that there are no limit cycles in the first quadrant. Look for a Dulac function of the form $x^r y^s$. After some algebra the problem reduces to showing that a certain two-parameter family of lines always has a member that does not pass through the (open) first quadrant.

Exercise 1.199. (a) Suppose that the system $\dot{x} = f(x, y)$, $\dot{y} = g(x, y)$ has a periodic orbit Γ with period T and B is a positive real valued function defined on some open neighborhood of Γ (as in Dulac's Criterion). Prove that Γ is a periodic orbit of the system $\dot{x} = B(x, y)f(x, y)$, $\dot{y} = B(x, y)g(x, y)$ with period

$$\tau = \int_0^T \frac{1}{B(x(t), y(t))} \, ds$$

where $t \mapsto (x(t), y(t))$ is a periodic solution of the original system whose orbit is Γ. (b) How does the period of the limit cycle of system (1.46) change if its vector field is multiplied by $(1 + x^2 + y^2)^\alpha$? Hint: The solution ρ of the initial value problem

$$\dot{\rho} = B((x(\rho), y(\rho)), \quad \rho(0) = 0$$

satisfies the identity $\rho(t + \tau) = \rho(t) + T$.

Exercise 1.200. [Uniqueness of Limit Cycles] (a) Prove the following proposition: If the divergence of a plane vector field is of fixed sign in an annular region Ω of the plane, then the associated differential equation has at most one periodic orbit in Ω. Hint: Use Green's theorem. (b) Recall Dulac's criterion from Exercise 1.198 and note that if the divergence of the plane vector field F is not of fixed sign in Ω, then it might be possible to find a nonnegative function $B : \Omega \to \mathbb{R}$ such that the divergence of BF does have fixed sign in Ω. As an example, consider the van der Pol oscillator,

$$\dot{x} = y, \qquad \dot{y} = -x + \lambda(1 - x^2)y$$

and the "Dulac function" $B(x,y) = (x^2 + y^2 - 1)^{-1/2}$. Show that van der Pol's system has at most one limit cycle in the plane. (The remarkable Dulac function B was discovered by L. A. Cherkas.) (c) Can you prove that the van der Pol oscillator has at least one limit cycle in the plane? Hint: Change coordinates using the Liénard transformation

$$u = x, \qquad v = y - \lambda(x - \frac{1}{3}x^3)$$

to obtain the Liénard system

$$\dot{u} = v + \lambda(u - \frac{1}{3}u^3), \qquad \dot{v} = -u.$$

In Chapter 5 we will prove that the van der Pol system has a limit cycle if $\lambda > 0$ is sufficiently small. In fact, this system has a limit cycle for each $\lambda > 0$. For this result, and for more general results about limit cycles of the important class of planar systems of the form

$$\dot{x} = y - F(x), \qquad \dot{y} = -g(x),$$

see [101, p. 154], [123, p. 215], [141, p. 267], and [183, p. 250].

Exercise 1.201. (a) Prove that the system

$$\dot{x} = x - y - x^3, \qquad \dot{y} = x + y - y^3$$

has a unique globally attracting limit cycle on the punctured plane. (b) Find all rest points of the system

$$\dot{x} = x - y - x^n, \qquad \dot{y} = x + y - y^n,$$

where n is a positive odd integer and determine their stability. (c) Prove that the system has a unique stable limit cycle. (d) What is the limiting shape of the limit cycle as $n \to \infty$?

Exercise 1.202. Show there is a unique limit cycle for system (1.48) with $|\mu| > 1$ by proving the existence of a fixed point for a Poincaré map and by proving that every limit cycle is stable. Hint: Recall the analysis of system (1.44) and consider $dv/d\theta$.

Exercise 1.203. Can a system of the form

$$\dot{x} = y, \qquad \dot{y} = f(x) - af'(x)y,$$

where f is a smooth function and a is a parameter, have a limit cycle? Hint: Consider a Liénard transformation.

Exercise 1.204. Draw the phase portrait of the system

$$\dot{x} = y + 2x(1 - x^2 - y^2), \qquad \dot{y} = -x.$$

Exercise 1.205. [Rigid Body Motion] The Euler equations for rigid body motion are presented in Exercise 1.77. Recall that the momentum vector is given by $M = A\Omega$ where A is a symmetric matrix and Ω is the angular velocity vector, and Euler's equation is given by $\dot{M} = M \times \Omega$. For ν a positive definite symmetric matrix and F a constant vector, consider the differential equation

$$\dot{M} = M \times \Omega + F - \nu M.$$

Here, the function $M \mapsto \nu M$ represents viscous friction and F is the external force (see [16]). Prove that all orbits of the differential equation are bounded, and therefore every orbit has a compact ω-limit set.

Exercise 1.206. (a) Prove that the origin is a center for the system $\ddot{x} + \dot{x}^2 + x = 0$. (b) Show that this system has unbounded orbits. (c) Describe the boundary between the bounded and unbounded orbits?

Exercise 1.207. Draw the phase portrait for the system $\ddot{x} = x^2 - x^3$. Is the solution with initial conditions $x(0) = \frac{1}{2}$ and $\dot{x}(0) = 0$ periodic?

Exercise 1.208. Draw the phase portrait of the Hamiltonian system $\ddot{x} + x - x^2 = 0$. Give an explicit formula for the Hamiltonian and use it to justify the features of the phase portrait.

Exercise 1.209. Let $t \mapsto x(t)$ denote the solution of the initial value problem

$$\ddot{x} + \dot{x} + x + x^3 = 0, \qquad x(0) = 1, \quad \dot{x}(0) = 0.$$

Determine $\lim\limits_{t \to \infty} x(t)$.

Exercise 1.210. Show that the system

$$\dot{x} = x - y - (x^2 + \frac{3}{2}y^2)x, \qquad \dot{y} = x + y - (x^2 + \frac{1}{2}y^2)y$$

has a unique limit cycle.

Exercise 1.211. Find the rest points in the phase plane of the differential equation $\ddot{x} + (\dot{x}^2 + x^2 - 1)\dot{x} + x = 0$ and determine their stability. Also, show that the system has a unique stable limit cycle.

Exercise 1.212. Determine the ω-limit set of the solution of the system

$$\dot{x} = 1 - x + y^3, \qquad \dot{y} = y(1 - x + y)$$

with initial condition $x(0) = 10$, $y(0) = 0$.

Exercise 1.213. Show that the system

$$\dot{x} = -y + xy, \qquad \dot{y} = x + \frac{1}{2}(x^2 - y^2)$$

has periodic solutions, but no limit cycles.

Exercise 1.214. Consider the van der Pol equation

$$\ddot{x} + (x^2 - \epsilon)\dot{x} + x = 0,$$

where ϵ is a real parameter. How does the stability of the trivial solution change with ϵ. Show that the van der Pol equation has a unique stable limit cycle for $\epsilon = 1$. What would you expect to happen to this limit cycle as ϵ shrinks to $\epsilon = 0$. What happens for $\epsilon < 0$?

Exercise 1.215. Find an explicit nonzero solution of the differential equation

$$t^2 x^2 \ddot{x} + \dot{x} = 0.$$

Define new variables $u = 2(3tx^2)^{-1/2}$, $v = -4\dot{x}(3x^3)^{-1/2}$ and show that

$$\frac{dv}{du} = \frac{3v(v - u^2)}{2u(v - u)}.$$

Draw the phase portrait of the corresponding first order system

$$\dot{u} = 2u(v - u), \qquad \dot{v} = 3v(v - u^2).$$

Exercise 1.216. [Yorke's Theorem] A theorem of James Yorke states that *if* $f : U \subseteq \mathbb{R}^n \to \mathbb{R}^n$ *is Lipschitz on the open set* U *with Lipschitz constant* L *and* Γ *is a periodic orbit of* $\dot{x} = f(x)$ *contained in* U, *then the period of* Γ *is larger than* $2\pi/L$ (see [238]). Use Yorke's theorem to estimate a lower bound for the period of the limit cycle solution of the system in Exercise 1.201 part (a). Note: The period of the periodic orbit is approximately 7.5. Hint: Use the mean value theorem and note that the norm of a matrix (with respect to the usual Euclidean norm) is the square root of the spectral radius of the matrix transpose times the matrix (that is; $\|A\| = \sqrt{\rho(A^T A)}$).

Exercise 1.217. [Poincaré index] Let C be a simple closed curve not passing through a rest point of the vector field X in the plane with components (f, g). Define the Poincaré index of X with respect to C to be

$$I(X, C) = \frac{1}{2\pi} \int_C d\arctan\left(\frac{g}{f}\right);$$

it is the total change the angle $(f(x, y), g(x, y))$ makes with respect to the (positive) x-axis as (x, y) traverses C exactly once counter clockwise (see, for example, [59] or [141]). (a) Prove: The index is an integer. (b) Prove: The index does not change with a deformation of C (as long as the deformed curve does not pass through a rest point). (c) Prove: If C is smooth and T is a continuous choice of the tangent vector along this curve, then $I(T, C) = 1$. In particular, the index of a vector field with respect to one of its closed orbits is unity. (d) The index of a point with respect to X is defined to be the index of X with respect to an admissible curve C that surrounds this point and no other rest point of X. Prove: The index of a regular point (a point that is not a rest point) is zero. (e) Prove: A periodic orbit surrounds at least one rest point.

1.10   Review of Calculus

The basic definitions of the calculus extend easily to multidimensional spaces. In fact, these definitions are essentially the same when extended to infinite dimensional spaces. Thus, we will begin our review with the definition of differentiation in a Banach space.

Definition 1.218. Let U be an open subset of a Banach space X, let Y denote a Banach space, and let the symbol $\| \ \|$ denote the norm in both Banach spaces. A function $f : U \to Y$ is called *(Fréchet) differentiable* at $a \in U$ if there is a bounded linear operator $Df(a) : X \to Y$, called the derivative of f, such that

$$\lim_{h \to 0} \frac{1}{\|h\|} \|f(a + h) - f(a) - Df(a)h\| = 0.$$

If f is differentiable at each point in U, then the function f is called differentiable.

Using the notation of Definition 1.218, let $L(X,Y)$ denote the Banach space of bounded linear transformations from X to Y, and note that the derivative of $f : U \to Y$ is the function $Df : U \to L(X,Y)$ given by $x \mapsto Df(x)$.

The following proposition is a special case of the chain rule.

Proposition 1.219. *Suppose that U is an open subset of a Banach space and $f : U \to Y$. If f is differentiable at $a \in U$ and $v \in U$, then*

$$\frac{d}{dt} f(a + tv)\big|_{t=0} = Df(a)v.$$

Proof. The proof is obvious for $v = 0$. Assume that $v \neq 0$ and consider the scalar function given by

$$\alpha(t) := \|\frac{1}{t}(f(a + tv) - f(a)) - Df(a)v)\|$$

$$= \frac{1}{|t|} \|f(a + tv) - f(a) - Df(a)tv\|$$

for $t \neq 0$. It suffices to show that $\lim_{t \to 0} \alpha(t) = 0$.

Choose $\epsilon > 0$. Since f is differentiable at a, there is some $\delta > 0$ such that

$$\frac{1}{\|h\|} \|f(a + h) - f(a) - Df(a)h\| < \epsilon$$

whenever $0 < \|h\| < \delta$. If $|t| < \delta \|v\|^{-1}$, then $\|tv\| < \delta$ and

$$\frac{1}{|t|\|v\|} \|f(a + tv) - f(a) - Df(a)tv\| < \epsilon.$$

In particular, we have that $\alpha(t) \leq \|v\|\epsilon$ whenever $|t| < \delta \|v\|^{-1}$, as required.

$\square$

The following is a list of standard facts about the derivative; the proofs are left as exercises. For the statements in the list, the symbols X, Y, X_i, and Y_i denote Banach spaces.

(i) If $f : X \to Y$ is differentiable at $a \in X$, then f is continuous at a.

(ii) If $f : X \to Y$ and $g : Y \to Z$ are both differentiable, then $h = g \circ f$ is differentiable, and its derivative is given by the chain rule

$$Dh(x) = Dg(f(x))Df(x).$$

(iii) If $f : X \to Y_1 \times \cdots \times Y_n$ is given by $f(x) = (f_1(x), \dots, f_n(x))$, and if f_i is differentiable for each i, then so is f and, in fact,

$$Df(x) = (Df_1(x), \dots, Df_n(x)).$$

(iv) If the function $f : X_1 \times X_2 \times \cdots \times X_n \to Y$ is given by $(x_1, \dots, x_n) \mapsto f(x_1, \dots, x_n)$, then the ith partial derivative of f at $a_1, \dots, a_n \in X_1 \times \cdots \times X_n$ is the derivative of the function $g : X_i \to Y$ defined by $g(x_i) = f(a_1, \dots, a_{i-1}, x_i, a_{i+1}, \dots, a_n)$. This derivative is denoted $D_i f(a)$. Of course, if f is differentiable, then its partial derivatives all exist and, if we define $h = (h_1, \dots, h_n)$, we have

$$Df(x)h = \sum_{i=1}^{n} D_i f(x)h_i.$$

Conversely, if all the partial derivatives of f exist and are continuous in an open set

$$U \subset X_1 \times X_2 \times \cdots \times X_n,$$

then f is continuously differentiable in U.

(v) If $f : X \to Y$ is a bounded linear map, then $Df(x) = f$ for all $x \in X$.

The C^r-norm of an r-times continuously differentiable function $f : U \to Y$, defined on an open subset U of X, is defined by

$$\|f\|_r = \|f\|_0 + \|Df\|_0 + \cdots \|D^r f\|_0$$

where $\| \ \|_0$ denotes the usual supremum norm, as well as the operator norms over U; for example,

$$\|f\|_0 = \sup_{u \in U} \|f(u)\|$$

and

$$\|Df\|_0 = \sup_{u \in U} \left(\sup_{\|x\|=1} \|Df(u)x\| \right).$$

Also, let us use $C^r(U, Y)$ to denote the set of all functions $f : U \to Y$ such that $\|f\|_r < \infty$. Of course, the set $C^r(U, Y)$ is a Banach space of functions with respect to the C^r-norm.

Although the basic definitions of differential calculus extend unchanged to the Banach space setting, this does not mean that there are no new phenomena in infinite dimensional spaces. The following examples and exercises illustrate some of the richness of the theory. The basic idea is that functions can be defined on function spaces in ways that are not available in the finite dimensional context. If such a function is defined, then its differentiability class often depends on the topology of the Banach space in a subtle manner.

Example 1.220. Let $X = C([0, 1])$ and define $F : X \to X$ by

$$F(g)(t) := \sin g(t)$$

(see [57]). We have the following proposition: The function F is continuously differentiable and

$$(DF(g)h)(t) = (\cos g(t))h(t).$$

To prove it, let us first compute

$$
\begin{aligned}
&|F(g + h)(t) - F(g)(t) - DF(g)h(t)| \\
&= |\sin(g(t) + h(t)) - \sin g(t) - (\cos g(t))h(t)| \\
&= |\sin g(t) \cos h(t) + \cos g(t) \sin h(t) - \sin g(t) - (\cos g(t))h(t)| \\
&= |(-1 + \cos h(t)) \sin g(t) + (-h(t) + \sin h(t)) \cos g(t)| \\
&\le \|F(g)\| |-1 + \cos h(t)| + \| \cos \circ g \| |-h(t) + \sin h(t)| \\
&\le \frac{1}{2} \left(\|F(g)\| \, \|h\|^2 + \| \cos \circ g \| \, \|h\|^2 \right).
\end{aligned}
$$

This proves that F is differentiable.

The function $DF : X \to L(X, X)$ given by $g \mapsto DF(g)$ is clearly continuous, in fact,

$$
\begin{aligned}
\|DF(g_1) - DF(g_2)\| &= \sup_{\|h\|=1} \|DF(g_1)h - DF(g_2)h\| \\
&= \sup_{\|h\|=1} \sup_{t} |(\cos g_1(t))h(t) - (\cos g_2(t))h(t)| \\
&\le \sup_{\|h\|=1} \sup_{t} |h(t)||g_1(t) - g_2(t)| \\
&= \|g_1 - g_2\|.
\end{aligned}
$$

Thus F is continuously differentiable, as required.

Example 1.221. Let $X := L^2([0, 1])$ and define $F : X \to X$ by

$$F(g)(t) = \sin g(t).$$

The function F is Lipschitz, but not differentiable.

To prove that F is Lipschitz, simply recall that $|\sin x - \sin y| \le |x - y|$ and estimate as follows:

$$\|F(g_1) - F(g_2)\|^2 = \int_0^1 |\sin g_1(t) - \sin g_2(t)|^2 \, dt$$

$$\le \int_0^1 |g_1(t) - g_2(t)|^2 \, dt$$

$$\le \|g_1 - g_2\|^2.$$

We will show that F is *not* differentiable at the origin. To this end, let us suppose that F is differentiable at the origin with derivative $DF(0)$. We have that $F(0) = 0$, and, by Proposition (1.219), all directional derivatives of F at the origin exist. Therefore, it follows that

$$\lim_{s \to 0} \frac{F(sg) - F(0)}{s} = \lim_{s \to 0} \frac{F(sg)}{s} = DF(0)g$$

for all $g \in L^2([0, 1])$.

To reach a contradiction, we will first prove that $DF(0)$ is the identity map on $L^2([0,1])$. To do this, it suffices to show that $DF(0)g = g$ for every *continuous* function $g \in L^2([0, 1])$. Indeed, this reduction follows because the (equivalence classes of) continuous functions are dense in $L^2([0,1])$.

Let us assume that g is continuous and square integrable. We will show that the directional derivative of F at the origin in the direction g exists and is equal to g. In other words, we will show that

$$\lim_{s \to 0} \frac{F(sg)}{s} = g;$$

that is,

$$\lim_{s \to 0} \int_0^1 \left| \frac{\sin(sg(t))}{s} - g(t) \right|^2 \, ds = 0. \tag{1.50}$$

Indeed, let us define

$$\psi_s(t) := \left| \frac{\sin(sg(t))}{s} - g(t) \right|^2, \qquad s > 0$$

and note that

$$\psi_s(t) \le \left(\left| \frac{\sin(sg(t))}{s} \right| + |g(t)| \right)^2.$$

Because $|\sin x| \le |x|$ for all $x \in \mathbb{R}$, we have the estimates

$$\psi_s(t) \le \left(\frac{|sg(t)|}{|s|} + |g(t)| \right)^2 \le 4|g(t)|^2.$$

Moreover, the function $t \mapsto 4|g(t)|^2$ is integrable, and therefore the function $t \mapsto \psi_s(t)$ is dominated by an integrable function.

If t is fixed, then

$$\lim_{s \to 0} \psi_s(t) = 0.$$

To prove this fact, let us observe that $|g(t)| < \infty$. If $g(t) = 0$, then $\psi_s(t) = 0$ for all s and the result is clear. If $g(t) \neq 0$, then

$$\psi_s(t) = \left| g(t) \left(\frac{\sin(sg(t))}{sg(t)} - 1 \right) \right|^2$$

$$= |g(t)|^2 \left| \frac{\sin(sg(t))}{sg(t)} - 1 \right|^2$$

and again $\psi_s(t) \to 0$ as $s \to 0$.

We have proved that the integrand of the integral in display (1.50) is dominated by an integrable function and converges to zero. Hence, the required limit follows from the dominated convergence theorem and, moreover, $DF(0)g = g$ for all $g \in L^2([0,1])$.

Because $DF(0)$ is the identity map, it follows that

$$\lim_{h \to 0} \frac{\|F(h) - h\|}{\|h\|} = 0.$$

But let us consider the sequence of functions $\{h_n\}_{n=1}^{\infty} \subset L^2([0,1])$ defined by

$$h_n(t) := \begin{cases} \pi/2, & 0 \leq t \leq 1/n, \\ 0, & t > 1/n. \end{cases}$$

Since

$$\|h_n\| = \left(\int_0^1 (h_n(t))^2 dt \right)^{1/2} = \left(\frac{1}{n} \frac{\pi^2}{4} \right)^{1/2} = \frac{1}{\sqrt{n}} \frac{\pi}{2},$$

it follows that $h_n \to 0$ as $n \to \infty$. Also, let us note that

$$\|F(h_n) - h_n\| = \left(\int_0^1 |\sin h_n(t) - h_n(t)|^2 dt \right)^{1/2}$$

$$= \left(\frac{1}{n} \left| 1 - \frac{\pi}{2} \right|^2 \right)^{1/2}$$

and therefore

$$\lim_{n \to \infty} \frac{\|F(h_n) - h_n\|}{\|h_n\|} = \lim_{n \to \infty} \frac{\frac{1}{\sqrt{n}}(1 - \frac{\pi}{2})}{\frac{1}{\sqrt{n}} \frac{\pi}{2}} = \frac{1 - \frac{\pi}{2}}{\frac{\pi}{2}} \neq 0.$$

This contradiction proves that F is not differentiable at the origin. Is F differentiable at any other point?

Exercise 1.222. Let $\mathrm{GL}(\mathbb{R}^n)$ denote the set of invertible linear transformations of $\mathbb{R}^n$ and let $f : \mathrm{GL}(\mathbb{R}^n) \to \mathrm{GL}(\mathbb{R}^n)$ be the function given by $f(A) = A^{-1}$. Prove that f is differentiable and compute its derivative.

Exercise 1.223. Consider the evaluation map

$$\mathrm{eval} : C^r(U, Y) \times U \to Y$$

defined by $(f, u) \mapsto f(u)$. Prove that eval is a C^r map. Also, compute its derivative.

Exercise 1.224. [Omega Lemma] (a) Suppose that $f : \mathbb{R} \to \mathbb{R}$ is a C^2 function such that the quantity $\sup_{x \in \mathbb{R}} |f''(x)|$ is bounded. Prove that $F : X \to X$ as in Example 1.221 is C^1. (b) The assumption that f is C^2 can be replaced by the weaker hypothesis that f is C^1. This is a special case of the omega lemma (see [2, p. 101]). *If M is a compact topological space, U is an open subset of a Banach space X, and g is in $C^r(U, Y)$ where Y is a Banach space and $r \geq 1$, then the map $\Omega_g : C^0(M, U) \to C^0(M, Y)$ given by $\Omega_g(f) = g \circ f$ is C^r and its derivative is given by*

$$(D\Omega_g(f)h)(m) = Dg(f(m))h(m).$$

Prove the omega lemma.

1.10.1   The Mean Value Theorem

The mean value theorem for functions of several variables is important. Let us begin with a special case.

Theorem 1.225. *Suppose that $[a, b]$ is a closed interval, Y is a Banach space, and $f : [a, b] \to Y$ is a continuous function. If f is differentiable on the open interval (a, b) and there is some number $M > 0$ such that $\|f'(t)\| \leq M$ for all $t \in (a, b)$, then*

$$\|f(b) - f(a)\| \leq M(b - a).$$

Proof. Let $\epsilon > 0$ be given and define $\phi : [a, b] \to \mathbb{R}$ by

$$\phi(t) = \|f(t) - f(a)\| - (M + \epsilon)(t - a).$$

Clearly, ϕ is a continuous function such that $\phi(a) = 0$. We will show that $\phi(b) \leq \epsilon$.

Define $S := \{t \in [a, b] : \phi(t) \leq \epsilon\}$. Since $\phi(a) = 0$, we have that $a \in S$. In particular $S \neq \emptyset$. By the continuity of ϕ, there is some number c such that $a < c < b$ and $[a, c] \subseteq S$. Moreover, since ϕ is continuous $\phi(t) \to \phi(c)$ as $t \to c$. Thus, since $\phi(t) \leq \epsilon$ for $a \leq t < c$, we must have $\phi(c) \leq \epsilon$ and, in fact, $[a, c] \subseteq S$.

Consider the supremum c^* of the set of all c such that $a \leq c \leq b$ and $[a, c] \subseteq S$. Let us show that $c^* = b$. If $c^* < b$, then consider the derivative of f at c^* and note that because

$$\lim_{\|h\|\to 0} \frac{\|f(c^* + h) - f(c^*) - f'(c^*)h\|}{\|h\|} = 0,$$

there is some h such that $c^* < c^* + h < b$ and

$$\|f(c^* + h) - f(c^*) - f'(c^*)h\| \leq \epsilon\|h\|.$$

Set $d = c^* + h$ and note that

$$\begin{aligned}
\|f(d) - f(c^*)\| &\leq \|f(c^* + h) - f(c^*) - f'(c^*)h\| + \|f'(c^*)h\| \\
&\leq \epsilon\|h\| + M\|h\| \\
&\leq (\epsilon + M)(d - c^*).
\end{aligned}$$

Moreover, since

$$\begin{aligned}
\|f(d) - f(a)\| &\leq \|f(d) - f(c^*)\| + \|f(c^*) - f(a)\| \\
&\leq (\epsilon + M)(d - c^*) + (M + \epsilon)(c^* - a) + \epsilon \\
&\leq (\epsilon + M)(d - a) + \epsilon,
\end{aligned}$$

we have that

$$\|f(d) - f(a)\| - (\epsilon + M)(d - a) \leq \epsilon,$$

and, as a result, $d \in S$, in contradiction to the fact that c^* is the supremum. Thus, $c^* = b$, as required.

Use the equality $c^* = b$ to conclude that

$$\begin{aligned}
\|f(b) - f(a)\| &\leq (\epsilon + M)(b - a) + \epsilon \\
&\leq M(b - a) + \epsilon(1 + (b - a))
\end{aligned}$$

for all $\epsilon > 0$. By passing to the limit as $\epsilon \to 0$, we obtain the inequality

$$\|f(b) - f(a)\| \leq M(b - a),$$

as required. $\qquad\qquad\qquad\qquad\qquad\qquad\qquad\qquad\qquad\qquad\qquad\qquad$ □

Theorem 1.226 (Mean Value Theorem). *Suppose that $f : X \to Y$ is differentiable on an open set $U \subseteq X$ with $a, b \in U$ and $a + t(b - a) \in U$ for $0 \leq t \leq 1$. If there is some $M > 0$ such that*

$$\sup_{0 \leq t \leq 1} \|Df(a + t(b - a))\| \leq M,$$

then

$$\|f(b) - f(a)\| \leq M\|b - a\|.$$

Proof. Define $g(t) := f(a + t(b - a))$. Clearly, g is differentiable on $[0, 1]$ and, by the chain rule, $g'(t) = Df(a + t(b - a))(b - a)$. In particular,

$$\|g'(t)\| \leq \|Df(a + t(b - a))\|\|b - a\| \leq M\|b - a\|.$$

Here, $g : [0, 1] \to Y$ and $\|g'(t)\| \leq M\|b - a\|$ for $0 \leq t \leq 1$. By the previous theorem,

$$\|g(1) - g(0)\| \leq M\|b - a\|,$$

that is,

$$\|f(b) - f(a)\| \leq M\|b - a\|. \qquad \qquad \Box$$

1.10.2 Integration in Banach Spaces

This section is a brief introduction to integration on Banach spaces following the presentation in [140]. As an application, we will give an alternative proof of the mean value theorem and a proof of a version of Taylor's theorem.

Let I denote a closed interval of real numbers and X a Banach space with norm $\| \ \|$. A *simple function* $f : I \to X$ is a function with the following property: There is a finite cover of I consisting of disjoint subintervals such that f restricted to each subinterval is constant. Here, each subinterval can be open, closed, or half open.

A sequence $\{f_n\}_{n=1}^{\infty}$ of not necessarily simple functions, each mapping I to X, *converges uniformly* to a function $f : I \to X$ if for each $\epsilon > 0$ there is an integer $N > 0$ such that $\|f_n(t) - f_m(t)\| < \epsilon$ whenever $n, m > N$ and $t \in I$.

Definition 1.227. A *regulated function* is a uniform limit of simple functions.

Lemma 1.228. *Every continuous function $f : I \to X$ is regulated.*

Proof. The function f is uniformly continuous. To see this, consider $F : I \times I \to X$ defined by $F(x, y) = f(y) - f(x)$ and note that F is continuous. Since the diagonal $D = \{(x, y) \in I \times I : x = y\}$ is a compact subset of $I \times I$ (Why?), its image $F(D)$ is compact in X. Hence, for each $\epsilon > 0$, a finite number of ϵ-balls in X cover the image of D. Taking the inverse images of the elements of some such covering, we see that there is an open cover $V_1, \ldots, V_n$ of the diagonal in $I \times I$ such that if $(x, y) \in V_i$, then $\|F(x, y)\| < \epsilon$. For each point $(x, x) \in D$, there is a ball centered at (x, x) and contained in $I \times I$ that is contained in some V_i. By compactness, a finite number of such balls cover D. Let δ denote the minimum radius of the balls in this finite subcover. If $|x - y| < \delta$, then $(x, y) \in B_\delta(x, x)$ and in fact $\|(x, y) - (x, x)\| = |y - x| < \delta$. Thus, $(x, y) \in V_i$ for some i in the set $\{1, \ldots, n\}$, and, as a result, we have that $\|F(x, y)\| < \epsilon$; that is, $\|f(y) - f(x)\| = \|F(x, y)\| < \epsilon$, as required.

Let us suppose that $I = \{x \in \mathbb{R} : a \leq x \leq b\}$. For each natural number n, there is some $\delta > 0$ such that if $|x - y| < \delta$, then $\|f(x) - f(y)\| < \frac{1}{n}$. Let us define a corresponding simple function f_n by $f_n(x) = f(a)$ for $a \leq x \leq a + \frac{\delta}{2}$, $f_n(x) = f(a + \frac{\delta}{2})$ for $a + \frac{\delta}{2} < x \leq a + \delta$, $f_n(x) = f(a + \delta)$ for $a + \delta < x \leq a + \frac{3\delta}{2}$, and so on until $a + k\frac{\delta}{2} \geq b$. This process terminates after a finite number of steps because I has finite length. Also, we have the inequality $\|f_n(x) - f(x)\| < \frac{1}{n}$ for all $x \in I$. Thus, the sequence of simple functions $\{f_n\}_{n=1}^{\infty}$ converges uniformly to f. $\qquad\square$

Definition 1.229. The *integral of a simple function* $f : I \to X$ over the interval $I = [a, b]$ is defined to be

$$\int_a^b f(t)\, dt := \sum_{j=1}^n \mu(I_j) v_j$$

where $I_1, \ldots, I_n$ is a partition of I, $f|_{I_j}(t) \equiv v_j$, and $\mu(I_j)$ denotes the length of the interval I_j.

Proposition 1.230. *If f is a simple function on I, then the integral of f over I is independent of the choice of the partition of I.*

Proof. The proof is left as an exercise. $\qquad\square$

Proposition 1.231. *If f is a regulated function defined on the interval $I = [a, b]$, and if $\{f_n\}_{n=1}^{\infty}$ is a sequence of simple functions converging uniformly to f, then the sequence defined by $n \mapsto \int_a^b f_n(t)\, dt$ converges in X. Moreover, if in addition $\{g_n\}_{n=1}^{\infty}$ is a sequence of simple functions converging uniformly to f, then*

$$\lim_{n \to \infty} \int_a^b f_n(t)\, dt = \lim_{n \to \infty} \int_a^b g_n(t)\, dt.$$

Proof. We will show that the sequence $n \mapsto \int_a^b f_n(t)\, dt$ is Cauchy. For this, consider the quantity

$$\left\| \int_a^b f_n(t)\, dt - \int_a^b f_m(t)\, dt \right\|.$$

Using χ_L to denote the characteristic function on the interval L, we have that, for some partitions of I and vectors $\{v_i\}$ and $\{w_i\}$,

$$f_n(x) = \sum_{i=1}^k \chi_{I_i}(x) v_i, \quad f_m(x) = \sum_{i=1}^l \chi_{J_i}(x) w_i.$$

The partitions $I_1, \ldots, I_k$ and $J_1, \ldots, J_l$ have a common refinement; that is, there is a partition of the interval I such that each subinterval in the new partition is contained in one of the subintervals $I_1, \ldots, I_k, J_1, \ldots, J_l$. Let this refinement be denoted by $K_1, \ldots, K_p$ and note that

$$f_n(x) = \sum_{i=1}^{p} \chi_{K_i}(x)\alpha_i, \qquad f_m(x) = \sum_{i=1}^{p} \chi_{K_i}(x)\beta_i.$$

Also, we have the inequality

$$\| \int_a^b f_n(t)\, dt - \int_a^b f_m(t)\, dt \| = \| \sum_{i=1}^{p} \mu(K_i)\alpha_i - \sum_{i=1}^{p} \mu(K_i)\beta_i \|$$

$$\leq \sum_{i=1}^{p} \mu(K_i)\|\alpha_i - \beta_i\|.$$

There are points $t_i \in K_i$ so that

$$\sum_{i=1}^{p} \mu(K_i)\|\alpha_i - \beta_i\| = \sum_{i=1}^{p} \mu(K_i)\|f_n(t_i) - f_m(t_i)\|$$

and, because $\sum_{i=1}^{p} \mu(K_i) = b - a$,

$$\sum_{i=1}^{p} \mu(K_i)\|f_n(t_i) - f_m(t_i)\| \leq (b-a)\max_i \|f_n(t_i) - f_m(t_i)\|$$

$$\leq (b-a)\max_{x \in I} \|f_n(x) - f_m(x)\|.$$

By combining the previous inequalities and using the fact that the sequence $\{f_n\}_{n=1}^{\infty}$ converges uniformly, it follows that the sequence $n \mapsto \int_a^b f_n(t)\, dt$ is a Cauchy sequence and thus converges to an element of X.

Suppose that $\{g_n\}_{n=1}^{\infty}$ is a sequence of simple functions that converges uniformly to f, and let us suppose that

$$\int_a^b f_n(t)\, dt \to F, \qquad \int_a^b g_n(t)\, dt \to G.$$

We have the estimates

$$\|F - G\| \leq \|F - \int_a^b f_n(t)\, dt\| + \| \int_a^b f_n\, dt - \int_a^b g_n\, dt\| + \| \int_a^b g_n\, dt - G\|$$

and

$$\| \int_a^b f_n\, dt - \int_a^b g_n\, dt\| \leq (b-a)\max_{x \in I} \|f_n(x) - g_n(x)\|$$

$$\leq (b-a)\max_{x \in I}(\|f_n(x) - f(x)\| + \|f(x) - g_n(x)\|).$$

The desired result, the equality $F = G$, follows by passing to the limit on both sides of the previous inequality. $\square$

In view of the last proposition, we have the following basic definition:

Definition 1.232. Let f be a regulated function on the interval $[a, b]$ and $\{f_n\}_{n=1}^{\infty}$ a sequence of simple functions converging uniformly to f in X. The *integral of f* denoted $\int_a^b f(t)\, dt$ is defined to be the limit of the sequence $n \mapsto \int_a^b f_n\, dt$ in X.

Proposition 1.233. *The functional $f \mapsto \int_a^b f(t)\, dt$, defined on the space of regulated functions, is linear.*

Proof. If f and g are regulated on the interval $[a, b]$, with sequences of simple functions $f_n \to f$ and $g_n \to g$, then $cf_n + dg_n \to cf + dg$ and

$$\int_a^b (cf + dg)(t)\, dt = \lim_{n \to \infty} \int_a^b (cf_n + dg_n)(t)\, dt.$$

But, for these simple functions, after a common refinement,

$$\int_a^b cf_n + dg_n\, dt = \sum_{i=1}^n \mu(I_i)(cv_i + dw_i) = c \sum_{i=1}^n \mu(I_i)v_i + d \sum_{i=1}^n \mu(I_i)w_i. \quad \square$$

Proposition 1.234. *If $\lambda : X \to \mathbb{R}$ is a continuous linear functional and if $f : I \to X$ is regulated, then the composition $\lambda f := \lambda \circ f : I \to \mathbb{R}$ is regulated, and*

$$\lambda \int_a^b f(t)\, dt = \int_a^b (\lambda f)(t)\, dt.$$

Proof. If $\{f_n\}_{n=1}^{\infty}$ is a sequence of simple functions converging uniformly to f and

$$f_n(x) = \sum_i \chi_{I_i}(x)v_i,$$

then

$$\lambda(f_n(x)) = \sum_i \chi_{I_i}(x)\lambda(v_i)$$

and, in particular, λf_n is a simple function for each n. Moreover, $\lambda \circ f$ is regulated by λf_n.

A continuous linear functional, by definition, has a bounded operator norm. Therefore, we have that

$$|\lambda f_n(x) - \lambda f(x)| = |\lambda(f_n(x) - f(x))|$$
$$\leq \|\lambda\| \|f_n(x) - f(x)\|$$

and

$$\left| \lambda \int_a^b f(t)\, dt - \int_a^b \lambda f(t)\, dt \right|$$

$$\leq \left| \lambda \int_a^b f(t)\, dt - \lambda \int_a^b f_n(t)\, dt \right| + \left| \lambda \int_a^b f_n(t)\, dt - \int_a^b \lambda f(t)\, dt \right|$$

$$\leq \|\lambda\| \left\| \int_a^b f(t)\, dt - \int_a^b f_n(t)\, dt \right\| + \left| \int_a^b \lambda f_n(t)\, dt - \int_a^b \lambda f(t)\, dt \right|.$$

The result follows by passing to the limit as $n \to \infty$.   □

Proposition 1.235. *If $f : [a, b] \to X$ is regulated, then*

$$\left\| \int_a^b f(t)\, dt \right\| \leq (b - a) \sup_{t \in [a,b]} \|f(t)\|. \qquad (1.51)$$

Proof. Note that the estimate (1.51) is true for simple functions; in fact, we have

$$\left\| \sum \mu(I_i) v_i \right\| \leq \sum \mu(I_i) \sup |v_i| \leq (b - a) \sup |v_i|.$$

Because f is regulated, there is a sequence $\{f_n\}_{n=1}^\infty$ of simple functions converging to f and, using this sequence, we have the following estimates:

$$\left\| \int_a^b f(t)\, dt \right\| \leq \left\| \int_a^b f(t)\, dt - \int_a^b f_n(t)\, dt \right\| + \left\| \int_a^b f_n(t)\, dt \right\|$$

$$\leq \left\| \int_a^b f(t)\, dt - \int_a^b f_n(t)\, dt \right\| + (b - a) \sup_x \|f_n(x)\|$$

$$\leq \left\| \int_a^b f(t)\, dt - \int_a^b f_n(t)\, dt \right\|$$
$$+ (b - a) \sup_x \|f_n(x) - f(x)\| + (b - a) \sup_x \|f(x)\|.$$

The desired result is obtained by passing to the limit as $n \to \infty$.   □

Let us now apply integration theory to prove the mean value theorem. We will use the following proposition.

Proposition 1.236. *Suppose that U is an open subset of X. If $f : U \to Y$ is a smooth function, and $x + ty \in U$ for $0 \leq t \leq 1$, then*

$$f(x + y) - f(x) = \int_0^1 Df(x + ty) y \, dt. \qquad (1.52)$$

Proof. Let $\lambda : Y \to \mathbb{R}$ be a continuous linear functional and consider the function $F : [0, 1] \to \mathbb{R}$ given by

$$F(t) = \lambda(f(x + ty)) =: \lambda f(x + ty).$$

The functional λ is C^1 because it is linear. Also, the composition of smooth maps is smooth. Thus, F is C^1.

By the fundamental theorem of calculus, we have that

$$F(1) - F(0) = \int_0^1 F'(t)\,dt,$$

or, equivalently,

$$\begin{aligned}
\lambda(f(x+y) - f(x)) &= \lambda f(x+y) - \lambda f(x) \\
&= \int_0^1 \lambda(Df(x+ty)y)\,dt \\
&= \lambda \int_0^1 Df(x+ty)y\,dt.
\end{aligned}$$

Here, $f(x+y) - f(x)$ and $\int_0^1 Df(x+ty)y\,dt$ are elements of Y, and λ has the same value on these two points. Moreover, by our construction, this is true for *all* continuous linear functionals. Thus, it suffices to prove the following claim: If u,v are in X and $\lambda(u) = \lambda(v)$ for all continuous linear functionals, then $u = v$. To prove the claim, set $w = u - v$ and note that $Z := \{tw : t \in \mathbb{R}\}$ is a closed subspace of Y. Moreover, $\lambda_0 : Z \to \mathbb{R}$ defined by $\lambda_0(tw) = t\|w\|$ is a linear functional on Z such that $\|\lambda_0(tw)\| = |t|\|w\| = \|tw\|$. Thus, $\|\lambda_0\| = 1$, and λ_0 is continuous. By the Hahn–Banach theorem, λ_0 extends to a continuous linear functional λ on all of Y. But for this extension we have, $\lambda(w) = \lambda(1 \cdot w) = \|w\| = 0$. Thus, we have $w = 0$, and $u = v$. $\qquad\square$

With the same hypotheses as in Proposition 1.236, the mean value theorem (Theorem 1.226) states that if $x + t(z - x) \in U$ for $0 \le t \le 1$, then

$$\|f(z) - f(x)\| \le \|z - x\| \sup_{t \in [0,1]} \|Df(x + t(z - x))\|. \tag{1.53}$$

Proof. By Proposition 1.236 we have that

$$\|f(z) - f(x)\| = \left\| \int_0^1 Df(x + t(z - x))(z - x)\,dt \right\|.$$

Also, the function $t \mapsto Df(x + t(z - x))(z - x)$ is continuous. Thus, the desired result is an immediate consequence of Lemma 1.228 and Proposition 1.235. $\qquad\square$

The next theorem is a special case of Taylor's theorem (see [2, p. 93] and Exercise 1.238).

Theorem 1.237 (Taylor's Theorem). *Suppose that U is an open subset of X. If $f : U \to Y$ is C^1 and $x + th \in U$ for $0 \le t \le 1$, then*

$$f(x + h) = f(x) + Df(x)h + \int_0^1 (Df(x + th)h - Df(x)h)\,dt.$$

Proof. By Proposition 1.236 we have

$$f(x + h) = f(x) + \int_0^1 Df(x + th)h\, dt$$

$$= f(x) + \int_0^1 ((Df(x + th)h - Df(x)h) + Df(x)h)\, dt$$

$$= f(x) + Df(x)h + \int_0^1 (Df(x + th)h - Df(x)h)\, dt,$$

as required.   □

Exercise 1.238. Prove the following generalization of Theorem 1.237. Suppose that U is an open subset of X. If $f : U \to Y$ is C^r and $x + th \in U$ for $0 \leq t \leq 1$, then

$$f(x + h) = f(x) + Df(x)h + D^2 f(x)h^2 + \cdots + D^r f(x)h^r$$
$$+ \int_0^1 \frac{(1 - t)^{r-1}}{(r - 1)!} (D^r f(x + th)h^r - D^r f(x)h^r)\, dt.$$

1.11   Contraction

A map transforming a complete metric space into itself that moves each pair of points closer together has a fixed point. This contraction principle has far reaching consequences including the existence and uniqueness of solutions of differential equations and the existence and smoothness of invariant manifolds. The basic theory is introduced in this section.

1.11.1   The Contraction Mapping Theorem

In this section, let us suppose that (X, d) is a metric space. A point $x_0 \in X$ is a *fixed point* of a function $T : X \to X$ if $T(x_0) = x_0$. The fixed point x_0 is called *globally attracting* if $\lim_{n \to \infty} T^n(x) = x_0$ for each $x \in X$.

Definition 1.239. Suppose that $T : X \to X$, and λ is a real number such that $0 \leq \lambda < 1$. The function T is called a *contraction* (with contraction constant λ) if

$$d(T(x), T(y)) \leq \lambda d(x, y)$$

whenever $x, y \in X$.

The next theorem is fundamental; it states that a contraction, viewed as a dynamical system, has a globally attracting fixed point.

Theorem 1.240 (Contraction Mapping Theorem). *If the function T is a contraction on the complete metric space (X, d) with contraction constant λ, then T has a unique fixed point $x_0 \in X$. Moreover, if $x \in X$, then the sequence $\{T^n(x)\}_{n=0}^{\infty}$ converges to x_0 as $n \to \infty$ and*

$$d(T^n(x), x_0) \leq \frac{\lambda^n}{1 - \lambda} d(T(x), x).$$

Proof. Let us prove first that fixed points of T are unique. Suppose that $T(x_0) = x_0$ and $T(x_1) = x_1$. Because T is a contraction, $d(T(x_0), T(x_1)) \leq \lambda d(x_0, x_1)$, and, because x_0 and x_1 are fixed points, $d(T(x_0), T(x_1)) = d(x_0, x_1)$. Thus, we have that

$$d(x_0, x_1) \leq \lambda d(x_0, x_1).$$

If $x_0 \neq x_1$, then $d(x_0, x_1) \neq 0$ and therefore $\lambda \geq 1$, in contradiction.

To prove the existence of a fixed point, let $x \in X$ and consider the corresponding sequence of iterates $\{T^n(x)\}_{n=1}^{\infty}$. By repeated applications of the contraction property, it follows that

$$d(T^{n+1}(x), T^n(x)) \leq \lambda d(T^n(x), T^{n-1}(x)) \leq \cdots \leq \lambda^n d(T(x), x).$$

Also, by using the triangle inequality together with this result, we obtain the inequalities

$$\begin{aligned}
d(T^{n+p}(x), T^n(x)) &\leq d(T^{n+p}(x), T^{n+p-1}(x)) + \cdots + d(T^{n+1}(x), T^n(x)) \\
&\leq (\lambda^{n+p-1} + \cdots + \lambda^n) d(T(x), x) \\
&\leq \lambda^n (1 + \lambda + \cdots + \lambda^{p-1}) d(T(x), x) \\
&\leq \frac{\lambda^n}{1 - \lambda} d(T(x), x). \qquad (1.54)
\end{aligned}$$

Since $0 \leq \lambda < 1$, the sequence $\{\lambda^n\}_{n=1}^{\infty}$ converges to zero, and therefore $\{T^n(x)\}_{n=1}^{\infty}$ is a Cauchy sequence. Thus, this sequence converges to some point $x_0 \in X$.

We will prove that x_0 is a fixed point of the map T. Let us first note that, because the sequences $\{T^{n+1}(x)\}_{n=0}^{\infty}$ and $\{T^n(x)\}_{n=1}^{\infty}$ are identical, $\lim_{n \to \infty} T^{n+1}(x) = x_0$. Also, by the contraction property, it follows that T is continuous and

$$d(T^{n+1}(x), T(x_0)) = d(T(T^n(x)), T(x_0)) \leq \lambda d(T^n(x), x_0).$$

Therefore, using the continuity of T, we have the required limit

$$\lim_{n \to \infty} T^{n+1}(x) = \lim_{n \to \infty} T(T^n(x)) = T(x_0).$$

To prove the estimate in the theorem, pass to the limit as $p \to \infty$ in the inequality (1.54) to obtain

$$d(x_0, T^n(x)) \leq \frac{\lambda^n}{1 - \lambda} d(T(x), x). \qquad \square$$

Exercise 1.241. Suppose that X is a set and n is a positive integer. Prove : If T is a function, $T : X \to X$, and if T^n has a unique fixed point, then T has a unique fixed point.

Exercise 1.242. Suppose that $g \in C[0, \infty)$, where $C[\alpha, \infty)$ denotes the Banach space of continuous functions that are bounded in the supremum norm on the interval $[\alpha, \infty)$ and $\int_1^\infty t|g(t)| \, dt < \infty$. Prove that if $\alpha > 0$ is sufficiently large, then the integral equation

$$x(t) = 1 + \int_t^\infty (t - s)g(s)x(s) \, ds$$

has a unique solution in $C[\alpha, \infty)$. (b) Relate the result of part (a) to the differential equation $\ddot{x} = -g(t)x$ and the existence of its solutions with specified limits as $t \to \infty$ (cf. [157, p. 132]).

1.11.2 Uniform Contraction

In this section we will consider contractions depending on parameters and prove a uniform version of the contraction mapping theorem.

Definition 1.243. Suppose that A is a set, $T : X \times A \to X$, and $\lambda \in \mathbb{R}$ is such that $0 \le \lambda < 1$. The function T is a *uniform contraction* if

$$d(T(x, a), T(y, a)) \le \lambda d(x, y)$$

whenever $x, y \in X$ and $a \in A$.

For uniform contractions in a Banach space where the metric is defined in terms of the Banach space norm by $d(x, y) = \|x - y\|$, we have the following result (see [57]).

Theorem 1.244 (Uniform Contraction Theorem). *Suppose that X and Y are Banach spaces, $U \subseteq X$ and $V \subseteq Y$ are open subsets, $\bar{U}$ denotes the closure of U, the function $T : \bar{U} \times V \to \bar{U}$ is a uniform contraction with contraction constant λ, and, for each $y \in V$, let $g(y)$ denote the unique fixed point of the contraction $x \mapsto T(x, y)$ in $\bar{U}$. If k is a non-negative integer and $T \in C^k(\bar{U} \times V, X)$, then $g : V \to X$ is in $C^k(V, X)$. Also, if T is real analytic, then so is g.*

Proof. We will prove the theorem for $k = 0, 1$. The proof for $k > 1$ uses an induction argument. The analytic case requires a proof of the convergence of the Taylor series of g.

By the definition of g given in the statement of the theorem, the identity $T(g(y), y) = g(y)$ holds for all $y \in V$. If $k = 0$, then

$$\|g(y + h) - g(y)\| = \|T(g(y + h), y + h) - T(g(y), y)\|$$
$$\le \|T(g(y + h), y + h) - T(g(y), y + h)\|$$
$$+ \|T(g(y), y + h) - T(g(y), y)\|$$
$$\le \lambda\|g(y + h) - g(y)\| + \|T(g(y), y + h) - T(g(y), y)\|,$$

and therefore

$$\|g(y+h) - g(y)\| \leq \frac{1}{1-\lambda} \|T(g(y), y+h) - T(g(y), y)\|.$$

But T is continuous at the point $(g(y), y)$. Thus, if $\epsilon > 0$ is given, there is some $\delta > 0$ such that

$$\|T(g(y), y+h) - T(g(y), y)\| < \epsilon \quad \text{whenever} \quad \|h\| < \delta.$$

In other words, g is continuous, as required.

Suppose that $k = 1$ and consider the function $g : V \to \bar{U}$ given by $g(y) = T(g(y), y)$. We will prove that g is C^1.

The first observation is simple. If g is C^1, then, by the chain rule,

$$Dg(y) = T_x(g(y), y)Dg(y) + T_y(g(y), y).$$

In other words, if $Dg(y)$ exists, we expect it to be a solution of the equation

$$z = T_x(g(y), y)z + T_y(g(y), y). \tag{1.55}$$

We will prove that, for each $y \in V$, the mapping

$$z \mapsto T_x(g(y), y)z + T_y(g(y), y),$$

on the Banach space of bounded linear transformations from Y to X, is a contraction. In fact, if z_1 and z_2 are bounded linear transformations from Y to X, then

$$\|T_x(g(y), y)z_1 + T_y(g(y), y) - (T_x(g(y), y)z_2 + T_y(g(y), y))\|$$
$$\leq \|T_x(g(y), y)\| \|z_1 - z_2\|.$$

Thus, the map is a contraction whenever $\|T_x(g(y), y)\| < 1$. In fact, as we will soon see, $\|T_x(g(y), y)\| \leq \lambda$. Once this inequality is proved, it follows from the contraction principle that for each $y \in V$ the equation (1.55) has a unique solution $z(y)$. The differentiability of the the function $y \mapsto g(y)$ is then proved by verifying the limit

$$\lim_{\|h\| \to 0} \frac{\|g(y+h) - g(y) - z(y)h\|}{\|h\|} = 0. \tag{1.56}$$

To obtain the required inequality $\|T_x(g(y), y)\| \leq \lambda$, note that T is C^1. In particular, the partial derivative T_x is a continuous function and

$$\lim_{\|h\| \to 0} \frac{\|T(x+h, y) - T(x, y) - T_x(x, y)h\|}{\|h\|} = 0.$$

Let $\xi \in X$ be such that $\|\xi\| = 1$ and note that for each $\epsilon > 0$, if we set $h = \epsilon \xi$, then we have

$$
\begin{aligned}
\|T_x(x,y)\xi\| &= \|\frac{1}{\epsilon} T_x(x,y)h\| \\
&\leq \frac{1}{\epsilon}\big(\|T(x+h,y) - T(x,y) - T_x(x,y)h\| \\
&\quad + \|T(x+h,y) - T(x,y)\|\big) \\
&\leq \frac{\|T(x+h,y) - T(x,y) - T_x(x,y)h\|}{\|h\|} + \frac{\lambda\|h\|}{\|h\|}.
\end{aligned}
$$

Passing to the limit as $\epsilon \to 0$, we obtain $\|T_x(x,y)\xi\| \leq \lambda$, as required.

To prove (1.56), set $\gamma = \gamma(h) := g(y+h) - g(y)$. Since, $g(y)$ is a fixed point of the contraction mapping T, we have

$$
\gamma = T(g(y) + \gamma, y + h) - T(g(y), y).
$$

Set

$$
\Delta := T(g(y) + \gamma, y + h) - T(g(y), y) - T_x(g(y), y)\gamma - T_y(g(y), y)h
$$

and note that

$$
\begin{aligned}
\gamma &= T(g(y) + \gamma, y + h) - T(g(y), y) - T_x(g(y), y)\gamma \\
&\quad -T_y(g(y), y)h + T_x(g(y), y)\gamma + T_y(g(y), y)h \\
&= T_x(g(y), y)\gamma + T_y(g(y), y)h + \Delta.
\end{aligned}
$$

Also, since T is C^1, we have for each $\epsilon > 0$ a $\delta > 0$ such that $\|\Delta\| < \epsilon(\|\gamma\| + \|h\|)$ whenever $\|\gamma\| < \delta$ and $\|h\| < \delta$.

The function $h \mapsto \gamma(h)$ is continuous. This follows from the first part of the proof since $T \in C^0$. Thus, we can find $\delta_1 > 0$ so small that $\delta_1 < \delta$ and $\|\gamma(h)\| < \delta$ whenever $\|h\| < \delta_1$, and therefore

$$
\|\Delta(\gamma(h), h)\| \leq \epsilon(\|\gamma(h)\| + \|h\|) \quad \text{whenever} \quad \|h\| < \delta_1.
$$

For $\|h\| < \delta_1$, we have

$$
\begin{aligned}
\|\gamma(h)\| &= \|T_x(g(y), y)\gamma + T_y(g(y), y)h + \Delta(\gamma, h)\| \\
&\leq \lambda\|\gamma\| + \|T_y(g(y), y)\|\|h\| + \epsilon(\|\gamma(h)\| + \|h\|)
\end{aligned}
$$

and, as a result,

$$
(1 - \lambda - \epsilon)\|\gamma(h)\| \leq (\|T_y(g(y), y)\| + \epsilon)\|h\|.
$$

If we take $\epsilon < 1 - \lambda$, then

$$
\|\gamma(h)\| \leq \frac{1}{1 - \lambda - \epsilon}(\|T_y(g(y), y)\| + \epsilon)\|h\| := \psi\|h\|,
$$

and it follows that

$$\|\Delta(\gamma(h), h)\| \le \epsilon(1 + \psi)\|h\|, \qquad \|h\| < \delta_1, \quad 0 < \epsilon < 1 - \lambda.$$

Finally, recall equation (1.55),

$$z = T_x(g(y), y)z + T_y(g(y), y),$$

and note that

$$(I - T_x(g(y), y))(\gamma(h) - z(y)h) = \gamma(h) - T_x(g(y), y)\gamma(h) - T_y(g(y), y)h$$
$$= \Delta(\gamma(h), h).$$

Also, since $\|T_x(g(y), y)\| < \lambda < 1$, we have

$$(I - T_x(g(y), y))^{-1} = I + \sum_{j=1}^{\infty} T_x^j$$

and

$$\|(I - T_x(g(y), y))^{-1}\| \le \frac{1}{1 - \|T_x\|} \le \frac{1}{1 - \lambda}.$$

This implies the inequality

$$\|\gamma(h) - z(y)h\| \le \frac{\epsilon}{1 - \lambda}(1 + \psi)\|h\|,$$

and the limit (1.56) follows.

By an application of the first part of the proof about solutions of contractions being *continuously* dependent on parameters, $y \mapsto z(y)$ is continuous. This completes the proof of the theorem for the case $k = 1$. □

Exercise 1.245. Let $C[0, \infty)$ denote the Banach space of continuous functions on the interval $[0, \infty)$ that are bounded in the supremum norm and $\mathcal{E}[0, \infty)$ the space of continuous functions bounded with respect to the norm $|f|_{\mathcal{E}} := \sup_{t \ge 0} |e^t f(t)|$. (a) Prove that $\mathcal{E}[0, \infty)$ is a Banach space. (b) For $f \in \mathcal{E}[0, \infty)$ and $\phi \in C[0, \infty)$ let

$$T(\phi, f)(t) := 1 + \int_t^{\infty} (t - s)f(s)\phi(s)\, ds.$$

Prove that $T : C[0, \infty) \times \mathcal{E}[0, \infty) \to C[0, \infty)$ is C^1. (c) Let $V \subset \mathcal{E}$ be the metric ball centered at the origin with radius $1/2$. Prove that there is a C^1 function $g : V \to C[0, \infty)$ such that $T(g(f), f) = g(f)$. (d) Show that the space $\mathcal{E}[0, \infty)$ can be replaced by other Banach spaces to obtain similar results. Can it be replaced by $C[0, \infty)$?

1.11.3   Fiber Contraction

In this section we will extend the contraction principle to bundles. The result of this extension, called the fiber contraction theorem [121], is useful in proving the smoothness of functions that are defined as fixed points of contractions.

Let X and Y be metric spaces. A map $\Gamma : X \times Y \to X \times Y$ of the form

$$\Gamma(x, y) = (\Lambda(x), \Psi(x, y)),$$

where $\Lambda : X \to X$ and $\Psi : X \times Y \to Y$, is called a *bundle map* over the base Λ with principal part Ψ. Here, the triple $(X \times Y, X, \pi)$, where $\pi : X \times Y \to X$ is the projection $\pi(x, y) = x$, is called the *trivial bundle* over X with fiber Y.

Definition 1.246. Suppose that $\mu \in \mathbb{R}$ is such that $0 \leq \mu < 1$. The bundle map $\Gamma : X \times Y \to X \times Y$ is called a *fiber contraction* if the function $y \mapsto \Psi(x, y)$ is a contraction with contraction constant μ for every $x \in X$.

Theorem 1.247 (Fiber Contraction Theorem). *Suppose that X and Y denote metric spaces, and that $\Gamma : X \times Y \to X \times Y$ is a continuous fiber contraction over $\Lambda : X \to X$ with principal part $\Psi : X \times Y \to Y$. If Λ has a globally attracting fixed point x_∞, and if y_∞ is a fixed point of the map $y \mapsto \Psi(x_\infty, y)$, then (x_∞, y_∞) is a globally attracting fixed point of Γ.*

Remark: The proof does not require the metric spaces X or Y to be complete.

Proof. Let d_X denote the metric for X, let d_Y denote the metric for Y, and let the metric on $X \times Y$ be defined by $d := d_X + d_Y$. We must show that for each $(x, y) \in X \times Y$ we have $\lim_{n \to \infty} \Gamma^n(x, y) = (x_\infty, y_\infty)$ where the limit is taken with respect to the metric d.

For notational convenience, let us denote the map $y \mapsto \Psi(x, y)$ by Ψ_x. Then, for example, we have

$$\Gamma^{n+1}(x, y) = (\Lambda^{n+1}(x), \Psi_{\Lambda^n(x)} \circ \Psi_{\Lambda^{n-1}(x)} \circ \cdots \circ \Psi_x(y))$$

and, using the triangle inequality, the estimate

$$d(\Gamma^{n+1}(x, y), (x_\infty, y_\infty)) \leq d(\Gamma^{n+1}(x, y), \Gamma^{n+1}(x, y_\infty))$$
$$+ d(\Gamma^{n+1}(x, y_\infty), (x_\infty, y_\infty)). \quad (1.57)$$

Note that

$$d(\Gamma^{n+1}(x, y), \Gamma^{n+1}(x, y_\infty)) = d_Y(\Psi_{\Lambda^n(x)} \circ \Psi_{\Lambda^{n-1}(x)} \circ \cdots \circ \Psi_x(y),$$
$$\Psi_{\Lambda^n(x)} \circ \Psi_{\Lambda^{n-1}(x)} \circ \cdots \circ \Psi_x(y_\infty)).$$

Moreover, if μ is the contraction constant for the fiber contraction Γ, then we have

$$d(\Gamma^{n+1}(x, y), \Gamma^{n+1}(x, y_\infty)) \leq \mu^{n+1} d_Y(y, y_\infty),$$

and therefore, $\lim_{n\to\infty} d(\Gamma^n(x,y), \Gamma^n(x,y_\infty)) = 0$.

For the second summand of (1.57), we have

$$d(\Gamma^{n+1}(x,y_\infty), (x_\infty, y_\infty)) = d_X(\Lambda^{n+1}(x), x_\infty)$$
$$+ d_Y(\Psi_{\Lambda^n(x)} \circ \cdots \circ \Psi_x(y_\infty), y_\infty).$$

By the hypothesis that x_∞ is a global attractor, the first summand on the right hand side of the last equality converges to zero as $n \to \infty$. Thus, to complete the proof, it suffices to verify the limit

$$\lim_{n\to\infty} d_Y(\Psi_{\Lambda^n(x)} \circ \Psi_{\Lambda^{n-1}(x)} \circ \cdots \circ \Psi_x(y_\infty), y_\infty) = 0. \qquad (1.58)$$

Let us observe that

$$d_Y(\Psi_{\Lambda^n(x)} \circ \cdots \circ \Psi_x(y_\infty), y_\infty) \le d_Y(\Psi_{\Lambda^n(x)} \circ \cdots \circ \Psi_x(y_\infty), \Psi_{\Lambda^n(x)}(y_\infty))$$
$$+ d_Y(\Psi_{\Lambda^n(x)}(y_\infty), y_\infty)$$
$$\le \mu d_Y(\Psi_{\Lambda^{n-1}(x)} \circ \cdots \circ \Psi_x(y_\infty), y_\infty)$$
$$+ d_Y(\Psi_{\Lambda^n(x)}(y_\infty), y_\infty),$$

and by induction that

$$d_Y(\Psi_{\Lambda^n(x)} \circ \Psi_{\Lambda^{n-1}(x)} \circ \cdots \circ \Psi_x(y_\infty), y_\infty) \le \sum_{j=0}^{n} \mu^{n-j} d_Y(\Psi_{\Lambda^j(x)}(y_\infty), y_\infty).$$

For each nonnegative integer m, define $a_m := d_Y(\Psi_{\Lambda^m(x)}(y_\infty), y_\infty)$. Each a_m is nonnegative and

$$a_m = d_Y(\Psi(\Lambda^m(x), y_\infty), \Psi(x_\infty, y_\infty)).$$

Using the continuity of Ψ and the hypothesis that x_∞ is a globally attracting fixed point, it follows that the sequence $\{a_m\}_{m=0}^{\infty}$ converges to zero and is therefore bounded. If A is an upper bound for the elements of this sequence, then for each $m = 0, 1, \ldots, \infty$ we have $0 \le a_m < A$.

Let $\epsilon > 0$ be given. There is some $K > 0$ so large that

$$0 \le a_k < \frac{1}{2}(1 - \mu)\epsilon$$

whenever $k \ge K$. Hence, if $n \ge K$, then

$$\sum_{j=0}^{n} \mu^{n-j} a_j = \sum_{j=0}^{K-1} \mu^{n-j} a_j + \sum_{j=K}^{n} \mu^{n-j} a_j$$
$$\le A \sum_{j=0}^{K-1} \mu^{n-j} + \frac{1}{2}(1 - \mu)\epsilon \sum_{j=K}^{n} \mu^{n-j}$$
$$\le A \frac{\mu^{n-K+1}}{1 - \mu} + \frac{1}{2}\epsilon.$$

Moreover, there is some $N \geq K$ such that

$$\mu^{n-K+1} < \frac{(1-\mu)\epsilon}{2A}$$

whenever $n \geq N$. In other words, $\lim_{n\to\infty} \sum_{j=0}^{n} \mu^{n-j} a_j = 0$. $\square$

As mentioned above, the fiber contraction principle is often used to prove that functions obtained as fixed points of contractions are smooth. We will use this technique as one method to prove that the flow defined by a smooth differential equation is smooth, and we will use a similar argument again when we discuss the smoothness of invariant manifolds. We will codify some of the ideas that are used in applications of the fiber contraction principle, and we will discuss a simple application to illustrate the procedure.

The setting for our analysis is given by a contraction $\Lambda : \mathcal{C} \to \mathcal{C}$, where $\mathcal{C}$ denotes a closed subset of a Banach space of continuous functions that map a Banach space X to a Banach space Y. Let $\alpha_\infty \in \mathcal{C}$ denote the unique fixed point of Λ, and recall that α_∞ is globally attracting; that is, if $\alpha \in \mathcal{C}$, then $\Lambda^n(\alpha) \to \alpha_\infty$ as $n \to \infty$.

Define the Banach space of all (supremum norm) bounded continuous functions from X to the linear maps from X to Y and denote this space by $C(X, L(X, Y))$. Elements of $C(X, L(X, Y))$ are the candidates for the derivatives of functions in $\mathcal{C}$. Also, let $\mathcal{C}^1$ denote the subset of $\mathcal{C}$ consisting of all continuously differentiable functions with bounded derivatives.

The first step of the method is to show that if $\alpha \in \mathcal{C}^1$, then the derivative of $\Lambda(\alpha)$ has the form

$$(D(\Lambda(\alpha)))(\xi) = \Psi(\alpha, D\alpha)(\xi)$$

where $\xi \in X$ and where Ψ is a map

$$\Psi : \mathcal{C} \times C(X, L(X, Y)) \to C(X, L(X, Y)).$$

Next, define the bundle map

$$\Gamma : \mathcal{C} \times C(X, L(X, Y)) \to \mathcal{C} \times C(X, L(X, Y))$$

by

$$(\alpha, \Phi) \mapsto (\Lambda(\alpha), \Psi(\alpha, \Phi))$$

and prove that Γ is a *continuous* fiber contraction. Warning: In the applications, the continuity of $\alpha \mapsto \Psi(\alpha, \Phi)$ is usually not obvious. It is easy to be deceived into believing that a map of the form $g \mapsto g \circ \alpha$ is continuous when g and α are continuous because the definition of the mapping involves the composition of two continuous functions. To prove that the map is continuous requires the omega lemma (Exercise 1.224), one of its relatives, or a new idea (see Theorem 1.249 and [80]).

Finally, pick a point $\alpha_0 \in \mathcal{C}^1$ so that $D\alpha_0 \in C(X, L(X, Y))$, let $(\phi_0, \Phi_0) = (\alpha_0, D\alpha_0)$, and define, for all positive integers n,

$$(\phi_{n+1}, \Phi_{n+1}) = \Gamma(\phi_n, \Phi_n).$$

By the fiber contraction principle, there is some $\Phi_\infty \in C(X, L(X, Y))$ such that $\lim_{n\to\infty}(\phi_n, \Phi_n) = (\alpha_\infty, \Phi_\infty)$. By the construction of Ψ, if $n \geq 0$, then $D(\phi_n) = \Phi_n$. If the convergence is uniform (or at least uniform on compact subsets of X), then we obtain the desired result, $D(\alpha_\infty) = \Phi_\infty$, as an application of the following theorem from advanced calculus (see Exercise 1.252).

Theorem 1.248. *If a sequence of differentiable functions is uniformly convergent and if the corresponding sequence of their derivatives is uniformly convergent, then the limit function of the original sequence is differentiable and its derivative is the limit of the corresponding sequence of derivatives.*

Moreover, we have $\Phi_\infty \in C(X, L(X, Y))$, and therefore Φ_∞ is continuous. In particular, the fixed point α_∞ is continuously differentiable.

We will formulate and prove a simple result to illustrate a typical application of the fiber contraction principle. For this, let us consider specifically the linear space $C^0(\mathbb{R}^M, \mathbb{R}^N)$ consisting of all continuous functions $f : \mathbb{R}^M \to \mathbb{R}^N$ and let $\mathcal{C}^0(\mathbb{R}^M, \mathbb{R}^N)$ denote the subspace consisting of all $f \in C^0(\mathbb{R}^M, \mathbb{R}^N)$ such that the supremum norm is finite; that is,

$$\|f\| := \sup_{\xi \in \mathbb{R}^M} |f(\xi)| < \infty.$$

Of course, $\mathcal{C}^0(\mathbb{R}^M, \mathbb{R}^N)$ is a Banach space with the supremum norm. Also, let $\mathcal{B}_\rho^0(\mathbb{R}^M, \mathbb{R}^N)$ denote the subset of $\mathcal{C}^0(\mathbb{R}^M, \mathbb{R}^N)$ such that, for

$$f \in \mathcal{B}_\rho^0(\mathbb{R}^M, \mathbb{R}^N),$$

the Lipschitz constant of f is bounded by ρ; that is,

$$\mathrm{Lip}(f) := \sup_{\xi_1 \neq \xi_2} \frac{|f(\xi_1) - f(\xi_2)|}{|\xi_1 - \xi_2|} \leq \rho.$$

The set $\mathcal{B}_\rho^0(\mathbb{R}^M, \mathbb{R}^N)$ is a closed subset of $\mathcal{C}^0(\mathbb{R}^M, \mathbb{R}^N)$ (see Exercise 1.251). Hence, $\mathcal{B}_\rho^0(\mathbb{R}^M, \mathbb{R}^N)$ is a complete metric space with respect to the supremum norm.

In case $f \in C^0(\mathbb{R}^M, \mathbb{R}^N)$ is continuously differentiable, its derivative Df is an element of $C^0(\mathbb{R}^M, L(\mathbb{R}^M, \mathbb{R}^N))$, the space of continuous functions from $\mathbb{R}^M$ to the linear maps from $\mathbb{R}^M$ to $\mathbb{R}^N$. The subset $\mathcal{F}$ of $C^0(\mathbb{R}^M, L(\mathbb{R}^M, \mathbb{R}^N))$ consisting of all elements that are bounded with respect to the norm

$$\|\Phi\| := \sup_{\xi \in \mathbb{R}^M} \left(\sup_{|v|=1} |\Phi(\xi)v| \right),$$

is a Banach space. The closed metric ball $\mathcal{F}_\rho$ of radius $\rho > 0$ centered at the origin of $\mathcal{F}$ (that is, all Φ such that $\|\Phi\| \leq \rho$) is a complete metric space relative to the norm on $\mathcal{F}$.

Theorem 1.249. *If $F : \mathbb{R}^N \to \mathbb{R}^N$ and $G : \mathbb{R}^M \to \mathbb{R}^N$ are continuously differentiable functions, $\|F\| + \|G\| < \infty$, $\|DF\| < 1$, and $\|DG\| < \infty$, then the functional equation $F \circ \phi - \phi = G$ has a unique solution α in $\mathcal{B}_\rho^0(\mathbb{R}^M, \mathbb{R}^N)$ for every $\rho > \|DG\|/(1 - \|DF\|)$. Moreover, α is continuously differentiable and $\|D\alpha\| < \rho$.*

Proof. Suppose that $\rho > \|DG\|/(1 - \|DF\|)$ or $\|DF\|\rho + \|G\| < \rho$. If $\phi \in \mathcal{B}_\rho^0(\mathbb{R}^M, \mathbb{R}^N)$, then the function $F \circ \phi - G$ is continuous. Also, we have that

$$\|F \circ \phi - G\| \leq \sup_{\xi \in \mathbb{R}^M} |F(\phi(\xi))| + \|G\| \leq \sup_{\zeta \in \mathbb{R}^N} |F(\zeta)| + \|G\| < \infty,$$

and, by the mean value theorem,

$$|F(\phi(\xi)) - G(\xi) - (F(\phi(\eta)) - G(\eta))| \leq \|DF\|\,|\phi(\xi) - \phi(\eta)|$$
$$+ \|DG\|\,|\xi - \eta|$$
$$\leq (\|DF\|\,\mathrm{Lip}(\phi) + \|DG\|)|\xi - \eta|,$$

where $\|DF\|\,\mathrm{Lip}(\phi) + \|DG\| < \rho$. It follows that the function $F \circ \phi - G$ is an element of the space $\mathcal{B}_\rho^0(\mathbb{R}^M, \mathbb{R}^N)$.

Let us define $\Lambda : \mathcal{B}_\rho^0(\mathbb{R}^M, \mathbb{R}^N) \to \mathcal{B}_\rho^0(\mathbb{R}^M, \mathbb{R}^N)$ by

$$\Lambda(\phi)(\xi) := F(\phi(\xi)) - G(\xi).$$

Also, note that by the hypothesis $\|DF\| < 1$. If ϕ_1 and ϕ_2 are in the space $\mathcal{B}_\rho^0(\mathbb{R}^M, \mathbb{R}^N)$, then

$$|\Lambda(\phi_1)(\xi) - \Lambda(\phi_2)(\xi)| < \|DF\|\|\phi_1 - \phi_2\|;$$

that is, Λ is a contraction on the complete metric space $\mathcal{B}_\rho^0(\mathbb{R}^M, \mathbb{R}^N)$. Therefore, there is a unique function $\alpha \in \mathcal{B}_\rho^0(\mathbb{R}^M, \mathbb{R}^N)$ such that $F \circ \alpha - \alpha = G$. Moreover, if $\phi \in \mathcal{B}_\rho^0(\mathbb{R}^M, \mathbb{R}^N)$, then $\lim_{n \to \infty} \Lambda^n(\phi) = \alpha$.

We will prove that the function α is continuously differentiable. To this end, note that for $\phi \in \mathcal{B}_\rho^0(\mathbb{R}^M, \mathbb{R}^N)$ and $\Phi \in \mathcal{F}_\rho$ we have

$$\|DF(\phi(\xi))\Phi(\xi) - DG(\xi)\| \leq \|DF\|\|\Phi\| + \|DG\| < \rho;$$

and, using this result, define $\Psi : \mathcal{B}_\rho^0(\mathbb{R}^M, \mathbb{R}^N) \times \mathcal{F}_\rho \to \mathcal{F}_\rho$ by

$$\Psi(\phi, \Phi)(\xi) := DF(\phi(\xi))\Phi(\xi) - DG(\xi).$$

The function $\Phi \mapsto \Psi(\phi, \Phi)$ is a contraction on $\mathcal{F}_\rho$ whose contraction constant is less than one and uniform over $\mathcal{B}_\rho^0(\mathbb{R}^M, \mathbb{R}^N)$; in fact, we have that

$$\|\Psi(\phi, \Phi_1)(\xi) - \Psi(\phi, \Phi_2)(\xi)\| \leq \|DF\|\|\Phi_1 - \Phi_2\|.$$

Thus, we have defined a bundle map

$$\Gamma : \mathcal{B}_\rho^0(\mathbb{R}^M, \mathbb{R}^N) \times \mathcal{F}_\rho \to \mathcal{B}_\rho(\mathbb{R}^M, \mathbb{R}^N) \times \mathcal{F}_\rho$$

given by

$$\Gamma(\phi, \Phi) := (\Lambda(\phi), \Psi(\phi, \Phi)),$$

which is a uniform contraction on fibers. To prove that Γ is a fiber contraction, it suffices to show that Γ is continuous.

Note that Λ is a contraction, $\Phi \mapsto \Psi(\phi, \Phi)$ is an affine function, and

$$|\Psi(\phi, \Phi)(\xi) - \Psi(\phi_0, \Phi_0)(\xi)| \leq \|DF\|\|\Phi - \Phi_0\| + \|\Phi_0\|\|DF \circ \phi - DF \circ \phi_0\|.$$

Thus, to prove the continuity of Γ, it suffices to show that the function $\phi \mapsto DF \circ \phi$ is continuous.

Fix $\phi_0 \in \mathcal{B}_\rho^0(\mathbb{R}^M, \mathbb{R}^N)$ and note that the function DF is *uniformly* continuous on the closed ball B centered at the origin in $\mathbb{R}^N$ with radius $\|\phi_0\| + 1$. Also, the image of ϕ_0 is in B. By the uniform continuity of DF, for each $\epsilon > 0$ there is a positive number $\delta < 1$ such that $|DF(\xi) - DF(\eta)| < \epsilon$ whenever $\xi, \eta \in B$ and $|\xi - \eta| < \delta$. If $\|\phi - \phi_0\| < \delta$ and $\xi \in \mathbb{R}^M$, then

$$|\phi(\xi)| \leq \|\phi - \phi_0\| + \|\phi_0\| < 1 + \|\phi_0\|;$$

and therefore, the image of ϕ is in the ball B. Thus, we have that

$$\|DF \circ \phi - DF \circ \phi_0\| < \epsilon \text{ whenever } \|\phi - \phi_0\| < \delta;$$

that is, DF is continuous at ϕ_0 (cf. Exercise 1.258).

Let Φ_∞ denote the unique fixed point of the contraction $\Phi \mapsto \Psi(\alpha, \Phi)$ over the fixed point α. Also, let us define a sequence in $\mathcal{B}_\rho^0(\mathbb{R}^M, \mathbb{R}^N) \times \mathcal{F}_\rho$ as follows: $(\phi_0, \Phi_0) = (0, 0)$ and, for each positive integer n,

$$(\phi_{n+1}, \Phi_{n+1}) := \Gamma(\phi_n, \Phi_n).$$

Note that $D\phi_0 = \Phi_0$ and, proceeding by induction, if $D\phi_n = \Phi_n$, then

$$D\phi_{n+1} = D(\Lambda(\phi_n)) = DF \circ \phi_n D\phi_n - DG$$
$$= \Psi(\phi_n, D\phi_n) = \Psi(\phi_n, \Phi_n) = \Phi_{n+1};$$

that is, $D\phi_n = \Phi_n$ for all integers $n \geq 0$.

By the fiber contraction theorem, we have that

$$\lim_{n \to \infty} \phi_n = \alpha, \qquad \lim_{n \to \infty} D\phi_n = \Phi_\infty.$$

The sequence $\{\phi_n\}_{n=0}^\infty$ converges uniformly to α and the sequence of its derivatives converges uniformly to Φ_∞. By Theorem 1.248 we have that α is differentiable with derivative Φ_∞. Thus, α is continuously differentiable. $\square$

Theorem 1.249 is an almost immediate corollary of the uniform contraction principle (see Exercise 1.254). For more sophisticated applications, there are at least three approaches to proving the smoothness of a map obtained by contraction: the uniform contraction principle, the fiber contraction principle, and the definition of the derivative. While the main difficulty in applying the uniform contraction principle is the proof of the smoothness of the uniform contraction, the main difficulty in applying the fiber contraction principle is the proof of the continuity of the fiber contraction, especially the continuity of the principle part of the fiber contraction with respect to the base point. The best choice is not always clear; it depends on the nature of the application and the skill of the applied mathematician.

Exercise 1.250. Let U denote an open ball in $\mathbb{R}^n$ or the entire space, and V an open ball in $\mathbb{R}^m$. Prove that the set of bounded continuous functions from U to $\mathbb{R}^n$ is a Banach space, hence a complete metric space. Also, prove that the set of continuous functions from U into $\bar{V}$ as well as the set of continuous functions from $\bar{V}$ to $\mathbb{R}^n$ are Banach spaces.

Exercise 1.251. Prove that $\mathcal{B}_\rho^0(\mathbb{R}^M, \mathbb{R}^N)$ is a closed subset of the Banach space $\mathcal{C}^0(\mathbb{R}^M, \mathbb{R}^N)$.

Exercise 1.252. Prove Theorem 1.248.

Exercise 1.253. Give a direct proof of Theorem 1.249 in case F is linear.

Exercise 1.254. Prove Theorem 1.249 using the uniform contraction principle. Hint: Construct a uniform contraction in $\mathbb{R}^N$ with parameter space $\mathbb{R}^M$ such that the solution of the functional equation is the map that assigns to each point in the parameter space the corresponding fixed point of the contraction.

Exercise 1.255. Suppose that $F : \mathbb{R} \to \mathbb{R}$ is continuous, and consider the function Λ given by $\phi \mapsto F \circ \phi$ on $C^0(\mathbb{R}, \mathbb{R})$. Show that $C^0(\mathbb{R}, \mathbb{R})$ is a metric space with metric $d(f, g) := \sup_{\xi \in \mathbb{R}} |f(\xi) - g(\xi)|$ and $\Lambda : C^0(\mathbb{R}, \mathbb{R}) \to C^0(\mathbb{R}, \mathbb{R})$. Construct a bounded continuous function F such that Λ is not continuous. Show that your Λ is continuous when restricted to $\mathcal{C}^0(\mathbb{R}, \mathbb{R})$.

Exercise 1.256. Derive the smoothness statement in the uniform contraction theorem as a corollary of the fiber contraction theorem.

Exercise 1.257. In the context of Theorem 1.249, prove that the solution of the functional equation $F \circ \phi - \phi = G$ is C^r (r-times continuously differentiable) if F and G are C^r, their C^r-norms are finite, and $\|DF\| < 1$. What condition would be required to prove that the solution is C^∞ or C^ω (real analytic).

Exercise 1.258. Prove the following lemma: If (X, d_X) and (Y, d_Y) are metric spaces, $f : X \to Y$ is continuous, and $K \subset X$ is compact, then f is uniformly continuous on K. Moreover, for every $\epsilon > 0$ there is a $\delta > 0$ such that $d_Y(f(x_1), f(x_2)) < \epsilon$ whenever $d_X(x_1, x_2) < \delta$ and $x_2 \in K$. Use this lemma to prove the continuity of the map $\phi \mapsto DF \circ \phi$ in the proof of Theorem 1.249. The lemma leads to only a slight improvement in the proof. On the other hand, this lemma simplifies the proof of continuity for many other fiber contractions.

1.11.4   The Implicit Function Theorem

The implicit function theorem is one of the most useful theorems in analysis. We will prove it as a corollary of the uniform contraction theorem.

Theorem 1.259 (Implicit Function Theorem). *Suppose that X, Y, and Z are Banach spaces, $U \subseteq X$, $V \subseteq Y$ are open sets, $F : U \times V \to Z$ is a C^1 function, and $(x_0, y_0) \in U \times V$ with $F(x_0, y_0) = 0$. If $F_x(x_0, y_0) : X \to Z$ has a bounded inverse, then there is a product neighborhood $U_0 \times V_0 \subseteq U \times V$ with $(x_0, y_0) \in U_0 \times V_0$ and a C^1 function $\beta : V_0 \to U_0$ such that $\beta(y_0) = x_0$ and $F(\beta(y), y) \equiv 0$. Moreover, if $F(x, y) = 0$ for $(x, y) \in U_0 \times V_0$, then $x = \beta(y)$.*

Proof. Define $L : Z \to X$ by $Lz = [F_x(x_0, y_0)]^{-1}z$ and $G : U \times V \to X$ by $G(x, y) = x - LF(x, y)$. Note that G is C^1 on $U \times V$ and $F(x, y) = 0$ if and only if $G(x, y) = x$. Moreover, we have that $G(x_0, y_0) = x_0$ and $G_x(x_0, y_0) = I - LF_x(x_0, y_0) = 0$.

Since G is C^1, there is a product neighborhood $U_0 \times V_1$ whose factors are two metric balls, $U_0 \subseteq U$ centered at x_0 and $V_1 \subseteq V$ centered at y_0, such that

$$\|G_x(x, y)\| < \frac{1}{2}$$

whenever $(x, y) \in U_0 \times V_1$.

Let us suppose that the ball U_0 has radius $\delta > 0$. Note that the function given by $y \mapsto F(x_0, y)$ is continuous and vanishes at y_0. Thus, there is a metric ball $V_0 \subseteq V_1$ centered at y_0 such that

$$\|L\|\|F(x_0, y)\| < \frac{\delta}{2}$$

for every $y \in V_0$. With this choice of V_0, if $(x, y) \in U_0 \times V_0$, then, by the mean value theorem,

$$
\begin{aligned}
\|G(x, y) - x_0\| &= \|G(x, y) - G(x_0, y) + G(x_0, y) - x_0\| \\
&\leq \|G(x, y) - G(x_0, y)\| + \|LF(x_0, y)\| \\
&\leq \sup_{u \in U_0} \|G_x(u, y)\|\|x - x_0\| + \frac{\delta}{2} \leq \delta.
\end{aligned}
$$

In other words, $G(x, y) \in \bar{U}_0$; that is, $G : \bar{U}_0 \times V_0 \to \bar{U}_0$.

Again, by the mean value theorem, it is easy to see that G is a uniform contraction; in fact,

$$
\begin{aligned}
\|G(x_1, y) - G(x_2, y)\| &\leq \sup_{u \in U_0} \|G_x(u, y)\|\|x_1 - x_2\| \\
&\leq \frac{1}{2}\|x_1 - x_2\|.
\end{aligned}
$$

Thus, there is a unique smooth function $y \mapsto \beta(y)$ defined on the open ball V_0 such that $\beta(y_0) = x_0$ and $G(\beta(y), y) \equiv \beta(y)$. In particular,

$$\beta(y) = \beta(y) - LF(\beta(y), y)$$

and therefore $F(\beta(y), y) \equiv 0$, as required. $\square$

1.12   Existence, Uniqueness, and Extension

In this section we will prove the basic existence and uniqueness theorems for differential equations. We will also prove a theorem on extension of solutions. While the theorems on existence, uniqueness, and extension are the foundation for theoretical study of ordinary differential equations, there is another reason to study their proofs. In fact, the techniques used in this section are very important in the modern development of our subject. In particular, the implicit function theorem is used extensively in perturbation theory, and the various extensions of the contraction principle are fundamental techniques used to prove the existence and smoothness of invariant manifolds. We will demonstrate these tools by proving the fundamental existence theorem for differential equations in two different ways.

Suppose that $J \subseteq \mathbb{R}$, $\Omega \subseteq \mathbb{R}^n$, and $\Lambda \subseteq \mathbb{R}^m$ are all open sets, and

$$f : J \times \Omega \times \Lambda \to \mathbb{R}^n$$

given by $(t, x, \lambda) \mapsto f(t, x, \lambda)$ is a continuous function. Recall that if $\lambda \in \Lambda$, then a *solution* of the ordinary differential equation

$$\dot{x} = f(t, x, \lambda) \tag{1.59}$$

is a differentiable function $\sigma : J_0 \to \Omega$ defined on some open subinterval $J_0 \subseteq J$ such that

$$\frac{d\sigma}{dt}(t) = f(t, \sigma(t), \lambda)$$

for all $t \in J_0$. For $t_0 \in J$, $x_0 \in \Omega$, and $\lambda_0 \in \Lambda$, the *initial value problem* associated with the differential equation (1.59) is given by the differential equation together with an initial value for the solution as follows:

$$\dot{x} = f(t, x, \lambda_0), \qquad x(t_0) = x_0. \tag{1.60}$$

If σ is a solution of the differential equation as defined above such that in addition $\sigma(t_0) = x_0$, then we say that σ is a solution of the initial value problem (1.60).

Theorem 1.260. *If the function $f : J \times \Omega \times \Lambda \to \mathbb{R}^n$ in the differential equation (1.59) is continuously differentiable, $t_0 \in J$, $x_0 \in \Omega$, and $\lambda_0 \in$*

Λ, *then there are open sets* $J_0 \subseteq J$, $\Omega_0 \subseteq \Omega$, *and* $\Lambda_0 \subseteq \Lambda$ *such that* $(t_0, x_0, \lambda_0) \in J_0 \times \Omega_0 \times \Lambda_0$, *and a unique* C^1 *function* $\sigma : J_0 \times \Omega_0 \times \Lambda_0 \to \mathbb{R}^n$ *given by* $(t, x, \lambda) \to \sigma(t, x, \lambda)$ *such that* $t \mapsto \sigma(t, x, \lambda)$ *is a solution of the differential equation (1.59) and* $\sigma(0, x, \lambda) = x$. *In particular,* $t \mapsto \sigma(t, x_0, \lambda_0)$ *is a solution of the initial value problem (1.60).*

Proof. The proof we will give is due to Joel Robbin [190]. Suppose that σ is a solution of the initial value problem (1.60), $\delta > 0$, and σ is defined on the interval $[t_0 - \delta, t_0 + \delta]$. In this case, if we define $\tau := (t - t_0)/\delta$ and $z(\tau) = \sigma(\delta\tau + t_0) - x_0$, then $z(0) = 0$ and for $-1 \le \tau \le 1$,

$$\frac{dz}{d\tau}(\tau) = \delta\dot{\sigma}(\delta\tau + t_0) = \delta f(\delta\tau + t_0, z + x_0, \lambda_0), \qquad (1.61)$$

at least if $z + x_0 \in \Omega$. Conversely, if the differential equation (1.61) has a solution defined on a subinterval of $-1 \le \tau \le 1$, then the differential equation (1.59) has a solution. Thus, it suffices to show the following proposition: If $\delta > 0$ is sufficiently small, then the differential equation (1.61) has a solution defined on the interval $-1 \le \tau \le 1$.

Choose an open ball centered at the origin with radius r that is in Ω and let U denote the open ball centered at the origin with radius $r/2$. Define the Banach spaces

$$X := \{\phi \in C^1([-1, 1], \mathbb{R}^n) : \phi(0) = 0\}, \quad Y := C([-1, 1], \mathbb{R}^n)$$

where the norm on Y is the usual supremum norm, the norm on X is given by

$$\|\phi\|_1 = \|\phi\| + \|\phi'\|,$$

and ϕ' denotes the first derivative of ϕ. Also, let X_0 denote the open subset of X consisting of those elements of X whose ranges are in the open ball at the origin with radius $r/2$.

Consider the function

$$F : (-1, 1) \times J \times U \times \Lambda \times X_0 \to Y$$

by

$$F(\delta, t, x, \lambda, \phi)(\tau) = \phi'(\tau) - \delta f(\delta\tau + t, \phi(\tau) + x, \lambda).$$

We will apply the implicit function theorem to F.

We will show that the function F is C^1. Since the second summand in the definition of F is C^1 by the omega lemma (see Exercise 1.224), it suffices to show that the map d given by $\phi \mapsto \phi'$ is a C^1 map from X to Y.

Note that $\phi' \in Y$ for each $\phi \in X$ and d is a linear transformation. Because

$$\|d\phi\| \le \|d\phi\| + \|\phi\| = \|\phi\|_1,$$

the linear transformation d is continuous. Since the map $d : X \to Y$ is linear *and bounded*, it is its own derivative. In particular, d is continuously differentiable.

If $(t_0, x_0, \lambda_0) \in J \times U \times \Lambda$, then $F(0, t_0, x_0, \lambda_0, 0)(\tau) = 0$. Also, if we set $\delta = 0$ before the partial derivative is computed, then it is easy to see that

$$F_\phi(0, t_0, x_0, \lambda_0, 0) = d.$$

In order to show that $F_\phi(0, t_0, x_0, \lambda_0, 0)$ has a bounded inverse, it suffices to show that d has a bounded inverse. To this end, define $L : Y \to X$ by

$$(Ly)(\tau) = \int_0^\tau y(s)\,ds.$$

Clearly,

$$(d \circ L)(y) = y \quad \text{and} \quad (L \circ d)(\psi) = \psi.$$

Thus, L is an inverse for d. Moreover, since

$$\|Ly\|_1 = \|Ly\| + \|(d \circ L)y\|$$
$$\leq \|y\| + \|y\| \leq 2\|y\|,$$

it follows that L is bounded.

By an application of the implicit function theorem to F, we have proved the existence of a unique smooth function $(\delta, t, x, \lambda) \mapsto \beta(\delta, t, x, \lambda)$, with domain an open set $K_0 \times J_0 \times \Omega_0 \times \Lambda_0$ containing the point $(0, t_0, x_0, \lambda_0)$ and range in X_0 such that $\beta(0, t_0, x_0, \lambda_0) = 0$ and

$$F(\delta, t, x, \lambda, \beta(\delta, t, x, \lambda)) \equiv 0.$$

Thus, there is some $\delta > 0$ such that

$$\tau \mapsto z(\tau, t_0, x_0, \lambda_0) := \beta(\delta, t_0, x_0, \lambda_0)(\tau)$$

is the required solution of the differential equation (1.61). Of course, this solution depends smoothly on τ and all of its parameters. $\square$

We will now consider a proof of Theorem 1.260 that uses the contraction principle and the fiber contraction theorem. For this, it is convenient to make a minor change in notation and to introduce a few new concepts.

Instead of working directly with the initial value problem (1.60), we will study the solutions of initial value problems of the form

$$\dot{x} = F(t, x), \qquad x(t_0) = x_0 \tag{1.62}$$

where there is no dependence on parameters. In fact, there is no loss of generality in doing so. Note that the initial value problem (1.60) is "equivalent" to the following system of differential equations:

$$\dot{y} = f(t, y, \lambda), \quad \dot{\lambda} = 0, \qquad y(t_0) = y_0. \tag{1.63}$$

In particular, if we define $x = (y, \lambda)$ and $F(t, (y, \lambda)) := (f(t, y, \lambda), 0)$, then solutions of the initial value problem (1.60) can be obtained from solutions of the corresponding initial value problem (1.62) in the obvious manner. Moreover, smoothness is preserved. Thus, it suffices to work with the initial value problem (1.62).

Although the existence of a local solution for the initial value problem (1.62) can be proved using only the continuity of the function F, if F is merely continuous, then a solution of the initial value problem may not be unique. A sufficient condition for uniqueness is the requirement that F is Lipschitz with respect to its second argument; that is, there is a constant $\lambda > 0$ such that for each $t \in J$ and for all $x_1, x_2 \in \Omega$,

$$|F(t, x_1) - F(t, x_2)| \le \lambda |x_1 - x_2|$$

where $|x|$ is the usual norm of $x \in \mathbb{R}^n$. We will not prove the most general possible result; rather we will prove the following version of Theorem 1.260.

Theorem 1.261. *If the function $F : J \times \Omega \to \mathbb{R}^n$ in the initial value problem (1.62) is continuous and Lipschitz (with respect to its second argument), $t_0 \in J$, and $x_0 \in \Omega$, then there are open sets $J_0 \subseteq J$ and $\Omega_0 \subseteq \Omega$ such that $(t_0, x_0) \in J_0 \times \Omega_0$ and a unique continuous function $\sigma : J_0 \times \Omega_0 \to \mathbb{R}^n$ given by $(t, x) \to \sigma(t, x)$ such that $t \mapsto \sigma(t, x)$ is a solution of the differential equation $\dot{x} = F(t, x)$ with $\sigma(t_0, x) = x$. In particular, $t \mapsto \sigma(t, x_0)$ is the unique solution of the initial value problem (1.62). If, in addition, F is C^1, then so is the function σ.*

Proof. The function $t \mapsto x(t)$ is a solution of the initial value problem if and only if it is a solution of the integral equation

$$x(t) = x_0 + \int_{t_0}^{t} F(s, x(s)) \, ds.$$

In fact, if $dx/dt = F(t, x)$, then, by integration, we obtain the integral equation. On the other hand, if $t \mapsto x(t)$ satisfies the integral equation, then, by the fundamental theorem of calculus

$$\frac{dx}{dt} = F(t, x(t)).$$

Fix $(t_0, x_0) \in J \times \Omega$. Let $b(t_0, \delta)$ and $B(x_0, \nu)$ denote metric balls centered at t_0 and x_0 with positive radii, respectively δ and ν, such that

$$\bar{b}(t_0, \delta) \times \bar{B}(x_0, \nu) \subseteq J \times \Omega.$$

Since F is continuous on $J \times \Omega$, there is some number $M > 0$ such that

$$\sup_{(t,x) \in b(t_0, \delta) \times B(x_0, \nu)} |F(t, x)| \le M.$$

Since F is Lipschitz on $J \times \Omega$, there is some number $\lambda > 0$ such that, for each $t \in J$ and all $x_1, x_2 \in \Omega$,

$$|F(t, x_1) - F(t, x_2)| \le \lambda |x_1 - x_2|.$$

If $F \in C^1$ on $J \times \Omega$, then there is some number $K > 0$ such that

$$\sup_{(t,x)\in b(t_0,\delta)\times B(x_0,\nu)} \|DF(t, x)\| \le K, \tag{1.64}$$

where, recall, $DF(t, x)$ is the derivative of the map $x \mapsto F(t, x)$ and

$$\|DF(t, x)\| := \sup_{\{v\in\mathbb{R}^n:|v|=1\}} |DF(t, x)v|$$

with $|x|$ the usual norm of $x \in \mathbb{R}^n$.

Choose $\delta > 0$ so that $\delta\lambda < \min(1, \frac{\nu}{2})$ and $\delta M < \frac{\nu}{2}$, define the Banach space of bounded continuous $\mathbb{R}^n$-valued functions on $b(t_0, \delta) \times B(x_0, \frac{\nu}{2})$ with the norm

$$\|\phi\| = \sup_{(t,x)\in b(t_0,\delta)\times B(x_0,\frac{\nu}{2})} |\phi(t, x)|,$$

and let X denote the complete metric space consisting of all functions in this Banach space with range in $\bar{B}(x_0, \nu)$. In case F is C^1, let us agree to choose δ as above, but with the additional restriction that $\delta K < 1$. Finally, define the operator Λ on X by

$$\Lambda(\phi)(t, x) = x + \int_{t_0}^t F(s, \phi(s, x))\, ds. \tag{1.65}$$

Let us prove that $\Lambda : X \to X$. Clearly, we have $\Lambda(\phi) \in C(b(t_0, \delta) \times B(x_0, \frac{\nu}{2}), \mathbb{R}^n)$. In view of the inequality

$$|\Lambda(\phi)(t, x) - x_0| \le |x - x_0| + \left|\int_{t_0}^t |F(s, \phi(s, x))|\, ds\right|$$
$$\le |x - x_0| + \delta M$$
$$< \frac{1}{2}\nu + \frac{1}{2}\nu,$$

the range of the operator Λ is in $\bar{B}(x_0, \nu)$, as required.

The operator Λ is a contraction. In fact, if $\phi_1, \phi_2 \in X$, then

$$|\Lambda(\phi_1)(t, x) - \Lambda(\phi_2)(t, x)| \le \left|\int_{t_0}^t |F(s, \phi_1(s, x)) - F(s, \phi_2(s, x))|\, ds\right|$$
$$\le \delta\lambda\|\phi_1 - \phi_2\|,$$

and therefore

$$\|\Lambda(\phi_1) - \Lambda(\phi_2)\| \le \delta\lambda\|\phi_1 - \phi_2\|,$$

as required. By the contraction principle, Λ has a unique fixed point. This function is a solution of the initial value problem (1.62) and it is continuously dependent on the initial condition.

If ϕ_∞ denotes the fixed point of Λ, then $\frac{d}{dt}\phi_\infty(t,x) = F(t,\phi_\infty(t,x))$. Because the functions ϕ_∞ and F are continuous, it follows that, for each fixed $x \in B(x_0, \frac{\nu}{2})$, the function $t \mapsto \phi_\infty(t,x)$ is C^1. To show that ϕ_∞ is C^1, it suffices to show that for each fixed $t \in b(t_0, \delta)$ the function $x \mapsto \phi_\infty(t,x)$ is C^1. We will prove this fact using the fiber contraction principle. The idea for this part of the proof is due to Jorge Sotomayor [211].

Let us define a Banach space consisting of the "candidates" for the derivatives of functions in X with respect to their second arguments. To this end, let $L(\mathbb{R}^n, \mathbb{R}^n)$ denote the set of linear transformations of $\mathbb{R}^n$ and define the Banach space

$$Y := C(b(t_0, \delta) \times B(x_0, \frac{\nu}{2}), L(\mathbb{R}^n, \mathbb{R}^n))$$

consisting of all indicated functions that are bounded with respect to the norm on Y given by

$$\|\Phi\| := \sup_{(t,x) \in b(t_0,\delta) \times B(x_0, \frac{\nu}{2})} \|\Phi(t,x)\|,$$

where, as defined above,

$$\|\Phi(t,x)\| := \sup_{\{v \in \mathbb{R}^n : |v| = 1\}} |\Phi(t,x)v|.$$

Let I denote the identity transformation on $\mathbb{R}^n$, $DF(t,x)$ the derivative of the map $x \mapsto F(t,x)$, and define $\Psi : X \times Y \to Y$ by

$$\Psi(\phi, \Phi)(t,x) := I + \int_{t_0}^{t} DF(s, \phi(s,x))\Phi(s,x)\,ds.$$

Also, define $\Gamma : X \times Y \to X \times Y$ by

$$\Gamma(\phi, \Phi) := (\Lambda(\phi), \Psi(\phi, \Phi)).$$

The function Γ is a continuous bundle map. The proof of this fact is left to the reader. Let us note, however, that the key result required here is the continuity of the function $\phi \mapsto \Psi(\phi, \Phi)$. The proof of this fact uses the continuity of DF and ϕ, and the compactness of the interval $\bar{b}(t_0, \delta)$ (see Exercise 1.258 and the proof of Theorem 1.249).

Let us prove that Γ is a fiber contraction. Recall that we have chosen the radius of the time interval, $\delta > 0$, so small that $\delta K < 1$, where the number K is defined in equation (1.64). Using this fact, we have

$$\|\Psi(\phi, \Phi_1)(t,x) - \Psi(\phi, \Phi_2)(t,x)\|$$
$$= \|\int_{t_0}^{t} DF(s, \phi(s,x))(\Phi_1(s,x) - \Phi_2(s,x))\,ds\|$$
$$< \delta K \|\Phi_1 - \Phi_2\|,$$

as required.

Let $\phi_0(t,x) \equiv x$ and note that $(\phi_0, I) \in X \times Y$. By the fiber contraction theorem (Theorem 1.247), the iterates of the point (ϕ_0, I) under Γ converge to a globally attracting fixed point, namely, $(\phi_\infty, \Phi_\infty)$, where in this case ϕ_∞ is the solution of the initial value problem (the fixed point of Λ) and Φ_∞ is the unique fixed point of the contraction $\Phi \mapsto \Psi(\phi_\infty, \Phi)$ on Y.

We will prove that $D\phi_\infty(t, \cdot) = \Phi_\infty(t, \cdot)$. (The derivative denoted by D is the partial derivative with respect to the second variable.) Let us start with the equation $D\phi_0(t,x) = I$, and for each integer $n > 1$ define $(\phi_n, \Phi_n) := \Gamma^n(\phi_0, I)$ so that

$$\Phi_{n+1}(t,x) = \Psi(\phi_n, \Phi_n)(t,x) := I + \int_{t_0}^{t} DF(s, \phi_n(s,x))\Phi_n(s,x)\, ds,$$

$$\phi_{n+1}(t,x) = x + \int_{t_0}^{t} F(s, \phi_n(s,x))\, ds.$$

Let us show the identity $D\phi_n(t, \cdot) = \Phi_n(t, \cdot)$ for each integer $n \geq 0$. The equation is true for $n = 0$. Proceeding by induction on n, let us assume that the equation is true for some fixed integer $n \geq 0$. Because we can "differentiate under the integral," the derivative

$$D\phi_{n+1}(t,x) = \frac{\partial}{\partial x}\left(x + \int_{t_0}^{t} F(s, \phi_n(s,x))\, ds\right)$$

is clearly equal to

$$I + \int_{t_0}^{t} DF(s, \phi_n(s,x))\Phi_n(s,x)\, ds = \Phi_{n+1}(t,x),$$

as required. Thus, we have proved that the sequence $\{D\phi_n(t, \cdot)\}_{n=0}^{\infty}$ converges to $\Phi_\infty(t, \cdot)$. Finally, by Theorem 1.248 we have that $D\phi_\infty(t, \cdot) = \Phi_\infty(t, \cdot)$. $\qquad\square$

Exercise 1.262. It is very easy to show that a C^2 differential equation has a C^1 flow. Why? We have proved above the stronger result that a C^1 differential equation has a C^1 flow. Show that a C^r differential equation has a C^r flow for $r = 2, 3, \ldots, \infty$. Also, show that a real analytic differential equation has a real analytic flow.

So far we have proved that initial value problems have unique solutions that exist on some (perhaps small) interval containing the initial time. If we wish to find a larger interval on which the solution is defined, the following problem arises. Suppose that the initial value problem

$$\dot{x} = f(t,x), \qquad x(t_0) = x_0$$

has a solution $t \mapsto \phi(t)$ defined on some interval J containing t_0. Maybe the solution is actually defined on some larger time interval $J_1 \supseteq J$. If we have a second solution $\psi(t)$ defined on J_1, then, by our local uniqueness result, $\psi(t) = \phi(t)$ on J. But, we may ask, does $\psi(t) = \phi(t)$ on J_1? The answer is *yes*.

To prove this fact, consider all the open intervals containing J. The union of all such intervals on which $\phi(t) = \psi(t)$ is again an open interval J^*; it is the largest open interval on which ϕ and ψ agree. Let us prove that $J^* \supseteq J_1$. If not, then the interval J^* has an end point $t_1 \in J_1$ that is not an endpoint of J_1. Suppose that t_1 is the right hand endpoint of J^*. By continuity,

$$\phi(t_1) = \psi(t_1).$$

Thus, by our local existence theorem, there is a unique solution of the initial value problem

$$\dot{x} = f(t, x), \qquad x(t_1) = \phi(t_1)$$

defined in some neighborhood of t_1. It follows that $\phi(t) = \psi(t)$ on some larger interval. This contradiction implies that $J^* \supseteq J_1$, as required. In particular, if a solution extends, then it extends uniquely.

Our existence theorem for solutions of initial value problems gives no information about the length of the maximal interval of existence. For example, recall that even if the vector field associated with a differential equation has no singularities, solutions of the differential equation may not exist for all $t \in \mathbb{R}$. The classic example (already mentioned) is the initial value problem

$$\dot{x} = x^2, \qquad x(0) = 1.$$

The maximal interval of existence of the solution $x(t) = (1 - t)^{-1}$ is the interval $(-\infty, 1)$. Moreover, this solution blows up in finite time, that is, $x(t) \to \infty$ as $t \to 1^-$. Of course, the maximal interval of existence need not be an open set (see Exercise 1.9).

In general, it is difficult to determine the exact domain on which a solution is defined. But, following the presentation in [123], the next theorem shows that our example illustrates the typical behavior.

Theorem 1.263. *Let $U \subseteq \mathbb{R}^n$ and $J \subseteq \mathbb{R}$ be open sets such that the open interval (α, β) is contained in J. Also, let $x_0 \in U$. If $f : J \times U \to \mathbb{R}^n$ is a C^1 function and the maximal interval of existence of the solution $t \to \phi(t)$ of the initial value problem $\dot{x} = f(t, x)$, $x(t_0) = x_0$ is $\alpha < t_0 < \beta$ with $\beta < \infty$, then for each compact set $K \subset U$ there is some $t \in (\alpha, \beta)$ such that $\phi(t) \notin K$. In particular, either $|\phi(t)|$ becomes unbounded or $\phi(t)$ approaches the boundary of U as $t \to \beta$.*

Proof. Suppose that the solution ϕ has maximal interval of existence (α, β) with $\beta < \infty$ and K is a compact subset of U such that $\phi(t) \in K$ for all

$t \in (\alpha, \beta)$. We will show that under these assumptions the interval (α, β) is not maximal.

The set $[t_0, \beta] \times K$ is compact. Thus, there is some $M > 0$ such that $|f(t,x)| < M$ for each $(t,x) \in [t_0, \beta] \times K$. Moreover, the function $\phi : [t_0, \beta) \to K$ is continuous.

We will show that the function ϕ extends continuously to the interval $[t_0, \beta]$. Note first that ϕ is uniformly continuous on $[t_0, \beta)$. In fact, if $s_1, s_2 \in [t_0, \beta)$ and $s_1 < s_2$, then

$$|\phi(s_2) - \phi(s_1)| = \left| \int_{s_1}^{s_2} f(t, \phi(t)) \, dt \right| \le M |s_2 - s_1|. \qquad (1.66)$$

A standard theorem from advanced calculus states that ϕ extends continuously to $[t_0, \beta]$. For completeness we will prove this fact for our special case.

Construct a sequence $\{t_n\}_{n=1}^{\infty}$ of numbers in the interval $[t_0, \beta)$ that converges to β, and recall that a convergent sequence is Cauchy. By inequality (1.66), the sequence $\{\phi(t_n)\}_{n=1}^{\infty}$ is also Cauchy. Hence, there is some $\omega \in \mathbb{R}$ such that $\phi(t_n) \to \omega$ as $n \to \infty$.

Let us extend the function ϕ to the closed interval $[t_0, \beta]$ by defining $\phi(\beta) = \omega$. We will prove that this extension is continuous. For this, it suffices to show that if $\{s_j\}_{j=1}^{\infty}$ is a sequence in $[t_0, \beta)$ that converges to β, then $\lim_{j \to \infty} \phi(s_j) = \omega$. (Why?)

We have that

$$|\phi(s_j) - \omega| \le |\phi(s_j) - \phi(t_j)| + |\phi(t_j) - \omega|.$$

Let $\epsilon > 0$ be given. If $\delta = \epsilon/(2M)$, then $|\phi(s) - \phi(t)| < \epsilon/2$ whenever $s, t \in [t_0, \beta)$ and $|s - t| < \delta$. Also, because

$$|s_j - t_j| \le |s_j - \beta| + |t_j - \beta|,$$

there is some integer N such that $|s_j - t_j| < \delta$ whenever $j \ge N$, and therefore

$$|\phi(s_j) - \omega| \le \frac{\epsilon}{2} + |\phi(t_j) - \omega|$$

whenever $j \ge N$. Moreover, since $\phi(t_j) \to \omega$ as $j \to \infty$, there is some $N_1 \ge N$ such that $|\phi(t_j) - \omega| < \epsilon/2$ whenever $j \ge N_1$. In particular, for $j \ge N_1$, we have $|\phi(s_j) - \omega| < \epsilon$, and it follows that $\phi(s_j) \to \omega$. In other words, ϕ extends continuously to β.

For $t_0 \le t < \beta$, the function ϕ is a solution of the differential equation. In particular, ϕ is continuously differentiable on $[t_0, \beta)$ and, on this interval,

$$\phi(t) = \phi(t_0) + \int_{t_0}^{t} f(s, \phi(s)) \, ds.$$

Moreover, since f is continuous and ϕ has a continuous extension, the map $s \mapsto f(s, \phi(s))$ is continuous on $[t_0, \beta]$. Thus, if follows that

$$
\begin{aligned}
\phi(\beta) &= \phi(t_0) + \lim_{t \to \beta^-} \int_{t_0}^t f(s, \phi(s))\, ds \\
&= \phi(t_0) + \int_{t_0}^\beta f(s, \phi(s))\, ds.
\end{aligned}
\tag{1.67}
$$

By the existence theorem for differential equations, there is a number $\delta > 0$ such that the initial value problem

$$
\dot{x} = f(t, x), \qquad x(\beta) = \phi(\beta)
$$

has a solution $t \mapsto \psi(t)$ defined on the interval $(\beta - \delta, \beta + \delta) \subseteq J$. Let us use this fact to define the continuous function $\gamma : [t_0, \beta + \delta) \to \mathbb{R}^n$ by

$$
\gamma(t) = \begin{cases} \phi(t), & \text{if } t_0 \le t \le \beta, \\ \psi(t), & \text{if } \beta < t < \beta + \delta. \end{cases}
$$

For $t_0 \le t \le \beta$, we have that

$$
\gamma(t) = \phi(t_0) + \int_{t_0}^t f(s, \gamma(s))\, ds.
\tag{1.68}
$$

Also, in view of equation (1.67), if $\beta < t < \beta + \delta$, then

$$
\begin{aligned}
\gamma(t) &= \phi(\beta) + \int_\beta^t f(s, \gamma(s))\, ds \\
&= \phi(t_0) + \int_{t_0}^t f(s, \gamma(s))\, ds.
\end{aligned}
$$

In other words, the equality (1.68) is valid on the interval $[t_0, \beta + \delta)$. It follows that γ is a solution of the differential equation that extends the solution ϕ. This violates the maximality of β—there is some t such that $\phi(t)$ is not in K. $\qquad \square$

2
Linear Systems and Stability of Nonlinear Systems

In this chapter we will study the differential equation

$$\dot{x} = A(t)x + f(x,t), \qquad x \in \mathbb{R}^n$$

where A is a smooth $n \times n$ matrix-valued function and f is a smooth function such that $f(0,t) = f_x(0,t) \equiv 0$. Note that if f has this form, then the associated *homogeneous linear system* $\dot{x} = A(t)x$ is the linearization of the differential equation along the *zero solution* $t \mapsto \phi(t) \equiv 0$.

One of the main objectives of the chapter is the proof of the basic results related to the principle of linearized stability. For example, we will prove that if the matrix A is constant and all of its eigenvalues have negative real parts, then the zero solution (also called the *trivial solution*) is asymptotically stable. Much of the chapter, however, is devoted to the general theory of homogeneous linear systems; that is, systems of the form $\dot{x} = A(t)x$. In particular, we will study the important special cases where A is a constant or periodic function.

In case $t \mapsto A(t)$ is a constant function, we will show how to reduce the solution of the system $\dot{x} = Ax$ to a problem in linear algebra. Also, by defining the matrix exponential, we will discuss the flow of this autonomous system as a one-parameter group with generator A.

Although the behavior of the general nonautonomous system $\dot{x} = A(t)x$ is not completely understood, the special case where $t \mapsto A(t)$ is a periodic matrix-valued function is reducible to the constant matrix case. We will develop a useful theory of periodic matrix systems, called Floquet theory, and use it to prove this basic result. The Floquet theory will appear again later when we discuss the stability of periodic nonhomogeneous systems. In particular, we will use Floquet theory in a stability analysis of the inverted pendulum (see Section 3.5).

Because linear systems theory is so well developed, it is used extensively in many areas of applied science. For example, linear systems theory is an essential tool for electromagnetics, circuit theory, and the theory of vibration. In addition, the results of this chapter are a fundamental component of control theory.

2.1   Homogeneous Linear Differential Equations

This section is devoted to a general discussion of the homogeneous linear system

$$\dot{x} = A(t)x, \qquad x \in \mathbb{R}^n$$

where $t \mapsto A(t)$ is a smooth function from some open interval $J \subseteq \mathbb{R}$ to the space of $n \times n$ matrices. Here, the continuity properties of matrix-valued functions are determined by viewing the space of $n \times n$ matrices as $\mathbb{R}^{n^2}$; that is, every matrix is viewed as an element in the Cartesian space by simply listing the rows of the matrix consecutively to form a row vector of length n^2. We will prove an important general inequality and then use it to show that solutions of linear systems cannot blow up in finite time. We will discuss the basic result that the set of solutions of a linear system is a vector space, and we will exploit this fact by showing how to construct the general solution of a linear homogeneous system with constant coefficients.

2.1.1   Gronwall's Inequality

The important theorem proved in this section does not belong to the theory of linear differential equations per se, but it is presented here because it will be used to prove the global existence of solutions of homogeneous linear systems.

Theorem 2.1 (Gronwall's Inequality). *Suppose that $a < b$ and let α, ϕ, and ψ be nonnegative continuous functions defined on the interval $[a, b]$. Moreover, suppose that α is differentiable on (a, b) with nonnegative continuous derivative $\dot{\alpha}$. If, for all $t \in [a, b]$,*

$$\phi(t) \leq \alpha(t) + \int_a^t \psi(s)\phi(s)\, ds, \tag{2.1}$$

then

$$\phi(t) \leq \alpha(t)e^{\int_a^t \psi(s)\, ds} \tag{2.2}$$

for all $t \in [a, b]$.

Proof. Assume for the moment that $\alpha(a) > 0$. In this case $\alpha(t) \geq \alpha(a) > 0$ on the interval $[a, b]$.

The function on the interval $[a, b]$ defined by $t \mapsto \alpha(t) + \int_a^t \psi(s)\phi(s)\,ds$ is positive and exceeds ϕ. Thus, we have that

$$\frac{\phi(t)}{\alpha(t) + \int_a^t \psi(s)\phi(s)\,ds} \leq 1.$$

Multiply both sides of this inequality by $\psi(t)$, add and subtract $\dot{\alpha}(t)$ in the numerator of the resulting fraction, rearrange the inequality, and use the obvious estimate to obtain the inequality

$$\frac{\dot{\alpha}(t) + \psi(t)\phi(t)}{\alpha(t) + \int_a^t \psi(s)\phi(s)\,ds} \leq \frac{\dot{\alpha}(t)}{\alpha(t)} + \psi(t),$$

which, when integrated over the interval $[a, t]$, yields the inequality

$$\ln\left(\alpha(t) + \int_a^t \psi(s)\phi(s)\,ds\right) - \ln(\alpha(a)) \leq \int_a^t \psi(s)\,ds + \ln(\alpha(t)) - \ln(\alpha(a)).$$

After we exponentiate both sides of this last inequality and use hypothesis (2.1), we find that, for each t in the interval $[a, b]$,

$$\phi(t) \leq \alpha(t)e^{\int_a^t \psi(s)\,ds} \leq \alpha(t)e^{\int_a^t \psi(s)\,ds}. \tag{2.3}$$

Finally, for the case $\alpha(a) = 0$, we have the inequality

$$\phi(t) \leq (\alpha(t) + \epsilon) + \int_a^t \psi(s)\phi(s)\,ds$$

for each $\epsilon > 0$. As a result of what we have just proved, we have the estimate

$$\phi(t) \leq (\alpha(t) + \epsilon)e^{\int_a^t \psi(s)\,ds}.$$

The desired inequality follows by passing to the limit (for each fixed $t \in [a, b]$) as $\epsilon \to 0$. $\qquad\square$

Exercise 2.2. What can you say about a continuous function $f : \mathbb{R} \to [0, \infty)$ if

$$f(x) \leq \int_0^x f(t)\,dt?$$

Exercise 2.3. Prove the "specific Gronwall lemma" [201]: If, for $t \in [a, b]$,

$$\phi(t) \leq \delta_2(t - a) + \delta_1\int_a^t \phi(s)\,ds + \delta_3,$$

where ϕ is a nonnegative continuous function on $[a, b]$, and $\delta_1 > 0$, $\delta_2 \geq 0$, and $\delta_3 \geq 0$ are constants, then

$$\phi(t) \leq \left(\frac{\delta_2}{\delta_1} + \delta_3\right)e^{\delta_1(t-a)} - \frac{\delta_2}{\delta_1}.$$

2.1.2 Homogeneous Linear Systems: General Theory

Consider the *homogeneous* linear system

$$\dot{x} = A(t)x, \qquad x \in \mathbb{R}^n. \tag{2.4}$$

By our general existence theory, the initial value problem

$$\dot{x} = A(t)x, \qquad x(t_0) = x_0 \tag{2.5}$$

has a unique solution that exists on some open interval containing t_0. The following theorem states that this open interval can be extended to the domain of A.

Theorem 2.4. *If $t \mapsto A(t)$ is continuous on the interval $\alpha < t < \beta$ and if $\alpha < t_0 < \beta$ (maybe $\alpha = -\infty$ or $\beta = \infty$), then the solution of the initial value problem (2.5) is defined on the open interval (α, β).*

Proof. Because the continuous function $t \mapsto A(t)$ is bounded on each compact subinterval of (α, β), it is easy to see that the function $(t, x) \mapsto A(t)x$ is locally Lipschitz with respect to its second argument. Consider the solution $t \mapsto \phi(t)$ of the initial value problem (2.5) given by the general existence theorem (Theorem 1.261) and let J_0 denote its maximal interval of existence. Suppose that J_0 does not contain (α, β). For example, suppose that the right hand end point b of J_0 is less than β. We will show that this assumption leads to a contradiction. The proof for the left hand end point is similar.

If $t \in J_0$, then we have

$$\phi(t) - \phi(t_0) = \int_{t_0}^{t} A(s)\phi(s)\, ds.$$

By the continuity of A and the compactness of $[t_0, b]$, there is some $M > 0$ such that $\|A(t)\| \le M$ for all $t \in [t_0, b]$. (The notation $\|\ \|$ is used for the matrix norm corresponding to some norm $|\ |$ on $\mathbb{R}^n$.) Thus, for $t \in J_0$, we have the following inequality:

$$|\phi(t)| \le |x_0| + \int_{t_0}^{t} \|A(s)\|\,|\phi(s)|\, ds$$

$$\le |x_0| + \int_{t_0}^{t} M|\phi(s)|\, ds.$$

In addition, by Gronwall's inequality, with $\psi(t) := M$, we have

$$|\phi(t)| \le |x_0| e^{M \int_{t_0}^{t} ds} = |x_0| e^{M(t - t_0)}.$$

Thus, $|\phi(t)|$ is uniformly bounded on $[t_0, b)$.

Because the boundary of $\mathbb{R}^n$ is empty, it follows from the extension theorem that $|\phi(t)| \to \infty$ as $t \to b^-$, in contradiction to the existence of the uniform bound. $\qquad\square$

Exercise 2.5. Use Gronwall's inequality to prove the following important inequality: If $t \mapsto \beta(t)$ and $t \mapsto \gamma(t)$ are solutions of the smooth differential equation $\dot{x} = f(x)$ and both are defined on the time interval $[0, T]$, then there is a constant $L > 0$ such that

$$|\beta(t) - \alpha(t)| \leq |\beta(0) - \alpha(0)|e^{Lt}.$$

Thus, two solutions diverge from each other at most exponentially fast. Also, if the solutions have the same initial condition, then they coincide. Therefore, the result of this exercise provides an alternative proof of the general uniqueness theorem for differential equations.

Exercise 2.6. Prove that if A is a linear transformation of $\mathbb{R}^n$ and $f : \mathbb{R}^n \to \mathbb{R}^n$ is a (smooth) function such that $|f(x)| \leq M|x| + N$ for positive constants M and N, then the differential equation $\dot{x} = Ax + f(x)$ has a complete flow.

Exercise 2.7. Suppose that $X(\cdot, \lambda)$ and $Y(\cdot, \lambda)$ are two vector fields with parameter $\lambda \in \mathbb{R}$, and the two vector fields agree to order N in λ; that is, $X(x, \lambda) = Y(x, \lambda) + O(\lambda^{N+1})$. If $x(t, \lambda)$ and $y(t, \lambda)$ are corresponding solutions defined on the interval $[0, T]$ with initial conditions at $t = 0$ that agree to order N in λ, prove that $x(T, \lambda)$ and $y(T, \lambda)$ agree to order N in λ. Hint: First prove the result for $N = 0$.

2.1.3 Principle of Superposition

The foundational result about linear homogeneous systems is the principle of superposition: The sum of two solutions is again a solution. A precise statement of this principle is the content of the next proposition.

Proposition 2.8. *If the homogeneous system (2.4) has two solutions $\phi_1(t)$ and $\phi_2(t)$, each defined on some interval (a, b), and if λ_1 and λ_2 are numbers, then $t \to \lambda_1\phi_1(t) + \lambda_2\phi_2(t)$ is also a solution defined on the same interval.*

Proof. To prove the proposition, we use the *linearity* of the differential equation. In fact, we have

$$\frac{d}{dt}(\lambda_1\phi_1(t) + \lambda_2\phi_2(t)) = \lambda_1\dot{\phi}_1(t) + \lambda_2\dot{\phi}_2(t)$$
$$= \lambda_1 A(t)\phi_1(t) + \lambda_2 A(t)\phi_2(t)$$
$$= A(t)(\lambda_1\phi_1(t) + \lambda_2\phi_2(t)).$$

$\square$

As a natural extension of the principle of superposition, we will prove that the set of solutions of the homogeneous linear system (2.4) is a finite dimensional vector space of dimension n.

Definition 2.9. A set of n solutions of the homogeneous linear differential equation (2.4), all defined on the same open interval J, is called a *fundamental set* of solutions on J if the solutions are linearly independent functions on J.

Proposition 2.10. *If $t \to A(t)$ is defined on the interval (a, b), then the system (2.4) has a fundamental set of solutions defined on (a, b).*

Proof. If $c \in (a, b)$ and $\mathbf{e}_1, \ldots, \mathbf{e}_n$ denote the usual basis vectors in $\mathbb{R}^n$, then there is a unique solution $t \mapsto \phi_i(t)$ such that $\phi_i(c) = \mathbf{e}_i$ for $i = 1, \ldots, n$. Moreover, by Theorem 2.4, each function ϕ_i is defined on the interval (a, b). Let us assume that the set of *functions* $\{\phi_i : i = 1, \ldots, n\}$ is linearly dependent and derive a contradiction. In fact, if there are scalars α_i, $i = 1, \ldots, n$, not all zero, such that $\sum_{i=1}^{n} \alpha_i \phi_i(t) \equiv 0$, then $\sum_{i=1}^{n} \alpha_i \mathbf{e}_i \equiv 0$. In view of the linear independence of the usual basis, this is the desired contradiction.   □

Proposition 2.11. *If $\mathcal{F}$ is a fundamental set of solutions of the linear system (2.4) on the interval (a, b), then every solution defined on (a, b) can be expressed as a linear combination of the elements of $\mathcal{F}$.*

Proof. Suppose that $\mathcal{F} = \{\phi_1, \ldots, \phi_n\}$. Pick $c \in (a, b)$. If $t \mapsto \phi(t)$ is a solution defined on (a, b), then $\phi(c)$ and $\phi_i(c)$, for $i = 1, \ldots, n$, are all vectors in $\mathbb{R}^n$. We will show that the set $B := \{\phi_i(c) : i = 1, \ldots, n\}$ is a basis for $\mathbb{R}^n$. If not, then there are scalars α_i, $i = 1, \ldots, n$, not all zero, such that $\sum_{i=1}^{n} \alpha_i \phi_i(c) = 0$. Thus, $y(t) := \sum_{i=1}^{n} \alpha_i \phi_i(t)$ is a solution with initial condition $y(c) = 0$. But the zero solution has the same initial condition. Thus, $y(t) \equiv 0$, and therefore $\sum_{i=1}^{n} \alpha_i \phi_i(t) \equiv 0$. This contradicts the hypothesis that $\mathcal{F}$ is a linearly independent set, as required.

Using the basis B, there are scalars $\beta_1, \ldots, \beta_n \in \mathbb{R}$ such that $\phi(c) = \sum_{i=1}^{n} \beta_i \phi_i(c)$. It follows that both ϕ and $\sum_{i=1}^{n} \beta_i \phi_i$ are solutions with the same initial condition, and, by uniqueness, $\phi = \sum_{i=1}^{n} \beta_i \phi_i$.   □

Definition 2.12. An $n \times n$ matrix function $t \mapsto \Psi(t)$, defined on an open interval J, is called a *matrix solution* of the homogeneous linear system (2.4) if each of its columns is a (vector) solution. A matrix solution is called a *fundamental matrix solution* if its columns form a fundamental set of solutions. In addition, a fundamental matrix solution $t \mapsto \Psi(t)$ is called the *principal fundamental matrix solution* at $t_0 \in J$ if $\Psi(t_0) = I$.

If $t \mapsto \Psi(t)$ is a matrix solution of the system (2.4) on the interval J, then $\dot{\Psi}(t) = A(t)\Psi(t)$ on J. By Proposition 2.10, there is a fundamental matrix solution. Moreover, if $t_0 \in J$ and $t \mapsto \Phi(t)$ is a fundamental matrix solution on J, then (by the linear independence of its columns) the matrix $\Phi(t_0)$ is invertible. It is easy to see that the matrix solution defined by $\Psi(t) := \Phi(t)\Phi^{-1}(t_0)$ is the principal fundamental matrix solution at t_0. Thus, system (2.4) has a principal fundamental matrix solution at each point in J.

Definition 2.13. The *state transition matrix* for the homogeneous linear system (2.4) on the open interval J is the family of fundamental matrix solutions $t \mapsto \Psi(t, \tau)$ parametrized by $\tau \in J$ such that $\Psi(\tau, \tau) = I$, where I denotes the $n \times n$ identity matrix.

Proposition 2.14. *If $t \mapsto \Phi(t)$ is a fundamental matrix solution for the system (2.4) on J, then $\Psi(t, \tau) := \Phi(t)\Phi^{-1}(\tau)$ is the state transition matrix. Also, the state transition matrix satisfies the Chapman–Kolmogorov identities*

$$\Psi(\tau, \tau) = I, \quad \Psi(t, s)\Psi(s, \tau) = \Psi(t, \tau)$$

and the identities

$$\Psi(t, s)^{-1} = \Psi(s, t), \qquad \frac{\partial \Psi}{\partial s}(t, s) = -\Psi(t, s)A(s).$$

Proof. See Exercise 2.15. □

A two-parameter family of operator-valued functions that satisfies the Chapman–Kolmogorov identities is called an *evolution family*.

In the case of constant coefficients, that is, in case $t \mapsto A(t)$ is a constant function, the corresponding homogeneous linear system is autonomous, and therefore its solutions define a flow. This result also follows from the Chapman–Kolmogorov identities.

To prove the flow properties, let us show first that if $t \mapsto A(t)$ is a constant function, then the state transition matrix $\Psi(t, t_0)$ depends only on the difference $t - t_0$. In fact, since $t \mapsto \Psi(t, t_0)$ and $t \mapsto \Psi(t+s, t_0+s)$ are both solutions satisfying the same initial condition at t_0, they are identical. In particular, with $s = -t_0$, we see that $\Psi(t, t_0) = \Psi(t - t_0, 0)$. If we define $\phi_t := \Psi(t, 0)$, then using the last identity together with the Chapman–Kolmogorov identities we find that

$$\Psi(t + s, 0) = \Psi(t, -s) = \Psi(t, 0)\Psi(0, -s) = \Psi(t, 0)\Psi(s, 0).$$

Thus, we recover the group property $\phi_{t+s} = \phi_t \phi_s$. Since, in addition, $\phi_0 = \Psi(0, 0) = I$, the family of operators ϕ_t defines a flow. In this context, ϕ_t is also called an *evolution group*.

If $t \mapsto \Phi(t)$ is a fundamental matrix solution for the linear system (2.4) and $v \in \mathbb{R}^n$, then $t \mapsto \Phi(t)v$ is a (vector) solution. Moreover, every solution is obtained in this way. In fact, if $t \mapsto \phi(t)$ is a solution, then there is some v such that $\Phi(t_0)v = \phi(t_0)$. (Why?) By uniqueness, we must have $\Phi(t)v = \phi(t)$. Also, note that $\Psi(t, t_0)v$ has the property that $\Psi(t_0, t_0)v = v$. In other words, Ψ "transfers" the initial state v to the final state $\Psi(t, t_0)v$. Hence, the name "state transition matrix."

Exercise 2.15. Prove Proposition 2.14.

Exercise 2.16. [Cocycles] A cocycle is a family of functions, each mapping from $\mathbb{R} \times \mathbb{R}^n$ to the set of linear transformations of $\mathbb{R}^n$ such that $\Phi(0, u) = I$ and $\Phi(t+s, u) = \Phi(t, \phi_s(u))\Phi(s, u)$. (To learn more about why cocycles are important, see [45].) Suppose $\dot{u} = f(u)$ is a differential equation on $\mathbb{R}^n$ with flow ϕ_t. Show that the family of principal fundamental matrix solutions $\Phi(t, u)$ of the family of variational equations $\dot{w} = Df(\phi_t(u))w$ is a cocycle over the flow ϕ_t.

Exercise 2.17. [Time-dependent vector fields] The solution of a nonautonomous differential equation $\dot{x} = X(t, x)$ is an evolution family $\phi^{t,s}$; that is, $t \mapsto \phi^{t,s}(\xi)$ is the solution with initial condition $\phi^{s,s}(\xi) = \xi$. (a) Prove that $\phi_\tau^{-t,t}(\xi) + \phi_x^{-t,t}(\xi)X(t, \xi) = 0$, where the subscripts denote partial derivatives—the reason why superscripts are used to denote the evolution variables. Hint: Differentiate the left-hand side of the Chapman-Kolmogorov identity $\phi^{-t,\tau}\phi^{\tau,t} = \phi^{-t,t}$ with respect to τ. (b) Define $F(t, \xi) = \phi^{-t,t}(\xi)$. Prove that $F(0, \xi) = \xi$ and

$$F_t(t, \xi) + F_x(t, x)X(t, x) = -X(-t, F(t, \xi)).$$

(c) Suppose that X is $2T$-periodic in t. Prove that ξ lies on a periodic orbit if and only if $F(T, \xi) = \xi$. (d) Suppose that X is $2T$-periodic in t and $X(t, x) = -X(-t, x)$. Show that every orbit is periodic. (For more on this subject see [163].)

The linear independence of a set of solutions of a homogeneous linear differential equation can be determined by checking the independence of a set of vectors obtained by evaluating the solutions at just one point. This useful fact is perhaps most clearly expressed by Liouville's formula, which has many other implications.

Proposition 2.18 (Liouville's Formula). *Suppose that $t \mapsto \Phi(t)$ is a matrix solution of the homogeneous linear system (2.4) on the open interval J. If $t_0 \in J$, then*

$$\det \Phi(t) = \det \Phi(t_0)e^{\int_{t_0}^t \operatorname{tr} A(s)\,ds}$$

where det *denotes determinant and* tr *denotes trace. In particular, $\Phi(t)$ is a fundamental matrix solution if and only if the columns of $\Phi(t_0)$ are linearly independent.*

Proof. The matrix solution $t \mapsto \Phi(t)$ is a differentiable function. Thus, we have that

$$\lim_{h \to 0} \frac{1}{h}[\Phi(t+h) - (I + hA(t))\Phi(t)] = 0.$$

In other words, using the "little oh" notation,

$$\Phi(t+h) = (I + hA(t))\Phi(t) + o(h). \qquad (2.6)$$

(The little oh has the following meaning: $f(x) = g(x) + o(h(x))$ if

$$\lim_{x \to 0+} \frac{|f(x) - g(x)|}{h(x)} = 0.$$

Thus, we should write $o(\pm h)$ in equation (2.6), but this technicality is not important in this proof.)

By the definition of the determinant of an $n \times n$ matrix, that is, if $B :=(b_{ij})$, then

$$\det B = \sum_{\sigma} \text{sgn}(\sigma) \prod_{i=1}^{n} b_{i,\sigma(i)},$$

and the following result: The determinant of a product of matrices is the product of their determinants, we have that

$$\det \Phi(t+h) = \det(I + hA(t)) \det \Phi(t) + o(h)$$
$$= (1 + h \, \text{tr} \, A(t)) \det \Phi(t) + o(h),$$

and therefore

$$\frac{d}{dt} \det \Phi(t) = \text{tr} \, A(t) \det \Phi(t).$$

Integration of this last differential equation gives the desired result.   □

Exercise 2.19. Find a fundamental matrix solution of the system

$$\dot{x} = \begin{pmatrix} 1 & -1/t \\ 1+t & -1 \end{pmatrix} x, \qquad t > 0.$$

Hint: $x(t) = \begin{pmatrix} 1 \\ t \end{pmatrix}$ is a solution.

Exercise 2.20. Suppose that every solution of $\dot{x} = A(t)x$ is bounded for $t \geq 0$ and let $\Phi(t)$ be a fundamental matrix solution. Prove that $\Phi^{-1}(t)$ is bounded for $t \geq 0$ if and only if the function $t \mapsto \int_0^t \text{tr} \, A(s) \, ds$ is bounded below. Hint: The inverse of a matrix is the adjugate of the matrix divided by its determinant.

Exercise 2.21. Suppose that the linear system $\dot{x} = A(t)x$ is defined on an open interval containing the origin whose right-hand end point is $\omega \leq \infty$ and the norm of every solution has a finite limit as $t \to \omega$. Show that there is a solution converging to zero as $t \to \omega$ if and only if $\int_0^\omega \text{tr} \, A(s) \, ds = -\infty$. Hint: A matrix has a nontrivial kernel if and only if its determinant is zero (cf. [115]).

Exercise 2.22. [Transport Theorem] Let ϕ_t denote the flow of the system $\dot{x} = f(x)$, $x \in \mathbb{R}^n$, and let Ω be a bounded region in $\mathbb{R}^n$. Define

$$V(t) = \int_{\phi_t(\Omega)} dx_1 dx_2 \cdots dx_n$$

and recall that the divergence of a vector field $f = (f_1, f_2, \ldots, f_n)$ on $\mathbb{R}^n$ with the usual Euclidean structure is

$$\text{div} \, f = \sum_{i=1}^{n} \frac{\partial f_i}{\partial x_i}.$$

(a) Use Liouville's theorem and the change of variables formula for multiple integrals to prove that

$$\dot{V}(t) = \int_{\phi_t(\Omega)} \operatorname{div} f(x) dx_1 dx_2 \cdots dx_n.$$

(b) Prove: The flow of a vector field whose divergence is everywhere negative contracts volume. (c) Suppose that $g : \mathbb{R}^n \times \mathbb{R} \to \mathbb{R}$ and, for notational convenience, let $dx = dx_1 dx_2 \cdots dx_n$. Prove the transport theorem:

$$\frac{d}{dt} \int_{\phi_t(\Omega)} g(x, t) \, dx = \int_{\phi_t(\Omega)} g_t(x, t) + \operatorname{div}(gf)(x, t) \, dx.$$

(d) Suppose that the mass in every open set remains unchanged as it is moved by the flow (that is, mass is conserved) and let ρ denote the corresponding mass-density. Prove that the density satisfies the equation of continuity

$$\frac{\partial \rho}{\partial t} + \operatorname{div}(\rho f) = 0.$$

(e) The flow of the system $\dot{x} = y$, $\dot{y} = x$ is area preserving. Show directly that the area of the unit disk is unchanged when it is moved forward two time units by the flow.

Exercise 2.23. Construct an alternate proof of Liouville's formula for the n-dimensional linear system $\dot{x} = A(t)x$ with fundamental matrix $\det \Phi(t)$ by differentiation of the function $t \mapsto \det \Phi(t)$ using the chain rule.   Hint: Compute $\frac{d}{dt} \det \Phi(t)$ directly as a sum of n determinants of matrices whose components are the components of $\Phi(t)$ and their derivatives with respect to t. For this computation note that the determinant is a multilinear function of its rows (or columns). Use the multilinearity with respect to rows. Substitute for the derivatives of components of Φ using the differential equation and use elementary row operations to reduce each determinant in the sum to a diagonal component of $A(t)$ times $\det \Phi(t)$.

2.1.4   Linear Equations with Constant Coefficients

In this section we will consider the homogeneous linear system

$$\dot{x} = Ax, \qquad x \in \mathbb{R}^n \tag{2.7}$$

where A is a real $n \times n$ (constant) matrix. We will show how to reduce the problem of constructing a fundamental set of solutions of system (2.7) to a problem in linear algebra. In addition, we will see that the principal fundamental matrix solution at $t = 0$ is given by the exponential of the matrix tA just as the fundamental scalar solution at $t = 0$ of the scalar differential equation $\dot{x} = ax$ is given by $t \mapsto e^{at}$.

Let us begin with the essential observation of the subject: The solutions of system (2.7) are intimately connected with the eigenvalues and eigenvectors of the matrix A. To make this statement precise, let us recall that a complex number λ is an eigenvalue of A if there is a complex *nonzero* vector v such that $Av = \lambda v$. In general, the vector v is called an *eigenvector associated with the eigenvalue* λ if $Av = \lambda v$. Moreover, the set of all eigenvectors associated with an eigenvalue forms a vector space. Because a real matrix can have complex eigenvalues, it is convenient to allow for complex solutions of the differential equation (2.7). Indeed, if $t \mapsto u(t)$ and $t \mapsto v(t)$ are real functions, and if $t \mapsto \phi(t)$ is defined by $\phi(t) := u(t) + iv(t)$, then ϕ is called a complex solution of system (2.7) provided that $\dot{u} + i\dot{v} = Au + iAv$. Of course, if ϕ is a complex solution, then we must have $\dot{u} = Au$ and $\dot{v} = Av$. Thus, it is clear that ϕ is a complex solution if and only if its real and imaginary parts are real solutions. This observation is used in the next proposition.

Proposition 2.24. *Let A be a real $n \times n$ matrix and consider the ordinary differential equation (2.7).*

(1) *The function given by $t \mapsto e^{\lambda t}v$ is a real solution if and only if $\lambda \in \mathbb{R}$, $v \in \mathbb{R}^n$, and $Av = \lambda v$.*

(2) *If $v \neq 0$ is an eigenvector for A with eigenvalue $\lambda = \alpha + i\beta$ such that $\beta \neq 0$, then the imaginary part of v is not zero. In this case, if $v = u + iw \in \mathbb{C}^n$, then there are two real solutions*

$$t \to e^{\alpha t}[(\cos \beta t)u - (\sin \beta t)w],$$

$$t \to e^{\alpha t}[(\sin \beta t)u + (\cos \beta t)w].$$

Moreover, these solutions are linearly independent.

Proof. If $Av = \lambda v$, then

$$\frac{d}{dt}(e^{\lambda t}v) = \lambda e^{\lambda t}v = e^{\lambda t}Av = Ae^{\lambda t}v.$$

In particular, the function $t \to e^{\lambda t}v$ is a solution.

If $\lambda = \alpha + i\beta$ and $\beta \neq 0$, then, because A is real, v must be of the form $v = u + iw$ for some $u, w \in \mathbb{R}^n$ with $w \neq 0$. The real and imaginary parts of the corresponding solution

$$
\begin{aligned}
e^{\lambda t}v &= e^{(\alpha + i\beta)t}(u + iw) \\
&= e^{\alpha t}(\cos \beta t + i \sin \beta t)(u + iw) \\
&= e^{\alpha t}[(\cos \beta t)u - (\sin \beta t)w + i((\sin \beta t)u + (\cos \beta t)w)]
\end{aligned}
$$

are real solutions of the system (2.7). To show that these real solutions are linearly independent, suppose that some linear combination of them with coefficients c_1 and c_2 is identically zero. Evaluation at $t = 0$ and at $t = \pi/(2\beta)$ yields the equations

$$c_1 u + c_2 w = 0, \qquad c_2 u - c_1 w = 0.$$

By elimination of u we find that $(c_1^2 + c_2^2)w = 0$. Since $w \neq 0$, both coefficients must vanish. This proves (2).

Finally, we will complete the proof of (1). Suppose that $\lambda = \alpha + i\beta$ and $v = u + iw$. If $e^{\lambda t}v$ is real, then $\beta = 0$ and $w = 0$. Thus, in fact, λ and v are real. On the other hand, if λ and v are real, then $e^{\lambda t}v$ is a real solution. In this case,

$$\lambda e^{\lambda t}v = A e^{\lambda t}v,$$

and we have that $\lambda v = Av$. $\square$

A fundamental matrix solution of system (2.7) can be constructed explicitly if the eigenvalues of A and their multiplicities are known. To illustrate the basic idea, let us suppose that $\mathbb{C}^n$ has a basis $\mathcal{B} := \{v_1, \dots, v_n\}$ consisting of eigenvectors of A, and let $\{\lambda_1, \dots, \lambda_n\}$ denote the corresponding eigenvalues. For example, if A has n distinct eigenvalues, then the set consisting of one eigenvector corresponding to each eigenvalue is a basis of $\mathbb{C}^n$. At any rate, if $\mathcal{B}$ is a basis of eigenvectors, then there are n corresponding solutions given by

$$t \mapsto e^{\lambda_i t}v_i, \quad i = 1, \dots, n,$$

and the matrix

$$\Phi(t) = [e^{\lambda_1 t}v_1, \dots, e^{\lambda_n t}v_n],$$

which is partitioned by columns, is a matrix solution. Because $\det \Phi(0) \neq 0$, this solution is a fundamental matrix solution, and moreover $\Psi(t) := \Phi(t)\Phi^{-1}(0)$ is the principal fundamental matrix solution of (2.7) at $t = 0$.

A principal fundamental matrix for a real system is necessarily real. To see this, let us suppose that $\Lambda(t)$ is the imaginary part of the principal fundamental matrix solution $\Psi(t)$ at $t = 0$. Since, $\Psi(0) = I$, we must have $\Lambda(0) = 0$. Also, $t \mapsto \Lambda(t)$ is a solution of the linear system. By the uniqueness of solutions of initial value problems, $\Lambda(t) \equiv 0$. Thus, even if some of the eigenvalues of A are complex, the principal fundamental matrix solution is real.

Continuing under the assumption that A has a basis $\mathcal{B}$ of eigenvectors, let us show that there is a change of coordinates that transforms the system $\dot{x} = Ax$, $x \in \mathbb{R}^n$, to a decoupled system of n scalar differential equations. To prove this result, let us first define the matrix $B := [v_1, \dots, v_n]$ whose columns are the eigenvectors in $\mathcal{B}$. The matrix B is invertible. Indeed, consider the action of B on the usual basis vectors and recall that the vector obtained by multiplication of a vector by a matrix is a linear combination of the columns of the matrix; that is, if $w = (w_1, \dots, w_n)$ is (the transpose of) a vector in $\mathbb{C}^n$, then the product Bw is equal to $\sum_{i=1}^n w_i v_i$. In particular, we have $Be_i = v_i$, $i = 1, \dots, n$. This proves that B is invertible. In fact, B^{-1} is the unique linear map such that $B^{-1}v_i = \mathbf{e}_i$.

Using the same idea, let us compute

$$
\begin{aligned}
B^{-1}AB &= B^{-1}A[v_1, \ldots, v_n] \\
&= B^{-1}[\lambda_1 v_1, \ldots, \lambda_n v_n] \\
&= [\lambda_1 \mathbf{e}_1, \ldots, \lambda_n \mathbf{e}_n] \\
&= \begin{pmatrix} \lambda_1 & & 0 \\ & \ddots & \\ 0 & & \lambda_n \end{pmatrix}.
\end{aligned}
$$

In other words, $D := B^{-1}AB$ is a diagonal matrix with the eigenvalues of A as its diagonal elements. The diffeomorphism of $\mathbb{C}^n$ given by the linear transformation $x = By$ transforms the system (2.7) to $\dot{y} = Dy$, as required. Or, using our language for general coordinate transformations, the push forward of the vector field with principal part $x \mapsto Ax$ by the diffeomorphism B^{-1} is the vector field with principal part $y \mapsto Dy$. In particular, the system $\dot{y} = Dy$ is given in components by

$$
\dot{y}_1 = \lambda_1 y_1, \ \ldots, \dot{y}_n = \lambda_n y_n.
$$

Note that if we consider the original system in the new coordinates, then it is obvious that the functions

$$
y_i(t) := e^{\lambda_i t} \mathbf{e}_i, \qquad i = 1, \ldots, n
$$

are a fundamental set of solutions for the differential equation $\dot{y} = Dy$. Moreover, by transforming back to the original coordinates, it is clear that the solutions

$$
x_i(t) := e^{\lambda_i t} B \mathbf{e}_i = e^{\lambda_i t} v_i, \qquad i = 1, \ldots, n
$$

form a fundamental set of solutions for the original system (2.7). Thus, we have an alternative method to construct a fundamental matrix solution: Change coordinates to obtain a new differential equation, construct a fundamental set of solutions for the new differential equation, and then transform these new solutions back to the original coordinates. Even if A is not diagonalizable, a fundamental matrix solution of the associated differential equation can still be constructed using this procedure. Indeed, we can use a basic fact from linear algebra: If A is a real matrix, then there is a nonsingular matrix B such that $D := B^{-1}AB$ is in (real) Jordan canonical form (see [59], [123], and Exercise 2.38). Then, as before, the system (2.7) is transformed by the change of coordinates $x = By$ into the linear system $\dot{y} = Dy$.

We will eventually give a detailed description of the Jordan form and also show that the corresponding canonical system of differential equations can be solved explicitly. This solution can be transformed back to the original coordinates to construct a fundamental matrix solution of $\dot{x} = Ax$.

Exercise 2.25. (a) Find the principal fundamental matrix solutions at $t = 0$ for the matrix systems:

1. $\dot{x} = \begin{pmatrix} 0 & 1 \\ 1 & 0 \end{pmatrix} x.$

2. $\dot{x} = \begin{pmatrix} 2 & -1 \\ 1 & 2 \end{pmatrix} x.$

3. $\dot{x} = \begin{pmatrix} 0 & 1 \\ 0 & 0 \end{pmatrix} x.$

4. $\dot{x} = \begin{pmatrix} 7 & -8 \\ 4 & -5 \end{pmatrix} x.$

(b) Solve the initial value problem for system 2 with initial value $x(0) = (1,0)$. (c) Find a change of coordinates (given by a matrix) that diagonalizes the system matrix of system 4. (d) Find the principal fundamental matrix solution at $t = 3$ for system 3.

Exercise 2.26. (a) Determine the flow of the first order system that is equivalent to the second order linear differential equation

$$\ddot{x} + \dot{x} + 4x = 0.$$

(b) Draw the phase portrait.

Exercise 2.27. [Euler's Equation] Euler's equation is the second order linear equation

$$t^2 \ddot{x} + bt\dot{x} + cx = 0, \qquad t > 0$$

with the parameters b and c. (a) Show that there are three different solution types according to the sign of $(b-1)^2 - 4c$. Hint: Guess a solution of the form $x = r^t$ for some number r. (b) Discuss, for each of the cases in part (a), the behavior of the solution as $t \to 0^+$. (c) Write a time-dependent linear first order system that is equivalent to Euler's equation. (d) Determine the principal fundamental matrix solution for the first order system in part (c) in case $b = 1$ and $c = -1$.

Instead of writing out the explicit, perhaps complicated, formulas for the components of the fundamental matrix solution of an $n \times n$ linear system of differential equations, it is often more useful, at least for theoretical considerations, to treat the situation from a more abstract point of view. In fact, we will show that there is a natural generalization of the exponential function to a function defined on the set of square matrices. Using this matrix exponential function, the solution of a linear homogeneous system with constant coefficients is given in a form that is analogous to the solution $t \mapsto e^{ta} x_0$ of the scalar differential equation $\dot{x} = ax$.

Recall that the set of linear transformations $\mathcal{L}(\mathbb{R}^n)$ (respectively $\mathcal{L}(\mathbb{C}^n)$) on $\mathbb{R}^n$ (respectively $\mathbb{C}^n$) is an n^2-dimensional Banach space with respect to the operator norm

$$\|A\| = \sup_{|v|=1} |Av|.$$

Most of the theory we will develop is equally valid for either of the vector spaces $\mathbb{R}^n$ or $\mathbb{C}^n$. When the space is not at issue, we will denote the Banach space of linear transformations by $\mathcal{L}(E)$ where E may be taken as either $\mathbb{R}^n$ or $\mathbb{C}^n$. The theory is also valid for the set of (operator norm) bounded linear transformations of an arbitrary Banach space.

Exercise 2.28. Prove: $\mathcal{L}(E)$ is a finite dimensional Banach space with respect to the operator norm.

Exercise 2.29. Prove: (a) If $A, B \in \mathcal{L}(E)$, then $\|AB\| \leq \|A\|\|B\|$. (b) If $A \in \mathcal{L}(E)$ and k is a nonnegative integer, then $\|A^k\| \leq \|A\|^k$.

Exercise 2.30. The space of $n \times n$ matrices is a topological space with respect to the operator topology. Prove that the set of matrices with n distinct eigenvalues is open and dense. A property that is defined on the countable intersection of open dense sets is called *generic* (see [123, p. 153–157]).

Proposition 2.31. *If $A \in \mathcal{L}(E)$, then the series $I + \sum_{k=1}^{\infty} \frac{1}{k!} A^k$ is absolutely convergent.*

Proof. Define

$$S_N := 1 + \|A\| + \frac{1}{2!}\|A^2\| + \cdots + \frac{1}{N!}\|A^N\|$$

and note that $\{S_N\}_{N=1}^{\infty}$ is a monotone increasing sequence of real numbers. Since (by Exercise 2.29) $\|A^k\| \leq \|A\|^k$ for every integer $k \geq 0$, it follows that $\{S_N\}_{N=1}^{\infty}$ is bounded above. In fact,

$$S_N \leq e^{\|A\|}$$

for every $N \geq 1$. $\square$

Define the *exponential map* $\exp : \mathcal{L}(E) \to \mathcal{L}(E)$ by

$$\exp(A) := I + \sum_{k=1}^{\infty} \frac{1}{k!} A^k.$$

Also, let us use the notation $e^A := \exp(A)$.

The main properties of the exponential map are summarized in the following proposition.

Proposition 2.32. *Suppose that $A, B \in \mathcal{L}(E)$.*

(0) If $A \in \mathcal{L}(\mathbb{R}^n)$, then $e^A \in \mathcal{L}(\mathbb{R}^n)$.

(1) If B is nonsingular, then $B^{-1}e^A B = e^{B^{-1}AB}$.

(2) If $AB = BA$, then $e^{A+B} = e^A e^B$.

(3) $e^{-A} = (e^A)^{-1}$. In particular, the image of exp is in the general linear group $GL(E)$ consisting of the invertible elements of $\mathcal{L}(E)$.

(4) $\frac{d}{dt}(e^{tA}) = Ae^{tA} = e^{tA}A$. In particular, $t \mapsto e^{tA}$ is the principal fundamental matrix solution of the system (2.7) at $t = 0$.

(5) $\|e^A\| \leq e^{\|A\|}$.

Proof. The proof of (0) is obvious.

To prove (1), define

$$S_N := I + A + \frac{1}{2!}A^2 + \cdots + \frac{1}{N!}A^N,$$

and note that if B is nonsingular, then $B^{-1}A^n B = (B^{-1}AB)^n$. Thus, we have that

$$B^{-1}S_N B = I + B^{-1}AB + \frac{1}{2!}(B^{-1}AB)^2 + \cdots + \frac{1}{N!}(B^{-1}AB)^N,$$

and, by the definition of the exponential map,

$$\lim_{N \to \infty} B^{-1}S_N B = e^{B^{-1}AB}.$$

Using the continuity of the linear map on $\mathcal{L}(E)$ defined by $C \mapsto B^{-1}CB$, it follows that

$$\lim_{N \to \infty} B^{-1}S_N B = B^{-1}e^A B,$$

as required.

While the proof of (4) given here has the advantage of being self contained, there are conceptually simpler alternatives (see Exercises 2.33–2.34). As the first step in the proof of (4), consider the following proposition: If $s, t \in \mathbb{R}$, then $e^{(s+t)A} = e^{sA}e^{tA}$. To prove it, let us denote the partial sums for the series representation of e^{tA} by

$$S_N(t) := I + tA + \frac{1}{2!}(tA)^2 + \cdots + \frac{1}{N!}(tA)^N$$

$$= I + tA + \frac{1}{2!}t^2 A^2 + \cdots + \frac{1}{N!}t^N A^N.$$

We claim that

$$S_N(s)S_N(t) = S_N(s+t) + \sum_{k=N+1}^{2N} P_k(s,t)A^k \qquad (2.8)$$

where $P_k(s,t)$ is a homogeneous polynomial of degree k such that

$$|P_k(s,t)| \leq \frac{(|s|+|t|)^k}{k!}.$$

To obtain this identity, note that the kth order term of the product, at least for $0 \leq k \leq N$, is given by

$$\Big(\sum_{j=0}^{k} \frac{1}{(k-j)!j!} s^{k-j}t^j\Big)A^k = \Big(\frac{1}{k!}\sum_{j=0}^{k} \frac{k!}{(k-j)!j!} s^{k-j}t^j\Big)A^k = \frac{1}{k!}(s+t)^k A^k.$$

Also, for $N+1 \leq k \leq 2N$, the kth order term is essentially the same, only some of the summands are missing. In fact, these terms all have the form

$$\Big(\sum_{j=0}^{k} \frac{\delta(j)}{(k-j)!j!} s^{k-j}t^j\Big)A^k$$

where $\delta(j)$ has value zero or one. Each such term is the product of A^k and a homogeneous polynomial in two variables of degree k. Moreover, because $|\delta(j)| \leq 1$, we obtain the required estimate for the polynomial. This proves the claim.

Using equation (2.8), we have the following inequality

$$\|S_N(s)S_N(t) - S_N(s+t)\| \leq \sum_{k=N+1}^{2N} |P_k(s,t)|\, \|A\|^k$$

$$\leq \sum_{k=N+1}^{2N} \frac{(|s|+|t|)^k}{k!} \|A\|^k.$$

Also, because the series

$$\sum_{k=0}^{\infty} \frac{(|s|+|t|)^k}{k!} \|A\|^k$$

is convergent, it follows that its partial sums, denoted Q_N, form a Cauchy sequence. In particular, if $\epsilon > 0$ is given, then for sufficiently large N we have

$$|Q_{2N} - Q_N| < \epsilon.$$

Moreover, since

$$Q_{2N} - Q_N = \sum_{k=N+1}^{2N} \frac{(|s| + |t|)^k}{k!} \|A\|^k,$$

it follows that

$$\lim_{N \to \infty} \|S_N(s) S_N(t) - S_N(s+t)\| = 0.$$

Using this fact and passing to the limit as $N \to \infty$ on both sides of the inequality

$$\|e^{sA} e^{tA} - e^{(s+t)A}\| \leq \|e^{sA} e^{tA} - S_N(s) S_N(t)\|$$
$$+ \|S_N(s) S_N(t) - S_N(s+t)\|$$
$$+ \|S_N(s+t) - e^{(s+t)A}\|,$$

we see that

$$e^{sA} e^{tA} = e^{(s+t)A}, \tag{2.9}$$

as required.

In view of the identity (2.9), the derivative of the function $t \mapsto e^{tA}$ is given by

$$\frac{d}{dt} e^{tA} = \lim_{s \to 0} \frac{1}{s} (e^{(t+s)A} - e^{tA})$$
$$= \lim_{s \to 0} \frac{1}{s} (e^{sA} - I) e^{tA}$$
$$= \left(\lim_{s \to 0} \frac{1}{s} (e^{sA} - I) \right) e^{tA}$$
$$= \left(\lim_{s \to 0} (A + R(s)) \right) e^{tA}$$

where

$$\|R(s)\| \leq \frac{1}{|s|} \sum_{k=2}^{\infty} \frac{|s|^k}{k!} \|A\|^k \leq |s| \sum_{k=2}^{\infty} \frac{|s|^{k-2}}{k!} \|A\|^k.$$

Moreover, if $|s| < 1$, then $\|R(s)\| \leq |s| e^{\|A\|}$. In particular, $R(s) \to 0$ as $s \to 0$ and as a result,

$$\frac{d}{dt} e^{tA} = A e^{tA}.$$

Since $A S_N(t) = S_N(t) A$, it follows that $A e^{tA} = e^{tA} A$. This proves the first statement of part (4). In particular $t \mapsto e^{tA}$ is a matrix solution of the system (2.7). Clearly, $e^0 = I$. Thus, the columns of e^0 are linearly independent. It follows that $t \mapsto e^{tA}$ is the principal fundamental matrix solution at $t = 0$, as required.

To prove (2), suppose that $AB = BA$ and consider the function $t \mapsto e^{t(A+B)}$. By (4), this function is a matrix solution of the initial value problem

$$\dot{x} = (A + B)x, \qquad x(0) = I.$$

The function $t \mapsto e^{tA}e^{tB}$ is a solution of the same initial value problem. To see this, use the product rule to compute the derivative

$$\frac{d}{dt}e^{tA}e^{tB} = Ae^{tA}e^{tB} + e^{tA}Be^{tB},$$

and use the identity $AB = BA$ to show that $e^{tA}B = Be^{tA}$. The desired result is obtained by inserting this last identity into the formula for the derivative. By the uniqueness of the solution of the initial value problem, the two solutions are identical.

To prove (3), we use (2) to obtain $I = e^{A-A} = e^{A}e^{-A}$ or, in other words, $(e^{A})^{-1} = e^{-A}$.

The result (5) follows from the inequality

$$\left\| I + A + \frac{1}{2!}A^2 + \cdots + \frac{1}{N!}A^N \right\| \leq \|I\| + \|A\| + \frac{1}{2!}\|A\|^2 + \cdots + \frac{1}{N!}\|A\|^N. \quad \square$$

We have defined the exponential of a matrix as an infinite series and used this definition to prove that the homogeneous linear system $\dot{x} = Ax$ has a fundamental matrix solution, namely, $t \mapsto e^{tA}$. This is a strong result because it does not use the existence theorem for differential equations. Granted, the uniqueness theorem is used. But it is an easy corollary of Gronwall's inequality (see Exercise 2.5). An alternative approach to the exponential map is to use the existence theorem and define the function $t \mapsto e^{tA}$ to be the principal fundamental matrix solution at $t = 0$. Proposition 2.32 can then be proved by using properties of the solutions of homogeneous linear differential equations.

Exercise 2.33. Show that the partial sums of the series representation of e^{tA} converge uniformly on compact subsets of $\mathbb{R}$. Use Theorem 1.248 to prove part (4) of Proposition 2.32.

Exercise 2.34. (a) Show that $\exp : \mathcal{L}(E) \to \mathcal{L}(E)$ is continuous. Hint: For $r > 0$, the sequence of partial sums of the series representation of $\exp(X)$ converges uniformly on $B_r(0) := \{X \in \mathcal{L}(E) : \|X\| < r\}$. (b) By Exercise 2.30, matrices with distinct eigenvalues are dense in $\mathcal{L}(E)$. Such matrices are diagonalizable (over the complex numbers). Show that if $A \in \mathcal{L}(E)$ is diagonalizable, then part (4) of Proposition 2.32 holds for A. (c) Use parts (a) and (b) to prove part (4) of Proposition 2.32. (d) Prove that $\exp : \mathcal{L}(E) \to \mathcal{L}(E)$ is differentiable and compute $D\exp(I)$.

Exercise 2.35. Define $\exp(A) = \Phi(1)$ where $\Phi(t)$ is the principal fundamental matrix at $t = 0$ for the system $\dot{x} = Ax$. (a) Prove that $\exp(tA) = \Phi(t)$. (b) Prove that $\exp(-A) = (\exp(A))^{-1}$.

Exercise 2.36. [Laplace Transform] (a) Prove that if A is an $n \times n$-matrix, then

$$e^{tA} - I = \int_0^t Ae^{\tau A}\, d\tau.$$

(b) Prove that if all eigenvalues of A have negative real parts, then

$$-A^{-1} = \int_0^\infty e^{\tau A}\, d\tau.$$

(c) Prove that if $s \in \mathbb{R}$ is sufficiently large, then

$$(sI - A)^{-1} = \int_0^\infty e^{-s\tau} e^{\tau A}\, d\tau;$$

that is, the Laplace transform of e^{tA} is $(sI - A)^{-1}$. (d) Solve the initial value problem $\dot{x} = Ax$, $x(0) = x_0$ using the method of the Laplace transform; that is, take the Laplace transform of both sides of the equation, solve the resulting algebraic equation, and then invert the transform to obtain the solution in the original variables. By definition, the Laplace transform of the (perhaps matrix valued) function f is

$$\mathcal{L}\{f\}s = \int_0^\infty e^{-s\tau} f(\tau)\, d\tau.$$

To obtain a matrix representation for e^{tA}, let us recall that there is a real matrix B that transforms A to real Jordan canonical form. Of course, to construct the matrix B, we must at least be able to find the eigenvalues of A, a task that is equivalent to finding the roots of a polynomial of degree n. Thus, for $n \geq 5$, it is generally impossible to construct the matrix B explicitly. But if B is known, then by using part (1) of Proposition 2.32, we have that

$$B^{-1} e^{tA} B = e^{tB^{-1}AB}.$$

Thus, the problem of constructing a principal fundamental matrix is solved as soon as we find a matrix representation for $e^{tB^{-1}AB}$.

The Jordan canonical matrix $B^{-1}AB$ is block diagonal, where each block corresponding to a real eigenvalue has the form "diagonal + nilpotent," and, each block corresponding to a complex eigenvalue with nonzero imaginary part has the form "block diagonal + block nilpotent." In view of this block structure, it suffices to determine the matrix representation for e^{tJ} where J denotes a single Jordan block.

Consider a block of the form

$$J = \lambda I + N$$

where N is the nilpotent matrix with zero components except on the super diagonal, where each component is unity and note that $N^k = 0$. We have that

$$e^{tJ} = e^{t(\lambda I + N)} = e^{t\lambda I} e^{tN} = e^{t\lambda}\left(I + tN + \frac{t^2}{2!}N^2 + \cdots + \frac{t^{k-1}}{(k-1)!}N^{k-1}\right)$$

where k is the dimension of the block.

If J is a Jordan block with diagonal 2×2 subblocks given by

$$R = \begin{pmatrix} \alpha & -\beta \\ \beta & \alpha \end{pmatrix} \tag{2.10}$$

with $\beta \neq 0$, then e^{tJ} is block diagonal with each block given by e^{tR}. To obtain an explicit matrix representation for e^{tR}, define

$$P := \begin{pmatrix} 0 & -\beta \\ \beta & 0 \end{pmatrix}, \qquad Q(t) := \begin{pmatrix} \cos\beta t & -\sin\beta t \\ \sin\beta t & \cos\beta t \end{pmatrix},$$

and note that $t \mapsto e^{tP}$ and $t \mapsto Q(t)$ are both solutions of the initial value problem

$$\dot{x} = \begin{pmatrix} 0 & -\beta \\ \beta & 0 \end{pmatrix} x, \qquad x(0) = I.$$

Thus, we have that $e^{tP} = Q(t)$ and

$$e^{tR} = e^{\alpha t} e^{tP} = e^{\alpha t} Q(t).$$

Finally, if the Jordan block J has the 2×2 block matrix R along its block diagonal and the 2×2 identity along its super block diagonal, then

$$e^{tJ} = e^{\alpha t} S(t) e^{tN} \tag{2.11}$$

where $S(t)$ is block diagonal with each block given by $Q(t)$, and N is the nilpotent matrix with 2×2 identity blocks on its super block diagonal. To prove this fact, note that J can be written as a sum $J = \alpha I + K$ where K has diagonal blocks given by P and super diagonal blocks given by the 2×2 identity matrix. Since the $n \times n$ matrix αI commutes with every matrix, we have that

$$e^{tJ} = e^{\alpha t} e^{tK}.$$

The proof is completed by observing that the matrix K can also be written as a sum of commuting matrices; namely, the block diagonal matrix with each diagonal block equal to P and the nilpotent matrix N.

We have outlined a procedure to find a matrix representation for e^{tA}. In addition, we have proved the following result.

Proposition 2.37. *If A is an $n \times n$ matrix, then e^{tA} is a matrix whose components are (finite) sums of terms of the form*

$$p(t)e^{\alpha t}\sin\beta t \ \text{ and } \ p(t)e^{\alpha t}\cos\beta t$$

where α and β are real numbers such that $\alpha + i\beta$ is an eigenvalue of A, and $p(t)$ is a polynomial of degree at most $n - 1$.

Exercise 2.38. [Jordan Form] Show that every real 2×2-matrix can be transformed to real Jordan canonical form and find the fundamental matrix solutions for the corresponding 2×2 real homogeneous linear systems of differential equations. Draw the phase portrait for each canonical system. Hint: For the case of a double eigenvalue suppose that $(A - \lambda I)V = 0$ and every eigenvector is parallel to V. Choose a vector W that is not parallel to V and note that $(A - \lambda I)W = Y \neq 0$. Since V and W are linearly independent, $Y = aV + bW$ for some real numbers a and b. Use this fact to argue that Y is parallel to V. Hence, there is a (nonzero) vector Z such that $(A - \lambda I)Z = V$. Define $B = [V, W]$ to be the indicated 2×2-matrix partitioned by columns and show that $B^{-1}AB$ is in Jordan form. To solve $\dot{x} = Ax$, use the change of variables $x = By$.

Exercise 2.39. Find the Jordan canonical form for the matrix

$$\begin{pmatrix} 1 & 1 & 2 \\ 0 & -2 & 1 \\ 0 & 0 & -2 \end{pmatrix}.$$

Exercise 2.40. Find the principal fundamental matrix solution at $t = 0$ for the linear differential equation whose system matrix is

$$\begin{pmatrix} 0 & 0 & 1 & 0 \\ 0 & 0 & 0 & 1 \\ 0 & 0 & 0 & -2 \\ 0 & a & 2 & 0 \end{pmatrix},$$

where $a := 4 - \omega^2$ and $0 \leq \omega \leq 1$, by changing variables so that the system matrix is in Jordan canonical form, computing the exponential, and changing back to the original variables.

Exercise 2.41. Suppose that $J = \lambda I + N$ is a $k \times k$-Jordan block and let B denote the diagonal matrix with main diagonal $1, \epsilon, \epsilon^2, \ldots, \epsilon^{k-1}$. (a) Show that $B^{-1}JB = \lambda I + \epsilon N$. (b) Prove: Given $\epsilon > 0$ and a matrix A, there is a diagonalizable matrix B such that $\|A - B\| < \epsilon$ (cf. Exercise 2.30). (c) Discuss the statement: A numerical algorithm for finding the Jordan canonical form will be ill conditioned.

Exercise 2.42. (a) Suppose that A is an $n \times n$-matrix such that $A^2 = I$. Find an explicit formula for e^{tA}. (b) Repeat part (a) in case $A^2 = -I$. (c) Solve the initial value problem

$$\dot{x} = \begin{pmatrix} 2 & -5 & 8 & -12 \\ 1 & -2 & 4 & -8 \\ 0 & 0 & 2 & -5 \\ 0 & 0 & 1 & -2 \end{pmatrix} x, \qquad x(0) = \begin{pmatrix} 1 \\ 0 \\ 0 \\ 1 \end{pmatrix}.$$

(d) Specify the stable manifold for the rest point at the origin of the linear system

$$\dot{x} = \begin{pmatrix} 2 & -3 & 4 & -4 \\ 1 & -2 & 4 & -4 \\ 0 & 0 & 2 & -3 \\ 0 & 0 & 1 & -2 \end{pmatrix} x.$$

Exercise 2.43. Prove that $\det e^A = e^{\operatorname{tr} A}$ for every $n \times n$ matrix A. Hint: Use Liouville's formula 2.18.

The scalar autonomous differential equation $\dot{x} = ax$ has the principal fundamental solution $t \mapsto e^{at}$ at $t = 0$. We have defined the exponential map on bounded linear operators and used this function to construct the analogous fundamental matrix solution $t \mapsto e^{tA}$ of the homogeneous autonomous system $\dot{x} = Ax$. The scalar nonautonomous homogeneous linear differential equation $\dot{x} = a(t)x$ has the principal fundamental solution

$$t \mapsto e^{\int_0^t a(s)\, ds}.$$

But, in the matrix case, the same formula with $a(s)$ replaced by $A(s)$ is not always a matrix solution of the linear system $\dot{x} = A(t)x$ (cf. [130] and see Exercise 2.50).

As an application of the methods developed in this section we will formulate and prove a special case of the Lie–Trotter product formula for the exponential of a sum of two $k \times k$-matrices when the matrices do not necessarily commute (see [221] for the general case).

Theorem 2.44. *If* $\gamma : \mathbb{R} \to \mathcal{L}(E)$ *is a* C^1-*function with* $\gamma(0) = I$ *and* $\dot{\gamma}(0) = A$, *then the sequence* $\{\gamma^n(t/n)\}_{n=1}^{\infty}$ *converges to* $\exp(tA)$. *In particular, if* A *and* B *are* $k \times k$-*matrices, then*

$$e^{t(A+B)} = \lim_{n \to \infty} \left(e^{\frac{t}{n}A} e^{\frac{t}{n}B} \right)^n.$$

Proof. Fix $T > 0$ and assume that $|t| < T$. We will first prove the following proposition: There is a number $M > 0$ such that $\|\gamma^j(t/n)\| \leq M$ whenever j and n are integers and $0 \leq j \leq n$. Using Taylor's theorem, we have the estimate

$$\|\gamma(t/n)\| \leq 1 + \frac{T}{n}\|A\| + \frac{T}{n}\int_0^1 \|\dot{\gamma}(st/n) - A\|\, ds.$$

Since $\sigma \mapsto \|\dot{\gamma}(\sigma) - A\|$ is a continuous function on the compact set $S := \{\sigma \in \mathbb{R} : |\sigma| \leq T\}$, we also have that $K := \sup\{\|\dot{\gamma}(\sigma) - A\| : \sigma \in S\} < \infty$, and therefore,

$$\|\gamma^j(t/n)\| \leq \|\gamma(t/n)\|^j \leq (1 + \frac{T}{n}(\|A\| + K))^n.$$

To finish the proof of the proposition, note that the sequence $\{(1 + \frac{T}{n}(\|A\| + K))^n\}_{n=1}^{\infty}$ is bounded—it converges to $\exp(T(\|A\| + K))$.

Using the (telescoping) identity

$$e^{tA} - \gamma^n(t/n) = \sum_{j=1}^{n} \left((e^{\frac{t}{n}A})^{n-j+1}\gamma^{j-1}(t/n) - (e^{\frac{t}{n}A})^{n-j}\gamma^j(t/n) \right)$$

$$= \sum_{j=1}^{n} \left((e^{\frac{t}{n}A})^{n-j}e^{\frac{t}{n}A}\gamma^{j-1}(t/n) - (e^{\frac{t}{n}A})^{n-j}\gamma(t/n)\gamma^{j-1}(t/n) \right),$$

we have the estimate

$$\|e^{tA} - \gamma^n(t/n)\| \le \sum_{j=1}^{n} e^{\frac{n-j}{n}T\|A\|}\|e^{\frac{t}{n}A} - \gamma(t/n)\|\|\gamma^{j-1}(t/n)\|$$

$$\le M\|e^{\frac{t}{n}A} - \gamma(t/n)\| \sum_{j=1}^{n} e^{(n-j)/nT\|A\|}$$

$$\le Mne^{T\|A\|}\|e^{\frac{t}{n}A} - \gamma(t/n)\|.$$

By Taylor's theorem (applied to each of the functions $\sigma \mapsto e^{\sigma A}$ and $\sigma \mapsto \gamma(\sigma)$), we obtain the inequality

$$\|e^{\frac{t}{n}A} - \gamma(t/n)\| \le \frac{T}{n}J(n)$$

where

$$J(n) := \int_0^1 \|\dot\gamma(st/n) - A\|\, ds + \int_0^1 \|A\|\|e^{\frac{st}{n}A} - I\|\, ds$$

is such that $\lim_{n\to\infty} J(n) = 0$. Since

$$\|e^{tA} - \gamma^n(t/n)\| \le MTe^{T\|A\|}J(n),$$

it follows that $\lim_{n\to\infty}\|e^{tA} - \gamma^n(t/n)\| = 0$, as required.

The second statement of the theorem follows from the first with A replaced by $A + B$ and $\gamma(t) := e^{tA}e^{tB}$. $\square$

The product formula in Theorem 2.44 gives a method to compute the solution of the differential equation $\dot x = (A + B)x$ from the solutions of the equations $\dot x = Ax$ and $\dot x = Bx$. Of course, if A and B happen to commute (that is, $[A, B] := AB - BA = 0$), then the product formula reduces to $e^{t(A+B)} = e^{tA}e^{tB}$ by part (2) of Proposition 2.32. It turns out that $[A, B] = 0$ is also a necessary condition for this reduction. Indeed, let us note first that $t \mapsto e^{tA}e^{tB}$ is a solution of the initial value problem

$$\dot W = AW + WB, \qquad W(0) = I. \tag{2.12}$$

If $t \mapsto e^{t(A+B)}$ is also a solution, then by substitution and a rearrangement of the resulting equality, we have the identity

$$A = e^{-t(A+B)}Ae^{t(A+B)}.$$

By computing the derivative with respect to t of both sides of this identity and simplifying the resulting equation, it follows that $[A, B] = 0$ (cf. Exercise 2.53).

What can we say about the product $e^{tA}e^{tB}$ in case $[A, B] \neq 0$? The answer is provided by (a special case of) the Baker-Campbell-Hausdorff formula

$$e^{tA}e^{tB} = e^{t(A+B)+(t^2/2)[A,B]+R(t,A,B)} \tag{2.13}$$

where $R(0, A, B) = R_t(0, A, B) = R_{tt}(0, A, B) = 0$ (see, for example, [225]).

To obtain formula (2.13), note that the curve $\gamma : \mathbb{R} \to \mathcal{L}(E)$ given by $t \mapsto e^{tA}e^{tB}$ is such that $\gamma(0) = I$. Also, the function $\exp : \mathcal{L}(E) \to \mathcal{L}(E)$ is such that $\exp(0) = I$ and $D\exp(0) = I$. Hence, by the inverse function theorem, there is a unique smooth curve $\Omega(t)$ in $\mathcal{L}(E)$ such that $\Omega(0) = 0$ and $e^{\Omega(t)} = e^{tA}e^{tB}$. Hence, the function $t \mapsto e^{\Omega(t)}$ is a solution of the initial value problem (2.12), that is,

$$D\exp(\Omega)\dot{\Omega} = Ae^{\Omega} + e^{\Omega}B. \tag{2.14}$$

By evaluation at $t = 0$, we have that $\dot{\Omega}(0) = A + B$. The equality $\ddot{\Omega}(0) = [A, B]$ is obtained by differentiating both sides of equation (2.14) with respect to t at $t = 0$. This computation requires the second derivative of $\exp$ at the origin in $\mathcal{L}(E)$. To determine this derivative, use the power series definition of $\exp$ to show that it suffices to compute the second derivative of the function $h : \mathcal{L}(E) \to \mathcal{L}(E)$ given by $h(X) = \frac{1}{2}X^2$. Since h is smooth, its derivatives can be determined by computing directional derivatives; in fact, we have that

$$Dh(X)Y = \frac{d}{dt}\frac{1}{2}(X + tY)^2\Big|_{t=0} = \frac{1}{2}(XY + YX),$$

$$D^2h(X)(Y, Z) = \frac{d}{dt}Dh(X + tZ)Y\Big|_{t=0} = \frac{1}{2}(YZ + ZY),$$

and $D^2\exp(0)(Y, Z) = \frac{1}{2}(YZ + ZY)$. The proof of formula (2.13) is completed by applying Taylor's theorem to the function Ω.

Exercise 2.45. Compute the principal fundamental matrix solution at $t = 0$ for the system $\dot{x} = Ax$ where

$$A := \begin{pmatrix} 1 & 2 & 3 \\ 0 & 1 & 4 \\ 0 & 0 & 1 \end{pmatrix}.$$

Exercise 2.46. Reduction to Jordan form is only one of many computational methods that can be used to determine the exponential of a matrix. Repeat Exercise 2.45 using the method presented in [112].

Exercise 2.47. Determine the phase portrait for the system

$$\begin{pmatrix} \dot{x} \\ \dot{y} \end{pmatrix} = \begin{pmatrix} 0 & 1 \\ -1 & -\mu \end{pmatrix} \begin{pmatrix} x \\ y \end{pmatrix}.$$

Make sure you distinguish the cases $\mu < -2$, $\mu > 2$, $\mu = 0$, $0 < \mu < 2$, and $-2 < \mu < 0$. For each case, find the principal fundamental matrix solution at $t = 0$.

Exercise 2.48. (a) Show that the general 2×2 linear system with constant coefficients decouples in polar coordinates, and the first-order differential equation for the angular coordinate θ can be viewed as a differential equation on the unit circle $\mathbb{T}^1$. (b) Consider the first-order differential equation

$$\dot{\theta} = \alpha \cos^2 \theta + \beta \cos \theta \sin \theta + \gamma \sin^2 \theta.$$

For $4\alpha\gamma - \beta^2 > 0$, prove that all orbits on the circle are periodic with period $4\pi(4\alpha\gamma - \beta^2)^{-1/2}$, and use this result to determine the period of the periodic orbits of the differential equation $\dot{\theta} = \eta + \cos\theta \sin\theta$ as a function of the parameter $\eta > 1$. Describe the behavior of this function as $\eta \to 1^+$ and give a qualitative explanation of the behavior. (c) Repeat the last part of the exercise for the differential equation $\dot{\theta} = \eta - \sin\theta$ where $\eta > 1$. (d) Show that an n-dimensional homogeneous linear differential equation induces a differential equation on the real projective space of dimension $n - 1$. (e) There is an intimate connection between the linear second-order differential equation

$$\ddot{y} - (q(t) + \dot{p}(t)/p(t))\dot{y} + r(t)p(t)y = 0$$

and the Riccati equation

$$\dot{x} = p(t)x^2 + q(t)x + r(t).$$

In fact, these equations are related by $x = -\dot{y}/(p(t)y)$. For example $\ddot{y} + y = 0$ is related to the Riccati equation $\dot{u} = -1 - u^2$, where in this case the change of variables is $x = \dot{y}/y$. Note that the unit circle in $\mathbb{R}^2$, with coordinates $(y, \dot{y})$, has coordinate charts given by $(y, \dot{y}) \mapsto \dot{y}/y$ and $(y, \dot{y}) \mapsto y/\dot{y}$. Thus, the transformation from the linear second-order equation to the Riccati equation is a local coordinate representation of the differential equation induced by the second-order linear differential equation on the circle. Explore and explain the relation between this coordinate representation and the polar coordinate representation of the first-order linear system. (f) Prove the cross-ratio property for Riccati equations: If x_i, $i = 1, 2, 3, 4$, are four linearly independent solutions of a Riccati equation, then the quantity

$$\frac{(x_1 - x_3)(x_2 - x_4)}{(x_1 - x_4)(x_2 - x_3)}$$

is constant. (g) Show that if one solution $t \mapsto z(t)$ of the Riccati equation is known, then the general solution can always be found by solving a linear equation after the substitution $x = z + 1/u$. (h) Solve the initial value problem

$$\dot{x} + x^2 + (2t + 1)x + t^2 + t + 1 = 0, \qquad x(1) = 1.$$

(see [200, p. 30] for this equation, and [71] for more properties of Riccati equations).

Exercise 2.49. The linearized Hill's equations for the relative motion of two satellites with respect to a circular reference orbit about the earth are given by

$$\ddot{x} - 2n\dot{y} - 3n^2 x = 0,$$
$$\ddot{y} + 2n\dot{x} = 0,$$
$$\ddot{z} + n^2 z = 0$$

where n is a constant related to the radius of the reference orbit and the gravitational constant. There is a five-dimensional manifold in the phase space corresponding to periodic orbits. An orbit with an initial condition not on this manifold contains a secular drift term. Determine the manifold of periodic orbits and explain what is meant by a secular drift term. Answer: The manifold of periodic orbits is the hyperplane given by $\dot{y} + 2nx = 0$.

Exercise 2.50. Find a matrix function $t \mapsto A(t)$ such that

$$t \mapsto \exp\left(\int_0^t A(s)\, ds \right)$$

is *not* a matrix solution of the system $\dot{x} = A(t)x$. Show that the given exponential formula is a solution in the scalar case. When is it a solution for the matrix case?

Exercise 2.51. In the Baker-Campbell-Hausdorff formula (2.13), the second-order correction term is $(t^2/2)[A, B]$. Prove that the third-order correction is $(t^3/12)([A, [A, B]] - [B, [A, B]])$.

Exercise 2.52. Show that the commutator relations

$$[A, [A, B]] = 0, \qquad [B, [A, B]] = 0$$

imply the identity

$$e^{tA}e^{tB} = e^{\Omega(t)} \tag{2.15}$$

where $\Omega(t) := t(A+B) + (t^2/2)[A, B]$. Is the converse statement true? Find (3×3) matrices A and B such that $[A, B] \neq 0$, $[A, [A, B]] = 0$, and $[B, [A, B]] = 0$. Verify identity (2.15) for your A and B. Hint: Suppose that $[A, [A, B]] = 0$ and $[B, [A, B]] = 0$. Use these relations to prove in turn $[\Omega(t), \dot{\Omega}(t)] = 0$, $D\exp(\Omega)\dot{\Omega} = \exp(\Omega)\dot{\Omega}$, and $[\exp(\Omega), \dot{\Omega}] = 0$. To prove the identity (2.15), it suffices to show that $t \mapsto \exp(\Omega(t))$ is a solution of the initial value problem (2.12). By substitution into the differential equation and some manipulation, prove that this function is a solution if and only if

$$\frac{d}{dt}(e^{-\Omega(t)} A e^{\Omega(t)} - t[A, B]) = 0.$$

Compute the indicated derivative and use the hypotheses to show that it vanishes.

Exercise 2.53. Find $n \times n$ matrices A and B such that $[A, B] \neq 0$ and $e^A e^B = e^{A+B}$ (see Problem 88-1 in *SIAM Review*, **31**(1), (1989), 125–126).

Exercise 2.54. Let A be an $n \times n$ matrix with components $\{a_{ij}\}$. Prove: Every component of e^A is nonnegative if and only if the off diagonal components of A are all nonnegative (that is, $a_{ij} \geq 0$ whenever $i \neq j$). Hint: The 'if' direction is an easy corollary of the Trotter product formula. But, this is not the best proof. To prove both directions, consider the positive invariance of the positive orthant in n-dimensional space under the flow of the system $\dot{x} = Ax$.

Exercise 2.55. [Lie Groups and Lax Pairs] Is the map

$$\exp : \mathcal{L}(E) \to GL(E)$$

injective? Is this map surjective? Do the answers to these questions depend on the choice of E as $\mathbb{R}^n$ or $\mathbb{C}^n$? Prove that the general linear group is a submanifold of $\mathbb{R}^N$ with $N = n^2$ in case $E = \mathbb{R}^n$, and $N = 2n^2$ in case $E = \mathbb{C}^n$. Show that the general linear group is a Lie group; that is, the group operation (matrix product), is a differentiable map from $GL(E) \times GL(E) \to GL(E)$. Consider the tangent space at the identity element of $GL(E)$. Note that, for each $A \in \mathcal{L}(E)$, the map $t \mapsto \exp(tA)$ is a curve in $GL(E)$ passing through the origin at time $t = 0$. Use this fact to prove that the tangent space can be identified with $\mathcal{L}(E)$. It turns out that $\mathcal{L}(E)$ is a Lie algebra. More generally, a vector space is called a Lie algebra if for each pair of vectors A and B, a product, denoted by $[A, B]$, is defined on the vector space such that the product is bilinear and also satisfies the following algebraic identities: (skew-symmetry) $[A, B] = -[B, A]$, and (the Jacobi identity)

$$[[A, B], C] + [[B, C], A] + [[C, A], B] = 0.$$

Show that $\mathcal{L}(E)$ is a Lie algebra with respect to the product $[A, B] := AB - BA$. For an elementary introduction to the properties of these structures, see [119].

The delicate interplay between Lie groups and Lie algebras leads to a far-reaching theory. To give a flavor of the relationship between these structures, consider the map $\mathrm{Ad} : GL(E) \to \mathcal{L}(\mathcal{L}(E))$ defined by $\mathrm{Ad}(A)(B) = ABA^{-1}$. This map defines the adjoint representation of the Lie group into the automorphisms of the Lie algebra. Prove this. Also, Ad is a homomorphism of groups: $\mathrm{Ad}(AB) = \mathrm{Ad}(A)\,\mathrm{Ad}(B)$. Note that we may as well denote the automorphism group of $\mathcal{L}(E)$ by $GL(\mathcal{L}(E))$. Also, define $\mathrm{ad} : \mathcal{L}(E) \to \mathcal{L}(\mathcal{L}(E))$ by $\mathrm{ad}(A)(B) = [A, B]$. The map ad is a homomorphism of Lie algebras. Now, $\varphi_t := \mathrm{Ad}(e^{tA})$ defines a flow in $\mathcal{L}(E)$. The associated differential equation is obtained by differentiation. Show that φ_t is the flow of the differential equation

$$\dot{x} = Ax - xA = \mathrm{ad}(A)x. \tag{2.16}$$

This differential equation is linear; thus, it has the solution $t \mapsto e^{t\,\mathrm{ad}(A)}$. By the usual argument it now follows that $e^{t\,\mathrm{ad}(A)} = \mathrm{Ad}(e^{tA})$. In particular, we have the commutative diagram

$$
\begin{array}{ccc}
\mathcal{L}(E) & \xrightarrow{\ \mathrm{ad}\ } & \mathcal{L}(\mathcal{L}(E)) \\
\downarrow{\scriptstyle\exp} & & \downarrow{\scriptstyle\exp} \\
GL(E) & \xrightarrow{\ \mathrm{Ad}\ } & GL(\mathcal{L}(E)).
\end{array}
$$

The adjoint representation of $GL(E)$ is useful in the study of the subgroups of $GL(E)$, and it is also used to identify the Lie group that is associated with a given Lie algebra. But consider instead the following application to spectral theory. A curve $t \mapsto L(t)$ in $\mathcal{L}(E)$ is called *isospectral* if the spectrum of $L(t)$ is the same as the spectrum of $L(0)$ for all $t \in \mathbb{R}$. We have the following proposition: Suppose that $A \in \mathcal{L}(E)$. If $t \mapsto L(t)$ is a solution of the differential equation (2.16), then the solution is isospectral. The proof is just a restatement of the content of the commutative diagram. In fact, $L(t)$ is similar to $L(0)$ because

$$L(t) = \text{Ad}(e^{tA})L(0) = e^{tA}L(0)e^{-tA}.$$

A pair of curves $t \mapsto L(t)$ and $t \mapsto M(t)$ is called a *Lax pair* if

$$\dot{L} = LM - ML.$$

The sign convention aside, the above proposition shows that if (L, M) is a Lax pair and if M is constant, then L is isospectral. Prove the more general result: If (L, M) is a Lax pair, then L is isospectral.

Finally, prove that

$$\frac{d}{dt}\left(e^{tA}e^{tB}e^{-tA}e^{-tB}\right)\Big|_{t=0} = 0$$

and

$$\frac{d}{dt}\left(e^{\sqrt{t}A}e^{\sqrt{t}B}e^{-\sqrt{t}A}e^{-\sqrt{t}B}\right)\Big|_{t=0} = AB - BA. \qquad (2.17)$$

As mentioned above, $[A, B]$ is in the tangent space at the identity of $GL(E)$. Thus, there is a curve $\gamma(t)$ in $GL(E)$ such that $\gamma(0) = I$ and $\dot{\gamma}(0) = [A, B]$. One such curve is $t \mapsto e^{t[A,B]}$. Since the Lie bracket $[A, B]$ is an algebraic object computed from the tangent vectors A and B, it is satisfying that there is another such curve formed from the curves $t \mapsto e^{tA}$ and $t \mapsto e^{tB}$ whose respective tangent vectors at $t = 0$ are A and B.

Exercise 2.56. Prove that if α is a real number and A is an $n \times n$ real matrix such that $\langle Av, v \rangle \leq \alpha|v|^2$ for all $v \in \mathbb{R}^n$, then $\|e^{tA}\| \leq e^{\alpha t}$ for all $t \geq 0$. Hint: Consider the differential equation $\dot{x} = Ax$ and the inner product $\langle \dot{x}, x \rangle$. Prove the following more general result suggested by Weishi Liu. Suppose that $t \mapsto A(t)$ and $t \mapsto B(t)$ are smooth $n \times n$ matrix valued functions defined on $\mathbb{R}$ such that $\langle A(t)v, v \rangle \leq \alpha(t)|v|^2$ and $\langle B(t)v, v \rangle \leq 0$ for all $t \geq 0$ and all $v \in \mathbb{R}^n$. If $t \mapsto x(t)$ is a solution of the differential equation $\dot{x} = A(t)x + B(t)x$, then

$$|x(t)| \leq e^{\int_0^t \alpha(s)\,ds}|x(0)|$$

for all $t \geq 0$.

Exercise 2.57. Let $v \in \mathbb{R}^3$, assume $v \neq 0$, and consider the differential equation

$$\dot{x} = v \times x, \quad x(0) = x_0$$

where $\times$ denotes the cross product in $\mathbb{R}^3$. Show that the solution of the differential equation is a rigid rotation of the initial vector x_0 about the direction v. If the differential equation is written as a matrix system

$$\dot{x} = Sx$$

where S is a 3×3 matrix, show that S is skew symmetric and that the flow $\phi_t(x) = e^{tS}x$ of the system is a group of orthogonal transformations. Show that every solution of the system is periodic and relate the period to the length of v.

Exercise 2.58. Consider the linear system $\dot{x} = A(t)x$ where $A(t)$ is a skew-symmetric $n \times n$-matrix for each $t \in \mathbb{R}$ with respect to some inner product on $\mathbb{R}^n$, and let $|\,|$ denote the corresponding norm. Show that $|\phi(t)| = |\phi(0)|$ for every solution $t \mapsto \phi(t)$.

Exercise 2.59. [An Infinite Dimensional ODE] Let E denote the Banach space $C([0,1])$ given by the set of all continuous functions $f : [0,1] \to \mathbb{R}$ with the supremum norm

$$\|f\| = \sup_{s \in [0,1]} |f(s)|$$

and consider the operator $U : E \to E$ given by $(Uf)(s) = f(as)$ where $0 \le a \le 1$. Also, let $g \in E$ denote the function given by $s \to bs$ where b is a fixed real number. Find the solution of the initial value problem

$$\dot{x} = Ux, \qquad x(0) = g.$$

This is a simple example of an ordinary differential equation on an infinite dimensional Banach space (see Section 3.6).

Exercise 2.60. Write a report on the application of the Lie-Trotter formula to obtain numerical approximations of the solution of the initial value problem $\dot{x} = (A + B)x$, $x(0) = v$ with expressions of the form

$$T(t,n)v = (e^{(t/n)A}e^{(t/n)B})^n v.$$

For example, approximate $x(1)$ for such systems where

$$A := \begin{pmatrix} a & 0 \\ 0 & b \end{pmatrix}, \qquad B := \begin{pmatrix} c & -d \\ d & c \end{pmatrix}.$$

Compare the results of numerical experiments using your implementation(s) of the "Lie-Trotter method" and your favorite choice of alternative method(s) to compute $x(1)$. Note that e^{tA} and e^{tB} can be input explicitly for the suggested example. Can you estimate the error $|T(1,n)v - e^{A+B}v|$? Generalizations of this scheme are sometimes used to approximate differential equations where the "vector field" can be split into two easily solved summands. Try the same idea to solve nonlinear ODE of the form $\dot{x} = f(x) + g(x)$ where e^{tA} is replaced by the flow of $\dot{x} = f(x)$ and e^{tB} is replaced by the flow of $\dot{x} = g(x)$.

2.2   Stability of Linear Systems

A linear homogeneous differential equation has a rest point at the origin. We will use our results about the solutions of constant coefficient homogeneous linear differential equations to study the stability of this rest point. The next result is fundamental.

Theorem 2.61. *Suppose that A is an $n \times n$ (real) matrix. The following statements are equivalent:*

(1) There is a norm $||_a$ on $\mathbb{R}^n$ and a real number $\lambda > 0$ such that for all $v \in \mathbb{R}^n$ and all $t \geq 0$,

$$|e^{tA}v|_a \leq e^{-\lambda t}|v|_a.$$

(2) If $||_g$ is an arbitrary norm on $\mathbb{R}^n$, then there is a constant $C > 0$ and a real number $\lambda > 0$ such that for all $v \in \mathbb{R}^n$ and all $t \geq 0$,

$$|e^{tA}v|_g \leq Ce^{-\lambda t}|v|_g.$$

(3) Every eigenvalue of A has negative real part.

Moreover, if $-\lambda$ exceeds the largest of all the real parts of the eigenvalues of A, then λ can be taken to be the decay constant in (1) or (2).

Corollary 2.62. *If every eigenvalue of A has negative real part, then the zero solution of $\dot{x} = Ax$ is asymptotically stable.*

Proof. We will show that $(1) \Rightarrow (2) \Rightarrow (3) \Rightarrow (1)$.

To show $(1) \Rightarrow (2)$, let $||_a$ be the norm in statement (1) and $||_g$ the norm in statement (2). Because these norms are defined on the finite dimensional vector space $\mathbb{R}^n$, they are equivalent; that is, there are constants $K_1 > 0$ and $K_2 > 0$ such that for all $x \in \mathbb{R}^n$ we have

$$K_1|x|_g \leq |x|_a \leq K_2|x|_g.$$

(Prove this!) Hence, if $t \geq 0$ and $v \in \mathbb{R}^n$, then

$$|e^{tA}v|_g \leq \frac{1}{K_1}|e^{tA}v|_a \leq \frac{1}{K_1}e^{-\lambda t}|v|_a \leq \frac{K_2}{K_1}e^{-\lambda t}|v|_g.$$

To show $(2) \Rightarrow (3)$, suppose that statement (2) holds but statement (3) does not. In particular, A has an eigenvalue $\mu \in \mathbb{C}$, say $\mu = \alpha + i\beta$ with $\alpha \geq 0$. Moreover, there is at least one eigenvector $v \neq 0$ corresponding to this eigenvalue. By Proposition 2.24, the system $\dot{x} = Ax$ has a solution $t \mapsto \gamma(t)$ of the form $t \to e^{\alpha t}((\cos \beta t)u - (\sin \beta t)w)$ where $v = u + iw$, $u \in \mathbb{R}^n$ and $w \in \mathbb{R}^n$. By inspection, $\lim_{t \to \infty} \gamma(t) \neq 0$. But if statement (2) holds, then $\lim_{t \to \infty} \gamma(t) = 0$, in contradiction.

To finish the proof we will show $(3) \Rightarrow (1)$. Let us assume that statement (3) holds. Since A has a finite set of eigenvalues and each of its eigenvalues has negative real part, there is a number $\lambda > 0$ such that the real part of each eigenvalue of A is less than $-\lambda$.

By Proposition 2.37, the components of e^{tA} are finite sums of terms of the form $p(t)e^{\alpha t}\sin \beta t$ or $p(t)e^{\alpha t}\cos \beta t$ where α is the real part of an eigenvalue of A and $p(t)$ is a polynomial of degree at most $n - 1$. In particular, if the matrix e^{tA}, partitioned by columns, is given by $[c_1(t), \ldots, c_n(t)]$, then each component of each vector $c_i(t)$ is a sum of such terms.

Let us denote the usual norm of a vector $v = (v_1, \ldots, v_n)$ in $\mathbb{R}^n$ by $|v|$. Also, $|v_i|$ is the absolute value of the real number v_i, or (if you like) the norm of the vector $v_i \in \mathbb{R}$. With this notation we have

$$|e^{tA}v| \leq \sum_{i=1}^{n} |c_i(t)||v_i|.$$

Because

$$|v_i| \leq \left(\sum_{j=1}^{n} |v_j|^2\right)^{1/2} = |v|,$$

it follows that

$$|e^{tA}v| \leq |v| \sum_{i=1}^{n} |c_i(t)|.$$

If $\beta_1, \ldots, \beta_\ell$ are the nonzero imaginary parts of the eigenvalues of A and if α denotes the largest real part of an eigenvalue of A, then using the structure of the components of the vector $c_i(t)$ it follows that

$$|c_i(t)|^2 \leq e^{2\alpha t} \sum_{k=0}^{2n-2} |d_{ki}(t)||t|^k$$

where each coefficient $d_{ki}(t)$ is a quadratic form in

$$\sin \beta_1 t, \ldots, \sin \beta_\ell t, \cos \beta_1 t, \ldots, \cos \beta_\ell t.$$

There is a constant $M > 0$ that does not depend on i or k such that the supremum of $|d_{ki}(t)|$ for $t \in \mathbb{R}$ does not exceed M^2. In particular, for each $i = 1, \ldots, n$, we have

$$|c_i(t)|^2 \leq e^{2\alpha t} M^2 \sum_{k=0}^{2n-2} |t|^k,$$

and as a result

$$|e^{tA}v| \leq |v| \sum_{i=1}^{n} |c_i(t)| \leq e^{\alpha t} nM|v| \left(\sum_{k=0}^{2n-2} |t|^k\right)^{1/2}.$$

Because $\alpha < -\lambda < 0$, there is some $\tau > 0$ such that for $t \geq \tau$, we have the inequality

$$e^{(\lambda+\alpha)t} nM \left(\sum_{k=0}^{2n-2} |t|^k\right)^{1/2} \leq 1,$$

or equivalently

$$e^{\alpha t} nM \left(\sum_{k=0}^{2n-2} |t|^k\right)^{1/2} \leq e^{-\lambda t}.$$

In particular, if $t \geq \tau$, then for each $v \in \mathbb{R}^n$ we have

$$|e^{tA}v| \leq e^{-\lambda t}|v|. \tag{2.18}$$

To finish the proof, we will construct a new norm for which the same inequality is valid for all $t \geq 0$. In fact, we will prove that

$$|v|_a := \int_0^\tau e^{\lambda s}|e^{sA}v|\, ds$$

is the required norm.

The easy proof required to show that $|\ |_a$ is a norm on $\mathbb{R}^n$ is left to the reader. To obtain the norm estimate, note that for each $t \geq 0$ there is a nonnegative integer m and a number T such that $0 \leq T < \tau$ and $t = m\tau + T$. Using this decomposition of t, we find that

$$
\begin{aligned}
|e^{tA}v|_a &= \int_0^\tau e^{\lambda s}|e^{sA}e^{tA}v|\, ds \\
&= \int_0^{\tau-T} e^{\lambda s}|e^{(s+t)A}v|\, ds + \int_{\tau-T}^\tau e^{\lambda s}|e^{(s+t)A}|\, ds \\
&= \int_0^{\tau-T} e^{\lambda s}|e^{m\tau A}e^{(s+T)A}v|\, ds \\
&\quad + \int_{\tau-T}^\tau e^{\lambda s}|e^{(m+1)\tau A}e^{(T-\tau+s)A}v|\, ds.
\end{aligned}
$$

Let $u = T + s$ in the first integral, let $u = T - \tau + s$ in the second integral, use the inequality (2.18), and, for $m = 0$, use the inequality $|e^{m\tau A}e^{uA}v| \leq e^{-\lambda m\tau}|v|$, to obtain the estimates

$$
\begin{aligned}
|e^{tA}v|_a &= \int_T^\tau e^{\lambda(u-T)}|e^{(m\tau+u)A}v|\, du + \int_0^T e^{\lambda(u+\tau-T)}|e^{((m+1)\tau+u)A}v|\, du \\
&\leq \int_T^\tau e^{\lambda(u-T)}e^{-\lambda(m\tau)}|e^{uA}v|\, du \\
&\quad + \int_0^T e^{\lambda(u+\tau-T)}e^{-\lambda(m+1)\tau}|e^{uA}v|\, du \\
&\leq \int_0^\tau e^{\lambda u}e^{-\lambda(m\tau+T)}|e^{uA}v|\, du \\
&= e^{-\lambda t}\int_0^\tau e^{\lambda u}|e^{uA}v|\, du \\
&\leq e^{-\lambda t}|v|_a,
\end{aligned}
$$

as required.   □

Recall that a matrix is *infinitesimally hyperbolic* if all of its eigenvalues have nonzero real parts. The following corollary of Theorem 2.61 is the basic result about the dynamics of hyperbolic linear systems.

Corollary 2.63. *If A is an $n \times n$ (real) infinitesimally hyperbolic matrix, then there are two A-invariant subspaces E^s and E^u of $\mathbb{R}^n$ such that $\mathbb{R}^n = E^s \oplus E^u$. Moreover, if $|\ \ |_g$ is a norm on $\mathbb{R}^n$, then there are constants $\lambda > 0$, $\mu > 0$, $C > 0$, and $K > 0$ such that for all $v \in E^s$ and all $t \geq 0$*

$$|e^{tA}v|_g \leq Ce^{-\lambda t}|v|_g,$$

and for all $v \in E^u$ and all $t \leq 0$

$$|e^{tA}v|_g \leq Ke^{\mu t}|v|_g.$$

Also, there exists a norm on $\mathbb{R}^n$ such that the above inequalities hold for $C = K = 1$ and $\lambda = \mu$.

Proof. The details of the proof are left as an exercise. But let us note that if A is infinitesimally hyperbolic, then we can arrange for the Jordan form J of A to be a block matrix

$$J = \begin{pmatrix} A_s & 0 \\ 0 & A_u \end{pmatrix}$$

where the eigenvalues of A_s all have negative real parts and the eigenvalues of A_u have positive real parts. Thus, there is an obvious J-invariant splitting of the vector space $\mathbb{R}^n$ into a stable space and an unstable space. By changing back to the original coordinates, it follows that there is a corresponding A-invariant splitting. The hyperbolic estimate on the stable space follows from Theorem 2.61 applied to the restriction of A to its stable subspace; the estimate on the unstable space follows from Theorem 2.61 applied to the restriction of $-A$ to the unstable subspace of A. Finally, an adapted norm on the entire space is obtained as follows:

$$|(v_s, v_u)|_a^2 = |v_s|_a^2 + |v_u|_a^2. \qquad \square$$

The basic result of this section—if all eigenvalues of the matrix A are in the left half plane, then the zero solution of the corresponding homogeneous system is asymptotically stable—is a special case of the principle of linearized stability. This result provides a method to determine the stability of the zero solution that does not require knowing other solutions of the system. As we will see, the same idea works in more general contexts. But, additional hypotheses are required for most generalizations.

Exercise 2.64. Find E^s, E^u, C, K, λ, and μ as in Corollary 2.63 (relative to the usual norm) for the matrix

$$A := \begin{pmatrix} 2 & 1 \\ 0 & -3 \end{pmatrix}.$$

Exercise 2.65. As a continuation of Exercise 2.56, suppose that A is an $n \times n$ matrix and that there is a number $\lambda > 0$ such that every eigenvalue of A has real part less than $-\lambda$. Prove that there is an inner product and associated norm such that $\langle Ax, x \rangle \leq -\lambda |x|^2$ for all $x \in \mathbb{R}^n$ and conclude that $|e^{tA}x| \leq e^{-\lambda t}|x|$. This gives an alternative method of constructing an adapted norm (see [123, p. 146]). Show that there is a constant $C > 0$ such that $|e^{tA}x| \leq Ce^{-\lambda t}|x|$ with respect to the usual norm. Moreover, show that there is a constant $k > 0$ such that if B is an $n \times n$ matrix, then $|e^{tB}x| \leq Ce^{k\|B-A\|-\lambda t}|x|$. In particular, if $\|B - A\|$ is sufficiency small, then there is some $\mu > 0$ such that $|e^{tB}x| \leq Ce^{-\mu t}|x|$.

Exercise 2.66. Suppose that A and B are $n \times n$-matrices and all the eigenvalues of B are positive real numbers. Also, let B^T denote the transpose of B. Show that there is a value μ_* of the parameter μ such that the rest point at the origin of the system

$$\dot{X} = AX - \mu X B^T$$

is asymptotically stable whenever $\mu > \mu_*$. Hint: X is a matrix valued variable. Show that the eigenvalues of the linear operator $X \mapsto AX - XB^T$ are given by differences of the eigenvalues of A and B. Prove this first in case A and B are diagonalizable and then use the density of the diagonalizable matrices (cf. [207, p. 331]).

2.3   Stability of Nonlinear Systems

Theorem 2.61 states that the zero solution of a constant coefficient homogeneous linear system is asymptotically stable if the spectrum of the coefficient matrix lies in the left half of the complex plane. The principle of linearized stability states that the same result is true for steady state solutions of nonlinear equations provided that the system matrix of the linearized system along the steady state solution has its spectrum in the left half plane. As stated, this principle is not a theorem. In this section, however, we will formulate and prove a theorem on linearized stability which is strong enough for most applications. In particular, we will prove that a rest point of an autonomous differential equation $\dot{x} = f(x)$ in $\mathbb{R}^n$ is asymptotically stable if all eigenvalues of the Jacobian matrix at the rest point have negative real parts. Our stability result is also valid for some nonhomogeneous nonautonomous differential equations of the form

$$\dot{x} = A(t)x + g(x, t), \qquad x \in \mathbb{R}^n \tag{2.19}$$

where $g : \mathbb{R}^n \times \mathbb{R} \to \mathbb{R}^n$ is a smooth function.

A fundamental tool used in our stability analysis is the formula, called the *variation of parameters formula,* given in the next proposition.

Proposition 2.67 (Variation of Parameters Formula). *Consider the initial value problem*

$$\dot{x} = A(t)x + g(x, t), \qquad x(t_0) = x_0 \tag{2.20}$$

and let $t \mapsto \Phi(t)$ be a fundamental matrix solution for the homogeneous system $\dot{x} = A(t)x$ that is defined on some interval J_0 containing t_0. If $t \mapsto \phi(t)$ is the solution of the initial value problem defined on some subinterval of J_0, then

$$\phi(t) = \Phi(t)\Phi^{-1}(t_0)x_0 + \Phi(t) \int_{t_0}^{t} \Phi^{-1}(s)g(\phi(s), s)\, ds. \tag{2.21}$$

Proof. Define a new function z by $z(t) = \Phi^{-1}(t)\phi(t)$. We have

$$\dot{\phi}(t) = A(t)\Phi(t)z(t) + \Phi(t)\dot{z}(t).$$

Thus,

$$A(t)\phi(t) + g(\phi(t), t) = A(t)\phi(t) + \Phi(t)\dot{z}(t)$$

and

$$\dot{z}(t) = \Phi^{-1}(t)g(\phi(t), t).$$

Also note that $z(t_0) = \Phi^{-1}(t_0)x_0$.

By integration,

$$z(t) - z(t_0) = \int_{t_0}^{t} \Phi^{-1}(s)g(\phi(s), s)\, ds,$$

or, in other words,

$$\phi(t) = \Phi(t)\Phi^{-1}(t_0)x_0 + \Phi(t) \int_{t_0}^{t} \Phi^{-1}(s)g(\phi(s), s)\, ds. \qquad \square$$

Let us note that in the special case where the function g in the differential equation (2.20) is a constant with respect to its first variable, the variation of parameters formula solves the initial value problem once a fundamental matrix solution of the associated homogeneous system is determined.

Exercise 2.68. Consider the linear system

$$\dot{u} = -\delta^2 u + v + \delta w, \quad \dot{v} = -u - \delta^2 v + \delta w, \quad \dot{w} = -\delta w$$

where δ is a parameter. Find the general solution of this system using matrix algebra and also by using the substitution $z = u + iv$. Describe the phase portrait for the system for each value of δ. Find an invariant line and determine the rate of change with respect to δ of the angle this line makes with the positive w-axis. Also, find the angular velocity of the "twist" around the invariant line.

Exercise 2.69. (a) Use variation of parameters to solve the system

$$\dot{x} = x - y + e^{-t}, \qquad \dot{y} = x + y + e^{-t}.$$

(b) Find the set of initial conditions at $t = 0$ so that $\lim_{t \to \infty}(x(t), y(t)) = (0, 0)$ whenever $t \mapsto (x(t), y(t))$ satisfies one of these initial conditions.

Exercise 2.70. Suppose that $g : \mathbb{R}^n \to \mathbb{R}^n$ is smooth and consider the family of solutions $t \mapsto \phi(t, \xi, \epsilon)$ of the family of differential equations

$$\dot{x} = Ax + \epsilon x + \epsilon^2 g(x)$$

with parameter ϵ such that $\phi(0, \xi, \epsilon) = \xi$. Compute the derivative $\phi_\epsilon(1, \xi, 0)$. Hint: Solve an appropriate variational equation using variation of parameters.

Exercise 2.71. The product $\Phi(t)\Phi^{-1}(s)$ appears in the variation of parameters formula where $\Phi(t)$ is the principal fundamental matrix for the system $\dot{x} = A(t)x$. Show that if A is a constant matrix or A is 1×1, then $\Phi(t)\Phi^{-1}(s) = \Phi(t - s)$. Prove that this formula does *not* hold in general for homogeneous linear systems.

Exercise 2.72. Give an alternative proof of Proposition 2.67 by verifying directly that the variation of parameters formula (2.21) is a solution of the initial value problem (2.20)

Exercise 2.73. Suppose that A is an $n \times n$-matrix all of whose eigenvalues have negative real parts. (a) Find a (smooth) function $f : \mathbb{R} \to \mathbb{R}$ so that a solution of the scalar equation $\dot{x} = -x + f(t)$ is not bounded for $t \geq 0$. (b) Show that there is a (smooth) function $f : \mathbb{R} \to \mathbb{R}^n$ so that a solution of the system $\dot{x} = Ax + f(t)$ is not bounded for $t \geq 0$. (c) Show that if the system $\dot{x} = Ax + f(t)$ does have a bounded solution, then all solutions are bounded.

Exercise 2.74. [Nonlinear Variation of Parameters] Consider the differential equations $\dot{y} = F(y, t)$ and $\dot{x} = f(t, x)$ and let $t \mapsto y(t, \tau, \xi)$ and $t \mapsto x(t, \tau, \xi)$ be the corresponding solutions such that $y(\tau, \tau, \xi) = \xi$ and $x(\tau, \tau, \xi) = \xi$. (a) Prove the nonlinear variation of parameters formula

$$x(t, \tau, \xi) = y(t, \tau, \xi) + \int_\tau^t [y_\tau(t, s, x(s, \tau, \xi)) + y_\xi(t, s, x(s, \tau, \xi))f(s, x(s, \tau, \xi))] \, ds.$$

Hint: Define $z(s) = y(t, s, x(s, \tau, \xi))$, differentiate z with respect to s, integrate the resulting formula over the interval $[\tau, t]$, and note that $z(t) = x(t, \tau, \xi)$ and $z(\tau) = y(t, \tau, \xi)$. (b) Derive the variation of parameters formula from the nonlinear variation of parameters formula. Hint: Consider $\dot{y} = A(t)y$ and $\dot{x} = A(t)x + h(t, x)$. Also, let $\Phi(t)$ denote a fundamental matrix for $\dot{y} = A(t)y$ and note that $d/dt \Phi^{-1}(t) = -\Phi^{-1}(t)A(t)$. (c) Consider the differential equation $\dot{x} = -x^3$ and prove that $x(t, \xi)$ (the solution such that $x(0, \xi) = \xi$) is $O(1/\sqrt{t})$ as $t \to \infty$; that is, there is a constant $C > 0$ such that $|x(t, \xi)| \leq C/\sqrt{t}$ as $t \to \infty$. Next suppose that M and δ are positive constants and $g : \mathbb{R} \to \mathbb{R}$ is such that $|g(x)| \leq Mx^4$ whenever $|x| < \delta$. Prove that if $t \mapsto x(t, \xi)$ is the solution of the differential equation $\dot{x} = -x^3 + g(x)$ such that $x(0, \xi) = \xi$ and $|\xi|$ is sufficiently small, then $|x(t, \xi)| \leq C/\sqrt{t}$. Hint: First show that the origin is asymptotically stable using a Lyapunov function. Write out the nonlinear variation of parameters formula, make an estimate, and use Gronwall's inequality.

The next proposition states an important continuity result for the solutions of nonautonomous systems with respect to initial conditions. To prove it, we will use the following lemma.

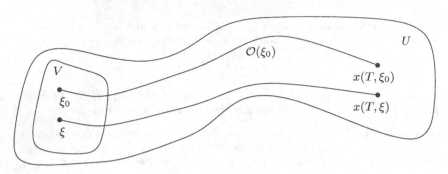

Figure 2.1: Local stability as in Proposition 2.76. For every open set U containing the orbit segment $\mathcal{O}(\xi_0)$, there is an open set V containing ξ_0 such that orbits starting in V stay in U on the time interval $0 \le t \le T$.

Lemma 2.75. *Consider a smooth function $f : \mathbb{R}^n \times \mathbb{R} \to \mathbb{R}^n$. If $K \subseteq \mathbb{R}^n$ and $A \subseteq \mathbb{R}$ are compact sets, then there is a number $L > 0$ such that*

$$|f(x,t) - f(y,t)| \le L|x - y|$$

for all $(x,t),(y,t) \in K \times A$.

Proof. The proof of the lemma uses compactness, continuity, and the mean value theorem. The details are left as an exercise. $\square$

Recall that a function f as in the lemma is called Lipschitz with respect to its first argument on $K \times A$ with Lipschitz constant L.

Proposition 2.76. *Consider, for each $\xi \in \mathbb{R}^n$, the solution $t \mapsto \phi(t, \xi)$ of the differential equation $\dot{x} = f(x,t)$ such that $\phi(0, \xi) = \xi$. If $\xi_0 \in \mathbb{R}^n$ is such that the solution $t \mapsto \phi(t, \xi_0)$ is defined for $0 \le t \le T$, and if $U \subseteq \mathbb{R}^n$ is an open set containing the orbit segment $\mathcal{O}(\xi_0) = \{\phi(t, \xi_0) : 0 \le t \le T\}$, then there is an open set $V \subseteq U$, as in Figure 2.1, such that $\xi_0 \in V$ and $\{\phi(t, \xi) : \xi \in V, \ 0 \le t \le T\} \subseteq U$; that is, the solution starting at each $\xi \in V$ exists on the interval $[0, T]$, and its values on this interval are in U.*

Proof. Let $\xi \in \mathbb{R}^n$, and consider the two solutions of the differential equation given by $t \mapsto \phi(t, \xi_0)$ and $t \mapsto \phi(t, \xi)$. For t in the intersection of the intervals of existence of these solutions, we have that

$$\phi(t, \xi) - \phi(t, \xi_0) = \xi - \xi_0 + \int_0^t f(\phi(s, \xi), s) - f(\phi(s, \xi_0), s)\, ds$$

and

$$|\phi(t, \xi) - \phi(t, \xi_0)| \le |\xi - \xi_0| + \int_0^t |f(\phi(s, \xi), s) - f(\phi(s, \xi_0), s)|\, ds.$$

We can assume without loss of generality that U is bounded, hence its closure is compact. It follows from the lemma that the smooth function f is Lipschitz on $U \times [0, T]$ with a Lipschitz constant $L > 0$. Thus, as long as $(\phi(t, \xi), t) \in U \times [0, T]$, we have

$$|\phi(t, \xi) - \phi(t, \xi_0)| \le |\xi - \xi_0| + \int_0^t L|\phi(s, \xi) - \phi(s, \xi_0)| \, ds$$

and by Gronwall's inequality

$$|\phi(t, \xi) - \phi(t, \xi_0)| \le |\xi - \xi_0| e^{Lt}.$$

Let $\delta > 0$ be such that δe^{LT} is less than the distance from $\mathcal{O}(\xi_0)$ to the boundary of U. Since, on the intersection J of the domain of definition of the solution $t \mapsto \phi(t, \xi)$ with $[0, T]$ we have

$$|\phi(t, \xi) - \phi(t, \xi_0)| \le |\xi - \xi_0| e^{LT},$$

the vector $\phi(t, \xi)$ is in the bounded set U as long as $t \in J$ and $|\xi - \xi_0| < \delta$. By the extension theorem, the solution $t \mapsto \phi(t, \xi)$ is defined at least on the interval $[0, T]$. Thus, the desired set V is $\{\xi \in U : |\xi - \xi_0| < \delta\}$.   □

We are now ready to formulate a theoretical foundation for Lyapunov's indirect method, that is, the method of linearization. The idea should be familiar: If the system has a rest point at the origin, the linearization of the system has an asymptotically stable rest point at the origin, and the nonlinear part is appropriately bounded, then the nonlinear system also has an asymptotically stable rest point at the origin.

Theorem 2.77. *Consider the initial value problem (2.20) for the case where $A := A(t)$ is a (real) matrix of constants. If all eigenvalues of A have negative real parts and there are positive constants $a > 0$ and $k > 0$ such that $|g(x, t)| \le k|x|^2$ whenever $|x| < a$, then there are positive constants C, b, and α that are independent of the choice of the initial time t_0 such that the solution $t \mapsto \phi(t)$ of the initial value problem satisfies*

$$|\phi(t)| \le C|x_0| e^{-\alpha(t - t_0)} \qquad (2.22)$$

for $t \ge t_0$ whenever $|x_0| \le b$. In particular, the function $t \mapsto \phi(t)$ is defined for all $t \ge t_0$, and the zero solution (the solution with initial value $\phi(t_0) = 0$), is asymptotically stable.

Proof. By Theorem 2.61 and the hypothesis on the eigenvalues of A, there are constants $C > 1$ and $\lambda > 0$ such that

$$\|e^{tA}\| \le C e^{-\lambda t} \qquad (2.23)$$

for $t \ge 0$. Fix $\delta > 0$ such that $\delta < a$ and $Ck\delta - \lambda < 0$, define $\alpha := \lambda - Ck\delta$ and $b := \delta/C$, and note that $\alpha > 0$ and $0 < b < \delta < a$.

If $|x_0| < b$, then there is a maximal half-open interval $J = \{t \in \mathbb{R} : t_0 \leq t < \tau\}$ such that the solution $t \to \phi(t)$ of the differential equation with initial condition $\phi(t_0) = x_0$ exists and satisfies the inequality

$$|\phi(t)| < \delta \tag{2.24}$$

on the interval J.

For $t \in J$, use the estimate

$$|g(\phi(t), t)| \leq k\delta|\phi(t)|,$$

the estimate (2.23), and the variation of parameters formula

$$\phi(t) = e^{(t-t_0)A}x_0 + e^{tA}\int_{t_0}^t e^{-sA}g(\phi(s), s)\, ds$$

to obtain the inequality

$$|\phi(t)| \leq Ce^{-\lambda(t-t_0)}|x_0| + \int_{t_0}^t Ce^{-\lambda(t-s)}k\delta|\phi(s)|\, ds.$$

Rearrange the inequality to the form

$$e^{\lambda(t-t_0)}|\phi(t)| \leq C|x_0| + Ck\delta\int_{t_0}^t e^{\lambda(s-t_0)}|\phi(s)|\, ds$$

and apply Gronwall's inequality to obtain the estimate

$$e^{\lambda(t-t_0)}|\phi(t)| \leq C|x_0|e^{Ck\delta(t-t_0)};$$

or equivalently

$$|\phi(t)| \leq C|x_0|e^{(Ck\delta-\lambda)(t-t_0)} \leq C|x_0|e^{-\alpha(t-t_0)}. \tag{2.25}$$

Thus, if $|x_0| < b$ and $|\phi(t)| < \delta$ for $t \in J$, then the required inequality (2.22) is satisfied for $t \in J$.

If J is not the interval $[t_0, \infty)$, then $\tau < \infty$. Because $|x_0| < \delta/C$ and in view of the inequality (2.25), we have that

$$|\phi(t)| < \delta e^{-\alpha(t-t_0)} \tag{2.26}$$

for $t_0 \leq t < \tau$. In particular, the solution is bounded by δ on the interval $[t_0, \tau)$. Therefore, by the extension theorem there is some number $\epsilon > 0$ such that the solution is defined on the interval $K := [t_0, \tau + \epsilon)$. Using the continuity of the function $t \mapsto |\phi(t)|$ on K and the inequality (2.26), it follows that

$$|\phi\tau)| \leq \delta e^{-\alpha(\tau-t_0)} < \delta.$$

By using this inequality and again using the continuity of the function $t \mapsto |\phi(t)|$ on K, there is a number $\eta > 0$ such that $t \mapsto \phi(t)$ is defined on the interval $[t_0, \tau + \eta)$, and, on this interval, $|\phi(t)| < \delta$. This contradicts the maximality of τ.   □

Corollary 2.78. *If $f : \mathbb{R}^n \to \mathbb{R}^n$ is smooth (at least class C^2), $f(\xi) = 0$, and all eigenvalues of $Df(\xi)$ have negative real parts, then the differential equation $\dot{x} = f(x)$ has an asymptotically stable rest point at ξ. Moreover, if $-\alpha$ is a number larger than every real part of an eigenvalue of $Df(\xi)$, and ϕ_t is the flow of the differential equation, then there is a neighborhood U of ξ and a constant $C > 0$ such that*

$$|\phi_t(x) - \xi| \le C|x - \xi|e^{-\alpha t}$$

whenever $x \in U$ and $t \ge 0$.

Proof. It suffices to prove the corollary for the case $\xi = 0$. By Taylor's theorem (Theorem 1.237), we can rewrite the differential equation in the form $\dot{x} = Df(0)x + g(x)$ where

$$g(x) := \int_0^1 (Df(sx) - Df(0))x \, ds.$$

The function $\xi \mapsto Df(\xi)$ is class C^1. Thus, by the mean value theorem (Theorem 1.53),

$$\|Df(sx) - Df(0)\| \le |sx| \sup_{\tau \in [0,1]} \|D^2 f(\tau s x)\|$$

$$\le |x| \sup_{\tau \in [0,1]} \|D^2 f(\tau x)\|.$$

Again, by the smoothness of f, there is an open ball B centered at the origin and a constant $k > 0$ such that

$$\sup_{\tau \in [0,1]} \|D^2 f(\tau x)\| < k$$

for all $x \in B$. Moreover, by an application of Proposition 1.235 and the above estimates we have that

$$|g(x)| \le \sup_{s \in [0,1]} |x| \|Df(sx) - Df(0)\| \le k|x|^2$$

whenever $x \in B$. The desired result now follows directly from Theorem 2.77.

$\square$

Exercise 2.79. Generalize the previous result to the Poincaré–Lyapunov Theorem: Let

$$\dot{x} = Ax + B(t)x + g(x,t), \quad x(t_0) = x_0, \quad x \in \mathbb{R}^n$$

be a smooth initial value problem. If

(1) A is a constant matrix with spectrum in the left half plane,

(2) $B(t)$ is the $n \times n$ matrix, continuously dependent on t such that $\|B(t)\| \to 0$ as $t \to \infty$,

(3) $g(x, t)$ is smooth and there are constants $a > 0$ and $k > 0$ such that

$$|g(x, t)| \leq k|x|^2$$

for all $t \geq 0$ and $|x| < a$,

then there are constants $C > 1$, $\delta > 0$, $\lambda > 0$ such that

$$|x(t)| \leq C|x_0|e^{-\lambda(t-t_0)}, \qquad t \geq t_0$$

whenever $|x_0| \leq \delta/C$. In particular, the zero solution is asymptotically stable.

Exercise 2.80. This exercise gives an alternative proof of the principle of linearized stability for autonomous systems using Lyapunov's direct method. (a) Consider the system

$$\dot{x} = Ax + g(x), \qquad x \in \mathbb{R}^n$$

where A is a real $n \times n$ matrix and $g : \mathbb{R}^n \to \mathbb{R}^n$ is a smooth function. Suppose that every eigenvalue of A has negative real part, and that for some $a > 0$, there is a constant $k > 0$ such that, using the usual norm on $\mathbb{R}^n$,

$$|g(x)| \leq k|x|^2$$

whenever $|x| < a$. Prove that the origin is an asymptotically stable rest point by constructing a quadratic Lyapunov function. Hint: Let $\langle \cdot, \cdot \rangle$ denote the usual inner product on $\mathbb{R}^n$, and let A^* denote the transpose of the real matrix A. Suppose that there is a real symmetric positive definite $n \times n$ matrix that also satisfies *Lyapunov's equation*

$$A^*B + BA = -I$$

and define $V : \mathbb{R}^n \to \mathbb{R}$ by

$$V(x) = \langle x, Bx \rangle.$$

Show that the restriction of V to a sufficiently small neighborhood of the origin is a strict Lyapunov function. To do this, you will have to estimate a certain inner product using the Schwarz inequality. Finish the proof by showing that

$$B := \int_0^\infty e^{tA^*} e^{tA} \, dt$$

is a symmetric positive definite $n \times n$ matrix which satisfies Lyapunov's equation. To do this, prove that A^* and A have the same eigenvalues. Then use the exponential estimates for hyperbolic linear systems to prove that the integral converges. (b) Give an alternative method to compute solutions of Lyapunov's equation using the following outline: Show that Lyapunov's equation in the form $A^*B + BA = S$, where A is diagonal, S is symmetric and positive definite, and all pairs of eigenvalues of A have nonzero sums, has a symmetric positive-definite solution B. In particular, under these hypotheses, the operator $B \mapsto A^*B + BA$ is invertible. Show that the same result is true without the hypothesis that A

is diagonal. Hint: Use the density of the diagonalizable matrices and the continuity of the eigenvalues of a matrix with respect to its components (see Exercises 2.66 and 8.1). (c) Prove that the origin is asymptotically stable for the system $\dot{x} = Ax + g(x)$ where

$$A := \begin{pmatrix} -1 & 2 & 0 \\ -2 & -1 & 0 \\ 0 & 0 & -3 \end{pmatrix}, \qquad g(u,v,w) := \begin{pmatrix} u^2 + uv + v^2 + wv^2 \\ w^2 + uvw \\ w^3 \end{pmatrix}$$

and construct the corresponding matrix B that solves Lyapunov's equation.

Exercise 2.81. Suppose that $f : \mathbb{R}^n \to \mathbb{R}^n$ is *conservative;* that is, there is some function $g : \mathbb{R}^n \to \mathbb{R}$ such that $f(x) = \operatorname{grad} g(x)$. Also, suppose that M and Λ are symmetric positive definite $n \times n$ matrices. Consider the differential equation

$$M\ddot{x} + \Lambda\dot{x} + f(x) = 0, \qquad x \in \mathbb{R}^n$$

and note that, in case M and Λ are diagonal, the differential equation can be viewed as a model of n particles each moving according to Newton's second law in a conservative force field with viscous damping. (a) Prove that the function $V : \mathbb{R}^n \to \mathbb{R}$ defined by

$$V(x,y) := \frac{1}{2}\langle My, y\rangle + \int_0^1 \langle f(sx), x\rangle\, ds$$

decreases along orbits of the associated first-order system

$$\dot{x} = y, \qquad M\dot{y} = -\Lambda y - f(x);$$

in fact, $\dot{V} = -\langle \Lambda y, y\rangle$. Conclude that the system has no periodic orbits. (b) Prove that if $f(0) = 0$ and $Df(0)$ is positive definite, then the system has an asymptotically stable rest point at the origin. Prove this fact in two ways: using the function V and by the method of linearization. Hint: To use the function V see Exercise 1.171. To use the method of linearization, note that M is invertible, compute the system matrix for the linearization in block form, suppose there is an eigenvalue λ, and look for a corresponding eigenvector in block form, that is the transpose of a vector (x, y). This leads to two equations corresponding to the block components corresponding to x and y. Reduce to one equation for x and then take the inner product with respect to x.

2.4   Floquet Theory

We will study linear systems of the form

$$\dot{x} = A(t)x, \qquad x \in \mathbb{R}^n \tag{2.27}$$

where $t \to A(t)$ is a T-periodic continuous matrix-valued function. The main theorem in this section, Floquet's theorem, gives a canonical form for fundamental matrix solutions. This result will be used to show that there is a periodic time-dependent change of coordinates that transforms system (2.27) into a homogeneous linear system with constant coefficients.

Floquet's theorem is a corollary of the following result about the range of the exponential map.

Theorem 2.82. *If C is a nonsingular $n \times n$ matrix, then there is an $n \times n$ matrix B (which may be complex) such that $e^B = C$. If C is a nonsingular real $n \times n$ matrix, then there is a real $n \times n$ matrix B such that $e^B = C^2$.*

Proof. If S is a nonsingular $n \times n$ matrix such that $S^{-1}CS = J$ is in Jordan canonical form, and if $e^K = J$, then $Se^K S^{-1} = C$. As a result, $e^{SKS^{-1}} = C$ and $B = SKS^{-1}$ is the desired matrix. Thus, it suffices to consider the nonsingular matrix C or C^2 to be a Jordan block.

For the first statement of the theorem, assume that $C = \lambda I + N$ where N is nilpotent; that is, $N^m = 0$ for some integer m with $0 \leq m < n$. Because C is nonsingular, $\lambda \neq 0$ and we can write $C = \lambda(I + (1/\lambda)N)$. A computation using the series representation of the function $t \mapsto \ln(1+t)$ at $t = 0$ shows that, formally (that is, without regard to the convergence of the series), if $B = (\ln \lambda)I + M$ where

$$M = \sum_{j=1}^{m-1} \frac{(-1)^{j+1}}{j\lambda^j} N^j,$$

then $e^B = C$. But because N is nilpotent, the series are finite. Thus, the formal series identity is an identity. This proves the first statement of the theorem.

The Jordan blocks of C^2 correspond to the Jordan blocks of C. The blocks of C^2 corresponding to real eigenvalues of C are all of the type $rI + N$ where $r > 0$ and N is real nilpotent. For a real matrix C all the complex eigenvalues with nonzero imaginary parts occur in complex conjugate pairs; therefore, the corresponding real Jordan blocks of C^2 are block diagonal or "block diagonal plus block nilpotent" with 2×2 diagonal subblocks of the form

$$\begin{pmatrix} \alpha & -\beta \\ \beta & \alpha \end{pmatrix}$$

as in equation (2.10). Some of the corresponding real Jordan blocks for the matrix C^2 might have real eigenvalues, but these blocks are again all block diagonal or "block diagonal plus block nilpotent" with 2×2 subblocks.

For the case where a block of C^2 is $rI + N$ where $r > 0$ and N is real nilpotent a real "logarithm" is obtained by the matrix formula given above. For block diagonal real Jordan block, write

$$R = r \begin{pmatrix} \cos\theta & -\sin\theta \\ \sin\theta & \cos\theta \end{pmatrix}$$

where $r > 0$, and note that a real logarithm is given by

$$\ln rI + \begin{pmatrix} 0 & -\theta \\ \theta & 0 \end{pmatrix}.$$

Finally, for a "block diagonal plus block nilpotent" Jordan block, factor the Jordan block as follows:

$$\mathcal{R}(I + \mathcal{N})$$

where $\mathcal{R}$ is block diagonal with R along the diagonal and $\mathcal{N}$ has 2×2 blocks on its super diagonal all given by R^{-1}. Note that we have already obtained logarithms for each of these factors. Moreover, it is not difficult to check that the two logarithms commute. Thus, a real logarithm of the Jordan block is obtained as the sum of real logarithms of the factors. □

Theorem 2.82 can be proved without reference to the Jordan canonical form (see [5]).

Theorem 2.83 (Floquet's Theorem). *If $\Phi(t)$ is a fundamental matrix solution of the T-periodic system (2.27), then, for all $t \in \mathbb{R}$,*

$$\Phi(t + T) = \Phi(t)\Phi^{-1}(0)\Phi(T).$$

In addition, there is a matrix B (which may be complex) such that

$$e^{TB} = \Phi^{-1}(0)\Phi(T)$$

and a T-periodic matrix function $t \mapsto P(t)$ (which may be complex valued) such that $\Phi(t) = P(t)e^{tB}$ for all $t \in \mathbb{R}$. Also, there is a real matrix R and a real $2T$-periodic matrix function $t \to Q(t)$ such that $\Phi(t) = Q(t)e^{tR}$ for all $t \in \mathbb{R}$.

Proof. Since the function $t \mapsto A(t)$ is periodic, it is defined for all $t \in \mathbb{R}$. Thus, by Theorem 2.4, all solutions of the system are defined for $t \in \mathbb{R}$.

If $\Psi(t) := \Phi(t + T)$, then $\Psi(t)$ is a matrix solution. Indeed, we have that

$$\dot{\Psi}(t) = \dot{\Phi}(t + T) = A(t + T)\Phi(t + T) = A(t)\Psi(t),$$

as required.

Define

$$C := \Phi^{-1}(0)\Phi(T) = \Phi^{-1}(0)\Psi(0),$$

and note that C is nonsingular. The matrix function $t \mapsto \Phi(t)C$ is clearly a matrix solution of the linear system with initial value $\Phi(0)C = \Psi(0)$. By the uniqueness of solutions, $\Psi(t) = \Phi(t)C$ for all $t \in \mathbb{R}$. In particular, we have that

$$\Phi(t + T) = \Phi(t)C = \Phi(t)\Phi^{-1}(0)\Phi(T),$$
$$\Phi(t + 2T) = \Phi((t + T) + T) = \Phi(t + T)C = \Phi(t)C^2.$$

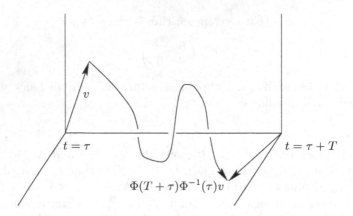

Figure 2.2: The figure depicts the geometry of the monodromy operator for the system $\dot{x} = A(t)x$ in the extended phase space. The vector v in $\mathbb{R}^n$ at $t = \tau$ is advanced to the vector $\Phi(T+\tau)\Phi^{-1}(\tau)v$ at $t = \tau + T$.

By Theorem 2.82, there is a matrix B, possibly complex, such that

$$e^{TB} = C.$$

Also, there is a real matrix R such that

$$e^{2TR} = C^2.$$

If $P(t) := \Phi(t)e^{-tB}$ and $Q(t) := \Phi(t)e^{-tR}$, then

$$P(t+T) = \Phi(t+T)e^{-TB}e^{-tB} = \Phi(t)Ce^{-TB}e^{-tB} = \Phi(t)e^{-tB} = P(t),$$

$$Q(t+2T) = \Phi(t+2T)e^{-2TR}e^{-tR} = \Phi(t)e^{-tR} = Q(t).$$

Thus, we have $P(t+T) = P(t)$, $Q(t+2T) = Q(t)$, and

$$\Phi(t) = P(t)e^{tB} = Q(t)e^{tR},$$

as required. $\square$

The representation $\Phi(t) = P(t)e^{tB}$ in Floquet's theorem is called a *Floquet normal form* for the fundamental matrix $\Phi(t)$. We will use this normal form to study the stability of the zero solution of periodic homogeneous linear systems.

Let us consider a fundamental matrix solution $\Phi(t)$ for the periodic system (2.27) and a vector $v \in \mathbb{R}^n$. The vector solution of the system starting at time $t = \tau$ with initial condition $x(\tau) = v$ is given by

$$t \mapsto \Phi(t)\Phi^{-1}(\tau)v.$$

If the initial vector is moved forward over one period of the system, then we again obtain a vector in $\mathbb{R}^n$ given by $\Phi(T+\tau)\Phi^{-1}(\tau)v$. The operator

$$v \mapsto \Phi(T+\tau)\Phi^{-1}(\tau)v$$

is called a *monodromy operator* (see Figure 2.2). Moreover, if we view the periodic differential equation (2.27) as the autonomous system

$$\dot{x} = A(\psi)x, \qquad \dot{\psi} = 1$$

on the phase cylinder $\mathbb{R}^n \times \mathbb{T}$ where ψ is an angular variable modulo T, then each monodromy operator is a (stroboscopic) Poincaré map for our periodic system. For example, if $\tau = 0$, then the Poincaré section is the fiber $\mathbb{R}^n$ on the cylinder at $\psi = 0$. Of course, each fiber $\mathbb{R}^n$ at $\psi = mT$ where m is an integer is identified with the fiber at $\psi = 0$, and the corresponding Poincaré map is given by

$$v \mapsto \Phi(T)\Phi^{-1}(0)v.$$

The eigenvalues of a monodromy operator are called *characteristic multipliers* of the corresponding time-periodic homogeneous system (2.27). The next proposition states that characteristic multipliers are nonzero complex numbers that are intrinsic to the periodic system—they do not depend on the choice of the fundamental matrix or the initial time.

Proposition 2.84. *The following statements are valid for the periodic linear homogeneous system (2.27).*

(1) *Every monodromy operator is invertible. Equivalently, every characteristic multiplier is nonzero.*

(2) *All monodromy operators have the same eigenvalues. In particular, there are exactly n characteristic multipliers, counting multiplicities.*

Proof. The first statement of the proposition is obvious from the definitions.

To prove statement (2), let us consider the principal fundamental matrix $\Phi(t)$ at $t = 0$. If $\Psi(t)$ is a fundamental matrix, then $\Psi(t) = \Phi(t)\Psi(0)$. Also, by Floquet's theorem,

$$\Phi(t+T) = \Phi(t)\Phi^{-1}(0)\Phi(T) = \Phi(t)\Phi(T).$$

Consider the monodromy operator $\mathcal{M}$ given by

$$v \mapsto \Psi(T+\tau)\Psi^{-1}(\tau)v$$

and note that

$$\begin{aligned}
\Psi(T+\tau)\Psi^{-1}(\tau) &= \Phi(T+\tau)\Psi(0)\Psi^{-1}(0)\Phi^{-1}(\tau) \\
&= \Phi(T+\tau)\Phi^{-1}(\tau) \\
&= \Phi(\tau)\Phi(T)\Phi^{-1}(\tau).
\end{aligned}$$

In particular, the eigenvalues of the operator $\Phi(T)$ are the same as the eigenvalues of the monodromy operator $\mathcal{M}$. Thus, all monodromy operators have the same eigenvalues.   □

Because

$$\Phi(t+T) = \Phi(t)\Phi^{-1}(0)\Phi(T),$$

some authors define characteristic multipliers to be the eigenvalues of the matrices defined by $\Phi^{-1}(0)\Phi(T)$ where $\Phi(t)$ is a fundamental matrix. Of course, both definitions gives the same characteristic multipliers. To prove this fact, let us consider the Floquet normal form $\Phi(t) = P(t)e^{tB}$ and note that $\Phi(0) = P(0) = P(T)$. Thus, we have that

$$\Phi^{-1}(0)\Phi(T) = e^{TB}.$$

Also, by using the Floquet normal form,

$$\begin{aligned}
\Phi(T)\Phi^{-1}(0) &= P(T)e^{TB}\Phi^{-1}(0) \\
&= \Phi(0)e^{TB}\Phi^{-1}(0) \\
&= \Phi(0)(\Phi^{-1}(0)\Phi(T))\Phi^{-1}(0),
\end{aligned}$$

and therefore $\Phi^{-1}(0)\Phi(T)$ has the same eigenvalues as the monodromy operator given by

$$v \mapsto \Phi(T)\Phi^{-1}(0)v.$$

In particular, the traditional definition agrees with our geometrically motivated definition.

Returning to consideration of the Floquet normal form $P(t)e^{tB}$ for the fundamental matrix $\Phi(t)$ and the monodromy operator

$$v \mapsto \Phi(T+\tau)\Phi^{-1}(\tau)v,$$

note that $P(t)$ is invertible and

$$\Phi(T+\tau)\Phi^{-1}(\tau) = P(\tau)e^{TB}P^{-1}(\tau).$$

Thus, the characteristic multipliers of the system are the eigenvalues of e^{TB}. A complex number μ is called a *characteristic exponent* (or a *Floquet exponent*) of the system, if ρ is a characteristic multiplier and $e^{\mu T} = \rho$. Note that if $e^{\mu T} = \rho$, then $\mu + 2\pi i k/T$ is also a Floquet exponent for each integer k. Thus, while there are exactly n characteristic multipliers for the periodic linear system (2.27), there are infinitely many Floquet exponents.

Exercise 2.85. Suppose that $a : \mathbb{R} \to \mathbb{R}$ is a T-periodic function. Find the characteristic multiplier and a Floquet exponent of the T-periodic system $\dot{x} = a(t)x$. Also, find the Floquet normal form for the principal fundamental matrix solution of this system at $t = t_0$.

Exercise 2.86. For the autonomous linear system $\dot{x} = Ax$ a fundamental matrix solution $t \mapsto \Phi(t)$ satisfies the identity $\Phi(T - t) = \Phi(T)\Phi^{-1}(t)$. Show that, in general, this identity does not hold for nonautonomous homogeneous linear systems. Hint: Write down a Floquet normal form matrix $\Phi(t) = P(t)e^{tB}$ that does not satisfy the identity and then show that it is the solution of a (periodic) nonautonomous homogeneous linear system.

Exercise 2.87. Suppose as usual that $A(t)$ is T-periodic and the Floquet normal form of a fundamental matrix solution of the system $\dot{x} = A(t)x$ has the form $P(t)e^{tB}$. (a) Prove that

$$\operatorname{tr} B = \frac{1}{T} \int_0^T \operatorname{tr} A(t)\, dt.$$

Hint: Use Liouville's formula 2.18. (b) By (a), the sum of the characteristic exponents is given by the right-hand side of the formula for the trace of B. Prove that the product of the characteristic multipliers is given by $\exp(\int_0^T \operatorname{tr} A(t)\, dt)$.

Let us suppose that a fundamental matrix for the system (2.27) is represented in Floquet normal form by $P(t)e^{tB}$. We have seen that the characteristic multipliers of the system are the eigenvalues of e^{TB}, but the definition of the Floquet exponents does not mention the eigenvalues of B. Are the eigenvalues of B Floquet exponents? This question is answered affirmatively by the following general theorem about the exponential map.

Theorem 2.88. *If A is an $n \times n$ matrix and if $\lambda_1, \ldots, \lambda_n$ are the eigenvalues of A repeated according to their algebraic multiplicity, then $\lambda_1^k, \ldots, \lambda_n^k$ are the eigenvalues of A^k and $e^{\lambda_1}, \ldots, e^{\lambda_n}$ are the eigenvalues of e^A.*

Proof. We will prove the theorem by induction on the dimension n.

The theorem is clearly valid for 1×1 matrices. Suppose that it is true for all $(n-1) \times (n-1)$ matrices. Define $\lambda := \lambda_1$, and let $v \neq 0$ denote a corresponding eigenvector so that $Av = \lambda v$. Also, let $\mathbf{e}_1, \ldots, \mathbf{e}_n$ denote the usual basis of $\mathbb{C}^n$. There is a nonsingular $n \times n$ matrix S such that $Sv = \mathbf{e}_1$. (Why?) Thus,

$$SAS^{-1}\mathbf{e}_1 = \lambda \mathbf{e}_1,$$

and it follows that the matrix SAS^{-1} has the block form

$$SAS^{-1} = \begin{pmatrix} \lambda & * \\ 0 & \tilde{A} \end{pmatrix}.$$

The matrix SA^kS^{-1} has the same block form, only with the block diagonal elements λ^k and $\tilde{A}^k$. Clearly the eigenvalues of this block matrix are λ^k together with the eigenvalues of $\widetilde{A}^k$. By induction, the eigenvalues of $\widetilde{A}^k$ are the kth powers of the eigenvalues of $\widetilde{A}$. This proves the second statement of the theorem.

Using the power series definition of exp, we see that $e^{SAS^{-1}}$ has block form, with block diagonal elements e^λ and $e^{\widetilde{A}}$. Clearly, the eigenvalues of this block matrix are e^λ together with the eigenvalues of $e^{\widetilde{A}}$. Again using induction, it follows that the eigenvalues of $e^{\widetilde{A}}$ are $e^{\lambda_2}, \ldots, e^{\lambda_n}$. Thus, the eigenvalues of $e^{SAS^{-1}} = Se^A S^{-1}$ are $e^{\lambda_1}, \ldots, e^{\lambda_n}$.   □

Theorem 2.88 is an example of a spectral mapping theorem. If we let $\sigma(A)$ denote the spectrum of the matrix A, that is, the set of all $\lambda \in \mathbb{C}$ such that $\lambda I - A$ is not invertible, then, for our finite dimensional matrix, $\sigma(A)$ coincides with the set of eigenvalues of A. Theorem 2.88 can be restated as follows: $e^{\sigma(A)} = \sigma(e^A)$.

The next result uses Floquet theory to show that the differential equation (2.27) is equivalent to a homogeneous linear system with constant coefficients. This result demonstrates that the stability of the zero solution can often be determined by the Floquet multipliers.

Theorem 2.89. *If the principal fundamental matrix solution of the T-periodic differential equation $\dot{x} = A(t)x$ (system (2.27)) at $t = 0$ is given by $Q(t)e^{tR}$ where Q and R are real, then the time-dependent change of coordinates $x = Q(t)y$ transforms this system to the (real) constant co-efficient linear system $\dot{y} = Ry$. In particular, there is a time-dependent (2T-periodic) change of coordinates that transforms the T-periodic system to a (real) constant coefficient linear system.*

(1) If the characteristic multipliers of the periodic system (2.27) all have modulus less than one; equivalently, if all characteristic exponents have negative real part, then the zero solution is asymptotically stable.

(2) If the characteristic multipliers of the periodic system (2.27) all have modulus less than or equal to one; equivalently, if all characteristic exponents have nonpositive real part, and if the algebraic multiplicity equals the geometric multiplicity of each characteristic multiplier with modulus one; equivalently, if the algebraic multiplicity equals the geometric multiplicity of each characteristic exponent with real part zero, then the zero solution is Lyapunov stable.

(3) If at least one characteristic multiplier of the periodic system (2.27) has modulus greater than one; equivalently, if a characteristic exponent has positive real part, then the zero solution is unstable.

Proof. We will prove the first statement of the theorem and part (1). The proof of the remaining two parts is left as an exercise. For part (2), note that since the differential equation is linear, the Lyapunov stability may reasonably be determined from the eigenvalues of a linearization.

By Floquet's theorem, there is a real matrix R and a real $2T$-periodic matrix $Q(t)$ such that the *principal* fundamental matrix solution $\Phi(t)$ of the system at $t = 0$ is represented by

$$\Phi(t) = Q(t)e^{tR}.$$

Also, there is a matrix B and a T-periodic matrix P such that

$$\Phi(t) = P(t)e^{tB}.$$

The characteristic multipliers are the eigenvalues of e^{TB}. Because $\Phi(0)$ is the identity matrix, we have that

$$\Phi(2T) = e^{2TR} = e^{2TB},$$

and in particular

$$(e^{TB})^2 = e^{2TR}.$$

By Theorem 2.88, the eigenvalues of e^{2TR} are the squares of the characteristic multipliers. These all have modulus less than one. Thus, by another application of Theorem 2.88, all eigenvalues of the real matrix R have negative real parts.

Consider the change of variables $x = Q(t)y$. Because

$$x(t) = Q(t)e^{tR}x(0)$$

and $Q(t)$ is invertible, we have that $y(t) = e^{tR}x(0)$; and therefore,

$$\dot{y} = Ry.$$

By our previous result about linearization (Lyapunov's indirect method), the zero solution of $\dot{y} = Ry$ is asymptotically stable. In fact, by Theorem 2.61, there are numbers $\lambda > 0$ and $C > 0$ such that

$$|y(t)| \leq Ce^{-\lambda t}|y(0)|$$

for all $t \geq 0$ and all $y(0) \in \mathbb{R}^n$. Because Q is periodic, it is bounded. Thus, by the relation $x = Q(t)y$, the zero solution of $\dot{x} = A(t)x$ is also asymptotically stable.   □

While the stability theorem just presented is very elegant, in applied problems it is usually impossible to compute the eigenvalues of e^{TB} explicitly. In fact, because $e^{TB} = \Phi(T)$, it is not at all clear that the eigenvalues can be found without solving the system, that is, without an explicit representation of a fundamental matrix. Note, however, that we only have to approximate *finitely* many numbers (the Floquet multipliers) to determine the stability of the system. This fact is important! For example, the stability can often be determined by applying a numerical method to approximate the Floquet multipliers.

Exercise 2.90. If the planar system $\dot{u} = f(u)$ has a limit cycle, then it is possible to find coordinates in a neighborhood of the limit cycle so that the differential equation has the form

$$\dot{\rho} = h(\rho, \varphi)\rho, \qquad \dot{\varphi} = \omega$$

where ω is a constant and for each ρ the function $\varphi \mapsto h(\rho, \varphi)$ is $2\pi/\omega$-periodic. Prove: If the partial derivative of h with respect to ρ is identically zero, then there is a coordinate system such that the differential equation in the new coordinates has the form

$$\dot{r} = cr, \qquad \dot{\phi} = \omega.$$

Hint: Use Exercise 2.85 and Theorem 2.89.

Exercise 2.91. View the damped periodically-forced Duffing equation $\ddot{x} + \dot{x} - x + x^3 = \epsilon \sin \omega t$ on the phase cylinder. The unperturbed system ($\epsilon = 0$) has a periodic orbit on the phase cylinder with period $2\pi/\omega$ corresponding to its rest point at the origin of the phase plane. Determine the Floquet multipliers associated with this periodic orbit of the unperturbed system; that is, the Floquet multipliers of the linearized system along the periodic orbit.

Exercise 2.92. Consider the system of two coupled oscillators with periodic parametric excitation

$$\ddot{x} + (1 + a \cos \omega t)x = y - x, \quad \ddot{y} + (1 + a \cos \omega t)y = x - y$$

where a and ω are nonnegative parameters. (See Section 3.3 for a derivation of the coupled oscillator model.) (a) Prove that if $a = 0$, then the zero solution is Lyapunov stable. (b) Using a numerical method (or otherwise), determine the Lyapunov stability of the zero solution for fixed but arbitrary values of the parameters. (c) What happens if viscous damping is introduced into the system? Hint: A possible numerical experiment might be designed as follows. For each point in a region of (ω, a)-space, mark the point green if the corresponding system has a Lyapunov stable zero solution; otherwise, mark it red. To decide which regions of parameter space might contain interesting phenomena, recall from your experience with second-order scalar differential equations with constant coefficients (mathematical models of springs) that resonance is expected when the frequency of the periodic excitation is rationally related to the natural frequency of the system. Consider resonances between the frequency ω of the excitation and frequency of periodic motions of the system with $a = 0$, and explore the region of parameter space near these parameter values. Although interesting behavior does occur at resonances, this is not the whole story. Because the monodromy matrix is symplectic (see [11, Sec. 42]), the characteristic multipliers have two symmetries: If λ is a characteristic multiplier, then so is its complex conjugate and its reciprocal. It follows that on the boundary between the stable and unstable regions a pair of characteristic exponents coalesce on the unit circle. Thus, it is instructive to determine the values of ω, with $a = 0$, for those characteristic multipliers that coalesce. These values of ω determine the points where unstable regions have boundary points on the ω-axis.

Is there a method to determine the characteristic exponents without finding the solutions of the differential equation (2.27) explicitly? An example of Lawrence Marcus and Hidehiko Yamabe shows no such method can be constructed in any obvious way from the eigenvalues of $A(t)$. Consider the π-periodic system $\dot{x} = A(t)x$ where

$$A(t) = \begin{pmatrix} -1 + \frac{3}{2}\cos^2 t & 1 - \frac{3}{2}\sin t \cos t \\ -1 - \frac{3}{2}\sin t \cos t & -1 + \frac{3}{2}\sin^2 t \end{pmatrix}. \tag{2.28}$$

It turns out that $A(t)$ has the (time independent) eigenvalues $\frac{1}{4}(-1 \pm \sqrt{7}\,i)$. In particular, the real part of each eigenvalue is negative. On the other hand,

$$x(t) = e^{t/2}\begin{pmatrix} -\cos t \\ \sin t \end{pmatrix}$$

is a solution, and therefore the zero solution is unstable!

The situation is not hopeless. An important example (Hill's equation) where the stability of the zero solution of the differential equation (2.27) can be determined in some cases is discussed in the next section.

Exercise 2.93. (a) Find the principal fundamental matrix solution $\Phi(t)$ at $t = 0$ for the Marcus–Yamabe system; its system matrix $A(t)$ is given in display (2.28). (b) Find the Floquet normal form for $\Phi(t)$ and its "real" Floquet normal form. (c) Determine the characteristic multipliers for the system. (d) The matrix function $t \mapsto A(t)$ is isospectral. Find a matrix function $t \mapsto M(t)$ such that $(A(t), M(t))$ is a Lax pair (see Exercise 2.55). Is every isospectral matrix function the first component of a Lax pair?

The Floquet normal form can be used to obtain detailed information about the solutions of the differential equation (2.27). For example, if we use the fact that the Floquet normal form decomposes a fundamental matrix into a periodic part and an exponential part, then it should be clear that for some systems there are periodic solutions and for others there are no nontrivial periodic solutions. It is also possible to have "quasi-periodic" solutions. The next lemma will be used to prove these facts.

Lemma 2.94. *If μ is a characteristic exponent for the homogeneous linear T-periodic differential equation (2.27) and $\Phi(t)$ is the principal fundamental matrix solution at $t = 0$, then $\Phi(t)$ has a Floquet normal form $P(t)e^{tB}$ such that μ is an eigenvalue of B.*

Proof. Let $\mathcal{P}(t)e^{tB}$ be a Floquet normal form for $\Phi(t)$. By the definition of characteristic exponents, there is a characteristic multiplier λ such that $\lambda = e^{\mu T}$, and, by Theorem 2.88, there is an eigenvalue ν of B such that $e^{\nu T} = \lambda$. Also, there is some integer $k \neq 0$ such that $\nu = \mu + 2\pi i k / T$.

Define $B := \mathcal{B} - (2\pi i k/T)I$ and $P(t) = \mathcal{P}(t)e^{(2\pi i k t/T)I}$. Note that μ is an eigenvalue of B, the function P is T-periodic, and

$$P(t)e^{tB} = \mathcal{P}(t)e^{t\mathcal{B}}.$$

It follows that $\Phi(t) = P(t)e^{tB}$ is a representation in Floquet normal form where μ is an eigenvalue of B.   □

A basic result that is used to classify the possible types of solutions that can arise is the content of the following theorem.

Theorem 2.95. *If λ is a characteristic multiplier of the homogeneous linear T-periodic differential equation (2.27) and $e^{T\mu} = \lambda$, then there is a (possibly complex) nontrivial solution of the form*

$$x(t) = e^{\mu t}p(t)$$

where p is a T-periodic function. Moreover, for this solution $x(t + T) = \lambda x(t)$.

Proof. Consider the principal fundamental matrix solution $\Phi(t)$ at $t = 0$. By Lemma 2.94, there is a Floquet normal form representation $\Phi(t) = P(t)e^{tB}$ such that μ is an eigenvalue of B. Hence, there is a vector $v \neq 0$ such that $Bv = \mu v$. Clearly, it follows that $e^{tB}v = e^{\mu t}v$, and therefore the solution $x(t) := \Phi(t)v$ is also represented in the form

$$x(t) = P(t)e^{tB}v = e^{\mu t}P(t)v.$$

The solution required by the first statement of the theorem is obtained by defining $p(t) := P(t)v$. The second statement of the theorem is proved as follows:

$$x(t + T) = e^{\mu(t+T)}p(t + T) = e^{\mu T}e^{\mu t}p(t) = \lambda x(t).   \square$$

Theorem 2.96. *Suppose that λ_1 and λ_2 are characteristic multipliers of the homogeneous linear T-periodic differential equation (2.27) and μ_1 and μ_2 are characteristic exponents such that $e^{T\mu_1} = \lambda_1$ and $e^{T\mu_2} = \lambda_2$. If $\lambda_1 \neq \lambda_2$, then there are T-periodic functions p_1 and p_2 such that*

$$x_1(t) = e^{\mu_1 t}p_1(t) \quad and \quad x_2(t) = e^{\mu_2 t}p_2(t)$$

are linearly independent solutions.

Proof. Let $\Phi(t) = P(t)e^{tB}$ (as in Lemma 2.94) be such that μ_1 is an eigenvalue of B. Also, let v_1 be a nonzero eigenvector corresponding to the eigenvalue μ_1. Since λ_2 is an eigenvalue of the monodromy matrix $\Phi(T)$, by Theorem 2.88 there is an eigenvalue μ of B such that $e^{T\mu} = \lambda_2 = e^{T\mu_2}$. It follows that there is an integer k such that $\mu_2 = \mu + 2\pi i k/T$. Also, because $\lambda_1 \neq \lambda_2$, we have that $\mu \neq \mu_1$. Hence, if v_2 is a nonzero eigenvector of

B corresponding to the eigenvalue μ, then the eigenvectors v_1 and v_2 are linearly independent.

As in the proof of Theorem 2.95, there are solutions of the form

$$x_1(t) = e^{\mu_1 t} P(t) v_1, \qquad x_2(t) = e^{\mu t} P(t) v_2.$$

Moreover, because $x_1(0) = v_1$ and $x_2(0) = v_2$, these solutions are linearly independent. Finally, let us note that x_2 can be written in the required form

$$x_2(t) = \left(e^{\mu t} e^{2\pi k i/T}\right)\left(e^{-2\pi k i/T} P(t) v_2\right). \qquad \square$$

The T-periodic system (2.27) has the Floquet normal form

$$t \mapsto Q(t) e^{tR}$$

where Q is a real $2T$-periodic function and R is real matrix. By Theorem 2.37 and 2.89, all solutions of the system are represented as finite sums of real solutions of the two types

$$q(t) r(t) e^{\alpha t} \sin \beta t \quad \text{and} \quad q(t) r(t) e^{\alpha t} \cos \beta t,$$

where q is $2T$-periodic, r is a polynomial of degree at most $n-1$, and $\alpha + i\beta$ is an eigenvalue of R. We will use Theorem 2.95 to give a more detailed description of the nature of these real solutions.

If the characteristic multiplier λ is a positive real number, then there is a corresponding real characteristic exponent μ. In this case, if the periodic function p in Theorem 2.95 is complex, then it can be represented as $p = r + is$ where both r and s are real T-periodic functions. Because our T-periodic system is real, both the real and the imaginary parts of a solution are themselves solutions. Hence, there is a real nontrivial solution of the form $x(t) = e^{\mu t} r(t)$ or $x(t) = e^{\mu t} s(t)$. Such a solution is periodic if and only if $\lambda = 1$ or equivalently if $\mu = 0$. On the other hand, if $\lambda \neq 1$ or $\mu \neq 0$, then the solution is unbounded either as $t \to \infty$ or as $t \to -\infty$. If $\lambda < 1$ (equivalently, $\mu < 0$), then the solution is asymptotic to the zero solution as $t \to \infty$. On the other hand, if $\lambda > 1$ (equivalently, $\mu > 0$), then the solution is unbounded as $t \to \infty$.

If the characteristic multiplier λ is a negative real number, then μ can be chosen to have the form $\nu + \pi i/T$ where ν is real and $e^{T\mu} = \lambda$. Hence, if we again take $p = r + is$, then we have the solution

$$e^{\mu t} p(t) = e^{\nu t} e^{\pi i t/T} (r(t) + is(t))$$

from which real nontrivial solutions are easily constructed. For example, if the real part of the complex solution is nonzero, then the real solution has the form

$$x(t) = e^{\nu t}(r(t) \cos(\pi t/T) - s(t) \sin(\pi t/T)).$$

Such a solution is periodic if and only if $\lambda = -1$ or equivalently if $\nu = 0$. In this case the solution is $2T$-periodic. If $\nu \neq 0$, then the solution is unbounded. If $\nu < 0$, then the solution is asymptotic to zero as $t \to \infty$. On the other hand, if $\nu > 0$, then the solution is unbounded as $t \to \infty$.

If λ is complex, then we have $\mu = \alpha + i\beta$ and there is a solution given by

$$x(t) = e^{\alpha t}(\cos \beta t + i \sin \beta t)(r(t) + is(t)).$$

Thus, there are real solutions

$$x_1(t) = e^{\alpha t}(r(t) \cos \beta t - s(t) \sin \beta t),$$
$$x_2(t) = e^{\alpha t}(r(t) \sin \beta t + s(t) \cos \beta t).$$

If $\alpha \neq 0$, then both solutions are unbounded. But, if $\alpha < 0$, then these solutions are asymptotic to zero as $t \to \infty$. On the other hand, if $\alpha > 0$, then these solutions are unbounded as $t \to \infty$. If $\alpha = 0$ and there are relatively prime positive integers m and n such that $2\pi m/\beta = nT$, then the solution is nT-periodic. If no such integers exist, then the solution is called *quasi-periodic*.

We will prove in Section 2.4.4 that the stability of a periodic orbit is determined by the stability of the corresponding fixed point of a Poincaré map defined on a Poincaré section that meets the periodic orbit. Generically, the stability of the fixed point of the Poincaré map is determined by the eigenvalues of its derivative at the fixed point. For example, if the eigenvalues of the derivative of the Poincaré map at the fixed point corresponding to the periodic orbit are all inside the unit circle, then the periodic orbit is asymptotically stable. It turns out that the eigenvalues of the derivative of the Poincaré map are closely related to the characteristic multipliers of a time-periodic system, namely, the variational equation along the periodic orbit. We will have much more to say about the general case later. Here we will illustrate the idea for an example where the Poincaré map is easy to compute.

Suppose that

$$\dot{u} = f(u, t), \qquad u \in \mathbb{R}^n \tag{2.29}$$

is a smooth nonautonomous differential equation. If there is some $T > 0$ such that $f(u, t + T) = f(u, t)$ for all $u \in \mathbb{R}^n$ and all $t \in \mathbb{R}$, then the system (2.29) is called T-periodic.

The nonautonomous system (2.29) is made "artificially" autonomous by the addition of a new equation as follows:

$$\dot{u} = f(u, \psi), \qquad \dot{\psi} = 1 \tag{2.30}$$

where ψ may be viewed as an angular variable modulo T. In other words, we can consider $\psi + nT = \psi$ whenever n is an integer. The phase cylinder for system (2.30) is $\mathbb{R}^n \times \mathbb{T}$, where $\mathbb{T}$ (topologically the unit circle) is defined to be $\mathbb{R}$ modulo T. This autonomous system provides the correct geometry with which to define a Poincaré map.

For each $\xi \in \mathbb{R}^n$, let $t \mapsto u(t, \xi)$ denote the solution of the differential equation (2.29) such that $u(0, \xi) = \xi$, and note that $t \mapsto (u(t, \xi), t)$ is the corresponding solution of the system (2.30). The set $\Sigma := \{(\xi, \psi) : \psi = 0\}$ is a Poincaré section, and the corresponding Poincaré map is given by $\xi \mapsto u(T, \xi)$.

If there is a point $p \in \mathbb{R}^n$ such that $f(p, t) = 0$ for all $t \in \mathbb{R}$, then the function $t \mapsto (p, t)$, or equivalently $t \mapsto (u(t, p), t)$, is a periodic solution of the system (2.30) with period T. Moreover, let us note that $u(T, p) = p$. Thus, the periodic solution corresponds to a fixed point of the Poincaré map as it should.

The derivative of the Poincaré map at p is the linear transformation of $\mathbb{R}^n$ given by the partial derivative $u_\xi(T, p)$. Moreover, by differentiating both the differential equation (2.29) and the initial condition $u(0, \xi) = \xi$ with respect to ξ, it is easy to see that the matrix function $t \mapsto u_\xi(t, p)$ is the principal fundamental matrix solution at $t = 0$ of the (T-periodic linear) variational initial value problem

$$\dot{W} = f_u(u(t, p), t)W, \qquad W(0) = I. \tag{2.31}$$

If the solution of system (2.31) is represented in the Floquet normal form $u_\xi(t, p) = P(t)e^{tB}$, then the derivative of the Poincaré map is given by $u_\xi(T, p) = e^{TB}$. In particular, the characteristic multipliers of the variational equation (2.31) coincide with the eigenvalues of the derivative of the Poincaré map. Thus, whenever the principle of linearized stability is valid, the stability of the periodic orbit is determined by the characteristic multipliers of the periodic variational equation (2.31).

As an example, consider the pendulum with oscillating support

$$\ddot{\theta} + (1 + a\cos \omega t)\sin \theta = 0.$$

The zero solution, given by $\theta(t) \equiv 0$, corresponds to a $2\pi/\omega$-periodic solution of the associated autonomous system. A calculation shows that the variational equation along this periodic solution is equivalent to the second order differential equation

$$\ddot{x} + (1 + a\cos \omega t)x = 0,$$

called a *Mathieu equation*. The normal form for the Mathieu equation is

$$\ddot{x} + (a - 2q\cos 2t)x = 0,$$

where a and q are parameters.

Since, as we have just seen (see also Exercise 2.92), equations of Mathieu type arise frequently in applications, the stability analysis of such equations is important (see, for example, [12], [18], [101], [127], [149], and [237]). In Section 2.4.2 we will show how the stability of the zero solution of the Mathieu equation, and, in turn, the stability of the zero solution of the pendulum with oscillating support, is related in a delicate manner to the amplitude a and the frequency ω of the periodic displacement.

Exercise 2.97. This is a continuation of Exercise 2.57. Suppose that $v : \mathbb{R} \to \mathbb{R}^3$ is a periodic function. Consider the differential equation

$$\dot{x} = v(t) \times x$$

and discuss the stability of its periodic solutions.

Exercise 2.98. Determine the stability type of the periodic orbit discussed in Exercise 2.91.

Exercise 2.99. (a) Prove that the system

$$\dot{x} = x - y - x(x^2 + y^2),$$
$$\dot{y} = x + y - y(x^2 + y^2),$$
$$\dot{z} = z + xz - z^3$$

has periodic orbits. Hint: Change to cylindrical coordinates, show that the cylinder (with radius one whose axis of symmetry is the z-axis) is invariant, and recall the analysis of equation (1.43). (b) Prove that there is a stable periodic orbit. (c) The stable periodic orbit has three Floquet multipliers. Of course, one of them is unity. Find (exactly) a vector v such that $\Phi(T)v = v$, where T is the period of the periodic orbit and $\Phi(t)$ is the principal fundamental matrix solution at $t = 0$ of the variational equation along the stable periodic solution. (d) Approximate the remaining two multipliers. Note: It is possible to represent these multipliers with integrals, but they are easier to approximate using a numerical method.

2.4.1 Lyapunov Exponents

An important generalization of Floquet exponents, called Lyapunov exponents, are introduced in this section. This concept is used extensively in the theory of dynamical systems (see, for example, [103], [144], [176], and [233]).

Consider a (nonlinear) differential equation

$$\dot{u} = f(u), \qquad u \in \mathbb{R}^n \tag{2.32}$$

with flow φ_t. If $\epsilon \in \mathbb{R}$, ξ, $v \in \mathbb{R}^n$, and $\eta := \xi + \epsilon v$, then the two solutions

$$t \mapsto \varphi_t(\xi), \qquad t \mapsto \varphi_t(\xi + \epsilon v)$$

start at points that are $O(\epsilon)$ close; that is, the absolute value of the difference of the two points in $\mathbb{R}^n$ is bounded by the usual norm of v times ϵ. Moreover, by Taylor expansion at $\epsilon = 0$, we have that

$$\varphi_t(\xi + \epsilon v) - \varphi_t(\xi) = \epsilon D\varphi_t(\xi)v + O(\epsilon^2)$$

where $D\varphi_t(\xi)$ denotes the derivative of the function $u \mapsto \varphi_t(u)$ evaluated at $u = \xi$. Thus, the first order approximation of the difference of the solutions at time t is $\epsilon D\varphi_t(\xi)v$ where $t \mapsto D\varphi_t(\xi)$ is the principal fundamental matrix solution at $t = 0$ of the linearized equation

$$\dot{W} = Df(\varphi_t(\xi))W$$

along the solution of the original system (2.32) starting at ξ. To see this fact, just note that

$$\dot{\varphi}_t(u) = f(\varphi_t(u))$$

and differentiate both sides of this identity with respect to u at $u = \xi$.

If we view v as a vector in the tangent space to $\mathbb{R}^n$ at ξ, denoted $T_\xi\mathbb{R}^n$, then $D\varphi_t(\xi)v$ is a vector in the tangent space $T_{\varphi_t(\xi)}\mathbb{R}^n$. For each such v, if $v \neq 0$, then it is natural to define a corresponding linear operator L, from the linear subspace of $T_\xi\mathbb{R}^n$ generated by v to the linear subspace of $T_{\varphi_t(\xi)}\mathbb{R}^n$ generated by $D\varphi_t(\xi)v$, defined by $L(av) = D\varphi_t(\xi)av$ where $a \in \mathbb{R}$. Let us note that the norm of this operator measures the relative "expansion" or "contraction" of the vector v; that is,

$$\|L\| = \sup_{a \neq 0} \frac{|D\phi_t(\xi)av|}{|av|} = \frac{|D\phi_t(\xi)v|}{|v|}.$$

Our two solutions can be expressed in integral form; that is,

$$\varphi_t(\xi) = \xi + \int_0^t f(\varphi_s(\xi))\, ds,$$

$$\varphi_t(\xi + \epsilon v) = \xi + \epsilon v + \int_0^t f(\varphi_s(\xi + \epsilon v))\, ds.$$

Hence, as long as we consider a finite time interval or a solution that is contained in a compact subset of $\mathbb{R}^n$, there is a Lipschitz constant $\mathrm{Lip}(f) > 0$ for the function f, and we have the inequality

$$|\varphi_t(\xi + \epsilon v) - \varphi_t(\xi)| \leq \epsilon|v| + \mathrm{Lip}(f)\int_0^t |\varphi_s(\xi + \epsilon v) - \varphi_s(\xi)|\, ds.$$

By Gronwall's inequality, the separation distance between the solutions is bounded by an exponential function of time. In fact, we have the estimate

$$|\varphi_t(\xi + \epsilon v) - \varphi_t(\xi)| \leq \epsilon|v|e^{t\,\mathrm{Lip}(f)}.$$

The above computation for the norm of L and the exponential bound for the separation rate between two solutions motivates the following definition (see [144]).

Definition 2.100. Suppose that $\xi \in \mathbb{R}^n$ and the solution $t \mapsto \varphi_t(\xi)$ of the differential equation (2.32) is defined for all $t \geq 0$. Also, let $v \in \mathbb{R}^n$ be a nonzero vector. The *Lyapunov exponent* at ξ in the direction v for the flow φ_t is defined to be

$$\chi(p, v) := \limsup_{t \to \infty} \frac{1}{t} \ln \left(\frac{|D\phi_t(\xi)v|}{|v|} \right).$$

As a simple example, let us consider the planar system

$$\dot{x} = -ax, \qquad \dot{y} = by$$

where a and b are positive parameters, and let us note that its flow is given by

$$\varphi_t(x, y) = (e^{-at}x, e^{bt}y).$$

By an easy computation using the definition of the Lyapunov exponents, it follows that if v is given by $v = (w, z)$ and $z \neq 0$, then $\chi(\xi, v) = b$. If $z = 0$ and $w \neq 0$, then $\chi(\xi, v) = -a$. In particular, there are exactly two Lyapunov exponents for this system. Of course, the Lyapunov exponents in this case correspond to the eigenvalues of the system matrix.

Although our definition of Lyapunov exponents is for autonomous systems, it should be clear that since the definition only depends on the fundamental matrix solutions of the associated variational equations along orbits of the system, we can define the same notion for solutions of abstract time-dependent linear systems. Indeed, for a T-periodic linear system

$$\dot{u} = A(t)u, \qquad u \in \mathbb{R}^n \tag{2.33}$$

with principal fundamental matrix $\Phi(t)$ at $t = 0$, the Lyapunov exponent defined with respect to the nonzero vector $v \in \mathbb{R}^n$ is

$$\chi(v) := \limsup_{t \to \infty} \frac{1}{t} \ln \left(\frac{|\Phi(t)v|}{|v|} \right).$$

Proposition 2.101. *If μ is a Floquet exponent of the system (2.33), then the real part of μ is a Lyapunov exponent.*

Proof. Let us suppose that the principal fundamental matrix $\Phi(t)$ is given in Floquet normal form by

$$\Phi(t) = P(t)e^{tB}.$$

If $\mu = a + bi$ is a Floquet exponent, then there is a corresponding vector v such that $e^{TB}v = e^{\mu T}v$. Hence, using the Floquet normal form, we have that

$$\Phi(T)v = e^{\mu T}v.$$

If $t \geq 0$, then there is a nonnegative integer n and a number r such that $0 \leq r < T$ and

$$\frac{1}{t} \ln \left(\frac{|\Phi(t)v|}{|v|} \right) = \frac{1}{T} \left(\frac{nT}{nT + r} \right) \left(\frac{1}{n} \ln \left(\frac{|P(nT + r)e^{rB}e^{n\mu T}v|}{|v|} \right) \right)$$

$$= \frac{1}{T} \left(\frac{nT}{nT + r} \right) \left(\frac{1}{n} \ln |e^{nTa}| + \frac{1}{n} \ln \left(\frac{|P(r)e^{rB}v|}{|v|} \right) \right).$$

Clearly, $n \to \infty$ as $t \to \infty$. Thus, it is easy to see that

$$\lim_{t \to \infty} \frac{1}{T} \left(\frac{nT}{nT + r} \right) \left(\frac{1}{n} \ln |e^{nTa}| + \frac{1}{n} \ln \left(\frac{|P(r)e^{rB}v|}{|v|} \right) \right) = a. \qquad \square$$

Let us suppose that a differential equation has a compact invariant set that contains an orbit whose closure is dense in the invariant set. Then, the existence of a positive Lyapunov exponent for this orbit ensures that nearby orbits tend to separate exponentially fast from the dense orbit. But, since these orbits are confined to a compact invariant set, they must also be bounded. This suggests that each small neighborhood in the invariant set undergoes both stretching and folding as it evolves with the flow. The subsequent kneading of the invariant set due to this stretching and folding would tend to mix the evolving neighborhoods so that they eventually intertwine in a complicated manner. For this reason, the existence of a positive Lyapunov exponent is often taken as a signature of "chaos." While this criterion is not always valid, the underlying idea that the stretching implied by a positive Lyapunov exponent is associated with complex motions is important in the modern theory of dynamical systems.

Exercise 2.102. Show that if two points are on the same orbit, then the corresponding Lyapunov exponents are the same.

Exercise 2.103. Prove the "converse" of Proposition 2.101; that is, every Lyapunov exponent for a time-periodic system is a Floquet exponent.

Exercise 2.104. If $\dot{x} = f(x)$, determine the Lyapunov exponent $\chi(\xi, f(\xi))$.

Exercise 2.105. How many Lyapunov exponents are associated with an orbit of a differential equation in an n-dimensional phase space.

Exercise 2.106. Suppose that x is in the omega limit set of an orbit. Are the Lyapunov exponents associated with x the same as those associated with the original orbit?

Exercise 2.107. In all the examples in this section, the lim sup can be replaced by lim. Are there examples where the superior limit is a finite number, but the limit does not exist? This is (probably) a challenging exercise! For an answer see [144] and [176].

2.4.2   Hill's Equation

A famous example where Floquet theory applies to give good stability results is Hill's equation,

$$\ddot{u} + a(t)u = 0, \qquad a(t+T) = a(t).$$

It was introduced by George W. Hill in his study of the motions of the moon. Roughly speaking, the motion of the moon can be viewed as a harmonic oscillator in a periodic gravitational field. But this model equation arises in many areas of applied mathematics where the stability of periodic motions is an issue. A prime example, mentioned in the previous section, is the stability analysis of small oscillations of a pendulum whose length varies with time.

If we define

$$x := \begin{pmatrix} u \\ \dot{u} \end{pmatrix},$$

then Hill's equation is equivalent to the first order system $\dot{x} = A(t)x$ where

$$A(t) = \begin{pmatrix} 0 & 1 \\ -a(t) & 0 \end{pmatrix}.$$

We will apply linear systems theory, especially Floquet theory, to analyze the stability of the zero solution of this linear T-periodic system.

The first step in the stability analysis is an application of Liouville's formula (2.18). In this regard, you may recall from your study of scalar second order linear differential equations that if $\ddot{u} + p(t)\dot{u} + q(t)u = 0$ and the Wronskian of the two solutions u_1 and u_2 is defined by

$$W(t) := \det \begin{pmatrix} u_1(t) & u_2(t) \\ \dot{u}_1(t) & \dot{u}_2(t) \end{pmatrix},$$

then

$$W(t) = W(0)e^{-\int_0^t p(s)\,ds}. \tag{2.34}$$

Note that for the equivalent first order system

$$\dot{x} = \begin{pmatrix} 0 & 1 \\ -q(t) & -p(t) \end{pmatrix} x = B(t)x$$

with fundamental matrix $\Psi(t)$, formula (2.34) is a special case of Liouville's formula

$$\det \Psi(t) = \det \Psi(0)e^{\int_0^t \operatorname{tr} B(s)ds}.$$

At any rate, let us apply Liouville's formula to the principal fundamental matrix $\Phi(t)$ at $t = 0$ for Hill's system to obtain the identity $\det \Phi(t) \equiv 1$. Since the determinant of a matrix is the product of the eigenvalues of the matrix, we have an important fact: The product of the characteristic multipliers of the monodromy matrix, $\Phi(T)$, is 1.

Let the characteristic multipliers for Hill's equation be denoted by λ_1 and λ_2 and note that they are roots of the characteristic equation

$$\lambda^2 - (\operatorname{tr}\Phi(T))\lambda + \det\Phi(T) = 0.$$

For notational convenience let us set $2\phi = \operatorname{tr}\Phi(T)$ to obtain the equivalent characteristic equation

$$\lambda^2 - 2\phi\lambda + 1 = 0$$

whose solutions are given by

$$\lambda = \phi \pm \sqrt{\phi^2 - 1}.$$

There are several cases to consider depending on the value of ϕ.

Case 1: If $\phi > 1$, then λ_1 and λ_2 are distinct positive real numbers such that $\lambda_1\lambda_2 = 1$. Thus, we may assume that $0 < \lambda_1 < 1 < \lambda_2$ with $\lambda_1 = 1/\lambda_2$ and there is a real number $\mu > 0$ (a characteristic exponent) such that $e^{T\mu} = \lambda_2$ and $e^{-T\mu} = \lambda_1$. By Theorem 2.95 and Theorem 2.96, there is a fundamental set of solutions of the form

$$e^{-\mu t}p_1(t), \qquad e^{\mu t}p_2(t)$$

where the real functions p_1 and p_2 are T-periodic. In this case, the zero solution is unstable.

Case 2: If $\phi < -1$, then λ_1 and λ_2 are both real and both negative. Also, since $\lambda_1\lambda_2 = 1$, we may assume that $\lambda_1 < -1 < \lambda_2 < 0$ with $\lambda_1 = 1/\lambda_2$. Thus, there is a real number $\mu > 0$ (a characteristic exponent) such that $e^{2T\mu} = \lambda_1^2$ and $e^{-2T\mu} = \lambda_2^2$. As in Case 1, there is a fundamental set of solutions of the form

$$e^{\mu t}q_1(t), \qquad e^{-\mu t}q_2(t)$$

where the real functions q_1 and q_2 are $2T$-periodic. Again, the zero solution is unstable.

Case 3: If $-1 < \phi < 1$, then λ_1 and λ_2 are complex conjugates each with nonzero imaginary part. Since $\lambda_1\bar{\lambda}_1 = 1$, we have that $|\lambda_1| = 1$, and therefore both characteristic multipliers lie on the unit circle in the complex plane. Because both λ_1 and λ_2 have nonzero imaginary parts, one of these characteristic multipliers, say λ_1, lies in the upper half plane. Thus, there is a real number θ with $0 < \theta T < \pi$ and $e^{i\theta T} = \lambda_1$. In fact, there is a solution of the form $e^{i\theta t}(r(t) + is(t))$ with r and s both T-periodic functions. Hence, there is a fundamental set of solutions of the form

$$r(t)\cos\theta t - s(t)\sin\theta t, \qquad r(t)\sin\theta t + s(t)\cos\theta t.$$

In particular, the zero solution is stable (see Exercise 2.113) but not asymptotically stable. Also, the solutions are periodic if and only if there are relatively prime positive integers m and n such that $2\pi m/\theta = nT$. If such integers exist, all solutions have period nT. If not, then these solutions are quasi-periodic.

We have just proved the following facts for Hill's equation: *Suppose that $\Phi(t)$ is the principal fundamental matrix solution of Hill's equation at $t = 0$. If $|\operatorname{tr}\Phi(T)| < 2$, then the zero solution is stable. If $|\operatorname{tr}\Phi(T)| > 2$, then the zero solution is unstable.*

Case 4: If $\phi = 1$, then $\lambda_1 = \lambda_2 = 1$. The nature of the solutions depends on the canonical form of $\Phi(T)$. If $\Phi(T)$ is the identity, then $e^0 = \Phi(T)$ and there is a Floquet normal form $\Phi(t) = P(t)$ where $P(t)$ is T-periodic and invertible. Thus, there is a fundamental set of periodic solutions and the zero solution is stable. If $\Phi(T)$ is not the identity, then there is a nonsingular matrix C such that

$$C\Phi(T)C^{-1} = I + N = e^N$$

where $N \neq 0$ is nilpotent. Thus, $\Phi(t)$ has a Floquet normal form $\Phi(t) = P(t)e^{tB}$ where $B := C^{-1}(\frac{1}{T}N)C$. Because

$$e^{tB} = C^{-1}(I + \frac{t}{T}N)C,$$

the matrix function $t \mapsto e^{tB}$ is unbounded, and therefore the zero solution is unstable.

Case 5: If $\phi = -1$, then the situation is similar to Case 4, except the fundamental matrix is represented by $Q(t)e^{tB}$ where $Q(t)$ is a $2T$-periodic matrix function.

By the results just presented, the stability of Hill's equation is reduced, in most cases, to a determination of the absolute value of the trace of its principal fundamental matrix evaluated after one period. While this is a useful fact, it leaves open an important question: Can the stability be determined without imposing a condition on the solutions of the equation? It turns out that in some special cases this is possible (see [149] and [237]). A theorem of Lyapunov [144] in this direction follows.

Theorem 2.108. *If $a : \mathbb{R} \to \mathbb{R}$ is a positive T-periodic function such that*

$$T \int_0^T a(t)\, dt \leq 4,$$

then all solutions of the Hill's equation $\ddot{x} + a(t)x = 0$ are bounded. In particular, the trivial solution is stable.

The proof of Theorem 2.108 is outlined in Exercises 2.113 and 2.116.

Exercise 2.109. Consider the second order system

$$\ddot{u} + \dot{u} + \cos(t)\, u = 0.$$

Prove: (a) If ρ_1 and ρ_2 are the characteristic multipliers of the corresponding first order system, then $\rho_1\rho_2 = \exp(-2\pi)$. (b) The Poincaré map for the system is dissipative; that is, it contracts area.

Exercise 2.110. Prove: The equation

$$\ddot{u} - (2\sin^2 t)\dot{u} + (1 + \sin 2t)u = 0.$$

does not have a fundamental set of periodic solutions. Does it have a nonzero periodic solution? Is the zero solution stable?

Exercise 2.111. Discuss the stability of the trivial solution of the scalar time-periodic system $\dot{x} = (\cos^2 t)x$.

Exercise 2.112. Prove: The zero solution is unstable for the system $\dot{x} = A(t)x$ where

$$A(t) := \begin{pmatrix} 1/2 - \cos t & 12 \\ 147 & 3/2 + \sin t \end{pmatrix}.$$

Exercise 2.113. Prove: If all solutions of the T-periodic system $\dot{x} = A(t)x$ are bounded, then the trivial solution is Lyapunov stable.

Exercise 2.114. For Hill's equation with period T, if the absolute value of the trace of $\Phi(T)$, where $\Phi(t)$ is the principal fundamental matrix at $t = 0$, is strictly less than two, show that there are no solutions of period T or $2T$. On the other hand, if the absolute value of the trace of $\Phi(T)$ is two, show that there is such a solution. Note that this property characterizes the boundary between the stable and unstable solutions.

Exercise 2.115. Prove: If $a(t)$ is an even T-periodic function, then Hill's equation has a fundamental set of solutions such that one solution is even and one is odd.

Exercise 2.116. Prove Theorem 2.108. Hint: If Hill's equation has an unbounded solution, then there is a real solution $t \mapsto x(t)$ and a real Floquet multiplier such that $x(t + T) = \lambda x(t)$. Define a new function $t \mapsto u(t)$ by

$$u(t) := \frac{\dot{x}(t)}{x(t)},$$

and show that u is a solution of the Riccati equation

$$\dot{u} = -a(t) - u^2.$$

Use the Riccati equation to prove that the solution x has at least one zero in the interval $[0, T]$. Also, show that x has two distinct zeros on some interval whose length does not exceed T. Finally, use the following proposition to finish the proof. If f is a smooth function on the finite interval $[\alpha, \beta]$ such that $f(\alpha) = 0$, $f(\beta) = 0$, and such that f is positive on the open interval (α, β), then

$$(\beta - \alpha) \int_\alpha^\beta \frac{|f''(t)|}{f(t)}\, dt > 4.$$

To prove this proposition, first suppose that f attains its maximum at γ and show that

$$\frac{4}{\beta - \alpha} \le \frac{1}{\gamma - \alpha} + \frac{1}{\beta - \gamma} = \frac{1}{f(\gamma)}\left(\frac{f(\gamma) - f(\alpha)}{\gamma - \alpha} - \frac{f(\beta) - f(\gamma)}{\beta - \gamma}\right).$$

Then, use the mean value theorem and the fundamental theorem of calculus to complete the proof.

Exercise 2.117. Prove: If $t \mapsto a(t)$ is negative, then the Hill's equation $\ddot{x} + a(t)x = 0$ has an unbounded solution. Hint: Multiply by x and integrate by parts.

2.4.3 Periodic Orbits of Linear Systems

In this section we will consider the existence and stability of periodic solutions of the time-periodic system

$$\dot{x} = A(t)x + b(t), \qquad x \in \mathbb{R}^n \tag{2.35}$$

where $t \mapsto A(t)$ is a T-periodic matrix function and $t \mapsto b(t)$ is a T-periodic vector function.

Theorem 2.118. *If the number one is not a characteristic multiplier of the T-periodic homogeneous system $\dot{x} = A(t)x$, then (2.35) has at least one T-periodic solution.*

Proof. Let us show first that if $t \mapsto x(t)$ is a solution of system (2.35) and $x(0) = x(T)$, then this solution is T-periodic. Define $y(t) := x(t + T)$. Note that $t \mapsto y(t)$ is a solution of (2.35) and $y(0) = x(0)$. Thus, by the uniqueness theorem $x(t + T) = x(t)$ for all $t \in \mathbb{R}$.

If $\Phi(t)$ is the principal fundamental matrix solution of the homogeneous system at $t = 0$, then, by the variation of parameters formula,

$$x(T) = \Phi(T)x(0) + \Phi(T)\int_0^T \Phi^{-1}(s)b(s)\,ds.$$

Therefore, $x(T) = x(0)$ if and only if

$$(I - \Phi(T))x(0) = \Phi(T)\int_0^T \Phi^{-1}(s)b(s)\,ds.$$

This equation for $x(0)$ has a solution whenever the number one is not an eigenvalue of $\Phi(T)$. (Note that the map $x(0) \mapsto x(T)$ is the Poincaré map. Thus, our periodic solution corresponds to a fixed point of the Poincaré map).

By Floquet's theorem, there is a matrix B such that the monodromy matrix is given by

$$\Phi(T) = e^{TB}.$$

In other words, by the hypothesis, the number one is not an eigenvalue of $\Phi(T)$. ☐

Corollary 2.119. *If $A(t) = A$, a constant matrix such that A is infinitesimally hyperbolic (no eigenvalues on the imaginary axis), then the differential equation (2.35) has at least one T-periodic solution.*

Proof. The monodromy matrix e^{TA} does not have 1 as an eigenvalue. $\square$

Exercise 2.120. Discuss the uniqueness of the T-periodic solutions of the system (2.35). Also, using Theorem 2.89, discuss the stability of the T-periodic solutions.

In system (2.35) if $b = 0$, then the trivial solution is a T-periodic solution. The next theorem states a general sufficient condition for the existence of a T-periodic solution.

Theorem 2.121. *If the T-periodic system (2.35) has a bounded solution, then it has a T-periodic solution.*

Proof. Consider the principal fundamental matrix solution $\Phi(t)$ at $t = 0$ of the homogeneous system corresponding to the differential equation (2.35). By the variation of parameters formula, we have the equation

$$x(T) = \Phi(T)x(0) + \Phi(T) \int_0^T \Phi^{-1}(s)b(s)\, ds.$$

Also, by Theorem 2.82, there is a constant matrix B such that $\Phi(T) = e^{TB}$. Thus, the stroboscopic Poincaré map P is given by

$$P(\xi) := \Phi(T)\xi + \Phi(T) \int_0^T \Phi^{-1}(s)b(s)\, ds$$

$$= e^{TB}\left(\xi + \int_0^T \Phi^{-1}(s)b(s)\, ds\right).$$

If the solution with initial condition $x(0) = \xi_0$ is bounded, then the sequence $\{P^j(\xi_0)\}_{j=0}^\infty$ is bounded. Also, P is an affine map; that is, $P(\xi) = L\xi + y$ where $L = e^{TB} = \Phi(T)$ is a real invertible linear map and y is an element of $\mathbb{R}^n$.

Note that if there is a point $x \in \mathbb{R}^n$ such that $P(x) = x$, then the system (2.35) has a periodic orbit. Thus, if we assume that there are no periodic orbits, then the equation

$$(I - L)\xi = y$$

has no solution ξ. In other words, y is not in the range $\mathcal{R}$ of the operator $I - L$.

There is some vector $v \in \mathbb{R}^n$ such that v is orthogonal to $\mathcal{R}$ and the inner product $\langle v, y \rangle$ does not vanish. Moreover, because v is orthogonal to the range, we have

$$\langle (I - L)\xi, v \rangle = 0$$

for each $\xi \in \mathbb{R}^n$, and therefore

$$\langle \xi, v \rangle = \langle L\xi, v \rangle. \tag{2.36}$$

Using the representation $P(\xi) = L\xi + y$ and an induction argument, it is easy to prove that if j is a nonnegative integer, then $P^j(\xi_0) = L^j\xi_0 + \sum_{k=0}^{j-1} L^k y$. By taking the inner product with v and repeatedly applying the reduction formula (2.36), we have

$$\langle P^j(\xi_0), v \rangle = \langle \xi_0, v \rangle + (j-1)\langle y, v \rangle.$$

Moreover, because $\langle v, y \rangle \neq 0$, it follows immediately that

$$\lim_{j \to \infty} \langle P^j(\xi_0), v \rangle = \infty,$$

and therefore the sequence $\{P^j(\xi_0)\}_{j=0}^{\infty}$ is unbounded, in contradiction. $\square$

2.4.4 Stability of Periodic Orbits

Consider a (nonlinear) autonomous system of differential equations on $\mathbb{R}^n$ given by $\dot{u} = f(u)$ with a periodic orbit Γ. Also, for each $\xi \in \mathbb{R}^n$, define the vector function $t \mapsto u(t, \xi)$ to be the solution of this system with the initial condition $u(0, \xi) = \xi$.

If $p \in \Gamma$ and $\Sigma' \subset \mathbb{R}^n$ is a section transverse to $f(p)$ at p, then, as a corollary of the implicit function theorem, there is an open set $\Sigma \subseteq \Sigma'$ and a function $T : \Sigma \to \mathbb{R}$, the time of first return to Σ', such that for each $\sigma \in \Sigma$, we have $u(T(\sigma), \sigma) \in \Sigma'$. The map $\mathcal{P}$, given by $\sigma \mapsto u(T(\sigma), \sigma)$, is the Poincaré map corresponding to the Poincaré section Σ.

The Poincaré map is defined only on Σ, a manifold contained in $\mathbb{R}^n$. It is convenient to avoid choosing local coordinates on Σ. Thus, we will view the elements in Σ also as points in the ambient space $\mathbb{R}^n$. In particular, if $v \in \mathbb{R}^n$ is tangent to Σ at p, then the derivative of $\mathcal{P}$ in the direction v is given by

$$D\mathcal{P}(p)v = (dT(p)v)f(p) + u_\xi(T(p), p)v. \tag{2.37}$$

The next proposition relates the spectrum of $D\mathcal{P}(p)$ to the Floquet multipliers of the first variational equation

$$\dot{W} = Df(u(t, p))W.$$

Proposition 2.122. *If Γ is a periodic orbit and $p \in \Gamma$, then the union of the set of eigenvalues of the derivative of a Poincaré map at $p \in \Gamma$ and the singleton set $\{1\}$ is the same as the set of characteristic multipliers of the first variational equation along Γ. In particular, zero is not an eigenvalue.*

Proof. Recall that $t \mapsto u_\xi(t, \xi)$ is the principal fundamental matrix solution at $t = 0$ of the first variational equation and, since

$$\frac{d}{dt} f(u(t, \xi)) = Df(u(t, \xi))u_t(t, \xi) = Df(u(t, \xi))f(u(t, \xi)),$$

the vector function $t \mapsto f(u(t, \xi))$ is the solution of the variational equation with the initial condition $W(0) = f(\xi)$. In particular,

$$u_\xi(T(p), p)f(p) = f(u(T(p), p)) = f(p),$$

and therefore $f(p)$ is an eigenvector of the linear transformation $u_\xi(T(p), p)$ with eigenvalue the number one.

Since Σ is transverse to $f(p)$, there is a basis of $\mathbb{R}^n$ of the form

$$f(p), s_1, \ldots, s_{n-1}$$

with s_i tangent to Σ at p for each $i = 1, \ldots, n-1$. It follows that the matrix $u_\xi(T(p), p)$ has block form, relative to this basis, given by

$$\begin{pmatrix} 1 & a \\ 0 & b \end{pmatrix}$$

where a is $1 \times (n-1)$ and b is $(n-1) \times (n-1)$. Moreover, each $v \in \mathbb{R}^n$ that is tangent to Σ at p has block form (the transpose of) $(0, v_\Sigma)$. As a result, we have the equality

$$u_\xi(T(p), p)v = \begin{pmatrix} 1 & a \\ 0 & b \end{pmatrix} \begin{pmatrix} 0 \\ v_\Sigma \end{pmatrix}.$$

The range of $D\mathcal{P}(p)$ is tangent to Σ at p. Thus, using equation (2.37) and the block form of $u_\xi(T(p), p)$, it follows that

$$D\mathcal{P}(p)v = \begin{pmatrix} dT(p)v + av_\Sigma \\ bv_\Sigma \end{pmatrix} = \begin{pmatrix} 0 \\ bv_\Sigma \end{pmatrix}.$$

In other words, the derivative of the Poincaré map may be identified with b and the differential of the return time map with $-a$. In particular, the eigenvalues of the derivative of the Poincaré map coincide with the eigenvalues of b. $\square$

Exercise 2.123. Prove that the characteristic multipliers of the first variational equation along a periodic orbit do not depend on the choice of $p \in \Gamma$.

Most of the rest of this section is devoted to a proof of the following fundamental theorem.

Theorem 2.124. *Suppose that Γ is a periodic orbit for the autonomous differential equation $\dot{u} = f(u)$ and $\mathcal{P}$ is a corresponding Poincaré map defined on a Poincaré section Σ such that $p \in \Gamma \cap \Sigma$. If the eigenvalues of the derivative $D\mathcal{P}(p)$ are inside the unit circle in the complex plane, then Γ is asymptotically stable.*

There are several possible proofs of this theorem. The approach used here is adapted from [123].

To give a complete proof of Theorem 2.124, we will require several preliminary results. Our first objective is to show that the point p is an asymptotically stable fixed point of the dynamical system defined by the Poincaré map on Σ.

Let us begin with a useful simple replacement of the Jordan normal form theorem that is adequate for our purposes here (see [129]).

Proposition 2.125. *An $n \times n$ (possibly complex) matrix A is similar to an upper triangular matrix whose diagonal elements are the eigenvalues of A.*

Proof. Let v be a nonzero eigenvector of A corresponding to the eigenvalue λ. The vector v can be completed to a basis of $\mathbb{C}^n$ that defines a matrix Q partitioned by the corresponding column vectors $Q := [v, y_1, \ldots, y_{n-1}]$. Moreover, Q is invertible and

$$[Q^{-1}v, Q^{-1}y_1, \ldots, Q^{-1}y_{n-1}] = [\mathbf{e}_1, \ldots, \mathbf{e}_n]$$

where $\mathbf{e}_1, \ldots, \mathbf{e}_n$ denote the usual basis elements.

Note that

$$
\begin{aligned}
Q^{-1}AQ &= Q^{-1}[\lambda v, Ay_1, \ldots, Ay_{n-1}] \\
&= [\lambda \mathbf{e}_1, Q^{-1}Ay_1, \ldots, Q^{-1}Ay_{n-1}].
\end{aligned}
$$

In other words, the matrix $Q^{-1}AQ$ is given in block form by

$$Q^{-1}AQ = \begin{pmatrix} \lambda & * \\ 0 & \tilde{A} \end{pmatrix}$$

where $\tilde{A}$ is an $(n-1) \times (n-1)$ matrix. In particular, this proves the theorem for all 2×2 matrices.

By induction, there is an $(n-1) \times (n-1)$ matrix $\tilde{R}$ such that $\tilde{R}^{-1}\tilde{A}\tilde{R}$ is upper triangular. The matrix $(QR)^{-1}AQR$ where

$$R = \begin{pmatrix} 1 & 0 \\ 0 & \tilde{R} \end{pmatrix}$$

is an upper triangular matrix with the eigenvalues of A as its diagonal elements, as required. $\square$

Let $\rho(A)$ denote the *spectral radius* of A, that is, the maximum modulus of the eigenvalues of A.

Proposition 2.126. *Suppose that A is an $n \times n$ matrix. If $\epsilon > 0$, then there is a norm on $\mathbb{C}^n$ such that $\|A\|_\epsilon < \rho(A) + \epsilon$. If A is a real matrix, then the restriction of the "ϵ-norm" to $\mathbb{R}^n$ is a norm on $\mathbb{R}^n$ with the same property.*

Proof. The following proof is adapted from [129]. By Proposition 2.125, there is a matrix Q such that

$$QAQ^{-1} = D + N$$

where D is diagonal with the eigenvalues of A as its diagonal elements, and N is upper triangular with each of its diagonal elements equal to zero.

Let $\mu > 0$, and define a new diagonal matrix S with diagonal elements

$$1, \mu^{-1}, \mu^{-2}, \dots, \mu^{1-n}.$$

A computation shows that

$$S(D + N)S^{-1} = D + SNS^{-1}.$$

Also, it is easy to show—by writing out the formulas for the components— that every element of the matrix SNS^{-1} is $O(\mu)$.

Define a norm on $\mathbb{C}^n$, by the formula

$$|v|_\mu := |SQv| = \langle SQv, SQv \rangle$$

where the angle brackets on the right hand side denote the usual Euclidean inner product on $\mathbb{C}^n$. It is easy to verify that this procedure indeed defines a norm on $\mathbb{C}^n$ that depends on the parameter μ.

Post multiplication by SQ of both sides of the equation

$$SQAQ^{-1}S^{-1} = D + SNS^{-1}$$

yields the formula

$$SQA = (D + SNS^{-1})SQ.$$

Using this last identity we have that

$$|Av|_\mu^2 = |SQAv|^2 = |(D + SNS^{-1})SQv|^2.$$

Let us define $w := SQv$ and then expand the last norm into inner products to obtain

$$|Av|_\mu^2 = \langle Dw, Dw \rangle + \langle SNS^{-1}w, Dw \rangle$$
$$+ \langle Dw, SNS^{-1}w \rangle + \langle SNS^{-1}w, SNS^{-1}w \rangle.$$

A direct estimate of the first inner product together with an application of the Schwarz inequality to each of the other inner products yields the following estimate:

$$|Av|_\mu^2 \leq (\rho^2(A) + O(\mu))|w|^2.$$

Moreover, we have that $|v|_\mu = |w|$. In particular, if $|v|_\mu = 1$ then $|w| = 1$, and it follows that

$$\|A\|_\mu^2 \leq \rho^2(A) + O(\mu).$$

Thus, if $\mu > 0$ is sufficiently small, then $\|A\|_\mu < \rho(A) + \epsilon$, as required. □

Corollary 2.127. *If all the eigenvalues of the $n \times n$ matrix A are inside the unit circle in the complex plane, then there is an "adapted norm" and a number λ, with $0 < \lambda < 1$, such that $|Av|_a < \lambda|v|_a$ for all vectors v, real or complex. In particular A is a contraction with respect to the adapted norm. Moreover, for each norm on $\mathbb{R}^n$ or $\mathbb{C}^n$, there is a positive number C such that $|A^n v| \leq C\lambda^n|v|$ for all nonnegative integers n.*

Proof. Under the hypothesis, we have $\rho(A) < 1$; thus, there is a number λ such that $\rho(A) < \lambda < 1$. Using Proposition 2.126, there is an adapted norm so that $\|A\|_a < \lambda$. This proves the first part of the corollary. To prove the second part, recall that all norms on a finite dimensional space are equivalent. In particular, there are positive numbers C_1 and C_2 such that

$$C_1|v| \leq |v|_a \leq C_2|v|$$

for all vectors v. Thus, we have

$$C_1|A^n v| \leq |A^n v|_a \leq |A|_a^n|v|_a \leq C_2\lambda^n|v|.$$

After dividing both sides of the last inequality by $C_1 > 0$, we obtain the desired estimate. □

We are now ready to return to the dynamics of the Poincaré map $\mathcal{P}$ defined above. Recall that Γ is a periodic orbit for the differential equation $\dot{u} = f(u)$ and $\mathcal{P} : \Sigma \to \Sigma'$ is defined by $\mathcal{P}(\sigma) = u(T(\sigma), \sigma)$ where T is the return time function. Also, we have that $p \in \Gamma \cap \Sigma$.

Lemma 2.128. *Suppose that $V \subseteq \mathbb{R}^n$ is an open set with compact closure $\bar{V}$ such that $\Gamma \subset V$ and $\bar{V}$ is contained in the domain of the function f. If $t_* \geq 0$, then there is an open set $W \subseteq V$ that contains Γ and is such that, for each point $\xi \in W$, the solution $t \mapsto u(t, \xi)$ is defined and stays in V on the interval $0 \leq t \leq t_*$. Moreover, if ξ and ν are both in W and $0 \leq t \leq t_*$, then there is a number $L > 0$ such that*

$$|u(t, \xi) - u(t, \nu)| < |\xi - \nu|e^{Lt_*}.$$

Proof. Note that $\bar{V}$ is a compact subset of the domain of the function f. By Lemma 2.75, f is globally Lipschitz on V with a Lipschitz constant $L > 0$. Also, there is a minimum *positive* distance m from the boundary of V to Γ.

An easy application of Gronwall's inequality can be used to show that if $\xi, \nu \in V$, then

$$|u(t, \xi) - u(t, \nu)| \leq |\xi - \nu| e^{Lt} \tag{2.38}$$

for all t such that both solutions are defined on the interval $[0, t]$.

Define the set

$$W_q := \{\xi \in \mathbb{R}^n : |\xi - q| e^{Lt_*} < m\}$$

and note that W_q is open. If $\xi \in W_q$, then

$$|\xi - q| < m e^{-Lt_*} < m.$$

Thus, it follows that $W_q \subseteq V$.

Using the extension theorem (Theorem 1.263), it follows that if $\xi \in W_q$, then the interval of existence of the solution $t \mapsto u(t, \xi)$ can be extended as long as the orbit stays in the compact set $\bar{V}$. The point q is on the periodic orbit Γ. Thus, the solution $t \to u(t, q)$ is defined for all $t \geq 0$. Using the definition of W_q and an application of the inequality (2.38) to the solutions starting at ξ and q, it follows that the solution $t \mapsto u(t, \xi)$ is defined and stays in V on the interval $0 \leq t \leq t_*$.

The union $W := \bigcup_{q \in \Gamma} W_q$ is an open set in V containing Γ with the property that all solutions starting in W remain in V at least on the time interval $0 \leq t \leq t_*$. □

Define the distance of a point $q \in \mathbb{R}^n$ to a set $S \subseteq \mathbb{R}^n$ by

$$\text{dist}(q, S) = \inf_{x \in S} |q - x|$$

where the norm on the right hand side is the usual Euclidean norm. Similarly, the (minimum) distance between two sets is defined as

$$\text{dist}(A, B) = \inf\{|a - b| : a \in A, b \in B\}.$$

(Warning: dist is not a metric.)

Proposition 2.129. *If $\sigma \in \Sigma$ and if $\lim_{n \to \infty} \mathcal{P}^n(\sigma) = p$, then*

$$\lim_{t \to \infty} \text{dist}(u(t, \sigma), \Gamma) = 0.$$

Proof. Let $\epsilon > 0$ be given and let Σ_0 be an open subset of Σ such that $p \in \Sigma_0$ and such that $\bar{\Sigma}_0$, the closure of Σ_0, is a compact subset of Σ. The return time map T is continuous; hence, it is uniformly bounded on the set $\bar{\Sigma}_0$, that is,

$$\sup\{T(\eta) : \eta \in \bar{\Sigma}_0\} = T^* < \infty.$$

Let V be an open subset of $\mathbb{R}^n$ with compact closure $\bar{V}$ such that $\Gamma \subset V$ and $\bar{V}$ is contained in the domain of f. By Lemma 2.128, there is an open set $W \subseteq V$ such that $\Gamma \subset W$ and such that, for each $\xi \in W$, the solution starting at ξ remains in V on the interval $0 \leq s \leq T^*$.

Choose $\delta > 0$ so small that the set

$$\Sigma_\delta := \{\eta \in \Sigma : |\eta - p| < \delta\}$$

is contained in $W \cap \Sigma_0$, and such that

$$|\eta - p|e^{LT^*} < \min\{m, \epsilon\}$$

for all $\eta \in \Sigma_\delta$. By Lemma 2.128, if $\eta \in \Sigma_\delta$, then, for $0 \leq s \leq T^*$, we have that

$$|u(s, \eta) - u(s, p)| < \epsilon.$$

By the hypothesis, there is some integer $N > 0$ such that $\mathcal{P}^n(\sigma) \in \Sigma_\delta$ whenever $n \geq N$.

Using the group property of the flow, let us note that

$$\mathcal{P}^n(\sigma) = u(\sum_{j=0}^{n-1} T(\mathcal{P}^j(\sigma)), \sigma).$$

Moreover, if $t \geq \sum_{j=0}^{N-1} T(\mathcal{P}^j(\sigma))$, then there is some integer $n \geq N$ and some number s such that $0 \leq s \leq T^*$ and

$$t = \sum_{j=0}^{n-1} T(\mathcal{P}^j(\sigma)) + s.$$

For this t, we have $\mathcal{P}^n(\sigma) \in \Sigma_\delta$ and

$$\begin{aligned}
\text{dist}(u(t, \sigma), \Gamma) &= \min_{q \in \Gamma} |u(t, \sigma) - q| \\
&\leq |u(t, \sigma) - u(s, p)| \\
&= |u(s, u(\sum_{j=0}^{n-1} T(\mathcal{P}^j(\sigma)), \sigma)) - u(s, p)| \\
&= |u(s, P^n(\sigma)) - u(s, p)|.
\end{aligned}$$

It follows that $\text{dist}(u(t, \sigma), \Gamma) < \epsilon$ whenever $t \geq \sum_{j=0}^{N-1} T(\mathcal{P}^j(\sigma))$. In other words,

$$\lim_{t \to \infty} \text{dist}(u(t, \sigma), \Gamma) = 0,$$

as required.   □

We are now ready for the proof of Theorem 2.124.

Proof. Suppose that V is a neighborhood of Γ. We must prove that there is a neighborhood U of Γ such that $U \subseteq V$ with the additional property that every solution of $\dot{u} = f(u)$ that starts in U stays in V and is asymptotic to Γ.

We may as well assume that V has compact closure $\bar{V}$ and $\bar{V}$ is contained in the domain of f. Then, by Lemma 2.128, there is an open set W that contains Γ and is contained in the closure of V with the additional property that every solution starting in W exists and stay in V on the time interval $0 \leq t \leq 2\tau$ where $\tau := T(p)$ is the period of Γ.

Also, let us assume without loss of generality that our Poincaré section Σ is a subset of a hyperplane Σ' and that the coordinates on Σ' are chosen so that p lies at the origin. By our hypothesis, the linear transformation $D\mathcal{P}(0) : \Sigma' \to \Sigma'$ has its spectrum inside the unit circle in the complex plane. Thus, by Corollary 2.127, there is an adapted norm on Σ' and a number λ with $0 < \lambda < 1$ such that $\|D\mathcal{P}(0)\| < \lambda$.

Using the continuity of the map $\sigma \to D\mathcal{P}(\sigma)$, the return time map, and the adapted norm, there is an open ball $\Sigma_0 \subseteq \Sigma$ centered at the origin such that $\Sigma_0 \subset W$, the return time map T restricted to Σ_0 is bounded by 2τ, and $\|D\mathcal{P}(\sigma)\| < \lambda$ whenever $\sigma \in \Sigma_0$. Moreover, using the mean value theorem, it follows that

$$|\mathcal{P}(\sigma)| = |\mathcal{P}(\sigma) - \mathcal{P}(0)| < \lambda|\sigma|,$$

whenever $\sigma \in \Sigma_0$. In particular, if $\sigma \in \Sigma_0$, then $\mathcal{P}(\sigma) \in \Sigma_0$.

Let us show that all solutions starting in Σ_0 are defined for all positive time. To see this, consider $\sigma \in \Sigma_0$ and note that, by our construction, the solution $t \mapsto u(t, \sigma)$ is defined for $0 \leq t \leq T(\sigma)$ because $T(\sigma) < 2\tau$. We also have that $u(T(\sigma), \sigma) = \mathcal{P}(\sigma) \in \Sigma_0$. Thus, the solution $t \mapsto u(t, \sigma)$ can be extended beyond the time $T(\sigma)$ by applying the same reasoning to the solution $t \to u(t, \mathcal{P}(\sigma)) = u(t + u(T\sigma), \sigma))$. This procedure can be extended indefinitely, and thus the solution $t \to u(t, \sigma)$ can be extended for all positive time.

Define $U := \{u(t, \sigma) : \sigma \in \Sigma_0 \text{ and } t > 0\}$. Clearly, $\Gamma \subset U$ and also every solution that starts in U stays in U for all $t \geq 0$. We will show that U is open. To prove this fact, let $\xi := u(t, \sigma) \in U$ with $\sigma \in \Sigma_0$. If we consider the restriction of the flow given by $u : (0, \infty) \times \Sigma_0 \to U$, then, using the same idea as in the proof of the rectification lemma (Lemma 1.120), it is easy to see that the derivative $Du(t, \sigma)$ is invertible. Thus, by the inverse function theorem (Theorem 1.121), there is an open set in U at ξ diffeomorphic to a product neighborhood of (t, σ) in $(0, \infty) \times \Sigma_0$. Thus, U is open.

To show that $U \subseteq V$, let $\xi := u(t, \sigma) \in U$ with $\sigma \in \Sigma_0$. There is some integer $n \geq 0$ and some number s such that

$$t = \sum_{j=0}^{n-1} T(\mathcal{P}^j(\sigma)) + s$$

where $0 \leq s < T(\mathcal{P}^n(\sigma)) < 2\tau$. In particular, we have that $\xi = u(s, \mathcal{P}^n(\sigma))$. But since $0 \leq s < 2\tau$ and $\mathcal{P}^n(\sigma) \in W$ it follows that $\xi \in V$.

Finally, for this same $\xi \in U$, we have as an immediate consequence of Proposition 2.129 that $\lim_{t \to \infty} \text{dist}(u(t, \mathcal{P}^n(\xi)), \Gamma) = 0$. Moreover, for each $t \geq 0$, we also have that

$$\text{dist}(u(t, \xi), \Gamma) = \text{dist}(u(t, u(s, \mathcal{P}^n(\xi))), \Gamma) = \text{dist}(u(s + t, \mathcal{P}^n(\xi)), \Gamma).$$

It follows that $\lim_{t \to \infty} \text{dist}(u(t, \xi), \Gamma) = 0$, as required. $\square$

A useful application of our results can be made for a periodic orbit Γ of a differential equation defined on the plane. In fact, there are exactly two characteristic multipliers of the first variational equation along Γ. Since one of these characteristic multipliers must be the number one, the product of the characteristic multipliers is the eigenvalue of the derivative of every Poincaré map defined on a section transverse to Γ. Because the determinant of a matrix is the product of its eigenvalues, an application of Liouville's formula proves the following proposition.

Proposition 2.130. *If Γ is a periodic orbit of period ν of the autonomous differential equation $\dot{u} = f(u)$ on the plane, and if $\mathcal{P}$ is a Poincaré map defined at $p \in \Gamma$, then, using the notation of this section, the eigenvalue λ_Γ of the derivative of $\mathcal{P}$ at p is given by*

$$\lambda_\Gamma = \det u_\xi(T(p), p) = e^{\int_0^\nu \text{div} f(u(t,p))\, dt}.$$

In particular, if $\lambda_\Gamma < 1$ then Γ is asymptotically stable, whereas if $\lambda_\Gamma > 1$ then Γ is unstable.

The flow near an attracting limit cycle is very well understood. A next proposition states that the orbits of points in the basin of attraction of the limit cycle are "asymptotically periodic."

Proposition 2.131. *Suppose that Γ is an asymptotically stable periodic orbit with period T. There is a neighborhood V of Γ such that if $\xi \in V$, then $\lim_{t \to \infty} |u(t + T, \xi) - u(t, \xi)| = 0$ where $| \; |$ is an arbitrary norm on $\mathbb{R}^n$. (In this case, the point ξ is said to have asymptotic period T.)*

Proof. By Lemma 2.128, there is an open set W such that $\Gamma \subset W$ and the function $\xi \mapsto u(T, \xi)$ is defined for each $\xi \in W$. Using the continuity of this function, there is a number $\delta > 0$ such that $\delta < \epsilon/2$ and

$$|u(T, \xi) - u(T, \eta)| < \frac{\epsilon}{2}$$

whenever $\xi, \eta \in W$ and $|\xi - \eta| < \delta$.

By the hypothesis, there is a number T^* so large that $\text{dist}(u(t,\xi), \Gamma) < \delta$ whenever $t \geq T^*$. In particular, for each $t \geq T^*$, there is some $q \in \Gamma$ such that $|u(t,\xi) - q| < \delta$. Using this fact and the group property of the flow, we have that

$$|u(t+T,\xi) - u(t,\xi)| \leq |u(T, u(t,\xi)) - u(T,q)| + |q - u(t,\xi)|$$
$$\leq \frac{\epsilon}{2} + \delta < \epsilon$$

whenever $t \geq T^*$. Thus, $\lim_{t\to\infty} |u(t+T,\xi) - u(t,\xi)| = 0$, as required. $\quad\square$

A periodic orbit can be asymptotically stable without being hyperbolic. In fact, it is easy to construct a limit cycle in the plane that is asymptotically stable whose Floquet multiplier is the number one. By the last proposition, points in the basin of attraction of such an attracting limit cycle have asymptotic periods equal to the period of the limit cycle. But, if the periodic orbit is hyperbolic, then a stronger result is true: Not only does each point in the basin of attraction have an asymptotic period, each such point has an asymptotic phase. This is the content of the next result.

Theorem 2.132. *If Γ is an attracting hyperbolic periodic orbit, then there is a neighborhood V of Γ such that for each $\xi \in V$ there is some $q \in \Gamma$ such that $\lim_{t\to\infty} |u(t,\xi) - u(t,q)| = 0$. (In this case, ξ is said to have asymptotic phase q.)*

Proof. Let Σ be a Poincaré section at $p \in \Gamma$ with compact closure, return map $\mathcal{P}$, and return-time map T. Without loss of generality, we will assume that for each $\sigma \in \Sigma$ we have (1) $\lim_{n\to\infty} \mathcal{P}^n(\sigma) = p$, (2) $T(\sigma) < 2T(p)$, and (3) $\|DT(\sigma)\| < 2\|DT(p)\|$.

By the hyperbolicity hypothesis, the spectrum of $D\mathcal{P}(p)$ is inside the unit circle; therefore, there are numbers C and λ such that $C > 0$, $0 < \lambda < 1$ and

$$|p - \mathcal{P}^n(\sigma)| < C\lambda^n \|p - \sigma\|.$$

Let

$$K := \frac{2C\|DT(p)\|}{1-\lambda} \sup_{\sigma\in\bar{\Sigma}} \|p - \sigma\| + 3T(p).$$

Using the implicit function theorem, it is easy to construct a neighborhood V of Γ such that for each $\xi \in V$, there is a number $t_\xi \geq 0$ with $\sigma_\xi := u(t_\xi, \xi) \in \Sigma$. Moreover, using Lemma 2.128, we can choose V such that every solution with initial point in V is defined at least on the time interval $-K \leq t \leq K$. Indeed, by the asymptotic stability of Γ, there is a neighborhood V of Γ such that every solution starting in V is defined for all positive time. If we redefine V to be the image of V under the flow for time K, then every solution starting in V is defined at least on the time interval $-K \le t \le K$.

We will show that if $\sigma_\xi \in \Sigma$, then there is a point $q_\xi \in \Gamma$ such that

$$\lim_{t \to \infty} |u(t, \sigma_\xi) - u(t, q_\xi)| = 0.$$

Using this fact, it follows that if $r := u(-t_\xi, q_\xi)$, then

$$\lim_{t \to \infty} |u(t, \xi) - u(t, r)| = \lim_{t \to \infty} |u(t - t_\xi, u(t_\xi, \xi)) - u(t - t_\xi, q_\xi)|$$

$$= \lim_{t \to \infty} |u(t - t_\xi, \sigma_\xi) - u(t - t_\xi, q_\xi)| = 0.$$

Thus, it suffices to prove the theorem for a point $\sigma \in \Sigma$.

Given $\sigma \in \Sigma$, define

$$s_n := nT(p) - \sum_{j=0}^{n-1} T(\mathcal{P}^j(\sigma)).$$

Note that

$$(n+1)T(p) - nT(p) = T(\mathcal{P}^n(\sigma)) + s_{n+1} - s_n,$$

and, as a result,

$$|s_{n+1} - s_n| = |T(p) - T(\mathcal{P}^n(\sigma))| \le 2\|DT(p)\| \|p - \mathcal{P}^n(\sigma)\|.$$

Hence,

$$|s_{n+1} - s_n| < 2\|DT(p)\| \|C\| \|p - \sigma\| \lambda^n$$

whenever $n \ge 0$.

Because $s_n = s_1 + \sum_{j=1}^{n-1}(s_{j+1} - s_j)$ and

$$\sum_{j=1}^{n-1} |s_{j+1} - s_j| < 2C\|DT(p)\| \|p - \sigma\| \sum_{j=1}^{n-1} \lambda^j < 2C\|DT(p)\| \frac{\|p - \sigma\|}{1 - \lambda},$$

the series $\sum_{j=1}^{\infty}(s_{j+1} - s_j)$ is absolutely convergent—its absolute partial sums form an increasing sequence that is bounded above. Thus, in fact, there is a number s such that $\lim_{n \to \infty} s_n = s$. Also, the sequence $\{s_n\}_{n=1}^{\infty}$ is uniformly bounded; that is,

$$|s_n| \le |s_1| + 2C\|DT(p)\| \frac{\|p - \sigma\|}{1 - \lambda} \le K.$$

Hence, the absolute value of its limit $|s|$ is bounded by the same quantity.

Let $\epsilon > 0$ be given. By the compactness of its domain, the function

$$u : [-K, K] \times \bar{\Sigma} \to \mathbb{R}^n$$

is uniformly continuous. In particular, there is a number $\delta > 0$ such that if (t_1, σ_1) and (t_2, σ_2) are both in the domain and if $|t_1 - t_2| + |\sigma_1 - \sigma_2| < \delta$, then

$$|u(t_1, \sigma_1) - u(t_2, \sigma_2)| < \epsilon.$$

In view of the equality,

$$u(nT(p), \sigma) = u(s_n, \mathcal{P}^n(\sigma)),$$

which follows from the definition of s_n, we have

$$|u(nT(p), \sigma) - u(s, p)| = |u(s_n, \mathcal{P}^n(\sigma)) - u(s, p)|.$$

Since for sufficiently large n,

$$|s_n - s| + |\mathcal{P}^n(\sigma) - p| < \epsilon,$$

it follows that

$$\lim_{n \to \infty} |u(nT(p), \sigma) - u(s, p)| = 0.$$

Also, for each $t \geq 0$, there is an integer $n \geq 0$ and a number $s(t)$ such that $0 \leq s(t) < T(p)$ and $t = nT(p) + s(t)$. Using this fact, we have the equation

$$|u(t, \sigma) - u(t, u(s, p))| = |u(s(t), u(nT(p), \sigma)) - u(s(t), u(nT(p), u(s, p)))|.$$

Also, because $q := u(s, p) \in \Gamma$ and Lemma 2.128, there is a constant $L > 0$ such that

$$|u(t, \sigma) - u(t, q)| = |u(s(t), u(nT(p), \sigma)) - u(s(t), q))|$$
$$\leq |u(nT(p), \sigma) - q|e^{LT(p)}.$$

By passing to the limit as $n \to \infty$, we obtain the desired result. $\qquad\square$

Necessary and sufficient conditions for the existence of asymptotic phase are known (see [47, 77]). An alternate proof of Theorem 2.132 is given in [47].

Exercise 2.133. Find a periodic solution of the system

$$\dot{x} = x - y - x(x^2 + y^2),$$
$$\dot{y} = x + y - y(x^2 + y^2),$$
$$\dot{z} = -z,$$

and determine its stability type. In particular, compute the Floquet multipliers for the monodromy matrix associated with the periodic orbit [128, p. 120].

Exercise 2.134. (a) Find an example of a planar system with a limit cycle such that some nearby solutions do not have an asymptotic phase. (b) Contrast and compare the asymptotic phase concept for the following planar systems that are defined in the punctured plane in polar coordinates:

1. $\dot{r} = r(1 - r)$, $\dot{\theta} = r$,
2. $\dot{r} = r(1 - r)^2$, $\dot{\theta} = r$,
3. $\dot{r} = r(1 - r)^n$, $\dot{\theta} = r$.

Exercise 2.135. Suppose that $v \neq 0$ is an eigenvector for the monodromy operator with associated eigenvalue λ_Γ as in Proposition 2.130. If $\lambda_\Gamma \neq 1$, then v and $f(p)$ are independent vectors that form a basis for $\mathbb{R}^2$. The monodromy operator expressed in this basis is diagonal. (a) Express the operators a and b defined in the proof of Proposition 2.122 in this basis. (b) What can you say about the derivative of the transit time map along a section that is tangent to v at p?

Exercise 2.136. This exercise is adapted from [235]. Suppose that $f : \mathbb{R}^2 \to \mathbb{R}$ is a smooth function and $A := \{(x,y) \in \mathbb{R}^2 : f(x,y) = 0\}$ is a regular level set of f. (a) Prove that each bounded component of A is an attracting hyperbolic limit cycle for the differential equation

$$\dot{x} = -f_y - f f_x, \qquad \dot{y} = f_x - f f_y.$$

(b) Prove that the bounded components of A are the only periodic orbits of the system. (c) Draw and explain the phase portrait of the system for the case where

$$f(x,y) = ((x - \epsilon)^2 + y^2 - 1)(x^2 + y^2 - 9).$$

Exercise 2.137. Consider an attracting hyperbolic periodic orbit Γ for an autonomous system $\dot{u} = f(u)$ with flow φ_t, and for each point $p \in \Gamma$, let Γ_p denote the set of all points in the phase space with asymptotic phase p. (a) Construct Γ_p for each p on the limit cycle in the planar system

$$\dot{x} = -y + x(1 - x^2 - y^2), \quad \dot{y} = x + y(1 - x^2 - y^2).$$

(b) Repeat the construction for the planar systems of Exercise 2.134. (c) Prove that $\mathcal{F} := \bigcup_{p \in \Gamma} \Gamma_p$ is an invariant foliation of the phase space in a neighborhood U of Γ. Let us take this to mean that every point in U is in one of the sets in the union $\mathcal{F}$ and the following invariance property is satisfied: If $\xi \in \Gamma_p$ and $s \in \mathbb{R}$, then $\varphi_s(\xi) \in \Gamma_{\varphi_s(p)}$. The second condition states that the flow moves fibers of the foliation (Γ_p is the fiber over p) to fibers of the foliation. (d) Are the fibers of the foliation smooth manifolds?

3

Applications

Is the subject of ordinary differential equations important? The ultimate answer to this question is certainly beyond the scope of this book. But two main points of evidence for an affirmative answer are provided in this chapter:

- Ordinary differential equations arise naturally from the foundations of physical science.
- Ordinary differential equations are useful tools for solving physical problems.

You will have to decide if the evidence is sufficient. Warning: If you pay too much attention to philosophical issues concerning the value of a mathematical subject, then you might stop producing mathematics. On the other hand, if you pay no attention to the value of a subject, then how will you know that it is worthy of study?

3.1 Origins of ODE: The Euler–Lagrange Equation

Let us consider a smooth function $L : \mathbb{R}^k \times \mathbb{R}^k \times \mathbb{R} \to \mathbb{R}$, a pair of points p_1, $p_2 \in \mathbb{R}^k$, two real numbers $t_1 < t_2$, and the set $C := C(p_1, p_2, t_1, t_2)$ of all smooth curves $q : \mathbb{R} \to \mathbb{R}^k$ such that $q(t_1) = p_1$ and $q(t_2) = p_2$. Using this data, there is a function $\Phi : C \to \mathbb{R}$ given by

$$\Phi(q) = \int_{t_1}^{t_2} L(q(t), \dot{q}(t), t)\, dt. \tag{3.1}$$

The Euler–Lagrange equation, an ordinary differential equation associated with the function L—called the *Lagrangian*—arises from the following problem: Find the extreme points of the function Φ. This variational problem is the basis for Lagrangian mechanics.

Recall from calculus that an extreme point of a smooth function is a point at which its derivative vanishes. To use this definition directly for the function Φ, we would have to show that C is a manifold and Φ is differentiable. This can be done. Instead, we will bypass these requirements by redefining the notion of extreme point. In effect, we will define the concept of *directional derivative* for a scalar function on a space of curves. Then, an extreme point is defined to be a point where all directional derivatives vanish.

Recall our geometric interpretation of the derivative of a smooth function on a manifold: For a tangent vector at a point in the domain of the function, choose a curve whose tangent at time $t = 0$ is the given vector, move the curve to the range of the function by composing it with the function, and then differentiate the resulting curve at $t = 0$ to produce a tangent vector on the range. This tangent vector is the image of the original vector under the derivative of the function. In the context of the function Φ on the space of curves C, let us consider a curve $\gamma : \mathbb{R} \to C$. Note that for $s \in \mathbb{R}$, the point $\gamma(s) \in C$ is itself a curve $\gamma(s) : \mathbb{R} \to \mathbb{R}^k$. So, in particular, if $t \in \mathbb{R}$, then $\gamma(s)(t) \in \mathbb{R}^k$. Rather than use the cumbersome notation $\gamma(s)(t)$, it is customary to interpret our curve of curves as a "variation of curves" in C, that is, as a smooth function $Q : \mathbb{R} \times \mathbb{R} \to \mathbb{R}^k$ with the "end conditions"

$$Q(s,t_1) \equiv p_1, \qquad Q(s,t_2) \equiv p_2.$$

In this interpretation, $\gamma(s)(t) = Q(s,t)$.

Fix a point $q \in C$ and suppose that $\gamma(0) = q$, or equivalently that $Q(0,t) = q(t)$. Then, as s varies we obtain a family of curves called a *variation* of the curve q. The tangent vector to γ at q is, by definition, the curve $V : \mathbb{R} \to \mathbb{R}^k \times \mathbb{R}^k$ given by $t \mapsto (q(t), v(t))$ where

$$v(t) := \left. \frac{\partial}{\partial s} Q(s,t) \right|_{s=0}.$$

Of course, v is usually not in C because it does not satisfy the required end conditions. On the other hand, v does satisfy a perhaps different pair of end conditions, namely,

$$v(t_1) = \left. \frac{\partial}{\partial s} Q(s,t_1) \right|_{s=0} = 0, \qquad v(t_2) = \left. \frac{\partial}{\partial s} Q(s,t_2) \right|_{s=0} = 0.$$

Let us view the vector V as an element in the "tangent space of C at q."

What is the directional derivative $D\Phi(q)V$ of Φ at q in the direction V? Following the prescription given above, we have the definition

$$D\Phi(q)V := \frac{\partial}{\partial s}\Phi(Q(s, \cdot))\Big|_{s=0}$$

$$= \int_{t_1}^{t_2} \frac{\partial}{\partial s} L\Big(Q(s, t), \frac{\partial Q}{\partial t}(s, t), t\Big)\Big|_{s=0} dt$$

$$= \int_{t_1}^{t_2} \Big(\frac{\partial L}{\partial q}\frac{\partial Q}{\partial s} + \frac{\partial L}{\partial \dot{q}}\frac{\partial^2 Q}{\partial s \partial t}\Big)\Big|_{s=0} dt. \qquad (3.2)$$

After evaluation at $s = 0$ and an integration by parts, we can rewrite the last integral to obtain

$$D\Phi(q)V = \int_{t_1}^{t_2}\Big[\frac{\partial L}{\partial q}(q(t), \dot{q}(t), t) - \frac{d}{dt}\Big(\frac{\partial L}{\partial \dot{q}}(q(t), \dot{q}(t), t)\Big)\Big]\frac{\partial Q}{\partial s}(0, t)\, dt$$

$$= \int_{t_1}^{t_2}\Big[\frac{\partial L}{\partial q}(q(t), \dot{q}(t), t) - \frac{d}{dt}\Big(\frac{\partial L}{\partial \dot{q}}(q(t), \dot{q}(t), t)\Big)\Big]v(t)\, dt. \quad (3.3)$$

The curve q is called an *extremal* if $D\Phi(q)V = 0$ for all tangent vectors V. Since for a given v we can construct $Q(s, t) := q(t) + sv(t)$ so that $\partial Q/\partial s = v$, the curve q is an extremal if the last integral in equation (3.3) vanishes for all smooth functions v that vanish at the points t_1 and t_2.

Proposition 3.1. *The curve q is an extremal if and only if it is a solution of the Euler–Lagrange equation*

$$\frac{d}{dt}\Big(\frac{\partial L}{\partial \dot{q}}\Big) - \frac{\partial L}{\partial q} = 0.$$

Proof. Clearly, if the curve q is a solution of the Euler–Lagrange equation, then, by formula (3.3), we have that $D\Phi(q) = 0$. Conversely, if $D\Phi(q) = 0$, we will show that q is a solution of the Euler–Lagrange equation. If not, then there is some time τ in the open interval (t_1, t_2) such that the quantity

$$\frac{\partial L}{\partial q}(q(t), \dot{q}(t), t) - \frac{d}{dt}\Big(\frac{\partial L}{\partial \dot{q}}(q(t), \dot{q}(t), t)\Big)$$

appearing in the formula (3.3) does not vanish when evaluated at τ. In this case, this quantity has constant sign on a closed interval containing the point τ. Moreover, there is a smooth nonnegative function v that vanishes outside this closed interval and is such that $v(\tau) = 1$ (see Exercise 3.15). Hence, $D\Phi(q)V \neq 0$ with respect to the variation $Q(s, t) := q(t) + sv(t)$, in contradiction. $\qquad\square$

When we search for the extreme points of a function, we are usually interested in its maxima or minima. The same is true for the function Φ defined above. In fact, the theory for determining the maxima and minima of Φ is similar to the usual finite-dimensional theory, but it is complicated by the technical problems of working in infinite-dimensional function spaces. The general theory is explained in books on the calculus of variations (see for example [82]).

In mechanics, the Lagrangian L is taken to be the difference between the kinetic energy and the potential energy of a particle, the corresponding function Φ is called the *action*, and a curve $q : \mathbb{R} \to \mathbb{R}^k$ is called a *motion*. Hamilton's principle states: *Every motion of a physical particle is an extremal of its action.* Of course, the motions of a particle as predicted by Newton's second law are the same as the motions predicted by Hamilton's principle (see Exercise 3.2).

Exercise 3.2. Prove: The motions of a particle determined by Newton's second law are the same as the motions determined by Hamilton's principle. In this context, Newton's law states that the time rate of change of the momentum (mass×velocity) is equal to the negative gradient of the potential energy.

One beautiful feature of Lagrangian mechanics, which is evident from the definition of extremals, is the following fact: Lagrangian mechanics is coordinate free. In particular, the *form* of the Euler–Lagrange equation does not depend on the choice of the coordinate system. Thus, to describe the motion of a particle, we are free to choose the coordinates $q := (q_1, \cdots, q_k)$ as we please and still use the same form of the Euler–Lagrange equation.

As an illustration, consider the prototypical example in mechanics: a free particle. Let (x, y, z) denote the usual Cartesian coordinates in space and $t \mapsto q(t) := (x(t), y(t), z(t))$ the position of the particle as time evolves. The kinetic energy of a particle with mass m *in Cartesian coordinates* is $\frac{m}{2}(\dot{x}^2(t) + \dot{y}^2(t) + \dot{z}^2(t))$. Thus, the "action functional" is given by

$$\Phi(q) = \int_{t_1}^{t_2} \frac{m}{2}(\dot{x}^2(t) + \dot{y}^2(t) + \dot{z}^2(t))\, dt,$$

the Euler–Lagrange equations are simply

$$m\ddot{x} = 0, \qquad m\ddot{y} = 0, \qquad m\ddot{z} = 0, \tag{3.4}$$

and each motion is along a straight line, as expected.

As an example of the Euler–Lagrange equations in a non-Cartesian coordinate system, let us consider the motion of a free particle in cylindrical coordinates (r, θ, z). To determine the Lagrangian, note that the *kinetic energy depends on the Euclidean structure of space*, that is, on the usual inner product. A simple computation shows that the kinetic energy of the motion $t \to (r(t), \theta(t), z(t))$ expressed in cylindrical coordinates is $\frac{m}{2}(\dot{r}^2(t) + r^2\dot{\theta}^2(t) + \dot{z}^2(t))$. For example, to compute the inner product of two tangent vectors relative to a cylindrical coordinate chart, move them to the Cartesian coordinate chart by the derivative of the cylindrical co-ordinate wrapping function (Section 1.8.5), and then compute the usual inner product of their images. The Euler–Lagrange equations are

$$m\ddot{r} - mr\dot{\theta}^2 = 0, \qquad m\frac{d}{dt}(r^2\dot{\theta}) = 0, \qquad m\ddot{z} = 0. \tag{3.5}$$

Clearly, cylindrical coordinates are not the best choice to determine the motion of a free particle. But it does indeed turn out that all solutions of system (3.5) lie on straight lines (see Exercise 3.5).

We will discuss some of the properties enjoyed by the Euler-Lagrange dynamical system. For simplicity we will consider only the case of autonomous Lagrangians, $L : \mathbb{R}^k \times \mathbb{R}^k \to \mathbb{R}$.

Define a new variable

$$p := \frac{\partial L}{\partial \dot{q}}(q, \dot{q}). \tag{3.6}$$

We will assume the Lagrangian is regular; that is, $\dot{q}$ is defined implicitly as a function $\alpha : \mathbb{R}^k \times \mathbb{R}^k \to \mathbb{R}^k$ in equation (3.6) so that $\dot{q} = \alpha(q, p)$. In this case, the Hamiltonian $H : \mathbb{R}^k \times \mathbb{R}^k \to \mathbb{R}$ is defined by

$$H(q, p) = p\dot{q} - L(q, \dot{q}) = p\alpha(q, p) - L(q, \alpha(q, p)).$$

This function is often written in the simple form

$$H = \frac{\partial L}{\partial \dot{q}}\dot{q} - L.$$

The transformation from the Lagrangian to the Hamiltonian is called the *Legendre transformation*.

Proposition 3.3. *If the Lagrangian is regular, then the Hamiltonian is a first integral of the corresponding Lagrangian dynamical system. Moreover, the Lagrangian equations of motion are equivalent to the Hamiltonian system*

$$\dot{q} = \frac{\partial H}{\partial p}(q, p), \qquad \dot{p} = -\frac{\partial H}{\partial q}(q, p).$$

Proof. Using the definition of p and α, and the Euler-Lagrange equation, we have that

$$\frac{d}{dt}H(q, p) = \dot{p}\alpha(q, p) + p\frac{d}{dt}\alpha(q, p) - \frac{\partial L}{\partial q}(q, \alpha(q, p))\alpha(q, p)$$

$$- \frac{\partial L}{\partial \dot{q}}(q, \alpha(q, p))\frac{d}{dt}\alpha(q, p)$$

$$= (\frac{d}{dt}\frac{\partial L}{\partial \dot{q}}(q, \alpha(q, p)) - \frac{\partial L}{\partial q}(q, \alpha(q, p)))\alpha(q, p)$$

$$= 0;$$

that is, H is constant along solutions of the Euler-Lagrange equation. We also have

$$\frac{\partial H}{\partial p} = \alpha + p\frac{\partial \alpha}{\partial p} - \frac{\partial L}{\partial \dot{q}}\frac{\partial \alpha}{\partial p}$$
$$= \dot{q}$$

and

$$\frac{\partial H}{\partial q} = p\frac{\partial \alpha}{\partial q} - \frac{\partial L}{\partial q} - \frac{\partial L}{\partial \dot{q}}\frac{\partial \alpha}{\partial p}$$
$$= -\frac{\partial L}{\partial q}$$
$$= -\frac{d}{dt}\frac{\partial L}{\partial \dot{q}}$$
$$= -\dot{p},$$

as required.   □

A first-order system equivalent to the Euler-Lagrange equation has dimension $2k$. Here, k is called the number of degrees of freedom. The space $\mathbb{R}^k$ with coordinate q is called the configuration space; its dimension is the same as the number of degrees of freedom. The space $\mathbb{R}^k \times \mathbb{R}^k$ with coordinates $(q, \dot{q})$ is called the state space; it corresponds to the tangent bundle of the configuration space. The space $\mathbb{R}^k \times \mathbb{R}^k$ with coordinates (q, p) is called the phase space; it corresponds to the cotangent bundle of the configuration space. A regular level set of the Hamiltonian is a $(2k-1)$-dimensional invariant manifold for the dynamical system. Hence, the existence of this first integral allows us to reduce the dimension of the dynamical system. In effect, we can consider the $(2k-1)$-dimensional first-order system on each regular level set of H. Since the intersection of two invariant sets is an invariant set, the dimension of the effective first-order system can be reduced further if there are additional first integrals.

There is an important connection between symmetries and first integrals of the Lagrangian equations of motion. The basic result in this direction is stated in the following theorem.

Theorem 3.4 (Noether's Theorem). *Suppose that ϕ_t is the flow of the differential equation $\dot{q} = f(q)$ for $q \in \mathbb{R}^k$ and $L : \mathbb{R}^k \times \mathbb{R}^k \to \mathbb{R}$ is an (autonomous) Lagrangian given by $(q, \dot{q}) \mapsto L(q, \dot{q})$. If*

$$L(\phi_s(q), D\phi_s(q)\dot{q}) = L(q, \dot{q})$$

for all $q \in \mathbb{R}^k$ and all s in some open interval that contains the origin in $\mathbb{R}$, then

$$I(q, \dot{q}) := \frac{\partial L}{\partial \dot{q}}(q, \dot{q})f(q)$$

is a first integral of the corresponding Euler-Lagrange equation.

Proof. Let q be a solution of the Euler-Lagrange equation and note that

$$\frac{d}{dt}I(q,\dot{q}) = \frac{d}{dt}\left(\frac{\partial L}{\partial \dot{q}}\right)f(q) + \frac{\partial L}{\partial \dot{q}}Df(q)\dot{q}$$

$$= \frac{\partial L}{\partial q}f(q) + \frac{\partial L}{\partial \dot{q}}Df(q)\dot{q}.$$

By the hypothesis,

$$\frac{d}{ds}L(\phi_s(q), D\phi_s(q)\dot{q})\Big|_{s=0} = 0.$$

Hence, we have that

$$\frac{\partial L}{\partial q}f(q) + \frac{\partial L}{\partial \dot{q}}\frac{d}{ds}D\phi_s(q)\dot{q}\Big|_{s=0} = 0.$$

To complete the proof of the desired equality $dI/dt = 0$, it suffices to show that

$$\frac{d}{ds}D\phi_s(q)\Big|_{s=0} = Df(q).$$

Because

$$\frac{d}{ds}\phi_s(q) = f(\phi_s(q)),$$

we have

$$\frac{d}{ds}D\phi_s(q) = Df(\phi_s(q))D\phi_s(q),$$

and therefore,

$$\frac{d}{ds}D\phi_s(q)|_{s=0} = Df(q)D\phi_0(q).$$

Since $\phi_0(q) \equiv q$, it follows that $D\phi_0(q)$ is the identity transformation of $\mathbb{R}^k$. □

By Proposition 3.3, the Lagrangian

$$L(x,y,\dot{x},\dot{y}) = \frac{1}{2}(\dot{x}^2 + \dot{y}^2) - U(x^2 + y^2), \tag{3.7}$$

where $U : \mathbb{R} \to \mathbb{R}$, has the Hamiltonian

$$H(x,y,\dot{x},\dot{y}) = \frac{1}{2}(\dot{x}^2 + \dot{y}^2) + U(x^2 + y^2)$$

as a first integral. Note that this Hamiltonian corresponds to the total energy of the system, which can be viewed as a model for the motion of a particle with unit mass in a potential field. Also, the Lagrangian of this system is invariant under the linear flow

$$\phi_t(x,y) = \begin{pmatrix} \cos t & -\sin t \\ \sin t & \cos t \end{pmatrix} \begin{pmatrix} x \\ y \end{pmatrix}.$$

By an application of Noether's theorem,

$$I(x,y,\dot{x},\dot{y}) = x\dot{y} - y\dot{x}$$

(the angular momentum of the particle) is a first integral of the corresponding Lagrangian dynamical system.

The existence of the integral I is a consequence of the circular symmetry of the Lagrangian, a feature that suggests the introduction of polar coordinates. In fact, the Lagrangian in polar coordinates is given by

$$L = \frac{1}{2}(\dot{r}^2 + r^2\dot{\theta}^2) - U(r^2),$$

and the Euler-Lagrange equations take the form

$$\ddot{r} = r\dot{\theta}^2 - 2rU'(r^2), \qquad \frac{d}{dt}(r^2\dot{\theta}) = 0. \tag{3.8}$$

By integration of the second equation, which is equivalent to the statement $\dot{I} = 0$, there is a constant k such that $\dot{\theta} = k/r^2$. Thus, the Euler-Lagrange equations decouple, and the reduced system is

$$\ddot{r} = \frac{k^2}{r^3} - 2rU'(r^2). \tag{3.9}$$

It is no accident that this system is itself an Euler-Lagrange equation (see [1, Ch.4] and Exercise 3.7).

We also have the equivalent Hamiltonian

$$H(r,\theta,p_r,p_\theta) = \frac{1}{2}\left(p_r^2 + \frac{p_\theta^2}{r^2}\right) + U(r^2).$$

The introduction of the new variables p_r and p_θ is required here because, by the definition of p,

$$p = (\dot{r}, r^2\dot{\theta}).$$

Hamilton's equations are given by

$$\dot{r} = p_r,$$
$$\dot{\theta} = p_\theta/r^2,$$
$$\dot{p}_r = \frac{p_\theta^2}{r^3} - 2rU'(r^2),$$
$$\dot{p}_\theta = 0.$$

Due to the existence of the second first-integral, these equations decouple. Note that the existence of this integral actually reduces the dimension of the first-order system by two dimensions. The effective Hamiltonian system

$$\dot{r} = p_r,$$
$$\dot{p}_r = \frac{k^2}{r^3} - 2rU'(r^2) \qquad (3.10)$$

with Hamiltonian

$$H = \frac{1}{2}(p_r^2 + \frac{k^2}{r^2}) + U(r^2)$$

is obtained by fixing an angular momentum $k := p_\theta$. This one degree-of-freedom system can be solved by quadrature. In particular, by fixing a value of the reduced Hamiltonian corresponding to a regular level set, we obtain a one-dimensional curve in the two-dimensional reduced space that corresponds to an orbit of the original Lagrangian system. Here, the existence of two independent integrals for our Lagrangian, which has two degrees of freedom, ensures that we can reduce the problem of finding solutions of the original system to a quadrature; that is, we can reduce the problem to the integration

$$t - t_0 = \int_{r(t_0)}^{r(t)} (2h - k^2/r^2 - 2U(r^2))^{-1/2} \, dr,$$

where h is the fixed total energy, the value of the (reduced) Hamiltonian along the orbit (see [1, Ch.4]). Some features of the solutions of the reduced system are discussed in the exercises.

For a system with k degrees of freedom, the problem of finding solutions can often be reduced to a quadrature if there are k independent integrals (see [12, Ch. 10]).

Exercise 3.5. Show that all solutions of system (3.5) lie on straight lines. Compare the parametrization of the solutions of system (3.5) with the solutions of the system (3.4). Hint: If necessary, read Section 3.2.2, especially, the discussion on the integration of the equations of motion for a particle in a central force field—Kepler's problem.

Exercise 3.6. Repeat Exercise 3.5 for spherical coordinates.

Exercise 3.7. Find a Lagrangian with Euler-Lagrange equation (3.9).

Exercise 3.8. (a) Discuss the qualitative features of the extremals for the Lagrangian (3.7) with $U(\xi) := \xi$ by studying the Euler-Lagrange equations (3.8) and the reduced system (3.10). In particular, show that the reduced system has periodic solutions. (b) Find a sufficient condition for the full Lagrangian system to have periodic solutions. (c) Show that these periodic solutions are restricted to invariant two-dimensional tori in the phase space. While the introduction of polar coordinates is viable, it is better to use symplectic polar coordinates: $x = \sqrt{2r}\cos\theta$ and $x = \sqrt{2r}\sin\theta$. To see one reason for this, repeat the exercise using these coordinates in the configuration space and recall Exercise 1.60. The underlying reason why symplectic polar coordinates are preferable is more subtle: they preserve the natural volume element in the configuration space. More generally, they preserve the symplectic form (see [12, Ch. 8]). (d) Repeat the exercise for the Kepler potential $U(\xi) = -\xi^{-1/2}$. Hint: See Section 3.2.2.

Exercise 3.9. [Geodesics] (a) Show that *images* in $\mathbb{R}^3$ of extremals of the Lagrangians $L(\dot{x}, \dot{y}, \dot{z}) = \dot{x}^2(t) + \dot{y}^2(t) + \dot{z}^2(t)$ and $\sqrt{L}$ are the same. Hint: For the second Lagrangian, consider curves parametrized by arc length. (b) Generalize to the case of Riemannian metrics, that is, compare the extremals of the functionals

$$\Phi(q) = \int_{t_1}^{t_2} (g_{k\ell} \dot{q}^k \dot{q}^\ell)^{1/2} \, dt$$

and

$$\Psi(q) = \int_{t_1}^{t_2} g_{k\ell} \dot{q}^k \dot{q}^\ell \, dt$$

where $(q, \dot{q}) = (q_1, q_2, \ldots, q_n, \dot{q}^1, \dot{q}^2, \ldots, \dot{q}^n)$, the Riemannian metric g is given in components, the matrix (g_{ij}) is positive definite and symmetric (see Exercise 1.139), and the Einstein summation convention is observed: sum on repeated indices over $\{1, 2, \ldots, n\}$. The images of the extremals of Φ are called geodesics of the Riemannian metric g. Show that the images of the extremals of Φ and Ψ are the same. (c) Show that the Euler-Lagrange equations for Ψ are equivalent to the geodesic equations

$$\ddot{q}^i + \Gamma^i_{k\ell} \dot{q}^k \dot{q}^\ell = 0$$

where

$$\Gamma^i_{k\ell} = \frac{1}{2} g^{ij} \left(\frac{\partial g_{\ell j}}{\partial q^k} + \frac{\partial g_{kj}}{\partial q^\ell} - \frac{\partial g_{\ell k}}{\partial q^j} \right)$$

are the Christoffel symbols and (g^{ij}) is the inverse of the matrix (g_{ij}). The Euler-Lagrange equations for Φ are the same provided that the extremals are parametrized by arc length.

Exercise 3.10. [Surfaces of Revolution] (a) Determine the geodesics on the cylinder $\{(x, y, z) \in \mathbb{R}^3 : x^2 + y^2 = 1\}$ with respect to the Riemannian metric obtained by restricting the usual inner product on $\mathbb{R}^3$. Hint: Use Exercise 3.9. To determine the metric, use the local coordinates $(r, \theta) \mapsto (\cos\theta, \sin\theta, r)$, where $q = (r, \theta)$ and $\dot{q} = (\dot{r}, \dot{\theta})$. Compute the tangent vector v_r in the direction of r at (r, θ) from the curve $(r + t, \theta)$ and compute the tangent vector v_θ in the direction θ from the curve $(r, \theta + t)$. The vector $\dot{q}$ on the cylinder is given by $\dot{r} v_r + \dot{\theta} v_\theta$, and the square of its length with respect to the usual metric on $\mathbb{R}^3$ is the desired Lagrangian; in fact, $L(r, \theta, \dot{r}, \dot{\theta}) = \dot{r}^2 + \dot{\theta}^2$. The geodesics on the cylinder are orbits of solutions of the corresponding Euler-Lagrange equations. (b) Repeat part (a) for the unit sphere in $\mathbb{R}^3$ and for the hyperboloid of one sheet $\{(x, y, z) \in \mathbb{R}^3 : z = x^2 + y^2\}$. (c) Generalize your results to determine the geodesics on a surface of revolution. In this case, prove that every meridian is a geodesic and find a necessary and sufficient condition for a latitude to be a geodesic. Hint: Use the parametrization

$$(r, \theta) \mapsto (\rho(r) \cos \theta, \rho(r) \sin \theta, r),$$

where ρ is the radial distance from the axis of revolution to the surface. (d) Prove Clairaut's relation: The radial distance from the axis of revolution to a point on a geodesic multiplied by the sine of the angle between the tangent to the geodesic at this point and the *tangent vector* at the same point in the direction of the axis of revolution is constant along the geodesic. Hint: Compute the cosine of $\pi/2$ minus the "Clairaut angle" using a dot product. Clairaut's relation is a geometric interpretation of the first integral obtained from the Euler-Lagrange equation with respect to the angle of rotation around the axis of revolution. (e) Give an example of a curve with the Clairaut relation that is not a geodesic.

Exercise 3.11. [Poincaré plane] Consider the Riemannian metric (Poincaré metric) on the upper half-plane given by $(dx^2 + dy^2)/y^2$ and the corresponding Lagrangian

$$L(x, y, \dot{x}, \dot{y}) = \frac{1}{2} \frac{\dot{x}^2 + \dot{y}^2}{y^2}.$$

The solutions of the Euler-Lagrange equations are geodesics by Exercise 3.9. (a) Show that every geodesic lies on a vertical line or a circle with center on the x-axis. Hint: The Lagrangian L is a first integral of the motion; that is, the length of the velocity vector along an extremal for L is constant. (b) A more sophisticated approach to this problem opens a rich mathematical theory. Show that the Euler-Lagrange equations for L are invariant under all linear fractional transformations of the upper half-plane given by

$$z \mapsto \frac{az + b}{cz + d}$$

where z is a complex variable and the coefficients $a, b, c, d \in \mathbb{R}$ are such that $ad - bc = 1$ and recall that linear fractional transformations preserve the set of all lines and circles. Hint: Write the Euler-Lagrange equations in complex form using $z = x + iy$ and show that the resulting equation is invariant under the transformations. Or, write the metric in complex form and show that it is invariant under the transformations.

Exercise 3.12. [Strain Energy] Strain is defined to be the relative change in length in some direction. For $f : \mathbb{R}^n \to \mathbb{R}^n$, relative change in length at $\xi \in \mathbb{R}^n$ is given by $(|f(\xi) - f(x)| - |x - \xi|)/|x - \xi|$. For a unit vector $v \in T_\xi \mathbb{R}^n$, the strain at ξ in the direction v is $|Df(\xi)v| - 1$. This formula is obtained by replacing x with $\xi + tv$ and passing to the limit at $t \to 0$. Consider the class $\mathcal{C}$ of all continuously differentiable invertible functions $f : [0, 1] \to [\alpha, \beta]$ with $f(0) = \alpha$ and $f(1) = \beta$ and define the total strain energy to be $E(f) = \int_0^1 (|f'(s)| - 1)^2 \, ds$. (a) Show that $g(s) = (\beta - \alpha)s + \alpha$ is an extremal and (b) g is the minimum of E over $\mathcal{C}$. Hints: For part (b), compute $E(g)$ and show directly that $E(f) \geq E(g)$ for every $f \in \mathcal{C}$. Use the fundamental theorem of calculus and the Cauchy-Schwarz inequality.

Exercise 3.13. [Minimal Surface Area] The determination of surfaces of revolution with minimal surface areas is a classic and mathematically rich problem in the calculus of variations (see [27]). (a) Consider all smooth functions $q : \mathbb{R} \to (0, \infty)$ whose graphs pass through the two point $(0, a)$ and (ℓ, b) in the plane, where a, b, and ℓ are positive. Prove that if the surface area of the surface of revolution obtained by revolving the graph of $q : [a, b] \to (0, \infty)$ about the horizontal axis $\{(t, y) : y = 0\}$ is an extremal, then $q(t) = c \cosh((t - d)/c)$ for suitable constants c and d; that is, this two-parameter family of functions solves the Euler-Lagrange equation for the functional whose Lagrangian is $L(q, \dot{q}) = 2\pi q \sqrt{1 + \dot{q}^2}$. Hint: To solve the Euler-Lagrange equation, use the first integral given by Proposition 3.3. (b) Show that the solution of the Euler-Lagrange equation with the boundary conditions imposed can have no solutions, one solution, or two solutions. Hint: To show that there can be two solutions, consider (for example) the case where $a = b = 2$ and $\ell = 1$. In this case q is a solution provided that $2 = c \cosh(1/(2c))$, and this equation for c has two positive roots. Which of these two solutions corresponds to the smallest surface area? Caution: If a surface with minimal surface area exists among the surfaces of revolution *generated by sufficiently smooth curves*, then the minimizing curve is an extremal. But, of course, there may be no such minimizer. Sufficient conditions for minima are known (see [27]). In the case $a = b = 2$ and $\ell = 1$, one of the two cantenaries is in fact the minimizer. Nature also knows the minimum solution. To demonstrate this fact, dip two hoops in a soap solution and pull them apart to form a surface of revolution.

Exercise 3.14. A derivative moved across an integral sign in display (3.2). Justify the result.

Exercise 3.15. [Bump functions] Prove that if $B_1 \subset B_2$ are open balls in $\mathbb{R}^n$, then there is a C^∞ function that has value one on B_1 and vanishes on the complement of B_2. Hint: Show that the function f defined to be $\exp(-1/(1 - |x|^2)$ for $|x| < 1$ and zero for $|x| \geq 1$ is C^∞; then, consider the function $x \mapsto \int_{-\infty}^{x} f(s) \, ds / \int_{-\infty}^{\infty} f(s) \, ds$, etc.

3.2 Origins of ODE: Classical Physics

What is classical physics? Look at Section 18–2 in Richard Feynman's lecture notes [87]; you might be in for a surprise. The fundamental laws of all of classical physics can be reduced to a few formulas. For example, a complete theory of electromagnetics is given by Maxwell's laws

$$\operatorname{div} \mathbf{E} = \rho/\epsilon_0,$$
$$\operatorname{curl} \mathbf{E} = -\frac{\partial \mathbf{B}}{\partial t},$$
$$\operatorname{div} \mathbf{B} = 0,$$
$$c^2 \operatorname{curl} \mathbf{B} = \frac{\mathbf{j}}{\epsilon_0} + \frac{\partial \mathbf{E}}{\partial t}$$

and the conservation of charge

$$\text{div}(\mathbf{j}) = -\frac{\partial \rho}{\partial t}.$$

Here $\mathbf{E}$ is the called electric field, $\mathbf{B}$ is the magnetic field, ρ is the charge density, $\mathbf{j}$ is the current, ϵ_0 is a constant, and c is the speed of light. The fundamental law of motion is Newton's law

$$\frac{d\mathbf{p}}{dt} = F$$

"the rate of change of the momentum is equal to the sum of the forces." The (relativistic) momentum of a particle is given by

$$\mathbf{p} := \frac{m}{\sqrt{1 - v^2/c^2}}\mathbf{v}$$

where, as is usual in the physics literature, $v := |\mathbf{v}|$ and the norm is the Euclidean norm. For a classical particle (velocity much less than the speed of light), the momentum is approximated by $\mathbf{p} = m\mathbf{v}$. There are two fundamental forces: The gravitational force

$$F = -\frac{GMm}{r^2}\mathbf{e}_r$$

on a particle of mass m due to a second mass M where G is the universal gravitational constant and $\mathbf{e}_r$ is the unit vector at M pointing in the direction of m; and the Lorentz force

$$F = q(\mathbf{E} + \mathbf{v} \times \mathbf{B})$$

where q is the charge on a particle in an electromagnetic field. That's it!

The laws of classical physics seem simple enough. Why then is physics, not to mention engineering, so complicated? The answer is that in almost all realistic applications there are *lots* of particles and the fundamental laws act together on all of the particles.

Rather than trying to model the motions of all the particles that are involved in an experiment or a physical phenomenon which we wish to explain, it is often more fruitful to develop constitutive force laws that are meant to approximate the true situation. The resulting equations of motion contain new "forces" that are not fundamental laws of nature. But, by using well conceived constitutive laws, the predictions we make from the corresponding equations of motion will agree with experiments or observations within some operating envelope. While the constitutive laws themselves may lead to complicated differential equations, these equations are supposed to be simpler and more useful than the equations of motion that would result from the fundamental force laws. Also, it is important to realize that in most circumstances no one knows how to write down the equations of motion from the fundamental force laws.

Let us consider a familiar example: the spring equation

$$m\ddot{x} = -\omega_0 x$$

for the displacement x of a mass (attached to the end of a spring) from its equilibrium position. This model uses Newton's law of motion, together with Hooke's restoring force law. Newton's law is fundamental (at least if we ignore relativistic effects); Hooke's law is not. It is meant to replace the law of motion that would result from modeling the motions of all the particles that make up the spring. When Hooke's (linear) law is not sufficiently accurate (for example, when we stretch a spring too far from its equilibrium position), we may refine the model to obtain a nonlinear equation of motion such as

$$m\ddot{x} = -\omega_0 x + \alpha x^3,$$

which is already a complicated differential equation.

Frictional forces are always modeled by constitutive laws. What law of nature are we using when we add a viscous damping force to a Hookian spring to obtain the model

$$m\ddot{x} = -\alpha \dot{x} - \omega_0 x?$$

The damping term is supposed to model a force due to friction. But what is friction? There are only two fundamental forces in classical physics and only four known forces in modern physics. Friction is not a nuclear force and it is not due to gravitation. Thus, at a fundamental level it must be a manifestation of electromagnetism. Is it possible to derive the linear form of viscous damping from Maxwell's laws? This discussion could become very philosophical!

These models introduce constitutive laws—in our example, restoring force and friction forces laws—that are not fundamental laws of nature. In reality, the particles (atoms) that constitute the spring obey the electromagnetic force law and the law of universal gravitation. But, to account for their motions using these fundamental force laws would result in a model with an enormous number of coupled equations that would be very difficult to analyze even if we knew how to write it down.

The most important point for us to appreciate is that Newton's law of motion—so basic for our understanding of the way the universe works—is expressed as an ordinary differential equation. Newton's law, the classical force laws, and constitutive laws are *the* origin of ordinary differential equations. It should be clear what Newton meant when he said "Solving differential equations is useful."

In the following subsections some applications of the theory of differential equations to problems that arise in classical physics are presented. The first section briefly describes the motion of a charged particle in a constant electromagnetic field. The second section is an introduction to the two-body problem, including Kepler motion, Delaunay elements, and perturbation forces. The analysis of two-body motion is used as a vehicle to explore a realistic important physical problem where it is not at all obvious how to obtain useful predictions from the complicated model system of differential equations obtained from Newton's law. Perturbations of two-body motion are considered in the final sections: Satellite motion about an oblate planet is used to illustrate the "method of averaging," and the diamagnetic Kepler problem—the motion of an electron of a hydrogen atom in a constant magnetic field—is used to illustrate some important transformation methods for the analysis of models of mechanical systems.

3.2.1 Motion of a Charged Particle

Let us consider a few simple exercises to "feel" the Lorentz force (for more see [87] and [138]). The equation of motion for a charged particle is

$$\frac{d\mathbf{p}}{dt} = q(\mathbf{E} + \mathbf{v} \times \mathbf{B})$$

where $\mathbf{p}$ is the momentum vector, q is a measure of the charge, and $\mathbf{v}$ is the velocity. We will consider the motion of a charged particle (classical and relativistic) in a constant electromagnetic field; that is, the electric field $\mathbf{E}$ and the magnetic field $\mathbf{B}$ are *constant* vector fields on $\mathbb{R}^3$.

In case $\mathbf{E} = 0$, let us consider the relativistic motion for a charged particle. Because the momentum is a nonlinear function of the velocity, it is useful to notice that the motion is "integrable." In fact, the two functions $\mathbf{p} \mapsto \langle \mathbf{p}, \mathbf{p} \rangle$ and $\mathbf{p} \mapsto \langle \mathbf{p}, \mathbf{B} \rangle$ are constant along orbits. Use this fact to conclude that $\mathbf{v} \mapsto \langle \mathbf{v}, \mathbf{v} \rangle$ is constant along orbits, and therefore the energy

$$\mathcal{E} := \frac{mc^2}{\sqrt{1 - v^2/c^2}}$$

is constant along orbits. It follows that the equation of motion can be recast in the form

$$\frac{\mathcal{E}}{c^2} \dot{\mathbf{v}} = q\mathbf{v} \times B,$$

and the solution can be found as in Exercise 2.57. The solution of this differential equation is important. For example, the solution can be used to place magnetic fields in an experiment so that charged particles are moved to a detector (see [87]).

Another important problem is to determine the drift velocity of a charged particle moving in a constant electromagnetic field (see Exercise 3.16).

Exercise 3.16. Consider a classical particle moving in space with a constant magnetic field pointing along the z-axis and a constant electric field parallel to the yz-plane. The equations of motion are

$$m\ddot{x} = qB_3\dot{y}, \quad m\ddot{y} = q(E_2 - B_3\dot{x}), \quad m\ddot{z} = qE_3. \tag{3.11}$$

(a) Solve the system (3.11). (b) Note that the first two components of the solution are periodic in time. Their average over one period gives a constant vector field, called the drift velocity field. Find this vector field. (c) Describe the motion of the charged particle in space.

Exercise 3.17. Use the theory of linear differential equations with constant coefficients to determine the motion for a "spatial oscillator" (see [138]) in the presence of a constant magnetic field. The equation of motion is

$$\dot{\mathbf{v}} = -\omega_0^2\mathbf{r} + \frac{q}{m}\mathbf{v} \times \mathbf{B}$$

where $\mathbf{r} = (x, y, z)$ is the position vector, the velocity is $\mathbf{v} = (\dot{x}, \dot{y}, \dot{z})$, and $\mathbf{B} = (0, 0, B_3)$. (This model uses Hooke's law). By rewriting the equations of motion in components, note that this model is a linear system with constant coefficients. (a) Find the general solution of the system. (b) Determine the frequency of the motion in the plane perpendicular to the magnetic field and the frequency in the direction of the magnetic field.

3.2.2 Motion of a Binary System

Let us consider two point masses, m_1 and m_2, moving in three-dimensional Euclidean space with corresponding position vectors R_1 and R_2. Also, let us define the relative position vector $R := R_2 - R_1$ and its length $r := |R|$. According to Newton's law (using of course the usual approximation for small velocities) and the gravitational force law, we have the equations of motion

$$m_1\ddot{R}_1 = \frac{G_0 m_1 m_2}{r^3}R + F_1, \quad m_2\ddot{R}_2 = -\frac{G_0 m_1 m_2}{r^3}R + F_2$$

where F_1 and F_2 are additional forces acting on m_1 and m_2 respectively. The relative motion of the masses is governed by the single vector equation

$$\ddot{R} = -\frac{G_0(m_1 + m_2)}{r^3}R + \frac{1}{m_2}F_2 - \frac{1}{m_1}F_1.$$

By rescaling distance and time such that $R = \alpha\bar{R}$ and $t = \beta\bar{t}$ with $G_0(m_1 + m_2)\beta^2 = \alpha^3$, we can recast the equations of motion in the simpler form

$$\ddot{R} = -\frac{1}{r^3}R + F. \tag{3.12}$$

We will study this differential equation.

The analysis of two-body interaction plays a central role in the history of science. This is reason enough to study the dynamics of the differential equation (3.12) and the surrounding mathematical terrain. As you will see, the intrinsic beauty, rich texture, and wide applicability of this subject make it one of the most absorbing topics in all of mathematics.

The following glimpse into celestial mechanics is intended to introduce an important application of ordinary differential equations, to see some of the complexity of a real world application and to introduce a special form of the dynamical equations that will provide some motivation for the theory of averaging presented in Chapter 7.

There are many different approaches to celestial mechanics. For a mathematician, perhaps the most satisfactory foundation for mechanics is provided by the theory of Hamiltonian systems. Although we will use Hamilton's equations in our analysis, the geometric context (symplectic geometry) for a modern treatment of the transformation theory for Hamiltonian systems (see [1], [12], and [160]) is unfortunately beyond the scope of this book. To bypass this theory, we will present an expanded explanation of the direct change of coordinates to the Delaunay elements given in [52] (see also [48] and [51]). In the transformation theory for Hamiltonian systems it is proved that our transformations are special coordinate transformations called *canonical transformations*. They have a special property: Hamilton's equations for the *transformed Hamiltonian* are exactly the differential equations given by the push forward of the original Hamiltonian vector field to the new coordinates. In other words, to perform a canonical change of coordinates we need only transform the Hamiltonian, not the differential equations; the transformed differential equations are obtained by computing Hamilton's equations from the transformed Hamiltonian. The direct method is perhaps not as elegant as the canonical transformation approach—we will simply push forward the Hamiltonian vector field in the usual way—but the direct transformation method is effective and useful. Indeed, we will construct special coordinates (action-angle coordinates) and show that they transform the Kepler system to a very simple form. Moreover, the direct method applies even if a nonconservative force F acts on the system; that is, even if the equations of motion are not Hamiltonian.

Let us begin by rewriting the second order differential equation (3.12) as the first order system

$$\dot{R} = V, \qquad \dot{V} = -\frac{1}{r^3}R + F \qquad (3.13)$$

defined on $\mathbb{R}^3 \times \mathbb{R}^3$. Also, let us use angle brackets to denote the usual inner product on $\mathbb{R}^3$ so that if $X \in \mathbb{R}^3$, then $|X|^2 = \langle X, X \rangle$.

The most important feature of system (3.13) is the existence of conserved quantities for the Kepler motion, that is, the motion with $F = 0$. In fact, total energy and angular momentum are conserved. The total energy of the

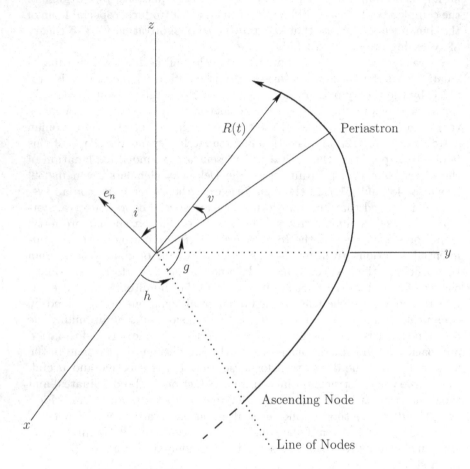

Figure 3.1: The osculating Kepler orbit in space.

Kepler motion $E : \mathbb{R}^3 \times \mathbb{R}^3 \to \mathbb{R}$ is given by

$$E(X, Y) := \frac{1}{2}\langle Y, Y \rangle - \frac{1}{\langle X, X \rangle^{1/2}}, \qquad (3.14)$$

the angular momentum by

$$A(X, Y) := X \times Y. \qquad (3.15)$$

Let us also define the total angular momentum

$$G(X, Y) := |A(X, Y)|. \qquad (3.16)$$

Note that we are using the term "angular momentum" in a nonstandard manner. In effect, we have defined the angular momentum to be the vector product of position and *velocity*; in physics, angular momentum is the vector product of position and *momentum*.

If $t \mapsto (R(t), V(t))$ is a solution of system (3.13), then

$$\dot{E} = \frac{d}{dt}E(R(t), V(t)) = \langle F, V \rangle, \qquad \dot{A} = R \times F. \qquad (3.17)$$

Thus, if $F = 0$, then E and A are constant on the corresponding orbit. In particular, in this case the projection of the Kepler orbit into physical space, corresponding to the curve $t \mapsto R(t)$, is contained in a fixed plane passing through the origin with normal vector given by the constant value of A along the orbit. In this case, the corresponding plane normal to A is called the *osculating plane*. At each instant of time, this plane contains the Kepler orbit that would result if the force F were not present thereafter. Refer to Figure 3.1 for a depiction of the curve $t \mapsto R(t)$ in space and the angles associated with the Delaunay elements, new coordinates that will be introduced as we proceed.

Let us define three functions $e_r : \mathbb{R}^3 \to \mathbb{R}^3$, and $e_b, e_n : \mathbb{R}^3 \times \mathbb{R}^3 \to \mathbb{R}^3$ by

$$e_r(X) = \frac{1}{|X|}X,$$

$$e_n(X, Y) = \frac{1}{G(X, Y)}X \times Y,$$

$$e_b(X, Y) = e_n(X, Y) \times e_r(X).$$

If $X, Y \in \mathbb{R}^3$ are linearly independent, then the vectors $e_r(X)$, $e_b(X, Y)$, and $e_n(X, Y)$ form an orthonormal frame in $\mathbb{R}^3$. Also, if these functions are evaluated along the (perturbed) solution $t \mapsto (R, \dot{R})$, then we have

$$e_r = \frac{1}{r}R, \quad e_n = \frac{1}{G}R \times \dot{R}, \quad e_b = \frac{1}{rG}(R \times \dot{R}) \times R = \frac{1}{rG}(r^2\dot{R} - \langle \dot{R}, R \rangle R).$$
$$(3.18)$$

(Note that subscripts are used in this section to denote coordinate directions, not partial derivatives.)

If e_x, e_y, e_z are the direction vectors for a fixed right-handed usual Cartesian coordinate system in $\mathbb{R}^3$, and if i denotes the inclination angle of the osculating plane relative to the z-axis, then

$$\cos i = \langle e_n, e_z \rangle. \qquad (3.19)$$

Of course, the angle i can also be viewed as a function on $\mathbb{R}^3 \times \mathbb{R}^3$ whose value at each point on the orbit is the inclination of the osculating plane. The idea is that we are defining new variables: They are all functions defined on $\mathbb{R}^3 \times \mathbb{R}^3$.

If the osculating plane is not coincident with the (x, y)-plane, then it meets this plane in a line, called the *line of nodes*. Of course, the line of nodes lies in the osculating plane and is orthogonal to the z-axis. Moreover, it is generated by the vector

$$e_{\mathrm{an}} := \langle e_b, e_z \rangle e_r - \langle e_r, e_z \rangle e_b.$$

The *angle of the ascending node h* between the x-axis and the line of nodes is given by

$$\cos h = \frac{1}{|e_{\mathrm{an}}|} \langle e_{\mathrm{an}}, e_x \rangle. \qquad (3.20)$$

If the osculating plane happens to coincide with the (x, y)-plane, then there is no natural definition of h. On the other hand, the angle $h(t)$ is continuous on each trajectory. Also, at a point where $i(t) = 0$, the angle h is defined whenever there is a continuous extension.

Let us compute the orthogonal transformation relative to the Euler angles i and h. This is accomplished in two steps: rotation about the z-axis through the angle h followed by rotation about the now rotated x-axis through the angle i. The rotation matrix about the z-axis is

$$M(h) := \begin{pmatrix} \cos h & -\sin h & 0 \\ \sin h & \cos h & 0 \\ 0 & 0 & 1 \end{pmatrix}.$$

To rotate about the new x-axis (after rotation by $M(h)$), let us rotate back to the original coordinates, rotate about the old x-axis through the angle i, and then rotate forward. The rotation about the x-axis is given by

$$M(i) := \begin{pmatrix} 1 & 0 & 0 \\ 0 & \cos i & -\sin i \\ 0 & \sin i & \cos i \end{pmatrix}.$$

Thus, the required rotation is

$$M := (M(h)M(i)M^{-1}(h))M(h) = M(h)M(i).$$

In components, if the new Cartesian coordinates are denoted x', y', z', then the transformation M is given by

$$x = x' \cos h - y' \sin h \cos i + z' \sin h \sin i,$$
$$y = x' \sin h + y' \cos h \cos i - z' \cos h \sin i,$$
$$z = y' \sin i + z' \cos i. \tag{3.21}$$

Also, by the construction, the normal e_n to the osculating plane is in the direction of the z'-axis.

If polar coordinates are introduced in the osculating plane

$$x' = r \cos \theta, \qquad y' = r \sin \theta,$$

then the position vector along our orbit in the osculating coordinates is given by

$$R(t) = (r(t) \cos \theta(t), r(t) \sin \theta(t), 0).$$

For a Kepler orbit, the osculating plane is fixed. Also, using the orthogonal transformation M, the vectors R and $\dot{R}$ are given in the original fixed coordinates by

$$R = M \begin{pmatrix} r \cos \theta \\ r \sin \theta \\ 0 \end{pmatrix}, \qquad \dot{R} = M \begin{pmatrix} \dot{r} \cos \theta - r\dot{\theta} \sin \theta \\ \dot{r} \sin \theta + r\dot{\theta} \cos \theta \\ 0 \end{pmatrix}. \tag{3.22}$$

If $X, Y \in \mathbb{R}^3$, then, because M is orthogonal, we have $\langle MX, MY \rangle = \langle X, Y \rangle$. As a consequence of this fact and definition (3.14), it follows that the total energy along the orbit is

$$E(R, \dot{R}) = \frac{1}{2}(\dot{r}^2 + r^2 \dot{\theta}^2) - \frac{1}{r}.$$

For arbitrary vectors $X, Y \in \mathbb{R}^3$, the coordinate-free definition of their vector product states that

$$X \times Y := (|X||Y| \sin \vartheta)\eta \tag{3.23}$$

where ϑ is the angle between X and Y and η is the unit vector orthogonal to X and Y such that the ordered triple (X, Y, η) has positive (right hand) orientation. Also, note that M is orientation preserving ($\det M > 0$). Using this fact and the definition the vector product, it follows that $MX \times MY = M(X \times Y)$ and the angular momentum along the orbit is

$$A(R, \dot{R}) = r^2 \dot{\theta}. \tag{3.24}$$

Thus, using equation (3.17), there is a constant (angular momentum) P_θ such that

$$r^2 \dot{\theta} = P_\theta, \qquad E = \frac{1}{2}\left(\dot{r}^2 + \frac{P_\theta^2}{r^2}\right) - \frac{1}{r}, \tag{3.25}$$

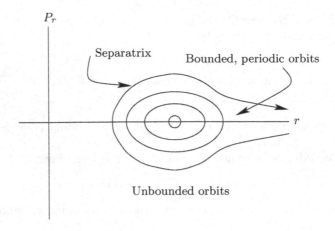

Figure 3.2: Schematic phase portrait of system (3.26). There is a center surrounded by a period annulus and "bounded" by an unbounded separatrix. The period orbits correspond to elliptical Keplerian motions.

and, because $\dot{E} = 0$, we also have

$$r^2\dot{\theta} = P_\theta, \qquad \dot{r}\left(\ddot{r} - \frac{P_\theta^2}{r^3} + \frac{1}{r^2}\right) = 0.$$

If $\dot{r} \equiv 0$, $r(0) \neq 0$, and $P_\theta \neq 0$, then the Kepler orbit is a circle; in fact, it is a solution of the system $\dot{r} = 0$, $\dot{\theta} = P_\theta r(0)^{-2}$. If $\dot{r}$ is not identically zero, then the motion is determined by Newton's equation

$$\ddot{r} = -\left(\frac{1}{r^2} - \frac{P_\theta^2}{r^3}\right).$$

The equivalent system in the phase plane,

$$\dot{r} = P_r, \qquad \dot{P}_r = -\frac{1}{r^2} + \frac{P_\theta^2}{r^3} \tag{3.26}$$

is Hamiltonian with energy

$$E = \frac{1}{2}\left(P_r^2 + \frac{P_\theta^2}{r^2}\right) - \frac{1}{r}.$$

It has a rest point with coordinates $(r, P_r) = (P_\theta^2, 0)$ and energy $-1/(2P_\theta^2)$. This rest point is a center surrounded by an annulus of periodic orbits, called a *period annulus,* that is "bounded" by an unbounded separatrix as depicted schematically in Figure 3.2. The separatrix crosses the r-axis at $r = \frac{1}{2}P_\theta^2$ and its energy is zero. Thus, if a Kepler orbit is bounded, it has negative energy.

Exercise 3.18. Prove all of the statements made about the phase portrait of system (3.26).

Exercise 3.19. The vector product is defined in display (3.23) in a coordinate-free manner. Suppose instead that $X = (x_1, x_2, x_3)$ and $Y = (y_1, y_2, y_3)$ in Cartesian coordinates, and their cross product is defined to be

$$X \times Y := \det \begin{pmatrix} e_1 & e_2 & e_3 \\ x_1 & x_2 & x_3 \\ y_1 & y_2 & y_3 \end{pmatrix}.$$

Discuss the relative utility of the coordinate-free versus coordinate-dependent definitions. Determine the vector product for vectors expressed in cylindrical or spherical coordinates. Think about the concept of coordinate free-definitions; it is important.

The rest of the discussion is restricted to orbits with negative energy and positive angular momentum $P_\theta > 0$.

The full set of differential equations for the Kepler motion is the first order system

$$\dot{r} = P_r, \quad \dot{\theta} = \frac{P_\theta}{r^2}, \quad \dot{P_r} = \frac{1}{r^2}\left(\frac{P_\theta^2}{r} - 1\right), \quad \dot{P_\theta} = 0. \tag{3.27}$$

Note that the angle θ increases along orbits because $\dot{\theta} = P_\theta/r^2 > 0$. Also, the bounded Kepler orbits project to periodic motions in the r–P_r phase plane. Thus, the bounded orbits can be described by two angles: the polar angle relative to the r-axis in the r–P_r plane and the angular variable θ. In other words, each bounded orbit lies on the (topological) cross product of two circles; that is, on a two-dimensional torus. Because the phase space for the Kepler motion is foliated by invariant two-dimensional tori, we will eventually be able to define special coordinates, called action-angle coordinates, that transform the Kepler system to a very simple form. In fact, the special class of Hamiltonian systems (called *integrable Hamiltonian systems*) that have a portion of their phase space foliated by invariant tori can all be transformed to a simple form by the introduction of action-angle coordinates. Roughly speaking, the action specifies a torus and the "angle" specifies the frequency on this torus. This is exactly the underlying idea for the construction of the action-angle coordinates for the Kepler system.

To integrate system (3.27), introduce a new variable $\rho = 1/r$ so that

$$\dot{\rho} = -\rho^2 P_r, \qquad \dot{P_r} = \rho^2(\rho P_\theta^2 - 1),$$

and then use θ as a time-like variable to obtain the *linear* system

$$\frac{d\rho}{d\theta} = -\frac{1}{P_\theta} P_r, \qquad \frac{dP_r}{d\theta} = \frac{1}{P_\theta}(\rho P_\theta^2 - 1). \tag{3.28}$$

Equivalently, we have the "harmonic oscillator model" for Kepler motion,

$$\frac{d^2\rho}{d\theta^2} + \rho = \frac{1}{P_\theta^2}. \tag{3.29}$$

It has the general solution

$$\rho = \frac{1}{P_\theta^2} + B\cos(\theta - g) \tag{3.30}$$

where the numbers B and g are determined by the initial conditions.

The Kepler orbit is an ellipse. In fact, by a rearrangement of equation (3.30) we have

$$r = \frac{P_\theta^2}{1 + P_\theta^2 B\cos(\theta - g)}.$$

If we introduce a new angle, the *true anomaly* $v := \theta - g$, and the usual quantities—the eccentricity e, and a, the semimajor axis of the ellipse—then

$$r = \frac{a(1 - e^2)}{1 + e\cos v}. \tag{3.31}$$

Exercise 3.20. Use the conservation of energy for system (3.27) to show that

$$\frac{dr}{d\theta} = \frac{r}{P_\theta}\left(2Er^2 + 2r - P_\theta^2\right)^{1/2}.$$

Solve this differential equation for $E < 0$ and show that the Kepler orbit is an ellipse.

Because the energy is constant on the Kepler orbit, if we compute the energy at $v = 0$ (corresponding to its *perifocus,* the point on the ellipse closest to the focus), then we have the corresponding values

$$r = a(1 - e), \qquad \dot{r} = 0, \qquad E = -\frac{1}{2a}. \tag{3.32}$$

Moreover, from the usual theory of conic sections, the semiminor axis is $b = a\sqrt{1 - e^2}$, and the area of the ellipse is

$$\pi ab = \pi a^2\sqrt{1 - e^2} = \pi P_\theta a^{3/2}. \tag{3.33}$$

The area element in polar coordinates is $\frac{1}{2}\rho^2\,d\theta$. Hence, if the period of the Kepler orbit is T, then

$$\int_0^{2\pi} \frac{r^2(\theta)}{2}\,d\theta = \int_0^T \frac{r^2(\theta(t))}{2}\dot{\theta}(t)\,dt = \pi P_\theta a^{3/2}.$$

Because $r^2\dot\theta = P_\theta$, the integral can be evaluated. The resulting equation is Kepler's third law

$$T^2 = 4\pi^2 a^3 \tag{3.34}$$

where, again, T is the orbital period and a is the semimajor axis of the corresponding elliptical orbit. For later reference, note that the frequency of the oscillation is

$$\omega := \frac{1}{a^{3/2}}. \tag{3.35}$$

Exercise 3.21. (a) Kepler's third law is given by equation (3.34) for *scaled* distance and time. Show that Kepler's third law in "unscaled" variables (with the same names as in (3.34)) is given by

$$T^2 = \frac{4\pi^2}{G_0(m_1 + m_2)} a^3.$$

(b) Show that the magnitude of the *physical* angular momentum in the unscaled variables for a Kepler orbit is

$$m_2 r^2 \dot\theta = m_2\Big(\frac{\alpha^3}{\beta^2} a(1 - e^2)\Big)^{1/2} = m_2\big(G_0(m_1 + m_2)a(1 - e^2)\big)^{1/2}$$

where r and t are unscaled, and with α and β as defined on page 240.

3.2.3 Perturbed Kepler Motion and Delaunay Elements

In this section we will begin an analysis of the influence of a force F on a Keplerian orbit by introducing new variables, called *Delaunay elements,* such that system (3.12) when recast in the new coordinates has a useful special form given below in display (3.59).

Recall the orthonormal frame $[e_r, e_b, e_n]$ in display (3.18), and let

$$F = F_r e_r + F_b e_b + F_n e_n.$$

The functions $L, G : \mathbb{R}^3 \times \mathbb{R}^3 \to \mathbb{R}$, two of the components of the Delaunay coordinate transformation, are defined by

$$L(X, Y) := (-2E(X, Y))^{-1/2}, \qquad G(X, Y) := |A(X, Y)|$$

where E is the energy and A the angular momentum. Using the results in display (3.17), we have that

$$\dot L = L^3 \langle F, \dot R \rangle, \qquad \dot G = \frac{1}{G} \langle R \times \dot R, R \times F \rangle.$$

Moreover, in view of the vector identities

$$\langle \alpha, \beta \times \gamma \rangle = \langle \gamma, \alpha \times \beta \rangle = \langle \beta, \gamma \times \alpha \rangle,$$
$$(\alpha \times \beta) \times \gamma = \langle \gamma, \alpha \rangle \beta - \langle \gamma, \beta \rangle \alpha,$$

and the results in display (3.18), it follows that

$$\dot{G} = \frac{1}{G} \langle F, (R \times \dot{R}) \times R \rangle = \frac{1}{G} \langle F, rGe_b \rangle = rF_b. \tag{3.36}$$

Also, using the formula for the triple vector product and the equality $r\dot{r} = \langle R, \dot{R} \rangle$, which is obtained by differentiating both sides of the identity $r^2 = \langle R, R \rangle$, we have

$$e_b = \frac{1}{rG}(R \times \dot{R}) \times R = \frac{1}{rG}(r^2 \dot{R} - \langle R, \dot{R} \rangle R)$$

and

$$\dot{R} = \dot{r}e_r + \frac{G}{r}e_b. \tag{3.37}$$

As a result, the Delaunay elements L and G satisfy the differential equations

$$\dot{L} = L^3 (\dot{r}F_r + \frac{G}{r}F_b), \qquad \dot{G} = rF_b. \tag{3.38}$$

If the force F does not vanish, then the relations found previously for the variables related to the osculating plane of the Kepler orbit are still valid, only now the quantities a, e, v, and $P_\theta > 0$ all depend on time. Thus, for example, if we use equation (3.24), then

$$G = P_\theta = \sqrt{a(1 - e^2)}. \tag{3.39}$$

Also, from equations (3.32), (3.37), and the definition of L we have

$$L = \sqrt{a}, \qquad e^2 = 1 - \frac{G^2}{L^2}, \tag{3.40}$$

and

$$-\frac{1}{2a} = \frac{1}{2}\langle \dot{R}, \dot{R} \rangle - \frac{1}{r} = \frac{1}{2}\left(\dot{r}^2 + \frac{a(1 - e^2)}{r^2}\right) - \frac{1}{r}. \tag{3.41}$$

Let us solve for $\dot{r}^2$ in equation (3.41), express the solution in the form

$$\dot{r}^2 = -\frac{1}{ar^2}(r - a(1 - e))(r - a(1 + e)),$$

and substitute for r from formula (3.31) to obtain

$$\dot{r} = \frac{e \sin v}{G}. \tag{3.42}$$

Hence, from equation (3.38),

$$\dot{L} = L^3 \left(F_r \frac{e}{G} \sin v + F_b \frac{G}{r} \right), \qquad \dot{G} = r F_b. \tag{3.43}$$

The Delaunay variable H is defined by

$$H := \langle A, e_z \rangle = G \langle \frac{1}{G} R \times \dot{R}, e_z \rangle = G \cos i \tag{3.44}$$

where i is the inclination angle of the osculating plane (see equation (3.19)). To find an expression for $\dot{H}$, let us first recall the transformation equations (3.21). Because e_n has "primed" coordinates $(0, 0, 1)$, it follows that e_n has original Cartesian coordinates

$$e_n = (\sin h \sin i, -\cos h \sin i, \cos i), \tag{3.45}$$

and, similarly, R is given by

$$R = r(\cos \theta \cos h - \sin \theta \sin h \cos i, \cos \theta \sin h + \sin \theta \cos h \cos i, \sin \theta \sin i). \tag{3.46}$$

Using equation (3.45) and equation (3.46), we have

$$\begin{aligned} e_b &= e_n \times e_r \\ &= \frac{1}{r} e_n \times R \\ &= (-\sin \theta \cos h - \cos \theta \sin h \cos i, \\ &\qquad -\sin \theta \sin h + \cos \theta \cos h \cos i, \cos \theta \sin i). \end{aligned} \tag{3.47}$$

By differentiating both sides of the identity $G e_n = A$, using the equation $e_r \times e_b = e_n$, and using the second identity of display (3.17), it follows that

$$\begin{aligned} G \dot{e}_n &= \dot{A} - \dot{G} e_n \\ &= r e_r \times F - r F_b e_n \\ &= r(F_b e_r \times e_b + F_n e_r \times e_n) - r F_b e_n \\ &= -r F_n e_b. \end{aligned} \tag{3.48}$$

The equations

$$\frac{di}{dt} = \frac{r F_n}{G} \cos \theta, \qquad \dot{h} = \frac{r F_n \sin \theta}{G \sin i} \tag{3.49}$$

are found by equating the components of the vectors obtained by substitution of the identities (3.45) and (3.47) into (3.48). Finally, using the definition (3.44) together with equations (3.43) and (3.49), we have the desired expression for the time derivative of H, namely,

$$\dot{H} = r(F_b \cos i - F_n \sin i \cos \theta). \tag{3.50}$$

Using the formula for $\dot{R}$ given in equation (3.37), the identity $\langle e_r, e_z \rangle = \sin \theta \sin i$ derived from equation (3.46), and the equation $\langle e_b, e_z \rangle = \cos \theta \sin i$ from (3.47), let us note that

$$\langle \dot{R}, e_z \rangle = \dot{r} \sin i \sin \theta + \frac{G}{r} \sin i \cos \theta.$$

A similar expression for $\langle \dot{R}, e_z \rangle$ is obtained by differentiation of both sides of the last equation in display (3.21) and substitution for di/dt from (3.49). The following formula for $\dot{\theta}$ is found by equating these two expressions:

$$\dot{\theta} = \frac{G}{r^2} - \frac{r F_n \cos i \sin \theta}{G \sin i}. \tag{3.51}$$

From equations (3.31) and (3.39),

$$r = \frac{G^2}{1 + e \cos v}. \tag{3.52}$$

Let us solve for $\dot{v}$ in the equation obtained by logarithmic differentiation of equation (3.52). Also, if we use the identity $1 - e^2 = G^2/L^2$ to find an expression for $\dot{e}$, substitute for $\dot{L}$ and $\dot{G}$ using the equation in display (3.43), and substitute for $\dot{r}$ using (3.42), then, after some simple algebraic manipulations,

$$\dot{v} = \frac{G}{r^2} + F_r \frac{G}{e} \cos v + F_b \frac{G^2}{re \sin v} \left(\frac{G \cos v}{e} - \frac{2r}{G} - \frac{r^2 \cos v}{eL^2G} \right).$$

A more useful expression for $\dot{v}$ is obtained by substituting for r from equation (3.52) to obtain

$$\dot{v} = \frac{G}{r^2} + F_r \frac{G}{e} \cos v - F_b \frac{G}{e} \left(1 + \frac{r}{G^2} \right) \sin v. \tag{3.53}$$

Recall equation (3.30), and define g, the *argument of periastron*, by $g := \theta - v$. Using equations (3.51) and (3.53), the time derivative $\dot{g}$ is

$$\dot{g} = -F_r \frac{G}{e} \cos v + F_b \frac{G}{e} \left(1 + \frac{r}{G^2} \right) \sin v - F_n \frac{r \cos i}{G \sin i} \sin(g + v). \tag{3.54}$$

The last Delaunay element, ℓ, called the *mean anomaly,* is defined with the aid of an auxiliary angle u, the *eccentric anomaly,* via *Kepler's equation*

$$\ell = u - e \sin u \tag{3.55}$$

where u is the unique angle such that

$$\cos u = \frac{e + \cos v}{1 + e \cos v}, \qquad \sin u = \frac{1 - e \cos u}{\sqrt{1 - e^2}} \sin v. \tag{3.56}$$

The lengthy algebraic computations required to obtain a useful expression for ℓ are carried out as follows: Differentiate both sides of Kepler's equation and solve for $\dot{\ell}$ in terms of $\dot{u}$ and $\dot{e}$. Use the relations (3.56) and equation (3.52) to prove the identity

$$r = L^2(1 - e\cos u) \tag{3.57}$$

and use this identity to find an expression for $\dot{u}$. After substitution using the previously obtained expressions for $\dot{r}$, $\dot{e}$, and $\dot{L}$, it is possible to show that

$$\dot{\ell} = \frac{1}{L^3} + \frac{r}{eL}\left[(-2e + \cos v + e\cos^2 v)F_r - (2 + e\cos v)\sin v F_b\right]. \tag{3.58}$$

In summary, the Delaunay elements (L, G, H, ℓ, g, h) for a Keplerian motion perturbed by a force F satisfy the following system of differential equations:

$$\dot{L} = L^3\left(F_r \frac{e}{G}\sin v + F_b \frac{G}{r}\right),$$

$$\dot{G} = rF_b,$$

$$\dot{H} = r(F_b\cos i - F_n\sin i\cos(g + v)),$$

$$\dot{\ell} = \frac{1}{L^3} + \frac{r}{eL}\left[(-2e + \cos v + e\cos^2 v)F_r - (2 + e\cos v)\sin v F_b\right],$$

$$\dot{g} = -F_r\frac{G}{e}\cos v + F_b\frac{G}{e}\left(1 + \frac{r}{G^2}\right)\sin v - F_n\frac{r\cos i}{G\sin i}\sin(g + v),$$

$$\dot{h} = rF_n\frac{\sin(g + v)}{G\sin i}. \tag{3.59}$$

Our transformation of the equations of motion for the perturbed Kepler problem to Delaunay elements—encoded in the differential equations in display (3.59)—is evidently not complete. Indeed, the components of the force F, as well as the functions

$$r, \quad e, \quad \cos v, \quad \sin v, \quad \cos i, \quad \sin i,$$

must be expressed in Delaunay elements. Assuming that this is done, it is still not at all clear how to extract useful information from system (3.59). Only one fact seems immediately apparent: If the force F is not present, then the Kepler motion relative to the Delaunay elements is a solution of the integrable system

$$\dot{L} = 0, \quad \dot{G} = 0, \quad \dot{H} = 0, \quad \dot{\ell} = \frac{1}{L^3}, \quad \dot{g} = 0, \quad \dot{h} = 0.$$

In fact, by inspection of this unperturbed system, it is clear that the Keplerian motion is very simple to describe in Delaunay coordinates: The three "action" variables L, G, and H remain constant and only *one* of the angular variables, namely ℓ, is not constant. In particular, for each initial condition, the motion is confined to a topological circle and corresponds to uniform rotation of the variable ℓ, that is, simple harmonic motion. The corresponding Keplerian orbit is periodic.

The result that two of the three angles that appear in system (3.59) remain constant for unperturbed motions is a special, perhaps magical, feature of the inverse square central force law. This special degeneracy of Kepler motion will eventually allow us to derive some rigorous results about the perturbed system, at least in the case where the force F is "small," that is, where $F = \epsilon F_\star$, the function $F_\star$ is bounded, and $\epsilon \in \mathbb{R}$ is regarded as a small parameter.

As we have just seen, a suitable change of coordinates can be used to transform the Kepler model equations to a very simple form. The underlying reason, mentioned previously, is that a region in the unperturbed phase space is foliated by invariant tori. Due to the special nature of the inverse square force law, each two-dimensional torus in this foliation is itself foliated by periodic orbits, that is, by one-dimensional tori. In other words, there is an open region of the unperturbed phase space filled with periodic orbits.

The foliation in the Kepler model is exceptional. In the generic case, we would not expect the flow on every invariant torus to be periodic. Instead, the flow on most of the tori would be quasi-periodic. To be more precise, consider a Poincaré section Σ and note that a two-dimensional torus in the foliation meets Σ in a one-dimensional torus $\mathbb{T}$. The associated Poincaré map is (up to a change of coordinates) a (linear) rotation on $\mathbb{T}$. The rotation angle is either rationally or irrationally related to 2π, and the angle of rotation changes continuously with respect to a parametrization of the tori in the foliation. In this scenario, the set of "quasi-periodic tori" and the set of "periodic tori" are both dense. But, with respect to Lebesgue measure, the set of quasi-periodic tori is larger. In fact, the set of quasi-periodic tori has measure one, whereas the set of periodic tori has measure zero (see Exercise 3.22). For the special case of Kepler motion, the flow is periodic on every torus.

Exercise 3.22. Consider the map $P : \mathbb{R}^2 \setminus \{(0,0)\} \to \mathbb{R}^2$ given by

$$P(x,y) = (x\cos(x^2 + y^2) - y\sin(x^2 + y^2), x\sin(x^2 + y^2) + y\cos(x^2 + y^2)).$$

(a) Prove that every circle C with radius r centered at the origin is an invariant set. (b) Prove that if m and n are integers and $r^2 = 2\pi m/n$, then every orbit of P on C is periodic. (c) Prove that if no such integers exist, then every orbit of P on C is dense. (d) Prove that the set of radii in a finite interval (for example the interval $[1, 2]$) corresponding to tori where all orbits of the Poincaré map are dense has measure one.

The origin of many important questions in the subject of differential equations arises from the problem of analyzing perturbations of integrable systems; that is, systems whose phase spaces are foliated by invariant tori. In fact, if the phase space of a system is foliated by k-dimensional tori, then there is a new coordinate system in which the equations of motion have the form

$$\dot{I} = \epsilon P(I, \theta), \qquad \dot{\theta} = \Omega(I) + \epsilon Q(I, \theta)$$

where I is a vector of "action variables," θ is a k-dimensional vector of "angle variables," and both P and Q are 2π-periodic functions of the angles. Poincaré called the analysis of this system the *fundamental problem of dynamics*. In other words, if we start with a "completely integrable" mechanical system in action-angle variables so that it has the form $\dot{I} = 0$, $\dot{\theta} = \Omega(I)$, and if we add a small force, then the problem is to describe the subsequent motion. This problem has been a central theme in mathematical research for over 100 years; it is still a prime source of important problems.

Let us outline a procedure to complete the transformation of the perturbed Kepler system (3.59) to Delaunay elements. Use equation (3.39) and the definition of L to obtain the formula

$$G^2 = L^2(1 - e^2),$$

and note that from the definition of H we have the identity

$$\cos i = \frac{H}{G}.$$

From our assumption that $G = P_\theta > 0$, and the inequality $0 \le i < \pi$, we can solve for e and i in terms of the Delaunay variables. Then, all the remaining expressions not yet transformed to Delaunay variables in system (3.59) are given by combinations of terms of the form $r^n \sin mv$ and $r^n \cos mv$ where n and m are integers. In theory we can use Kepler's equation to solve for u as a function of ℓ and e. Thus, if we invert the transformation (3.56) and also use equation (3.57), then we can express our combinations of r and the trigonometric functions of v in Delaunay variables.

The inversion of Kepler's equation is an essential element of the transformation to Delaunay variables. At a more fundamental level, the inversion of Kepler's equation is required to find the position of a planet on its elliptical orbit as a function of time. The rigorous treatment of the inversion problem seems to have played a very important role in the history of 19th century mathematics.

Problem 3.23. Write a report on the history and the mathematics related to Kepler's equation (see [61] and [91]). Include an account of the history of the theory of Bessel functions ([60], [228]) and complex analysis ([15], [31]).

To invert Kepler's equation formally, set $w := u - \ell$ so that

$$w = e\sin(w + \ell),$$

suppose that

$$w = \sum_{j=1}^{\infty} w_j e^j,$$

use the sum formula for the sine, and equate coefficients in Kepler's equation for w to obtain the w_j as trigonometric functions of ℓ. One method that can be used to make this inversion rigorous, the method used in Bessel's original treatment, is to expand in Fourier series.

It is easy to see, by an analysis of Kepler's equation, that the angle ℓ is an odd function of u. Thus, after inverting, u is an odd function of ℓ as is $e\sin u$. Thus,

$$e\sin u = \frac{2}{\pi} \sum_{\nu=1}^{\infty} \left(\int_0^{\pi} e\sin u \sin \nu\ell \, d\ell \right) \sin \nu\ell,$$

and, after integration by parts,

$$e\sin u = \frac{2}{\pi} \sum_{\nu=1}^{\infty} \left(\frac{1}{\nu} \int_0^{\pi} \cos \nu\ell (e\cos u \frac{du}{d\ell}) \, d\ell \right) \sin \nu\ell.$$

By Kepler's equation $e\cos u = 1 - d\ell/du$. Also, we have that

$$e\sin u = \frac{2}{\pi} \sum_{\nu=1}^{\infty} \left(\frac{1}{\nu} \int_0^{\pi} \cos(\nu(u - e\sin u)) \, du \right) \sin \nu\ell.$$

Bessel defined the Bessel function of the first kind

$$J_\nu(x) := \frac{1}{2\pi} \int_0^{2\pi} \cos(\nu s - x\sin s) \, ds = \frac{1}{\pi} \int_0^{\pi} \cos(\nu s - x\sin s) \, ds$$

so that

$$e\sin u = \sum_{\nu=1}^{\infty} \frac{2}{\nu} J_\nu(\nu e) \sin \nu\ell.$$

Hence, if we use the definition of the Bessel function and Kepler's equation, then

$$u = \ell + \sum_{\nu=1}^{\infty} \frac{2}{\nu} J_\nu(\nu e) \sin \nu\ell.$$

By similar, but increasingly more difficult calculations, all products of the form

$$r^n \sin mu, \quad r^n \cos mu, \quad r^n \sin mv, \quad r^n \cos mv,$$

where n and m are integers, can be expanded in Fourier series in ℓ whose νth coefficients are expressed as linear combinations of $J_\nu(\nu e)$ and $J'_\nu(\nu e)$ (see [135] and [228]). Thus, we have at least one method to transform system (3.59) to Delaunay elements.

3.2.4   Satellite Orbiting an Oblate Planet

The earth is not a sphere. In this section we will consider the perturbations of satellite motion caused by this imperfection.

The law of universal gravitation states that two particles (point masses) attract by the inverse square law. The earth is composed of lots of particles. But, if the earth is idealized as a sphere with uniformly distributed mass, then the gravitational force exerted on a satellite obeys the inverse square law for the earth considered as a point mass concentrated at the center of the sphere. But because the true shape of the earth is approximated by an oblate spheroid that "bulges" at the equator, the gravitational force exerted on a satellite depends on the position of the satellite relative to the equator. As we will see, the equations of motion of an earth satellite that take into account the oblateness of the earth are quite complex. At first sight it will probably not be at all clear how to derive useful predictions from the model. But, as an illustration of some of the ideas introduced so far in this chapter, we will see how to transform the model equations into action-angle variables. Classical perturbation theory can then be used to make predictions.

Introduce Cartesian coordinates so that the origin is at the center of mass of an idealized planet viewed as an axially symmetric oblate spheroid whose axis of symmetry is the z-axis. The "multipole" approximation of the corresponding gravitational potential has the form

$$-\frac{G_0 m_1}{r} + U(r, z) + O\left(\left(\frac{R_0}{r}\right)^3\right)$$

where

$$U = -\frac{1}{2}\frac{G_0 m_1 J_2 R_0^2}{r^3}\left(1 - 3\frac{z^2}{r^2}\right),$$

m_1 is the mass of the planet, R_0 is the equatorial radius, and J_2 is a constant related to the moments of inertia of the planet (see [94] and [152]). Note that the first term of the multipole expansion is just the point mass gravitational law that determines the Kepler motion.

The oblateness problem has been widely studied by many different methods, some more direct than our Delaunay variable approach (see [152], [184], [94], and [201]). But, our approach serves to illustrate some general methods that are widely applicable.

As an approximation to the gravitational potential of the planet, let us drop the higher order terms and consider Kepler motion perturbed by the force determined by the second term of the multipole expansion, that is, the perturbing force per unit of mass is the negative gradient of U. Since the satellite has negligible mass compared to the planet, we may as well assume that the motion of the planet is not affected by the satellite. Thus, under this assumption and in our notation, if we let m_2 denote the mass of the satellite, then $F_2 = -m_2 \operatorname{grad} U$ and the equation of motion for the satellite is given by

$$\ddot{R} = -\frac{G_0 m_1}{r^3} R - \operatorname{grad} U. \tag{3.60}$$

To use the general formulas for transformation to Delaunay variables given in display (3.59), we must first rescale system (3.60). For this, let β denote a constant measured in seconds and let $\alpha := (G_0 m_1)^{1/3} \beta^{2/3}$, so that α is measured in meters. Then, rescaling as in the derivation of equation (3.12), we obtain the equation of motion

$$\ddot{R} = -\frac{1}{r^3} R + F \tag{3.61}$$

where

$$F_x = -\frac{\epsilon}{r^5}\left(1 - 5\frac{z^2}{r^2}\right)x,$$

$$F_y = -\frac{\epsilon}{r^5}\left(1 - 5\frac{z^2}{r^2}\right)y,$$

$$F_z = -\frac{\epsilon}{r^5}\left(3 - 5\frac{z^2}{r^2}\right)z, \tag{3.62}$$

and

$$\epsilon := \frac{3}{2} J_2 \frac{R_0^2}{(G_0 m_1)^{2/3} \beta^{4/3}}$$

is a dimensionless parameter. It turns out that if we use parameter values for the earth of

$$G_0 m_1 \approx 4 \times 10^{14} m^3/\mathrm{sec}^2, \qquad R_0 \approx 6.3 \times 10^6 m, \qquad J_2 \approx 10^{-3},$$

then $\epsilon \approx 11\beta^{-4/3}$.

By adjusting our "artificial" scale parameter β, we can make the parameter ϵ as small as we like. But there is a cost: The unit of time in the scaled system is β seconds. In particular, if ϵ is small, then the unit of time is large. At any rate, the rescaling suggests that we can treat ϵ as a "small parameter."

We have arrived at a difficult issue in the analysis of our problem that often arises in applied mathematics. The perturbation parameter ϵ in our model system is a function of β. But we don't like this. So we will view β as *fixed*, and let ϵ be a *free* variable. Acting under this assumption, let us suppose that we are able to prove a theorem: If $\epsilon > 0$ is sufficiently small, then the system .... Does our theorem apply to the original unscaled system? Strictly speaking, the answer is "no"! Maybe our sufficiently small values of ϵ are *smaller* than the value of ϵ corresponding to the fixed value of β.

There are several ways to avoid the snag. For example, if we work harder, we might be able to prove a stronger theorem: There is a function $\beta \mapsto \epsilon_0(\beta)$ given by ... such that if $0 \leq \epsilon_0(\beta)$, then the corresponding system .... In this case, if β is fixed and the corresponding value of ϵ is smaller than $\epsilon_0(\beta)$, then all is well. But, in most realistic situations, the desired stronger version of our hypothetical theorem is going to be very difficult (if not impossible) for us to prove. Thus, we must often settle for the weaker version of our theorem and be pleased that the conclusion of our theorem holds for some choices of the parameters in the scaled system. This might be good. For example, we can forget about the original model, declare the scaled model to be *the* mathematical model, and use our theorems to make a prediction from the scaled model. If our predictions are verified by experiment, then we might be credited with an important discovery. At least we will be able to say with some confidence that we *understand* the associated phenomena mathematically, and we can be reasonably certain that we are studying a useful model. Of course, qualitative features of our scaled model might occur in the original model (even if we cannot prove that they do) even for physically realistic values of the parameters. This happens all the time. Otherwise, no one would be interested in perturbation theory. Thus, we have a reason to seek evidence that our original model predicts the same phenomena that are predicted by the scaled model with a small parameter. We can gather evidence by performing experiments with a numerical method, or we can try to prove another theorem.

Returning to our satellite, let us work with the scaled system and treat ϵ as a small parameter. To express the components of the force resolved relative to the frame $[e_r, e_b, e_n]$ in Delaunay elements, let us transform the vector field F to this frame using the transformation M defined in display (3.46) followed by a transformation to (e_r, e_b, e_n)-coordinates. In fact, the required transformation is

$$N := \begin{pmatrix} \cos h & -\sin h & 0 \\ \sin h & \cos h & 0 \\ 0 & 0 & 1 \end{pmatrix} \begin{pmatrix} 1 & 0 & 0 \\ 0 & \cos i & -\sin i \\ 0 & \sin i & \cos i \end{pmatrix} \begin{pmatrix} \cos \theta & -\sin \theta & 0 \\ \sin \theta & \cos \theta & 0 \\ 0 & 0 & 1 \end{pmatrix}.$$

Note that the angle θ is given by $\theta = g + v$ and the position vector is given by $R = (r, 0, 0)$ in the (e_r, e_b, e_n)-coordinates. Using the usual "push forward" change of coordinates formula

$$N^{-1}F(N(R)),$$

it follows that the transformed components of the force are

$$F_r = -\frac{\epsilon}{r^4}(1 - 3\sin^2(g + v)\sin^2 i),$$

$$F_b = -\frac{\epsilon}{r^4}\sin(2g + 2v)\sin^2 i,$$

$$F_n = -\frac{\epsilon}{r^4}\sin(g + v)\sin 2i. \tag{3.63}$$

Substitution of the force components (3.63) into system (3.59), followed by expansion in Fourier series in ℓ, gives the equations of motion for a satellite orbiting an oblate planet. While the resulting equations are quite complex, it turns out that the equation for H is very simple; in fact, $\dot{H} = 0$. This result provides a useful internal check for our formulas expressed in Delaunay elements because it can be proved directly from the definition of H as the z-component of the angular momentum in the original Cartesian coordinates: Simply differentiate in formula (3.44) and then use formula (3.17). Thus, we have extracted one prediction from the equations of motion: The z-component of the angular momentum remains constant as time evolves. Of course, the axial symmetry of the mass also suggests that H is a conserved quantity.

Recall that system (3.59) has a striking feature: while the angle ℓ is changing rapidly in the scaled time, the actions L, G, and H together with the other angles g and h change (relatively) slowly—their derivatives have order ϵ. As a rough measure of the time scale on which the model is valid, let us determine the time scale on which the satellite maintains "forward motion" (that is, $\dot{\ell} > 0$). Since $\dot{L} = O(\epsilon)$, we have that $L(t) \approx L(0) + C_1\epsilon t$ where $L(0) > 0$ and C_1 is a constant. By substitution of this expression into the right-hand side of the equation

$$\dot{\ell} = \frac{1}{L^3} + O(\epsilon)$$

and using $C_2\epsilon$ to approximate the $O(\epsilon)$ term in this equation, the approximation of the right-hand side can only vanish if t has order $\epsilon^{-4/3}$. As an approximation, we may as well say that the forward motion for the perturbed solution is maintained at least on a time scale of order $1/\epsilon$, that is, for $0 \le t \le C/\epsilon$, where C is some positive constant.

Because of the slow variation of the actions, it seems natural, at least since the time of Laplace, to study the *average* motion of the slow variables relative to ℓ. The idea is that all of the slow variables are undergoing rapid periodic oscillations due to the change in ℓ, at least over a long time scale.

Thus, if we average out these rapid oscillations, then the "drift" of the slow variables will be apparent. As mentioned before, we will make this idea precise in Chapter 7. Let us see what predictions can be made after this averaging is performed on the equations of motion of the satellite orbiting the oblate planet.

The averages that we wish to compute are the integral averages over ℓ on the interval $[0, 2\pi]$ of the right hand sides of the equations of motion in display (3.59). As we have seen, the variable ℓ appears when we change r, $\cos v$, and $\sin v$ to Delaunay variables. Let us consider the procedure for the variable G. Note that after substitution,

$$\dot{G} = -\epsilon \sin^2 i \frac{\sin(2g + 2v)}{r^3}.$$

Using the sum formula for the sine, we see that we must find the average

$$\langle \frac{\sin 2v}{r^3} \rangle := \frac{1}{2\pi} \int_0^{2\pi} \frac{\sin 2v}{r^3} d\ell$$

and the average $\langle (\cos 2v)/r^3 \rangle$. (Warning: Angle brackets are used to denote averages and inner products. But this practice should cause no confusion if the context is taken into account.) The procedure for computing these and all the other required averages for the Delaunay differential equations is evident from the following example. Differentiate in Kepler's equation (3.55) to obtain the identity

$$\frac{d\ell}{dv} = (1 - e \cos u) \frac{du}{dv}$$

and likewise in the expression for $\cos u$ in display (3.56) to obtain

$$\frac{du}{dv} = \frac{1 - e \cos u}{\sqrt{1 - e^2}}.$$

Combine these results to compute

$$\frac{d\ell}{dv} = \frac{r^2}{GL^3}$$

and use a change of variable in the original integral to find the average

$$\langle \frac{\sin 2v}{r^3} \rangle = \frac{1}{2\pi L^3 G} \int_0^{2\pi} \frac{\sin 2v}{r} dv.$$

Finally, substitution for r from equation (3.52) and an easy integration are used to prove that $\langle (\sin 2v)/r^3 \rangle = 0$ and $\langle (\cos 2v)/r^3 \rangle = 0$. As a result, it follows that $\dot{G} = 0$. Similarly, the complete set of *averaged* equations of motion are

$$\dot{L} = \dot{G} = \dot{H} = 0,$$

$$\dot{g} = -\epsilon \frac{1}{2L^3 G^4}(5\sin^2 i - 4),$$

$$\dot{h} = -\epsilon \frac{1}{L^3 G^4}\cos i \qquad (3.64)$$

where $\cos i = H/G$. Let us note that the dependent variables that appear in the averaged system (3.64) should perhaps be denoted by new symbols so that solutions of system (3.64) are not confused with solutions of the original system. This potential confusion will not arise here.

Finally, we have arrived at a system that we can analyze. In fact, the (square root of the) semimajor axis of the osculating ellipse, the total angular momentum, and the z-component of the angular momentum are constant on average. The argument of periastron g, or, if you like, the angle from the equatorial plane to the line corresponding to the perigee (closest approach of the satellite) is changing on average at a rate proportional to $4 - 5\sin^2 i$. If the inclination i of the osculating plane—an angle that is on average fixed—is less than the critical inclination where $\sin^2 i = \frac{4}{5}$, then the perigee of the orbit advances. If the inclination angle is larger than the critical inclination, then the perigee is retrograde. Similarly, the rate of regression of the ascending node—given by the angle h in the equatorial plane relative to the x-axis—is proportional to the quantity $-\cos i$. Thus, for example, if the orbit is polar $(i = \frac{\pi}{2})$, then the rate of regression is zero on average. These observations indicate the importance of the critical inclination, but a deeper analysis is required to predict the satellite's motion near the critical inclination (see [64]–[68]).

The averaging computation that we have just completed is typical of many "first order" approximations in mechanics. Averaging is, of course, only one of the basic methods that have been developed to make predictions from "realistic" systems of ordinary differential equations that originate in celestial mechanics.

Exercise 3.24. Because the perturbation force for the oblate planet comes from a potential, the force is conservative. In fact, the perturbed system in this case is Hamiltonian with the total energy, including the correction to the gravitational potential, as the Hamiltonian function. It turns out that the coordinate change to Delaunay variables is of a very special type, called canonical. For a canonical change of coordinates it is not necessary to change variables directly in the equations of motion. Instead, it suffices to change variables in the Hamiltonian function and then to derive the new equations of motion in the usual way from the transformed Hamiltonian (see Section 1.8.1). Show, by constructing an example, that a general change of coordinates is *not* canonical. Assume that the Delaunay coordinates are canonical, write the Hamiltonian in Delaunay variables, and derive from it the Hamiltonian equations of motion in Delaunay variables. In particular, show, using the form of the Hamiltonian differential equations, that the average of $\dot{L}$ over the angle ℓ must vanish. This provides an internal check for the formulas derived in this section. Do you see how one might obtain the averaged equations directly from the Hamiltonian? One reason why the Hamiltonian approach was not used in the derivation of the equations in Delaunay elements is that we have not developed the theory required to prove that the change to Delaunay variables is canonical. Another reason is that our approach works even if the perturbing force is not conservative.

3.2.5   The Diamagnetic Kepler Problem

In this section we will derive equations of motion for the electron of the hydrogen atom in a constant magnetic field. The purpose of the section is to apply Delaunay variables, averaging, and the transformation from Lagrangian to Hamiltonian mechanics. We will discuss a version of Larmor's theorem.

Consider the *classical* equations for the motion of an electron of an atom in the presence of a *constant* magnetic field. Let us assume that the electron is subjected to the Coulomb potential relative to the nucleus of the atom and to the Lorentz force due to the constant magnetic field B. If q is the charge of an electron and Z is the atomic number of the atom, and if, as usual, R is the position of the electron relative to Cartesian coordinates centered at the nucleus, V is the velocity of the electron, and $r := |R|$, then the Coulomb potential is

$$U := \frac{kZq(-q)}{r} = -\frac{kZq^2}{r}$$

where k is a constant. (Note the similarity to the gravitational potential!) In our choice of units, q is measured in coulombs and kq^2, often denoted e^2 in physics where of course e is not the eccentricity, has value $kq^2 \approx (1.52 \times 10^{-14})^2 \text{Nm}^2$ in mks units where N is used to denote newtons. For the rest of this section let us suppose that $Z = 1$, the atomic number of the hydrogen atom.

Let us assume that the constant magnetic field B is parallel to the z-axis and the electric field E vanishes. Then, as we have seen previously, the Lorentz force is given by

$$qV \times B.$$

According to Newton's law, the equations of motion are given in vector form by

$$\dot{p} = qV \times B - \frac{kq^2}{r^3}R \tag{3.65}$$

where p is the momentum. Because the speed of the electron of a hydrogen atom is about one percent of the speed of light ([87]), let us use the classical momentum $p = mV$.

Equation (3.65) is given in components by

$$m\ddot{x} = -\frac{kq^2}{r^3}x + qb\dot{y},$$

$$m\ddot{y} = -\frac{kq^2}{r^3}y - qb\dot{x},$$

$$m\ddot{z} = -\frac{kq^2}{r^3}z, \tag{3.66}$$

and after rescaling as in equation (3.61) it has the form

$$\ddot{R} = -\frac{1}{r^3}R + F$$

where $F := \epsilon(\dot{y}, -\dot{x}, 0)$, $\epsilon := 2\omega\beta$, and $\omega := \frac{1}{2m}qb$ is called the Larmor frequency. Using the formulas in Section 3.2.3, $V = \dot{R}$ expressed in the frame $[e_r, e_b, e_n]$ is given by

$$\dot{R} = \frac{e \sin v}{G}e_r + \frac{G}{r}e_b,$$

and, after some computation, it follows that

$$F_r = \epsilon \frac{G \cos i}{r},$$

$$F_b = -\epsilon \frac{e \cos i \sin v}{G},$$

$$F_n = -\epsilon \frac{\sin i}{G}(e \sin g + \sin(g + v)).$$

Although the equations of motion (3.59) for this choice of F are complicated, the corresponding averaged system is simple; in fact, we have

$$\dot{H} = \dot{G} = \dot{L} = \dot{g} = 0, \qquad \dot{h} = -\frac{\epsilon}{2}. \tag{3.67}$$

This result is a special case of Larmor's theorem: For a charged particle influenced by a centrally symmetric field (for example, the Coulomb field) and an axially symmetric weak magnetic field, there is a time-dependent coordinate system rotating with uniform velocity about the axis of symmetry of the magnetic field (at the Larmor frequency) such that, in the new coordinates, the motion is the same as for the charged particle influenced by the centrally symmetric field only.

To determine the Lagrangian and Hamiltonian formulation of the model equation (3.65), let us recast system (3.65) in the form

$$\frac{d}{dt}\left(p - \frac{1}{2}q(R \times B)\right) = \frac{1}{2}q(V \times B) - \operatorname{grad} U. \tag{3.68}$$

This first step may appear to arrive out of thin air. In fact, it is just the bridge from Newtonian to Lagrangian mechanics. As we will see in a moment, system (3.68) is in the form of an Euler–Lagrange equation.

In physics, a new vector field A, called the vector potential, is introduced so that $B = \operatorname{curl} A$. For our constant magnetic field, an easy computation shows that $A = \frac{1}{2} B \times R$. This vector field can be substituted into the left hand side of equation (3.68) and used in the rest of the computation. Let us, however, continue using the original fields.

If we define the Lagrangian

$$\mathcal{L} := \frac{1}{2m} \langle p, p \rangle + \frac{q}{2} \langle B \times R, V \rangle - U,$$

then, using a vector identity, we have

$$\frac{\partial \mathcal{L}}{\partial V} = p - \frac{q}{2} (R \times B),$$

$$\frac{\partial \mathcal{L}}{\partial R} = -\frac{q}{2} \frac{\partial}{\partial R} \langle R, B \times V \rangle - \operatorname{grad} U$$

$$= \frac{q}{2} V \times B - \operatorname{grad} U;$$

that is, equation (3.68) with $\mathcal{Q} := R = (x, y, z)$ is exactly

$$\frac{d}{dt} \left(\frac{\partial \mathcal{L}}{\partial \dot{\mathcal{Q}}} \right) = \frac{\partial \mathcal{L}}{\partial \mathcal{Q}}. \qquad (3.69)$$

In view of our derivation of the Euler–Lagrange equation in Section 3.1, we have reason to expect that there is a variational principle associated with the differential equation (3.69). This is indeed the case (see [138]).

The position variable $\mathcal{Q}$ and the velocity variable $\dot{\mathcal{Q}}$ define a coordinate system on $\mathbb{R}^3 \times \mathbb{R}^3$. Let us define a new variable

$$\mathcal{P} := \frac{\partial \mathcal{L}}{\partial \dot{\mathcal{Q}}} (\mathcal{Q}, \dot{\mathcal{Q}}) = p - \frac{q}{2} \mathcal{Q} \times B \qquad (3.70)$$

and note that the relation (3.70) can be inverted to obtain $\dot{\mathcal{Q}}$ as a function of $\mathcal{Q}$ and $\mathcal{P}$. In fact, there is a function α such that

$$\mathcal{P} \equiv \frac{\partial \mathcal{L}}{\partial \dot{\mathcal{Q}}} (\mathcal{Q}, \alpha(\mathcal{Q}, \mathcal{P})).$$

Thus, we have defined new coordinates $(\mathcal{P}, \mathcal{Q})$ on $\mathbb{R}^3 \times \mathbb{R}^3$.

The reason for introducing $\mathcal{P}$ is so that we can define the Hamiltonian

$$\mathcal{H} := \mathcal{P} \dot{\mathcal{Q}} - \mathcal{L}(\mathcal{Q}, \dot{\mathcal{Q}}) = \mathcal{P} \alpha(\mathcal{Q}, \mathcal{P}) - \mathcal{L}(\mathcal{Q}, \alpha(\mathcal{Q}, \mathcal{P})).$$

This terminology is justified by the following results:

$$\frac{\partial \mathcal{H}}{\partial \mathcal{P}} = \dot{\mathcal{Q}} + \left(\mathcal{P} - \frac{\partial \mathcal{L}}{\partial \dot{\mathcal{Q}}}\right)\frac{\partial \alpha}{\partial \mathcal{P}} = \dot{\mathcal{Q}},$$

$$\frac{\partial \mathcal{H}}{\partial \mathcal{Q}} = -\frac{\partial \mathcal{L}}{\partial \mathcal{Q}} = -\frac{d}{dt}\frac{\partial \mathcal{L}}{\partial \dot{\mathcal{Q}}} = -\dot{\mathcal{P}}.$$

Thus, the original system is equivalent to the Hamiltonian system with Hamiltonian $\mathcal{H}$. In particular, $\mathcal{H}$ is constant along orbits.

By the definition of $\mathcal{H}$, we have

$$\mathcal{H} = \langle p + \frac{q}{2}(B \times R), V \rangle - \frac{1}{2m}\langle p, p \rangle - \frac{q}{2}\langle B \times R, V \rangle + U(\mathcal{Q})$$

$$= \frac{1}{2m}\langle p, p \rangle + U(\mathcal{Q})$$

$$= \frac{1}{2m}\langle \mathcal{P} + \frac{q}{2}\mathcal{Q} \times B, \mathcal{P} + \frac{q}{2}\mathcal{Q} \times B \rangle + U(\mathcal{Q}).$$

For the constant magnetic field $B := (0, 0, b)$, the Hamiltonian is

$$\mathcal{H} = \frac{1}{2m}\left(\mathcal{P}_1^2 + \mathcal{P}_2^2 + \mathcal{P}_3^2\right) + \omega(y\mathcal{P}_1 - x\mathcal{P}_2) + \frac{m\omega^2}{2}(x^2 + y^2) - \frac{kq^2}{r} \quad (3.71)$$

where $\omega := \frac{1}{2m}qb$ is the Larmor frequency. Here, the magnitude b of the magnetic field has units $\mathrm{N\,sec/(coul\,m)}$ and the Larmor frequency ω has units $1/\mathrm{sec}$.

Recall that, in general, the Hamiltonian is constant along solutions of the corresponding Hamiltonian system. Thus, $\mathcal{H}$ is a first integral of its corresponding Hamiltonian system. An interesting feature of this system is the existence of another independent first integral. In fact, the angular momentum function

$$(x, y, z, \mathcal{P}_1, \mathcal{P}_2, \mathcal{P}_3) \mapsto y\mathcal{P}_1 - x\mathcal{P}_2 \quad (3.72)$$

is constant on orbits (see Exercise 3.26).

By Larmor's theorem we expect that there is a rotating coordinate system (rotating with the Larmor frequency about the axis of symmetry of the magnetic field) such that the transformed system is Hamiltonian but with the "angular momentum term" $\omega(y\mathcal{P}_1 - x\mathcal{P}_2)$ eliminated from the corresponding Hamiltonian. Indeed, an easy computation shows that the change of variables $(x', y', \mathcal{P}_1', \mathcal{P}_2') \to (x, y, \mathcal{P}_1, \mathcal{P}_2)$ given by

$$x = x'\cos\Omega t - y'\sin\Omega t, \qquad y = x'\sin\Omega t + y'\cos\Omega t,$$
$$\mathcal{P}_1 = \mathcal{P}_1'\cos\Omega t - \mathcal{P}_2'\sin\Omega t, \qquad \mathcal{P}_2 = \mathcal{P}_1'\sin\Omega t + \mathcal{P}_2'\cos\Omega t$$

transforms the Hamiltonian system to a new system whose Hamiltonian (after renaming the variables) is

$$\mathcal{H} = \frac{1}{2m}\left(\mathcal{P}_1^2 + \mathcal{P}_2^2 + \mathcal{P}_3^2\right) + (\omega + \Omega)(y\mathcal{P}_1 - x\mathcal{P}_2) + \frac{m\omega^2}{2}(x^2 + y^2) - \frac{kq^2}{r}.$$

Hence, if $\Omega := -\omega$, then the Hamiltonian for the transformed system is given by

$$\mathcal{H}^* = \frac{1}{2m}\left(\mathcal{P}_1^2 + \mathcal{P}_2^2 + \mathcal{P}_3^2\right) + \frac{m\omega^2}{2}(x^2 + y^2) - \frac{kq^2}{r}. \qquad (3.73)$$

Because of the analogy with the perturbed Kepler motion, this is called the diamagnetic Kepler problem ([105]). The corresponding Hamiltonian equations of motion are

$$\dot{x} = \frac{1}{m}\mathcal{P}_1, \quad \dot{y} = \frac{1}{m}\mathcal{P}_2, \quad \dot{z} = \frac{1}{m}\mathcal{P}_3,$$

$$\dot{\mathcal{P}}_1 = -\left(\frac{kq^2}{r^3}x + m\omega^2 x\right),$$

$$\dot{\mathcal{P}}_2 = -\left(\frac{kq^2}{r^3}y + m\omega^2 y\right),$$

$$\dot{\mathcal{P}}_3 = -\frac{kq^2}{r^3}z. \qquad (3.74)$$

Equivalently, we have the second order system

$$m\ddot{x} = -\frac{kq^2}{r^3}x - m\omega^2 x,$$

$$m\ddot{y} = -\frac{kq^2}{r^3}y - m\omega^2 y,$$

$$m\ddot{z} = -\frac{kq^2}{r^3}z \qquad (3.75)$$

which is given in vector form by

$$\ddot{R} = -\frac{kq^2 m^{-1}}{r^3}R + F_0 \qquad (3.76)$$

where $F_0 = -\omega^2(x, y, 0)$. This last system is again analogous to the equation for relative motion of the perturbed two-body problem.

Let us note that the Kepler orbit of the electron is rotating with the Larmor frequency ω about the axis of symmetry of the constant magnetic field. For a weak magnetic field, the rotation frequency is slow. According to our reduced Hamiltonian system in the rotating coordinates, the perturbation due to the magnetic field in the rotating system has order ω^2. Thus, the main effect is rotation of the electron about the axis of symmetry of the magnetic field. The second order effects, however, are important.

As an instructive project, rescale system (3.75) and use equations (3.59) to transform the diamagnetic Kepler problem to Delaunay elements. Also, average the transformed equations and discuss the average motion of the electron (see Exercise 3.28). It turns out that the diamagnetic Kepler problem has very complex (chaotic) motions; it is one of the model equations studied in the subject called quantum chaos, but that is another story ([93], [105]).

Exercise 3.25. Verify the equations (3.67).

Exercise 3.26. Prove that the function given in display (3.72) is constant on orbits of the Hamiltonian system with Hamiltonian (3.71). What corresponding quantity is conserved for system (3.65)?

Exercise 3.27. Transform equation (3.66) directly to equation (3.75) using a rotating coordinate system.

Exercise 3.28. Show that system (3.75) can be rescaled in space and time to the dimensionless form

$$\ddot{R} = -\frac{1}{r^3}R - \omega^2\beta^2 \begin{pmatrix} x \\ y \\ 0 \end{pmatrix} \tag{3.77}$$

where β is measured in seconds. Define $\epsilon := \omega^2\beta^2$ and show that the scaled system is equivalent to the first order system

$$\dot{x} = \mathcal{P}_1, \qquad \dot{y} = \mathcal{P}_2, \qquad \dot{z} = \mathcal{P}_3,$$
$$\dot{\mathcal{P}}_1 = -x/r^3 - \epsilon x, \qquad \dot{\mathcal{P}}_2 = -y/r^3 - \epsilon y, \qquad \dot{\mathcal{P}}_3 = -z/r^3.$$

Use the result of Exercise 3.26 and the new variables $(\rho, \theta, z, \mathcal{P}_\rho, \mathcal{P}_\theta, \mathcal{P}_z)$ given by

$$x = \rho\cos\theta, \qquad y = \rho\sin\theta,$$
$$\mathcal{P}_\rho = \cos\theta\,\mathcal{P}_1 + \sin\theta\,\mathcal{P}_2, \qquad \mathcal{P}_\theta = x\,\mathcal{P}_2 - y\,\mathcal{P}_1, \qquad \mathcal{P}_z = \mathcal{P}_3$$

to show that the differential equation expressed in the new variables decouples so that the set of orbits with zero angular momentum correspond to solutions of the subsystem

$$\dot{\rho} = \mathcal{P}_\rho, \quad \dot{z} = \mathcal{P}_z, \quad \dot{\mathcal{P}}_\rho = -\frac{\rho}{(\rho^2 + z^2)^{3/2}} - \epsilon\rho, \quad \dot{\mathcal{P}}_z = -\frac{z}{(\rho^2 + z^2)^{3/2}}. \tag{3.78}$$

Exercise 3.29. Consider system (3.77) to be in the form

$$\ddot{R} = -\frac{1}{r^3}R + F.$$

Show that

$$F_r = -\epsilon r\cos^2\theta, \qquad F_b = \epsilon r\sin\theta\cos\theta, \qquad F_n = 0,$$

and the averaged Delaunay system is given by

$$\dot{L} = 0, \qquad \dot{G} = \epsilon\frac{5}{4}L^2(L^2 - G^2)\sin 2g, \qquad \dot{g} = -\epsilon\frac{1}{4}L^2 G(3 + 5\cos 2g).$$

Draw the phase cylinder portrait of the (g, G)-subsystem. Find the rest points and also show that there is a homoclinic orbit.

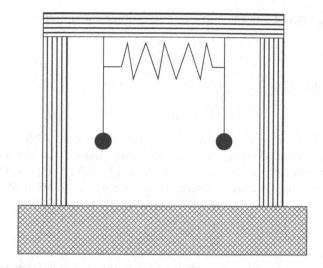

Figure 3.3: Two pendula connected by a spring. To build a simple bench model, consider suspending two lead fishing weights on "droppers" from a stretched horizontal section of monofilament.

Exercise 3.30. Consider the diamagnetic Kepler problem as a perturbation, by the Coulomb force, of the Hamiltonian system with Hamiltonian

$$\mathcal{H}_0^* = \frac{1}{2m}\left(\mathcal{P}_1^2 + \mathcal{P}_2^2 + \mathcal{P}_3^2\right) + \frac{m\omega^2}{2}(x^2 + y^2).$$

Write out Hamilton's equations, change coordinates so that the equations of motion corresponding to the Hamiltonian $\mathcal{H}_0^*$ are in action-angle form (use polar coordinates), and find the perturbation in the new coordinates. Is averaging reasonable for this system?

3.3   Coupled Pendula: Normal Modes and Beats

Consider a pendulum of length L and mass m where the angle with positive orientation with respect to the downward vertical is θ and let g denote the gravitational constant (near the surface of the earth). The kinetic energy of the mass is given by $K := \frac{m}{2}(L\dot{\theta})^2$ and the potential energy is given by $U := -mgL\cos\theta$. There are several equivalent ways to obtain the equations of motion; we will use the Lagrangian formulation. Recall that the Lagrangian of the system is

$$\mathcal{L} := K - U = \frac{m(L\dot{\theta})^2}{2} + mgL\cos\theta, \tag{3.79}$$

and the equation of motion is given by Lagrange's equation

$$\frac{d}{dt}\frac{\partial \mathcal{L}}{\partial \dot\theta} - \frac{\partial \mathcal{L}}{\partial \theta} = 0.$$

Thus, the equation of motion for the pendulum is

$$mL^2\ddot\theta + mgL\sin\theta = 0. \tag{3.80}$$

For two identical pendula coupled by a Hookian spring with spring constant k, the potential energy due to the spring depends on (at least) the placement of the spring and the choice of the plane for the pendulum motion. For small oscillations, a reasonable linear approximation of the true potential energy due to the spring is given by

$$\frac{1}{2}ka\ell^2(\theta_1 - \theta_2)^2$$

where a is a dimensionless constant and ℓ is the distance from the pivot point of the pendulum to the point where the spring is attached. Here, we do not model the physical attachment of the spring and the pendula. Perhaps a different expression for the potential energy of the spring would be required to model a laboratory experiment. Our model, however, is a reasonable approximation as long as the spring moves in a fixed plane where the spring is either stretched or twisted by the pendulum motion. The corresponding Lagrangian is

$$\mathcal{L} = \frac{m}{2}\left((L\dot\theta_1)^2 + (L\dot\theta_2)^2\right) + mgL(\cos\theta_1 + \cos\theta_2) - \frac{1}{2}ka\ell^2(\theta_1 - \theta_2)^2,$$

and the equations of motion are given by

$$mL^2\ddot\theta_1 + mgL\sin\theta_1 + ka\ell^2(\theta_1 - \theta_2) = 0,$$
$$mL^2\ddot\theta_2 + mgL\sin\theta_2 - ka\ell^2(\theta_1 - \theta_2) = 0.$$

Let us note that time (the independent variable) and the parameters of the system are rendered dimensionless by rescaling time via $t = \mu s$, where $\mu := (L/g)^{1/2}$, and introducing the dimensionless constant $\alpha := ka\ell^2/(mgL)$ to obtain the system

$$\theta_1'' + \sin\theta_1 + \alpha(\theta_1 - \theta_2) = 0,$$
$$\theta_2'' + \sin\theta_2 - \alpha(\theta_1 - \theta_2) = 0.$$

To study the motions of the system for "small" oscillations of the pendula, the approximation $\sin\theta \approx \theta$—corresponding to linearization of the system of differential equations at the origin—yields the model

$$\theta_1'' + (1 + \alpha)\theta_1 - \alpha\theta_2 = 0,$$
$$\theta_2'' - \alpha\theta_1 + (1 + \alpha)\theta_2 = 0. \tag{3.81}$$

Although this linear second order system can be expressed as a first order system and solved in the usual manner by finding the eigenvalues and eigenvectors of the system matrix, there is a simpler way to proceed. In fact, if Θ is defined to be the transpose of the state vector (θ_1, θ_2), then system (3.81) has the form

$$\Theta'' = A\Theta$$

where A is the matrix

$$\begin{pmatrix} -(1+\alpha) & \alpha \\ \alpha & -(1+\alpha) \end{pmatrix}.$$

The idea is to diagonalize the symmetric matrix A by a linear change of variables of the form $\Theta = BZ$, where B is an orthogonal matrix whose columns are unit-length eigenvectors of A, so that the transformed differential equation, $Z'' = B^{-1}ABZ$, decouples. For system (3.81), the component form of the inverse change of variables $Z = B^{-1}\Theta$ is

$$x = \frac{1}{\sqrt{2}}(\theta_1 + \theta_2), \qquad y = \frac{1}{\sqrt{2}}(\theta_1 - \theta_2),$$

and the decoupled system is

$$x'' = -x, \qquad y'' = -(1 + 2\alpha)y.$$

There are two *normal modes* of oscillation. If $y(s) \equiv 0$, then $\theta_1(s) - \theta_2(s) \equiv 0$ and the pendula swing "in phase" with unit frequency relative to the scaled time. If $x(s) \equiv 0$, then $\theta_1(s) + \theta_2(s) \equiv 0$ and the pendula swing "in opposing phase" with angular frequency $(1 + 2\alpha)^{1/2}$ in the scaled time. The frequency of the second normal mode is larger than the frequency of the first normal mode due to the action of the spring; the spring has no effect on the first normal mode.

Consider the following experiment. Displace the second pendulum by a small amount and then release it from rest. What happens?

In our mathematical model, the initial conditions corresponding to the experiment are

$$\theta_1 = 0, \quad \theta_1' = 0, \quad \theta_2 = a, \quad \theta_2' = 0.$$

The predicted motion of the system, with $\beta := \sqrt{1 + 2\alpha}$, is given by

$$x(s) = \frac{a}{\sqrt{2}}\cos s, \qquad y(s) = -\frac{a}{\sqrt{2}}\cos \beta s,$$

and

$$\theta_1(s) = \frac{a}{2}(\cos s - \cos \beta s), \qquad \theta_2(s) = \frac{a}{2}(\cos s + \cos \beta s).$$

Use the usual identities for $\cos(\phi \pm \psi)$ with

$$\phi := \frac{1+\beta}{2}s, \qquad \psi := \frac{1-\beta}{2}s,$$

to obtain

$$\theta_1(s) = \left(a\sin\frac{\beta-1}{2}s\right)\sin\frac{\beta+1}{2}s, \quad \theta_2(s) = \left(a\cos\frac{\beta-1}{2}s\right)\cos\frac{\beta+1}{2}s,$$

and note that each pendulum swings with quasi-frequency $\frac{1}{2}(\beta+1)$ and (relatively) slowly varying amplitude. Also, the "beats" of the two pendula are out of phase; that is, whereas the first pendulum is almost motionless and the second pendulum swings at maximum amplitude when s is approximately an integer multiple of $2\pi/(\beta-1)$, the second pendulum is almost motionless and the first pendulum swings at maximum amplitude when s is approximately an odd integer multiple of $\pi/(\beta-1)$. This interesting exchange-of-energy phenomenon can be observed even with very crude experimental apparatus—try it.

Exercise 3.31. Suppose that the kinetic energy of a mechanical system is given by $\frac{1}{2}\langle K\dot{\Theta}, \dot{\Theta}\rangle$ and its potential energy is given by $\frac{1}{2}\langle P\Theta, \Theta\rangle$, where Θ is the state vector, K and P are symmetric matrices, and the angle brackets denote the usual inner product. If both quadratic forms are positive definite, show that they can be simultaneously diagonalized. In this case, the resulting system decouples. Solutions corresponding to the oscillation of a single component while all other components are at rest are called *normal modes*. Determine the frequencies of the normal modes (see [12]).

Exercise 3.32. (a) Find the general solution of the system

$$\ddot{x} = a\sin t(x\sin t + y\cos t) - x, \quad \ddot{y} = a\cos t(x\sin t + y\cos t) - y.$$

Hint: Find a time-dependent transformation that makes the system autonomous. (b) Find the Floquet multipliers corresponding to the zero solution. (c) For which values of a is the zero solution stable; for which values is it stable?

Exercise 3.33. Build a bench top experiment with two "identical" coupled pendula (see Figure 3.3), and tune it until the beat phenomena are observed. Show that a measurement of the length of the pendula together with a measurement of the number of oscillations of one pendulum per second suffices to predict the time interval required for each pendulum to return to rest during a beating regime. Does your prediction agree with the experiment? How sensitive is the predicted value of this time scale relative to errors in the measurements of the lengths of the pendula and the timing observation? Approximate the spring constant in your physical model?

Problem 3.34. Consider small oscillations of the coupled pendula in case there are two different pendula, that is, pendula with different lengths or different masses. What happens if there are several pendula coupled together in a ring or

Figure 3.4: Representation of the Fermi–Ulam–Pasta coupled oscillator.

in series? What about oscillations that are not small? What predictions (if any) made from the linear model remain valid for the nonlinear model? What happens if there is damping in the system?

3.4   The Fermi–Ulam–Pasta Oscillator

The analysis of the small oscillations of coupled pendula in Section 3.3 can be generalized in many different directions. Here we will consider a famous example due to Enrico Fermi, Staniaław Ulam, and John R. Pasta [86] that can be viewed as a model for a series of masses coupled to their nearest neighbors by springs. The original model was obtained as the discretization of a partial differential equation model of a string—one of the ways that systems of ordinary differential equations arise in applied mathematics.

Let us consider N identical masses positioned on a line as in Figure 3.4, and let us suppose that the masses are constrained to move only on this line. Moreover, let us suppose that the masses are coupled by springs, but with the first and last masses pinned to fixed positions. If x_k denotes the displacement of the kth mass from its equilibrium position; then, using Newton's second law, the equations of motion are given by

$$m\ddot{x}_k = F(x_{k+1} - x_k) - F(x_k - x_{k-1}), \qquad k = 1, \ldots, N-2$$

where $F(x_{k+1} - x_k)$ is the force exerted on the kth mass from the right and $-F(x_k - x_{k-1})$ is the force exerted on the kth mass from the left.

One of the Fermi–Ulam–Pasta models uses the scalar force law

$$F(z) = \alpha(z + \beta z^2), \qquad \alpha > 0, \quad \beta \geq 0,$$

to model the restoring force of a nonlinear spring. This choice leads to the following equations of motion:

$$m\ddot{x}_k = \alpha(x_{k-1} - 2x_k + x_{k+1})(1 + \beta(x_{k+1} - x_{k-1})), \quad k = 1, \ldots N-2 \tag{3.82}$$

where we also impose the boundary conditions

$$x_0(t) \equiv 0, \qquad x_{N-1}(t) \equiv 0.$$

If we set $\beta = 0$ in the equations (3.82), then we obtain the linearization of system (3.82) at the point corresponding to the rest positions of the masses.

The first objective of this section is to determine the normal modes and the general solution of this linearization.

Let us define the state vector x with components $(x_1, \ldots, x_{N-2})$, and let us write the system in matrix form

$$\ddot{x} = c^2 Q x \qquad (3.83)$$

where $c^2 = \alpha/m$ and

$$Q = \begin{pmatrix} -2 & 1 & 0 & 0 & \cdots & & 0 \\ 1 & -2 & 1 & 0 & \cdots & & 0 \\ \vdots & & & & \ddots & & \\ 0 & \cdots & & & & 1 & -2 \end{pmatrix}.$$

Because Q is a real negative definite symmetric matrix, the matrix has a basis of eigenvectors corresponding to real negative eigenvalues. If v is an eigenvector corresponding to the eigenvalue λ, then $e^{c\sqrt{-\lambda}\,it}v$ is a solution of the matrix system. The corresponding normal mode is the family of real solutions of the form

$$x(t) = R\cos(c\sqrt{-\lambda}\,t + \rho)\,v$$

where R and ρ depend on the initial conditions.

If $v = (v_1, \ldots, v_{N-2})$ is an eigenvector of Q with eigenvalue λ, then

$$v_{k-1} - 2v_k + v_{k+1} = \lambda v_k, \qquad k = 1, \ldots, N-2.$$

To solve this linear three term recurrence, set $v_k = a^k$, and note that a^k gives a solution if and only if

$$a^2 - (2 + \lambda)a + 1 = 0. \qquad (3.84)$$

Also, note that the product of the roots of this equation is unity, and one root is given by

$$r = \frac{2 + \lambda + \sqrt{\lambda(4 + \lambda)}}{2}.$$

Thus, using the linearity of the recurrence, the general solution has the form

$$v_k = \mu r^k + \nu r^{-k}$$

where μ and ν are arbitrary scalars. Moreover, in view of the boundary conditions, $v_0 = 0$ and $v_{N-1} = 0$, we must take $\mu + \nu = 0$ and $r^{N-1} - 1/r^{N-1} = 0$. In particular, r must satisfy the equation $r^{2(N-1)} = 1$. Thus, the possible choices for r are the roots of unity

$$r_\ell = e^{\pi i \ell/(N-1)}, \qquad \ell = 0, 1, \ldots, 2N - 3.$$

We will show that the r_ℓ for $\ell = 1, \ldots, N-2$ correspond to $N-2$ distinct eigenvalues of the $(N-2) \times (N-2)$ matrix Q as follows: The eigenvalue

$$\lambda_\ell = -4 \sin^2\left(\frac{\pi\ell}{2(N-1)}\right)$$

corresponding to r_ℓ is obtained by solving equation (3.84) with $a = r_\ell$; that is, the equation

$$e^{2\pi i\ell/(N-1)} - (2 + \lambda)e^{\pi i\ell/(N-1)} + 1 = 0.$$

The remaining choices for r_ℓ of course cannot lead to new eigenvalues. But to see this directly consider the range of integers ℓ expressed in the form

$$0, 1, \ldots, N-2, N-1, (N-1)+1, \ldots, (N-1)+N-2$$

to check that the corresponding r_ℓ are given by

$$1, r_1, \ldots, r_{N-2}, -1, -r_1, \ldots, -r_{N-2},$$

and the corresponding λ_ℓ are

$$0, \lambda_1, \ldots, \lambda_{N-2}, -4, \lambda_{N-2}, \ldots, \lambda_1.$$

Here, the choices $r = 1$ and $r = -1$, corresponding to $\ell = 0$ and $\ell = N-1$, give $v_k \equiv 0$. Hence, they do not yield eigenvalues.

The components of the eigenvectors corresponding to the eigenvalue λ_ℓ are given by

$$v_k = \mu\left(e^{\pi i\ell k/(N-1)} - e^{-\pi i\ell k/(N-1)}\right) = 2i\mu \sin\left(\frac{\pi\ell k}{N-1}\right)$$

where μ is a scalar. If $\mu = 1/(2i)$, then we have, for each $\ell = 1, \ldots, N-2$, the associated eigenvector v^ℓ with components

$$v_k^\ell = \sin\left(\frac{\pi\ell k}{N-1}\right).$$

Because Q is symmetric, its eigenvectors corresponding to distinct eigenvalues are orthogonal with respect to the usual inner product. Moreover, we have that

$$\langle v^\ell, v^\ell \rangle = \sum_{k=1}^{N-2} \sin^2\left(\frac{\pi\ell k}{N-1}\right) = \frac{N-1}{2}$$

where the last equality can be proved by first applying the identity

$$\sin\theta = \frac{e^{i\theta} - e^{-i\theta}}{2i}$$

and then summing the resulting geometric series. Thus, the vectors

$$\left(\frac{2}{N-1}\right)^{1/2} v^1, \ldots, \left(\frac{2}{N-1}\right)^{1/2} v^{N-2}$$

form an orthonormal basis of $\mathbb{R}^{N-2}$.

The general solution of the system (3.82) with $\beta = 0$ is given by the vector solution $t \mapsto x(t)$ with components

$$x_k(t) = \left(\frac{2}{N-1}\right)^{1/2} \sum_{\ell=1}^{N-2} (\gamma_\ell p_\ell(t) + \eta_\ell q_\ell(t)) \sin(\frac{\pi \ell k}{N-1})$$

where $c^2 = \alpha/m$,

$$p_\ell(t) = \cos(2ct \sin(\frac{\pi \ell}{2(N-1)})), \qquad q_\ell(t) = \sin(2ct \sin(\frac{\pi \ell}{2(N-1)})),$$

and γ_ℓ, η_ℓ are real constants. In vector form, this solution is given by

$$x(t) = \left(\frac{2}{N-1}\right)^{1/2} \sum_{\ell=1}^{N-2} (\gamma_\ell p_\ell(t) + \eta_\ell q_\ell(t)) v^\ell;$$

it is the solution of the first-order system corresponding to the model (3.82) with initial value

$$x(0) = \left(\frac{2}{N-1}\right)^{1/2} \sum_{\ell=1}^{N-2} \gamma_\ell v^\ell,$$

$$\dot{x}(0) = 2c \left(\frac{2}{N-1}\right)^{1/2} \sum_{\ell=1}^{N-2} \eta_\ell \sin(\frac{\pi \ell}{2(N-1)}) v^\ell.$$

Moreover, let us note that if we use the orthonormality of the normalized eigenvectors, then the scalars γ_ℓ and η_ℓ can be recovered with the inversion formulas

$$\gamma_\ell = \langle x(0), \left(\frac{2}{N-1}\right)^{1/2} v^\ell \rangle,$$

$$\eta_\ell = \langle \dot{x}(0), \left(\frac{2}{N-1}\right)^{1/2} (2c \sin \frac{\pi \ell}{2(N-1)})^{-1} v^\ell \rangle.$$

Now that we have determined the normal modes and the general solution of the linearized system, let us use them to describe the Fermi–Ulam–Pasta experiments.

If B is the matrix whose columns are the ordered orthonormal eigenvectors of Q, then the linear coordinate transformation $x = By$ decouples the system of differential equations (3.82) with $\beta = 0$ into a system of the form

$$\ddot{y}_k = c^2 \lambda_k y_k, \qquad k = 1, \ldots, N-2$$

where λ_k is the eigenvalue corresponding to the kth column of B. Note that the total energy of this kth mode is given by

$$E_k := \frac{1}{2}\left(\dot{y}_k^2 - c^2 \lambda_k y_k^2\right),$$

and this energy can be easily computed from the vector solution $x(t)$ by using the identity

$$\left(\frac{2}{N-1}\right)^{1/2}\langle x(t), v^k\rangle = y_k.$$

Fermi, Ulam, and Pasta expected that after an initial excitation the averages over time of the linear mode energies $E_k(t)$ of the nonlinear ($\beta \neq 0$) oscillator (3.82) would tend to equalize after a sufficiently long time period. The process leading to this "equipartition of energy" is called *thermalization*. In fact, the purpose of their original experiment—numerical integration of the system starting with nonzero energy in only one normal mode—was to determine the length of time required for thermalization to occur. Contrary to their expectation, the results of the experiment suggested that thermalization does *not* occur. For example, for some choices of the system parameters, the energy becomes distributed among the various linear modes for a while, but eventually almost all the energy returns to the initial mode. Later, most of the energy might be in the second mode before returning again to the first mode, and so on. For other values of the system parameters the recurrence is not as pronounced, but none of their experiments suggested that thermalization does occur. The explanation for this "beat phenomenon" and for the nonexistence of thermalization leads to some very beautiful mathematics and mathematical physics (see the article by Richard Palais [180]). It is remarkable that the first numerical dynamics experiments performed on a computer (during 1954–55) turned out to be so important (see [230]).

Exercise 3.35. Solve the differential equation (3.83) by converting it to a first order system and then finding a basis of eigenvectors.

Exercise 3.36. Describe the geometry of the modes of oscillation of the masses in the Fermi–Ulam–Pasta model with respect to the linearized model (3.83). For example, is it possible that all the masses move so that the distances between adjacent masses stays fixed?

Exercise 3.37. Solve system (3.82) with $\beta = 0$ and periodic boundary conditions: $x_0(t) = x_{N-1}(t)$.

Exercise 3.38. Repeat the Fermi–Ulam–Pasta experiment. Begin with the parameters

$$N = 32, \qquad m = 1.0, \qquad \alpha = 1.0, \qquad \beta = 0.25,$$

and choose an initial condition so that the velocity of each mass is zero and all the energy is in the first mode; for example, take

$$x_k(0) = \left(\frac{2}{N-1}\right)^{1/2} \sin(\frac{\pi k}{N-1}).$$

Integrate system (3.82) numerically and output the mode energies for at least the first few modes. Discuss how the mode energies change over time.

Exercise 3.39. Consider system (3.82) with its first boundary condition replaced by $x_0(t) = A \sin \omega t$. What is the general behavior of this system? Note: This problem is open-ended; it does not seem to have a simple answer.

3.5   The Inverted Pendulum

Consider a pendulum with oscillating vertical displacement. We will outline a proof of the following amazing fact: The inverted pendulum can be made stable for certain rapid oscillations of small vertical displacement.

Historical comments and a very interesting description of the stabilization phenomenon based on topological methods is given by Mark Levi (see [142]). David Acheson's book [4] includes results on the stabilization of the *double* pendulum.

The equation of motion for the displaced inverted pendulum is obtained as a modification of the pendulum model (3.80). For this, let H be a smooth one-periodic function with unit amplitude where L is the length of the pendulum and g is the gravitational constant. Let us also incorporate two control parameters, the amplitude δ and the (relative) frequency Ω of the vertical displacement, so that the displacement function is given by

$$t \mapsto \delta H(\Omega t).$$

This displacement may be viewed as an external force with period $1/\Omega$ by taking the force to be

$$\mathcal{F} := mL\Omega^2 \delta H''(\Omega t) \sin \theta.$$

An alternative way to view the model is to imagine the pendulum inside an "Einstein elevator" that is being periodically accelerated. In this case, the external force is perceived as a time-dependent change in the gravitational field. The new gravitational "constant" measured in some units, say cm/sec/sec, is given by

$$g - \Omega^2 \delta H''(\Omega t),$$

and the equation of motion is obtained by replacing g by this difference in the model (3.80). The minus sign is not important, it is there to make this formulation compatible with the Lagrangian formulation. Indeed, with the choice for the force given above and the Lagrangian (3.79), the Lagrange equation

$$\frac{d}{dt}\frac{d\mathcal{L}}{d\dot{\theta}} - \frac{d\mathcal{L}}{d\theta} = \mathcal{F}$$

yields the following equation of motion:

$$\ddot{\theta} + \frac{g}{L}\sin\theta = \delta\frac{\Omega^2}{L}H''(\Omega t)\sin\theta.$$

Let us rescale time with the change of variable given by

$$t = \frac{1}{\Omega}s.$$

Also, after this change of time, let us use the scaled period and amplitude of the displacement

$$\alpha := \frac{g}{L\Omega^2}, \qquad \beta := \frac{\delta}{L},$$

and the function given by

$$a(s) := H''(s)$$

to construct the dimensionless equation of motion

$$\theta'' + (\alpha - \beta a(s))\sin\theta = 0 \qquad\qquad (3.85)$$

where the function $s \mapsto a(s)$ is periodic with period one.

To study the stability of the rest point at $\theta = \pi$ corresponding to the inverted pendulum, the equation of motion is linearized at $\theta = \pi$ to obtain the periodic linear system

$$w'' - (\alpha - \beta a(s))w = 0 \qquad\qquad (3.86)$$

with parameters α and β. While we will consider the two-parameter family of differential equations in the entire parameter plane, only those systems with parameters in the first quadrant correspond to the physical pendulum.

We will outline a proof of the following proposition:

Proposition 3.40. *If $a(s) = \sin 2\pi s$ in the differential equation (3.86), then the point $(\alpha, \beta) = (0,0)$ is a boundary point of an open subset in the first quadrant of the parameter plane such that the differential equation corresponding to each point of this open set has a stable trivial solution.*

The open set mentioned in Proposition 3.40 contains points close to the origin of the parameter plane that correspond to high-frequency small-amplitude displacements which stabilize the inverted pendulum. Proposition 3.40 is true if the normalized displacement $a(s) = \sin 2\pi s$ is replaced by certain other periodic functions that will also be determined.

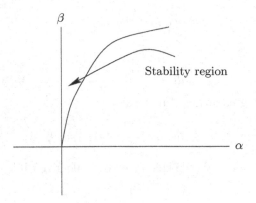

Figure 3.5: Stabilization region for the inverted pendulum.

According to Proposition 3.40, the (periodic) trivial solution of the linearized pendulum model equation is stable. Because such a periodic solution is not hyperbolic, we have no method that can be used to prove that the corresponding rest position of the original nonlinear model of the inverted pendulum is stable. In fact, the principle of linearized stability gives the correct insight: the rest position of the original nonlinear model of the inverted pendulum is stable whenever its linearization is stable. The *proof* of this fact seems to require an analysis (KAM theory) that is beyond the scope of this book (see [142]).

We will use Floquet theory, as in our analysis of Hill's equation, to prove Proposition 3.40.

Let $\Phi(s, \alpha, \beta)$ denote the principal fundamental matrix at $s = 0$ for the first order system

$$w' = z, \qquad z' = (\alpha - \beta a(s))w \tag{3.87}$$

corresponding to the differential equation (3.86). Recall from our study of Hill's equation that the trivial solution of the system (3.87) is stable provided that

$$|\operatorname{tr} \Phi(1, \alpha, \beta)| < 2.$$

If $(\alpha, \beta) = (0, 0)$, then

$$\Phi(1, 0, 0) = \exp \begin{pmatrix} 0 & 1 \\ 0 & 0 \end{pmatrix} = \begin{pmatrix} 1 & 1 \\ 0 & 1 \end{pmatrix}.$$

At this point of the parameter space $\operatorname{tr} \Phi(1, 0, 0) = 2$. Thus, it is reasonable to look for nearby points where $\operatorname{tr} \Phi(1, 0, 0) < 2$. The idea is simple enough: Under our assumption that H is smooth, so is the function $\tau : \mathbb{R}^2 \to \mathbb{R}$ given by $(\alpha, \beta) \mapsto \operatorname{tr} \Phi(1, \alpha, \beta)$. We will use the implicit function theorem to show that the boundary of the region of stability, depicted in Figure 3.5, is a smooth curve that passes through the origin of the parameter space and into the positive first quadrant.

To compute the partial derivative $\tau_\alpha(0,0)$, let $A(s)$ denote the system matrix for the system (3.87) and use the matrix equation $\Phi' = A(s)\Phi$ to obtain the variational initial value problem

$$\Phi'_\alpha = A(s)\Phi_\alpha + A_\alpha(s)\Phi, \qquad \Phi_\alpha(0) = 0.$$

At $(\alpha, \beta) = (0,0)$, the variational equation is given by the following (inhomogeneous) linear system

$$\begin{aligned} W' &= \begin{pmatrix} 0 & 1 \\ 0 & 0 \end{pmatrix} W + \begin{pmatrix} 0 & 0 \\ 1 & 0 \end{pmatrix} \Phi(s,0,0) \\ &= \begin{pmatrix} 0 & 1 \\ 0 & 0 \end{pmatrix} W + \begin{pmatrix} 0 & 0 \\ 1 & 0 \end{pmatrix} \begin{pmatrix} 1 & s \\ 0 & 1 \end{pmatrix} \\ &= \begin{pmatrix} 0 & 1 \\ 0 & 0 \end{pmatrix} W + \begin{pmatrix} 0 & 0 \\ 1 & s \end{pmatrix}. \end{aligned}$$

It can be solved by variation of parameters to obtain

$$\Phi_\alpha(1,0,0) = W(1) = \begin{pmatrix} \frac{1}{2} & * \\ * & \frac{1}{2} \end{pmatrix};$$

and therefore, $\tau_\alpha(0,0) = 1$.

Define the function $g(\alpha, \beta) := \tau(\alpha, \beta) - 2$ and note that we now have

$$g(0,0) = 0, \qquad g_\alpha(0,0) = 1.$$

By an application of the implicit function theorem, there is a function $\beta \mapsto \gamma(\beta)$, defined for sufficiently small β, such that $\gamma(0) = 0$ and $\tau(\gamma(\beta), \beta) \equiv 2$.

To determine which "side" of the curve $\Gamma := \{(\alpha, \beta) : \alpha = \gamma(\beta)\}$ corresponds to the region of stability, let us consider points on the positive α-axis. In this case, the linearized equation has constant coefficients:

$$w'' - \alpha w = 0.$$

Its principal fundamental matrix solution at $s = 0$ evaluated at $s = 1$ is given by

$$\begin{pmatrix} \cosh \sqrt{\alpha} & \frac{1}{\sqrt{\alpha}} \sinh \sqrt{\alpha} \\ \sqrt{\alpha} \sinh \sqrt{\alpha} & \cosh \sqrt{\alpha} \end{pmatrix},$$

and, for $\alpha > 0$, we have $\tau(\alpha, 0) = 2 \cosh \sqrt{\alpha} > 2$.

By the implicit function theorem, the curve Γ in the parameter space corresponding to $\operatorname{tr} \Phi(1, \alpha, \beta) = 2$ is unique. Also, by the computation above, the positive α-axis lies in the unstable region. Because $\tau_\alpha(0,0) = 1$, we must have $\tau_\alpha(\gamma(\beta), \beta) > 0$ as long as β is sufficiently small. Thus, it follows that the trace of the monodromy matrix increases through the value 2 as the curve Γ is crossed. In particular, the trace of the monodromy matrix is less than 2 on the left side of this curve; that is, Γ forms a boundary of the stable region as depicted in Figure 3.5.

Finally, to determine the conditions on the periodic displacement so that the restriction of Γ to $\beta > 0$ lies in the first quadrant, we will use the equality

$$\tau_\beta(0,0) = -\int_0^1 a(s) \, ds$$

(see Exercise 3.41).

Because the original external excitation of the pendulum is periodic, the average value of its second derivative (corresponding to the function $s \mapsto a(s)$) is zero. In this case, we have that $\tau_\beta(0,0) = 0$, or equivalently, $\gamma'(0) = 0$. A portion of the stability region will be as depicted in Figure 3.5 provided that $\gamma''(0) > 0$. In this case, if the amplitude, $\beta = \delta > 0$, of the periodic perturbation is sufficiently small, then there is a range of sufficiently high frequencies Ω (recall that $\alpha := 1/\Omega^2$) such that the linearized pendulum motion is stabilized. By differentiation of the implicit relation $\tau(\gamma(\beta), \beta) = 2$, it is easy to see that the required condition on the second derivative of γ is equivalent to the inequality $\tau_{\beta\beta}(0,0) < 0$. Of course, this requirement is not always satisfied (see Exercise 3.43), but it is satisfied for $a(s) = \sin(2\pi s)$ (see Exercise 3.42).

Exercise 3.41. Prove that

$$\tau_\beta(0,0) = -\int_0^1 a(s) \, ds.$$

Exercise 3.42. Prove that $\tau_{\beta\beta}(0,0) < 0$ for the case $a(s) = \sin(2\pi s)$. Hint: Compute the variational derivatives directly in terms of the second order equation (3.86).

Exercise 3.43. Find a condition on the function $a(s)$ so that $\tau_{\beta\beta}(0,0) < 0$. Also, if $a(s)$ is expressed as a convergent Fourier series, find the corresponding condition in terms of its Fourier coefficients. Hint: If

$$a(s) = \sum_{k=1}^{\infty} a_k \cos(2\pi k s) + b_k \sin(2\pi k s),$$

then

$$\tau_{\beta\beta}(1,0,0) = 2\Big(\sum_{k=1}^{\infty} \frac{1}{2\pi k} b_k\Big)^2 - \sum_{k=1}^{\infty} \Big(\frac{1}{2\pi k}\Big)^2 (a_k^2 + 3b_k^2).$$

Exercise 3.44. Find an example of a periodic function $s \mapsto a(s)$ with period one such that $\tau_{\beta\beta}(0,0) > 0$. For this choice of the displacement, the inverted pendulum is not stabilized for small $\beta > 0$.

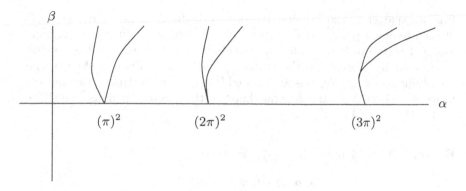

Figure 3.6: Regions of instability (Arnold tongues) for the linearized pendulum.

Exercise 3.45. What can you say about the stability of the inverted pendulum using Lyapunov's theorem (Theorem 2.108) or Exercise 2.114?

Let us consider the stability of the noninverted pendulum. Note that the linearization of the differential equation (3.85) at $\theta = 0$ is given by

$$w'' + (\alpha - \beta a(s))w = 0,$$

and let $\Psi(s, \alpha, \beta)$ denote the principal fundamental matrix solution at $s = 0$ of the corresponding homogeneous linear system. In this case, we have

$$\operatorname{tr} \Psi(1, \alpha, 0) = 2 \cos \sqrt{\alpha}.$$

Because the boundaries of the regions of instability are given by

$$|\operatorname{tr} \Psi(1, \alpha, \beta)| = 2,$$

they intersect the α-axis only if $\sqrt{\alpha}$ is an integer multiple of π. In view of the equation $\alpha = g/(L\Omega^2)$, these observations suggest the zero solution is unstable for small amplitude displacements whenever there is an integer n such that the period of the displacement is

$$\frac{1}{\Omega} = \frac{n}{2}\left(2\pi\left(\frac{L}{g}\right)^{1/2}\right);$$

that is, the period of the displacement is a half-integer multiple of the natural frequency of the pendulum. In fact, the instability of the pendulum for a small amplitude periodic displacement with $n = 1$ is demonstrated in every playground by children pumping up swings.

The proof that the instability boundaries do indeed cross the α-axis at the "resonant" points $(\alpha, \beta) = ((n\pi)^2, 0)$, for $n = 1, \dots, \infty$, is obtained from an analysis of the Taylor expansion of the function given by $\Psi(1, \alpha, \beta)$ at each resonant point (see Exercise 3.46). Typically, the instability regions are as depicted in Figure 3.6. The instability region with $n = 1$ is "open" at $\beta = 0$ (the tangents to the boundary curves have distinct slopes); the remaining instability regions are "closed." It is an interesting mathematical problem to determine the general shape of the stability regions (see [101], [153], [154], and [149]).

Exercise 3.46. Suppose that $a(s) = \sin(2\pi s)$ and set

$$g(\alpha, \beta) = \operatorname{tr} \Psi(1, \alpha, \beta) - 2.$$

Show that $g_\alpha((n\pi)^2, 0) = 0$ and $g_\beta((n\pi)^2, 0) = 0$. Thus, the implicit function theorem cannot be applied directly to obtain the boundaries of the regions of instability, the boundary curves are singular at the points where they meet the α-axis. By computing appropriate higher order derivatives and analyzing the resulting Taylor expansion of g, show that the regions near the α-axis are indeed as depicted in Figure 3.6. Also, show that the regions become "thinner" as n increases.

3.6 Origins of ODE: Partial Differential Equations

In this section there is an elementary discussion of three big ideas:

- Certain partial differential equations (PDE) can be viewed as ordinary differential equations with an infinite dimensional phase space.

- Finite dimensional approximations of some PDE are systems of ordinary differential equations.

- Traveling wave fronts in PDE can be determined by ordinary differential equations.

While these ideas are very important and therefore have been widely studied, only a few elementary illustrations will be given here. The objective of this section is to introduce these ideas as examples of how ordinary differential equations arise and to suggest some very important areas for further study (see [32], [108], [107], [118], [177], [182], [206], [220], and [239]). We will also discuss the solution of first order PDE as an application of the techniques of ordinary differential equations.

Most of the PDE mentioned in this section can be considered as models of "reaction-diffusion" processes. To see how these models are derived, imagine some substance distributed in a medium. The density of the substance is represented by a function $u : \mathbb{R}^n \times \mathbb{R} \to \mathbb{R}$ so that $(x, t) \mapsto u(x, t)$ gives its density at the site with coordinate x at time t.

If Ω is a region in space with boundary $\partial\Omega$, then the rate of change of the amount of the substance in Ω is given by the flux of the substance through the boundary of Ω plus the amount of the substance generated in Ω; that is,

$$\frac{d}{dt}\int_\Omega u\, d\mathcal{V} = -\int_{\partial\Omega} X\cdot\eta\, d\mathcal{S} + \int_\Omega f\, d\mathcal{V}$$

where X is the vector field representing the motion of the substance; $d\mathcal{V}$ is the volume element; $d\mathcal{S}$ is the surface element; the vector field η is the outer unit normal field on the boundary of Ω; and f, a function of density, position and time, represents the amount of the substance generated in Ω. The minus sign on the flux term is required because we are measuring the rate of change of the amount of substance *in* Ω. If, for example, the flow is all out of Ω, then $X\cdot\eta \geq 0$ and the minus sign is required because the rate of change of the amount of substance in Ω must be negative.

If Stokes's theorem is applied to rewrite the flux term and the time derivative is interchanged with the integral of the density, then

$$\int_\Omega u_t\, d\mathcal{V} = -\int_\Omega \operatorname{div} X\, d\mathcal{V} + \int_\Omega f\, d\mathcal{V}.$$

Moreover, because the region Ω is arbitrary in the integral identity, it is easy to prove the fundamental balance law

$$u_t = -\operatorname{div} X + f. \tag{3.88}$$

To obtain a useful dynamical equation for u from equation (3.88), we need a constitutive relation between the density u of the substance and the flow field X. It is not at all clear how to derive this relationship from the fundamental laws of physics. Thus, we have an excellent example of an important problem where physical intuition must be used to propose a constitutive law whose validity can only be tested by comparing the results of experiments with the predictions of the corresponding model. Problems of this type lie at the heart of applied mathematics and physics.

For equation (3.88), the classic constitutive relation—called Darcy's, Fick's, or Fourier's law depending on the physical context—is

$$X = -K\operatorname{grad} u + \mu V \tag{3.89}$$

where $K \geq 0$ and μ are functions of density, position, and time; and V denotes the flow field for the medium in which our substance is moving. The minus sign on the gradient term represents the assumption that the substance diffuses from higher to lower concentrations.

By inserting the relation (3.89) into the balance law (3.88), we obtain the dynamical equation

$$u_t = \operatorname{div}(K\operatorname{grad} u) - \operatorname{div}(\mu V) + f. \tag{3.90}$$

Also, if we assume that the diffusion coefficient K is equal to k^2 for some constant k, the function μ is given by $\mu(u, x, t) = \gamma u$ where γ is a constant, and V is an incompressible flow field (div $V = 0$); then we obtain the most often used reaction-diffusion-convection model equation

$$u_t + \gamma \operatorname{grad} u \cdot V = k^2 \Delta u + f. \tag{3.91}$$

In this equation, the gradient term is called the *convection term,* the Laplacian term is called the *diffusion term,* and f is the *source term.* Let us also note that if the diffusion coefficient is zero, the convection coefficient is given by $\gamma = 1$, the source function vanishes, and V is not necessarily incompressible, then the dynamical equation (3.90) reduces to the law of conservation of mass, also called the *continuity equation,* given by

$$u_t + \operatorname{div}(uV) = 0. \tag{3.92}$$

Because equation (3.91) is derived from general physical principles, this PDE can be used to model many different phenomena. As a result, there is a vast scientific literature devoted to its study. We will not be able to probe very deeply, but we will use equation (3.91) to illustrate a few aspects of the analysis of these models where ordinary differential equations arise naturally.

3.6.1 Infinite Dimensional ODE

A simple special case of the reaction-diffusion-convection model (3.91) is the linear diffusion equation (the heat equation) in one spatial dimension, namely, the PDE

$$u_t = k^2 u_{xx} \tag{3.93}$$

where k^2 is the *diffusivity constant.* This differential equation can be used to model heat flow in an insulated bar. In fact, let us suppose that the bar is idealized to be the interval $[0, \ell]$ on the x-axis so that $u(x, t)$ represents the temperature of the bar at the point with coordinate x at time t. Moreover, because the bar has finite length, let us model the heat flow at the ends of the bar where we will consider just two possibilities: The bar is insulated at both ends such that we have the Neumann boundary conditions

$$u_x(0, t) = 0, \qquad u_x(\ell, t) = 0;$$

or, heat is allowed to flow through the ends of the bar, but the temperature at the ends is held constant at zero (in some appropriate units) such that we have the Dirichlet boundary conditions

$$u(0, t) = 0, \qquad u(\ell, t) = 0.$$

If one set of boundary conditions is imposed and an initial temperature distribution, say $x \mapsto u_0(x)$, is given on the bar, then we would expect that there is a unique scalar function $(x, t) \mapsto u(x, t)$, defined on the set $[0, \ell] \times [0, \infty)$ that satisfies the PDE, the initial condition $u(x, 0) = u_0(x)$, and the boundary conditions. Of course, if such a solution exists, then for each $t > 0$, it predicts the corresponding temperature distribution $x \mapsto u(x, t)$ on the bar. In addition, if there is a solution of the boundary value problem corresponding to each initial temperature distribution, then we have a situation that is just like the phase flow of an ordinary differential equation. Indeed, let us consider a linear space $\mathcal{E}$ of temperature distributions on the rod and let us suppose that if a function $v : [0, \ell] \to \mathbb{R}$ is in $\mathcal{E}$, then there is a solution $(x, t) \mapsto u(x, t)$ of the boundary value problem with v as the initial temperature distribution such that $x \mapsto u(x, t)$ is a function in $\mathcal{E}$ whenever $t > 0$. In particular, all the functions in $\mathcal{E}$ must satisfy the boundary conditions. If this is the case, then we have defined a function $(0, \infty) \times \mathcal{E} \to \mathcal{E}$ given by $(t, v) \mapsto \varphi_t(v)$ such that $\varphi_0(v)(x) = v(x)$ and $(x, t) \mapsto \varphi_t(v)(x)$ is the solution of the boundary value problem with initial temperature distribution v. In other words, we have defined a dynamical system with (semi) flow φ_t whose phase space is the function space $\mathcal{E}$ of possible temperature distributions on the bar. For example, for the Dirichlet problem, we might take $\mathcal{E}$ to be the subset of $C^2[0, \ell]$ consisting of those functions that vanish at the ends of the interval $[0, \ell]$.

Taking our idea a step further, let us define the linear transformation A on $\mathcal{E}$ by

$$Au = k^2 u_{xx}.$$

Then, the PDE (3.93) can be rewritten as

$$\dot{u} = Au, \tag{3.94}$$

an ordinary differential equation on the infinite dimensional space $\mathcal{E}$. Also, to remind ourselves of the boundary conditions, let us write $A = A_{\mathcal{N}}$ if Neumann boundary conditions are imposed and $A = A_{\mathcal{D}}$ for Dirichlet boundary conditions.

Although the linear homogeneous differential equation (3.94) is so simple that its solutions can be given explicitly, we will see how the general solution of the PDE can be found by treating it as an ordinary differential equation.

Let us begin by determining the rest points of the system (3.94). In fact, a rest point is a function $v : [0, \ell] \to \mathbb{R}$ that satisfies the boundary conditions and the second order ordinary differential equation $v_{xx} = 0$. Clearly, the only possible choices are affine functions of the form $v = cx + d$ where c and d are real numbers. There are two cases: For $A_{\mathcal{N}}$ we must have $c = 0$, but d is a free variable. Thus, there is a line in the function space $\mathcal{E}$ corresponding to the constant functions in $\mathcal{E}$ that consists entirely of rest points. For the Dirichlet case, both c and d must vanish and there is a unique rest point at the origin of the phase space.

Having found the rest points for the differential equation (3.94), let us discuss their stability. By analogy with the finite dimensional case, let us recall that we have discussed two methods that can be used to determine the stability of rest points: linearization and Lyapunov's direct method. In particular, for the finite dimensional case, the method of linearization is valid as long as the rest point is hyperbolic, and, in this case, the eigenvalues of the system matrix for the linearized system at the rest point determine its stability type.

Working formally, let us apply the method of linearization at the rest points of the system (3.94). Since this differential equation is already linear, we might expect the stability of these rest points to be determined from an analysis of the position in the complex plane of the eigenvalues of the system operator A. By definition, if λ is an eigenvalue of the operator $A_{\mathcal{D}}$ or $A_{\mathcal{N}}$, then there must be a nonzero function v on the interval $[0, \ell]$ that satisfies the boundary conditions and the ordinary differential equation

$$k^2 v_{xx} = \lambda v.$$

If v is an eigenfunction with eigenvalue λ, then we have that

$$\int_0^\ell k^2 v_{xx} v \, dx = \int_0^\ell \lambda v^2 \, dx. \qquad (3.95)$$

Let us suppose that v is square integrable, that is,

$$\int_0^\ell v^2 \, dx < \infty$$

and also smooth enough so that integration by parts is valid. Then, equation (3.95) is equivalent to the equation

$$v_x v \Big|_0^\ell - \int_0^\ell v_x^2 \, dx = \frac{\lambda}{k^2} \int_0^\ell v^2 \, dx.$$

Thus, if either Dirichlet or Neumann boundary conditions are imposed, then the boundary term from the integration by parts vanishes, and therefore the eigenvalue λ must be a nonpositive real number.

For $A_{\mathcal{D}}$, if $\lambda = 0$, then there is no nonzero eigenfunction. If $\lambda < 0$, then the eigenvalue equation has the general solution

$$v(x) = c_1 \cos \alpha x + c_2 \sin \alpha x$$

where $\alpha := (-\lambda)^{1/2}/k$ and c_1 and c_2 are constants; and, in order to satisfy the Dirichlet boundary conditions, we must have

$$\begin{pmatrix} 1 & 0 \\ \cos \alpha \ell & \sin \alpha \ell \end{pmatrix} \begin{pmatrix} c_1 \\ c_2 \end{pmatrix} = \begin{pmatrix} 0 \\ 0 \end{pmatrix}$$

for some nonzero vector of constants. In fact, the determinant of the matrix must vanish, and we therefore have to impose the condition that $\alpha \ell$ is an integer multiple of π; or equivalently,

$$\lambda = -\left(\frac{n\pi k}{\ell}\right)^2$$

with a corresponding eigenfunction given by

$$x \mapsto \sin \frac{n\pi}{\ell} x$$

for each integer $n = 1, 2, \ldots, \infty$. By a similar calculation for $A_\mathcal{N}$, we have that $\lambda = 0$ is an eigenvalue with a corresponding eigenfunction $v \equiv 1$, and again the same real numbers

$$\lambda = -\left(\frac{n\pi k}{\ell}\right)^2$$

are eigenvalues, but this time with corresponding eigenfunctions

$$x \mapsto \cos \frac{n\pi}{\ell} x.$$

The nature of the real parts of the eigenvalues computed in the last paragraph and the principle of linearized stability suggest that the origin is an asymptotically stable rest point for the Dirichlet problem. On the other hand, the rest points of the Neumann problem seem to be of a different type: each of these rest points would appear to have a one-dimensional center manifold and an infinite dimensional stable manifold. All of these statements are true. But to prove them, certain modifications of the corresponding finite dimensional results are required. For example, the principle of linearized stability is valid for rest points of infinite dimensional ODE under the assumption that all points in the *spectrum* of the operator given by the linearized vector field at the rest point (in our case the operator A) have negative real parts that are bounded away from the imaginary axis in the complex plane (see, for example, [206, p. 114]). More precisely, the required hypothesis is that there is some number $\alpha > 0$ such that the real part of every point in the spectrum of the operator is less than $-\alpha$. In general, the principle of linearized stability fails for rest points of differential equations in infinite dimensional spaces (see Exercise 3.48).

Recall that a complex number λ is in the spectrum of the linear operator A if the operator $A - \lambda I$ does not have a *bounded* inverse. Of course, if $v \neq 0$ is an eigenfunction with eigenvalue λ, then the operator $A - \lambda I$ is not injective and indeed λ is in the spectrum. In a finite dimensional space, if an operator is injective, then it is invertible. Hence, the only complex numbers in the spectrum of a finite dimensional linear operator are eigenvalues. But, in an infinite dimensional space, there can be points in the spectrum that are not eigenvalues (see [78]). For example, let us define the space $L^2(0, \ell)$ to be all (real) functions $v : [0, \ell] \to \mathbb{R}$ such that

$$\int_0^\ell v^2(x) \, dx < \infty \tag{3.96}$$

where we consider two such functions v and w to be equal if

$$\int_0^\ell (v(x) - w(x))^2 \, dx = 0,$$

and consider the operator $B : L^2 \to L^2$ given by $(Bf)(x) \mapsto xf(x)$. This operator has no eigenvalues, yet the entire interval $[0, \ell]$ is in its spectrum. (Why?)

The operators $A_{\mathcal{D}}$ and $A_{\mathcal{N}}$, considered as operators defined in $L^2(0, \ell)$, have spectra that consist entirely of eigenvalues (pure point spectrum). To prove this claim, note first that these operators are not defined on all of L^2. After all, a square integrable function does not have to be differentiable. Instead, we can view our operators to be defined on the subset of L^2 consisting of those functions that have two derivatives both contained in L^2. Then, the claim about the spectra of $A_{\mathcal{D}}$ and $A_{\mathcal{N}}$ can be proved in two steps. First, if a complex number λ is not an eigenvalue, then for all $w \in L^2$ there is some function v that satisfies the boundary conditions and the differential equation

$$k^2 v_{xx} - \lambda v = w.$$

In other words, there is an operator $B : L^2 \to L^2$ given by $Bw = v$ such that $(A - \lambda I)Bw = w$. The boundedness of B is proved from the explicit construction of B as an integral operator. Also, it can be proved that $B(A - \lambda I)v = v$ for all v in the domain of A (see Exercise 3.47). Using these facts and the theorem on linearized stability mentioned above, it follows that the origin is an asymptotically stable rest point for the Dirichlet problem.

Exercise 3.47. Show that the spectrum of the operator in $L^2(0, \ell)$ given by $Av = v_{xx}$ with either Dirichlet or Neumann boundary conditions consists only of eigenvalues. Prove the same result for the operator $Av = av_{xx} + bv_x + cv$ where a, b, and c are real numbers.

Exercise 3.48. This exercise discusses an example (a slight modification of an example in [45, p. 32]) of an infinite dimensional linear differential equation such that the spectrum of the system matrix is in the left half-plane and the trivial solution is unstable. It requires some Hilbert space theory. For each $n \geq 1$ let A_n denote the $n \times n$-matrix with $a_{i,i+1} = 1$ and all other components zero (that is, the super diagonal is all ones, every other component is zero). Let $\mathcal{H}$ denote the (complex) Hilbert space $X = \oplus_{n \geq 1} \mathbb{C}^n$ with the ℓ^2-norm, and define the operator

A in $\mathcal{H}$ by $Ax = \{A_n x_n + 2\pi i n x_n\}_{n=1}^{\infty}$. (a) Prove that A_n is nilpotent and the spectrum of A_n is $\{0\}$. (b) Prove that A_n is an unbounded operator and it can be densely defined in $\mathcal{H}$. (c) Prove that $\{2\pi i n : n \geq 1\}$ are eigenvalues of A. (d) Prove that the differential equation $\dot{x} = (A - \frac{1}{2}I)x$ on $\mathcal{H}$ has the solution $T^t x = \{e^{-t/2} e^{2\pi i n t} e^{tA_n}\}_{n=1}^{\infty}$, which is a semi-group defined for $t \geq 0$. (e) Prove that the spectrum of $A - \frac{1}{2}I$ lies in the left half-plane. Hint: Show that the resolvent of A (that is, $R(A, \lambda) = (A - \lambda I)^{-1}$) is a bounded operator whenever the real part of λ is greater than zero using the following fact: The resolvent of A_n can be represented as a finite sum. (f) Let $p_n := (1, 1, \ldots, 1)/\sqrt{n} \in C^n$ and let x_n denote the element of $\mathcal{H}$ such that the nth component of x_n is p_n and all other components are zero. Prove that $\|x_n\| = 1$. (g) Prove that $\lim_{n \to \infty} |e^{tA_n} p_n - e^t p_n| = 0$ for each fixed $t \geq 0$. (h) Prove that T^t is not stable. Hint: It suffices to show that, for the semigroup S^t generated by A, if $C > 1$ is given, there is an element $x \in \mathcal{H}$ such that $\|x\| = 1$ and $\|S^t x\| \geq Ce^{t/2}$. Prove this inequality using part (g).

In view of our results for finite dimensional linear systems, we expect that if we have a linear evolution equation $\dot{v} = Av$, even in an infinite dimensional phase space, and if $Aw = \lambda w$, then $e^{t\lambda} w$ is a solution. This is indeed the case for the PDE (3.93). Moreover, for linear evolution equations, we can use the principle of superposition to deduce that every linear combination of solutions of this type is again a solution. If we work formally, that is, without proving convergence, and if we use the eigenvalues and eigenvectors computed above, then the "general solution" of the Dirichlet problem is given by

$$u(x, t) = \sum_{n=1}^{\infty} e^{-\left(\frac{n\pi k}{\ell}\right)^2 t} a_n \sin \frac{n\pi}{\ell} x,$$

and the general solution of the Neumann problem is given by

$$u(x, t) = \sum_{n=0}^{\infty} e^{-\left(\frac{n\pi k}{\ell}\right)^2 t} b_n \cos \frac{n\pi}{\ell} x$$

where a_n and b_n are real numbers.

If the initial condition $u(x, 0) = u_0(x)$ is given, then, for instance, for the Dirichlet problem we must have that

$$u_0(x) = \sum_{n=1}^{\infty} a_n \sin \frac{n\pi}{\ell} x.$$

In other words, the initial function u_0 must be represented by a Fourier sine series. What does this mean? The requirement is that the Fourier sine series converges to u_0 in the space L^2 endowed with its natural norm,

$$\|v\| := \left(\int_0^{\ell} v^2(x) \, dx \right)^{1/2}.$$

In fact, the inner product space L^2 is a Hilbert space; that is, with respect to this norm, every Cauchy sequence in L^2 converges (see [199]). The precise requirement for u_0 to be represented by a Fourier sine series is that there are real numbers a_n and corresponding L^2 partial sums

$$\sum_{n=1}^{N} a_n \sin \frac{n\pi}{\ell} x$$

such that

$$\lim_{N\to\infty} \|u_0 - u_N\| = 0.$$

If the initial function u_0 is continuous, then for our special case the corresponding solution obtained by Fourier series also converges pointwise to a C^2 function that satisfies the PDE in the classical sense. We will show in a moment that this solution is unique, and therefore the special solutions of the PDE obtained from the eigenvalues and corresponding eigenfunctions do indeed form a fundamental set of solutions for our boundary value problems.

There are several ways to prove that solutions of the diffusion equation with a given initial condition are unique. We will use the "energy method"; an alternative uniqueness proof is based on the maximum principle (see Exercise 3.49). To show the uniqueness result, let us note that if two solutions of either the Dirichlet or Neumann boundary value problem satisfy the same initial condition, then the difference u of these two solutions is a solution of the same boundary value problem but with initial value the zero function. Using an integration by parts, we also have the equality

$$\frac{d}{dt} \int_0^\ell \frac{1}{2} u^2 \, dx = \int_0^\ell u_t u \, dx = k^2 \int_0^\ell u_{xx} u \, dx = -k^2 \int_0^\ell u_x^2 \, dx.$$

It follows that the function

$$t \mapsto \int_0^\ell \frac{1}{2} u^2(x,t) \, dx$$

is not increasing, and therefore it is bounded above by its value at $t = 0$, namely,

$$\int_0^\ell \frac{1}{2} u^2(x,0) \, dx = 0.$$

The conclusion is that $u(x,t) \equiv 0$, as required.

Exercise 3.49. Prove the maximum principle: If $u_t(x,t) = k^2 u_{xx}(x,t)$ is a C^2 function on the open rectangle $(0,\ell) \times (0,T)$ and a continuous function on the closure of this rectangle, then the maximum of the function u is assumed either on the line $(0, \ell) \times \{0\}$ or on one of the lines

$$\{0\} \times [0, T], \qquad \{\ell\} \times [0, T].$$

Also, use the maximum principle to prove the uniqueness of solutions of the boundary value problem with initial condition for the diffusion equation. Hint: Use calculus (see [217, p. 41]).

Exercise 3.50. Solve the PDE (3.93) by the method of separation of variables; that is, assume that there is a solution of the form $u(x, t) = p(x)q(t)$, substitute this expression into the PDE, impose the boundary conditions, and determine the general form of the functions p and q.

Using the explicit form of the Fourier series representations of the general solutions of the heat equation with Dirichlet or Neumann boundary conditions, we can see that these solutions are very much like the solutions of a homogeneous linear ordinary differential equation: They are expressed as superpositions of fundamental solutions and they obviously satisfy the flow property $\varphi_s(\varphi_t(v)) = \varphi_{s+t}(v)$ as long as s and t are not negative (the series solutions do not necessarily converge for $t < 0$). Because of this restriction on the time variable, the solutions of our evolution equation are said to be *semi-flows* or *semi-groups*.

In the case of Dirichlet boundary conditions, if we look at the series solution, then we can see immediately that the origin is in fact globally asymptotically stable. For the Neumann problem there is a one-dimensional invariant manifold of rest points, and all other solutions are attracted exponentially fast to this manifold. Physically, if the temperature is held fixed at zero at the ends of the bar, then the temperature at each point of the bar approaches zero at an exponential rate, whereas if the bar is insulated at its ends, then the temperature at each point approaches the average value of the initial temperature distribution.

Our discussion of the scalar diffusion equation, PDE (3.93), has served to illustrate the view that a (parabolic) PDE is an ordinary differential equation on an infinite dimensional space. Moreover, as we have seen, if we choose to study a PDE from this viewpoint, then our experience with ordinary differential equations can be used to advantage as a faithful guide to its analysis.

Exercise 3.51. Verify the semi-flow property $\varphi_s(\varphi_t(v)) = \varphi_{s+t}(v)$ for the solutions of the scalar heat equation with Dirichlet or Neumann boundary conditions. Generalize this result to the equation $u_t = u_{xx} + f(u)$ under the assumption that every initial value problem for this equation has a local solution. Hint: How is the flow property proved for finite dimensional autonomous equations?

Let us now consider the nonlinear PDE

$$u_t = k^2 u_{xx} + f(u, x, t), \qquad 0 < x < \ell, \quad t > 0 \qquad (3.97)$$

where f is a smooth function that represents a heat source in our heat conduction model.

To illustrate the analysis of rest points for a nonlinear PDE, let us assume that the source term f for the PDE (3.97) depends only on its first argument, and let us impose, as usual, either Dirichlet or Neumann boundary conditions. In this situation, the rest points are given by those solutions of the ordinary differential equation

$$k^2 u_{xx} + f(u) = 0 \qquad (3.98)$$

that also satisfy the Dirichlet or Neumann boundary conditions.

The boundary value problem (3.98) is an interesting problem in ordinary differential equations. Let us note first that if we view the independent variable as "time," then the second order differential equation (3.98) is just Newton's equation for a particle of mass k^2 moving in a potential force field with force $-f(u)$. In addition, the corresponding first order system in the phase plane is the Hamiltonian system

$$\dot{u} = v, \qquad \dot{v} = -f(u)$$

whose total energy is given by

$$H(u, v) := \frac{k^2}{2} v^2 + F(u)$$

where F, the potential energy, can be taken to be

$$F(u) := \int_0^u f(w)\, dw,$$

and, as we know, the phase plane orbits all lie on curves of constant energy. We will use these facts below.

A rest point of the PDE (3.97) with our special form of f and Dirichlet boundary conditions corresponds to a trajectory in the phase plane that starts on the v-axis and returns to the v-axis again exactly at time $x = \ell$. On the other hand, a rest point for the PDE with Neumann boundary conditions corresponds to a trajectory in the phase plane that starts on the u-axis and returns to the u-axis at time $x = \ell$.

Though the nonlinear boundary value problems that have just been described are very difficult in general, they can be "solved" in some important special cases. As an example, let us consider the following Dirichlet boundary value problem

$$u_t = u_{xx} + u - u^3, \qquad u(0, t) = 0, \quad u(\ell, t) = 0 \qquad (3.99)$$

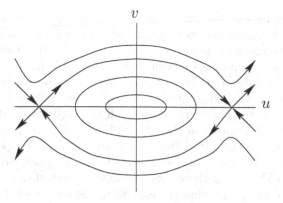

Figure 3.7: Phase portrait of the system $\dot{u} = v$ $\dot{v} = -u + u^3$.

(see Exercise 3.54 for Neumann boundary conditions). Note first that the constant functions with values 0 or ± 1 are solutions of the differential equation $u_{xx} + u - u^3 = 0$, but only the zero solution satisfies the Dirichlet boundary conditions. Thus, there is exactly one constant rest point. Let us determine if there are any nonconstant rest points.

The phase plane system corresponding to the steady state equation for the PDE (3.99) is given by

$$\dot{u} = v, \qquad \dot{v} = -u + u^3.$$

It has saddle points at $(\pm 1, 0)$ and a center at $(0, 0)$. Moreover, the period annulus surrounding the origin is bounded by a pair of heteroclinic orbits that lie on the curve

$$\frac{1}{2}v^2 + \frac{1}{2}u^2 - \frac{1}{4}u^4 = \frac{1}{4}$$

(see Figure 3.7). Using this fact, it is easy to see that the interval $(0, 1/\sqrt{2})$ on the v-axis is a Poincaré section for the annulus of periodic orbits. Also, a glance at the phase portrait of the system shows that only the solutions that lie on these periodic orbits are candidates for nonconstant steady states for the PDE; they are the only periodic orbits in the phase plane that meet the v-axis at more than one point. Also, let us notice that the phase portrait is symmetric with respect to each of the coordinate axes. In view of this symmetry, if we define the period function

$$T : \left(0, \frac{1}{\sqrt{2}}\right) \to \mathbb{R} \tag{3.100}$$

so that $T(a)$ is the minimum period of the periodic solution starting at $u(0) = 0$, $v(0) = a$, then

$$u\left(\frac{1}{2}T(a)\right) = 0, \qquad v\left(\frac{1}{2}T(a)\right) = -a.$$

Hence, solutions of our boundary value problem that correspond to rest points for the PDE also correspond to periodic solutions whose half-periods are exactly some integer submultiple of ℓ; equivalently, these solutions correspond to those real numbers a such that $0 < a < 1/\sqrt{2}$ and $T(a) = 2\ell/n$ for some positive integer n. In fact, each such a corresponds to exactly two rest points of the PDE; namely, $x \mapsto u(x)$ and $x \mapsto u(\ell - x)$ where $x \mapsto (u(x), v(x))$ is the phase trajectory such that $u(0) = 0$ and $v(0) = a$.

The number and position in the phase plane of all rest point solutions of the PDE can be determined from the following three propositions: (i) $T(a) \to 2\pi$ as $a \to 0^+$; (ii) $T(a) \to \infty$ as $a \to (1/\sqrt{2})^-$; and (iii) $T'(a) > 0$ (see Exercise 3.52). Using these facts, it follows that there is at most a finite number of rest points that correspond to the integers $1, 2, \ldots, n$ such that $n < \ell/\pi$.

Exercise 3.52. Prove that the period function T given in display (3.100) has a positive first derivative. Hint: Find the explicit time-dependent periodic solutions of the first order system $\dot{u} = v$, $\dot{v} = -u + u^3$ using Jacobi elliptic functions. For a different method, see [34] and [196].

Exercise 3.53. Find the rest points for the Dirichlet boundary value problem

$$u_t = u_{xx} + au - bu^2, \qquad u(x, 0) = 0, \quad u(x, \ell) = 0$$

(see [41]).

Are the rest points of the PDE (3.99) stable? It turns out that the stability problem for nonconstant rest points, even for our scalar PDE, is too difficult to describe here (see [206, p. 530]). On the other hand, we can say something about the stability of the constant rest point at the origin for the PDE (3.99). In fact, let us note that if $\ell < \pi$, then it is the only rest point. Moreover, its stability can be determined by linearization.

Let us first describe the linearization procedure for a PDE. The correct formulation is simple if we view the PDE as an ordinary differential equation on a function space. Indeed, we can just follow the recipe for linearizing an ordinary differential equation of the form $\dot{u} = g(u)$. Let us recall that if z is a rest point and g is a smooth function, then the linearization of the ordinary differential equation at z is

$$\dot{x} = Dg(z)(x - z),$$

or equivalently

$$\dot{w} = Dg(z)w$$

where $w := x - z$. Moreover, if the eigenvalues of $Dg(z)$ all have negative real parts, then the rest point z is asymptotically stable (see Section 2.3).

In order to linearize at a rest point of a PDE, let us suppose that the function $x \mapsto z(x)$ is a rest point for the PDE

$$u_t = g(u)$$

where $g(u) := u_{xx} + f(u)$ and $f : \mathbb{R} \to \mathbb{R}$ is a differentiable function. If the domain of $A_{\mathcal{D}}$ is viewed as the function space $C^2[0, \ell]$, then the function $g : C^2[0, \ell] \to C^0[0, \ell]$ is differentiable. This statement follows because the function $u \mapsto u_{xx}$ is linear and the function f is smooth. But there is a subtle point: in the definition of g we must view the notation $f(u)$ to mean $f \circ u$ where $u \in C^2[0, \ell]$. The difficulty is that the smoothness of the function $u \mapsto f \circ u$ depends on the topology of the function space to which u belongs (see Example 1.221).

Once we know that g is differentiable, its derivative can be easily computed as a directional derivative; in fact,

$$Dg(z)v = \frac{d}{dt}g(z + tv)\Big|_{t=0} = v_{xx} + Df(z)v.$$

Therefore, by definition, the linearized equation at the rest point z is given by

$$\dot{w} = w_{xx} + Df(z(x))w. \tag{3.101}$$

For a nonconstant rest point, the linearized equation (3.101) depends on the space variable x. The determination of stability in this case is often quite difficult—recall the stability analysis for periodic solutions of finite dimensional ordinary differential equations. For a constant rest point, the linearized equation has the form $\dot{w} = Aw$ where A is the linear operator given by $w \mapsto w_{xx} + Df(z)w$ for z a fixed number. In this case, as mentioned previously, it seems natural to expect the following result: If the spectrum of A is in the open left half-plane and bounded away from the imaginary axis, then the rest point is asymptotically stable. In fact, this result, when properly interpreted, is true for the PDE (3.99). But to prove it, we have to specify the function space on which the spectrum is to be computed and recast the arguments used for ordinary differential equations in an infinite dimensional setting. For the PDE (3.99) the idea—derived from our study of ordinary differential equations—of applying the principle of linearized stability is justified, but some functional analysis is required to carry it out (see [206, Chapter 11]).

Our example is perhaps too simple; there are PDE where the linearized stability of a steady state can be easily proved, but the stability of the rest point is an open question. The problem for a general PDE of the form

$$u_t = Au + f(u)$$

is that the linear operator A, the function f, and the linearized operator $A + Df(z)$ must all satisfy additional hypotheses before the ODE arguments for the validity of the principle of linearized stability can be verified in the infinite dimensional case. This fact is an important difference between the theory of ordinary differential equations and the theory of PDE.

Let us put aside the theoretical justification of linearized stability and reconsider the rest point at the origin for the PDE (3.99) where the linearized system is given by

$$w_t = w_{xx} + w, \qquad w(0) = 0, \quad w(\ell) = 0.$$

In this case, the spectrum of the differential operator defined by

$$Aw = w_{xx} + w$$

consists only of eigenvalues (see Exercise 3.47). In fact, using the analysis of the spectrum of the operator $w \to w_{xx}$ given above, the spectrum of A is easily obtained by a translation. In fact, the spectrum is

$$\left\{ 1 - \left(\frac{n\pi}{\ell} \right)^2 : n = 1, 2, \dots, \infty \right\}.$$

Because

$$1 - \left(\frac{n\pi}{\ell} \right)^2 \leq 1 - \left(\frac{\pi}{\ell} \right)^2,$$

the spectrum of A lies in the left half of the complex plane and is bounded away from the imaginary axis if and only if $1 < \pi^2/\ell^2$. Hence, using this fact and assuming the validity of the principle of linearized stability, we have the following proposition: If $\ell < \pi$, then the origin is the only steady state and it is asymptotically stable.

Let us go one step further in our qualitative analysis of the PDE $u_t = u_{xx} + f(u)$ by showing that there are no periodic solutions. In fact, this claim is true independent of the choice of $\ell > 0$ and for an arbitrary smooth source function f. The idea for the proof, following the presentation in [206], is to show that there is a function (essentially a Lyapunov function) that decreases on orbits. In fact, let us define

$$E(u) = -\int_0^\ell \left(\frac{1}{2} u(x) u_{xx}(x) + F(u(x)) \right) dx$$

where F is an antiderivative of f and note that

$$\dot{E} = -\int_0^\ell \left(\frac{1}{2} u_t u_{xx} + \frac{1}{2} u u_{txx} + f(u) u_t \right) dx.$$

After integration by parts twice for the integral of the second term in the integrand, and after imposing either Dirichlet or Neumann boundary conditions, it follows that

$$\dot{E} = -\int_0^\ell (u_{xx} + f(u)) u_t \, dx = -\int_0^\ell (u_{xx} + f(u))^2 \, dx.$$

Hence, except for the rest points, the time derivative of E is negative along orbits. In particular, there are no periodic orbits. Can the function E be used to give a proof of the stability of the rest point at the origin?

For the PDE (3.99) with $\ell < \pi$ we have now built up a rather complete picture of the phase portrait. In fact, we know enough to conjecture that there is a unique rest point that is globally asymptotically stable. Is this conjecture true?

Exercise 3.54. Analyze the existence of rest points, the stability types of constant rest points, and the phase portrait for the Neumann boundary value problem

$$u_t = u_{xx} + u - u^3, \qquad u_x(0, t) = 0, \quad u_x(\ell, t) = 0.$$

Note that there are three constant rest points. Use equation (3.101) to determine their stability types.

3.6.2 Galërkin Approximation

Since most differential equations, ODE or PDE, cannot be solved, it is natural to seek approximate solutions. For example, numerical methods are often used to obtain approximate values of state variables. But the utility of approximation methods goes far beyond number crunching: approximations are used to gain insight into the qualitative behavior of dynamical systems, to test computer codes, and to obtain existence proofs. Indeed, approximation methods are central elements of applied mathematics. In this section we will take a brief look at a special case of *Galërkin's method,* one of the classic approximation methods for PDE. It is one of an array of methods that are based on the idea of finding finite dimensional approximations of infinite dimensional dynamical systems.

As a remark, let us note that other approximation methods for PDE are based on the idea of finding finite dimensional invariant (or approximately invariant) submanifolds in the infinite dimensional phase space. Recall that rest points and periodic orbits are finite dimensional invariant submanifolds. But these are only the simplest examples. In fact, let us note that a rest point or a periodic orbit might have an infinite dimensional stable manifold and a finite dimensional center manifold. In this case, the local dynamical behavior is determined by the dynamics on the center manifold because nearby orbits are attracted to the center manifold. An important generalization of this basic situation is the concept of an inertial manifold. By definition, an *inertial manifold* M is a finite dimensional submanifold in the phase space that has two properties: M is positively invariant, and every solution is attracted to M at an exponential rate (see [220]).

In general, if there is an attracting finite dimensional invariant manifold, then the dynamical system restricted to this invariant set is an ordinary differential equation that models the asymptotic behavior of the full infinite dimensional PDE. In particular, the ω-limit set of every solution lies on this manifold. Thus, the existence of such an invariant manifold provides the theoretical basis for a complete understanding of the infinite dimensional dynamical system using the techniques of ordinary differential equations. Unfortunately, it is usually very difficult to prove the existence of attracting invariant manifolds. Even if an invariant manifold does exist, it is often equally difficult to obtain a specification of the manifold that would be required to reduce the original infinite dimensional dynamical system to an ordinary differential equation. As an alternative, an approximation method—such as Galërkin's method—that does not require the existence of an invariant manifold can often be employed with great success.

The following philosophical question seems to accompany all theoretical approximation methods for PDE "Is the set of reduced equations—presumably a system of nonlinear ordinary differential equations—easier to analyze than the original PDE?" In general, the answer to this question is clearly "no." If, however, the finite dimensional approximation is low dimensional or of some special form, then often qualitative analysis is possible, and useful insights into the dynamics of the original system can be obtained. Perhaps the best answer to the question is to avoid the implied choice between infinite dimensional and finite dimensional analysis. The best approach to an applied problem is with a mind free of prejudice. Often several different methods, including physical thinking and numerical analysis, are required to obtain consistent and useful predictions from a model.

Let us begin our discussion of the Galërkin approximation method with an elementary, but key idea. Recall that a (real) vector space H is an inner product space if there is a bilinear form (denoted here by angle brackets) such that if $h \in H$, then $\langle h, h \rangle \geq 0$ and $\langle h, h \rangle = 0$ if and only if $h = 0$. It follows immediately that if $v \in H$ and $\langle v, h \rangle = 0$ for all $h \in H$, then $v = 0$. We will use this fundamental result to solve equations in H. Indeed, suppose that we wish to find a solution of the (linear) equation

$$Au = b. \tag{3.102}$$

If there is a vector $u_0 \in H$ such that $\langle Au_0 - b, h \rangle = 0$ for all $h \in H$, then u_0 is a solution of the equation.

Definition 3.55. Suppose that S is a subspace of the Hilbert space H. A *Galërkin approximation* of a solution of equation (3.102) is an element $u_S \in S$ such that

$$\langle Au_S - b, s \rangle = 0$$

for all $s \in S$.

Of course, every $h \in H$ is an approximation of a solution. To obtain a useful approximation, we will consider a sequence of subspaces, $S_1 \subset$

$S_2 \subset \cdots$, whose union is dense in H together with corresponding Galërkin approximations $u_n \in S_n$ such that $\langle Au_n - b, s \rangle = 0$ for all $s \in S_n$. In this case, we might expect that the sequence $u_1, u_2, \ldots$ converges to a solution of the equation (3.102).

If H is a finite dimensional inner product space and the subspaces

$$S_1 \subset S_2 \subset S_3 \subset \cdots \subset H$$

are strictly nested, then a corresponding sequence of Galërkin approximations is finite. Thus, we do not have to worry about convergence. But, if H is an infinite dimensional Hilbert space, then the approximating subspaces must be chosen with care in order to ensure the convergence of the sequence of Galërkin approximations.

Let us recall that a sequence $B = \{\nu_i\}_{i=1}^{\infty}$ of linearly independent elements in H is called a *Hilbert space basis* if the linear manifold S spanned by B—all finite linear combinations of elements in B—is dense in H; that is, if $h \in H$, then there is a sequence in S that converges to h in the natural norm defined from the inner product. A Hilbert space that has such a basis is called *separable*.

Galërkin's principle. *Suppose that H is a Hilbert space, $B = \{\nu_i\}_{i=1}^{\infty}$ is a Hilbert space basis for H, and $A : H \to H$ is a linear operator. Also, for each positive integer n, let S_n denote the linear manifold spanned by the finite set $\{\nu_1, \ldots, \nu_n\}$. Then, for each positive integer n, there is some $u_n \in S_n$ such that $\langle Au_n - b, s \rangle = 0$ for all $s \in S_n$. Moreover, the sequence $\{u_n\}_{n=1}^{\infty}$ converges to a solution of the equation $Au = b$.*

The Galërkin principle is not a theorem. In fact, the Galërkin approximations may not exist or the sequence of approximations may not converge. The applicability of the method depends on the equation we propose to solve, the choice of the space H, and the choice of the basis B.

Existence of Weak Solutions

Let us consider the steady state equation

$$-u_{xx} + g(x)u - f(x) = 0, \qquad 0 < x < \ell, \tag{3.103}$$

with either Dirichlet or Neumann boundary conditions where f and g are smooth functions defined on the interval $[0, \ell]$. The basic idea is to look for a weak solution. To see what this means, note that if u is a solution of the differential equation (3.103), then

$$\int_0^{\ell} (-u_{xx} + gu - f)\phi \, dx = 0 \tag{3.104}$$

whenever ϕ is a square integrable function defined on $[0, \ell]$. In the Hilbert space $L^2(0, \ell)$ (see display (3.96)), the inner product of two functions v and

w is

$$\langle v, w \rangle := \int_0^\ell v(x)w(x)\, dx.$$

Therefore, if u is a solution of the equation (3.103), then equation (3.104) merely states that the inner product of ϕ with the zero function in L^2 vanishes. Moreover, if we define the operator $Au = -u_{xx} + gu$ and the function $b = f$, then $\langle Au - b, \phi \rangle = 0$ whenever ϕ is in the Hilbert space $L^2(0, \ell)$. Turning this analysis around, we can look for a function u such that $\langle Au - b, \phi \rangle = 0$ for all ϕ in L^2. In this case u is called a *weak solution* of the PDE.

Although L^2 is a natural Hilbert spaces of functions, the following problem arises if we try to apply the Galërkin method in L^2 to the PDE (3.103): the elements in L^2 are not all differentiable; therefore, the operator A is not defined on all of $L^2(0, \ell)$.

In which Hilbert space should we look for a solution? By asking this question, we free ourselves from the search for a *classical* or *strong* solution of the PDE (3.103), that is, a twice continuously differentiable function that satisfies the PDE and the boundary conditions. Instead, we will seek a *weak solution* by constructing a Hilbert space H whose elements are in L^2 such that a Galërkin formulation of our partial differential equation makes sense in H. If our boundary value problem has a classical solution, and we choose the Hilbert space H as well as the Galërkin formulation appropriately, then the L^2 equivalence class of the classical solution will also be in H. Moreover, if we are fortunate, then the weak solution of the boundary value problem obtained by applying the Galërkin principle in H will be exactly the equivalence class of the classical solution.

To construct the appropriate Hilbert space of candidate solutions for the equation (3.104), let us first formally apply the *fundamental method for PDE* (that is, integration by parts) to obtain the identity

$$\int_0^\ell (-u_{xx} + gu - f)\phi\, dx = \int_0^\ell (u_x\phi_x + gu\phi - f\phi)\, dx - u_x\phi\Big|_0^\ell. \quad (3.105)$$

If the functions ϕ and u are sufficiently smooth so that the integration by parts is valid, then equation (3.104) is equivalent to an equation involving only first derivatives with respect to the variable x, namely, the equation

$$\int_0^\ell (u_x\phi_x + gu\phi)\, dx - u_x\phi\Big|_0^\ell = \int_0^\ell f\phi\, dx. \quad (3.106)$$

In other words, to use equation (3.106) as a Galërkin formulation of our boundary value problem, we must define a Hilbert space H whose elements have one derivative with respect to x in L^2. Moreover, suppose that such a Hilbert space H exists. If we find a function $u \in H$ such that equation (3.106) holds for all $\phi \in H$ and u *happens to be smooth*, then the integration by parts is valid and we also have a solution of equation (3.104) for all smooth functions ϕ. Using this fact, it is easy to prove that u satisfies the PDE (3.103) pointwise, that is, u is a classical solution (see Exercise 3.56).

Exercise 3.56. Suppose that u is a C^2 function. If equation (3.104) holds for every $\phi \in C^\infty$, then prove that $-u_{xx} + g(x)u - f(x) = 0$.

If Dirichlet boundary conditions are imposed, then the boundary conditions must be built into the Hilbert space of test functions from which we select ϕ. In other words, we must impose the condition that the test functions satisfy the Dirichlet boundary conditions. The appropriate Hilbert space $H_D^1(0, \ell)$ is defined to be the completion with respect to the Sobolev norm of the set of smooth functions on $[0, \ell]$ that satisfy the Dirichlet boundary conditions. Here, the Sobolev norm of a function ϕ is given by

$$\|\phi\|_1 := \left(\int_0^\ell \phi^2(x)\, dx + \int_0^\ell \phi_x^2(x)\, dx \right)^{1/2}.$$

The Sobolev space $H_D^1(0, \ell)$ is a Hilbert space with respect to the inner product

$$\langle \phi, \psi \rangle_1 = \int_0^\ell \phi\psi\, dx + \int_0^\ell \phi_x\psi_x\, dx.$$

Informally, $H_D^1(0, \ell)$ is the space of functions that satisfy the Dirichlet boundary conditions and have one derivative in L^2.

We have the following Galërkin or weak formulation of our Dirichlet boundary value problem: Find $u \in H_D^1(0, \ell)$ such that

$$(u, \phi) := \int_0^\ell (u_x\phi_x + gu\phi)\, dx = \int_0^\ell f\phi\, dx = \langle f, \phi \rangle \qquad (3.107)$$

for all $\phi \in H_D^1(0, \ell)$. (Note: In equation (3.107) the inner product $\langle f, \phi \rangle$ is the L^2 inner product, not the H^1 inner product.) If u is a weak solution of the Dirichlet boundary value problem, then, using the definition of the Sobolev space, we can be sure that u is the limit of smooth functions that satisfy the boundary conditions. Of course, u itself is only defined abstractly as an equivalence class, thus it only satisfies the boundary conditions in the generalized sense, that is, u is the limit of a sequence of functions that satisfy the boundary conditions.

For the Neumann boundary value problem, again using equation (3.105), the appropriate space of test functions is $H^1(0, \ell)$, the space defined just like H_D^1 except that no boundary conditions are imposed. This requires a bit of explanation. First, we have the formal statement of the weak formulation of the Neumann problem: Find a function u in $H^1(0, \ell)$ such that, with the same notation as in display (3.107),

$$(u, \phi) = \langle f, \phi \rangle$$

for all $\phi \in H^1(0, \ell)$. We will show the following proposition: *If u is smooth enough so that the integration by parts in display (3.105) is valid and the equivalence class of u in $H^1(0, \ell)$ is a weak solution of the Neumann problem, then u satisfies the Neumann boundary conditions.* In fact, if $\phi \in H_D^1(0, \ell) \subset H^1(0, \ell)$, then ϕ is a limit of smooth functions that satisfy the Dirichlet boundary conditions. By using integration by parts for a sequence of smooth functions converging to ϕ in $H_D^1(0, \ell)$ and passing to the limit, we have the identity

$$\int_0^\ell (-u_{xx} + gu)\phi \, dx = \int_0^\ell f\phi \, dx \qquad (3.108)$$

for all $\phi \in H_D^1(0, \ell)$. In other words, $-u_{xx} + gu - f$ is the zero element of $H_D^1(0, \ell)$. By Exercise 3.57, the space $H_D^1(0, \ell)$ is a dense subspace of $H^1(0, \ell)$. Thus, it is easy to see that the identity (3.108) holds for all $\phi \in H^1(0, \ell)$. Finally, by this identity, the boundary term in the integration by parts formula in display (3.105) must vanish for each $\phi \in H^1(0, \ell)$; hence u satisfies the Neumann boundary conditions, as required. Our weak formulation is therefore consistent with the classical boundary value problem: If a weak solution of the Neumann boundary value problem happens to be smooth, then it will satisfy the Neumann boundary conditions.

Exercise 3.57. Prove that $H_D^1(0, \ell)$ is a dense subspace of $H^1(0, \ell)$.

Our analysis leads to the natural question "If a weak solution exists, then is it automatically a strong (classical) solution?" The answer is "yes" for the example problems that we have formulated here, but this important regularity result is beyond the scope of our discussion. Let us simply remark that the regularity of the weak solution depends on the form of the PDE. It is also natural to ask if our *weak* boundary value problems have solutions. While the existence theory for boundary value problems is difficult in general, we will formulate and prove an elementary result that implies the existence of solutions for some of the examples that we have considered. The proof of this result uses the contraction principle.

Let us suppose that H is a real Hilbert space, that $(\, , \,)$ is a bilinear form on H (it maps $H \times H \to \mathbb{R}$), $\langle \, , \, \rangle$ is the inner product on H, and $\| \, \| := \langle \, , \, \rangle^{1/2}$ is the natural norm. The bilinear form is called *continuous* if there is a constant $a > 0$ such that

$$|(u, v)| \le a\|u\|\|v\|$$

for all $u, v \in H$. The bilinear form is called *coercive* if there is a constant $b > 0$ such that

$$(u, u) \geq b\|u\|^2$$

for all $u \in H$.

Theorem 3.58 (Lax–Milgram Theorem). *If H is a real Hilbert space, $(\ ,\ )$ is a continuous and coercive bilinear form on H, and F is a bounded linear functional $F : H \to \mathbb{R}$, then there is a unique $u \in H$ such that*

$$(u, \phi) = F(\phi)$$

for every $\phi \in H$. Moreover,

$$\|u\| \leq \frac{1}{b}\|F\|.$$

Proof. The main tool of the proof is a standard result in Hilbert space theory, the Riesz representation theorem: If F is a bounded linear functional, then there is a unique $f \in H$ such that $F(\phi) = \langle f, \phi \rangle$ for every $\phi \in H$ (see [199]). In particular, this is true for the functional F in the statement of the theorem.

If $u \in H$, then the function given by $\phi \mapsto (u, \phi)$ is a linear functional on H. To see that this functional is bounded, use the continuity of the bilinear form to obtain the estimate

$$|(u, \phi)| \leq a\|u\|\|\phi\|$$

and note that $\|u\| < \infty$. The Riesz theorem now applies to each such functional. Therefore, there is a function $A : H \to H$ such that

$$(u, \phi) = \langle Au, \phi \rangle$$

for all $\phi \in H$. Moreover, using the linearity of the bilinear form, it follows that A is a linear transformation.

It is now clear that the equation in the statement of the theorem has a unique solution if and only if the equation $Au = f$ has a unique solution for each $f \in H$.

By the continuity and the coerciveness of the bilinear form, if $u, v, \phi \in H$, then

$$\langle A(u - v), \phi \rangle = (u - v, w) \leq a\|u - v\|\|\phi\|, \qquad (3.109)$$
$$\langle A(u - v), u - v \rangle = (u - v, u - v) \geq b\|u - v\|^2. \qquad (3.110)$$

Also, by the Schwarz inequality, we have that

$$\sup_{\|\phi\| \leq 1} |\langle v, \phi \rangle| \leq \|v\|,$$

and, for $\phi := (1/\|v\|)v$, this upper bound is attained. Thus, the norm of the linear functional $\phi \mapsto \langle w, \phi \rangle$ is $\|w\|$. In particular, using the inequality (3.109), we have

$$\|Au - Av\| = \sup_{\|w\| \leq 1} \langle A(u-v), \phi \rangle \leq a\|u-v\|. \qquad (3.111)$$

Define the family of operators $\mathcal{A}^\lambda : H \to H$ by

$$\mathcal{A}^\lambda \phi = \phi - \lambda(A\phi - f), \qquad \lambda > 0,$$

and note that $\mathcal{A}^\lambda u = u$ if and only if $Au = f$. Thus, to solve the equation $Au = f$, it suffices to show that for at least one choice of $\lambda > 0$, the operator $\mathcal{A}^\lambda$ has a unique fixed point.

By an easy computation using the definition of the norm, equation (3.109), the Schwarz inequality, and equation (3.111), we have that

$$\|\mathcal{A}^\lambda u - \mathcal{A}^\lambda v\|^2 = (1 - 2\lambda a + \lambda^2 a^2)\|u - v\|^2.$$

The polynomial in λ vanishes at $\lambda = 0$ and its derivative at this point is negative. It follows that there is some $\lambda > 0$ such that the corresponding operator is a contraction on the complete metric space H. By the contraction mapping theorem, there is a unique fixed point $u \in H$. Moreover, for this u we have proved that $(u, u) = F(u)$. Therefore,

$$\|u\|\|F\| \geq \langle f, u \rangle \geq b\|u\|^2,$$

and the last statement of the theorem follows.   □

Construction of Weak Solutions

The Lax–Milgram theorem is a classic result that gives us one way to prove the existence of weak solutions for our boundary value problems. One way to *construct* a solution, or at least a computable approximation of a solution, is to use the Galërkin method described above. In fact, with the previously defined notation, let us consider one of the finite dimensional Hilbert spaces S_n of H. Note that if the hypotheses of the Lax–Milgram theorem are satisfied, then there is a unique $u_n \in S_n$ such that

$$(u_n, s) = \langle f, s \rangle \qquad (3.112)$$

for all $s \in S_n$ with the additional property that

$$\|u_n\| \leq \frac{1}{b}\|f\| \qquad (3.113)$$

where b is the coercivity constant. The Galërkin principle is the statement that the sequence $\{u_n\}_{n=1}^\infty$ converges to the unique solution u of the weak boundary value problem. The approximation u_n can be expressed as a linear combination of the vectors $\nu_1, \ldots, \nu_n$ that, by our choice, form a basis of the subspace S_n. Thus, there are real numbers $c_1, \ldots, c_n$ such that

$$u_n = \sum_{j=1}^{n} c_j \nu_j.$$

Also, each element $s \in S_n$ is given in coordinates by

$$s = \sum_{i=1}^{n} s_i \nu_i.$$

Thus, the equation (3.112) is given in coordinates by the system of equations

$$\sum_{j=1}^{n} c_j (\nu_j, \nu_i) = \langle f, \nu_i \rangle, \qquad i = 1, \ldots n,$$

or, in the equivalent matrix form for the unknown vector $(c_1, \ldots c_n)$, we have the equation

$$\mathcal{S} \begin{pmatrix} c_1 \\ \vdots \\ c_n \end{pmatrix} = \begin{pmatrix} \langle f, \nu_1 \rangle \\ \vdots \\ \langle f, \nu_n \rangle \end{pmatrix}$$

where $\mathcal{S}$, called the *stiffness matrix*—the terminology comes from the theory of elasticity—is given by $\mathcal{S}_{ij} := (\nu_j, \nu_i)$. Of course, by the Lax–Milgram theorem, $\mathcal{S}$ is invertible and the matrix system can be solved to obtain the approximation u_n.

Does the sequence of approximations $\{u_n\}_{n=1}^{\infty}$ converge? The first observation is that, by the inequality (3.113), the sequence of approximates is bounded. Let u be the weak solution given by the Lax–Milgram theorem. Subtract the equality $(u_n, s) = \langle f, s \rangle$ from the equality $(u, s) = \langle f, s \rangle$ to see that

$$(u - u_n, s) = 0 \qquad (3.114)$$

for all $s \in S_n$. Also, using the coerciveness of the bilinear form, if $\phi \in S_n$, then

$$b \|u - u_n\|^2 \le (u - u_n, u - u_n) = (u - u_n, u - u_n + \phi - \phi)$$
$$= (u - u_n, \phi - u_n) + (u - u_n, u - \phi).$$

Moreover, with $u_n, \phi \in S_n$ and equation (3.114), we have the inequality

$$b \|u - u_n\|^2 \le (u - u_n, u - \phi) \le a \|u - u_n\| \|u - \phi\|.$$

It follows that

$$\|u - u_n\| \leq \frac{a}{b}\|u - \phi\| \tag{3.115}$$

for all $\phi \in S_n$.

Recall that the linear span of the sequence $\{\nu_j\}_{j=1}^{\infty}$ is assumed to be dense in H. Hence, for each $\epsilon > 0$ there is some integer m and constants $c_1, \ldots, c_m$ such that

$$\left\|u - \sum_{j=1}^{m} c_j \nu_j\right\| < \epsilon.$$

If we set $n = m$ and $v = \sum_{j=1}^{m} c_j \nu_j$ in the inequality (3.115), then

$$\|u - u_n\| \leq \frac{a}{b}\epsilon.$$

In other words, the sequence of Galërkin approximations converges to the weak solution, as required.

In the context of the steady state problem with which we started, namely, the PDE (3.103), the Lax–Milgram theorem applies if g is bounded above zero (see Exercise 3.59). For example, let g be a constant function given by $g(x) = \lambda > 0$, consider Dirichlet boundary conditions, and let S_n denote the span of the subset $\{\nu_1, \nu_2, \ldots, \nu_n\}$ of $H_D^1(0, \ell)$, where

$$\nu_j(x) := \sin \frac{j\pi}{\ell} x.$$

The smooth function f is represented by a Fourier (sine) series,

$$f(x) = \sum_{j=1}^{\infty} a_j \sin \frac{j\pi}{\ell} x,$$

on the interval $(0, \ell)$, and the corresponding Galërkin approximation is

$$u_n(x) = \sum_{j=1}^{n} \frac{a_j}{\lambda + (j\pi/\ell)^2} \sin \frac{j\pi}{\ell} x, \tag{3.116}$$

exactly the partial sum of the Fourier series approximation of the solution (see Exercise 3.60).

Exercise 3.59. Prove that if g is bounded above zero, then the bilinear form

$$(u, v) := \int_0^{\ell} (u_x v_x + guv) \, dx$$

is continuous and coercive on the spaces H_D^1 and H^1. Also, prove that if f is smooth, then $F(\phi) := \int_0^{\ell} f\phi \, dx$ is a continuous linear functional on H_D^1 and H^1.

Exercise 3.60. Suppose g is a negative constant. Find the stiffness matrix for the Galërkin approximation for the PDE (3.103) with Dirichlet boundary conditions using the basis given by

$$\nu_j(x) := \sin \frac{j\pi}{\ell} x, \qquad j = 1, 2, \dots, \infty$$

for H_0^1, and verify the approximation (3.116). Also, consider the PDE (3.103) with Neumann boundary conditions, and find the Galërkin approximations corresponding to the basis

$$1, \cos \frac{\pi x}{\ell}, \sin \frac{\pi x}{\ell}, \dots.$$

Galërkin Approximations and ODE

We have now seen one very simple example where the Galërkin principle can be turned into a theorem. Let us take this as a prototype argument to justify the Galërkin principle and proceed to our main objective in this section: to see how the Galërkin method leads to problems in ordinary differential equations.

Let us consider the PDE

$$u_t = u_{xx} + f(x, t), \qquad 0 < x < \ell, \quad t > 0 \tag{3.117}$$

with either Dirichlet or Neumann boundary conditions. The weak form of this boundary value problem is derived from the integration by parts formula

$$\int_0^\ell (u_t - u_{xx} - f(x,t))\phi \, dx = \int_0^\ell (u_t \phi + u_x \phi_x - f(x,t))\phi \, dx - u_x \phi \Big|_0^\ell.$$

Just as before, we can formulate two weak boundary value problems.

The Dirichlet Problem: Find $u(x, t)$, a family of functions in $H_D^1(0, \ell)$ such that

$$\int_0^\ell (u_t \phi + u_x \phi_x) \, dx = \int_0^\ell f\phi \, dx$$

for all $\phi \in H_D^1(0, \ell)$.

The Neumann Problem: Find $u(x, t)$, a family of functions in $H^1(0, \ell)$ with the same integral condition satisfied for all $\phi \in H^1(0, \ell)$.

To apply the Galërkin method, choose $\nu_1, \nu_2, \dots$ a linearly independent sequence whose span is dense in the Hilbert space $H_D^1(0, \ell)$ or $H^1(0, \ell)$, and define the finite dimensional spaces S_n as before. The new wrinkle is that we will look for an approximate solution in the subspace S_n of the form

$$u_n(x, t) = \sum_{j=1}^{n} c_j(t) \nu_j(x)$$

where the coefficients are differentiable functions of time. According to the Galërkin principle, let us search for the unknown functions $c_1, \ldots, c_n$ so that we have $(u_n, s) = \langle f, s \rangle$ for all $s \in S_n$. Expressed in coordinates, the requirement is that the unknown functions satisfy the system of n ordinary differential equations

$$\sum_{j=1}^{n} c_j'(t) \int_0^\ell \nu_j \nu_k \, dx + \sum_{j=1}^{n} c_j(t) \int_0^\ell (\nu_j)_x (\nu_k)_x \, dx = \int_0^\ell f \nu_k \, dx$$

indexed by $k = 1, \ldots, n$. In matrix form, we have the linear system of ordinary differential equations

$$\mathcal{M}C' + \mathcal{S}C = F(t)$$

where $\mathcal{M}$, given by

$$\mathcal{M}_{kj} := \int_0^\ell \nu_j \nu_k \, dx$$

is called the *mass matrix*, $\mathcal{S}$, given by

$$\mathcal{S}_{kj} := \int_0^\ell (\nu_j)_x (\nu_k)_x \, dx$$

is the stiffness matrix, and $C := (c_1, \ldots, c_n)$. If the initial condition for the PDE (3.117) is $u(x, 0) = u_0(x)$, then the usual choice for the initial condition for the approximate system of ordinary differential equations is the element $u_0^n \in S_n$ such that

$$\langle u_0^n, s \rangle = \langle u_0, s \rangle$$

for all $s \in S_n$. This "least squares" approximation always exists. (Why?)

We have, in effect, described some aspects of the theoretical foundations of the finite element method for obtaining numerical approximations of PDE (see [216]). But a discussion of the techniques that make the finite element method a practical computational tool is beyond the scope of this book.

The Galërkin method was originally developed to solve problems in elasticity. This application yields some interesting dynamical problems for the corresponding systems of ordinary differential equations. Let us consider, for instance, the PDE (more precisely the *integro-PDE*),

$$u_{xxxx} + \left(\alpha - \beta \int_0^1 u_x^2 \, dx \right) u_{xx} + \gamma u_x + \delta u_t + \epsilon u_{tt} = 0$$

that is derived in the theory of aeroelasticity as a model of panel flutter where $u(x, t)$ represents the deflection of the panel (see for example the book of Raymond L. Bisplinghoff and Holt Ashley [26, p. 428] where the physical interpretation of this equation and its parameters are given explicitly). The boundary conditions for "simply supported" panel edges are

$$u(0, t) = u(1, t) = 0, \qquad u_{xx}(0, t) = u_{xx}(1, t) = 0.$$

If we take just the first Fourier mode, that is, the Galërkin approximation with trial function

$$u_1(x, t) = c(t) \sin \pi x,$$

then we obtain the equation

$$\epsilon \ddot{c} + \delta \dot{c} + \pi^2 (\pi^2 - \alpha) c + \frac{\pi^4}{2} \beta c^3 = 0. \tag{3.118}$$

Let us note that if $\pi^2 - \alpha < 0$, then this Galërkin approximation is a form of Duffing's equation with damping. We have already developed some of the tools needed to analyze this equation. In fact, most solutions are damped oscillations whose ω-limits are one of two possible asymptotically stable rest points (see Exercise 3.61). On the other hand, if a periodic external force is added to this system, then very complex dynamics are possible (see [124] and Chapter 6).

Exercise 3.61. Draw representative phase portraits for the family of differential equations (3.118). How does the phase portrait depend on the choice of the parameters?

Exercise 3.62. Consider the basis functions

$$\nu_j(x) := \sin(j\pi x / \ell)$$

for $H_D^1(0, \ell)$. (a) Find the mass matrix and the stiffness matrix for the Galërkin approximations for the weak Dirichlet boundary value problem (3.117) with $f(x, t) := \sin(\pi x / \ell) \cos \omega t$. (b) Solve the corresponding system of linear differential equations for the nth approximation $u_n(x, t)$. (c) What can you say qualitatively about the solutions of the Galërkin approximations? What long term dynamical behavior of the PDE (3.117) is predicted by the Galërkin approximations? (d) Find a steady state solution? (e) Repeat the analysis for $f(x, t) = \cos \omega t$. (f) Do you see a problem with the validity of these formal computations? (g) Formulate and solve analogous problems for Neumann boundary conditions.

Exercise 3.63. Consider a two (Fourier) mode Galërkin approximation for the PDE

$$u_t = k^2 u_{xx} + u - u^3 + a \cos \omega t, \qquad 0 < x < \ell, \quad t > 0$$

with either Dirichlet or Neumann boundary conditions. (a) What is the general character of the solutions in the phase plane? Hint: Start with the case where there is a time-independent source term ($a = 0$) and consider the stability of the steady state solution of the PDE at $u \equiv 0$. (b) Is the (linearized) stability criterion for the PDE reflected in the stability of the corresponding rest point in the phase plane of the approximating ordinary differential equation? (c) Is the ω-limit set of every solution of the approximation a rest point?

3.6.3 Traveling Waves

The concept of traveling wave solutions will be introduced in this section for the classic reaction-diffusion model system

$$u_t = k^2 u_{xx} + au(1 - u), \qquad x \in \mathbb{R}, \quad t > 0 \tag{3.119}$$

where k and $a > 0$ are constants.

The PDE (3.119), often called *Fisher's equation*, can be used to model many different phenomena. For example, this equation is a model of logistic growth with diffusion ([88], [173]), and it is also a model of neutron flux in a nuclear reactor (see [182]). For a general description of this and many other models of this type see [173] and [182].

Let us begin with the observation that equation (3.119) can be rescaled to remove the explicit dependence on the system parameters. In fact, with respect to the new time and space variables

$$\tau = at, \qquad \xi = \frac{\sqrt{a}}{k}x,$$

equation (3.119) can be recast in the form

$$u_\tau = u_{\xi\xi} + u(1 - u). \tag{3.120}$$

Therefore, with no loss of generality, we will consider the original model equation (3.119) for the case $a = 1$ and $k = 1$.

In some applications, Fisher's equation is considered on the whole real line (that is, $(-\infty < x < \infty)$) and the physically relevant boundary conditions are

$$\lim_{x \to -\infty} u(x, t) = 1, \qquad \lim_{x \to \infty} u(x, t) = 0. \tag{3.121}$$

For example, u might measure the diseased fraction of a population, which is distributed in some spatial direction (with spatial position x). The infected portion of the population increases to unity in the negative spatial direction; it decreases to zero in the positive x direction.

The basic idea is to look for a solution of equation (3.119) in the form of a *traveling wave*, that is,

$$u(x, t) = U(x - ct)$$

where the wave form is given by the function $U : \mathbb{R} \to \mathbb{R}$ and the wave speed is $|c| \neq 0$. For definiteness and with respect to our disease model, let us assume that $c > 0$. By substituting the traveling wave *ansatz* into Fisher's equation, we obtain the second order nonlinear ordinary differential equation

$$\ddot{U} + c\dot{U} + U - U^2 = 0,$$

which is equivalent to the phase plane system

$$\dot{U} = V, \qquad \dot{V} = -U - cV + U^2. \tag{3.122}$$

All solutions of the system (3.122) correspond to traveling wave solutions of Fisher's equation. But, a meaningful solution of a physical model must satisfy additional properties. For example, in the case where u measures the infected fraction of a population, we must have $0 \leq u \leq 1$ and the boundary conditions (3.121).

Fisher's equation, for a population model in which we ignore diffusion, reduces to the one-dimensional ordinary differential equation $\dot{u} = u - u^2$ for logistic growth, whose dynamics can be completely determined. In particular, the phase space is $\mathbb{R}$, there is an unstable rest point at $u = 0$, a stable rest point at $u = 1$, and a connecting orbit, that is, an orbit with α-limit set $\{0\}$ and ω-limit set $\{1\}$. For our disease model, this result predicts that if some fraction of the population is infected, then the entire population is eventually infected.

This suggests a natural question: Is there a traveling wave solution $u(x,t) = U(x - ct)$ for the PDE (3.119) such that $0 < u(x,t) < 1$, u satisfies the boundary conditions (3.121), and

$$\lim_{t \to \infty} u(x,t) = 1, \qquad \lim_{t \to -\infty} u(x,t) = 0?$$

In other words, is there an orbit—for the PDE viewed as an infinite dimensional ordinary differential equation in a space of solutions that incorporates the boundary conditions—connecting the steady states $u \equiv 0$ and $u \equiv 1$ as in the case of the one-dimensional logistic model?

Let us note that all the required conditions are satisfied if $0 < U(s) < 1$ and

$$\lim_{s \to \infty} U(s) = 0, \qquad \lim_{s \to -\infty} U(s) = 1.$$

An answer to our question is given by the following proposition.

Proposition 3.64. *The PDE (3.120) with the boundary conditions (3.121) has a traveling wave solution $(x,t) \mapsto u(x,t) = U(x-ct)$, with $0 < u(x,t) < 1$, whose orbit connects the steady states $u \equiv 0$ and $u \equiv 1$ if and only if $c \geq 2$.*

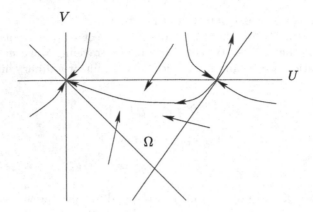

Figure 3.8: The invariant region Ω for the system (3.122) in case $c \geq 2$.

Proof. Note that the solution $u(x,t) = U(x - ct)$ is a connecting orbit if $0 < U(s) < 1$, and

$$\lim_{s \to \infty} U(s) = 0, \qquad \lim_{s \to -\infty} U(s) = 1.$$

The system matrix of the linearized phase plane equations (3.122) at the origin has eigenvalues

$$\frac{1}{2}(-c \pm \sqrt{c^2 - 4}),$$

and its eigenvalues at the point $(1,0)$ are given by

$$\frac{1}{2}(-c \pm \sqrt{c^2 + 4}).$$

Therefore, if $c > 0$, then there is a hyperbolic sink at the origin and a hyperbolic saddle at the point $(1,0)$. Moreover, if a connecting orbit exists, then the corresponding phase plane solution $s \mapsto (U(s), V(s))$ must be on the unstable manifold of the saddle and the stable manifold of the sink.

Note that if $c < 2$, then the sink at the origin is of *spiral* type. Hence, even if there is a connecting orbit in this case, the corresponding function U cannot remain positive.

Assume that $c \geq 2$ and consider the lines in the phase plane given by

$$V = \frac{1}{2}\left(-c + \sqrt{c^2 - 4}\right)U, \qquad V = \frac{1}{2}\left(-c + \sqrt{c^2 + 4}\right)(U - 1). \quad (3.123)$$

They correspond to eigenspaces at the rest points. In particular, the second line is tangent to the unstable manifold of the saddle point at $(U,V) = (1,0)$. The closed triangular region Ω (see Figure 3.8) in the phase plane bounded by the lines (3.123) and the line given by $V = 0$ is positively invariant. This result is easily checked by computing the dot product of the vector field corresponding to system (3.122) with the appropriate normal fields along these lines to see that the vector field points into Ω at every point on its boundary except for the rest points. Indeed, *along the lines* defined by the equations in display (3.123), we have

$$\dot{V} - \frac{1}{2}(-c + \sqrt{c^2 - 4})\dot{U} = U^2 \geq 0,$$

$$\dot{V} - \frac{1}{2}(-c + \sqrt{c^2 + 4})\dot{U} = (U - 1)^2 \geq 0, \tag{3.124}$$

and $\dot{V} = -U + U^2 < 0$ for $0 < U < 1$ along the line with equation $V = 0$.

Suppose (as we will soon see) that the unstable manifold at the saddle intersects the region Ω. Then a solution that starts on this portion of the unstable manifold must remain in the region Ω for all positive time. Thus, the ω-limit set of the corresponding orbit is also in Ω. Because $\dot{U} \leq 0$ in Ω, there are no periodic orbits in Ω and no rest points in the interior of Ω. By the Poincaré–Bendixson theorem, the ω-limit set must be contained in the boundary of Ω. In fact, this ω-limit set must be the origin.

To complete the proof, we will show that the unstable manifold at the saddle has nonempty intersection with the interior of Ω. To prove this fact, let us first recall that the unstable manifold is tangent to the line given by the second equation in display (3.123). We will show that the tangency is quadratic and the unstable manifold lies above this line.

In the new coordinates given by

$$Z = U - 1, \qquad W = V,$$

the saddle rest point is at the origin for the equivalent first order system

$$\dot{Z} = W, \qquad \dot{W} = Z - cW + Z^2.$$

The additional change of coordinates

$$Z = P, \qquad W = Q + \alpha P := Q + \frac{1}{2}(-c + \sqrt{c^2 + 4})P$$

transforms the system so that the unstable manifold of the saddle point is tangent to the horizontal P-axis. We will show that the unstable manifold is above this axis in some neighborhood of the origin; it then follows from the second formula in display (3.124) that the unstable manifold lies above the P-axis globally.

Note that the unstable manifold is given, locally at least, by the graph of a smooth function $Q = h(P)$ with $h(0) = h'(0) = 0$. Since this manifold is invariant, we must have that $\dot{Q} = h'(P)\dot{P}$, and therefore, by an easy computation,

$$P^2 - (c + \alpha)h(P) = h'(P)(h(P) + \alpha P). \tag{3.125}$$

The function h has the form $h(P) = \beta P^2 + O(P^3)$. By substitution of this expression into equation (3.125), we obtain the inequality

$$\beta = (3\alpha + c)^{-1} > 0,$$

as required. $\square$

Much more can be said about the traveling wave solutions that we have just found. For example, the ω-limit set of most solutions of the PDE (3.119), which start with physically realistic initial conditions, is the traveling wave solution with wave speed $c = 2$. This result was proved by Andrei N. Kolmogorov, Ivan G. Petrovskii, and Nikolai S. Piskunov [136] (see also [17] and [25]). For a detailed mathematical account of traveling wave solutions see the book of Paul C. Fife [88] and also [173] and [206].

Exercise 3.65. The reaction term for the PDE studied in Proposition 3.64 is $g(u) = u(1 - u)$. Prove a generalization of this proposition where the reaction term is given by a function g that is C^1 on $(0, 1)$ and such that g' is bounded on $(0, 1)$, $g(0) = g(1) = 0$, $g(u) > 0$ on $(0, 1)$, $g'(0) = 1$, and $g'(u) < 1$ on $(0, 1]$ (see [136]).

Exercise 3.66. Show that the PDE

$$u_t - u^2 u_x = u_{xx} + u, \qquad x \in \mathbb{R}, \quad t \geq 0$$

has a nonconstant solution that is periodic in both space and time.

Exercise 3.67. For positive constants α and β, find a traveling wave solution of the Korteweg-de Vries (KdV) equation

$$\eta_t + (1 + \frac{3}{2}\alpha\eta)\eta_x + \frac{1}{3}\beta\eta_{xxx} = 0, \qquad x \in \mathbb{R}, \quad t \geq 0$$

in the form $\eta(x, t) = N(x - ct)$ where N, N', and N'' all vanish as the argument of N approaches $\pm\infty$ and $c > 1$. The solution is an approximation of the "great wave of translation" observed by John Scott Russell (1808-1882). Further analysis of this equation led to the theory of solitons—nonlinear solitary waves that retain their basic shape after interaction with other solitary waves of the same type (see for example [217]).

3.6.4 First Order PDE

Consider the model equation (3.91) in case there is no diffusion, but the medium moves with velocity field V; that is, the differential equation

$$u_t + \gamma \operatorname{grad} u \cdot V = f. \tag{3.126}$$

This is an important example of a first order partial differential equation. Other examples are equations of the form

$$u_t + (f(u))_x = 0,$$

called *conservation laws* (see [206]), and equations of the form

$$S_t + H(S_q, q, t) = 0,$$

called *Hamilton–Jacobi equations* (see [12]). We will show how such PDE can be solved using ordinary differential equations in case the derivatives of the unknown function appear linearly and there is only one space dimension. The theory can be generalized to fully nonlinear equations with several space dimensions (see, for example, [81] and Exercises 3.73 and 3.74).

The equation (3.126), for $\gamma = 1$, is given by

$$u_t + v(x, t)u_x = f(u, x, t),$$

or, with a redefinition of the names of the functions, it has the more general form

$$f(x, y, u)u_x + g(x, y, u)u_y = h(x, y, u). \qquad (3.127)$$

We will solve the PDE (3.127) using the following basic idea: *If the graph G of a function $z = u(x, y)$ is an invariant manifold for the first order system*

$$\dot{x} = f(x, y, z), \quad \dot{y} = g(x, y, z), \quad \dot{z} = h(x, y, z), \qquad (3.128)$$

then u is a solution of the PDE (3.127). Indeed, because

$$(x, y) \mapsto (x, y, u(x, y), u_x(x, y), u_y(x, y), -1)$$

is a normal vector field on G, it follows from the results of Section 1.8.1 that the manifold G is invariant if and only if the dot product of the vector field associated with the system (3.128) and the normal field is identically zero; that is, if and only if equation (3.127) holds. The orbits of the system (3.128) are called *characteristics* of the PDE (3.127).

Perhaps it is possible to find an invariant manifold for the first order system (3.128) by an indirect method, but we will construct the invariant manifold directly from appropriate initial data. To see how this is done, let us suppose that we have a curve in space given by $\gamma : \mathbb{R} \to \mathbb{R}^3$ such that in coordinates

$$\gamma(s) = (\gamma_1(s), \gamma_2(s), \gamma_3(s)).$$

This curve is called *noncharacteristic* at $\gamma(0)$ (with respect to the PDE (3.127)) if

$$\dot{\gamma}_1(0)g(\gamma(0)) - \dot{\gamma}_2(0)f(\gamma(0)) \neq 0.$$

The geometric interpretation is clear: The projection of the tangent vector of the noncharacteristic curve at $\gamma(0)$ onto its first two components is transverse to the first two components of the tangent to the characteristic at this point.

Let φ_t denote the flow of the system (3.128), and define $\mathcal{H} : \mathbb{R}^2 \to \mathbb{R}^3$ by

$$(s, t) \mapsto \varphi_t(\gamma(s)). \tag{3.129}$$

Also, define $H : \mathbb{R}^2 \to \mathbb{R}^2$ by projection of the image of $\mathcal{H}$ onto its first two components. More precisely, let e_1, e_2, e_3 be the usual basis vectors for $\mathbb{R}^3$ and let the usual inner product be denoted by angle brackets. Then H is given by

$$(s, t) \mapsto (\langle \varphi_t(\gamma(s)), e_1 \rangle, \langle \varphi_t(\gamma(s)), e_2 \rangle).$$

The image of $\mathcal{H}$ is an invariant set for system (3.128). We will prove the following proposition: *if γ is a noncharacteristic curve at $\gamma(0)$ and both $|s|$ and $|t|$ are sufficiently small, then the image of $\mathcal{H}$ is an invariant two-dimensional manifold.*

Using this proposition, we will obtain the desired solution of our PDE by simply reparameterizing the invariant manifold with the coordinate transformation given by H.

The idea of the proof of the proposition is to show that $DH(0,0)$ is invertible and then apply the inverse function theorem. To show that H is locally invertible at $\gamma(0)$, note that

$$DH(0,0)e_1 = \frac{d}{ds} H(s,0) \Big|_{s=0}$$

$$= \frac{d}{ds} (\gamma_1(s), \gamma_2(s)) \Big|_{s=0}$$

$$= (\dot{\gamma}_1(0), \dot{\gamma}_2(0)),$$

and similarly

$$DH(0,0)e_2 = (f(\gamma(0)), g(\gamma(0))).$$

Because the curve γ is noncharacteristic at $\gamma(0)$, the matrix representation of $DH(0,0)$ has nonzero determinant and is therefore invertible. By the inverse function theorem, H is locally invertible at the origin.

Let $\mathcal{H} = (\mathcal{H}_1, \mathcal{H}_2, \mathcal{H}_3)$ and suppose that H is invertible. We have the identity

$$\mathcal{H}(H^{-1}(x,y)) = (x, y, \mathcal{H}_3(H^{-1}(x,y))).$$

In other words, the surface given by the range of $\mathcal{H}$ is locally the graph of the function u defined by

$$u(x,y) = \mathcal{H}_3(H^{-1}(x,y)).$$

Hence, u is a local solution of the PDE (3.127). Moreover, since $u(H(s,t)) = \mathcal{H}_3(s,t)$, we have that

$$u(\gamma_1(s), \gamma_2(s)) = \gamma_3(s).$$

This last equation is used to encode initial data for the PDE along a non-characteristic curve.

We have now proved that if we are given a noncharacteristic curve, then there is a corresponding local solution of the PDE (3.127). Also, we have obtained a method to construct such a solution.

The method is very simple: Choose a noncharacteristic curve γ, which encodes initial data via the formula $u(\gamma_1(s), \gamma_2(s)) = \gamma_3(s)$; determine the flow ϕ_t of the first order system (3.128) for the characteristics; define the function $\mathcal{H}$ given by $\mathcal{H}(s,t) = \phi_t(\gamma(s))$, and invert the associated function H given by $H(s,t) = (\mathcal{H}_1(s,t), \mathcal{H}_2(s,t))$. The corresponding solution of the PDE (3.127) is $u(x,y) = \mathcal{H}_3(H^{-1}(x,y))$.

As an example, let us consider the model equation

$$u_\tau + a\sin(\omega\tau)u_x = u - u^2, \qquad 0 \le x \le 1, \quad \tau \ge 0$$

with initial data $u(x,0) = u_0(x)$ defined on the unit interval. A phenomenological interpretation of this equation is that u is the density of a species with logistic growth in a moving medium that is changing direction with frequency ω and amplitude a. We have used τ to denote the time parameter so that we can write the first order system for the characteristics in the form

$$\dot\tau = 1, \qquad \dot x = a\sin(\omega\tau), \qquad \dot z = z - z^2. \tag{3.130}$$

To specify the initial data, define the noncharacteristic curve given by $s \mapsto (0, s, u_0(s))$. After solving system (3.130) and using the definition of $\mathcal{H}$ in display (3.129), we have that

$$\mathcal{H}(s,t) = \left(t, s + \frac{a}{\omega}(1 - \cos\omega t), \frac{e^t u_0(s)}{1 + u_0(s)(e^t - 1)}\right).$$

Also, because H^{-1} is given explicitly by

$$H^{-1}(\tau, x) = (\tau, x - \frac{a}{\omega}(1 - \cos\omega\tau)),$$

we have the solution

$$u(x,\tau) = \frac{e^\tau u_0(x - \frac{a}{\omega}(1 - \cos\omega\tau))}{1 + (e^\tau - 1)u_0(x - \frac{a}{\omega}(1 - \cos\omega\tau))}. \tag{3.131}$$

What does our model predict? For example, if the initial condition is given by a positive function u_0, then the ω-limit set of the corresponding solution of the PDE is the constant function $u \equiv 1$, the solution corresponding to no drift. On the other hand, if the initial population is distributed so that some regions have zero density, then the fate of the initial population is more complicated (see Exercise 3.68).

Exercise 3.68. What long term behavior for the corresponding model equation is predicted by the solution (3.131)? How does your answer depend on the choice of u_0, a, and ω?

Exercise 3.69. Find the general solution of the PDE $au_x + bu_y = h(x, y)$ in case a and b are constants and h is a given function. Hint: Solve the homogeneous equation ($h = 0$) first.

Exercise 3.70. (a) Find the general solution of the PDE $au_x + bu_y = cu$ in case a, b, and c are constants using the method developed in this section. (b) Find the general solution using the following alternative method: Find an integrating factor for the expression $u_y - (c/b)u$; that is, find a function μ so that $(\mu(y)u)_y = u_y - (c/b)u$.

Exercise 3.71. Solve the PDE $xu_x + yu_y = 2u$ with u prescribed on the unit circle. Hint: Define the noncharacteristic curve

$$s \mapsto (\cos s, \sin s, h(\cos s, \sin s)).$$

Exercise 3.72. (a) Find the general solution of the PDE $xu_x - yu_y = 2u$. (b) Give a geometric description of the noncharacteristic curves.

Exercise 3.73. [Several Space Dimensions] Solve the PDE $u_t = a(x, y)u_x + b(x, y)u_y$ with initial data $u(0, x, y) = g(x, y)$. Hint: This is an example of a first order PDE with two space variables. The method of solution illustrates the generalization of the theory presented in this section. (a) Write the PDE in the standard form $u_\tau - a(x, y)u_x - b(x, y)u_y = 0$, where t is replaced by τ for notational convenience, and let ψ_t denote the flow of the differential equation

$$\dot{x} = a(x, y), \qquad \dot{x} = b(x, y).$$

Show that $\tilde{\psi}_t = \psi_{-t}$ is the flow of the differential equation

$$\dot{x} = -a(x, y), \qquad \dot{x} = -b(x, y).$$

(b) The characteristics of the PDE are the solutions of the system

$$\dot{\tau} = 1, \quad \dot{x} = -a(x, y), \quad \dot{x} = -b(x, y), \quad \dot{z} = 0.$$

Show that the flow ϕ_t of this system has the form

$$\phi_t(\tau, x, y, z) = (t + \tau, \psi_{-t}(x, y), z).$$

(c) Because the solution u has three arguments, the initial data must be encoded into a noncharacteristic *surface* of the form

$$\gamma(r, s) = (\gamma_1(r, s), \gamma_2(r, s), \gamma_3(r, s), \gamma_4(r, s)).$$

Define the notion of a noncharacteristic surface and show that the parametrized surface given by the function $\gamma(r, s) = (0, r, s, g(r, s))$ is a noncharacteristic surface. (d) Define $\mathcal{H}(r, s, t) = \phi_t(\gamma(r, s))$. Show that $\mathcal{H}_4(r, s, t) = g(r, s)$. (e) Define H to be the function $\mathcal{H}$ composed with the linear projection onto its first three components. Show that $H^{-1}(\tau, x, y) = (\psi_\tau(x, y), \tau)$. (f) By the theory in this section, the solution of the PDE with the given initial data is $u(t, x, y) = g(\psi_t(x, y))$. Verify that this is the solution by direct substitution into the PDE. (g) Find the solution of the initial value problem

$$u_t = -yu_x + xu_y, \qquad u(0, x, y) = 2x^2 + y^2.$$

Exercise 3.74. [Several Space Dimensions (Continued)] (a) Let X denote a vector field on $\mathbb{R}^2$. Generalize the method outlined in Exercise 3.73 to cover the initial value problem for the equation of continuity

$$u_t + \operatorname{div}(uX) = 0, \qquad u(0, x, y) = g(x, y).$$

(b) Suppose that u denotes the density of some substance in a medium whose instantaneous direction of motion in the physical three-dimensional space is approximated by the vector field $X(x, y, z) = (-(x+y), x-y, 0)$. Suppose that the initial density is constant, say $u(x, y, z) = 1$, and determine the density at time $t > 0$. Give a physical interpretation of your answer. (c) Repeat part (b) for the initial density $u(x, y, z) = x^2 + y^2$.

Exercise 3.75. A function U that is constant along the orbits of an ordinary differential equation is called an *invariant function,* or a *first integral.* In symbols, if we have a differential equation $\dot{x} = f(x)$ with flow φ_t, then U is invariant provided that $U(\phi_t(x)) = U(x)$ for all x and t for which the flow is defined. Show that U is invariant if and only if $\langle \operatorname{grad} U(x), f(x) \rangle \equiv 0$. Equivalently, the directional derivative of U in the direction of the vector field given by f vanishes. Consider the differential equation

$$\dot{\theta} = 1, \qquad \dot{\phi} = \alpha$$

where $\alpha \in \mathbb{R}$. Also, consider both θ and ϕ as angular variables so that the differential equation can be viewed as an equation on the torus. Give necessary and sufficient conditions on α so that there is a smooth invariant function *defined on the torus.*

Exercise 3.76. A simple example of a conservation law is the (nonviscous) Burgers's equation $u_t + uu_x = 0$. Burgers's equation with viscosity is given by

$$u_t + uu_x = \frac{1}{Re} u_{xx}$$

where Re is called the *Reynold's number.* This is a simple model that incorporates two of the main features in fluid dynamics: convection and diffusion. Solve the nonviscous Burgers's equation with initial data $u(x, 0) = (1 - x)/2$ for $-1 < x < 1$. Note that the solution cannot be extended for all time. This is a general phenomenon that appears in the study of conservation laws that is related to the existence of *shock waves* (see [206]). Also, consider the viscous Burgers's equation on the same interval with the same initial data and with boundary conditions

$$u(-1, t) = 1, \qquad u(1, t) = 0.$$

How can we find Galërkin approximations? The problem is that with the nonhomogeneous boundary conditions, there is no vector space of functions that satisfy the boundary conditions. To overcome this problem, we can look for a solution of our problem in the form

$$u(x,t) = v(x,t) + \frac{1}{2}(1-x)$$

where v satisfies the equation

$$v_t + (v + \frac{1}{2}(1-x))(v_x - \frac{1}{2}) = v_{xx}$$

and Dirichlet boundary conditions. Determine the Galërkin approximations using trigonometric trial functions. Use a numerical method to solve the resulting differential equations, and thus approximate the solution of the PDE. For a numerical analyst's approach to this problem, consider the Galërkin approximations with respect to the "test function basis" of Chebyshev polynomials given by

$$T_0(x) = 1, \qquad T_1(x) = x, \qquad T_2(x) = 2x^2 - 1$$

and

$$T_{n+1}(x) = 2xT_n(x) - T_{n-1}(x).$$

The Chebyshev polynomials are orthogonal (but not orthonormal) with respect to the inner product defined by

$$\langle f, g \rangle := \int_{-1}^{1} f(x)g(x)(1-x^2)^{-1/2}\, dx.$$

Also, the Chebyshev polynomials do not satisfy the boundary conditions. In spite of these facts, proceed as follows: Look for a Galërkin approximation in the form

$$u_n(x,t) = \sum_{i=1}^{n} c_i(t)T_{n-1}(x),$$

but only construct the corresponding system of differential equations for

$$c_1, \ldots, c_{n-2}.$$

Then, define the last two coefficients so that the boundary conditions are satisfied (see [90]). Compare numerical results. Finally, note that Burgers's equation can, in principle, be solved explicitly by the Hopf–Cole transformation. In fact, if u is a solution of Burgers's equation and ψ is defined so that $\psi_x = u$, then ψ is defined up to a function that depends only on the time variable. An appropriate choice of the antiderivative satisfies the equation

$$\psi_t + \frac{1}{2}\psi_x^2 = \frac{1}{Re}\psi_{xx}.$$

If ϕ is defined by the equation $\psi = -(2/Re)\ln\phi$, then

$$\phi_t = \frac{1}{Re}\phi_{xx}.$$

Thus, solutions of the heat equation can be used to construct solutions of Burgers's equation. Because Burgers's equation can be solved explicitly, this PDE is a useful candidate for testing numerical codes.

4
Hyperbolic Theory

This chapter is an introduction to the theory of hyperbolic structures in differential equations. The basic idea that is discussed might be called the principle of hyperbolic linearization: If the system matrix of the linearized flow of a differential equation has no eigenvalue with zero real part, then the nonlinear flow behaves locally like the linear flow. This idea has far-reaching consequences that are the subject of many important and useful mathematical results. Here we will discuss two fundamental theorems: the center and stable manifold theorem for a rest point and Hartman's theorem.

4.1 Invariant Manifolds

The stable manifold theorem is a basic result in the theory of ordinary differential equations. It and many closely related results, for example, the center manifold theorem, form the foundation for analyzing the dynamical behavior of a dynamical system in the vicinity of an invariant set. In this section we will consider some of the theory that is used to prove such results, and we will prove the existence of invariant manifolds related to the simplest example of an invariant set, namely, a rest point. Nevertheless, the ideas that we will discuss can be used to prove much more general theorems. In fact, some of the same ideas can be used to prove the existence and properties of invariant manifolds for infinite dimensional dynamical systems.

The concept of an invariant manifold for a rest point arises from the study of linear systems. Recall that if A is the system matrix of a linear differential equation on $\mathbb{R}^n$, then the spectrum of A splits naturally—from the point of view of stability theory—into three subsets: the eigenvalues with negative, zero, and positive real parts. After a linear change of coordinates that transforms A to its real Jordan normal form, we find that the differential equation $\dot{u} = Au$ decouples into a system

$$\dot{x} = Sx, \quad \dot{y} = Uy, \quad \dot{z} = Cz$$

where $(x, y, z) \in \mathbb{R}^k \times \mathbb{R}^\ell \times \mathbb{R}^m$ with $k + \ell + m = n$, and where S, U and C are linear operators whose eigenvalues have all negative, positive, and zero real parts, respectively. The subspace $\mathbb{R}^k \subset \mathbb{R}^n$ is called the *stable manifold* of the rest point at the origin for the original system $\dot{u} = Au$, the subspace $\mathbb{R}^\ell$ is called the *unstable manifold,* and the subspace $\mathbb{R}^m$ is called the *center manifold.*

According to Theorem 2.61, there are constants $K > 0$, $a > 0$, and $b > 0$ such that if $\xi \in \mathbb{R}^k$ and $\zeta \in \mathbb{R}^n$, then

$$\|x(t, \xi)\| = \|e^{tS}\xi\| \leq Ke^{-at}\|\xi\|, \quad t \geq 0,$$

$$\|y(t, \zeta)\| = \|e^{tU}\zeta\| \leq Ke^{bt}\|\zeta\|, \quad t \leq 0, \tag{4.1}$$

where $t \mapsto x(t, \xi)$ is the solution of the differential equation $\dot{x} = Sx$ with the initial condition $x(0, \xi) = \xi$, and y is defined similarly. Here, $\| \ \|$ is an arbitrary norm on $\mathbb{R}^n$. There are no such exponential estimates on the center manifold.

An analysis of the dynamics on the center manifold, when it exists, is often one of the main reasons for finding a center manifold in the first place. In this regard, let us recall that the flow of a nonlinear system near a rest point where the linearization has an eigenvalue with zero real part is *not* determined by the linearized flow. For example, the linearization at a rest point of a planar system might have a center, whereas the corresponding rest point for the nonlinear system is a sink or a source. In this case the center manifold at the rest point is an open subset of the plane. As this example shows, we can expect the most complicated (and most interesting) dynamics near a nonhyperbolic rest point to occur on a corresponding center manifold. If a center manifold has dimension less than the dimension of the phase space, then the most important dynamics can be studied by considering the restriction of the original system to a center manifold. To illustrate, let us imagine a multidimensional system that has a rest point with a codimension two stable manifold and a two-dimensional center manifold. Then, as we will see, the orbits of the nonlinear system are all locally attracted to the center manifold, and therefore the nontrivial dynamics can be determined by studying a planar system. This "center manifold reduction" to a lower dimensional system is one of the main applications of the theory.

The stable manifold theorem states that if the linear system $\dot{u} = Au$ has no center manifold, then the nonlinear system $\dot{u} = Au + H(u)$, where $H : \mathbb{R}^n \to \mathbb{R}^n$ with $H(0) = 0$ and $DH(0) = 0$, has stable and unstable manifolds corresponding to the stable and unstable manifolds of the linear system. These manifolds are invariant sets that contain the rest point at the origin, and they have the same dimensions as the corresponding linear manifolds. In fact, the corresponding linear manifolds are their tangent spaces at the rest point. Moreover, the flow restricted to the stable and the unstable manifolds has exponential (hyperbolic) estimates similar to the inequalities in display (4.1)

There are several different methods available to prove the existence of invariant manifolds. Each of these methods has technical as well as conceptual advantages and disadvantages. Here we will use the *Lyapunov–Perron method*. The basic idea is to determine the invariant manifold as the graph of a function that is obtained as the fixed point of an integral operator on a Banach space. An alternative method based on the "graph transform" is also very important (see [122] and [194]). The Lyapunov–Perron method has wide applicability and it can be used to prove very general theorems. While the graph transform method is perhaps even more far-reaching, the main reason for using the Lyapunov–Perron method here is that the theory illustrates many useful ODE techniques.

For the invariant manifold theory that we will discuss, it is not necessary to assume the existence of an infinitesimally hyperbolic linearization. Instead, it suffices to assume that the spectrum of the linearization has a spectral gap; that is, the spectrum is separated into two vertical strips in the complex plane such that the maximum of the real parts of the eigenvalues in the left hand strip is strictly less than the minimum of the real parts of the eigenvalues in the right hand strip. This hypothesis is exactly the condition required to apply the Lyapunov–Perron method to obtain the existence of an invariant manifold. The stable, unstable, and center manifold theorems are easily obtained as corollary results.

We will use the notation C^1 as a prefix to denote spaces of continuously differentiable functions. If f is such a function, then let $\|f\|_1$ denote the C^1-norm given by the sum of the supremum norm of f and the supremum norm of its derivative Df, where the supremum is taken over the domain of definition of the function.

The next theorem is the main result of this section. It states the existence of a smooth global invariant manifold at a rest point of a nonlinear system, provided that the linearization of the system at the rest point has a spectral gap and the nonlinear remainder terms are sufficiently small. The proof of this theorem is quite long, but it is not too difficult to understand. The idea is to set up a contraction in an appropriate Banach space of continuous functions so that the fixed point of the contraction is a function whose graph is the desired invariant manifold. Then the fiber contraction principle is applied to show that this function is smooth. The proof uses many important ODE techniques that are well worth learning.

Theorem 4.1. *Suppose that a and b are real numbers with $a < b$, $S :$ $\mathbb{R}^k \to \mathbb{R}^k$ and $U : \mathbb{R}^\ell \to \mathbb{R}^\ell$ are linear transformations, each eigenvalue of S has real part less than a, and each eigenvalue of U has real part greater than b. If $F \in C^1(\mathbb{R}^k \times \mathbb{R}^\ell, \mathbb{R}^k)$ and $G \in C^1(\mathbb{R}^k \times \mathbb{R}^\ell, \mathbb{R}^\ell)$ are such that $F(0,0) = 0$, $DF(0,0) = 0$, $G(0,0) = 0$, $DG(0,0) = 0$, and $\|F\|_1$ and $\|G\|_1$ are sufficiently small, then there is a unique function $\alpha \in C^1(\mathbb{R}^k, \mathbb{R}^\ell)$ with the following properties*

$$\alpha(0) = 0, \quad D\alpha(0) = 0, \quad \sup_{\xi \in \mathbb{R}^k} \|D\alpha(\xi)\| < \infty,$$

whose graph, namely the set

$$W(0,0) = \{(x,y) \in \mathbb{R}^k \times \mathbb{R}^\ell : y = \alpha(x)\},$$

is an invariant manifold for the system of differential equations given by

$$\dot{x} = Sx + F(x,y), \qquad \dot{y} = Uy + G(x,y). \tag{4.2}$$

Moreover, if $(\xi, \alpha(\xi)) \in W(0,0)$, then for each $\lambda > a$ there is a constant $C > 0$ such that the solution $t \mapsto (x(t), y(t))$ of the system (4.2) with initial condition $(\xi, \alpha(\xi))$ satisfies the exponential estimate

$$\|x(t)\| + \|y(t)\| \leq Ce^{\lambda t}\|\xi\|.$$

Proof. We will use several Banach spaces and several different norms. The proofs that these spaces with the indicated norms are indeed Banach spaces are left to the reader. We will outline a proof for just one of the spaces.

Let $C^0(\mathbb{R}^N, \mathbb{R}^M)$ denote the linear space of all continuous functions

$$f : \mathbb{R}^N \to \mathbb{R}^M,$$

and let us use it to define the following Banach spaces: $\mathcal{C}^0(\mathbb{R}^N, \mathbb{R}^M)$, the set of all functions $f \in C^0(\mathbb{R}^N, \mathbb{R}^M)$ such that $f(0) = 0$ and

$$\|f\|_0 = \sup_{\xi \in \mathbb{R}^N} \|f(\xi)\| < \infty;$$

$\mathcal{C}^1(\mathbb{R}^N, \mathbb{R}^M)$, the set of all continuously differentiable functions

$$f \in \mathcal{C}^0(\mathbb{R}^N, \mathbb{R}^M)$$

such that $f(0) = 0$, $Df(0) = 0$, and

$$\|f\|_1 = \|f\|_0 + \|Df\|_0 < \infty;$$

and $\mathcal{E}^0(\mathbb{R}^N, \mathbb{R}^M)$, the set of all functions $f \in C^0(\mathbb{R}^N, \mathbb{R}^M)$ such that $f(0) = 0$ and

$$\|f\|_{\mathcal{E}} = \sup\left\{\frac{\|f(\xi)\|}{\|\xi\|} : \xi \in \mathbb{R}^N, \xi \neq 0\right\} < \infty.$$

Also, for $f \in C^0(\mathbb{R}^N, \mathbb{R}^M)$, let the Lipschitz constant of f be denoted by

$$\operatorname{Lip}(f) := \sup_{\xi \neq \eta} \frac{\|f(\xi) - f(\eta)\|}{\|\xi - \eta\|}$$

whenever the indicated supremum is finite.

Proposition A: The space $\mathcal{E}^0(\mathbb{R}^N, \mathbb{R}^M)$ with the $\mathcal{E}$-norm is a Banach space.

To prove the proposition, let us assume for the moment that if $\{f_n\}_{n=1}^{\infty}$ is a sequence in $\mathcal{E}^0(\mathbb{R}^N, \mathbb{R}^M)$ that converges in the $\mathcal{E}$-norm to a function $f : \mathbb{R}^N \to \mathbb{R}^M$, then the sequence converges uniformly on compact subsets of $\mathbb{R}^N$. Using the usual theorems on uniform convergence, it follows from this fact that the limit function f is continuous on $\mathbb{R}^N$. Also, there is a sufficiently large integer n such that $\|f - f_n\|_{\mathcal{E}} < 1$. Thus, for this choice of n, we have that

$$\|f\|_{\mathcal{E}} \leq \|f - f_n\|_{\mathcal{E}} + \|f_n\|_{\mathcal{E}} < 1 + \|f_n\|_{\mathcal{E}},$$

and as a result we see that the $\mathcal{E}$-norm of f is bounded.

To show that $\mathcal{E}^0(\mathbb{R}^N, \mathbb{R}^M)$ is a Banach space, we must show that it is complete. To this end, suppose that the above sequence is Cauchy. We will show that the sequence converges to a function $f : \mathbb{R}^N \to \mathbb{R}^M$ with $f(0) = 0$. By the facts claimed above, we must then have that $f \in \mathcal{E}^0(\mathbb{R}^N, \mathbb{R}^M)$, as required.

Let us define a function $f : \mathbb{R}^N \to \mathbb{R}^M$. First, set $f(0) = 0$. If $\xi \in \mathbb{R}^N$ is not the zero vector, let $\epsilon > 0$ be given and note that there is an integer J such that

$$\frac{\|f_m(\xi) - f_n(\xi)\|}{\|\xi\|} < \frac{\epsilon}{\|\xi\|}$$

whenever m and n exceed J. Thus, the sequence $\{f_n(\xi)\}_{n=1}^{\infty}$ is a Cauchy sequence in $\mathbb{R}^M$, and hence it has a limit that we define to be $f(\xi)$.

We claim that the sequence $\{f_n\}_{n=1}^{\infty}$ converges to the function f in the $\mathcal{E}$-norm. To prove the claim, let $\epsilon > 0$ be given. There is an integer J, as before, such that, if $\xi \neq 0$, then

$$\frac{\|f_n(\xi) - f_p(\xi)\|}{\|\xi\|} < \frac{\epsilon}{2}$$

whenever the integers n and p exceed J. It follows that if $\xi \in \mathbb{R}^N$, including $\xi = 0$, then the inequality

$$\|f_n(\xi) - f_p(\xi)\| \leq \frac{\epsilon}{2}\|\xi\|$$

holds whenever n and p exceed J. Using this fact, we have the following estimates

$$\|f_n(\xi) - f(\xi)\| \leq \|f_n(\xi) - f_p(\xi)\| + \|f_p(\xi) - f(\xi)\|$$
$$\leq \frac{\epsilon}{2}\|\xi\| + \|f_p(\xi) - f(\xi)\|.$$

If we now pass to the limit as $p \to \infty$, we find that, for all $\xi \in \mathbb{R}^N$,

$$\frac{\|f_n(\xi) - f(\xi)\|}{\|\xi\|} \leq \frac{\epsilon}{2} < \epsilon$$

whenever n exceeds J. Thus, we have that $\|f_n - f\|_{\mathcal{E}} < \epsilon$ whenever n exceeds J, and therefore the sequence converges to f in the $\mathcal{E}$-norm.

To finish the proof, we must show that convergence in the $\mathcal{E}$-norm is uniform on compact sets. To this end, suppose that $\{f_n\}_{n=1}^{\infty}$ converges to f in the $\mathcal{E}$-norm, let $\mathcal{K}$ be a compact subset of $\mathbb{R}^N$, and let $\epsilon > 0$ be given. Also, let us define $r := \sup_{\xi \in \mathcal{K}} \|\xi\|$. There is an integer J such that if $\xi \neq 0$, then

$$\frac{\|f_n(\xi) - f(\xi)\|}{\|\xi\|} < \frac{\epsilon}{r+1}$$

whenever n exceeds J. Hence, as before, if $\xi \in \mathcal{K}$, then

$$\|f_n(\xi) - f(\xi)\| \leq \frac{\epsilon}{r+1}\|\xi\| \leq \epsilon \frac{r}{r+1} < \epsilon$$

whenever n exceeds J. It follows that the convergence is uniform on the compact set $\mathcal{K}$. This completes the proof of Proposition A.

Let us define two subsets of the Banach spaces defined above as follows:

$$\mathcal{B}_\rho^0(\mathbb{R}^N, \mathbb{R}^M) := \{f \in \mathcal{E}^0(\mathbb{R}^N, \mathbb{R}^M) : \mathrm{Lip}(f) \leq \rho\},$$

$$\mathcal{B}_\delta^1(\mathbb{R}^N, \mathbb{R}^M) := \{f \in \mathcal{C}^1(\mathbb{R}^N, \mathbb{R}^M) : \|f\|_1 < \delta\}.$$

The set $\mathcal{B}_\rho^0(\mathbb{R}^N, \mathbb{R}^M)$ is a closed (in fact, compact) subset of $\mathcal{E}^0(\mathbb{R}^N, \mathbb{R}^M)$, while the set $\mathcal{B}_\delta^1(\mathbb{R}^N, \mathbb{R}^M)$ is an open subset of $\mathcal{C}^1(\mathbb{R}^N, \mathbb{R}^M)$. Moreover, the set $\mathcal{B}_\rho^0(\mathbb{R}^N, \mathbb{R}^M)$ is a complete metric space with respect to the metric given by the $\mathcal{E}$-norm.

Fix $\rho > 0$. If $\delta > 0$, $F \in \mathcal{B}_\delta^1(\mathbb{R}^k \times \mathbb{R}^\ell, \mathbb{R}^k)$, and $\alpha \in \mathcal{B}_\rho^0(\mathbb{R}^k, \mathbb{R}^\ell)$, then the differential equation

$$\dot{x} = Sx + F(x, \alpha(x)) \tag{4.3}$$

has a continuous flow. In fact, for each $\xi \in \mathbb{R}^k$, there is a solution $t \mapsto x(t, \xi, \alpha)$ such that $x(0, \xi, \alpha) = \xi$ and such that $(t, \xi, \alpha) \mapsto x(t, \xi, \alpha)$ is a continuous function.

To compress notation, let

$$\chi(t) := x(t, \xi, \alpha)$$

and note that the function $t \mapsto \chi(t)$ is defined for all $t \geq 0$.

By the hypotheses of the theorem, there is a constant $K > 0$ such that for all $\xi \in \mathbb{R}^k$ and for all $\nu \in \mathbb{R}^\ell$, we have the following exponential estimates

$$\|e^{tS}\xi\| \leq Ke^{at}\|\xi\|, \qquad \|e^{-tU}\nu\| \leq Ke^{-bt}\|\nu\|$$

for all $t \geq 0$. A direct proof of these estimates under the spectral gap condition is similar to the proof of Theorem 2.61. These estimates can also be obtained as a corollary to this theorem. Hint: Apply Theorem 2.61 to new operators $S + cI$ and $U + cI$, where c is a real number chosen so that there is a spectral gap containing the origin.

Using the inequality $\|DF\|_0 < \delta$ and the mean value theorem, we have that

$$\|F(x, y)\| = \|F(x, y) - F(0, 0)\| \leq \delta(\|x\| + \|y\|)$$

where we are using the sum of the norms on each factor for the norm on the product space $\mathbb{R}^k \times \mathbb{R}^\ell$. Also, after obtaining a similar estimate for α, and combining these estimates, it follows that

$$\|F(x, \alpha(x))\| \leq \delta(1 + \rho)\|x\|.$$

By an application of the variation of parameters formula given in Proposition (2.67), the function χ satisfies the integral equation

$$\chi(t) = e^{tS}\xi + \int_0^t e^{(t-\tau)S} F(\chi(\tau), \alpha(\chi(\tau))) \, d\tau \tag{4.4}$$

from which we obtain the estimate

$$\|\chi(t)\| \leq Ke^{at}\|\xi\| + \int_0^t K\delta(1 + \rho)e^{a(t-\tau)}\|\chi(\tau)\| \, d\tau.$$

Equivalently, we have

$$e^{-at}\|\chi(t)\| \leq K\|\xi\| + \int_0^t K\delta(1 + \rho)e^{-a\tau}\|\chi(\tau)\| \, d\tau,$$

and by an application of Gronwall's inequality, we obtain the estimate

$$\|x(t, \xi)\| = \|\chi(t)\| \leq K\|\xi\|e^{(K\delta(1+\rho)+a)t}. \tag{4.5}$$

In particular, the solution $t \mapsto x(t, \xi)$ does not blow up in finite time. Hence, it is defined for all $t \geq 0$.

For $\alpha \in \mathcal{B}_\rho^0(\mathbb{R}^k, \mathbb{R}^\ell)$ and $G \in \mathcal{B}_\delta^1(\mathbb{R}^k \times \mathbb{R}^\ell, \mathbb{R}^\ell)$, if the graph

$$\mathcal{M}_\alpha := \{(x, y) \in \mathbb{R}^k \times \mathbb{R}^\ell : y = \alpha(x)\}$$

is an invariant set for the system (4.2), then the function

$$t \mapsto y(t, \xi, \alpha) := \alpha(x(t, \xi, \alpha))$$

is a solution of the differential equation

$$\dot{y} = Uy + G(x, y) \tag{4.6}$$

with initial condition $y(0, \xi, \alpha) = \alpha(\xi)$. Equivalently, by variation of parameters and with the notational definition $\gamma(t) := y(t, \xi, \alpha)$, we have the equation

$$e^{-tU}\gamma(t) - \alpha(\xi) = \int_0^t e^{-\tau U} G(\chi(\tau), \alpha(\chi(\tau)))\, d\tau.$$

Note that

$$\|e^{-tU}\gamma(t)\| \le Ke^{-bt}\rho\|\chi(t)\| \le K^2\rho\|\xi\|e^{(K\delta(1+\rho)+a-b)t}. \tag{4.7}$$

Thus, using the inequality $a - b < 0$, if we choose δ so that

$$0 < \delta < \frac{b - a}{K(1 + \rho)},$$

then $\lim_{t\to\infty} \|e^{-tU}\gamma(t)\| = 0$ and

$$\alpha(\xi) = -\int_0^\infty e^{-\tau U} G(\chi(\tau), \alpha(\chi(\tau)))\, d\tau. \tag{4.8}$$

Conversely, if $\alpha \in \mathcal{B}_\rho^0(\mathbb{R}^k, \mathbb{R}^\ell)$ satisfies the integral equation (4.8), then the graph of α is an invariant manifold. To see this, consider a point $(\xi, \alpha(\xi))$ on the graph of α, and redefine $\chi(t, \xi) := x(t, \xi, \alpha)$ and $\gamma(t) := \alpha(\chi(t, \xi))$. We will show that γ is a solution of the differential equation (4.6). Indeed, from the integral equation (4.8), we have that

$$\frac{d}{dt}\left(e^{-tU}\gamma(t)\right) = -\frac{d}{dt}\int_0^\infty e^{-(t+\tau)U} G(\chi(\tau, \chi(t, \xi)), \alpha(\chi(\tau, \chi(t, \xi))))\, d\tau$$

$$= -\frac{d}{dt}\int_0^\infty e^{-(t+\tau)U} G(\chi(\tau + t, \xi), \alpha(\chi(\tau + t, \xi)))\, d\tau$$

$$= -\frac{d}{dt}\int_t^\infty e^{-sU} G(\chi(s, \xi), \alpha(\chi(s, \xi)))\, ds$$

$$= e^{-tU} G(\chi(t, \xi), \gamma(t)).$$

In other words,

$$e^{-tU}\dot{\gamma}(t) - e^{-tU}U\gamma(t) = e^{-tU}G(\chi(t,\xi),\gamma(t)),$$

and therefore γ is a solution of the differential equation (4.6), as required.

Proposition B: Suppose that $\rho > 0$. If $\delta > 0$ is sufficiently small, $F \in \mathcal{B}_\delta^1(\mathbb{R}^k \times \mathbb{R}^\ell, \mathbb{R}^k)$, and $G \in \mathcal{B}_\delta^1(\mathbb{R}^k \times \mathbb{R}^\ell, \mathbb{R}^\ell)$, then the *Lyapunov–Perron operator* Λ defined by

$$\Lambda(\alpha)(\xi) := -\int_0^\infty e^{-tU}G(x(t,\xi,\alpha), \alpha(x(t,\xi,\alpha)))\, dt$$

is a contraction on the complete metric space $\mathcal{B}_\rho^0(\mathbb{R}^k, \mathbb{R}^\ell)$.

Let us first prove that, for sufficiently small $\delta > 0$, the Lyapunov-Perron operator Λ maps the metric space $\mathcal{B}_\rho^0(\mathbb{R}^k, \mathbb{R}^\ell)$ into itself. To show that $\Lambda(\alpha)$ is Lipschitz with $\mathrm{Lip}(\Lambda(\alpha)) \le \rho$, consider $\xi, \eta \in \mathbb{R}^k$ and note that

$$\|\Lambda(\alpha)(\xi) - \Lambda(\alpha)(\eta)\| \le K(1+\rho)\|G\|_1 \int_0^\infty e^{-bt}\|x(t,\xi,\alpha) - x(t,\eta,\alpha)\|\, dt.$$

Using the integral equation (4.4), we have the estimate

$$\|x(t,\xi,\alpha) - x(t,\eta,\alpha)\|$$
$$\le Ke^{at}\|\xi - \eta\| + \int_0^t K\|F\|_1(1+\rho)e^{a(t-\tau)}\|x(\tau,\xi,\alpha) - x(\tau,\eta,\alpha)\|\, d\tau.$$

After multiplying both sides of this last inequality by e^{-at} and applying Gronwall's inequality, we have that

$$\|x(t,\xi,\alpha) - x(t,\eta,\alpha)\| \le K\|\xi - \eta\|e^{(K\|F\|_1(1+\rho)+a)t}. \tag{4.9}$$

Returning to the original estimate, let us substitute the inequality (4.9) and carry out the resulting integration to obtain the inequality

$$\|\Lambda(\alpha)(\xi) - \Lambda(\alpha)(\eta)\| \le \frac{K^2\delta(1+\rho)}{b - a - K\delta(1+\rho)}\|\xi - \eta\|. \tag{4.10}$$

If $\|F\|_1$ and $\|G\|_1$ are sufficiently small, that is, if $\delta > 0$ is sufficiently small, then it follows that $\Lambda(\alpha)$ is a Lipschitz continuous function with Lipschitz constant less than ρ. In fact, it suffices to take

$$0 < \delta < \min\left\{\frac{b-a}{K(1+\rho)}, \frac{(b-a)\rho}{K(1+\rho)(K+\rho)}\right\}.$$

If $\delta > 0$ is less than the first element in the brackets, then the denominator of the fraction in inequality (4.10) is positive. If $\delta > 0$ is less than the second element, then the fraction is less than ρ. Moreover, if we take $\xi = 0$, then

$x(t, 0, \alpha) \equiv 0$ is the corresponding solution of the differential equation (4.3), and it follows that $\Lambda(\alpha)(0) = 0$.

To show that $\|\Lambda(\alpha)\|_{\mathcal{E}} < \infty$, let us use the estimate (4.10) with $\eta = 0$ to get

$$\sup_{\xi \neq 0} \frac{\|\Lambda(\alpha)(\xi)\|}{\|\xi\|} \leq \frac{K^2 \delta (1 + \rho)}{b - a - K\delta(1 + \rho)} < \infty.$$

This completes the proof that Λ is a transformation of the complete metric space $\mathcal{B}_\rho^0(\mathbb{R}^k, \mathbb{R}^\ell)$ into itself.

It remains to show that Λ is a contraction. By definition, the norm on the product $\mathbb{R}^k \times \mathbb{R}^\ell$ is the sum of the Euclidean norms on the factors. Thus, because $\|G\|_1$ is finite and the Lipschitz constant for all functions in the space $\mathcal{B}_\rho^0(\mathbb{R}^k, \mathbb{R}^\ell)$ does not exceed ρ, we obtain the inequalities

$$\|\Lambda(\alpha)(\xi) - \Lambda(\beta)(\xi)\|$$
$$\leq K \int_0^\infty e^{-bt} \|G(x(t, \xi, \alpha), \alpha(x(t, \xi, \alpha))) - G(x(t, \xi, \beta), \beta(x(t, \xi, \beta)))\| \, dt$$
$$\leq K\|G\|_1 \int_0^\infty e^{-bt} (\|x(t, \xi, \alpha) - x(t, \xi, \beta)\|$$
$$\qquad + \|\alpha(x(t, \xi, \alpha)) - \alpha(x(t, \xi, \beta))\| + \|\alpha(x(t, \xi, \beta)) - \beta(x(t, \xi, \beta))\| \, dt$$
$$\leq K\|G\|_1 \int_0^\infty e^{-bt} ((1 + \rho)\|x(t, \xi, \alpha) - x(t, \xi, \beta)\|$$
$$\qquad + \|\alpha - \beta\|_{\mathcal{E}} \|x(t, \xi, \beta)\|) \, dt. \tag{4.11}$$

To estimate the terms in the integrand of the last integral in the display (4.11), let us use the integral equation (4.4) to obtain the estimate

$$\|x(t, \xi, \alpha) - x(t, \xi, \beta)\| \leq$$
$$K \int_0^t e^{a(t-\tau)} \|F(x(\tau, \xi, \alpha), \alpha(x(\tau, \xi, \alpha))) - F(x(\tau, \xi, \beta), \beta(x(\tau, \xi, \beta)))\| \, d\tau.$$

Then, by proceeding exactly as in the derivation of the estimate (4.11), it is easy to show that

$$\|x(t, \xi, \alpha) - x(t, \xi, \beta)\|$$
$$\leq K\|F\|_1 \int_0^t e^{a(t-\tau)}(1 + \rho)\|x(\tau, \xi, \alpha) - x(\tau, \xi, \beta)\| \, d\tau$$
$$\qquad + K\|F\|_1 \|\alpha - \beta\|_{\mathcal{E}} \int_0^t \|x(\tau, \xi, \beta)\| \, d\tau.$$

After inserting the inequality (4.5), integrating the second integral, and multiplying both sides of the resulting inequality by e^{-at}, we find that

$$e^{-at}\|x(t,\xi,\alpha) - x(t,\xi,\beta)\| \leq \int_0^t K(1+\rho)\|F\|_1 e^{-a\tau}\|x(\tau,\xi,\alpha) - x(\tau,\xi,\beta)\|\,d\tau$$
$$+ \frac{K\|F\|_1\|\xi\|}{\delta(1+\rho)}\|\alpha - \beta\|_\varepsilon \big(e^{K\delta(1+\rho)t} - 1\big).$$

Then, an application of Gronwall's inequality followed by some algebraic manipulations can be used to show the estimate

$$\|x(t,\xi,\alpha) - x(t,\xi,\beta)\| \leq \frac{K}{1+\rho}\|\alpha - \beta\|_\varepsilon\|\xi\|e^{(2K\delta(1+\rho)+a)t}.$$

Returning to the main estimate, if we insert the last inequality as well as the inequality (4.5), then an integration together with some obvious manipulations yields the estimate

$$\|\Lambda(\alpha) - \Lambda(\beta)\|_\varepsilon \leq \frac{2K^2\delta}{b - a - 2K\delta(1+\rho)}\|\alpha - \beta\|_\varepsilon.$$

Moreover, if

$$0 < \delta < \min\left\{\frac{b-a}{2K(1+\rho)}, \frac{b-a}{2K(K+1+\rho)}\right\},$$

then Λ has a contraction constant strictly less than one, as required.

Taking into account all the restrictions on δ, if

$$0 < \delta < \frac{b-a}{K}\min\left\{\frac{\rho}{(1+\rho)(K+\rho)}, \frac{1}{2(1+\rho)}, \frac{1}{2(K+1+\rho)}\right\},$$

then the Lyapunov–Perron operator has a fixed point whose graph is a Lipschitz continuous invariant manifold that passes through the origin. This completes the proof of Proposition B.

We will use the fiber contraction principle to prove the smoothness of the invariant manifold that is the graph of the function α obtained as the fixed point of the Lyapunov–Perron operator (see [191] for a similar proof for the case of diffeomorphisms). To this end, let us follow the prescription outlined after the proof of the fiber contraction theorem (Theorem 1.247).

The space of "candidates for the derivatives of functions in $\mathcal{B}_\rho^0(\mathbb{R}^k, \mathbb{R}^\ell)$" is, in the present case, the set $\mathcal{F} = \mathcal{C}^0(\mathbb{R}^k, L(\mathbb{R}^k, \mathbb{R}^\ell))$ of all bounded continuous functions Φ that map $\mathbb{R}^k$ into the linear maps from $\mathbb{R}^k$ into $\mathbb{R}^\ell$ with $\Phi(0) = 0$ and with the norm

$$\|\Phi\|_\mathcal{F} := \sup_{\xi\in\mathbb{R}^k}\|\Phi(\xi)\|,$$

where $\|\Phi(\xi)\|$ denotes the usual operator norm of the linear transformation $\Phi(\xi)$. Also, let $\mathcal{F}_\rho$ denote the closed ball in $\mathcal{F}$ with radius ρ, that is,

$$\mathcal{F}_\rho := \{\Phi \in \mathcal{F} : \|\Phi\| \le \rho\}$$

where $\rho > 0$ is the number chosen in the first part of the proof.

Proposition C: Suppose that $\beta \in \mathcal{B}_\rho^0(\mathbb{R}^k, \mathbb{R}^\ell)$, the function $t \mapsto x(t, \xi, \beta)$ is the solution of the differential equation (4.3) with parameter β and initial condition $x(0, \xi, \beta) = \xi$, and $\Phi \in \mathcal{F}_\rho$. If $\|F\|_1$ and $\|G\|_1$ are both sufficiently small, then Ψ given by

$$
\begin{aligned}
\Psi(\beta, \Phi)(\xi) := -\int_0^\infty e^{-tU} [& G_x(x(t, \xi, \beta), \beta(x(t, \xi, \beta)))W(t, \xi, \beta, \Phi) \\
& + G_y(x(t, \xi, \beta), \beta(x(t, \xi, \beta)))\Phi(x(t, \xi, \beta))W(t, \xi, \beta, \Phi)]\, dt,
\end{aligned}
$$
(4.12)

where $t \mapsto W(t, \xi, \beta, \Phi)$ is the solution of the initial value problem

$$
\begin{aligned}
\dot{w} = Sw &+ F_x(x(t, \xi, \beta), \beta(x(t, \xi, \beta)))w \\
&+ F_y(x(t, \xi, \beta), \beta(x(t, \xi, \beta)))\Phi(x(t, \xi, \beta))w, \\
w(0) &= I,
\end{aligned}
$$
(4.13)

defines a function from $\mathcal{B}_\rho^0 \times \mathcal{F}_\rho$ to $\mathcal{F}_\rho$. If, in addition, β is continuously differentiable and if Λ denotes the Lyapunov–Perron operator, then $D\Lambda(\beta) = \Psi(\beta, D\beta)$.

To prove Proposition C, let us note first that W is continuous, and it satisfies the integral equation

$$
\begin{aligned}
W(t, \xi, \beta, \Phi) := e^{tS} &+ \int_0^t e^{(t-s)S}[F_x(x(s, \xi, \beta), \beta(x(s, \xi, \beta)))W(s, \xi, \beta, \Phi) \\
&+ F_y(x(s, \xi, \beta), \beta(x(s, \xi, \beta)))\Phi(x(s, \xi, \beta))W(s, \xi, \beta, \Phi)]\, ds.
\end{aligned}
$$
(4.14)

The integral equation is simply obtained by applying the variation of parameters formula to the differential equation (4.13); the continuity of W follows from the continuity of the solutions of differential equations with respect to parameters because the right hand side of the differential equation is continuous in (t, ξ, β, Φ). This fact is not obvious. But, it can be proved from the observation that the terms on the right hand side of the differential equation can all be rewritten as compositions of continuous functions. For example, to show that

$$(t, \xi, \beta, \Phi) \mapsto \Phi(x(t, \xi, \beta))$$

is continuous, note that x is continuous and the function $\mathbb{R}^k \times \mathcal{F}_\rho \to L(\mathbb{R}^k, \mathbb{R}^\ell)$ given by $(\zeta, \Phi) \mapsto \Phi(\zeta)$ is continuous. The continuity of the last function follows from the continuity of the elements of $\mathcal{F}$ and the estimate

$$\|\Phi_1(\zeta) - \Phi_2(\eta)\| \le \|\Phi_1(\zeta) - \Phi_1(\eta)\| + \|\Phi_1 - \Phi_2\|_{\mathcal{F}}.$$

We will show that the improper integral in the definition of Ψ is convergent. Using the hypotheses of the theorem and estimating in the usual manner, we have that

$$\|\Psi(\beta, \Phi)(\xi)\| \leq \int_0^\infty Ke^{-bt}\|G\|_1(1+\rho)\|W(t, \xi, \beta, \Phi)\| \, dt. \qquad (4.15)$$

An upper bound for $\|W(t, \xi, \beta, \Phi)\|$ can be obtained from the integral equation (4.14). In fact, estimating once again in the usual manner, we have

$$\|W(t, \xi, \beta, \Phi)\| \leq Ke^{at} + \int_0^t K\|F\|_1(1+\rho)e^{(t-s)a}\|W(s, \xi, \beta, \Phi)\| \, ds.$$

After multiplying both sides of this last inequality by e^{-at} and then applying Gronwall's inequality, we obtain the estimate

$$\|W(t, \xi, \beta, \Phi)\| \leq Ke^{(K\|F\|_1(1+\rho)+a)t}. \qquad (4.16)$$

If the inequality (4.16) is inserted into the estimate (4.15), with

$$\|F\|_1 \leq \delta < \frac{b-a}{K(1+\rho)},$$

and the resulting integral is evaluated, then we have that

$$\|\Psi(\beta, \Phi)(\xi)\| \leq \frac{K^2\|G\|_1(1+\rho)}{b-a-K\delta(1+\rho)}. \qquad (4.17)$$

Thus, the original integral converges. Moreover, if the quantity $\|G\|_1$ is sufficiently small—the upper bound

$$\|G\|_1 \leq \delta \leq \frac{\rho(b-a-K\delta(1+\rho))}{K^2(1+\rho)}$$

will suffice—then

$$\|\Psi(\beta, \Phi)(\xi)\| \leq \rho.$$

Finally, it is easy to check that $\Psi(\beta, \Phi)(0) = 0$. Therefore, $\Psi(\beta, \Phi) \in \mathcal{F}_\rho$, as required.

If β is continuously differentiable, then the solution $t \mapsto x(t, \xi, \beta)$ of the differential equation (4.3) given by

$$\dot{x} = Sx + F(x, \beta(x))$$

is continuously differentiable by Theorem 1.261. Moreover, if we define $\Phi := D\beta$, then $W(t, \xi, \beta) := x_\xi(t, \xi, \beta)$ (the solution of the first variational equation of the differential equation (4.3)) is the corresponding solution of the integral equation (4.14). In this case, the integrand of the integral expression for $\Lambda(\beta)(\xi)$ is clearly a differentiable function of ξ with derivative exactly the integrand of the integral expression for $\Psi(\beta, D\beta)(\xi)$. As we have shown above, this integrand is bounded above by an integrable function. Thus, differentiation under the integral sign is justified, and in fact, $D\Lambda(\beta) = \Psi(\beta, D\beta)$, as required. This completes the proof of the proposition.

Let us show that

$$\Gamma : \mathcal{B}_\rho^0(\mathbb{R}^k, \mathbb{R}^\ell) \times \mathcal{F}_\rho \to \mathcal{B}_\rho^0(\mathbb{R}^k, \mathbb{R}^\ell) \times \mathcal{F}_\rho$$

given by $(\beta, \Phi) \mapsto (\Lambda(\beta), \Psi(\beta, \Phi))$, is a continuous fiber contraction. For this, let $\beta \in \mathcal{B}_\rho^0(\mathbb{R}^k, \mathbb{R}^\ell)$ and consider the estimates (analogous to those made previously) given by

$$\|\Psi(\beta, \Phi_1)(\xi) - \Psi(\beta, \Phi_2)(\xi)\|$$
$$\leq K \int_0^\infty e^{-bt} (\|G_x W + G_y \Phi_1 W - G_x W - G_y \Phi_2 W)\| \, dt$$
$$\leq K \int_0^\infty e^{-bt} \|G\|_1 \|\Phi_1 - \Phi_2\|_{\mathcal{F}} |W(t, \xi, \beta, \Phi_2)| \, dt, \qquad (4.18)$$

where, for notational convenience, the arguments of some of the functions in the integrands are suppressed.

Using the estimate (4.16), we have that

$$\|\Psi(\beta, \Phi_1)(\xi) - \Psi(\beta, \Phi_2)(\xi)\|$$
$$\leq K^2 \|G\|_1 \int_0^\infty e^{(K\|F\|_1(1+\rho) + a - b)t} \, dt \, \|\Phi_1 - \Phi_2\|_{\mathcal{F}}$$
$$\leq \frac{K^2 \|G\|_1}{b - a - K\|F\|_1(1 + \rho)} \|\Phi_1 - \Phi_2\|_{\mathcal{F}}. \qquad (4.19)$$

Thus, if

$$0 < \delta < \frac{b - a}{K(K + 1 + \rho)},$$

$\|F\|_1 \leq \delta$, and $\|G\|_1 \leq \delta$, then

$$0 < \frac{2K^2 \|F\|_1}{b - a - 2K\|G\|_1(1 + \rho)} < 1,$$

and therefore Γ is a fiber contraction.

We must show that Γ is continuous. As in the proof of Theorem 1.249, it suffices to show that the map Ψ is continuous with respect to its first argument. This result follows from the next proposition.

Proposition D: Suppose that $\Phi \in \mathcal{F}_\rho$, $H : \mathbb{R}^k \times \mathbb{R}^\ell \to L(\mathbb{R}^\ell, \mathbb{R}^\ell)$, and $K : \mathbb{R}^k \times \mathbb{R}^\ell \to L(\mathbb{R}^k, \mathbb{R}^\ell)$. If H and K are bounded continuous functions, then the functions $\Delta : \mathcal{B}_\rho^0(\mathbb{R}^k, \mathbb{R}^\ell) \to \mathcal{F}_\rho$, given by

$$\Delta(\beta)(\xi) = \int_0^\infty e^{-tU} H(x(t, \xi, \beta), \beta(x(t, \xi, \beta)))\Phi(x(t, \xi, \beta))W(t, \xi, \beta, \Phi))\, dt,$$

and $\Upsilon : \mathcal{B}_\rho^0(\mathbb{R}^k, \mathbb{R}^\ell) \to \mathcal{F}_\rho$, given by

$$\Upsilon(\beta)(\xi) = \int_0^\infty e^{-tU} K(x(t, \xi, \beta), \beta(x(t, \xi, \beta)))W(t, \xi, \beta, \Phi))\, dt,$$

where W is defined in Proposition C, are continuous.

The proof of Proposition D is rather long; we will outline the ideas of the proof for the function Δ.

It suffices to show that Δ is continuous at each point of its domain; that is, for each $\alpha \in \mathcal{B}_\rho^0(\mathbb{R}^k, \mathbb{R}^\ell)$ and each $\epsilon > 0$, there is some $\delta > 0$ such that $\|\Delta(\beta) - \Delta(\alpha)\|_\mathcal{F} < \epsilon$ whenever $\|\beta - \alpha\|_\varepsilon < \delta$.

For notational convenience, let

$$h(t, \xi, \beta) := H(x(t, \xi, \beta), \beta(x(t, \xi, \beta))).$$

By using the definition of Δ, the hyperbolic estimates, and the triangle inequality, the proof is reduced to showing that the supremum over $\xi \in \mathbb{R}^k$ of each of the three integrals

$$I_1 := K \int_0^\infty e^{-bt} \|H\|_0 \|\Phi\|_\mathcal{F} \|W(t, \xi, \beta) - W(t, \xi, \alpha)\|\, dt,$$

$$I_2 := K \int_0^\infty e^{-bt} \|H\|_0 \|\Phi(x(t, \xi, \beta)) - \Phi(x(t, \xi, \alpha))\| \|W(t, \xi, \alpha)\|\, dt,$$

$$I_3 := K \int_0^\infty e^{-bt} \|H(t, \xi, \beta) - H(t, \xi, \alpha)\| \|\Phi\|_\mathcal{F} \|W(t, \xi, \alpha)\|\, dt,$$

is less than $\epsilon/3$ whenever $\|\beta - \alpha\|_\varepsilon$ is sufficiently small.

The inequality (4.16) implies the convergence of the integrals I_1, I_2, and I_3. Thus, there is some $T > 0$ such that the supremum over $\xi \in \mathbb{R}^k$ of each corresponding integral over $[T, \infty]$ is less than $\epsilon/6$ uniformly with respect to β. With this estimate in hand, it suffices to show that the supremum over $\xi \in \mathbb{R}^k$ of each corresponding integral over the compact interval $[0, T]$ is bounded above by $\epsilon/6$ for sufficiently small $\|\beta - \alpha\|_\varepsilon$. In other words, it suffices to show the continuity with respect to β of integrals over a finite time interval. For example, for I_1 it suffices to show that the function $h : \mathcal{B}_\rho^0(\mathbb{R}^k, \mathbb{R}^\ell) \to \mathcal{F}$ given by

$$h(\beta)(\xi) := \int_0^T e^{-bt} W(t, \xi, \beta)\, dt$$

is continuous. This result requires a technical observation. From advanced calculus, an integral of the form

$$\int_0^T \omega(t, \beta)\, dt$$

is continuous with respect to β provided that ω is continuous. But, this theorem is usually stated for $\omega : \mathbb{R} \times \mathbb{R}^m \to \mathbb{R}^n$; that is, for β a variable in a *finite-dimensional* space. The usual proof is to argue that the corresponding function of β is continuous at each point $\alpha \in \mathbb{R}^m$ by using the uniform continuity of ω on a compact set of the form $[0, T] \times B$ where B is a closed ball in $\mathbb{R}^m$ with center at a point α. But, in our case, the set $\mathcal{B}_\rho^0(\mathbb{R}^k, \mathbb{R}^\ell)$ is not compact because it is a ball in an infinite-dimensional Banach space. A proof that includes infinite-dimensional parameters is only slightly more complicated; it uses the compactness of $[0, T]$. The idea is simple: Pick $\alpha \in \mathcal{B}_\rho^0(\mathbb{R}^k, \mathbb{R}^\ell)$ and $\epsilon > 0$; and, for each $t \in [0, T]$, use the continuity of ω to find a product neighborhood $\{s : |s - t| < \delta_t\} \times \{\beta : \|\beta - \alpha\|_\varepsilon < \delta_t\}$ in $\mathbb{R} \times \mathcal{B}_\rho(\mathbb{R}^k, \mathbb{R}^\ell)$ on which

$$\int_0^T \|\omega(s, \beta) - \omega(t, \alpha)\|\, dt < \epsilon/T.$$

Because $[0, T]$ is compact, there are finitely many such neighborhoods that cover $[0, T] \times \{\alpha\}$. For a positive δ less than the minimum of the corresponding $\{\delta_{t_1}, \delta_{t_2}, \dots, \delta_{t_N}\}$, we have that

$$\|\omega(t, \beta) - \omega(t, \alpha)\| < \epsilon/T$$

whenever $\|\beta - \alpha\|_\varepsilon < \delta$, as required.

To complete the proof of Proposition D, it suffices to show that the function $\omega : [0, T] \times \mathcal{B}_\rho^0(\mathbb{R}^k, \mathbb{R}^\ell) \to \mathcal{F}$, given by $\omega(t, \beta)(\xi) = W(t, \xi, \beta)$, the function $p : [0, T] \times \mathcal{B}_\rho^0(\mathbb{R}^k, \mathbb{R}^\ell) \to \mathcal{F}$, given by $p(t, \beta)(\xi) = \Phi(x(t, \xi, \beta))$, and the function $h : [0, T] \times \mathcal{B}_\rho^0(\mathbb{R}^k, \mathbb{R}^\ell) \to \mathcal{F}$, given by $h(t, \beta)(\xi) = H(t, \xi, \beta)$ are continuous. We have already indicated the idea for the proof of this fact for ω and p (see the discussion following display (4.14)); the proof for h is similar. This completes the outline of the proof of Proposition D.

Continuing with the proof of the theorem, let us define $(\phi_0, \Phi_0) = (0, 0) \in \mathcal{B}_\rho^0(\mathbb{R}^k, \mathbb{R}^\ell) \times \mathcal{F}_\rho$ and note that $D\phi_0 = \Phi_0$. Also, let us define recursively a sequence $\{(\phi_n, \Phi_n)\}_{n=0}^\infty$ by

$$(\phi_{n+1}, \Phi_{n+1}) := \Gamma(\phi_n, \Phi_n) = (\Lambda(\phi_n), \Psi(\phi_n, \Phi_n)).$$

If $D\phi_n = \Phi_n$, then, by Proposition C, $D\Lambda(\phi_n) = \Psi(\phi_n, \Phi_n)$ and $D\Lambda(\phi_n) \in \mathcal{F}_\rho$. Thus, $D\phi_{n+1} = D\Lambda(\phi_n) = \Psi(\phi_n, \Phi_n) = \Phi_{n+1}$. Moreover, if α is the fixed point of the Lyapunov–Perron operator, then by the fiber contraction theorem and the fact that $\mathcal{F}_\rho$ is a complete metric space, there is some $\Phi_\infty \in \mathcal{F}_\rho$ such that

$$\lim_{n \to \infty} \phi_n = \alpha, \qquad \lim_{n \to \infty} \Phi_n = \Phi_\infty.$$

The sequence $\{\phi_n\}_{n=0}^\infty$ converges in $\mathcal{E}^0(\mathbb{R}^k, \mathbb{R}^\ell)$ to α and its sequence of derivatives converges uniformly to a continuous function—an element of $\mathcal{F}_\rho$. By Theorem 1.248, α is continuously differentiable with derivative Φ_∞, provided that the convergence of the sequence $\{\phi_n\}_{n=0}^\infty$ is uniform. While the norm in $\mathcal{E}^0(\mathbb{R}^k, \mathbb{R}^\ell)$ is not the uniform norm, the convergence is uniform on compact subsets of $\mathbb{R}^k$. As differentiability and continuity are local properties, the uniform convergence on compact subsets of $\mathbb{R}^k$ of the sequence $\{\phi_n\}_{n=0}^\infty$ to α is sufficient to obtain the desired result: α is continuously differentiable with derivative Φ_∞.

For a direct proof that the function α is continuously differentiable, consider $\xi, h \in \mathbb{R}^k$ and note that, by the fundamental theorem of calculus, if n is a positive integer, then

$$\phi_n(\xi + h) - \phi_n(\xi) = \int_0^1 \frac{d}{dt}\phi_n(\xi + th)\, dt = \int_0^1 \Phi_n(\xi + th)h\, dt.$$

If we pass to the limit as $n \to \infty$ and use the uniform convergence of $\{\Phi_n\}_{n=0}^\infty$ to the continuous function Φ_∞, then we have the identity

$$\alpha(\xi + h) - \alpha(\xi) = \int_0^1 \Phi_\infty(\xi + th)h\, dt,$$

and consequently the estimate

$$\|\alpha(\xi + h) - \alpha(\xi) - \Phi_\infty(\xi)h\| \le \left\| \int_0^1 \Phi_\infty(\xi + th)h\, dt - \int_0^1 \Phi_\infty(\xi)h\, dt \right\|$$

$$\le \|h\| \int_0^1 \|\Phi_\infty(\xi + th) - \Phi_\infty(\xi)\|\, dt.$$

The Lebesgue dominated convergence theorem can be used to show that the last integral converges to zero as $h \to 0$. This proves that $D\alpha = \Phi_\infty$, as required. $\qquad \square$

As a remark, note that for the existence and smoothness of the invariant manifold in Theorem 4.1, both $\|F\|_1$ and $\|G\|_1$ are required to be less than the minimum of the numbers

$$\frac{(b - a)\rho}{K(1 + \rho)(K + \rho)}, \qquad \frac{b - a}{2K(1 + \rho)}, \qquad \frac{b - a}{2K(K + 1 + \rho)}.$$

Of course, if K is given, then there is an optimal value of ρ, namely, the value that makes the minimum of the three numbers as large as possible.

Theorem 4.1 requires that the nonlinear terms F and G in the differential equation (4.2) have sufficiently small C^1-norms over the entire product space $\mathbb{R}^k \times \mathbb{R}^\ell$. Clearly, we cannot expect a differential equation whose linearization at a rest point has a spectral gap to also have globally small corresponding nonlinear terms. To overcome this difficulty, we will use the following observation: By restricting attention to a sufficiently small open set containing the rest point, the C^1-norm of the nonlinear terms can be made as small as we like.

Suppose that the coordinates are already chosen so that the rest point is at the origin and the differential equation is given in a product neighborhood of the origin in the form

$$\dot{x} = Sx + f(x,y), \qquad \dot{y} = Uy + g(x,y) \tag{4.20}$$

where f and g, together with all their first partial derivatives, vanish at the origin.

Let $\delta > 0$ be given as in the proof of Theorem 4.1. Choose a ball B_r at the origin with radius $r > 0$ such that

$$\sup_{(x,y)\in B_r} \|Df(x,y)\| < \frac{\delta}{3}, \qquad \sup_{(x,y)\in B_r} \|Dg(x,y)\| < \frac{\delta}{3}.$$

Then, using the mean value theorem, we have that

$$\sup_{(x,y)\in B_r} \|f(x,y)\| < \frac{\delta r}{3}, \qquad \sup_{(x,y)\in B_r} \|g(x,y)\| < \frac{\delta r}{3}.$$

Moreover, there is a smooth "bump function" (cf. Exercise 4.17) $\gamma : \mathbb{R}^k \times \mathbb{R}^\ell \to \mathbb{R}$, also called in this context a "cut-off function," with the following properties:

(i) $\gamma(x,y) \equiv 1$ for $(x,y) \in B_{r/3}$;
(ii) the function γ vanishes on the complement of B_r in $\mathbb{R}^k \times \mathbb{R}^\ell$;
(iii) $\|\gamma\| = 1$ and $\|D\gamma\| \leq 2/r$.

With these constructions, it follows that

$$\|D(\gamma \cdot f)\| \leq \|D\gamma\|\|f\| + \|\gamma\|\|Df\| < \left(\frac{2}{r}\right)\left(\frac{\delta r}{3}\right) + \frac{\delta}{3} < \delta \tag{4.21}$$

with the same upper estimate for $\|D(\gamma \cdot G)\|$.

If we define $F(x,y) := \gamma(x,y)f(x,y)$ and $G(x,y) := \gamma(x,y)g(x,y)$, then the system

$$\dot{x} = Sx + F(x,y), \qquad \dot{y} = Uy + G(x,y)$$

has a global C^1 invariant manifold. The subset of this manifold that is contained in the ball $B_{r/3}$ is an invariant manifold for the system (4.20).

If the rest point is hyperbolic, so that $a < 0 < b$ in Theorem 4.1, then we have proved the existence and uniqueness of a stable manifold at the rest point. In particular, solutions starting on this invariant manifold converge to the origin as $t \to \infty$. To obtain the existence of an unstable manifold, simply reverse the direction of the independent variable, $t \to -t$, and apply Theorem 4.1 to the resulting differential equation.

Of course, the local invariant manifolds that are produced in the manner just described may very well be just small portions of the entire invariant manifolds at the rest point. It's just that one of the global invariant manifolds may not be the graph of a function. If $W^s_{loc}(0,0)$ denotes the local stable manifold for a rest point at the origin, and if ϕ_t denotes the flow of the corresponding differential equation, then we can define the stable manifold by

$$W^s(0,0) := \bigcup_{t \leq 0} \phi_t(W^s_{loc}(0,0)).$$

It can be shown that $W^s(0,0)$ is an immersed disk. A similar statement holds for the unstable manifold.

In case the rest point is not hyperbolic, let us consider the system

$$\dot{x} = Sx + f(x,y,z), \quad \dot{y} = Uy + g(x,y,z), \quad \dot{z} = Cz + h(x,y,z)$$

where we have already changed coordinates so that $(x,y,z) \in \mathbb{R}^k \times \mathbb{R}^\ell \times \mathbb{R}^m$ with the spectrum of S in the left half plane, the spectrum of U in the right half plane, and the spectrum of C on the imaginary axis. If, for example, we group the first and last equations so that the system is expressed in the form

$$\begin{pmatrix} \dot{x} \\ \dot{z} \end{pmatrix} = \begin{pmatrix} S & 0 \\ 0 & C \end{pmatrix} \begin{pmatrix} x \\ z \end{pmatrix} + \begin{pmatrix} f(x,y,z) \\ h(x,y,z) \end{pmatrix},$$

then we are in the situation of Theorem 4.1 where the corresponding spectral gap is bounded by $a = 0$ and some $b > 0$. An application of Theorem 4.1 produces a "center-stable manifold"—a manifold $W^{cs}(0,0)$ given as the graph of a smooth function $\alpha : \mathbb{R}^k \times \mathbb{R}^m \to \mathbb{R}^\ell$. Using a reversal of the independent variable and a second application of Theorem 4.1, let us produce a center-unstable manifold $W^{cu}(0,0)$ given as the graph of a smooth function $\omega : \mathbb{R}^\ell \times \mathbb{R}^m \to \mathbb{R}^k$. The intersection of these manifolds is denoted by $W^c(0,0)$ and is a center manifold for the original system (that is, an invariant manifold that is tangent to the eigenspace of the linear system corresponding to the eigenvalues with zero real parts). To prove this fact, we will show that $W^c(0,0)$ is given, at least locally, as the graph of a function $\mu : \mathbb{R}^m \to \mathbb{R}^k \times \mathbb{R}^\ell$ with $\mu(0) = 0$ and $D\mu(0) = 0$.

There seems to be a technical point here that depends on the choice of the number ρ. Recall that $\rho > 0$ was used in the proof of Theorem 4.1 as the bound on the Lipschitz constants for the functions considered as candidates for fixed points of the Lyapunov–Perron operator. If $0 < \rho < 1$, then we will show, as a corollary of Theorem 4.1, that there is a smooth global center manifold. If $\rho > 1$, then we will show that there is a local center manifold. Of course, in the proof of Theorem 4.1 we were free to choose $\rho < 1$ as long as we were willing to take the C^1-norms of the nonlinear terms sufficiently small, perhaps smaller than is required to prove the existence of the center-stable and the center-unstable manifolds.

Let us suppose that $0 < \rho < 1$. If there is a smooth function $\nu : \mathbb{R}^m \to \mathbb{R}^k$ with $\nu(0) = 0$ that satisfies the functional equation

$$\nu(z) = \omega(\alpha(\nu(z), z), z), \qquad (4.22)$$

then it is easy to check that $W^c(0,0)$ is the graph of the smooth function $\zeta \mapsto (\nu(\zeta), \alpha(\nu(\zeta), \zeta))$, as required.

In order to solve the functional equation, let us consider the Banach space $\mathcal{E}^0(\mathbb{R}^m, \mathbb{R}^k)$ with the $\mathcal{E}$-norm as defined in the proof of Theorem 4.1, the subset $\mathcal{B}_\rho^0(\mathbb{R}^m, \mathbb{R}^k)$ consisting of all elements of $\mathcal{E}^0(\mathbb{R}^m, \mathbb{R}^k)$ whose Lipschitz constants do not exceed ρ, and the operator Λ that is defined for functions in $\mathcal{B}_\rho^0(\mathbb{R}^m, \mathbb{R}^k)$ by

$$\Lambda(\nu)(\zeta) := \omega(\alpha(\nu(\zeta), \zeta), \zeta).$$

We are using the same symbol to denote this operator as we used to denote the Lyapunov–Perron operator because the proof that each of these operators has a smooth fixed function is essentially the same.

To show that Λ is a contraction on the complete metric space $\mathcal{B}_\rho^0(\mathbb{R}^m, \mathbb{R}^k)$, note that if $\nu \in \mathcal{B}_\rho^0(\mathbb{R}^m, \mathbb{R}^k)$, then $\Lambda(\nu)$ is continuous on $\mathbb{R}^m$. Moreover, it is easy to show the following inequality:

$$\|\Lambda(\nu)(\zeta_1) - \Lambda(\nu)(\zeta_2)\| \leq \mathrm{Lip}(\omega)\,\mathrm{Lip}(\alpha)\,\mathrm{Lip}(\nu)\|\zeta_1 - \zeta_2\| \leq \rho^3 \|\zeta_1 - \zeta_2\|.$$

In particular, we have that $\|\Lambda(\nu)(\zeta_1)\| \leq \rho^3 \|\zeta_1\|$. It now follows that $\|\Lambda(\nu)\|_\mathcal{E} < \infty$ and $\mathrm{Lip}(\Lambda(\nu)) \leq \rho$. Thus, Λ maps the complete metric space $\mathcal{B}_\rho^0(\mathbb{R}^m, \mathbb{R}^k)$ into itself. Also, we have that

$$\|\Lambda(\nu_1)(\zeta) - \Lambda(\nu_2)(\zeta)\| \leq \mathrm{Lip}(\omega)\,\mathrm{Lip}(\alpha)\|\nu_1(\zeta) - \nu_2(\zeta)\| \leq \rho^2 \|\nu_1 - \nu_2\|_\mathcal{E}\|\zeta\|,$$

and, as a result, Λ is a contraction on $\mathcal{B}_\rho^0(\mathbb{R}^m, \mathbb{R}^k)$. Therefore, Λ has a globally attracting fixed point $\nu \in \mathcal{B}_\rho^0(\mathbb{R}^m, \mathbb{R}^k)$.

To show that ν is smooth, we can again use the fiber contraction principle. In fact, the proof is completely analogous to the proof of the smoothness of the invariant manifold in Theorem 4.1 (see also the discussion after the fiber contraction theorem (Theorem (1.247))). We will outline the main steps of the proof.

Consider the set $\mathcal{F} = \mathcal{C}^0(\mathbb{R}^m, L(\mathbb{R}^m, \mathbb{R}^k))$ of all bounded continuous functions Φ that map $\mathbb{R}^m$ into the bounded linear maps from $\mathbb{R}^m$ into $\mathbb{R}^k$ with $\Phi(0) = 0$ and with the norm

$$\|\Phi\|_{\mathcal{F}} := \sup_{\xi \in \mathbb{R}^k} \|\Phi(\xi)\|$$

where, as before, $\|\Phi(\xi)\|$ denotes the operator norm of the transformation $\Phi(\xi)$. Also, let $\mathcal{F}_\rho$ denote the closed ball in $\mathcal{F}$ with radius ρ, that is,

$$\mathcal{F}_\rho := \{\Phi \in \mathcal{F} : \|\Phi\| \leq \rho\}.$$

For $\phi \in \mathcal{B}_\rho^0(\mathbb{R}^m, \mathbb{R}^k)$ and for $\Phi \in \mathcal{F}_\rho$, let us define

$$\Psi(\phi, \Phi)(\zeta) := \omega_y(\alpha(\phi(\zeta), \zeta), \zeta)[\alpha_x(\alpha(\phi(\zeta), \zeta), \zeta)\Phi(\zeta) + \alpha_z(\phi(\zeta), \zeta), \zeta)]$$
$$+ \omega_z(\alpha(\phi(\zeta), \zeta), \zeta).$$

It is easy to check that Ψ maps $\mathcal{B}_\rho^0(\mathbb{R}^m, \mathbb{R}^k) \times \mathcal{F}_\rho$ into $\mathcal{F}_\rho$. Moreover, if ϕ is continuously differentiable, then $D\Lambda(\phi) = \Psi(\phi, D\phi)$.

The transformation $\Gamma : \mathcal{B}_\rho^0(\mathbb{R}^m, \mathbb{R}^k) \times \mathcal{F}_\rho \to \mathcal{B}_\rho^0(\mathbb{R}^m, \mathbb{R}^k) \times \mathcal{F}_\rho$ given by $\Gamma(\phi, \Phi) = (\Lambda(\phi), \Psi(\phi, \Phi))$ is a fiber contraction. In fact, we have

$$\|\Psi(\phi, \Phi_1)(\zeta) - \Psi(\phi, \Phi_2)(\zeta)\| \leq \rho^2 \|\Phi_1 - \Phi_2\|.$$

Also, let us define Φ_∞ to be the unique fixed point of the map $\Phi \mapsto \Psi(\nu, \Phi)$ where, recall, ν is the unique fixed point of Λ.

Let $(\phi_0, \Phi_0) = (0, 0) \in \mathcal{B}_\rho^0(\mathbb{R}^m, \mathbb{R}^k) \times \mathcal{F}_\rho$ and define recursively the sequence $\{(\phi_0, \Phi_0)\}_{n=0}^\infty$ by

$$(\phi_{n+1}, \Phi_{n+1}) = \Gamma(\phi_n, \Phi_n).$$

It is easy to check that $\Phi_n = D\phi_n$ for each nonnegative integer n. By the fiber contraction principle, we have that $\lim_{n\to\infty} \phi_n = \nu$ and

$$\lim_{n\to\infty} \Phi_n = \Phi_\infty.$$

As before, by using the uniform convergence on compact subsets of $\mathbb{R}^m$ of the sequence $\{\phi_n\}_{n=0}^\infty$ and the uniform convergence of $\{\Phi_n\}_{n=0}^\infty$, we can conclude that ν is continuously differentiable with derivative Φ_∞.

If $\rho > 1$, let us consider the map $\Delta : \mathbb{R}^k \times \mathbb{R}^m \to \mathbb{R}^k$ defined by

$$\Delta(x, z) := x - \omega(\alpha(x, z), z).$$

An application of the implicit function theorem at the origin produces a local solution $z \mapsto \nu(z)$ that can be used as above to define a function whose graph is a subset $W_{loc}^c(0, 0)$ of $W^c(0, 0)$.

We have proved that a C^1 differential equation has C^1 *local* invariant manifolds at a rest point. It should be reasonably clear that the methods

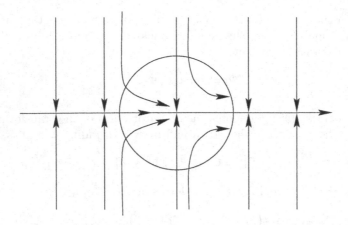

Figure 4.1: Schematic phase portrait for system (4.23) modified with a cut-off function that removes the nonlinearity outside of the indicated disk. Note that only the horizontal axis is a *global* center manifold for the modified differential equation.

of proof used in this section, together with an induction argument, can be used to show that if $1 \leq r < \infty$, then a C^r differential equation has C^r local invariant manifolds at a rest point. The case of C^∞, or analytic, differential equations is more difficult. For example, an analytic differential equation may not have a C^∞ center manifold (see [103, p. 126]).

Let us note that (local) center manifolds may not be unique. For example, the rest point at the origin for the planar differential equation

$$\dot{x} = x^2, \qquad \dot{y} = -y \qquad\qquad (4.23)$$

has infinitely many center manifolds (see Exercise 4.2). This fact may seem contrary to the uniqueness of the invariant manifolds proved in Theorem 4.1. The apparent contradiction arises because only one of the local center manifolds for the differential equation (4.23) is defined globally. More precisely, if this differential equation is modified by a cut-off function, then only one of the local center manifolds extends as the graph of a globally defined function (see Figure 4.1). Indeed, in the unbounded region where the cut-off function vanishes, the modified vector field is given by the linearized equations at the rest point, and for this linear system the only invariant one-dimensional manifold that is the graph of a function over the horizontal axis is the horizontal axis itself.

The local stable and unstable manifolds are unique. The key observation is that, unlike for the center manifold case, the linearization at a hyperbolic rest point, which defines the modified vector field in the region where the cut-off function vanishes, is such that local invariant manifolds for the original system would extend globally for the modified vector field as graphs of functions. Thus, the existence of more than one local stable or unstable manifold would violate Theorem 4.1.

Exercise 4.2. Show that the system (4.23) has infinitely many local center manifolds.

4.2 Applications of Invariant Manifolds

The most basic application of invariant manifold theory is the rigorous proof that the phase portraits of rest points of nonlinear systems have invariant manifolds akin to the (linear) invariant subspaces at the zero solution of a constant coefficient homogeneous linear system. But the applications of invariant manifold theory go far beyond this fact. It turns out that invariant sets (for example, periodic orbits, invariant tori, etc.) also have associated invariant manifolds. It is even possible to have a system (called a uniformly hyperbolic system or an Anosov system) where every orbit has associated nontrivial stable and unstable manifolds. The existence of invariant manifolds provides an important part of the analysis required to understand the dynamical behavior of a differential equation near an invariant set, for example a steady state.

Another important application of invariant manifold theory arises when we are interested in the qualitative changes in the phase portrait of a family of differential equations that depends on one or more parameters. For example, let us imagine that the phase portrait of a family at some parameter value has a rest point (more generally, an invariant set) that is not hyperbolic. In this case we expect that the qualitative dynamical behavior of the system will change—a bifurcation will occur—when the parameter is varied. Often, if there are qualitative changes, then they are confined to changes on a center manifold. After all, the dynamics on stable and unstable manifolds is very simple: asymptotic attraction in forward or backward time to the invariant manifold. This observation often allows the reduction of a multidimensional problem to a much lower dimensional differential equation restricted to the center manifold as we will now explain.

Let us consider a differential equation that depends on a parameter ν. Moreover, let us assume that the differential equation has a rest point whose position in space is a smooth function of ν near $\nu = 0$. In this case, there is a change of coordinates that fixes the rest point at the origin and transforms the system to the form

$$
\begin{aligned}
\dot{x} &= S(\nu)x + F(x, y, z, \nu), \\
\dot{y} &= U(\nu)y + G(x, y, z, \nu), \\
\dot{z} &= C(\nu)z + H(x, y, z, \nu)
\end{aligned}
$$

where S, U, and C are matrices that depend on the parameter. Moreover, $C(0)$ has eigenvalues with zero real parts, $S(0)$ has eigenvalues with negative real parts, and $U(0)$ has eigenvalues with positive real parts.

There is a standard "trick" that is quite important. Let us introduce ν as a new dependent variable; that is, let us consider the system

$$
\begin{aligned}
\dot{x} &= S(\nu)x + F(x, y, z, \nu), \\
\dot{y} &= U(\nu)y + G(x, y, z, \nu), \\
\dot{z} &= C(\nu)z + H(x, y, z, \nu), \\
\dot{\nu} &= 0.
\end{aligned}
$$

Also, note that if we expand the matrices S, U, and C in powers of ν at $\nu = 0$ to obtain, for example, $S(\nu) = S(0) + \nu \mathcal{S}(\nu)$, then the term $\nu \mathcal{S}(\nu)x$ is a nonlinear term with respect to our new differential equation, and therefore it can be grouped together with $F(x, y, z, \nu)$ in the first equation. Hence, by an obvious redefinition of the symbols, we lose no generality if we consider the system in the form

$$
\begin{aligned}
\dot{x} &= Sx + F(x, y, z, \nu), \\
\dot{y} &= Uy + G(x, y, z, \nu), \\
\dot{z} &= Cz + H(x, y, z, \nu) \\
\dot{\nu} &= 0
\end{aligned}
$$

where the matrices S, U and C do not depend on ν. Moreover, by grouping together the last two equations, let us view the system as having its center part augmented by one extra center direction corresponding to ν. If ν is a vector of parameters, then there may be several new center directions.

By our general theorem, there is a center manifold given as the graph of a function with components $x = \alpha(z, \nu)$ and $y = \beta(z, \nu)$ defined on the space with coordinates (z, ν). In particular, the center manifold depends smoothly on the coordinate ν in some open ball containing $\nu = 0$, and therefore the restriction of the original differential equation to this invariant center manifold—its *center manifold reduction*—depends smoothly on the parameter ν and has the form

$$
\dot{z} = Cz + H(\alpha(z, \nu), \beta(z, \nu), z, \nu). \tag{4.24}
$$

The "interesting" dynamics near the original rest point for ν near $\nu = 0$ is determined by analyzing the family of differential equations (4.24). In fact, this construction is one of the most important applications of center manifold theory.

The qualitative behavior for center manifold reduced systems is the same on all local center manifolds. Moreover, each bounded invariant set of the original system, sufficiently close to the rest point under consideration, is also an invariant set for each center manifold reduced system (see, for example, [58]).

Exercise 4.3. (a) Determine a center manifold for the rest point at the origin of the system

$$\dot{x} = -xy, \qquad \dot{y} = -y + x^2 - 2y^2$$

and a differential equation for the dynamics on this center manifold. (b) Show that every solution of the system is attracted to the center manifold (see the interesting article by A. J. Roberts [195]). (c) Determine the stability type of the rest point at the origin. Hint: Look for the center manifold as a graph of a function of the form

$$y = h(x) = -\alpha x^2 + \beta x^3 + \cdots.$$

Why does the expected h have $h(0) = 0$ and $h'(0) = 0$? The condition for invariance is $\dot{y} = h'(x)\dot{x}$ with $y = h(x)$. Find the first few terms of the series expansion for h, formulate a conjecture about the form of h, and then find h explicitly. Once h is known, the dynamical equation for the flow on the center manifold is given by $\dot{x} = -xh(x)$. (Why?) (d) Find an explicit equation for the unstable manifold of the saddle point at the origin for the system

$$\dot{x} = \epsilon x - xy, \qquad \dot{y} = -y + x^2 - 2y^2,$$

and find the differential equation that gives the dynamics on this invariant manifold. (e) How does the phase portrait change as ϵ passes through $\epsilon = 0$.

Exercise 4.4. Find the third order Taylor series approximation of the (scalar) center manifold reduced family at the origin, as in display (4.24), for the system

$$\dot{z} = \epsilon - z + w + \frac{1}{4}((1 + \epsilon)z^2 - 2\epsilon wz - (1 - \epsilon)w^2),$$

$$\dot{w} = \epsilon + z - w - \frac{1}{4}((1 + \epsilon)z^2 - (2\epsilon - 4)wz + (3 + \epsilon)w^2).$$

Exercise 4.5. The system

$$\dot{p} = -\frac{1}{5}p(3p^5 - 5q^4p^2 + 13q^2p - 3), \quad \dot{q} = -\frac{1}{5}q(p^5 + q^2p - 1).$$

derived in Exercise 1.145 has a semi-hyperbolic rest point at $(p, q) = (1, 0)$. Use a center manifold reduction to determine the phase portrait in a neighborhood of this rest point. Hint: Use the hint for part (c) of Exercise 4.3.

4.3 The Hartman–Grobman Theorem

In the last section we proved the existence of stable and unstable manifolds for hyperbolic rest points of differential equations using a classic idea that is worth repeating: The existence of the desired object, for example an invariant manifold, is equivalent to the existence of a fixed point for a properly defined map on a function space, and the hyperbolicity hypothesis is used to prove that this map is a contraction. This same idea is used in this section to prove the Hartman–Grobman theorem (Theorem 1.47). This result provides a perfect setting for our exploration of various aspects of hyperbolicity theory. See [11], [165], [186], [181], [192], and [54] for the origins of this marvelous proof, and for the original proofs see [102] and [113].

4.3.1 Diffeomorphisms

We will prove the Hartman–Grobman theorem for discrete dynamical systems, that is, dynamical systems defined by diffeomorphisms as follows: Suppose that $\Omega \subseteq \mathbb{R}^n$ and $F : \Omega \to \Omega$ is a diffeomorphism. The orbit of $\xi \in \Omega$ is the set of all iterates of ξ under transformation by F. More precisely, the orbit of ξ is the set $\{F^\ell(\xi) : \ell \in \mathbb{Z}\}$, where F^{-1} denotes the inverse of F, we define $F^0(\xi) = \xi$ and (by induction) $F^{\ell+1}(\xi) := F(F^\ell(\xi))$ for every integer ℓ. A *fixed point* of the dynamical system defined by F (that is, a point ξ such that $F(\xi) = \xi$) is analogous to a rest point for the dynamical system defined by a differential equation.

There is, of course, a very close connection between the dynamical systems defined by differential equations and those defined by diffeomorphisms. If, for example, φ_t is the flow of an autonomous differential equation, then for each fixed $t \in \mathbb{R}$ the *time t map* given by $\xi \mapsto \varphi_t(\xi)$ is a diffeomorphism on its domain that defines a dynamical system whose orbits are all subsets of the orbits of the flow. Also, a Poincaré map is a diffeomorphism whose orbits correspond to features of the phase portrait of its associated differential equation. In particular, a fixed point of a Poincaré map corresponds to a periodic orbit of the associated differential equation.

Let φ_t be the flow of the differential equation $\dot{x} = f(x)$ and recall that $t \mapsto D\phi_t(\zeta)$ is the solution of the variational initial value problem

$$\dot{W} = Df(\varphi_t(\zeta))W, \qquad W(0) = I.$$

In particular, if ζ is a rest point, then the solution of the initial value problem is

$$D\phi_t(\zeta) = e^{tDf(\zeta)}.$$

Thus, if ζ is a hyperbolic rest point and $t \neq 0$, then the linear transformation $D\phi_t(\zeta)$ has no eigenvalues on the unit circle of the complex plane. For this reason, a fixed point of a diffeomorphism is called *hyperbolic* if the derivative of the diffeomorphism at the fixed point has no eigenvalue on the unit circle.

The next theorem is a version of the Hartman–Grobman theorem for diffeomorphisms. Informally, it states that the phase portrait near a hyperbolic fixed point is the same, up to a *continuous* change of coordinates, as the phase portrait of the dynamical system induced by the derivative of the diffeomorphism evaluated at the fixed point.

Theorem 4.6 (Hartman–Grobman Theorem). *Let Ω be an open subset of $\mathbb{R}^n$. If ζ is a hyperbolic fixed point for the diffeomorphism $F : \Omega \to \Omega$, then there is an open set $U \in \mathbb{R}^n$ containing ζ and a homeomorphism H with domain U such that*

$$F(H(x)) = H(DF(\zeta)x)$$

whenever $x \in U$ and both sides of the equation are defined.

The proof of Theorem 4.6 is based on the idea that the conjugating homeomorphism is the solution of a functional equation. Sufficient conditions for the appropriate functional equation to have a unique solution are given in the following key lemma.

Lemma 4.7. *Suppose that $A : \mathbb{R}^n \to \mathbb{R}^n$ is an invertible hyperbolic linear transformation and $p : \mathbb{R}^n \to \mathbb{R}^n$ is a smooth function such that $p(0) = 0$. If $0 < \alpha < 1$ and the C^1-norm of the function p is sufficiently small, then there is a unique continuous function $h : \mathbb{R}^n \to \mathbb{R}^n$ such that $\|h\| \leq \alpha$, $h(0) = 0$, and*

$$h(Ax) - Ah(x) = p(x + h(x)) \tag{4.25}$$

for every x in $\mathbb{R}^n$.

Proof. For $h : \mathbb{R}^n \to \mathbb{R}^n$, define the linear operator L by

$$L(h)(x) = h(Ax) - Ah(x),$$

the (nonlinear) operator Φ by

$$\Phi(h)(x) = p(x + h(x)) - p(x),$$

and recast equation (4.25) in the form

$$L(h)(x) = \Phi(h)(x) + p(x). \tag{4.26}$$

The operator L is invertible on the Banach space $C(\mathbb{R}^n)$, the space of bounded continuous transformations of $\mathbb{R}^n$ with the supremum norm. To prove this fact, let us use the hyperbolicity of the linear transformation A to decompose the space $\mathbb{R}^n$ as the direct sum of the invariant linear eigenspaces E^s and E^u that correspond, respectively, to the subsets of the spectrum of A that lie inside and outside of the unit circle. Of course, if the fixed point is a sink or a source, then there is only one eigenspace.

By Corollary 2.127, there are adapted norms (both denoted by $|\ |$) on the eigenspaces such that the sum of these norms is equivalent to the norm on $\mathbb{R}^n$, and in addition there is a number λ, with $0 < \lambda < 1$, such that if $x = x_s + x_u \in E^s \oplus E^u$, then

$$|Ax_s| < \lambda |x_s|, \qquad |A^{-1}x_u| < \lambda |x_u|.$$

Also, if h is a transformation of $\mathbb{R}^n$, then h can be expressed uniquely as a sum of functions $h = h_s + h_u$ where $h_s : \mathbb{R}^n \to E^s$ and $h_u : \mathbb{R}^n \to E^u$.

Using the projections to E^s and E^u, let us note that

$$L(h)(x) = [h_s(Ax) - A(h_s(x))] + [h_u(Ax) - A(h_u(x))].$$

Because the eigenspaces are invariant sets for A, it follows that the equation

$$L(h)(x) = p(x),$$

where $p : \mathbb{R}^n \to \mathbb{R}^n$, has a solution h if and only if the "operator equations"

$$L_s(h_s)(x) := h_s(Ax) - Ah_s(x) = p_s(x),$$

$$L_u(h_u)(x) := h_u(Ax) - Ah_u(x) = p_u(x)$$

both have solutions. In particular, to prove that L is invertible, it suffices to prove that L_s and L_u are both invertible as operators on the respective spaces $C^0(\mathbb{R}^n, E^s)$ and $C^0(\mathbb{R}^n, E^u)$ where $C^0(\mathbb{R}^n, E^s)$, respectively $C^0(\mathbb{R}^n, E^u)$, denotes the space of continuous bounded functions from $\mathbb{R}^n$ to E^s, respectively E^u, with the adapted norm.

Let us define two additional operators S and U by

$$S(h_s)(x) := h_s(Ax), \qquad U(h_u)(x) := h_u(Ax)$$

so that

$$L_s(h_s) = (S - A)h_s, \qquad L_u(h_u) = (U - A)h_u.$$

Because A is invertible, both of the operators S and U are invertible; for example, we have that $S^{-1}(h_s)(x) = h_s(A^{-1}x)$. Moreover, it is easy to prove directly from the definition of the operator norm that these operators and their inverses all have norm one. Thus, we have

$$\|S^{-1}A\| \le \|S^{-1}\|\,\|A\| < \lambda < 1,$$

and therefore the operator $I - S^{-1}A$ is invertible. In fact, its inverse is given by

$$I - S^{-1}A = I + \sum_{\ell=1}^{\infty} (S^{-1}A)^{\ell}$$

where the Neumann series is easily proved to be absolutely convergent by comparison with the geometric series

$$\sum_{\ell=0}^{\infty} \lambda^{\ell} = \frac{1}{1 - \lambda}.$$

Because the operator L_s can be rewritten in the form

$$L_s = S - A = S(I - S^{-1}A),$$

it is invertible with inverse

$$L_s^{-1} = (I - S^{-1}A)^{-1}S^{-1}.$$

Moreover, we have the following norm estimate:

$$\|L_s^{-1}\| \leq \frac{1}{1-\lambda}.$$

Similarly, for the operator L_u, we have that

$$L_u = U - A = A(A^{-1}U - I) = -A(I - A^{-1}U)$$

with

$$\|A^{-1}U\| < \lambda < 1.$$

Therefore, the inverse of L_u is given by

$$L_u^{-1} = -(I - A^{-1}U)^{-1}A^{-1},$$

and, in addition, we have the norm estimate

$$\|L_u^{-1}\| \leq \frac{\lambda}{1-\lambda} < \frac{1}{1-\lambda}.$$

Using the norm estimates for the inverses of the operators L_s and L_u, it follows that L is invertible and

$$\|L^{-1}\| < \frac{2}{1-\lambda}.$$

Let us recast equation (4.26) in the form

$$h = L^{-1}\Phi(h) + L^{-1}p$$

and note that the solutions of equation (4.26) are exactly the fixed points of the operator T defined by

$$T(h) := L^{-1}\Phi(h) + L^{-1}p.$$

Also, the set

$$C_\alpha^0 := \{h \in C^0(\mathbb{R}^n) : \|h\| \leq \alpha, \ h(0) = 0\}$$

is a complete metric subspace of the Banach space $C^0(\mathbb{R}^n)$. Thus, to complete the proof of the lemma, it suffices to show the following proposition: $T : C_\alpha^0 \to C_\alpha^0$ and T is a contraction.

To prove the proposition, note that if $h(0) = 0$, then $T(h)(0) = 0$ and and

$$\begin{aligned}
\|T(h)\| &\leq \|L^{-1}\|(\|\Phi(h)\| + \|p\|) \\
&\leq \frac{2}{1-\lambda}\Big(\sup_{x \in \mathbb{R}^n} |p(x + h(x)) - p(x)| + \|p\| \Big) \\
&\leq \frac{2}{1-\lambda}\Big(\sup_{x \in \mathbb{R}^n} |Dp(x)| \, \|h\| + \|p\| \Big) \\
&\leq \frac{2}{1-\lambda}\big(\|p\|_1 \|h\| + \|p\| \big) \\
&\leq \frac{2}{1-\lambda}(1 + \alpha)\|p\|_1
\end{aligned}$$

where $\| \ \|_1$ denotes the C^1 norm. Hence, if

$$\|p\|_1 < \frac{\alpha}{1+\alpha}(1 - \lambda),$$

then T is a transformation of the space C_α^0. Moreover, because

$$\begin{aligned}
\|T(h_1) - T(h_2)\| &= \|L^{-1}(\Phi(h_1) - \Phi(h_2))\| \\
&\leq \frac{2}{1-\lambda} \sup_{x \in \mathbb{R}^n} \|p(x + h_1(x)) - p(x + h_2(x))\| \\
&\leq \frac{2}{1-\lambda} \|p\|_1 \|h_1 - h_2\|,
\end{aligned}$$

the same restriction on the size of $\|p\|_1$ ensures that T is a contraction. $\square$

Let us prove Theorem 4.6.

Proof. Assume, without loss of generality, that ζ is the origin of $\mathbb{R}^n$. Also, define $A := DF(0)$ and note that, because F is a diffeomorphism, A is an *invertible* hyperbolic linear transformation.

Choose $\alpha \in \mathbb{R}$ such that $0 < \alpha < 1$. If we define $f(x) := F(x) - Ax$, then $f(0) = 0$ and $Df(0) = 0$. Thus, using the continuity of f, there is an open neighborhood V of the origin such that the C^1-norm of the restriction of f to V is less than $\alpha/3$. This norm is defined as usual in terms of C^0-norms as follows

$$\|f\|_1 = \|f\| + \|Df\|.$$

By using an appropriate bump function, as in the derivation of the estimate (4.21), there is a smooth function f^* defined on all of $\mathbb{R}^n$ such that $f = f^*$ on V and the C^1-norm of f^* (with the supremum taken over $\mathbb{R}^n$) does not exceed three times the C^1-norm of the restriction of f to V; that is, since $\|f\|_1 \leq \alpha/3$, we have that $\|f^*\|_1 < \alpha$.

Apply Lemma 4.7 with $p = f^*$ and define a new continuous function $H : \mathbb{R}^n \to \mathbb{R}^n$ by

$$H(x) = x + h(x). \tag{4.27}$$

Using equation (4.25), it is easy to see that $F^*(H(x)) = H(A(x))$ for all $x \in \mathbb{R}^n$, where $F^*(x) := f^*(x) + Ax$. This function H, restricted to a suitably small neighborhood of the origin, is a candidate for the required local homeomorphism. Indeed, to complete the proof of the theorem, we will show that there is an open set U containing the origin and contained in V such that the restriction of H to U is a homeomorphism.

To prove that H is injective, let us suppose that $H(x) = H(y)$ for some points x and y in $\mathbb{R}^n$. Using the identities

$$H(Ax) = F^*(H(x)) = F^*(H(y)) = H(Ay),$$

we have that

$$H(A^\ell x) = H(A^\ell y)$$

for every integer ℓ. But then

$$A^\ell x + h(A^\ell x) = A^\ell y + h(A^\ell y)$$

and

$$\|A^\ell x - A^\ell y\| = \|h(A^\ell x) - h(A^\ell y)\| \leq 2\|h\|.$$

In particular, the set

$$\{\|A^\ell (x - y)\| : \ell \in \mathbb{Z}\}$$

is bounded. But this is a contradiction unless $x = y$. In fact, because A is a hyperbolic linear transformation on $\mathbb{R}^n$, if $z \neq 0$, then either

$$\lim_{\ell \to \infty} \|A^\ell z\| = \infty, \quad \text{or} \quad \lim_{\ell \to -\infty} \|A^\ell z\| = \infty.$$

Thus, H is injective.

There is an open neighborhood U of the origin such that its closure $\bar{U}$ is compact, contained in V, and $H(\bar{U}) \subset V$. Because H is a continuous injective function on the compact set $\bar{U} \subset \mathbb{R}^n$, an elementary argument using point set topology [168, p. 167] shows that H restricted to $\bar{U}$ is a homeomorphism onto its image. In particular, H has a continuous inverse defined on $H(\bar{U})$. This inverse restricted to $H(U)$ is still continuous. Thus, H restricted to U is a homeomorphism onto its image. Finally, by Brouwer's theorem on invariance of domain (see [169, p. 207]), $H(U)$ is open.   □

4.3.2   Differential Equations

In this section we will prove the following version of the Hartman–Grobman theorem for a hyperbolic rest point of an autonomous differential equation.

Theorem 4.8. *Suppose that ζ is a rest point of the differential equation $\dot{x} = f(x)$ on $\mathbb{R}^n$ with flow φ_t and ψ_t is the flow of the linearized system $\dot{x} = Df(\zeta)(x - \zeta)$. If ζ is a hyperbolic rest point, then there is an open subset U of $\mathbb{R}^n$ such that $\zeta \in U$ and a homeomorphism G with domain U such that $G(\varphi_t(x)) = \psi_t(G(x))$ whenever $x \in U$ and both sides of the equation are defined.*

While the proofs of Theorem 4.8 and the Hartman–Grobman theorem for diffeomorphisms are similar, there are some subtle differences. For example, note that whereas the conjugating homeomorphism H in the diffeomorphism case is a solution of the functional equation $F(G(x)) = G(Df(\zeta)x)$, the corresponding equation in Theorem 4.8 has the form

$$G(F(x)) = Df(\zeta)G(x).$$

If G is a homeomorphism, then these two equations are equivalent. But, the form of these equations is important for the method used here to *prove* the existence of G.

Let us begin with a lemma analogous to Lemma 4.7. For the statement of this lemma, recall that $C(\mathbb{R}^n)$ denotes the Banach space of bounded continuous transformations of $\mathbb{R}^n$ with the supremum norm.

Lemma 4.9. *If $A : \mathbb{R}^n \to \mathbb{R}^n$ is an invertible hyperbolic linear transformation and $F : \mathbb{R}^n \to \mathbb{R}^n$ is a homeomorphism, then the operator given by*

$$\Phi(g)(x) = Ag(x) - g(F(x))$$

is a bounded linear transformation with a bounded inverse on the Banach space $C(\mathbb{R}^n)$.

Proof. If $g \in C(\mathbb{R}^n)$, then clearly $x \mapsto \Phi(g)(x)$ is a continuous transformation of $\mathbb{R}^n$. Also, it is clear that Φ is a linear operator. The norm estimate

$$|\Phi(g)(x)| \le |Ag(x)| + |g(F(x))| \le \|A\| \, \|g\| + \|g\|$$

(where $\|A\|$ is the operator norm of the linear transformation A and $\|g\|$ is the supremum norm of the function g), shows that Φ is a bounded linear operator on $C(\mathbb{R}^n)$.

The proof that the operator Φ has a bounded inverse is similar to the proof of Lemma 4.7. In fact, relative to the splitting $\mathbb{R}^n = E^s \oplus E^u$, the operator Φ is given by $\Phi = \Phi_s + \Phi_u$ where

$$\Phi_s(g_s) := A \circ g_s - g_s \circ F, \qquad \Phi_u(g_u) := A \circ g_u - g_u \circ F.$$

The important point is that the operators S and U defined by

$$S(g_s) := g_s \circ F, \qquad U(g_u) := g_u \circ F$$

and their inverses are all bounded, and they all have operator norm one. The operators Φ_s and Φ_u are inverted using Neumann series, as in the proof of Lemma 4.7. □

Let us prove Theorem 4.8.

Proof. Assume that $\zeta = 0$ and define $B := Df(0)$. Also, note that

$$\psi_t(x) = e^{tB}x$$

and define $A := \psi_1$, the time-one map of the flow ψ_t.

By using an appropriate bump function γ defined on a neighborhood of the origin, the differential equation

$$\dot{x} = f^*(x), \tag{4.28}$$

where $f^* := \gamma f$, has a complete flow φ_t^* together with the following additional properties.

  (i) The function $f^* : \mathbb{R}^n \to \mathbb{R}^n$ has a finite Lipschitz constant.

  (ii) There is an open neighborhood V of the origin such that the time one map $F := \varphi_1^*$ agrees with the time one map φ_1 of the flow φ_1 on V.

  (iii) The function $p(x) := F(x) - Ax$ has finite C^1-norm that is sufficiently small so that, by Lemma 4.7, there is a function $h \in C(\mathbb{R}^n)$ with $h(0) = 0$, $\|h\| < 1$, and

$$h(Ax) - Ah(x) = p(x + h(x))$$

for all $x \in \mathbb{R}^n$.

Let us prove first that there is a continuous map $G : \mathbb{R}^n \to \mathbb{R}^n$ such that $G(F(x)) = A(G(x)$; that is, G conjugates the time one maps of the linear and nonlinear flows. In fact, because p has finite C^1-norm, it follows that $p \in C(\mathbb{R}^n)$ and by Lemma 4.9, there is a unique $g \in C(\mathbb{R}^n)$ such that

$$Ag(x) - g(F(x)) = p(x).$$

By defining $G(x) := x + g(x)$, we have that

$$G(F(x)) = AG(x). \tag{4.29}$$

To construct a conjugacy between the linear and the nonlinear flows, use the "time one conjugacy" G to define $\mathcal{G} : \mathbb{R}^n \to \mathbb{R}^n$ by

$$\mathcal{G}(x) := \int_0^1 \psi_{-s}(G(\varphi_s^*(x))) \, ds.$$

We will show that

$$\psi_t(\mathcal{G}(x)) = \mathcal{G}(\varphi_t^*(x)). \tag{4.30}$$

In fact, using the linearity of ψ_t, the change of variable $\tau = s + t - 1$, the flow property, and equation (4.29), we have

$$
\begin{aligned}
\psi_{-t}(\mathcal{G}(\varphi_t^*(x))) &= \int_0^1 \psi_{-t-s}(G(\varphi_{s+t}^*(x)))\, ds \\
&= \int_{t-1}^t \psi_{-\tau}(\psi_{-1}(G(\varphi_1^*(\varphi_\tau^*(x)))\, d\tau \\
&= \int_{t-1}^t \psi_{-\tau}(G(\varphi_\tau^*(x)))\, d\tau.
\end{aligned}
$$

By splitting the last integral into the two parts

$$\int_{t-1}^0 \psi_{-\tau}(G(\varphi_\tau^*(x)))\, d\tau + \int_0^t \psi_{-\tau}(G(\varphi_\tau^*(x)))\, d\tau,$$

changing the variable in the first integral to $\sigma := \tau + 1$, and using the flow property together with equation (4.29), we obtain the identity

$$
\begin{aligned}
\psi_{-t}(\mathcal{G}(\varphi_t^*(x))) &= \int_t^1 \psi_{-\sigma+1}(G(\varphi_{-1+\sigma}^*(x)))\, d\sigma + \int_0^t \psi_{-\tau}(G(\varphi_\tau^*(x)))\, d\tau \\
&= \mathcal{G}(x),
\end{aligned}
$$

as required.

Recall equation (4.29) and note that if we set $t = 1$ in equation (4.30), then we have the functional identities

$$\mathcal{G}(F(x)) = A\mathcal{G}(x), \qquad G(F(x)) = AG(x).$$

By Lemma 4.9, the function G is unique among all continuous transformations of $\mathbb{R}^n$ of the form $G(x) = x + g(x)$ that satisfy the same functional equation, provided that $g \in C(\mathbb{R}^n)$. Thus, to prove that $\mathcal{G} = G$, it suffices to show that the function $x \mapsto \mathcal{G}(x) - x$ is in $C(\mathbb{R}^n)$. To see this fact, let us note first that we have the identities

$$
\begin{aligned}
\mathcal{G}(x) - x &= \int_0^1 \psi_{-s}(G(\varphi_s^*(x)))\, ds - x \\
&= \int_0^1 \psi_{-s}[G(\varphi_s^*(x)) - \psi_s(x)]\, ds \\
&= \int_0^1 \psi_{-s}[G(\varphi_s^*(x)) - \varphi_s^*(x) + \varphi_s^*(x) - \psi_s(x)]\, ds,
\end{aligned}
$$

and the estimate

$$|\mathcal{G}(x) - x| \le e^{\|B\|}(\,|\mathcal{G}(\varphi_s^*(x)) - \varphi_s^*(x)| + |\varphi_s^*(x) - \psi_t(x)|\,)$$

$$\le e^{\|B\|}(\,\|g\| + \sup_{0 \le s \le 1} |\varphi_s^*(x) - \psi_s(x)|\,).$$

Also, for $0 \le s \le 1$, we have the inequalities

$$|\varphi_s^*(x) - \psi_s(x)| \le \int_0^s |f^*(\varphi_t^*(x)) - B\psi_t(x)|\, dt$$

$$\le \int_0^s |f^*(\varphi_t^*(x)) - f^*(\psi_t(x))| + |f^*(\psi_t(x)) - B\psi_t(x)|\, dt$$

$$\le \int_0^s \mathrm{Lip}(f^*)|\varphi_t^*(x) - \psi_t(x)|\, dt + (\|f^*\| + \|B\| e^{\|B\|}).$$

Thus, by Gronwall's inequality,

$$\sup_{0 \le s \le 1} |\varphi_s^*(x) - \psi_s(x)| \le (\|f^*\| + \|B\| e^{\|B\|}) e^{\mathrm{Lip}(f^*)},$$

and, as a result, the function $x \mapsto \mathcal{G}(x) - x$ is in $C(\mathbb{R}^n)$.

It remains to show that $G = \mathcal{G}$ is a homeomorphism when restricted to some neighborhood of the origin. Using property (iii) given above and the proof of the Hartman–Grobman theorem for diffeomorphisms, the function h given in property (iii) can be used to define a continuous function H by $H(x) = x + h(x)$ so that

$$F(H(x)) = H(Ax).$$

Thus,

$$G(H(Ax)) = G(F(H(x))) = AG(H(x))$$

and, with $K := G{\circ}H$, we have $K(A(x)) = A(K(x))$. Moreover, the function K has the form

$$K(x) = x + h(x) + g(x + h(x))$$

where, by the construction of G and H, the function

$$\alpha : x \mapsto h(x) + g(x + h(x))$$

is in $C(\mathbb{R}^n)$ and $A(\alpha(x)) - \alpha(Ax) = 0$. By Lemma 4.9, there is only one function α in $C(\mathbb{R}^n)$ that solves this functional equation. It follows that $\alpha(x) \equiv 0$. Therefore, K is the identity function and $G(H(x)) = x$ for all $x \in \mathbb{R}^n$. Because there is an open set U containing the origin such that the restriction of H to U is a homeomorphism onto its image, we must have that G restricted to $H(U)$ is the inverse of H. In particular, G restricted to $H(U)$ is a homeomorphism onto U. $\square$

Exercise 4.10. Suppose A is an invertible linear transformation of $\mathbb{R}^n$. Let $\mathcal{L}$ denote the set of all Lipschitz functions mapping $\mathbb{R}^n$ to $\mathbb{R}^n$ and, for $\alpha \in \mathcal{L}$, let $\mathrm{Lip}(\alpha)$ denote the (least) Lipschitz constant for α. Prove: There is an $\epsilon > 0$ such that if $\alpha \in \mathcal{L}$ and $\mathrm{Lip}(\alpha) < \epsilon$, then $A + \alpha : \mathbb{R}^n \to \mathbb{R}^n$ is continuous and bijective. Also, prove that the inverse map is Lipschitz, hence continuous. This result is a version of the Lipschitz inverse function theorem.

Exercise 4.11. [Toral Automorphisms] Consider the torus $\mathbb{T}^2 = \mathbb{R}^2/\mathbb{Z}^2$, that is, all equivalence classes of points in the plane where two points are equivalent if their difference is in the integer lattice. A unimodular matrix, for example

$$A := \begin{pmatrix} 2 & 1 \\ 1 & 1 \end{pmatrix},$$

induces a map on $\mathbb{T}^2$ called a *toral automorphism*. (a) Prove that A is a hyperbolic linear map (spectrum off the unit circle). (b) Prove that the map induced by A on $\mathbb{T}^2$ is invertible. (c) Determine all periodic points of the induced map. (d) Prove that the induced map has a dense orbit. (e) Show that every orbit of the induced map has a one-dimensional stable and a one-dimensional unstable manifold, the sets defined as the points in $\mathbb{T}^2$ that are asymptotic to the given orbit under forward, respectively backward, iteration. Hyperbolic toral automorphisms are the prototypical examples of Anosov (uniformly hyperbolic) dynamical systems and they enjoy many interesting dynamical properties; for example, they are "chaotic maps" where the entire phase space is a "chaotic attractor". Also, note that toral automorphisms are examples of area preserving dynamical systems: the measures of subsets of the phase space do not change under iteration by the map. (The flow of a Hamiltonian system has the same property.) (f) Prove that hyperbolic toral automorphisms are *ergodic*; that is, they are area preserving maps such that every one of their invariant sets has measure zero or measure one. Hint: See [133]. The first order system $\dot{x} = 1$, $\dot{y} = \alpha$ induces a flow on the torus (where x and y are viewed as angular variables modulo one). (g) Prove that the flow of this system is measure preserving. (h) Prove that the flow is ergodic if α is irrational.

4.3.3   Linearization via the Lie Derivative

A proof of the Hartman–Grobman theorem for differential equations is given in this section that does not require the Hartman–Grobman theorem for diffeomorphisms (see [54]).

We will use the concept of the Lie derivative for vector fields.

Definition 4.12. Suppose that X and Y are vector fields on $\mathbb{R}^n$ and ϕ_t is the flow of X. The *Lie derivative* of Y in the direction of X, denoted $L_X Y$, is the vector field given by

$$L_X Y(x) = \frac{d}{dt} D\phi_{-t}(\phi_t(x))Y(\phi_t(x))\big|_{t=0}.$$

Also, we will use the following elementary result.

Proposition 4.13. *If X, Y, and Z are vector fields, ϕ_t is the flow of X, and $L_X Y = Z$, then*

$$\frac{d}{dt} D\phi_{-t}(\phi_t(x)) Y(\phi_t(x)) = D\phi_{-t}(\phi_t(x)) Z(\phi_t(x)).$$

Proof. Note that

$$\frac{d}{dt} D\phi_{-t}(\phi_t(x)) Y(\phi_t(x)) = \frac{d}{ds} D\phi_{-(t+s)}(\phi_{(t+s)}(x)) Y(\phi_{(t+s)}(x))\big|_{s=0}.$$

Set $y = \phi_t(x)$. Using the cocycle property

$$D\phi_{-t-s}(z) = D\phi_{-t}(\phi_{-s}(z)) D\phi_{-s}(z),$$

where $z \in \mathbb{R}^n$ (see Exercise 2.16), and the identity $\phi_{t+s}(x) = \phi_s(y)$, we have that

$$\frac{d}{ds} D\phi_{-(t+s)}(\phi_{t+s}(x)) Y(\phi_{t+s}(x))\big|_{s=0} = D\phi_{-t}(y) \frac{d}{ds} D\phi_{-s}(\phi_s(y)) Y(\phi_s(y))\big|_{s=0}$$
$$= D\phi_{-t}(\phi_t(x)) Z(\phi_t(x)).$$

$\square$

A C^1 vector field f on $\mathbb{R}^n$ such that $f(0) = 0$ is called *locally topologically conjugate* to its linearization $A := Df(0)$ at the origin if there is a homeomorphism $G : U \to V$ of neighborhoods of the origin such that the flows of f and A are locally conjugated by G; that is,

$$G(e^{tA}x) = \phi_t(G(x))$$

whenever $x \in U$, $t \in \mathbb{R}^n$, and both sides of the conjugacy equation are defined. Recall that a matrix is *infinitesimally hyperbolic* if all of its eigenvalues have nonzero real parts.

Theorem 4.14 (Hartman–Grobman Theorem). *Let f be a C^1 vector field on $\mathbb{R}^n$ such that $f(0) = 0$. If the linearization A of f at the origin is infinitesimally hyperbolic, then f is locally topologically conjugate to A at the origin.*

Proof. For each $r > 0$ there is a smooth bump function $\rho : \mathbb{R}^n \to [0, 1]$ with the following properties: $\rho(x) \equiv 1$ for $|x| < r/2$, $\rho(x) \equiv 0$ for $|x| > r$, and $|d\rho(x)| < 4/r$ for $x \in \mathbb{R}^n$. The vector field $Y = A + \xi$ where $\xi(x) := \rho(x)(f(x) - Ax)$ is equal to f on the open ball of radius $r/2$ at the origin. Thus, it suffices to prove that Y is locally conjugate to A at the origin.

Suppose that φ_t is the flow of Y. We will seek a solution $G = id + \eta$ of the conjugacy equation

$$G(e^{tA}x) = \varphi_t(G(x)) \tag{4.31}$$

where $\eta : \mathbb{R}^n \to \mathbb{R}^n$ is continuous and differentiable in the direction A. Let us first note that equation (4.31) is equivalent to the equation

$$e^{-tA}G(e^{tA}x) = e^{-tA}\varphi_t(G(x)). \qquad (4.32)$$

By substituting $G = id + \eta$ and differentiating both sides of the resulting equation with respect to t at $t = 0$, we obtain the *infinitesimal conjugacy equation*

$$L_A\eta = \xi \circ (id + \eta) \qquad (4.33)$$

where

$$L_A\eta := \frac{d}{dt}(e^{-tA}\eta(e^{tA}))\big|_{t=0} \qquad (4.34)$$

is the Lie derivative of η in the direction of the vector field given by A. (If G is a conjugacy, then the right-hand side of equation (4.32) is differentiable; therefore, in this case, the Lie derivative of G in the direction A is defined.)

We will show that if $r > 0$ is sufficiently small, then the infinitesimal conjugacy equation has a bounded continuous solution $\eta : \mathbb{R}^n \to \mathbb{R}^n$ (differentiable along A) such that $G := id + \eta$ is a homeomorphism of $\mathbb{R}^n$ whose restriction to the ball of radius $r/2$ at the origin is a local conjugacy as in equation (4.31).

Since A is infinitesimally hyperbolic, $A = A^+ \oplus A^-$, where $A^+ : \mathbb{R}^n \to \mathbb{R}^n$ is a linear map whose spectrum is in the left half of the complex plane and $A^- : \mathbb{R}^n \to \mathbb{R}^n$ is a linear map whose spectrum is in the right half of the complex plane. Put $\mathbf{E}^+ = \text{Range}(A^+)$ and $\mathbf{E}^- = \text{Range}(A^-)$. By Corollary 2.63, there are positive constants C and λ such that

$$|e^{tA}v^+| \le Ce^{-\lambda t}|v^+|, \qquad |e^{-tA}v^-| \le Ce^{-\lambda t}|v^-| \qquad (4.35)$$

for $t \ge 0$, $v^+ \in \mathbf{E}^+$, and $v^- \in \mathbf{E}^-$. The Banach space $\mathcal{B}$ of bounded (in the supremum norm) continuous vector fields on $\mathbb{R}^n$, which we identity with bounded continuous functions from $\mathbb{R}^n$ to $\mathbb{R}^n$, splits into the complementary subspaces $\mathcal{B}^+$ and $\mathcal{B}^-$ of vector fields with ranges, respectively, in $\mathbf{E}^+$ or $\mathbf{E}^-$. In particular, a vector field $\eta \in \mathcal{B}$ has a unique representation $\eta = \eta^+ + \eta^-$, where $\eta^+ \in \mathcal{B}^+$ and $\eta^- \in \mathcal{B}^-$.

The function Υ on $\mathcal{B}$ defined by

$$(\Upsilon\eta)(x) = \int_0^\infty e^{tA}\eta^+(e^{-tA}x)\,dt - \int_0^\infty e^{-tA}\eta^-(e^{tA}x)\,dt \qquad (4.36)$$

is a bounded linear operator $\Upsilon : \mathcal{B} \to \mathcal{B}$. The boundedness of Υ follows from the hyperbolic estimates (4.35) and the boundedness of the projections $P^\pm : \mathbb{R}^n \to \mathbf{E}^\pm$; for instance,

$$|e^{tA}\eta^+(e^{-tA}x)| \le Ce^{-\lambda t}|P^+\eta(e^{-tA}x)|$$
$$\le Ce^{-\lambda t}\|P^+\|\|\eta\|. \qquad (4.37)$$

The continuity of the function $x \mapsto (\Upsilon\eta)(x)$ is an immediate consequence of the following lemma from advanced calculus—essentially the Weierstrass M-test—and the estimates of the integrands as in display (4.37).

Lemma 4.15. *Suppose that $h : [0,\infty) \times \mathbb{R}^n \to \mathbb{R}^m$, given by $(t,x) \mapsto h(t,x)$, is continuous (respectively, the partial derivative h_x is continuous). If for each $y \in \mathbb{R}^n$ there is an open set $S \subset \mathbb{R}^n$ with compact closure $\bar{S}$ and a function $M : [0,\infty) \to \mathbb{R}$ such that $y \in S$, the integral $\int_0^\infty M(t)\,dt$ converges, and $|h(t,x)| \leq M(t)$ (respectively, $|h_x(t,x)| \leq M(t)$) whenever $t \in [0,\infty)$ and x is in $\bar{S}$, then $H : \mathbb{R}^n \to \mathbb{R}^m$ given by $H(x) = \int_0^\infty h(t,x)\,dt$ is continuous (respectively, H is continuously differentiable and $DH(x) = \int_0^\infty h_x(t,x)\,dt$).*

Using the definition of L_A in display (4.34) and the fundamental theorem of calculus, it is easy to prove that Υ is a right inverse for L_A, that is, $L_A \Upsilon = \mathrm{id}_{\mathcal{B}}$. As a consequence, if

$$\eta = \Upsilon(\xi \circ (\mathrm{id} + \eta)) := F(\eta), \tag{4.38}$$

then η is a solution of the infinitesimal conjugacy equation (4.33).

The function F defined in display (4.38) maps $\mathcal{B}$ into $\mathcal{B}$. Also, if η_1 and η_2 are in $\mathcal{B}$, then (by the linearity of Υ and the mean value theorem applied to the function ξ)

$$\|F(\eta_1) - F(\eta_2)\| \leq \|\Upsilon\| \, \|\xi \circ (\mathrm{id} + \eta_1) - \xi \circ (\mathrm{id} + \eta_2)\|$$
$$\leq \|\Upsilon\| \, \|D\xi\| \, \|\eta_1 - \eta_2\|.$$

Using the definition of ξ and the properties of the bump function ρ, we have that

$$\|D\xi\| \leq \sup_{|x| \leq r} \|Df(x) - A\| + \frac{4}{r} \sup_{|x| \leq r} |f(x) - Ax|.$$

Because Df is continuous, there is some positive number r such that $\|Df(x) - A\| < 1/(10\|\Upsilon\|)$ whenever $|x| \leq r$. By Taylor's theorem (applied to the C^1 function f) and the obvious estimate of the integral form of the remainder, if $|x| \leq r$, then $|f(x) - Ax| < r/(10\|\Upsilon\|)$. For the number $r > 0$ just chosen, we have the estimate $\|\Upsilon\|\|D\xi\| < 1/2$; therefore, F is a contraction on $\mathcal{B}$. By the contraction mapping theorem applied to the restriction of F on the closed subspace $\mathcal{B}_0$ of $\mathcal{B}$ consisting of those elements that vanish at the origin, the equation (4.38) has a unique solution $\eta \in \mathcal{B}_0$, which also satisfies the infinitesimal conjugacy equation (4.33).

We will show that $G := \mathrm{id} + \eta$ is a local conjugacy. Apply Proposition 4.13 to the infinitesimal conjugacy equation (4.33) to obtain the identity

$$\frac{d}{dt}(e^{-tA}\eta(e^{tA}x)) = e^{-tA}\xi(G(e^{tA}x)).$$

Using the definitions of G and Y, it follows immediately that

$$\frac{d}{dt}(e^{-tA}G(e^{tA}x)) = -e^{-tA}AG(e^{tA}x) + e^{-tA}Y(G(e^{tA}x))$$

and (by the product rule)

$$e^{-tA}\frac{d}{dt}G(e^{tA}x) = e^{-tA}Y(G(e^{tA}x)).$$

Therefore, the function given by $t \mapsto G(e^{tA}x)$ is the integral curve of Y starting at the point $G(x)$. But, by the definition of the flow φ_t of Y, this integral curve is the function $t \mapsto \varphi_t(G(x))$. By uniqueness, $G(e^{tA}x) = \varphi_t(G(x))$. Because Y is linear on the complement of a compact set, Gronwall's inequality can be used to show that the flow of Y is complete. Hence, the conjugacy equation holds for all $t \in \mathbb{R}$.

It remains to show that the continuous function $G : \mathbb{R}^n \to \mathbb{R}^n$ given by $G(x) = x + \eta(x)$ is a homeomorphism. Since η is bounded on $\mathbb{R}^n$, the map $G = \text{id} + \eta$ is surjective. To prove this fact, choose $y \in \mathbb{R}^n$, note that the equation $G(x) = y$ has a solution of the form $x = y + z$ if $z = -\eta(y + z)$, and apply Brouwer's fixed point theorem to the map $z \mapsto -\eta(y + z)$ on the ball of radius $\|\eta\|$ centered at the origin. (Using similar ideas, it is also easy to prove that G is proper; that is, the inverse image under G of every compact subset of $\mathbb{R}^n$ is compact.) We will show next that G is injective. If x and y are in $\mathbb{R}^n$ and $G(x) = G(y)$, then $\varphi_t(G(x)) = \varphi_t(G(y))$ and, by the conjugacy relation, $e^{tA}x + \eta(e^{tA}x) = e^{tA}y + \eta(e^{tA}y)$. By the linearity of e^{tA}, we have that

$$|e^{tA}(x - y)| = |\eta(e^{tA}y) - \eta(e^{tA}x)| \le 2\|\eta\|.$$

But, if u is a nonzero vector in $\mathbb{R}^n$, then the function $t \mapsto |e^{tA}u|$ is unbounded on $\mathbb{R}$ (see Exercise 4.16). Hence, $x = y$, as required. By Brouwer's theorem on invariance of domain, the bijective continuous map G is a homeomorphism. (Brouwer's theorem can be avoided by using instead the following elementary fact: A continuous, proper, bijective map from $\mathbb{R}^n$ to $\mathbb{R}^n$ is a homeomorphism.) $\square$

Exercise 4.16. Suppose $A : \mathbb{R}^n \to \mathbb{R}^n$ is a linear map and $u \in \mathbb{R}^n$. Show that if A is infinitesimally hyperbolic and $u \ne 0$, then the function $t \mapsto |e^{tA}u|$ is unbounded for $t \in \mathbb{R}$.

Exercise 4.17. Prove that for each $r > 0$ there is a C^∞ function $\rho : \mathbb{R}^n \to [0,1]$ with the following properties: $\rho(x) \equiv 1$ for $|x| < r/2$, $\rho(x) \equiv 0$ for $|x| > r$, and $|d\rho(x)| < 4/r$ for $x \in \mathbb{R}^n$.

In the classic paper [114], Philip Hartman shows that if $a > b > 0$ and $c \ne 0$, then there is no C^1 linearizing conjugacy at the origin for the analytic differential equation

$$\dot{x} = ax, \quad \dot{y} = (a - b)y + cxz, \quad \dot{z} = -bz. \tag{4.39}$$

On the other hand, he proves the following two results. (1) If a C^2 vector field has a rest point such that either all eigenvalues of its linearization have negative real parts or all eigenvalues have positive real parts, then the vector field is locally C^1 conjugate to its linearization. (2) If a C^2 *planar* vector field has a hyperbolic rest point, then the vector field is locally C^1 conjugate to its linearization. Hartman proves the analogs of these theorems for diffeomorphisms and then derives the corresponding theorems for vector fields as corollaries (cf. [54]). We also note that Shlomo Sternberg proved that the analytic planar system

$$\dot{x} = -x, \quad \dot{y} = -2y + x^2 \tag{4.40}$$

is not C^2 linearizable. Thus, it should be clear that smooth linearization is a delicate issue.

In general, the conjugacy obtained as in the proof of Theorem 4.14 is not smooth. This fact is illustrated by linearizing the smooth scalar vector field given by $f(x) = -ax + \xi(x)$ where $a > 0$, $\xi(0) = 0$, and $\xi'(0) = 0$. Suppose that ξ vanishes outside a sufficiently small open subset of the origin with radius $r > 0$ so that, as in the proof of Theorem 4.14, the linearizing transformation is $G = \mathrm{id} + \eta$ and

$$\eta(x) = \int_0^\infty e^{-at} \xi(e^{at}x + \eta(e^{at}x)) \, dt. \tag{4.41}$$

Set $\Xi := \xi \circ (\mathrm{id} + \eta)$ and $R := r + \|\eta\|$. By an application of the (reverse) triangle inequality, $|x + \eta(x)| \geq r$ whenever $|x| \geq R$. Hence, $\Xi(x) = 0$ whenever $|x| > R$. For $x \neq 0$, the change of variable $u := e^{at}$ transforms equation (4.41) into

$$\eta(x) = \frac{1}{a} \int_1^{R/|x|} \frac{\Xi(ux)}{u^2} \, du.$$

Moreover, if $x > 0$, then (with $w = ux$)

$$\eta(x) = \frac{x}{a} \int_x^R \frac{\Xi(w)}{w^2} \, dw,$$

and if $x < 0$, then

$$\eta(x) = -\frac{x}{a} \int_{-R}^x \frac{\Xi(w)}{w^2} \, dw.$$

If η were continuously differentiable in a neighborhood of the origin, then we would have the identity

$$\eta'(x) = \frac{1}{a} \int_x^R \frac{\Xi(w)}{w^2} \, dw - \frac{\Xi(x)}{ax}$$

for $x > 0$ and the identity

$$\eta'(x) = -\frac{1}{a}\int_{-R}^{x}\frac{\Xi(w)}{w^2}\,dw - \frac{\Xi(x)}{ax}$$

for $x < 0$. Because the left-hand and right-hand derivatives agree at $x = 0$ and $\Xi(0) = \Xi'(0) = 0$, it would follow that

$$\int_{0}^{R}\frac{\Xi(w)}{w^2}\,dw = -\int_{-R}^{0}\frac{\Xi(w)}{w^2}\,dw.$$

But this equality is not true in general. For example, it is not true if $\xi(x) = \rho(x)x^2$ where ρ is a bump function as in the proof of Theorem 4.14. In this case, the integrands are nonnegative and not identically zero.

There are at least two ways to avoid the difficulty just described (cf. [54]). First, note that the operator L_A, for the case $Ax = -ax$, has the formal right inverse given by

$$(\Upsilon\eta)(x) := -\int_{0}^{\infty}e^{at}\eta(e^{-at}x)\,dt.$$

Thus, the formal conjugacy is $G = \mathrm{id} + \eta$, where

$$\eta(x) = -\int_{0}^{\infty}e^{at}\xi(e^{-at}x + \eta(e^{-at}x))\,dt.$$

In this case, no inconsistency arises from the assumption that $\eta'(0)$ exists, and this method does produce a smooth conjugacy for the scalar vector fields under consideration here.

Another idea that can be used to search for a smooth conjugacy is to differentiate both sides of the desired conjugacy relation

$$e^{tA}G(x) = G(\phi_t(x))$$

with respect to t at $t = 0$. Or, equivalently, we can seek a conjugacy $G = \mathrm{id} + \eta$ such that

$$DG(G^{-1}(y))f(G^{-1}(y)) = Ay,$$

where $y = x + \eta(x)$; that is, the push forward of f by G is the linearization of f at the origin. In this case, it is easy to see that η determines a linearizing transformation G if η is a (smooth) solution of the first-order partial differential equation

$$D\eta(x)f(x) + a\eta(x) = -\xi(x).$$

To determine η, let ϕ_t denote the flow of f, replace x by the integral curve $t \mapsto \phi_t(x)$, and note that along this characteristic curve

$$\frac{d}{dt}\eta(\phi_t(x)) + a\eta(\phi_t(x)) = -\xi(\phi_t(x))$$

(see Section 3.6.4). By variation of parameters, we have that

$$\frac{d}{dt} e^{at} \eta(\phi_t(x)) = -e^{at} \xi(\phi_t(x));$$

and, after integration, we have the identity

$$e^{at} \eta(\phi_t(x)) = \eta(x) - \int_0^t e^{as} \xi(\phi_s(x))\, ds.$$

Recall that a function $h : \Omega \subseteq \mathbb{R}^k \to \mathbb{R}^n$ is called Hölder continuous on Ω if there are positive constants $c > 0$ and $0 < \nu \le 1$ such that

$$|h(x) - h(y)| \le c|x - y|^{\nu}$$

for all $x, y \in \Omega$. The number ν is called the Hölder exponent. If η is continuously differentiable, $\eta(0) = 0$, $D\eta(0) = 0$, and $D\eta$ is Hölder continuous with exponent η, then (using Taylor's theorem)

$$|\eta(x)| \le \int_0^1 |D\eta(sx)x|\, ds = \frac{c}{1+\nu} |x|^{1+\nu}.$$

By Corollary 2.78, if x is near $x = 0$ and $\epsilon > 0$ is given, then there is some constant $C > 0$ such that $|\phi_t(x)| \le C e^{(\epsilon-a)t}$. Combining these estimates, it is easy to see that there is some constant $K > 0$ such that

$$|e^{at} \eta(\phi_t(x))| \le K e^{((1+\eta)\epsilon - \eta a)t}.$$

Hence, if $0 < \epsilon < a\eta/(1+\eta)$, then

$$\lim_{t\to\infty} |e^{at} \eta(\phi_t(x))| = 0$$

and we would have

$$\eta(x) = \int_0^\infty e^{at} \xi(\phi_t(x))\, dt; \qquad (4.42)$$

that is, the function η defined by the smooth linearizing transformation $G = \mathrm{id} + \eta$ is given by the formula (4.42).

If, for example, $\xi \in C^2$, then the function η defined in display (4.42) is C^1 and $G = \mathrm{id} + \eta$ is a smooth linearizing transformation (see Exercise 4.20). As a test of the validity of this method, consider the differential equation $\dot{x} = -ax + x^2$, where $a > 0$. In this case, the flow ϕ_t can be computed explicitly and the integral (4.42) can be evaluated to obtain the smooth near-identity linearizing transformation $G : (-a, a) \to \mathbb{R}$ given by

$$G(x) = x + \frac{x^2}{a - x}. \qquad (4.43)$$

Exercise 4.18. Why are Hölder exponents restricted to the interval $(0, 1]$?

Exercise 4.19. Verify that the function G defined in display (4.43) is a linearizing transformation (at the origin) for the differential equation $\dot{x} = -ax + x^2$, where $a > 0$.

Exercise 4.20. Suppose that $a > 0$, $r > 0$, and $\xi : \mathbb{R} \to [0, \infty)$ is a C^2 function which vanishes in the complement of the interval $(-r, r)$. Also, suppose that $\xi(0) = \xi'(0) = 0$. Show that if r is sufficiently small and ϕ_t is the flow of the differential equation $\dot{x} = -ax + \xi(x)$, then η defined in display (4.42) is a C^1 function.

5

Continuation of Periodic Solutions

A fundamental engineering problem is to determine the response of a physical system to an applied force. In this chapter some mathematical ideas are introduced that can be used to address a classic case of this problem where the physical system is an oscillator that is modeled by a differential equation with periodic orbits and the applied force is modeled as a small periodic perturbation. Partial answers to several important questions will be given. Which, if any, of the unperturbed periodic orbits persist under the perturbation? Are the perturbed periodic orbits stable? Can the perturbed periodic orbits be approximated by analytic formulas? Although we will restrict most of our discussion to planar systems—the case of most practical value, many of the results of the theory presented here can be easily generalized to multidimensional systems (see [189] for some results in the spirit of this chapter). On the other hand, the multidimensional case will be discussed in detail for systems with first integrals and in the exercises.

Continuation theory has a long history in applied science and mathematics; it is still an active area of mathematical research. Thus, there is a mathematical and scientific literature on this topic that is far too extensive to be reviewed here. Nevertheless, every student of the subject should be aware of the classic books by Aleksandr A. Andronov, Aleksandr A. Vitt, and Semen E. Khaiken [9], Nikolai N. Bogoliubov and Yuri A. Mitropolsky [28], Chihiro Hayashi [117], Nikolai Minorsky [162], and James J. Stoker [215]; and the more recent works of Miklós Farkas [84], John Guckenheimer and Philip Holmes [103], Jack K. Hale [106], Jirair K. Kevorkian and Julian D. Cole [134], James Murdock [170], Ali H. Nayfey [175], and Stephen W. Wiggins [233].

5.1   A Classic Example: van der Pol's Oscillator

An important mathematical model in the history of our subject is known as *van der Pol's equation*

$$\ddot{x} + \epsilon(x^2 - 1)\dot{x} + \omega^2 x = a \sin \Omega t. \tag{5.1}$$

After this differential equation was introduced by Lord Rayleigh [188] in 1883, it has been suggested as a model for many different physical phenomena. For example, Balthasar van der Pol [222] investigated it more extensively when he studied the equation in 1926 as a model of the voltage in a triode circuit. Then, just two years later, van der Pol and Johannes van der Mark [224] proposed the equation as a model for the human heartbeat. In this introduction, we will use the differential equation (5.1) to illustrate some of the ideas that will be explored more fully later in this chapter.

Let us observe some of the dynamical features of the van der Pol equation. If $a = 0$ and $\epsilon = 0$, then equation (5.1) is the familiar model of a linear spring; that is, a spring with restoring force modeled by Hooke's law. This equation is often referred to as the spring equation or the harmonic oscillator. The term $a \sin \Omega t$ represents a periodic external force with amplitude a, period $2\pi/\Omega$ and frequency Ω. The term $\epsilon(x^2 - 1)\dot{x}$ can be viewed as representing a nonlinear damping. The "direction" of this damping depends on the state $(x, \dot{x})$ of the system where x represents position and $\dot{x}$ represents velocity. In fact, the energy of the spring is given by

$$E := \frac{1}{2}\dot{x}^2 + \frac{1}{2}\omega^2 x^2,$$

and has time derivative

$$\dot{E} = a\dot{x} \sin \Omega t - \epsilon(x^2 - 1)\dot{x}^2.$$

Thus, the external forcing and the nonlinear damping cause energy fluctuations. Energy due to the damping leaves the system while $|x| > 1$ and is absorbed while $|x| < 1$.

Our subject is motivated by the following basic question: If the current state of the system is known, what does the model predict about its future states? Even though the van der Pol equation has been studied intensively, we cannot give a complete answer to this question. Nevertheless, as we will see, many useful predictions can be made. In particular, in this section we will show how to determine the steady state behavior of the system when there is no external force and the damping is small.

Let us consider the unforced, weakly damped, scaled van der Pol equation given by

$$\ddot{x} + \epsilon(x^2 - 1)\dot{x} + x = 0. \tag{5.2}$$

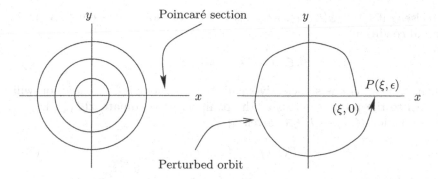

Figure 5.1: The left panel depicts the phase portrait for the harmonic oscillator. The right panel shows a perturbed orbit with initial state $(\xi, 0)$ on the positive x-axis that returns to the positive x-axis at the point $P(\xi, \epsilon)$.

The corresponding unperturbed $(\epsilon = 0)$ equation $\ddot{x} + x = 0$ is explicitly solvable. Indeed, the solution with initial state $(x_0, \dot{x}_0)$ is given by

$$t \mapsto x_0 \cos t + \dot{x}_0 \sin t.$$

In particular, all solutions of the unperturbed system, except for the solution corresponding to the rest point at $(0,0)$, are periodic with period 2π. Hence, there is no problem predicting the future states of the unperturbed system.

What happens when $\epsilon \neq 0$? Does the differential equation (5.2) have a periodic solution? If it does, then can we find a "formula" that represents the solution? Or, if this is not possible, how can we approximate the periodic solution? Is the periodic solution stable? We will approach such questions using the geometric interpretation of the differential equation as a system in the phase plane; that is, as the equivalent first order system given by

$$\dot{x} = -y,$$
$$\dot{y} = x - \epsilon(x^2 - 1)y. \tag{5.3}$$

Here, the choice $\dot{x} = y$ works just as well, but the minus sign ensures that trajectories move in the positive sense of the usual orientation of the Euclidean plane.

If $\epsilon = 0$, then all orbits of system (5.3), except the rest point at the origin, are circles that intersect the positive x-axis as shown in the left panel of Figure 5.1. To investigate the orbits of the system (5.3) for $\epsilon \neq 0$, we will consider the Poincaré map defined on the positive x-axis.

Let us note that if $\epsilon \neq 0$ is sufficiently small, then the orbit of the solution of system (5.3) with initial condition $(x(0), y(0)) = (\xi, 0)$ remains close to the circle with radius ξ at least until it returns to the x-axis after a finite time $T(\xi, \epsilon)$ that depends on the initial point and the value of ϵ. More precisely, if $t \mapsto (x(t,\xi,\epsilon), y(t,\xi,\epsilon))$ is the solution of system (5.3) with initial condition

$$x(0,\xi,\epsilon) = \xi, \qquad y(0,\xi,\epsilon) = 0,$$

then, as long as ϵ is sufficiently small, the trajectory of this solution will return to the positive x-axis at the point with coordinate $x(T(\xi,\epsilon),\xi,\epsilon)$. The function $(\xi,\epsilon) \mapsto P(\xi,\epsilon)$ given by

$$P(\xi,\epsilon) := x(T(\xi,\epsilon),\xi,\epsilon) \tag{5.4}$$

is called the *parametrized return map* (see the right panel of Figure 5.1).

If $P(\xi,\epsilon) = \xi$, then $t \mapsto (x(t,\xi,\epsilon), y(t,\xi,\epsilon))$ is a periodic solution of the system (5.3) with period $T(\xi,\epsilon)$. In other words, if ξ is a fixed point of the map $\xi \mapsto P(\xi,\epsilon)$ or a zero of the associated *displacement function* $\delta(\xi,\epsilon) = x(T(\xi,\epsilon),\xi,\epsilon) - \xi$, then $(\xi,0)$ is the initial point for a periodic orbit of the perturbed system.

Because $\delta(\xi,0) \equiv 0$, it is natural to look for the root ξ implicitly as a function β of ϵ such that, for $\epsilon \neq 0$, the point $\xi = \beta(\epsilon)$ is the initial point of a periodic solution of system (5.3). More precisely, we seek a function β defined on some neighborhood of $\epsilon = 0$ in $\mathbb{R}$ such that $\delta(\beta(\epsilon),\epsilon) \equiv 0$. The obvious way to find an implicit solution is to apply the implicit function theorem (Theorem 1.259).

In the present context, the displacement function is defined by $\delta : U \times V \to \mathbb{R}$ where U and V are both open subsets of $\mathbb{R}$. Moreover, we have that $\delta(\xi,0) \equiv 0$. If there were some point $(\xi,0)$ such that $\delta_\xi(\xi,0) \neq 0$, then by the implicit function theorem there would be an implicit solution and our problem would be solved. But it is clear that the hypothesis of the implicit function theorem is *not* satisfied. In fact, because $\delta(\xi,0) \equiv 0$, we have that $\delta_\xi(\xi,0) \equiv 0$. As we will see, however, the implicit function theorem does apply after a further reduction.

Let us use the Taylor series of δ at $\epsilon = 0$ to obtain the equation

$$\delta(\xi,\epsilon) = \epsilon\delta_\epsilon(\xi,0) + O(\epsilon^2)$$

where the $O(\epsilon^2)$ term denotes the remainder. This notation is used formally in the following way: The statement $f(\epsilon) = g(\epsilon) + O(\epsilon^2)$ means that there are constants $K > 0$ and $\epsilon_0 > 0$ such that the inequality

$$|f(\epsilon) - g(\epsilon)| < K\epsilon^2.$$

holds for $|\epsilon| < \epsilon_0$. The required reduction is accomplished by defining a new function

$$\Delta(\xi,\epsilon) := \delta_\epsilon(\xi,0) + O(\epsilon)$$

so that

$$\delta(\xi,\epsilon) = \epsilon(\delta_\epsilon(\xi,0) + O(\epsilon)) = \epsilon\Delta(\xi,\epsilon).$$

Clearly, if there is a function $\epsilon \mapsto \beta(\epsilon)$ such that $\Delta(\beta(\epsilon), \epsilon) \equiv 0$, then $\delta(\beta(\epsilon), \epsilon) \equiv 0$.

Even though the implicit function theorem does not apply to the displacement function δ, it might well apply to the function Δ. At any rate, we have reduced the original search for a periodic solution of the perturbed van der Pol equation to the problem of finding implicit solutions of the equation $\Delta(\xi, \epsilon) = 0$. Thus, by the implicit function theorem, we have the following proposition: If $\xi > 0$ is a *simple zero* of the function $\xi \mapsto \Delta(\xi, 0)$, that is, $\Delta(\xi, 0) = 0$ and $\Delta_\xi(\xi, 0) \neq 0$, or equivalently if $\delta_\epsilon(\xi, 0) = 0$ and $\delta_{\xi\epsilon}(\xi, 0) \neq 0$, then an implicit solution $\xi = \beta(\epsilon)$ exists. The function $\xi \mapsto \delta_\epsilon(\xi, 0)$ is called the *reduced displacement function,* and a simple zero of the reduced bifurcation function (respectively the corresponding unperturbed periodic orbit) is called a *continuation point* of periodic solutions of the system (5.3) (respectively a continuable periodic orbit). Also, a periodic orbit is said to *persist* if it is continuable. The ideas used to prove our proposition recur in every continuation problem that we will consider; their implementation constitutes the first part, called the *reduction step,* in the solution of the continuation problem.

The second part of the continuation method is the *identification step,* that is, the identification of the reduced displacement function in terms of the original differential equation. For system (5.3), perhaps the most direct route to the identification of the reduced displacement function is via a change to polar coordinates. But, as an illustration of a general method, let us work directly in the original variables and identify the reduced function by solving a variational equation derived from system (5.3).

To carry out the identification step, apply the chain rule to compute the partial derivative

$$\delta_\epsilon(\xi, 0) = \dot{x}(T(\xi, 0), \xi, 0)T_\epsilon(\xi, 0) + x_\epsilon(T(\xi, 0), \xi, 0)$$

and evaluate at $\epsilon = 0$ to obtain the equality

$$\dot{x}(T(\xi, 0), \xi, 0) = -y(0, \xi, 0) = 0.$$

In particular, the function $\xi \mapsto \dot{x}(T(\xi, 0), \xi, 0)T_\epsilon(\xi, 0)$ and all of its derivatives vanish. Thus, to complete the identification step it suffices to determine the partial derivative $x_\epsilon(T(\xi, 0), \xi, 0)$. To do this, let us compute the partial derivative with respect to ϵ at $\epsilon = 0$ of both sides of the differential equation (5.3) to obtain a *variational equation.* Also, let us compute the partial derivative with respect to ϵ of both sides of each of the initial conditions $x(0, \xi, \epsilon) = \xi$ and $y(0, \xi, \epsilon) = 0$ to obtain the corresponding (variational) initial value problem

$$\dot{x}_\epsilon = -y_\epsilon, \quad \dot{y}_\epsilon = x_\epsilon - (x^2 - 1)y, \quad x_\epsilon(0, \xi, 0) = 0, \quad y_\epsilon(0, \xi, 0) = 0 \quad (5.5)$$

whose solution is $t \mapsto (x_\epsilon(t, \xi, 0), y_\epsilon(t, \xi, 0))$.

The variational initial value problem (5.5) is expressed in matrix form by

$$\dot{W} = AW + b(t), \qquad W(0) = 0 \tag{5.6}$$

where

$$A = \begin{pmatrix} 0 & -1 \\ 1 & 0 \end{pmatrix}, \qquad b(t) = \begin{pmatrix} 0 \\ (1 - x^2(t, \xi, 0))y(t, \xi, 0) \end{pmatrix},$$

and this nonhomogeneous 2×2 linear system is readily solved by the variation of parameters formula (2.67). Indeed, let us recall that the principal fundamental matrix solution at $t = 0$ of the associated homogeneous linear system $\dot{W} = AW$ is the 2×2 matrix function $t \mapsto \Phi(t)$ with $\dot{\Phi} = A\Phi$ and $\Phi(0) = I$, and the solution $t \mapsto W(t)$ of the initial value problem (5.6) is given by

$$W(t) = \Phi(t)W(0) + \Phi(t) \int_0^t \Phi^{-1}(s)b(s) \, ds. \tag{5.7}$$

Moreover, for the system (5.3), we have that $W(0) = 0$, $T(\xi, 0) = 2\pi$, and

$$\Phi(t) = e^{tA} = \begin{pmatrix} \cos t & -\sin t \\ \sin t & \cos t \end{pmatrix}.$$

It follows that

$$x(t, \xi, 0) = \xi \cos t, \qquad y(t, \xi, 0) = \xi \sin t$$

and, in addition,

$$\begin{pmatrix} x_\epsilon(2\pi, \xi, 0) \\ y_\epsilon(2\pi, \xi, 0) \end{pmatrix} = \Phi(2\pi) \int_0^{2\pi} \Phi^{-1}(s)b(s) \, ds$$

$$= \begin{pmatrix} \int_0^{2\pi} \sin s[(1 - \xi^2 \cos^2 s)\xi \sin s] \, ds \\ \int_0^{2\pi} \cos s[(1 - \xi^2 \cos^2 s)\xi \sin s] \, ds \end{pmatrix}.$$

After an elementary integration, we have that

$$\delta_\epsilon(\xi, 0) = \frac{\pi}{4}\xi(4 - \xi^2), \qquad \xi > 0, \tag{5.8}$$

and therefore $\xi = 2$ is a simple zero of the reduced displacement function $\xi \mapsto \delta_\epsilon(\xi, 0)$. Hence, the unperturbed periodic orbit with radius 2 persists. But since $\xi = 2$ is the only zero of the displacement function, all other periodic orbits of the unperturbed system are destroyed by the perturbation. In particular, there is a function $\epsilon \mapsto \beta(\epsilon)$ defined on some neighborhood of $\epsilon = 0$ such that $\beta(0) = 2$, and for each ϵ in the domain of β the corresponding van der Pol system (5.3) has a periodic orbit with initial condition $(x(0), y(0)) = (\beta(\epsilon), 0)$.

The theory we have just developed to analyze the existence of continuations of periodic solutions of the van der Pol equation will be generalized in the next two sections of this chapter. In Sections 5.3.6 and 5.3.7 we will discuss a method that can be used to obtain analytical approximations of the perturbed periodic orbit. For an analysis of the stability of the perturbed periodic solution see Exercise 5.3.

Let us formalize what we have done so far by considering the weakly linear system

$$\dot{u} = Au + \epsilon g(u), \quad u \in \mathbb{R}^2 \tag{5.9}$$

where

$$u = \begin{pmatrix} x \\ y \end{pmatrix}, \quad A = \begin{pmatrix} 0 & -1 \\ 1 & 0 \end{pmatrix}, \quad g(u) = \begin{pmatrix} g_1(u) \\ g_2(u) \end{pmatrix}.$$

By repeating the steps of the argument made for system (5.3), it is easy to prove the following theorem.

Theorem 5.1. *A simple zero of the function $B : (0, \infty) \to \mathbb{R}$ given by*

$$\xi \mapsto \int_0^{2\pi} g_1(\xi \cos s, \xi \sin s) \cos s + g_2(\xi \cos s, \xi \sin s) \sin s \, ds$$

is a continuation point of periodic solutions of the system (5.9). Moreover, if ξ_0 is a continuation point, then $B(\xi_0) = 0$.

Exercise 5.2. Apply Theorem 5.1 to find the continuation points of periodic solutions for the system

$$\dot{x} = -y + \epsilon p(x, y), \quad \dot{y} = x + \epsilon q(x, y)$$

where p and q are entire functions with series representations given by

$$p = \sum p_{ij} x^i y^j, \quad q = \sum q_{ij} x^i y^j.$$

For example, give a complete analysis when p, q are quadratic polynomials and again when p, q are cubic polynomials.

Exercise 5.3. [Stability] Prove that for sufficiently small ϵ the stability of the perturbed periodic solution passing near the continuation point $(\xi, 0)$ is determined by the size of $P_\xi(\xi, \epsilon)$. In particular, show that $P_\xi(\xi, \epsilon) \geq 0$ and prove the following statements: If $P_\xi(\xi, \epsilon) < 1$, then the periodic solution is (asymptotically) stable; and if $P_\xi(\xi, \epsilon) > 1$, then the periodic solution is (asymptotically) unstable. Also, note that

$$P(\xi, \epsilon) = P(\xi, 0) + \epsilon P_\epsilon(\xi, 0) + O(\epsilon^2),$$

and therefore

$$P_\xi(\xi, \epsilon) - 1 = \epsilon(\delta_{\xi\epsilon}(\xi, 0) + O(\epsilon)).$$

If, for example, $\epsilon > 0$ is sufficiently small and $\delta_{\xi\epsilon}(\xi, 0) < 0$, then the periodic orbit is stable. Thus, if ϵ is sufficiently small, then to determine the stability, it suffices to compute the sign of the mixed partial derivative at the continuation point ξ. Apply your results to determine the stability of the perturbed periodic orbit for the van der Pol equation.

Exercise 5.4. The period of the perturbed periodic orbit for the van der Pol oscillator is given by the function

$$\epsilon \mapsto T(\beta(\epsilon), \epsilon)$$

where T is the return time function that appears in the definition of the Poincaré map (5.4) and β is the implicit solution of the corresponding displacement function. Determine the first two terms of the Taylor series at $\epsilon = 0$ of the period of the perturbed periodic orbit. Hint: Use the identity

$$y(T(\beta(\epsilon), \epsilon), \beta(\epsilon), \epsilon) \equiv 0.$$

We will learn a more efficient method for computing the period of the perturbed periodic orbit in Section 5.3.6 (see Exercise 5.50).

5.1.1   Continuation Theory and Applied Mathematics

Continuation theory, also called regular perturbation theory, is very useful in applied mathematics where we wish to make predictions from a differential equation model of a physical process. In most instances, our model is a family of differential equations; that is, the model depends on parameters. If a member of the family—obtained by fixing the parameters—has a dynamical feature (for example, a rest point, periodic orbit, or invariant manifold) that is relevant to the analysis of our applied problem, then there is a natural and fundamental question: Does this feature persist if we change the parameter values? Continuation theory is a diverse collection of tools that can be used to answer this question in some situations.

In the rest of this chapter, we will extend the continuation theory for periodic solutions introduced in Section 5.1 to cover more complex problems. But, as in the example provided by the van der Pol equation, we will always look for continuations of unperturbed periodic solutions in a family of differential equations with a small parameter. We will see that the underlying ideas for the general continuation analysis are the same as those introduced in this section: Construct an appropriate displacement function; reduce to a bifurcation function whose simple zeros correspond—by an application of the implicit function theorem—to continuation points; and identify the reduced bifurcation function in terms of the given differential equation.

Perhaps our analysis of the continuation of periodic solutions for the general weakly nonlinear system provides initial evidence for the notion that the proof of a general result such as Theorem 5.1 is often easy compared with the task of applying the result to a realistic model. For our example, where the perturbation term is a single harmonic, the bifurcation function is a quadratic polynomial (formula (5.8)) and its roots are therefore easy to determine. If, however, we consider a perturbation with several harmonics, as for example in Exercise 5.2, then the problem of finding the number and position of the persistent unperturbed periodic solutions becomes more difficult. This illustrates a maxim that lies at the heart of many problems in applied mathematics: The more realistic the model, the more difficult it is to apply general theorems.

Maxim number two: General theorems are always too weak. If you work hard and are fortunate, you might develop all of the ideas necessary to prove a classic and beautiful theorem such as Theorem 5.1. You may then go to your collaborator, a very good engineer, and proudly announce your result: "If ... and ϵ is sufficiently small, then there is a periodic solution." But you know what is coming! Your collaborator will say, "That's interesting, but *how small* do I have to make the perturbation so that I can be sure there is a periodic orbit?" You are now invited to find a computable number $\epsilon_0 > 0$ and a proof that periodic solutions exist at least for $|\epsilon| < \epsilon_0$. If you succeed in doing this for the model equation (5.2), then your collaborator will be happy for a moment. But before long she comes back to you with a new perturbation term in mind: "Does your method apply if we add ... ?"

When confronted with an applied problem, there is a natural tendency for a mathematician to try to prove a theorem. Perhaps by now you feel that your contribution to the applied project is not receiving enough credit. But in fact your results are enormously valuable. Because you have answered some basic questions, *new questions can be asked.* You have also provided a way to *understand* why a periodic orbit exists. After proving a few more theorems that apply to show the existence of periodic orbits for a few more basic model equations, your understanding of periodic orbits begins to coalesce into a theory that gives a conceptual framework, which can be used by you, and others, to discuss the existence of periodic orbits in systems that are too complex to analyze rigorously.

In general, the applied mathematician faces a highly nontrivial, perhaps impossible, task when trying to rigorously verify the hypotheses of general theorems for realistic models of physical systems. In fact, doing so might require the development of a new area of mathematics. Most often, we are left to face the realization that rigorous results can only be obtained for simplified models.

Do not be discouraged.

The analysis of a mathematical model, even a simple one, deepens our understanding, sharpens our formulation of results, forces us to seek new methods of analysis, and often reveals *new phenomena.* In addition, rigorous results for simple models provide test cases that can be used to debug implementations of numerical methods that we intend to use to obtain predictions from more realistic models.

When we return as mathematicians to confront a realistic model of our original physical problem (the understanding of which is the real object of the game), it is not always clear how to continue doing *mathematics*. Instead, we turn to computation and investigate numerical methods. Perhaps we become experts in computer algebra, or we investigate computer graphics in order to find useful visual representations of our data, and so on. But when our simulations are implemented, we are happy to have knowledge of the range of expected phenomena, we are happy to be able to test our code on the simplified models we have rigorously analyzed, and we are happy to verify numerically the hypotheses of a general theorem that we have proved. All of this helps us gain confidence in our predictions.

By running our simulations, we find evidence for an answer to our original physical question. But during the process, we might also see unexpected results or we conceive new ideas to improve our simulations. These experiences motivate us to find additional rigorous results. Thus, we are naturally led back to questions in mathematics. And so it goes—a natural cycle that will be repeated many times during our attempts to understand physical phenomena.

Our technical skills will improve and our depth of understanding will increase as we master more sophisticated mathematical methods and learn from the experience of doing applied mathematics. The remainder of this chapter is intended to help provide an example of an area of applicable mathematics as well as the opportunity to gain some useful experience with some types of differential equations that appear as mathematical models.

5.2   Autonomous Perturbations

In this section we will consider the periodic solutions of the system

$$\dot{u} = f(u) + \epsilon g(u), \quad u \in \mathbb{R}^2 \tag{5.10}$$

where ϵ is a small parameter and the unperturbed system

$$\dot{u} = f(u) \tag{5.11}$$

has periodic solutions. If the unperturbed differential equation (5.11) is nonlinear, then there are at least two cases to consider in our search for periodic solutions of system (5.10): system (5.11) has a limit cycle (see Definition 1.178); and system (5.11) has an (invariant) annulus of periodic solutions. In the limit cycle case, we wish to determine if the limit cycle persists after perturbation; in the case of an invariant annulus of periodic solutions, we wish to determine which, if any, of its constituent periodic solutions persist.

Let us begin with the general assumption that the unperturbed system (5.11) has a periodic solution Γ. To employ the method suggested in

Section 5.1, we must define a displacement function. To do this, let us choose a point $v \in \Gamma$ and a curve Σ_1 that is transverse to Γ at v. By an application of the implicit function theorem, there is an open segment $\Sigma \subseteq \Sigma_1$ with $v \in \Sigma$ and some $\epsilon_0 > 0$ such that for each $\sigma \in \Sigma$ the solution of the system (5.10) with $|\epsilon| < \epsilon_0$ that has initial value σ returns to Σ_1 after some finite positive time. More precisely, there is a return time function $\mathcal{T} : \Sigma \times (-\epsilon_0, \epsilon_0) \to \mathbb{R}$ and a (parametrized) Poincaré map $P : \Sigma \times (-\epsilon_0, \epsilon_0) \to \Sigma_1$. The subset $\Sigma \subseteq \Sigma_1$ is called a *Poincaré section*.

Although the usual definitions of Poincaré section and Poincaré map do not mention parametrized systems, the important idea in the definition of the Poincaré section Σ is that solutions starting in Σ return to Σ_1. Let us also note that for each ϵ in the interval $(-\epsilon_0, \epsilon_0)$ the corresponding Poincaré map $\sigma \mapsto P(\sigma, \epsilon)$ is defined on the fixed Poincaré section Σ.

In the example in Section 5.1, the Poincaré section is a line. Here, by allowing the Poincaré section Σ to be a curve, we create a new technical problem: What is the definition of displacement on the *manifold* Σ? There are at least two options. We could define $\Delta : \Sigma \times (-\epsilon_0, \epsilon_0) \to \mathbb{R}^2$ by $\Delta(\sigma, \epsilon) := P(\sigma, \epsilon) - \sigma$. If we do so, then the "displacement" is a *vector* in $\mathbb{R}^2$. Alternatively, if we view Σ_1 as a one-dimensional manifold, then we can define the displacement function $\delta : \mathbb{R} \times (-\epsilon_0, \epsilon_0) \to \mathbb{R}$ relative to a local coordinate representation of Σ. Indeed, let us choose a function $\sigma : \mathbb{R} \to \Sigma \subseteq \mathbb{R}^2$ such that $\sigma(0) = v$ and for each $\xi \in \mathbb{R}$ the vector $\dot{\sigma}(\xi)$ is a nonzero tangent vector to Σ at $\sigma(\xi)$. A displacement function is then defined by

$$\delta(\xi, \epsilon) := \sigma^{-1}(P(\sigma(\xi), \epsilon)) - \xi. \tag{5.12}$$

If we want to avoid local coordinates, then our naïve notion of distance will have to be replaced by some measure of distance on the manifold Σ. This could be a reason to study differential geometry! The introduction of manifolds might *seem* unnecessarily complex, and certainly, the mention of manifolds and local coordinates can be avoided as long as the discussion is about curves. But, for generalizations of our continuation theory to higher dimensional problems, these ideas are unavoidable. Even in the one-dimensional case, since we will have to compute partial derivatives of the displacement, we must ultimately make some choice of local coordinates. Hence, we may as well make this choice at the outset. Let us also note that our analysis is based on the implicit function theorem. For this reason, it is advantageous to study a function $\mathbb{R} \times \mathbb{R} \to \mathbb{R}$, the usual context for the implicit function theorem, rather than a function $\mathbb{R} \times \mathbb{R} \to \mathbb{R}^2$. Thus, we will work with the definition of displacement given by equation (5.12).

Consider the case where the unperturbed system (5.11) has a limit cycle Γ with period $2\pi/\omega$ and let δ be defined as in equation (5.12). We have $\delta(0, 0) = 0$. Also, because Γ is isolated among periodic solutions of the system (5.11), the function $\xi \mapsto \delta(\xi, 0)$ does not vanish in some punctured neighborhood of $\xi = 0$. Thus, in this case the function δ is already in a form where the implicit function theorem can be directly applied. In fact, we have the following proposition: *If $\delta_\xi(0,0) \neq 0$, then Γ persists.* The conclusion means that there is a continuous function $\epsilon \mapsto \beta(\epsilon)$ defined in some interval containing $\epsilon = 0$ with $\beta(0) = 0$ and $\delta(\beta(\epsilon), \epsilon) \equiv 0$. Also, it is easy to identify $\delta_\xi(0,0)$. By the definition given in equation (5.12), the number $\delta_\xi(0,0) + 1$ is the local representative of the derivative of the Poincaré map on Σ at $\{v\} = \Gamma \cap \Sigma$. In other words, $\delta_\xi(0,0) \neq 0$ if and only if the derivative of the Poincaré map is not the identity at v. A periodic orbit in the plane with this property is called *hyperbolic*. More generally, a periodic orbit Γ is *hyperbolic* if the derivative of the Poincaré map at v has no eigenvalue with modulus one.

To identify $\delta_\xi(0,0)$ in terms of the function f, let

$$t \mapsto u(t, \zeta, \epsilon), \quad \zeta \in \mathbb{R}^2, \ \epsilon \in \mathbb{R}$$

denote the solution of system (5.10) with initial condition $u(0, \zeta, \epsilon) = \zeta$, and define the local representation of the return time map $T : \mathbb{R} \times (-\epsilon_0, \epsilon_0) \to \mathbb{R}$ by $T(\xi, \epsilon) = \mathcal{T}(\sigma(\xi), \epsilon)$. From the definition of the displacement in display (5.12), we have

$$\sigma(\delta(\xi, \epsilon) + \xi) = P(\sigma(\xi), \epsilon) = u(T(\xi, \epsilon), \sigma(\xi), \epsilon). \tag{5.13}$$

Set $\epsilon = 0$ and note that $\xi \mapsto \sigma(\delta(\xi, \epsilon) + \xi)$ defines a curve in $\Sigma_1 \subseteq \mathbb{R}^2$. After differentiation with respect to ξ at $\xi = 0$, we obtain an equality between tangent vectors to Σ at v. In fact,

$$\begin{aligned}(\delta_\xi(0,0) + 1)\dot\sigma(0) &= \dot u(T(0,0), v, 0)T_\xi(0,0) + u_\zeta(T(0,0), v, 0)\dot\sigma(0) \\ &= T_\xi(0,0)f(v) + u_\zeta(2\pi/\omega, v, 0)\dot\sigma(0). \end{aligned} \tag{5.14}$$

To be (absolutely) precise, the left hand side is

$$\sigma_*(0)\Big[(\delta_\xi(0,0) + 1)\frac{\partial}{\partial\xi}\Big]$$

where $\frac{\partial}{\partial\xi}$ denotes the unit tangent vector to $\mathbb{R}$ at $\xi = 0$ and $\sigma_*(0)$ is the linear map given by the differential of σ. This differential is a linear map from the tangent space of $\mathbb{R}$ at $\xi = 0$ to the tangent space of Σ at v. We represent this quantity as a vector in $\mathbb{R}^2$ that is tangent to Σ at v:

$$\sigma_*(0)(\delta_\xi(0,0) + 1)\frac{\partial}{\partial\xi} = (\delta_\xi(0,0) + 1)\sigma_*(0)\frac{\partial}{\partial\xi} = (\delta_\xi(0,0) + 1)\dot\sigma(0).$$

Similar remarks apply to the identifications made on the right hand side.

An expression for $\delta_\xi(0,0)$ can be determined from equation (5.14) once we compute the derivative $u_\zeta(2\pi/\omega, v, 0)$. Let us note that by taking the partial derivative with respect to ζ in the equations

$$\dot u(t, \zeta, 0) = f(u(t, \zeta, 0)), \quad u(0, \zeta, 0) = \zeta,$$

it is easy to see that the function $t \mapsto u_{\zeta}(t, \zeta, 0)$ is the matrix solution of the homogeneous variational equation (also called the first variational equation) given by

$$\dot{W} = Df(u(t, \zeta, 0))W \qquad (5.15)$$

with initial condition $W(0) = I$ where Df denotes the derivative of the function f. In other words, $t \mapsto u_{\zeta}(t, \zeta, 0)$ is the principal fundamental matrix solution of the system (5.15) at $t = 0$ and the desired derivative $u_{\zeta}(2\pi/\omega, v, 0)$ is just the value of the solution of the variational initial value problem at $t = 2\pi/\omega$.

Let $\varphi_t(\zeta) := u(t, \zeta, 0)$ denote the flow of the differential equation (5.11) and $t \mapsto \Phi(t)$ the principal fundamental matrix solution of the system (5.15) at $t = 0$. The following proposition is simple but fundamental:

$$\Phi(t)f(\zeta) = f(\varphi_t(\zeta)).$$

To prove it, note that $\Phi(0)f(\zeta) = f(\zeta)$ and

$$\frac{d}{dt}f(\varphi_t(\zeta)) = Df(\varphi_t(\zeta))f(\varphi_t(\zeta)).$$

Thus, $t \mapsto f(\varphi_t(\zeta))$ and $t \mapsto \Phi(t)f(\zeta)$ are solutions of the same initial value problem, and therefore they must be equal.

Define $f^{\perp} = Rf$ where R is the rotation matrix $\left(\begin{smallmatrix} 0 & -1 \\ 1 & 0 \end{smallmatrix}\right)$ and note that f and $f^{\perp}$ are linearly independent at each point of the plane at which f is nonzero (for example at each point on Γ). If $f(\zeta) \neq 0$, then there are two real-valued functions $t \mapsto a(t, \zeta)$ and $t \mapsto b(t, \zeta)$ such that

$$\Phi(t)f^{\perp}(\zeta) = a(t, \zeta)f(\varphi_t(\zeta)) + b(t, \zeta)f^{\perp}(\varphi_t(\zeta)). \qquad (5.16)$$

We will soon find useful formulas for a and b. Before we do so, let us note that the fundamental matrix $\Phi(t)$ is represented as a linear transformation from $\mathbb{R}^2$, with the basis $\{f(\zeta), f^{\perp}(\zeta)\}$, to $\mathbb{R}^2$, with the basis $\{f(\varphi_t(\zeta)), f^{\perp}(\varphi_t(\zeta))\}$, by the matrix

$$\Phi(t) = \begin{pmatrix} 1 & a(t, \zeta) \\ 0 & b(t, \zeta) \end{pmatrix}. \qquad (5.17)$$

In equation (5.14), $\dot{\sigma}(0)$ is a tangent vector at $v \in \Sigma \subseteq \mathbb{R}^2$. Hence, there are real constants c_1 and c_2 such that

$$\dot{\sigma}(0) = c_1 f(v) + c_2 f^{\perp}(v),$$

and therefore

$$(\delta_{\xi}(0,0) + 1)(c_1 f(v) + c_2 f^{\perp}(v))$$
$$= T_{\xi}(0,0)f(v) + \Phi(2\pi/\omega)(c_1 f(v) + c_2 f^{\perp}(v))$$
$$= T_{\xi}(0,0)f(v) + c_1 f(v) + c_2 a(2\pi/\omega, v)f(v) + c_2 b(2\pi/\omega, v)f^{\perp}(v).$$

Moreover, because Σ is transverse to Γ, we have $c_2 \neq 0$. Using this fact and the linear independence of f and $f^\perp$, it follows that

$$\delta_\xi(0,0) = b(2\pi/\omega, v) - 1, \tag{5.18}$$

$$\begin{aligned} T_\xi(0,0) &= -c_2 a(2\pi/\omega, v) + c_1 \delta_\xi(0,0) \\ &= -c_2 a(2\pi/\omega, v) + c_1(b(2\pi/\omega, v) - 1). \end{aligned} \tag{5.19}$$

Let us identify the quantities $a(2\pi/\omega, v)$ and $b(2\pi/\omega, v)$ *geometrically*. From equation (5.18), it is clear that $b(2\pi/\omega, v)$ is the (local representative of the) derivative of the Poincaré map for the unperturbed system (5.11) at $\{v\} = \Gamma \cap \Sigma$. If $\dot\sigma(0) = -f^\perp(v)$ (for example, if we take $t \mapsto \sigma(t)$ to be the solution of the differential equation $\dot u = -f^\perp(u)$ with initial condition $u(0) = v$), then $c_1 = 0$, $c_2 = -1$, and $a(2\pi/\omega, \zeta)$ is the derivative of the (local representative of the) return time map for (5.11) on Σ at v.

Recall that the Euclidean divergence and curl of the vector function $f : \mathbb{R}^2 \to \mathbb{R}^2$ with $f(x,y) = (f_1(x,y), f_2(x,y))$ are defined as follows:

$$\operatorname{div} f(x,y) := \frac{\partial f_1}{\partial x}(x,y) + \frac{\partial f_2}{\partial y}(x,y),$$

$$\operatorname{curl} f(x,y) := \frac{\partial f_2}{\partial x}(x,y) - \frac{\partial f_1}{\partial y}(x,y).$$

Also, the scalar curvature function of the smooth curve $t \mapsto (x(t), y(t))$ is given by

$$\kappa := \frac{\dot x \ddot y - \dot y \ddot x}{(\dot x^2 + \dot y^2)^{3/2}}.$$

We will write $\kappa(t, \zeta)$ to denote the scalar curvature along the curve $t \mapsto \varphi_t(\zeta)$ given by the phase flow φ_t of an autonomous planar differential equation.

Theorem 5.5 (Diliberto's Theorem). *Let φ_t denote the flow of the differential equation $\dot u = f(u)$, $u \in \mathbb{R}^2$. If $f(\zeta) \neq 0$, then the principal fundamental matrix solution $t \mapsto \Phi(t)$ at $t = 0$ of the homogeneous variational equation*

$$\dot W = Df(\varphi_t(\zeta))W$$

is such that

$$\Phi(t)f(\zeta) = f(\varphi_t(\zeta)),$$
$$\Phi(t)f^\perp(\zeta) = a(t,\zeta)f(\varphi_t(\zeta)) + b(t,\zeta)f^\perp(\varphi_t(\zeta))$$

where

$$b(t,\zeta) = \frac{|f(\zeta)|^2}{|f(\varphi_t(\zeta))|^2} e^{\int_0^t \operatorname{div} f(\varphi_s(\zeta))\, ds}, \tag{5.20}$$

$$a(t,\zeta) = \int_0^t \left(2\kappa(s,\zeta)|f(\varphi_s(\zeta))| - \operatorname{curl} f(\varphi_s(\zeta))\right) b(s,\zeta)\, ds. \tag{5.21}$$

The integral formulas (5.20) and (5.21) for $a(t, \zeta)$ and $b(t, \zeta)$ seem to have been first obtained by Stephen P. Diliberto [74]. We note, however, that his formula for $a(t, \zeta)$ incorrectly omits the factor 2 of the curvature term.

Proof. By definition

$$t \mapsto a(t)f(\varphi_t(\zeta)) + b(t)f^\perp(\varphi_t(\zeta))$$

is the solution of the variational equation (5.15) with initial value $f^\perp(\zeta)$. In particular, $a(0) = 0$, $b(0) = 1$, and

$$
\begin{aligned}
a(t)Df(\varphi_t(\zeta))&f(\varphi_t(\zeta)) \\
&+ a'(t)f(\varphi_t(\zeta)) + b(t)Df^\perp(\varphi_t(\zeta))f(\varphi_t(\zeta)) + b'(t)f^\perp(\varphi_t(\zeta)) \\
&= a(t)Df(\varphi_t(\zeta))f(\varphi_t(\zeta)) + b(t)Df(\varphi_t(\zeta))f^\perp(\varphi_t(\zeta)). \quad (5.22)
\end{aligned}
$$

After taking the inner product with $f^\perp(\varphi_t(\zeta))$ and suppressing the arguments of various functions, we obtain the equation

$$b'|f|^2 = b(\langle Df \cdot f^\perp, f^\perp \rangle - \langle Df^\perp \cdot f, f^\perp \rangle).$$

Since $f^\perp = Rf$, where $R = \begin{pmatrix} 0 & -1 \\ 1 & 0 \end{pmatrix}$, we have

$$\langle Df^\perp \cdot f, f^\perp \rangle = \langle RDf \cdot f, Rf \rangle = \langle Df \cdot f, f \rangle$$

and

$$b'|f|^2 = b(\langle Df \cdot f^\perp, f^\perp \rangle + \langle Df \cdot f, f \rangle - 2\langle Df \cdot f, f \rangle).$$

By an easy (perhaps lengthy) computation, it follows that

$$b' = b \operatorname{div} f - b\frac{d}{dt} \ln |f|^2.$$

The solution of this differential equation with the initial condition $b(0) = 1$ is exactly formula (5.20).

From equation (5.22), taking the inner product this time with $f(\varphi_t(\zeta))$, we obtain

$$
\begin{aligned}
a'|f|^2 &= b(\langle Df \cdot f^\perp, f \rangle - \langle Df^\perp \cdot f, f \rangle) \\
&= b(\langle f^\perp, (Df)^* f \rangle - \langle RDf \cdot f, f \rangle) \\
&= b(\langle f^\perp, (Df)^* f \rangle + \langle f^\perp, Df \cdot f \rangle) \\
&= b(\langle f^\perp, 2Df \cdot f \rangle + \langle f^\perp, ((Df)^* - (Df))f \rangle)
\end{aligned}
\quad (5.23)
$$

where $*$ denotes the transpose. Also, by simple computations, we have

$$\langle f^\perp, 2Df \cdot f \rangle = 2\kappa |f|^3,$$
$$\langle f^\perp, ((Df)^* - (Df))f \rangle = -|f|^2 \operatorname{curl} f$$

where the scalar curvature κ, the curl, and the other functions are evaluated on the curve $t \mapsto \varphi_t(\zeta)$. After substitution of these formulas into equation (5.23), an integration yields formula (5.21).   □

Recall that the periodic orbit Γ is hyperbolic if the derivative of the Poincaré map on Σ at $v = \Gamma \cap \Sigma$ has no eigenvalue with modulus one. By our geometric identification, this derivative is just $b(2\pi/\omega, v)$. Using the equality $|f(\varphi_{2\pi/\omega}(v))| = |f(v)|$ and Diliberto's theorem, we have the identification

$$b(2\pi/\omega, v) = e^{\int_0^{2\pi/\omega} \operatorname{div} f(\varphi_t(v)) \, dt}.$$

Thus, the derivative of the Poincaré map is independent of the choice of section Σ. In addition, by a change of variables, it is easy to see that the derivative does not depend on $v \in \Gamma$. These remarks give an alternate proof of Proposition 2.130, which we restate here in a slightly different form.

Proposition 5.6. *A periodic solution* $t \mapsto \varphi_t(\zeta)$ *of* $\dot{u} = f(u)$ *with period* $2\pi/\omega$ *is hyperbolic if and only if*

$$\int_0^{2\pi/\omega} \operatorname{div} f(\varphi_t(\zeta)) \, dt \neq 0. \tag{5.24}$$

Also, using equation (5.18) together with the implicit function theorem, we have a theorem on persistence.

Theorem 5.7. *A hyperbolic periodic solution of the differential equation* $\dot{u} = f(u)$ *persists for autonomous perturbations.*

Exercise 5.8. Prove: If φ_t is the flow of the differential equation $\dot{x} = f(x)$ with the periodic orbit Γ, then $\int_0^{2\pi/\omega} \operatorname{div} f(\varphi_t(\zeta)) \, dt$ does not depend on the choice of $\zeta \in \Gamma$.

Exercise 5.9. With respect to Proposition 5.6, suppose that Γ is the periodic orbit corresponding to the periodic solution $t \mapsto \varphi_t(\zeta)$. Show that the inequality

$$\int_\Gamma \operatorname{div} f \, ds < 0$$

is *not* sufficient to prove that Γ is a stable limit cycle.

Exercise 5.10. Suppose that Γ is a hyperbolic periodic solution with period T of the planar system $\dot{u} = f(u)$ and $\zeta \in \Gamma$. Using the notation of Diliberto's theorem, define

$$g(\varphi_t(\zeta)) = \frac{1}{b(t,\zeta)}\left(\frac{a(T,\zeta)}{b(T,\zeta) - 1} + a(t,\zeta)\right) f(\varphi_t(\zeta)) + f^\perp(\varphi_t(\zeta)).$$

Prove the following facts: (a) $\Phi(t)g(\zeta) = b(t,\zeta)g(\varphi_t(\zeta))$. (b) $g(\varphi_T(\zeta)) = g(\zeta)$. (c) The vector g is nowhere parallel to f. (d) The vector field determined by g on Γ is invariant under the linearized flow.

Exercise 5.11. [Multidimensional Systems] Suppose that $\dot{u} = f(u)$, $u \in \mathbb{R}^n$ has a periodic orbit Γ cut transversely by an $(n-1)$-dimensional surface $\Sigma \subseteq \mathbb{R}^n$. Here, transversality means that $f(v)$ is not tangent to Σ for $v = \Gamma \cap \Sigma$. (a) Show that the analogue of Theorem 5.7 is valid in this context. Hint: Although there is no obvious substitute for Diliberto's formulas, the ideas of this section apply. Use the definition of hyperbolicity, that is, the derivative of the Poincaré map on Σ at v has its spectrum off the unit circle in the complex plane; and then proceed abstractly by following the same argument presented for Theorem 5.7. (b) Is the hyperbolicity hypothesis necessary when $n > 2$? Can you prove a stronger result?

Exercise 5.12. Obtain equation (5.20) using Liouville's formula (2.18). Warning: At first sight, in the context of equation (5.20), it might appear that the fundamental matrix for system (5.15) is given by $\left(\begin{smallmatrix} 1 & a(t) \\ 0 & b(t) \end{smallmatrix} \right)$ relative to the basis

$$\{f(\varphi_t(\zeta)), f^{\perp}(\varphi_t(\zeta))\}.$$

But, this matrix does not represent the fundamental matrix solution in any fixed basis. Rather, it represents a transition from the initial basis given by $\{f(\zeta), f^{\perp}(\zeta)\}$ to the basis $\{f(\varphi_t(\zeta)), f^{\perp}(\varphi_t(\zeta))\}$.

Exercise 5.13. Let $t \mapsto \alpha(t)$ be a nonconstant solution of the scalar second-order differential equation $\ddot{x} = f(x)$. (a) Show that the principal fundamental matrix solution $\Phi(t)$ at $t = 0$ of the first variational equation along α is defined by

$$\Phi(t) = \begin{pmatrix} \dot{\alpha}(t) & \beta(t) \\ f(\alpha(t)) & \dot{\beta}(t) \end{pmatrix}$$

where $\dot{\alpha}(0)\dot{\beta}(0) - \beta(0)\ddot{\alpha}(0) = 1$, β is given by $t \mapsto \dot{\alpha}(t)\gamma(t)$, γ is an antiderivative of $\dot{\alpha}(t)^{-2}$ defined on the time intervals where $\dot{x}$ does not vanish, and β, defined by removing singularities, is smooth on the domain of α (cf. [63]). Also, Compare this result with the fundamental matrix solution given by Diliberto's theorem. Hint: Use Abel's formula. Show that the second-order ODE is reversible. A planar system $\dot{x} = p(x,y)$, $\dot{y} = q(x,y)$ is called reversible if $t \mapsto (x(-t), -y(-t))$ is a solution whenever $t \mapsto (x(t), y(t))$ is a solution. For the second order ODE, $t \mapsto x(T-t)$ is a solution whenever $t \mapsto x(t)$ is a solution. Show that if $\dot{x}$ has two zeros, then $x(t)$ is periodic. In case $\dot{x}$ has no zeros, show that γ is an antiderivative such that $\dot{\alpha}(0)\dot{\beta}(0) - \beta(0)\ddot{\alpha}(0) = 1$. In case $\dot{x}$ has exactly one zero at $t = T$ and $T \geq 0$, show that the function γ is an antiderivative on the interval $(-\infty, T)$ (or on the maximal interval of time less than T on which the solution x is defined) such that $\dot{\alpha}(0)\dot{\beta}(0) - \beta(0)\ddot{\alpha}(0) = 1$, and $\gamma(T + s) := -\gamma(T - s)$ for $s > 0$. A similar construction is made if $T < 0$. In case $\dot{x}$ has two zeros, x is periodic and γ is an antiderivative defined periodically on the whole line such that the initial condition $\dot{\alpha}(0)\dot{\beta}(0) - \beta(0)\ddot{\alpha}(0) = 1$ is satisfied.

Problem 5.14. How can Diliberto's theorem be generalized to the case of variational equations for differential equations defined in $\mathbb{R}^n$ for $n > 2$? A solution of this exercise together with some examples would perhaps make a nice research article.

To determine the persistence of periodic orbits of the differential equation (5.11), our main hypothesis, $\delta_\xi(0,0) \neq 0$, is equivalent to requiring the unperturbed periodic solution to be hyperbolic. Let us consider the continuation problem for nonhyperbolic periodic orbits.

If an unperturbed planar periodic orbit is not hyperbolic, then we cannot determine an implicit solution of the equation $\delta(\xi, \epsilon) = 0$ by a direct application of the implicit function theorem. Instead, the main new tool for the analysis is the (Weierstrass) preparation theorem. The following statement is a special case of this important result (see [8], [29], and [57]).

Theorem 5.15 (Preparation Theorem). *If $\delta : \mathbb{R} \times \mathbb{R} \to \mathbb{R}$, given by $(\xi, \epsilon) \mapsto \delta(\xi, \epsilon)$, is analytic (or C^∞) and*

$$\delta(0,0) = \frac{\partial \delta}{\partial \xi}(0,0) = \frac{\partial^2 \delta}{\partial \xi^2}(0,0) = \cdots = \frac{\partial^{n-1}\delta}{\partial \xi^{n-1}}(0,0) = 0, \quad \frac{\partial^n \delta}{\partial \xi^n}(0,0) \neq 0,$$

then there are n smooth functions $a_i : \mathbb{R} \to \mathbb{R}$ defined near $\epsilon = 0$ and a function $U : \mathbb{R} \times \mathbb{R} \to \mathbb{R}$ defined near $(\xi, \epsilon) = (0,0)$ such that $a_i(0) = 0$, $i = 1, \ldots, n$, $U(0,0) \neq 0$, and

$$\delta(\xi, \epsilon) = (a_0(\epsilon) + a_1(\epsilon)\xi + \cdots + a_{n-1}(\epsilon)\xi^{n-1} + \xi^n)U(\xi, \epsilon).$$

The name "preparation theorem" is used because the function δ, written in the form given in the conclusion of the theorem, is *prepared* for a study of its zeros. Moreover, because $U(0,0) \neq 0$ (such a function U is called a *unit* in the algebra of functions defined in a neighborhood of the origin), the zeros of the function $\delta(\xi, \epsilon)$ near $(\xi, \epsilon) = (0,0)$ are exactly the zeros of the *Weierstrass polynomial*

$$a_0(\epsilon) + a_1(\epsilon)\xi + \cdots + a_{n-1}(\epsilon)\xi^{n-1} + \xi^n.$$

In particular, there are at most n zeros for each fixed ϵ near $\epsilon = 0$.

For the case where δ is the displacement function associated with a periodic orbit Γ, the *multiplicity* of Γ is defined to be the degree n of the Weierstrass polynomial. If $n = 1$, then Γ is hyperbolic and exactly one continuation point of periodic solutions exists for $|\epsilon| \neq 0$ sufficiently small. It follows from the preparation theorem that if Γ has multiplicity n, then there is some choice of the function g in the differential equation (5.10) such that n families of periodic solutions bifurcate from Γ at $\epsilon = 0$. But, for each specific perturbation, the actual number of continuations can only be determined by analyzing the coefficients of the Weierstrass polynomial.

Exercise 5.16. Show that the system

$$\dot{x} = -y + x(x^2 + y^2 - 1)^2,$$
$$\dot{y} = x + y(x^2 + y^2 - 1)^2 \tag{5.25}$$

has a limit cycle with multiplicity 2.

As an illustration of the ideas just presented, let us analyze the continuation problem for a periodic orbit Γ with multiplicity 2.

Using the displacement function δ associated with Γ, we have that

$$\delta(0,0) = \delta_\xi(0,0) = 0, \quad \delta_{\xi\xi}(0,0) \neq 0,$$

and, by the preparation theorem,

$$\delta(\xi,\epsilon) = (a_0(\epsilon) + a_1(\epsilon)\xi + \xi^2)U(\xi,\epsilon) \tag{5.26}$$

where $a_0(0) = 0$, $a_1(0) = 0$, but $U(0,0) \neq 0$. We will solve for ξ implicitly with respect to ϵ. But, in anticipation of a bifurcation at $\epsilon = 0$, we cannot expect to have a smooth continuation given by a function $\epsilon \mapsto \beta(\epsilon)$ such that $\beta(0) = 0$ and $\delta(\beta(\epsilon), \epsilon) \equiv 0$. More likely, there are implicit solutions defined for $\epsilon > 0$ or $\epsilon < 0$, but not both. For this reason, we say there are N *positive branches* at the bifurcation point $(0,0)$ if there is some $\epsilon_0 > 0$ and N continuous functions $\beta_1, \dots, \beta_N$, each defined for $0 \leq \epsilon < \epsilon_0$ such that for each $j = 1, \dots, N$, $\beta_j(0) = 0$, and $\delta(\beta_j(\epsilon), \epsilon) \equiv 0$. Negative branches are defined analogously for $-\epsilon_0 < \epsilon \leq 0$. Of course, the number and position of the branches is determined by the roots of the Weierstrass polynomial.

With respect to the Weierstrass polynomial in display (5.26), we have

$$a_0(\epsilon) = a_{01}\epsilon + O(\epsilon^2), \qquad a_1(\epsilon) = O(\epsilon),$$

and therefore the roots of this Weierstrass polynomial are given by

$$\xi = \beta(\epsilon) = \frac{-a_1(\epsilon) \pm \sqrt{-4a_{01}\epsilon + O(\epsilon^2)}}{2}.$$

If $\epsilon \neq 0$ has fixed sign and $a_{01}\epsilon > 0$, then there are no real branches. On the other hand, if $a_{01}\epsilon < 0$, then there are two real branches given by

$$\beta_1(\epsilon) = \sqrt{-a_{01}\epsilon} + O(\epsilon), \qquad \beta_2(\epsilon) = -\sqrt{-a_{01}\epsilon} + O(\epsilon).$$

To identify the coefficient a_{01}, compute the derivatives

$$\delta_\epsilon(0,0) = a_{01}U(0,0),$$
$$\delta_{\xi\xi}(0,0) = 2U(0,0),$$

and note that

$$a_{01} = 2\delta_\epsilon(0,0)/\delta_{\xi\xi}(0,0). \tag{5.27}$$

Of course $\delta_{\xi\xi}(0,0)$ is just the second derivative of the unperturbed Poincaré map $\xi \mapsto \sigma^{-1}P(\sigma(\xi),0)$. A formula for the derivative $\delta_\epsilon(0,0)$ will be computed below.

Let us apply the result in equation (5.27) to the bifurcation of limit cycles for the system

$$
\begin{aligned}
\dot{x} &= -\, y + x(x^2 + y^2 - 1)^2, \\
\dot{y} &= x + y(x^2 + y^2 - 1)^2 + \epsilon(x^2 - 1)y.
\end{aligned} \tag{5.28}
$$

By a change to polar coordinates, we have the equivalent system

$$
\begin{aligned}
\dot{r} &= r(r^2 - 1)^2 + \epsilon r \sin^2 \theta (r^2 \cos^2 \theta - 1), \\
\dot{\theta} &= 1 + \epsilon \cos \theta \sin \theta (r^2 \cos^2 \theta - 1).
\end{aligned}
$$

Note that, for r near $r = 1$, if ϵ is sufficiently small, then we can treat θ as a time-like variable and obtain the following differential equation for r:

$$
\frac{dr}{d\theta} = F(r, \theta, \epsilon) := \frac{r(r^2 - 1)^2 + \epsilon r \sin^2 \theta (r^2 \cos^2 \theta - 1)}{1 + \epsilon \cos \theta \sin \theta (r^2 \cos^2 \theta - 1)}. \tag{5.29}
$$

Also, for each ξ near $\xi = 1$, let us define the function $\theta \mapsto r(\theta, \xi, \epsilon)$ to be the unique solution of the differential equation (5.29) with the initial condition $r(0, \xi, \epsilon) = \xi$.

Note that the displacement function is given by $\delta(\xi, \epsilon) = r(2\pi, \xi, \epsilon) - \xi$. Thus, to compute the partial derivative $\delta_\xi(\xi, \epsilon)$, it suffices to solve the variational initial value problem

$$
\frac{d}{d\theta} r_\xi = F_r(r(\theta, \xi, \epsilon), \xi, \epsilon) r_\xi, \quad r_\xi(0, \xi, \epsilon) = 1
$$

to obtain the useful formula

$$
r_\xi(\theta, \xi, \epsilon) = e^{\int_0^\theta F_r(r(s,\xi,\epsilon),\xi,\epsilon)\, ds}.
$$

By Exercise 5.16, the point $\xi = 1$ corresponds to the unperturbed limit cycle. Thus, if we view ξ as a coordinate on the positive x-axis, then $\delta(1, 0) = r(2\pi, 1, 0) - 1 = 0$. Moreover, we have

$$
r_\xi(2\pi, \xi, 0) = e^{\int_0^{2\pi} (r^2 - 1)(5r^2 - 1)\, d\theta},
$$

and therefore $\delta_\xi(1, 0) = 0$. By taking one more derivative with respect to ξ, let us note that

$$
\delta_{\xi\xi}(1, 0) = r_{\xi\xi}(2\pi, 1, 0) = \int_0^{2\pi} 8 r_\xi \, d\theta = 16\pi
$$

is positive. To compute $\delta_\epsilon(1, 0)$, solve the variational initial value problem

$$
\frac{d}{d\theta} r_\epsilon = \sin^2 \theta (\cos^2 \theta - 1), \quad r_\epsilon(0, 1, 0) = 0
$$

to obtain

$$\delta_\epsilon(1,0) = r_\epsilon(2\pi, 1, 0) = -\int_0^{2\pi} \sin^4 \theta \, d\theta < 0.$$

By our analysis of the Weierstrass polynomial, there are two branches of periodic solutions for small $\epsilon > 0$. One branch consists of stable limit cycles; the other branch consists of unstable limit cycles. We will outline a method for proving this fact, but the details are left to the reader.

The stability of the perturbed limit cycles is determined by $\delta_\xi(\beta(\epsilon), \epsilon)$. In fact, the orbit is unstable if $\delta_\xi(\beta(\epsilon), \epsilon) > 0$ and stable if $\delta_\xi(\beta(\epsilon), \epsilon) < 0$. To prove this claim, recall that $\delta(\xi, \epsilon) = \sigma^{-1}(P(\sigma(\xi), \epsilon)) - \xi$ and the stability type is determined by the derivative of the Poincaré map. Since $\delta_\xi(0,0) = 0$, the stability type for small ϵ is determined by the sign of $\delta_{\epsilon\xi}(0,0)$. If $\delta_{\epsilon\xi}(0,0) > 0$ and a branch of continued periodic solutions exists for $\epsilon > 0$, then the branch is unstable. If the branch exists for $\epsilon < 0$, then the branch is stable. If $\delta_{\epsilon\xi}(0,0) < 0$ and the branch exists for $\epsilon > 0$, then it is stable, whereas, if the branch exists for $\epsilon < 0$, then it is unstable.

We have discussed a complete analysis for autonomous perturbations in case Γ is hyperbolic, and we have just indicated how to approach the problem when Γ has finite multiplicity. Let us consider the case where Γ has infinite multiplicity; that is, when $\delta(\xi, 0) \equiv 0$. This, of course, is not quite correct if by infinite multiplicity we mean that all partial derivatives of the displacement function δ with respect to ξ vanish at $(\xi, \epsilon) = (0, 0)$; maybe δ is infinitely flat but still $\delta(\xi, 0) \neq 0$ for $\xi \neq 0$. On the other hand, if δ is analytic (it will be if the differential equation (5.10) is analytic), then infinite multiplicity at the point $\xi = 0$ does imply that $\delta(\xi, 0) \equiv 0$.

Exercise 5.17. Give an example of an *infinitely flat limit cycle*: The periodic orbit is isolated but $\partial^k \delta / \partial \xi^k (0,0) = 0$ for $k = 1, 2, 3, \ldots$.

Suppose that $\delta(\xi, 0) \equiv 0$ and consider the perturbation series

$$\delta(\xi, \epsilon) = \delta_\epsilon(\xi, 0)\epsilon + \frac{1}{2!}\delta_{\epsilon\epsilon}(\xi, 0)\epsilon^2 + O(\epsilon^3).$$

Note that

$$\delta(\xi, \epsilon) = \epsilon(\delta_\epsilon(\xi, 0) + O(\epsilon)). \tag{5.30}$$

Here, since $\delta(\xi, 0) \equiv 0$, the periodic orbit Γ is contained in a *period annulus*; that is, an annulus in the plane consisting entirely of periodic orbits of the unperturbed differential equation (5.11) (see, for example, Figure 3.2).

Although we could consider continuations from the fixed periodic orbit Γ, it is traditional to consider all of the periodic orbits in the period annulus together. Let us determine if any of the periodic orbits in the annulus persist. For this problem, if we recall equation (5.30) and use the implicit function theorem, then the reduction step is easy: *A simple zero of the function $\xi \mapsto \delta_\epsilon(\xi, 0)$ is a continuation point of periodic solutions.* Equivalently, if $\delta_\epsilon(\xi_0, 0) = 0$ and $\delta_{\xi\epsilon}(\xi_0, 0) \neq 0$, then the periodic solution Γ_{ξ_0} of the unperturbed system (5.11) with initial value $\sigma(\xi_0) \in \Sigma$ persists.

For the identification step, we will find a useful formula for $\delta_\epsilon(\xi, 0)$. Let us first compute the partial derivative with respect to ϵ in equation (5.13) to obtain the identity

$$\delta_\epsilon(\xi, 0)\dot\sigma(\xi) = T_\epsilon(\xi, 0)f(\sigma(\xi)) + u_\epsilon(T(\xi, 0), \sigma(\xi), 0), \qquad (5.31)$$

and note that $t \mapsto u_\epsilon(t, \sigma(\xi), 0)$ is the solution of the inhomogeneous variational initial value problem

$$\dot W = Df(\varphi_t(\sigma(\xi)))W + g(\varphi_t(\sigma(\xi))), \qquad W(0) = 0, \qquad (5.32)$$

where the initial condition follows from the identity $u(0, \sigma(\xi), 0) \equiv \sigma(\xi)$. (The differential equation in display (5.32) is also called the *second variational equation.*)

By the variation of constants formula,

$$u_\epsilon(T(\xi, 0), \sigma(\xi), 0) = \Phi(T(\xi, 0)) \int_0^{T(\xi, 0)} \Phi^{-1}(s)g(\varphi_s(\sigma(\xi)))\, ds$$

where $\Phi(t)$ denotes the principal fundamental matrix solution of the system (5.15) at $t = 0$.

Let us use the identifications given in equations (5.20) and (5.21) by first expressing the function g in the form

$$g(\varphi_t(\sigma(\xi))) = c_1(t, \sigma(\xi))f(\varphi_t(\sigma(\xi))) + c_2(t, \sigma(\xi))f^\perp(\varphi_t(\sigma(\xi)))$$

with

$$c_1(t, \sigma(\xi)) = \frac{1}{|f(\varphi_t(\sigma(\xi)))|^2}\langle g(\varphi_t(\sigma(\xi))), f(\varphi_t(\sigma(\xi)))\rangle,$$

$$c_2(t, \sigma(\xi)) = \frac{1}{|f(\varphi_t(\sigma(\xi)))|^2}\langle g(\varphi_t(\sigma(\xi))), f^\perp(\varphi_t(\sigma(\xi)))\rangle$$

$$:= \frac{1}{|f(\varphi_t(\sigma(\xi)))|^2}f(\varphi_t(\sigma(\xi))) \wedge g(\varphi_t(\sigma(\xi))).$$

Also, note that the inverse of the matrix (5.17) represents the action of the inverse of the principal fundamental matrix at $t = 0$ from the span of $\{f, f^\perp\}$ at $u(t, \sigma(\xi), 0)$ to the span of $\{f, f^\perp\}$ at $\sigma(\xi)$. Likewise, the matrix in equation (5.17) evaluated at $T(\xi, 0)$ is the matrix representation of the fundamental matrix with respect to the basis $\{f, f^\perp\}$ at $\sigma(\xi)$. Thus, we have that

$$\Phi(T(\xi, 0)) = \begin{pmatrix} 1 & a(T(\xi, 0), \sigma(\xi)) \\ 0 & b(T(\xi, 0), \sigma(\xi)) \end{pmatrix},$$

$$\Phi^{-1}(s)g(\varphi_s(\sigma(\xi))) = \frac{1}{b(s, \sigma(\xi))} \begin{pmatrix} b(s, \sigma(\xi)) & -a(s, \sigma(\xi)) \\ 0 & 1 \end{pmatrix} \begin{pmatrix} c_1(s, \sigma(\xi)) \\ c_2(s, \sigma(\xi)) \end{pmatrix},$$

and

$$u_\epsilon(T(\xi,0),\xi,0) = (\mathcal{N}(\xi) + a(T(\xi,0),\sigma(\xi))\mathcal{M}(\xi))f(\sigma(\xi))$$
$$+ b(T(\xi,0),\sigma(\xi))\mathcal{M}(\xi)f^\perp(\sigma(\xi)) \qquad (5.33)$$

where

$$\mathcal{M}(\xi) := \int_0^{T(\xi,0)} \frac{1}{b(t,\sigma(\xi))|f(\varphi_t(\sigma(\xi)))|^2} f(\varphi_t(\sigma(\xi))) \wedge g(\varphi_t(\sigma(\xi)))\, dt$$

$$= \frac{1}{|f(\sigma(\xi))|^2}$$
$$\times \int_0^{T(\xi,0)} e^{-\int_0^t \operatorname{div} f(\varphi_s(\sigma(\xi)))\, ds} f(\varphi_t(\sigma(\xi))) \wedge g(\varphi_t(\sigma(\xi)))\, dt,$$

$$\mathcal{N}(\xi) := \int_0^{T(\xi,0)} \frac{1}{|f(\varphi_t(\sigma(\xi)))|^2} \langle g(\varphi_t(\sigma(\xi))), f(\varphi_t(\sigma(\xi)))\rangle\, dt$$
$$- \int_0^{T(\xi,0)} \frac{a(t,\sigma(\xi))}{b(t,\sigma(\xi))|f(\varphi_t(\sigma(\xi)))|^2} f(\varphi_t(\sigma(\xi))) \wedge g(\varphi_t(\sigma(\xi)))\, dt.$$

After taking the inner product of both sides of the equation (5.31) with the vector $f^\perp(\sigma(\xi))$, and using the formulas for $\mathcal{M}$ and $\mathcal{N}$, the quantity $\delta_\epsilon(\xi,0)$ is seen to be given by

$$\delta_\epsilon(\xi,0) = \frac{b(T(\xi,0),\sigma(\xi))|f(\sigma(\xi))|^2}{\langle \dot\sigma(\xi), f^\perp(\sigma(\xi))\rangle} \mathcal{M}(\xi). \qquad (5.34)$$

In this formula, $\langle \dot\sigma, f^\perp\rangle \neq 0$ because Σ is transverse to the unperturbed periodic solutions, and $b(t,\zeta) \neq 0$ because $|f|$ does not vanish along the unperturbed periodic orbit.

The autonomous Poincaré–Andronov–Melnikov function is defined by

$$M(\xi) := \int_0^{T(\xi,0)} e^{-\int_0^t \operatorname{div} f(\varphi_s(\sigma(\xi)))\, ds} f(\varphi_t(\sigma(\xi))) \wedge g(\varphi_t(\sigma(\xi)))\, dt. \qquad (5.35)$$

Here, $\xi \mapsto T(\xi,0)$ is a local representation of the *period function* associated with the period annulus of the differential equation (5.11); the number $T(\xi,0)$ is the minimum period of the periodic orbit labeled by ξ, that is, the orbit passing through the point in the plane with coordinates $(\xi,0)$. It should be clear that values of the function M are independent of the choice of Poincaré section. In fact, as long as ξ is a smooth parameter for the periodic solutions in our period annulus, the value of M at a particular periodic solution is not altered by the choice of the parametrization.

Theorem 5.18. *Suppose that the differential equation (5.11) has a period annulus $\mathcal{A}$ whose periodic solutions are parametrized by a smooth function $\sigma : \mathbb{R} \to \mathcal{A}$ given by $\xi \mapsto \sigma(\xi)$. If ξ_0 is a simple zero of the function*

$\xi \mapsto M(\xi)$ *given by the formula (5.35) for the perturbed system (5.10),* *then the periodic solution of the unperturbed system (5.11) passing through* ξ_0 *is continuable.*

Proof. This result follows immediately from the formula (5.34) and the persistence of the simple zeros of $\xi \mapsto \delta_\epsilon(\xi, 0)$. We only remark that, in general, if $\alpha(\xi) = \beta(\xi)\gamma(\xi)$ with $\beta(\xi)$ nonvanishing, then the simple zeros of α and γ coincide. $\qquad\square$

Exercise 5.19. Find the continuable periodic solutions of the perturbed harmonic oscillator in each of the following systems:

1. weakly damped van der Pol equation:

$$\ddot{x} + \epsilon(x^2 - 1)\dot{x} + \omega^2 x = 0;$$

2. nonlinear weakly damped van der Pol equation:

$$\ddot{x} + \epsilon(x^2 - 1)\dot{x} + \omega^2 x - \epsilon\lambda x^3 = 0;$$

3. modified van der Pol equation:

$$\ddot{x} + \epsilon(x^2 + \dot{x}^2 - 1)\dot{x} + x = 0.$$

5.3   Nonautonomous Perturbations

Let us consider the periodic solutions of the nonautonomous periodically perturbed system

$$\dot{u} = f(u) + \epsilon g(u, t, \epsilon), \quad u \in \mathbb{R}^2. \tag{5.36}$$

More precisely, let us suppose that the unperturbed system has a periodic solution Γ whose period is $2\pi/\omega$ and that $t \mapsto g(u, t, \epsilon)$ is periodic with period

$$\eta := \eta(\epsilon) = \frac{n}{m}\frac{2\pi}{\omega} + k\epsilon + O(\epsilon^2) \tag{5.37}$$

where n, m are relatively prime positive integers and $k \in \mathbb{R}$ is the "detuning parameter." In particular, at $\epsilon = 0$ we have

$$m\eta(0) = n\frac{2\pi}{\omega} \tag{5.38}$$

and we say that the periodic solution Γ is in $(m : n)$ *resonance* with the perturbation g. Equation (5.38) is called a *resonance relation*. If, as before, we let $t \mapsto u(t, \zeta, \epsilon)$ denote the solution of the differential equation (5.36)

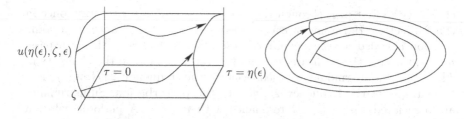

Figure 5.2: The left panel depicts an orbit on a invariant cylinder starting at ζ and returning to the Poincaré section at $\tau = 0$ for the Poincaré
map (5.40). The right panel is a schematic depiction of the same orbit on
the torus formed by identifying the Poincaré section at $\tau = 0$ with the plane
at $\tau = \eta(\epsilon)$. If the orbit were to close on the mth return to the section, it
would be an $(m:1)$ subharmonic.

with initial condition $u(0, \zeta, \epsilon) = \zeta$ in $\mathbb{R}^2$, then we have that $t \mapsto u(t, \zeta, 0)$
defines a $2\pi/\omega$-periodic function for each $\zeta \in \Gamma$.

   The nonautonomous differential equation (5.36) is equivalent to the first
order system

$$\begin{aligned} \dot{u} &= f(u) + \epsilon g(u, \tau, \epsilon), \\ \dot{\tau} &= 1 \end{aligned} \tag{5.39}$$

in the extended phase plane. Because g is a periodic function of time, it is
customary to view τ as an angular variable modulo $\eta(\epsilon)$. This leads to the
very useful geometric interpretation of the system (5.39) as a differential
system on the *phase cylinder* $\mathbb{R}^2 \times \mathbb{T}$ where

$$\mathbb{T} := \{e^{2\pi i\tau/\eta(\epsilon)} : \tau \in \mathbb{R}\}.$$

   There is an annular region $A \subseteq \mathbb{R}^2$ containing Γ and some $\epsilon_0 > 0$ such
that $\Sigma = A \times \{1\} \subseteq \mathbb{R}^2 \times \mathbb{T}$ is a Poincaré section for the system (5.39) with
associated (parametrized) Poincaré map $P : \Sigma \times (-\epsilon_0, \epsilon_0) \to \mathbb{R}^2$ defined by

$$(\zeta, 1, \epsilon) \mapsto u(\eta(\epsilon), \zeta, \epsilon).$$

But, because the set $A \times \{1\}$ is naturally identified with A, we will view
the Poincaré map as the map $P : A \times (-\epsilon_0, \epsilon_0) \to \mathbb{R}^2$ given by

$$(\zeta, \epsilon) \mapsto u(\eta(\epsilon), \zeta, \epsilon). \tag{5.40}$$

   We are going to look for $(m:n)$ subharmonic solutions of the perturbed
system (5.36), that is, periodic solutions of the differential equation (5.36)
with period $m\eta(\epsilon)$. They correspond to periodic points of period m for
the Poincaré map. Actually, there is a finer classification of such solutions
that is often made as follows: A periodic solution is called a *harmonic* if it

closes at the first pass through the Poincaré section after rotating once in the $\mathbb{T}$ direction. Harmonics are associated with $(1:1)$ resonance. A periodic solution is called a *subharmonic of order* m if it closes at the mth pass, $m > 1$, through the Poincaré section after rotating once in the $\mathbb{T}$ direction. The name "subharmonic" is used because the frequency $2\pi/(m\eta(\epsilon))$ is a submultiple of the frequency $2\pi/\eta(\epsilon)$ of the perturbation. Subharmonics are associated with $(m:1)$ resonance with $m > 1$. A periodic solution is called an (m, n) *ultrasubharmonic* if it closes at the mth pass through the Poincaré section after rotating n times, $n > 1$, in the $\mathbb{T}$ direction. Ultrasubharmonics are associated with $(m:n)$ resonance with $n > 1$. The geometry of subharmonic orbits in the extended phase plane is depicted in Figure 5.2.

The key point derived from our geometric interpretation of the perturbation problem is the following: A periodic point of period m for the Poincaré map is a periodic solution with period $m\eta(\epsilon)$ for the system (5.36). To see this, let ζ be a periodic point of period m so that

$$u(m\eta(\epsilon), \zeta, t) = \zeta.$$

Consider the solution $t \mapsto u(t, \zeta, \epsilon)$ of the system (5.36) and the function given by

$$v(t) := u(t + m\eta(\epsilon), \zeta, \epsilon),$$

and note that

$$\dot{v} = f(v) + \epsilon g(v, t + m\eta(\epsilon), \epsilon).$$

Using the periodicity of g, this last equation simplifies to yield

$$\dot{v} = f(v) + \epsilon g(v, t, \epsilon),$$

and therefore $t \mapsto v(t)$ is a solution of the differential equation (5.36). As $v(0) = \zeta$ and $u(0, \zeta, \epsilon) = \zeta$, the solutions $t \mapsto u(t, \zeta, \epsilon)$ and $t \mapsto v(t)$ must be the same; that is, $u(t + m\eta(\epsilon), \zeta, \epsilon) = u(t, \zeta, \epsilon)$ and the function $t \mapsto u(t, \zeta, \epsilon)$ is $m\eta(\epsilon)$-periodic.

As before, let us define the *(parametrized) displacement function* δ : $A \times (-\epsilon_0, \epsilon_0) \to \mathbb{R}^2$ by

$$\delta(\zeta, \epsilon) = u(m\eta(\epsilon), \zeta, \epsilon) - \zeta. \tag{5.41}$$

Here there is no need for a local coordinate representation via a coordinate chart: Points in the domain $A \times (\epsilon_0, \epsilon_0) \subset \mathbb{R}^2 \times \mathbb{R}$ are already expressed in local coordinates.

Clearly, if $\zeta \in \Gamma$, where Γ is a resonant periodic solution of the differential equation (5.11), then $\delta(\zeta, 0) = 0$; in effect,

$$\delta(\zeta, 0) = u(m\eta(0), \zeta, 0) - \zeta = u(m\frac{n}{m}\frac{2\pi}{\omega}, \zeta, 0) - \zeta = 0.$$

To see if Γ persists, we would like to apply the implicit function theorem to the function δ at the point $(\zeta, 0)$ where $\delta(\zeta, 0) = 0$. Thus, we would like to show that the linear map $\delta_\zeta(\zeta, 0) : \mathbb{R}^2 \to \mathbb{R}^2$ is invertible. But, for a point ζ that lies on a resonant periodic solution Γ, this map always has a nontrivial kernel. In fact, we have that $\delta_\zeta(\zeta, 0)f(\zeta) \equiv 0$ for $\zeta \in \Gamma$. This result is geometrically obvious. But to construct an analytic proof, let us use the definition of the directional derivative and the group property of the unperturbed flow to obtain the identity

$$
\begin{aligned}
\delta_\zeta(\zeta, 0)f(\zeta) &= \frac{d}{dt}\delta(u(t, \zeta, 0), 0)\big|_{t=0} \\
&= \frac{d}{dt}(u(2\pi n/\omega, u(t, \zeta, 0), 0) - u(t, \zeta, 0))\big|_{t=0} \\
&= \frac{d}{dt}(u(2\pi n/\omega + t, \zeta, 0) - u(t, \zeta, 0))\big|_{t=0} \\
&= f(u(2\pi n/\omega, \zeta, 0)) - f(\zeta) = 0.
\end{aligned}
\tag{5.42}
$$

We have just proved that the kernel of the linear transformation $\delta_\zeta(\zeta, 0)$ contains the subspace generated by $f(\zeta)$. Here and hereafter we will let $[v]$ denote the subspace spanned by the enclosed vector. In particular, we have $[f(\zeta)] \subseteq \text{Kernel}\,\delta_\zeta(\zeta, 0)$. The analysis to follow later in this chapter falls naturally into two cases: $[f(\zeta)] = \text{Kernel}\,\delta_\zeta(\zeta, 0)$ and $\text{Kernel}\,\delta_\zeta(\zeta, 0) = \mathbb{R}^2$. After a short section devoted to the continuation of periodic orbits from unperturbed rest points where the kernel of the derivative of the displacement can be trivial, we will develop some of the theory required to determine the continuable periodic orbits in each of these two cases.

Exercise 5.20. [Multidimensional Oscillators] Suppose that the system $\dot{u} = f(u)$, for the vector case $u \in \mathbb{R}^n$, has a T-periodic orbit Γ given by the solution $t \mapsto \gamma(t)$. (a) Show that the number one is a Floquet multiplier of the (periodic) variational equation $\dot{w} = Df(\gamma(t))w$. (b) Prove that if the Floquet multiplier one has algebraic multiplicity one and if $g : \mathbb{R}^n \times \mathbb{R} \times \mathbb{R} \to \mathbb{R}^n$ is a smooth function given by $(u, t, \epsilon) \mapsto g(u, t, \epsilon)$ such that the corresponding map given by $t \mapsto g(u, t, \epsilon)$ is T-periodic for each u and ϵ, then Γ persists in the family $\dot{u} = f(u) + \epsilon g(u, t, \epsilon)$. (c) Is the same result true if the geometric multiplicity is one?

5.3.1 Rest Points

Let us suppose that the unperturbed system

$$\dot{u} = f(u),$$

derived from the system (5.36) by setting $\epsilon = 0$, has a rest point $u = \zeta$. This point is a fixed point of the unperturbed Poincaré map and a zero of the unperturbed displacement function. In particular, the rest point corresponds to a periodic solution of the artificially autonomous system

$$\dot{u} = f(u), \qquad \dot{\tau} = 1, \tag{5.43}$$

where τ is considered as an angular variable modulo $\eta(0)$. To determine if the corresponding periodic solution continues, we have the following theorem.

Theorem 5.21. *If ζ is a rest point for the unperturbed system $\dot{u} = f(u)$ derived from the system (5.36), and the Jacobian matrix $Df(\zeta)$ has no eigenvalue of the form $2\pi Ni/\eta$ where N is an integer, then the periodic orbit with period $\eta(0)$ for system (5.43) corresponding to ζ persists as an $\eta(\epsilon)$-periodic solution of equation (5.36).*

Proof. The partial derivative

$$\delta_\zeta(\zeta, 0) = u_\zeta(\eta(0), \zeta, 0) - I$$

is easily computed by solving the variation initial value problem

$$\dot{W} = Df(\zeta)W, \qquad W(0) = I$$

to obtain

$$\delta_\zeta(\zeta, 0) = e^{\eta Df(\zeta)} - I.$$

The matrix $\delta_\zeta(\zeta, 0)$ is invertible if and only if the number one is not an eigenvalue of $e^{\eta Df(\zeta)}$. Thus, the desired result follows from Theorem 2.88 and the implicit function theorem. $\square$

Exercise 5.22. Describe the bifurcations of rest points that may occur in case $2\pi Ni/\eta$ is an eigenvalue of $Df(\zeta)$ for some integer N.

5.3.2   Isochronous Period Annulus

If the coordinate neighborhood $A \subset \mathbb{R}^2$ containing the unperturbed periodic orbit Γ is a period annulus $\mathcal{A}$, it is possible that every periodic solution in $\mathcal{A}$ has the same period, that is, the period annulus is *isochronous*. In this case, if a resonance relation holds for one periodic solution in $\mathcal{A}$, then it holds for all of the periodic solutions in $\mathcal{A}$. We will determine the continuable periodic solutions for an unperturbed system with an isochronous period annulus.

Let us note that a period annulus for a linear system is necessarily isochronous. Although there are nonlinear systems with isochronous period annuli (just transform a linear system with a period annulus by a nonlinear change of coordinates), they can be difficult to recognize.

Exercise 5.23. Prove that the following systems have isochronous period annuli.

1. $\ddot{x} + 1 - \sqrt{1 + 2x} = 0$.

2. (Loud's system) $\dot{x} = -y + Bxy$, $\dot{y} = x + Dx^2 + Fy^2$ in case $(D/B, F/B)$ is one of the following:

$$(0, 1), \quad \left(-\frac{1}{2}, 2\right), \quad \left(0, \frac{1}{4}\right), \quad \left(-\frac{1}{2}, \frac{1}{2}\right)$$

(see [42] and [148]).

Loud's theorem states that every quadratic system with an isochronous period annulus can be transformed by a linear change of coordinates to one of the four systems mentioned above. An interesting unsolved pure mathematics problem is to determine the number and positions of critical points for the period functions of the period annuli of Loud's system as the parameters B, D, and F are varied. For example, there are some period functions with two critical points. It is not known if this is the maximum number (see [42]).

For the rest of this subsection, let us assume that the unperturbed system (5.11) has an isochronous period annulus $\mathcal{A}$ where every periodic orbit has period $2\pi/\omega$. In this case, $\delta(\zeta, 0) \equiv 0$ and $\delta_\zeta(\zeta, 0) \equiv 0$ for $\zeta \in \mathcal{A}$.

Because the perturbation series for the displacement function (see display (5.41)) has the form

$$\delta(\zeta, \epsilon) = \epsilon(\delta_\epsilon(\zeta, 0) + O(\epsilon)),$$

we have the following proposition: *A simple zero ζ of the function $\zeta \mapsto \delta_\epsilon(\zeta, 0)$ is an (ultra)subharmonic continuation point.* In other words, there is a number $\epsilon_0 > 0$ and a continuous function $\beta : (-\epsilon_0, \epsilon_0) \to \mathbb{R}^2$ given by $\epsilon \mapsto \beta(\epsilon)$ such that $\beta(0) = \zeta$ and $\delta(\beta(\epsilon), \epsilon) \equiv 0$. Of course, $\beta(\epsilon)$ is the initial value of a subharmonic solution of the differential equation (5.36). This result is the now familiar reduction step of our analysis.

To identify the function $\zeta \mapsto \delta_\epsilon(\zeta, 0)$, we simply compute this partial derivative from the definition of the displacement (5.41) to obtain

$$\delta_\epsilon(\zeta, 0) = m\eta'(0)f(\zeta) + u_\epsilon(m\eta(0), \xi, 0)$$
$$= mkf(\zeta) + u_\epsilon(2\pi n/\omega, \zeta, 0).$$

As before, $t \mapsto u_\epsilon(t, \zeta, 0)$ is the solution of a variational initial value problem, namely,

$$\dot{W} = Df(\varphi_t(\zeta))W + g(\varphi_t(\zeta), t, 0), \qquad W(0) = 0$$

where φ_t is the flow of the unperturbed system. The solution of the initial value problem is obtained just as in the derivation of equation (5.33). The only difference is the "nonautonomous" nature of g, but this does

not change any of the formal calculations. In fact, with the notation as in
equation (5.33), we obtain

$$u_\epsilon(2\pi n/\omega, \zeta, 0) = (\mathcal{N}(\zeta) + a(2\pi n/\omega, \zeta)\mathcal{M}(\zeta))f(\zeta)$$
$$+ b(2\pi n/\omega, \zeta)\mathcal{M}(\zeta)f^\perp(\zeta). \tag{5.44}$$

By the geometric interpretation of the functions a and b given following
equation (5.19), these functions are readily reinterpreted in the present
context. In fact, since every orbit of our isochronous period annulus is not
hyperbolic, we must have $b(2\pi/\omega, \zeta) = 1$, and, since the period function is
constant, we also have $a(2\pi/\omega, \zeta) = 0$. Thus, we obtain the identity

$$\delta_\epsilon(\zeta, 0) = (mk + \mathcal{N}(\zeta))f(\zeta) + \mathcal{M}(\zeta)f^\perp(\zeta). \tag{5.45}$$

Exercise 5.24. Show that $b(n2\pi/\omega, \zeta) = b^n(2\pi/\omega, \zeta)$ and

$$a(2\pi n/\omega, \zeta) = a(2\pi/\omega, \zeta) \sum_{j=0}^{n-1} b^j(2\pi/\omega, \zeta).$$

Theorem 5.25. *Suppose the differential equation (5.36) is such that the
unperturbed system has an isochronous period annulus $\mathcal{A}$ with period $2\pi/\omega$
and the perturbation $g(u, t, \epsilon)$ has period $\nu(\epsilon) = (n/m)2\pi/\omega + k\epsilon + O(\epsilon^2)$
where n and m are relatively prime positive integers. If the bifurcation
function $B : \mathcal{A} \to \mathbb{R}^2$ given by $\zeta \mapsto (mk + \mathcal{N}(\zeta), \mathcal{M}(\zeta))$ has a simple zero
ζ, then ζ is a continuation point of $(m : n)$ (ultra)subharmonics for the
system (5.36).*

Proof. The theorem follows from equation (5.45). Indeed, if

$$F(\zeta) := \begin{pmatrix} f_1(\zeta) & -f_2(\zeta) \\ f_2(\zeta) & f_1(\zeta) \end{pmatrix} \quad \text{and} \quad \mathcal{B}(\zeta) := \begin{pmatrix} mk + \mathcal{N}(\zeta) \\ \mathcal{M}(\zeta) \end{pmatrix},$$

then

$$B(\zeta) = F(\zeta) \cdot \mathcal{B}(\zeta),$$

and the simple zeros of B coincide with the simple zeros of $\mathcal{B}$.   □

Theorem 5.25, specialized to the case where the unperturbed system is
linear, is slightly more general than Theorem 5.1. For example, suppose
that $f(u) = Au$ where

$$A = \begin{pmatrix} 0 & -\omega \\ \omega & 0 \end{pmatrix}, \quad \omega > 0.$$

Since $\operatorname{div} f \equiv 0$ and $|f|$ is constant on orbits, $b(t, \zeta) \equiv 1$. Also, let us note
that

$$2\kappa(t, \zeta)|f(\varphi_t(\zeta))| - \operatorname{curl} f(\varphi_t(\zeta)) = 2\frac{1}{|\zeta|}\omega|\zeta| - 2\omega = 0,$$

and therefore $a(t, \zeta) \equiv 0$. (This is a good internal check that the formula for a is correct!) Thus, in this special case,

$$\mathcal{N}(\zeta) = \int_0^{n2\pi/\omega} \frac{1}{|f(\varphi_t(\zeta))|^2} \langle f(\varphi_t(\zeta)), g(\varphi_t(\zeta), t, 0) \rangle \, dt,$$

$$\mathcal{M}(\zeta) = \frac{1}{|f(\zeta)|^2} \int_0^{n2\pi/\omega} f(\varphi_t(\zeta)) \wedge g(\varphi_t(\zeta), t, 0) \, dt.$$

More explicitly, we have that

$$\mathcal{N}(\zeta) = \frac{1}{\omega|\zeta|^2} \int_0^{n2\pi/\omega} x g_2(x, y, t, 0) - y g_1(x, y, t, 0) \, dt,$$

$$\mathcal{M}(\zeta) = -\frac{1}{\omega|\zeta|^2} \int_0^{n2\pi/\omega} x g_1(x, y, t, 0) + y g_2(x, y, t, 0) \, dt \qquad (5.46)$$

where

$$x := x(t, \zeta) = \zeta_1 \cos \omega t - \zeta_2 \sin \omega t, \quad y := y(t, \zeta) = \zeta_1 \sin \omega t + \zeta_2 \cos \omega t.$$

Let us consider the stability of the perturbed (ultra)subharmonics. Note that the "perturbation series" for the Poincaré map is given by

$$P(\zeta, \epsilon) = \zeta + \epsilon P_\epsilon(\zeta, 0) + O(\epsilon^2),$$

and $P_\epsilon(\zeta, 0) = \delta_\epsilon(\zeta, 0)$. Thus, the formula for the partial derivative of the Poincaré map with respect to ϵ is given by equation (5.45) and

$$P(\zeta, \epsilon) = \begin{pmatrix} \zeta_1 \\ \zeta_2 \end{pmatrix}$$
$$+ \epsilon \left(km\omega \begin{pmatrix} -\zeta_2 \\ \zeta_1 \end{pmatrix} + \begin{pmatrix} \int_0^{n2\pi/\omega} g_1 \cos \omega t + g_2 \sin wt \, dt \\ \int_0^{n2\pi/\omega} g_2 \cos \omega t - g_1 \sin wt \, dt \end{pmatrix} \right)$$
$$+ O(\epsilon^2) \qquad (5.47)$$

where g_1 and g_2 are evaluated at $(x, y, t, 0)$.

It should be clear that the stability of the perturbed (ultra)subharmonics is determined by the eigenvalues of the matrix $P_\zeta(\zeta, \epsilon)$, called the *linearized Poincaré map* evaluated at the fixed point of $\zeta \mapsto P(\zeta, \epsilon)$ corresponding to the subharmonic. The subharmonic is stable if both eigenvalues lie inside the unit circle in the complex plane. Of course, if the linearized Poincaré map is hyperbolic, then the local behavior near the periodic orbit is determined—stability is just a special case of this more general fact.

It is not too difficult to show that if $\epsilon > 0$ is sufficiently small, then the matrix $P_\zeta(\zeta, \epsilon)$ *evaluated at the perturbed fixed point* has both of its eigenvalues inside the unit circle in the complex plane provided that each eigenvalue of the matrix $P_{\zeta\epsilon}(\zeta, 0)$ has negative real part. For $\epsilon < 0$, each eigenvalue of the matrix $P_{\zeta\epsilon}(\zeta, 0)$ must have positive real part. Equivalently, it suffices to have

$$\det P_{\zeta\epsilon}(\zeta, 0) > 0, \quad \epsilon \operatorname{tr} P_{\zeta\epsilon}(\zeta, 0) < 0.$$

The proof of this fact contains a pleasant surprise.

The perturbation series for the Poincaré map evaluated along the curve $\epsilon \mapsto (\beta(\epsilon), \epsilon)$ has the form

$$P_\zeta(\beta(\epsilon), \epsilon) = I + \epsilon A + \epsilon^2 B + O(\epsilon^3)$$

where $A = P_{\zeta\epsilon}(\beta(0), 0)$. In particular, we have used the equality $P_{\zeta\zeta}(\zeta, 0) = 0$. The characteristic polynomial of the first order approximation of this matrix, namely, $I + \epsilon A$, has coefficients that contain terms of second order in ϵ. Thus, it appears that second order terms in the perturbation series are required for computing the eigenvalues to first order. Fortunately, there is an unexpected cancellation, and the eigenvalues, for $\epsilon > 0$, are given by

$$1 + \epsilon \frac{1}{2}\left(\operatorname{tr} A \pm \sqrt{\operatorname{tr}^2 A - 4 \det A}\right) + O(\epsilon^2). \tag{5.48}$$

Using formula (5.48), it is easy to show that if the eigenvalues of A have nonzero real parts, then the first order terms of the expansion determine the stability. If A has an eigenvalue with zero real part, then higher order terms in the perturbation expansion must be considered (see [172]).

General formulas for the eigenvalues of $P_{\zeta\epsilon}(\zeta, 0)$ can be obtained in terms of certain partial derivatives of $\mathcal{M}$ and $\mathcal{N}$. But, such formulas are usually not useful. A better approach is to use the special properties of the system under investigation.

Exercise 5.26. Prove the statements following equation (5.48) concerning the eigenvalues of the matrix $P_\zeta(\zeta, \epsilon)$.

5.3.3   The Forced van der Pol Oscillator

In this subsection we will outline, by formulating a series of exercises, some applications of the continuation theory (developed so far in this chapter) to the classic case of the van der Pol oscillator. Also, we mention briefly some of the additional structures that can be studied using our first order methods.

Exercise 5.27. Find the (ultra)subharmonics for the periodically forced van der Pol oscillator

$$\ddot{x} + \epsilon(x^2 - 1)\dot{x} + x = \epsilon a \sin \Omega t.$$

In particular, for fixed $a \neq 0$, find the regions in (Ω, ϵ) space near the line $\epsilon = 0$ where (ultra)subharmonics exist.

The regions mentioned in Exercise 5.27 are called *entrainment domains* or, in some of the electrical engineering literature, they are called *synchronization domains*. We cannot determine the entire extent of the entrainment domains because our first order theory is only valid for sufficiently small $|\epsilon|$. Higher order methods can be used to obtain more information (see, for example, [161] and [109], and for some classic numerical experiments [117]).

To use the formulas for $\mathcal{M}$ and $\mathcal{N}$ in display (5.46), let us consider the first order system in the phase plane given by

$$\dot{x} = -y, \qquad \dot{y} = x + \epsilon(-(x^2 - 1)y - a\sin\Omega t).$$

Also, let us consider curves in the (Ω, ϵ) parameter space of the form $\epsilon \mapsto (\Omega(\epsilon), \epsilon)$ where

$$\Omega(\epsilon) = \frac{m}{n} - k\left(\frac{m}{n}\right)^2 \epsilon + O(\epsilon^2),$$

$$\eta(\epsilon) = 2\pi\frac{n}{m} + k\epsilon + O(\epsilon^2).$$

To complete Exercise 5.27, start by looking for *harmonics;* that is, look for periodic solutions of the perturbed system with periods close to 2π for Ω near $\Omega = 1$. Set $m = n = 1$. To help debug the computations for this example, first try the case $k = 0$ where k is the detuning, and show that there is a harmonic at the point (ζ_1, ζ_2) provided that $\zeta_2 = 0$ and ζ_1 is a root of the equation $\zeta_1^3 - 4\zeta_1 + 4a = 0$. This corresponds to perturbation in the vertical direction in the parameter space. Show that the harmonic will be stable if $|\zeta_1| > 2$ and that there is a unique (stable) harmonic in case $a = 1$.

There is a very interesting difference between the $(1:1)$ resonance and the $(m:n)$ resonance with $m/n \neq 1$. To glimpse into this structure, consider the $(m:n)$ resonance where $m/n \neq 1$ and use equation (5.47) to compute the following first order approximation of the associated Poincaré map:

$$\zeta_1 \mapsto \zeta_1 + \epsilon\left(-km\zeta_2 + n\pi\zeta_1 - \frac{n\pi}{4}\zeta_1(\zeta_1^2 + \zeta_2^2)\right),$$

$$\zeta_2 \mapsto \zeta_2 + \epsilon\left(km\zeta_1 + n\pi\zeta_2 - \frac{n\pi}{4}\zeta_2(\zeta_1^2 + \zeta_2^2)\right). \qquad (5.49)$$

This map preserves the origin. Thus, it is natural to study the map in polar coordinates where it is represented to first order by

$$r \mapsto r + \epsilon n\pi r\left(1 - \frac{r^2}{4}\right),$$

$$\theta \mapsto \theta + \epsilon mk. \qquad (5.50)$$

Here the first order formula in the rectangular coordinates goes over to a formula in polar coordinates that contains higher order terms in ϵ that have been deleted. Can we safely ignore these higher order terms?

For the $(1 : 1)$ resonance with k included as a parameter, a similar first order computation yields the map

$$\zeta_1 \mapsto \zeta_1 + \epsilon\left(-k\zeta_2 + \pi\zeta_1 - a\pi - \frac{\pi}{4}\zeta_1(\zeta_1^2 + \zeta_2^2)\right),$$
$$\zeta_2 \mapsto \zeta_2 + \epsilon\left(k\zeta_1 + \pi\zeta_2 - \frac{\pi}{4}\zeta_2(\zeta_1^2 + \zeta_2^2)\right). \tag{5.51}$$

Note that if this map preserves the origin, then $a = 0$. Thus, polar coordinates are not the natural coordinates for studying this map. Instead, a useful representation of the map is obtained by changing to the complex coordinate $z = \zeta_1 + i\zeta_2$ where the map is represented in the form

$$z \mapsto z + \epsilon\left((\pi + ki)z - \frac{1}{4}\pi z^2 \bar{z} - a\pi\right). \tag{5.52}$$

We will that show the dynamics of the map defined in display (5.49) are quite different from the dynamics of the map in display (5.51).

For the map (5.50), the circle $r = 2$ is an invariant set and every point in the plane except the origin is attracted to this circle under iteration. On the circle, the map gives a rational or an irrational rotation depending on whether or not k is rational. In other words, an analysis of the dynamics at this approximation suggests that there is an invariant torus in the phase space of the differential equation and solutions of the differential equation that do not start on the torus are attracted at an exponential rate to this torus in positive time. Roughly speaking, such an invariant torus is called *normally hyperbolic;* for the precise definition of normal hyperbolicity see [85] and [122].

Solutions of the differential equation on the invariant torus may wind around the torus, as in the case of irrational rotation, or they may be attracted to a subharmonic on the torus as in the case of rational rotation. There are general theorems that can be used to show that a normally hyperbolic torus will persist with the addition of a small perturbation (see, for example, [85], [106], and [122], and also [39] and [46]). Thus, we see that there is a second possible type of entrainment. It is possible that solutions are entrained to the torus when there are no periodic solutions on the torus. In this case, corresponding to an irrational rotation, every solution on the torus is dense; that is, every solution on the torus has the entire torus as its omega limit set.

Exercise 5.28. View the circle as the set $\{e^{i\theta} : \theta \in \mathbb{R}\}$ and define the (linear) rotation on the circle through angle α as the map $e^{i\theta} \mapsto e^{i(\theta+\alpha)}$ for some fixed $k \in \mathbb{R}$. Prove the following classic result of Jacobi: If α is a rational multiple of π, then every point on the circle is periodic under the rotation map. If α is an irrational multiple of π, then the orbit of each point on the circle is dense in the circle. In the irrational case, the solutions are called *quasi-periodic.*

For the forced van der Pol oscillator, we cannot determine the quasi-periodicity of the flow by looking at the first order approximation of the Poincaré map—the flow on the torus is nonlinear. There are actually three possibilities. The nonlinear flow can have all its orbits dense, all its orbits periodic, or it can have isolated periodic solutions. We have to be careful here because the nonlinear Poincaré map on the invariant torus is not, in general, a rotation as defined above. Rather, it is likely to be conjugate to a rotation by a nonlinear change of coordinates.

The Poincaré map will have a stable subharmonic, or at least an isolated subharmonic, on the invariant torus provided that the bifurcation function has *simple* zeros on this torus. We will have more to say about this topic below.

For the case $m/n \neq 1$, an examination of the map (5.50) shows that a necessary condition for the existence of subharmonics on the invariant torus near $r = 1$ is that $k = 0$. In the $(m : n)$ entrainment domain in the (Ω, ϵ) (frequency-amplitude) parameter space the curves corresponding to the subharmonics would have to be expressible as series

$$\Omega = \frac{m\omega}{n} - \left(\frac{m}{n}\right)^2 \left(\frac{k}{2\pi}\right) \omega^2 \epsilon + \sum_{j=2}^{\infty} \Omega_j \epsilon^j$$

(see equation (5.37)). But because $k = 0$, it follows that they are all of the form $\Omega = m\omega/n + O(\epsilon^2)$. Thus, all such curves have the same tangent line at $\epsilon = 0$, namely, the line given by $\Omega = m\omega/n$. The portion of the entrainment domain near $\epsilon = 0$ that is filled by such curves is called an *Arnold tongue.*

For the map (5.51), there are fixed points corresponding to harmonics but not necessarily an invariant torus. In case $k = 0$, there is a fixed point only if $\zeta_2 = 0$ and

$$\zeta_1^3 - 4\zeta_1 + 4a = 0.$$

In case $k \neq 0$, the computations are more complicated. There are many different ways to proceed. One effective method is "Gröbner basis reduction" (see [70]). Without going into the definition of a Gröbner basis for a polynomial ideal, the reduction method is an algorithm that takes as input a set of polynomials (with rational coefficients) and produces a new set of polynomials with the same zero set. The reduced set is in a good normal form for further study of its zero set. The output depends on the ordering of the variables. In particular, the Gröbner basis is not unique.

For example, by using the MAPLE V command gbasis with the lexicographic ordering of the variables ζ_1 and ζ_2, the equations

$$-k\zeta_2 + \pi\zeta_1 - a\pi - \frac{\pi}{4}\zeta_1 \left(\zeta_1^2 + \zeta_2^2\right) = 0,$$

$$k\zeta_1 + \pi\zeta_2 - \frac{\pi}{4}\zeta_2 \left(\zeta_1^2 + \zeta_2^2\right) = 0,$$

can be reduced to

$$4k^2\zeta_1 + 4k\pi\zeta_2 + a\pi^2\zeta_2^2 = 0,$$
$$16k^3a\pi + \left(16k^4 + 16k^2\pi^2\right)\zeta_2 + 8ak\pi^3\zeta_2^2 + a^2\pi^4\zeta_2^3 = 0. \tag{5.53}$$

By an inspection of the equations (5.53), it is clear that there are either one, two, or three fixed points in the Poincaré section. If there is exactly one solution, then (for sufficiently small $\epsilon > 0$), either it corresponds to a stable harmonic that attracts the entire phase space, and, as a result, there is no invariant torus, or, it corresponds to an unstable harmonic and there is an invariant torus. The first order approximation of the Poincaré map restricted to the invariant circle corresponding to this invariant torus may be conjugate to either a rational or an irrational rotation. In case it is rational, each point on the invariant circle is periodic. On the other hand, if it is irrational, then each point has a dense orbit. Are these properties present in the perturbed Poincaré map?

If there are three harmonics, several different phase portraits are possible, but generally the Poincaré map has a sink, a source, and a saddle. The "most likely" possibility in this case is to have the unstable separatrices of the saddle attracted to the sink. In this case, the separatrices together with the saddle and the sink form an invariant "circle" that corresponds to an invariant torus for the flow of the differential equation. We may ask if this set is a manifold. The answer is not obvious. For example, if the linearization of the Poincaré map at the sink happens to have complex eigenvalues, then the separatrices will "roll up" at the sink and the invariant "circle" will not be smooth. In our case, however, if ϵ is small, then the linearization of the Poincaré map is near the identity; and therefore, this roll up phenomenon does not occur. Does this mean the invariant circle *is* smooth?

The case where there is an "invariant torus"—consisting of a saddle, its unstable manifold, and a sink—is particularly interesting from the point of view of applications. For example, a trajectory starting near the *stable* manifold of the saddle will be "entrained" by the harmonic corresponding to the saddle on perhaps a very long time scale. But, unless the orbit stays on the stable manifold, a very unlikely possibility, it will eventually leave the vicinity of the saddle along the unstable manifold. Ultimately, the orbit will be entrained by the sink. But, because of the possibility of a long sojourn time near the saddle, it is often not clear in practice, for example, in a numerical experiment, when a trajectory has become entrained (phase locked) to the input frequency with a definite phase. This phenomenon might be the cause of some difficulties if we wish to control the response of the oscillator.

Which regions in the (k, a) parameter space correspond to the existence of three harmonics? Answer: the region of the parameter space where the cubic polynomial (5.53) has three distinct real roots. To find this region,

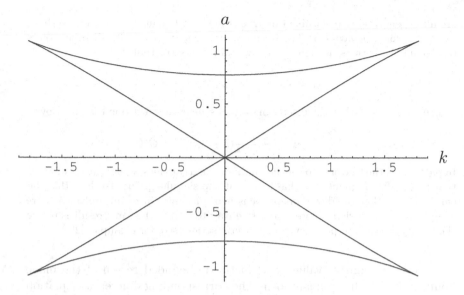

Figure 5.3: Discriminant locus for the polynomial (5.53). The bounded region corresponds to the existence of three harmonics for the periodically forced van der Pol equation.

let us first compute the *discriminant locus* of the polynomial, that is, the set of points in the parameter space where the cubic polynomial has a double root (see [29]). Of course, the discriminant locus is the zero set of the discriminant of the polynomial. Equivalently, the discriminant locus is given by the set of points in the parameter space where the polynomial and its first derivative have a simultaneous solution. This set is also the zero set of the resultant of the polynomial and its first derivative. In our case, a computation shows that the discriminant locus of the cubic polynomial in display (5.53) is the zero set of the polynomial

$$\Delta(k, a) := 27\pi^6 a^4 - 16\pi^6 a^2 - 144\pi^4 a^2 k^2 + 64\pi^4 k^2 + 128\pi^2 k^4 + 64k^6.$$
(5.54)

The discriminant locus is also the boundary of the region corresponding to the existence of three real roots. This region is the bounded region depicted in Figure 5.3.

Exercise 5.29. The discriminant locus corresponds to an invariant curve for the Hamiltonian system

$$\dot{k} = -\frac{\partial \Delta}{\partial a}(k, a), \quad \dot{a} = \frac{\partial \Delta}{\partial k}(k, a)$$
(5.55)

with Hamiltonian Δ. Show that the invariant set consists of six trajectories and five rest points (zeros of the vector field). The four rest points not at the origin are all degenerate—the Jacobian matrix at each rest point has zero eigenvalues. Study the local behavior of the discriminant locus at each of its singular points to explain the corners in Figure 5.3. For example, show that

$$k_0 = -\frac{\sqrt{3}}{3}\pi, \qquad a_0 = -\frac{4}{9}\sqrt{6}$$

is a rest point and that the discriminant locus near this rest point is given by

$$a - a_0 = \frac{\sqrt{2}}{\pi}(k - k_0) \pm \frac{2}{3}\frac{3^{1/4}}{\pi^{3/2}}(k - k_0)^{3/2} + O((k - k_0)^2).$$

In particular, the tangents to the discriminant locus at the singular point coincide; that is, the discriminant locus has a cusp at the singular point. To show this you can just note that the discriminant locus is a quadratic in a^2 and solve. A more complicated but perhaps more instructive way to obtain the same result is to use the theory of Newton polygons and Puiseux series (see, for example, [29]).

For each parameter value (k, a) in the unbounded region of the plane bounded by the discriminant locus, the corresponding differential equation has one subharmonic solution. We can determine the stability of this subharmonic using the formulas given in the preceding section following formula (5.47). In particular, there are curves in the parameter space starting near each cusp of the discriminant locus that separates the regions corresponding to stable and unstable harmonics. These curves are exactly the curves in the (k, a) parameter space given by the parameter values where the following conditions (see formula (5.48)) are met at some fixed point of the first order linearized Poincaré map: The trace of the linearization of the $O(\epsilon)$ term of the first order Poincaré map vanishes and its determinant is positive. We call these the *PAH curves* in honor of Poincaré, Andronov, and Hopf.

To determine the PAH curve, note first that the trace of the $O(\epsilon)$ term of the linearization (5.51) is given by $\pi(2 - z\bar{z})$ and use fact if there is a fixed point, then the $O(\epsilon)$ term of the map (5.52) vanishes. Thus, (k, a) lies on the PAH curve when the determinant is positive and the following two equations have a simultaneous solution:

$$2 - z\bar{z} = 0,$$

$$(\pi + ki)z - \frac{1}{4}\pi z^2\bar{z} - a\pi = 0.$$

All three conditions are satisfied provided that (k, a) lies on one of the curves given by

$$a^2\pi^2 = \frac{\pi^2}{2} + 2k^2, \qquad |k| \le \frac{\pi}{2}. \tag{5.56}$$

The portion of the PAH curve in the region where $k > 0$ and $a > 0$ is depicted in Figure 5.4. Note that the PAH curve does not pass through the cusp on the discriminant locus; rather, it "stops" on the discriminant locus at the point $(k, a) = (\pi/2, 1)$. This suggests there are more bifurcations for parameter values in the region corresponding to three harmonics—inside the bounded region cut off by the discriminant locus. This is indeed the case. A more detailed bifurcation diagram and references to the literature on these bifurcations can be found in [103, p. 71] where the first order approximation is obtained by the method of averaging, a topic that will be covered in Chapter 7.

Exercise 5.30. Compare and contrast our computation of the first order approximation of the Poincaré map with the first order approximation obtained by the method of averaging, see Chapter 7 and [103], [233], or [201].

Exercise 5.31. Find the points where the PAH curve intersects the discriminant locus. Show that the determinant of the linearized Poincaré map vanishes at a fixed point of the first order Poincaré map exactly when the parameter value defining the map is on the discriminant locus. Study the bifurcations at the point $(k, a) = (\pi/2, 1)$ on the hyperbola (5.56) to account for the end point of the PAH curve. The set of (k, a) points where the determinant of the $O(\epsilon)$ term of the linearized Poincaré map vanishes at the fixed point is determined by finding the parameters (k, a) where the equations

$$\pi^2 - \pi^2 z\bar{z} + \frac{3}{16}\pi^2 (z\bar{z})^2 + k^2 = 0,$$

$$(\pi + ki)z - \frac{1}{4}\pi z^2 \bar{z} - a\pi = 0.$$

have a simultaneous solution for z.

The eigenvalues of the linearized Poincaré map are generally complex conjugates $\lambda(k, a)$ and $\bar{\lambda}(k, a)$; they lie on the unit circle in the complex plane when (k, a) is on one of the PAH curves. In other words, the stability of a corresponding harmonic is not determined by the first order terms of the perturbation series at this point. If we consider a second curve $\mu \mapsto (k(\mu), a(\mu))$ that crosses one of the boundary curves at the parameter value $\mu = 0$, then we will see that the fixed point of the Poincaré map changes its stability as we cross from $\mu < 0$ to $\mu > 0$. For example, the real parts of the eigenvalues of the linearized map may change from negative to positive values—the stability changes in this case from stable to unstable. The bifurcation corresponding to this loss of stability is called *Hopf bifurcation*. The theory for this bifurcation is quite subtle; it will be discussed in detail in Chapter 8. But roughly speaking if the parameter value $\mu > 0$ is sufficiently small, then the Poincaré map has an invariant circle with "radius" approximately $\sqrt{\mu}$ and "center" approximately at the unstable harmonic.

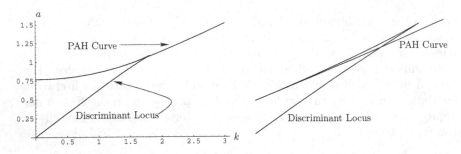

Figure 5.4: The left panel depicts the PAH curve in the region in the (k, a) parameter space with $k > 0$ and $a > 0$ together with the discriminant locus. The right panel is a blowup of the figure near the cusp on the discriminant locus.

Exercise 5.32. Show (numerically) that the Hopf bifurcation occurs as described in this section for the forced van der Pol oscillator. See Figure 5.4. For example, fix $k = 2$ and $\epsilon = .001$, then compute phase portraits of the Poincaré map for several choices of a in the range $a = 1.2$ to $a = 1.1$.

Exercise 5.33. Determine the "phase portrait" of the Poincaré map for the forced van der Pol oscillator near $(1 : 1)$ resonance for the case when the parameters (k, a) lie on the discriminant locus. In particular, determine the phase portrait in case (k, a) is a singular point of the discriminant locus. How does the phase portrait change on a curve of parameter values that passes through the discriminant locus?

Exercise 5.34. Code a numerical simulation of the Poincaré map for the forced van der Pol oscillator and verify that the first order analysis of this section predicts the dynamical behavior of the iterated Poincaré map.

Exercise 5.35. Consider the forced van der Pol oscillator near $(1 : 1)$ resonance for fixed input amplitude a, for example, $a = \frac{3}{4}$. Determine the value of the detuning for which the amplitude of the response is maximum.

5.3.4 Regular Period Annulus

In this section we will discuss a continuation theory for periodic solutions of the periodically perturbed oscillator

$$\dot{u} = f(u) + \epsilon g(u, t, \epsilon), \quad u \in \mathbb{R}^2, \tag{5.57}$$

in case the unperturbed system has a resonant periodic orbit that is contained in a nonisochronous period annulus.

Consider an unperturbed resonant periodic orbit Γ for system (5.57) that is contained in a period annulus $\mathcal{A}$, and recall that if $\mathcal{A}$ is isochronous, then all of the orbits in $\mathcal{A}$ are resonant and the unperturbed displacement function $\zeta \mapsto \delta(\zeta, 0)$, defined as in display (5.41) by

$$\delta(\zeta, \epsilon) = u(m\eta(\epsilon), \zeta, \epsilon) - \zeta, \tag{5.58}$$

vanishes identically. If $\mathcal{A}$ is not isochronous, then we expect that although the unperturbed displacement function does vanish on Γ, it does not vanish on nearby periodic orbits.

What happens when we attempt to apply the implicit function theorem? For each $z \in \Gamma$ we have $\delta(z, 0) = 0$. If, in addition, the linear transformation $\delta_\zeta(z, 0) : \mathbb{R}^2 \to \mathbb{R}^2$ were invertible, then z would be a continuation point. But, as demonstrated by the result in display (5.42), this linear transformation is not invertible. In particular, all vectors tangent to Γ are in its kernel.

In this section, we will consider the case where the kernel of the derivative of the displacement function at each point $z \in \Gamma$ is exactly the one-dimensional tangent space to Γ at z. In other words, we will assume that Kernel $\delta_\zeta(\zeta, 0) = [f(\zeta)]$. If this condition is met, then Γ, as well as the corresponding invariant torus for the system

$$\dot{u} = f(u), \qquad \dot{\tau} = 1,$$

is called *normally nondegenerate*.

Before proceeding to the continuation analysis, let us consider a geometrical interpretation of our assumption about the kernel of the derivative of the displacement function. For this, we do not need to assume that the periodic orbit Γ is contained in a period annulus. Instead, we may assume more generally that there is a region $R \subseteq \mathbb{R}^2$ and an $\epsilon_0 > 0$ such that the displacement function $\delta : R \times (-\epsilon_0, \epsilon_0) \to \mathbb{R}^2$ is defined. Also, let us assume that there is a curve $\Sigma \subseteq R$ transverse to Γ—a Poincaré section—such that the return time map $\mathcal{T} : \Sigma \times (-\epsilon_0, \epsilon_0) \to \mathbb{R}$ is defined. The following proposition gives the geometrical conditions we seek.

Proposition 5.36. *Suppose Γ is an $(m : n)$ resonant unperturbed periodic solution of the periodically perturbed oscillator (5.57); $\mathcal{T} : \Sigma \times (-\epsilon_0, \epsilon_0) \to \mathbb{R}$ is the return time map defined on a Poincaré section Σ with $\{v\} = \Gamma \cap \Sigma$; and $R \subseteq \mathbb{R}^2$ is a region containing Γ such that for some $\epsilon_0 > 0$ the displacement $\delta : R \times (-\epsilon_0, \epsilon_0) \to \mathbb{R}^2$ given in equation (5.58) is defined. If Γ is contained in a period annulus $\mathcal{A} \subseteq R$ such that the differential $\mathcal{T}_*(v, 0)$ of $\sigma \mapsto \mathcal{T}(\sigma, 0)$ at $\sigma = v$ is nonsingular, then Kernel $\delta_\zeta(\zeta, 0) = [f(\zeta)]$ and Range $\delta_\zeta(\zeta, 0) = [f(\zeta)]$ for each $\zeta \in \Gamma$. If Γ is a hyperbolic limit cycle, or if $\mathcal{T}_*(v, 0)$ is nonsingular, then Kernel $\delta_\zeta(\zeta, 0) = [f(\zeta)]$ for all $\zeta \in \Gamma$. Moreover, if Σ is orthogonal to Γ at v, then*

$$\text{Range}\,\delta_\zeta(\zeta, 0) = [r_1 f(\zeta) + r_2 f^\perp(\zeta)]$$

for each $\zeta \in \Gamma$ where, for a and b as in Diliberto's theorem (Theorem 5.5),

$$r_1 = a(2\pi n/\omega, v) = -\sum_{j=0}^{n-1} b^j (2\pi/\omega, v)(\mathcal{T}_*(\zeta, 0) f^\perp(\zeta)),$$

$$r_2 = b(2\pi n/\omega, v) - 1 = b^n(2\pi/\omega, v) - 1.$$

Proof. By equation (5.42), we have $[f(\zeta)] \subseteq \text{Kernel} \, \delta_\zeta(\zeta, 0)$. Consider the vector field $f^\perp$ and the solution $t \mapsto u^\perp(t, \zeta)$ of the initial value problem

$$\dot{u} = f^\perp(u), \quad u(0) = \zeta.$$

In other words $u^\perp(t, \zeta)$ is the flow of $f^\perp$. We have

$$\delta_\zeta(\zeta, 0) f^\perp(\zeta) = \frac{d}{dt}(u(m\eta(0), u^\perp(t, \zeta), 0) - u^\perp(t, \zeta))\Big|_{t=0}$$

$$= u_\zeta(2\pi n/\omega, \zeta, 0) f^\perp(\zeta) - f^\perp(\zeta).$$

Here $t \mapsto u_\zeta(t, \zeta, 0)$ is the principal fundamental matrix at $t = 0$ for the variational equation (5.15). Thus, from equation (5.16) and Exercise 5.24 we have

$$\delta_\zeta(\zeta, 0) f^\perp(\zeta) = \Big(\sum_{j=0}^{n-1} b^j(2\pi/\omega, v)\Big) a(2\pi/\omega, v) f(\zeta)$$

$$+ (b^n(2\pi/\omega, v) - 1) f^\perp(\zeta). \tag{5.59}$$

If Γ is contained in a period annulus, then $b(2\pi/\omega, v) = 1$ and, by equation (5.19),

$$a(2\pi/\omega, v) = -\frac{|f(\zeta)|^2}{\langle \dot{\sigma}(0), f^\perp(\zeta)\rangle} T'(0)$$

where $\dot{\sigma}(0)$ is the tangent vector to Σ at ζ determined by the parametrization of Σ. Thus,

$$\delta_\zeta(\zeta, 0) f^\perp(\zeta) = -n \frac{|f(\zeta)|^2}{\langle \dot{\sigma}(0), f^\perp(\zeta)\rangle} T'(0) f(\zeta) \neq 0$$

and $\text{Kernel} \, \delta_\zeta(\zeta, 0) = [f(\zeta)]$. Also, the range of $\delta_\zeta(\zeta, 0)$ is $[f(\zeta)]$.

On the other hand, if Γ is a hyperbolic limit cycle, then $b(2\pi/\omega, v) \neq 1$, the coefficient of $f^\perp(\zeta)$ in equation (5.59) does not vanish, and $f^\perp(\zeta) \notin \text{Kernel} \, \delta_\zeta(\zeta, 0)$. Hence, in this case we have that $\text{Kernel} \, \delta_\zeta(\zeta, 0) = [f(\zeta)]$. Moreover, if Σ is orthogonal to Γ at ζ, then

$$\delta_\zeta(\zeta, 0) f^\perp(\zeta) = \Big(\sum_{j=0}^{n-1} b^j(2\pi/\omega, v)\Big)\Big(-\frac{T'(0)|f(\zeta)|^2}{\langle \dot{\sigma}(0), f^\perp(\zeta)\rangle}\Big) f(\zeta)$$

$$+ (b^n(2\pi/\omega, v) - 1) f^\perp(\zeta).$$

$\square$

We say that a period annulus $\mathcal{A}$ is a *regular period annulus* if the differential $\mathcal{T}_*(\zeta, 0)$ of the return time, defined as in the previous theorem, is nonsingular at every point of $\mathcal{A}$. Let us note that the differential $\mathcal{T}_*(\zeta, 0)$ is nonsingular if and only if the corresponding period function for the period annulus $\mathcal{A}$ has a nonvanishing derivative; that is, the period function is regular.

Every resonant periodic orbit contained in a regular period annulus is normally nondegenerate. Also, by Proposition 5.36, if Γ is a resonant periodic orbit in $\mathcal{A}$ and $\zeta \in \Gamma$, then both the kernel and range of the partial derivative $\delta_\zeta(\zeta, 0)$ are given by $[f(\zeta)]$. In particular, if we restrict the linear map $\delta_\zeta(\zeta, 0)$ to $[f^\perp(\zeta)]$, then the map $\delta_\zeta(\zeta, 0) : [f^\perp(\zeta)] \to [f(\zeta)]$ is an isomorphism. We will use these facts in the analysis to follow.

Exercise 5.37. Prove that a linear map on a finite dimensional vector space, when restricted to a complement of its kernel, is an isomorphism onto its range. What happens in an infinite dimensional space?

Let us reiterate a basic fact: The partial derivative $\delta_\zeta(\zeta, 0)$ of the displacement function, when viewed as a map on all of $\mathbb{R}^2$, has a nontrivial kernel. Although this precludes a direct application of the implicit function theorem to solve the equation $\delta(\zeta, \epsilon) = 0$ on $\mathbb{R}^2 \times \mathbb{R}$, we can use the implicit function theorem to reduce our search for continuation points to the problem of solving a related equation on a lower dimensional space. This is accomplished by using an important technique called Lyapunov–Schmidt reduction. This method is very general. In fact, it works for equations defined on Banach spaces when the linear map playing the role of our derivative $\delta_\zeta(\zeta, 0)$ is a Fredholm operator. We will give a brief introduction to these simple but powerful ideas in an abstract setting. As we will demonstrate when we apply the method to our continuation problem, it is very fruitful to keep the *idea* of the method firmly in mind, but it may not be efficient to adapt all of the abstraction verbatim. Also, on a first reading and for the applications to be made later in this section, it is sufficient to consider only finite dimensional real Banach spaces, that is, $\mathbb{R}^n$ with the usual norm.

Let us suppose that $\mathcal{B}_1$ and $\mathcal{B}_2$ are Banach spaces and $L(\mathcal{B}_1, \mathcal{B}_2)$ denotes the bounded linear maps from $\mathcal{B}_1$ to $\mathcal{B}_2$. Also, with this notation, let us recall that a map $G : \mathcal{B}_1 \to \mathcal{B}_2$ is C^1 if there is a continuous map $\mathcal{L} : \mathcal{B}_1 \to L(\mathcal{B}_1, \mathcal{B}_2)$ such that

$$\lim_{n \to 0} \frac{\|G(x + h) - G(x) - \mathcal{L}(x) \cdot h\|}{\|h\|} = 0$$

for each $x \in \mathcal{B}_1$. A map $A \in L(\mathcal{B}_1, \mathcal{B}_2)$ is called *Fredholm* if it has a finite dimensional kernel and a closed range with a finite dimensional complement. The *index* of a Fredholm map is defined to be the difference of the dimensions of its corange and kernel.

Suppose that X and Y are open subsets of the Banach spaces $\mathcal{B}$ and $\mathcal{E}$ respectively, and $F : X \times Y \to \mathcal{B}$, given by $(x, y) \mapsto F(x, y)$, is a C^1 map such that $F(0, 0) = 0$. Since F is C^1, the partial derivative $F_x(0, 0)$ is a bounded linear map on $\mathcal{B}$. Let us assume in addition that $F_x(0, 0)$ is Fredholm with index zero.

If $\mathcal{B}$ is finite dimensional, as in our application where F is the displacement function and $\mathcal{B} = \mathbb{R}^2$, then every linear map is automatically Fredholm. Although we will use the hypothesis that our Fredholm map has index zero to ensure that the final reduced bifurcation function is a map between finite dimensional spaces *of the same dimension,* the general Lyapunov–Schmidt reduction technique does not require the Fredholm map to have index zero.

Let K denote the kernel of the Fredholm map $F_x(0, 0)$, and let R denote its range. There are subspaces KC and RC such that $\mathcal{B} = K \oplus KC$ and $\mathcal{B} = R \oplus RC$. The complement RC exists by the Fredholm hypothesis. The existence of a complement KC for the finite dimensional kernel K in an infinite dimensional Banach space is a consequence of the Hahn–Banach theorem (see [199, p. 105]). Indeed, choose a basis for the finite dimensional subspace, apply the Hahn–Banach theorem to extend the corresponding dual basis functionals to the entire Banach space, use the extended functionals to define a projection from the entire space to the finite dimensional subspace, and construct the desired complement as the range of the complementary projection.

The complementary subspaces KC and RC are not unique. In fact, in the applications, the correct choices for these spaces can be an important issue. On the other hand, there are always complementary linear projections $\mathcal{P} : \mathcal{B} \to R$ and $\mathcal{Q} : \mathcal{B} \to RC$ corresponding to the direct sum splitting of $\mathcal{B}$. Also, there is a product neighborhood of the origin in X of the form $U \times V$ where $U \subseteq K$ and $V \subseteq KC$.

Consider the map $H : U \times V \times Y \to R$ defined by $(u, v, y) \mapsto \mathcal{P}F(u+v, y)$. Its partial derivative with respect to v at $(0, 0, 0)$ is given by

$$\mathcal{P}F_x(0, 0)\big|_{KC} : KC \to R. \qquad (5.60)$$

The map $\mathcal{P}$ is the projection to the range of $F_x(0, 0)$. Thus, $\mathcal{P}F_x(0, 0)\big|_{KC} = F_x(0, 0)\big|_{KC}$. Note that in a finite dimensional space the map (5.60) is an isomorphism. The same result is true in an infinite dimensional space under the assumption that $F_x(0, 0)$ is Fredholm. In effect, the open mapping theorem (see [199, p. 99]) states that a continuous bijective linear map of Banach spaces is an isomorphism.

The main idea of the Lyapunov–Schmidt reduction results from the observation that, by the implicit function theorem applied to the map H, there are open sets $U_1 \subseteq U$ and $Y_1 \subseteq Y$, and a C^1 map $h : U_1 \times Y_1 \to KC$, with $h(0,0) = 0$ such that

$$\mathcal{P}F(u + h(u,y), y) \equiv 0.$$

The (Lyapunov–Schmidt) reduced function $\widetilde{F} : U_1 \times Y_1 \to RC$ associated with F is defined by

$$(u, y) \mapsto \mathcal{Q}F(u + h(u,y), y).$$

Clearly, $\widetilde{F}(0,0) = 0$. If there is a continuous function $y \mapsto \beta(y)$, with $\beta(y) \in U_1$ such that $\beta(0) = 0$ and $\widetilde{F}(\beta(y), y) \equiv 0$, then

$$\mathcal{Q}F(\beta(y) + h(\beta(y), y), y) \equiv 0,$$
$$\mathcal{P}F(\beta(y) + h(\beta(y), y), y) \equiv 0.$$

In particular, since $\mathcal{P}$ and $\mathcal{Q}$ are projections to complementary subspaces of $\mathcal{B}$, we must have

$$F(\beta(y) + h(\beta(y), y), y) \equiv 0,$$

that is, $y \mapsto \beta(y) + h(\beta(y), y)$ is an implicit solution of $F(x, y) = 0$ for x as a function of y near $(x, y) = (0, 0)$.

The implicit function theorem cannot be used directly to find u as an implicit function of y for the reduced equation $\widetilde{F}(u, y) = 0$. (If this were possible, then we would have been able to solve the original equation $F(x, y) = 0$ by an application of the implicit function theorem.) To prove this fact, let us consider the partial derivative

$$\widetilde{F}_u(0,0) = \mathcal{Q}F_x(0,0)(I + h_u(0,0)) : K \to RC.$$

Here $r := F_x(0,0)(I + h_u(0,0))u \in R$, so $\mathcal{Q}r = 0$. Thus, $\widetilde{F}_u(0,0)$ is not invertible; it is in fact the zero operator.

Although the implicit function theorem does not apply directly to the reduced function, we may be able to apply it after a further reduction. For example, in the applications to follow, we will have a situation where

$$\widetilde{F}(u, 0) \equiv 0. \tag{5.61}$$

In this case, under the assumption that $F \in C^2$, let us apply Taylor's theorem (Theorem 1.237) to obtain the representation $\widetilde{F}(u, y) = \widetilde{F}_y(u, 0)y + G(u, y)y$ where $(u, y) \mapsto G(u, y)$ is the C^1 function given by

$$G(u, y) = \int_0^1 (\widetilde{F}_y(u, ty) - \widetilde{F}_y(u, y))\, dt.$$

Thus, we also have

$$\widetilde{F}(u, y) = (\widetilde{F}_y(u, 0) + G(u, y))y$$

where $G(u, 0) = 0$ and

$$G_y(u, 0) = \int_0^1 (t\widetilde{F}_{yy}(u, 0) - \widetilde{F}_{yy}(u, 0))\, dt = -\frac{1}{2}\widetilde{F}_{yy}(u, 0).$$

In particular, let us note that the simple zeros of the *reduced bifurcation function* $B : U_1 \to RC$ defined by

$$B(u) = \widetilde{F}_y(u, 0)$$

are the same as the simple zeros of the function $u \mapsto \widetilde{F}_y(u, 0) + G(u, 0)$. Thus, by another application of the implicit function theorem, it follows that if the reduced bifurcation function B has a simple zero, then the equation

$$\widetilde{F}_y(u, 0) + G(u, y) = 0$$

has an implicit solution. Therefore, the simple zeros of the reduced bifurcation function B are continuation points.

Let us now apply the Lyapunov–Schmidt reduction to our continuation problem in case the resonant periodic orbit Γ is contained in a regular period annulus. For definiteness, let $\Gamma_{m/n} := \Gamma$ denote the unperturbed periodic solution that is in $(m : n)$ resonance with the periodic perturbation, and recall from Proposition 5.36 that $\operatorname{Kernel}\delta_\zeta(\zeta, 0) = [f(\zeta)]$ and $\operatorname{Range}\delta_\zeta(\zeta, 0) = [f(\zeta)]$. According to the Lyapunov–Schmidt reduction, we should choose coordinates and projections relative to the splitting $\mathbb{R}^2 = [f(\zeta)] \oplus [f^\perp(\zeta)]$. But, in keeping with the philosophy that the Lyapunov–Schmidt reduction is merely a guide to the analysis, we will consider instead a coordinate system that has the required splitting property "infinitesimally;" that is, we will choose coordinates tangent to the summands of the splitting rather than coordinates on the subspaces themselves.

Let φ_t denote the flow of the system $\dot{u} = f(u)$ and let ψ_t denote the flow of the system $\dot{u} = f^\perp(u)$. Of course, $\varphi_t(\zeta) = u(t, \zeta, 0)$. Define

$$\Upsilon(\rho, \phi) = \varphi_\phi \psi_\rho(v),$$

where $v \in \Gamma_{m/n}$ is viewed as arbitrary but fixed. Also, the subscripts on φ and ψ denote the temporal parameter in the respective flows, not partial derivatives. This should cause no confusion if the context is taken into account.

The (ρ, ϕ) coordinates are defined in some annulus containing $\Gamma_{m/n}$. They have the property that for ρ fixed, $\phi \mapsto \Upsilon(\rho, \phi)$ is tangent to f, whereas for ϕ fixed at $\phi = 0$, the map $\rho \mapsto \Upsilon(\rho, \phi)$ is tangent to $f^\perp$. More precisely, we have that

$$\Upsilon_\rho(\rho, \phi) = D\Upsilon(\rho, \phi)\frac{\partial}{\partial \rho} = D\varphi_\phi(\psi_\rho(v))f^\perp(\psi_\rho(v)),$$

$$\Upsilon_\phi(\rho, \phi) = D\Upsilon(\rho, \phi)\frac{\partial}{\partial \phi} = f(\Upsilon(\rho, \phi))$$

where $\partial/\partial\rho$ (respectively $\partial/\partial\phi$) denotes the unit vector field tangent to the ρ-axis (respectively the ϕ-axis) of the coordinate plane. Also, in the new (local) coordinates, the displacement is given by

$$\Delta(\rho, \phi, \epsilon) := \delta(\Upsilon(\rho, \phi), \epsilon) \qquad (5.62)$$

and we have that

$$\Delta(0, \phi, 0) = \delta(\varphi_\phi(v), 0) \equiv 0.$$

The first step of the Lyapunov–Schmidt reduction method is to find an implicit solution of the map H given by $H(\rho, \phi, \epsilon) := \mathcal{P} \cdot \Delta(\rho, \phi, \epsilon)$ where $\mathcal{P} := \mathcal{P}(\phi)$ is a projection onto the range $[f(\varphi_\phi(v))]$ of the linear map $\delta_\zeta(\varphi_\phi(v), 0)$. For definiteness, let $\langle \, , \, \rangle$ denote the usual inner product on $\mathbb{R}^2$ and define H by

$$(\rho, \phi, \epsilon) \mapsto \langle \Delta(\rho, \phi, \epsilon), f(\varphi_\phi(v)) \rangle.$$

The partial derivative of H with respect to ρ—the direction complementary to the kernel—evaluated at $(0, \phi, 0)$ is given by $\langle \Delta_\rho(0, \phi, 0), f(\varphi_\phi(v)) \rangle$. Using Diliberto's theorem, Proposition 5.36, and equation (5.59), we have

$$\begin{aligned}
\Delta_\rho(0, \phi, 0) &= \delta_\zeta(\varphi_\phi(v), 0) D\Upsilon(0, \phi)\frac{\partial}{\partial\rho} \\
&= \delta_\zeta(\varphi_\phi(v), 0) D\varphi_\phi(v) f^\perp(v) \\
&= \delta_\zeta(\varphi_\phi(v), 0)\big(a(\phi)f(\varphi_\phi(v)) + b(\phi)f^\perp(\varphi_\phi(v))\big) \\
&= b(\phi)\delta_\zeta(\varphi_\phi(v), 0)f^\perp(\varphi_\phi(v)) \\
&= b(\phi)a(2\pi n/\omega)f(\varphi_\phi(v)). \qquad (5.63)
\end{aligned}$$

Thus,

$$H_\rho(0, \phi, 0) = \langle \Delta_\rho(0, \phi, 0), f(\varphi_\phi(v)) \rangle = b(\phi)a(2\pi n/\omega)|f(\varphi_\phi(v))|^2 \neq 0,$$

and we can apply (as expected) the implicit function theorem to obtain an implicit function $(\phi, \epsilon) \mapsto h(\phi, \epsilon)$ such that $h(\phi, 0) = 0$ and

$$\langle \Delta(h(\phi, \epsilon), \phi, \epsilon), f(\varphi_\phi(v)) \rangle \equiv 0.$$

Also, because $\Delta(0, \phi, 0) \equiv 0$ and the implicit solution produced by an application of the implicit function theorem is unique, we have that $h(\phi, 0) \equiv 0$.

The second step of the Lyapunov–Schmidt reduction is to consider the zeros of the reduced displacement function $\widetilde{\Delta}$ given by

$$(\phi, \epsilon) \mapsto \mathcal{Q}(\phi)(\Delta(h(\phi, \epsilon), \phi, \epsilon)) = \langle \Delta(h(\phi, \epsilon), \phi, \epsilon), f^\perp(\varphi_\phi(v)) \rangle$$

where $\mathcal{Q}(\phi)$ is the indicated linear projection onto the complement of the range of the partial derivative $\delta_\zeta(\varphi_\phi(v), 0)$. Here, as mentioned previously, we can make a further reduction. In fact, because

$$\langle \Delta(h(\phi, 0), \phi, 0), f^\perp(\varphi_\phi(v)) \rangle \equiv 0,$$

it follows that
$$\widetilde{\Delta}(\phi, \epsilon) = \epsilon(\widetilde{\Delta}_\epsilon(\phi, 0) + O(\epsilon)).$$

Let us define the *bifurcation function* $B : \mathbb{R} \to \mathbb{R}$ by

$$B(\phi) := \widetilde{\Delta}_\epsilon(\phi, 0).$$

By the general remarks following equation (5.61), the simple zeros of B are (ultra)subharmonic continuation points. This ends the reduction phase of our analysis.

We will identify the bifurcation function B geometrically and analytically. As we will see in a moment,

$$B(\phi) = \widetilde{\Delta}_\epsilon(\phi, 0) = \mathcal{Q}(\phi)P_\epsilon(\varphi_\phi(v), 0) \qquad (5.64)$$

where P is the Poincaré map. Also, let us note that if we take $\mathcal{Q}$ to be an arbitrary projection to the complement of the range of $\delta_\zeta(\varphi_\phi(v), 0)$, then we will obtain an equivalent bifurcation function, that is, a bifurcation function with the same simple zeros. In any case, the bifurcation function is the projection onto the complement of the range of the partial derivative of the Poincaré map with respect to the bifurcation parameter.

To determine an analytic expression for the bifurcation function and to show that the representation (5.64) is valid, start with the definitions of B and $\widetilde{\Delta}$, and compute the derivative of $\epsilon\widetilde{\Delta}(\phi, \epsilon)$ at $\epsilon = 0$ to obtain the formula

$$B(\phi) = \langle \Delta_\rho(h(\phi, 0), \phi, 0)h_\epsilon(\phi, 0) + \Delta_\epsilon(h(\phi, 0), \phi, 0), \ f^\perp(\varphi_\phi(v)) \rangle$$
$$= \langle \Delta_\rho(0, \phi, 0)h_\epsilon(\phi, 0) + \Delta_\epsilon(0, \phi, 0), \ f^\perp(\varphi_\phi(v)) \rangle.$$

By using equation (5.63) and the identity $h(\phi, 0) \equiv 0$, it follows that

$$B(\phi) = \langle \Delta_\epsilon(0, \phi, 0), \ f^\perp(\varphi_\phi(v)) \rangle.$$

Here, $\Delta_\rho(0, \phi, 0)h_\epsilon(\phi, 0)$ is viewed as the vector $\Delta_\rho(0, \phi, 0)$ multiplied by the scalar $h_\epsilon(\phi, 0)$. Strictly speaking, $\Delta_\rho(0, \phi, 0)$ is a linear transformation $\mathbb{R} \to \mathbb{R}^2$ represented by a 2×1 matrix that we identify with a vector in $\mathbb{R}^2$ and $h_\epsilon(\phi, 0)$ is a linear transformation $\mathbb{R} \to \mathbb{R}$ that we identify with a scalar.

To find a formula for the partial derivative $\Delta_\epsilon(0, \phi, 0)$, first use the definition $\delta(\zeta, \epsilon) = u(m\eta(\epsilon), \zeta, \epsilon) - \zeta$ and compute the partial derivative with respect to ϵ to obtain the equation

$$\Delta_\epsilon(0, \phi, 0) = m\eta'(0)f(\varphi_\phi(v)) + u_\epsilon(2\pi n/\omega, \varphi_\phi(v), 0). \qquad (5.65)$$

Then, by equation (5.33), we have

$$u_\epsilon(2\pi n/\omega, \zeta, 0) = (\mathcal{N} + a\mathcal{M})f(\zeta) + b\mathcal{M}f^\perp(\zeta), \qquad (5.66)$$

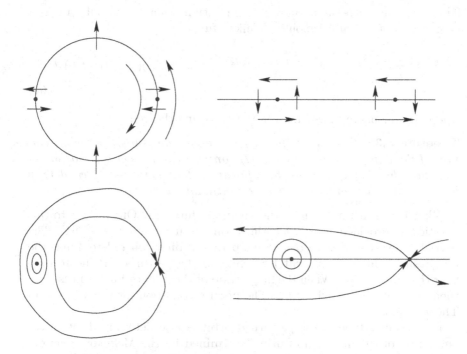

Figure 5.5: The top left panel depicts a resonant periodic orbit Γ viewed as a manifold of fixed points for the Poincaré map. The "twist" in the tangential directions along Γ due to the changing periods of the periodic orbits in the period annulus and the normal "push" directions due to the perturbations as detected by the Melnikov function are also depicted. In this illustration, the Melnikov function has two zeros. The top right panel shows the local directions of twist and push near the continuation points corresponding to these zeros. The bottom two panels depict the perturbed fixed points of the Poincaré map (subharmonics of the perturbed differential equation) and their stability types as would be expected by inspection of the directions of twist and push. The local phase portraits of the perturbed periodic orbits are seen to be saddles and rotation points that alternate in the direction of the unperturbed resonant orbit. The global depiction of the stable and unstable manifolds of the saddle point only illustrates one of many possibilities. Also, the rotation point is depicted as a center, but of course it can be a source or sink, depending on the nature of the perturbation.

and therefore
$$B(\phi) = b(2\pi n/\omega)|f(\varphi_\phi(v))|^2 \mathcal{M}(\phi).$$

Thus, ϕ is an (ultra)subharmonic continuation point if and only if ϕ is a simple zero of the subharmonic Melnikov function

$$M(\phi) := \int_0^{2\pi n/\omega} e^{-\int_0^t \operatorname{div} f(\varphi_{s+\phi}(v))\,ds} f(\varphi_{t+\phi}(v)) \wedge g(\varphi_{t+\phi}(v), t, 0)\,dt.$$

$$(5.67)$$

These arguments are formalized in the following theorem.

Theorem 5.38. *If Γ is an $(m : n)$ resonant unperturbed periodic solution of the differential equation (5.57) contained in a regular period annulus, then the simple zeros of the bifurcation function $\phi \mapsto M(\phi)$ defined by (5.67) are the (ultra)subharmonic continuation points.*

What is the real meaning of the Melnikov function? One answer to this question is provided by the identification given by equation (5.64). The partial derivative of the Poincaré map in the direction ϵ determines the infinitesimal direction of drift for orbits of the perturbed Poincaré map near the point $\varphi_\phi(v)$. When the magnitude of the infinitesimal drift is zero, then we expect a periodic orbit. The precise condition for this is given in Theorem 5.38.

The stability type of the perturbed orbit is also determined by an examination of the direction of drift determined by the Melnikov function. In fact, the resonant periodic orbit is fixed by the unperturbed Poincaré map. By the assumption that the resonant orbit is "normally nondegenerate," the drift of the unperturbed Poincaré map is in opposite directions on opposite sides of the resonant orbit. The sign of the Melnikov function determines the drift in the direction of the complement to the range of the infinitesimal displacement, a direction that is known to be transverse to the unperturbed orbit. A plot of these directions at a continuable point suggests the stability type as seen in Figure 5.5.

Exercise 5.39. Use the Lyapunov–Schmidt reduction to determine conditions on the triplet of functions (g_1, g_2, g_3) so that the system of equations

$$1 - x^2 - y^2 - z^2 + \epsilon g_1(x, y, z) = 0,$$
$$1 - x^2 - y^2 - z^2 + \epsilon g_2(x, y, z) = 0,$$
$$xyz + \epsilon g_3(x, y, z) = 0,$$

has solutions for small $\epsilon \neq 0$. What (additional) condition assures that roots found in this way are simple?

Exercise 5.40. Consider the forced rotor given by

$$\ddot{\theta} + \epsilon \sin \theta = \epsilon \sin t.$$

The associated first order system with $\dot{\phi} = v$ can be considered as a differential equation on the cylinder $\mathbb{R} \times \mathbb{T}$, where $\mathbb{T}$ denotes the unit circle. In this interpretation, all orbits of the unperturbed system are periodic. Moreover, the periodic orbit Γ corresponding to $v = 0$ is $(1 : 1)$ resonant. Show that Γ is normally nondegenerate, and determine the continuation points on Γ. What can you say about the $(m : n)$ resonant periodic orbits? Change the time scale in the differential equation to slow time $\tau = \sqrt{\epsilon}t$. What is the meaning of the continuable periodic solutions relative to the transformed differential equation? The slow time equation is a rapidly forced pendulum. Does it have subharmonics?

Exercise 5.41. Suppose that $F : \mathbb{R}^3 \times \mathbb{R} \to \mathbb{R}^3$ is a function given in the form

$$F(u, \epsilon) = g(u) + \epsilon h(u)$$

where $g, h : \mathbb{R}^3 \to \mathbb{R}^3$ are smooth vector functions with

$$g(u) = (g_1(u), g_2(u), g_3(u)), \qquad h(u) = (h_1(u), h_2(u), h_3(u)).$$

Prove the following theorem: If the slot functions g_1 and g_2 are identical and $v \in \mathbb{R}^3$ is such that $g(v) = 0$ and the vectors $\operatorname{grad} g_1(v)$ and $\operatorname{grad} g_3(v)$ are linearly independent, then there is a curve $s \mapsto \gamma(s)$ in $\mathbb{R}^3$ such that $\gamma(0) = 0$, $\dot{\gamma}(s) \neq 0$, and $F(\gamma(s), 0) \equiv 0$. If such a curve exists and $s = 0$ is a simple zero of the scalar function given by $s \mapsto h_2(\gamma(s)) - h_1(\gamma(s))$, then there is a curve $\epsilon \mapsto \beta(\epsilon)$ is $\mathbb{R}^3$ such that $\beta(0) = v$ and $F(\beta(\epsilon), \epsilon) \equiv 0$. Moreover, for each sufficiently small $\epsilon \neq 0$, the point $\beta(\epsilon)$ is a simple zero of the function $u \mapsto F(u, \epsilon)$ (see [49]).

Exercise 5.42. Prove that the roots of a (monic) polynomial depend continuously on its coefficients.

5.3.5 Limit Cycles–Entrainment–Resonance Zones

In Section 5.3.4 we considered the continuation of (ultra)subharmonics of the differential equation

$$\dot{u} = f(u) + \epsilon g(u, t, \epsilon), \quad u \in \mathbb{R}^2 \tag{5.68}$$

from a resonant unperturbed periodic orbit contained in a period annulus. Here, we will consider continuation of (ultra)subharmonics from a resonant unperturbed limit cycle.

If we view the differential equation (5.68) as an autonomous first order system on the phase cylinder $\mathbb{R}^2 \times \mathbb{T}$, then the unperturbed differential equation has an invariant torus $\Gamma \times \mathbb{T}$. For the theory in this section, it is not necessary to determine the fate of the invariant torus after perturbation. In fact, if Γ is a hyperbolic limit cycle, then the corresponding invariant torus is a normally hyperbolic invariant manifold. Roughly speaking, an invariant manifold is attracting and normally hyperbolic if the linearized flow for each orbit on the manifold contracts normal vectors at an exponential rate, and if the slowest such rate is faster than the fastest contraction rate for a vector that is tangent to the manifold. There is a similar definition if the manifold is repelling or if it has both attracting and repelling normal directions. In our case, if the limit cycle Γ is attracting, then its normal contraction rate is exponential and its tangential contraction rate is zero. Moreover, the invariant unperturbed torus corresponding to Γ inherits this behavior (see Exercise 5.44). Thus, this invariant torus is normally hyperbolic. In this case, by a powerful, important theorem (see [85] and [122]), the normally hyperbolic torus persists after perturbation. The continuation theory in this section describes the flow on this perturbed invariant torus. Typically, there is an even number of (ultra)subharmonics that alternate in stability around the perturbed torus.

By the above remarks, if our unperturbed system has a resonant, attracting hyperbolic limit cycle, then after perturbation there is an attracting invariant torus and nearby perturbed orbits are attracted to stable (ultra)subharmonic orbits on this torus. In the engineering literature this phenomenon is called *entrainment*: As nearby orbits are attracted to the perturbed invariant torus, their quasi-periods approach the periods of the (ultra)subharmonics on the perturbed torus. In particular, the asymptotic periods are entrained to a multiple of the period of the input perturbation. For a perturbation of small amplitude, this entrained period is close to the resonant period $m\eta(0)$ as in equation (5.38). We will determine a bifurcation function whose simple zeros are the continuation points of these (ultra)subharmonics.

For the remainder of this section let us consider the periodically perturbed oscillator (5.68) under the following assumptions:

(*i*)   There is an unperturbed periodic orbit Γ in $(m : n)$ resonance with the periodic perturbation as in the equation (5.38).

(*ii*) There is a region $R \subseteq \mathbb{R}^2$ with $\Gamma \subset R$ such that the displacement $\delta : R \times (-\epsilon_0, \epsilon_0) \to \mathbb{R}^2$ is defined for some $\epsilon_0 > 0$.

(*iii*) As in Proposition 5.36, the periodic orbit Γ is a hyperbolic limit cycle, or alternatively, the differential of the return time map $\sigma \mapsto \mathcal{T}(\sigma, 0)$ at $v = \Gamma \cap \Sigma$ defined on some curve Σ transverse to Γ, is nonsingular.

Let us also note that by the third hypothesis and Proposition 5.36, we have that $\mathrm{Kernel}\, \delta_\zeta(\zeta, 0) = [f(\zeta)]$ for $\zeta \in \Gamma$, and therefore the invariant torus $\Gamma \times \mathbb{T}$ is normally nondegenerate.

The analysis required to obtain the bifurcation function in case the unperturbed resonant periodic orbit is a limit cycle is analogous to the analysis carried out in the last section for the case of a regular period annulus. In particular, using the same notation as before, we can apply the Lyapunov–Schmidt reduction to the displacement function represented in the same (ρ, ϕ)-coordinates.

By the abstract theory of the Lyapunov–Schmidt reduction, ρ can be defined implicitly as a function of (ϕ, ϵ) when it is projected onto the range of the infinitesimal displacement, that is, the partial derivative of the displacement with respect to the space variable. It is easy and instructive to verify this directly. In fact, with respect to the (ρ, ϕ)-coordinates, the section Σ as in Proposition 5.36 is just an integral curve of $f^{\perp}$, and hence it is orthogonal to Γ. Thus, let us first consider the map

$$(\rho, \phi, \epsilon) \mapsto \langle \Delta(\rho, \phi, \epsilon), r_1 f(\varphi_\phi(v)) + r_2 f^{\perp}(\varphi_\phi(v)) \rangle$$

where r_1 and r_2 are defined in Proposition 5.36 and Δ is the local coordinate representation defined in display (5.62) of the displacement. By equation (5.63) and this proposition, its differential with respect to ρ at $(\rho, \epsilon) = (0, 0)$ is given by

$$\begin{aligned}
&\langle \Delta_\rho(0, \phi, 0),\ r_1 f(\varphi_\phi(v)) + r_2 f^{\perp}(\varphi_\phi(v)) \rangle \\
&= \langle \delta_\zeta(\varphi_\phi(v), 0) D\Upsilon(0, \phi) \partial/\partial\rho,\ r_1 f(\varphi_\phi(v)) + r_2 f^{\perp}(\varphi_\phi(v)) \rangle \\
&= \langle b(\phi) \delta_\zeta(\varphi_\phi(v), 0) f^{\perp}(\varphi_\phi(v)),\ r_1 f(\varphi_\phi(v)) + r_2 f^{\perp}(\varphi_\phi(v)) \rangle \\
&= b(\phi) \langle r_1 f(\varphi_\phi(v)) + r_2 f^{\perp}(\varphi_\phi(v)),\ r_1 f(\varphi_\phi(v)) + r_2 f^{\perp}(\varphi_\phi(v)) \rangle \\
&= b(\phi) |f(\varphi_\phi(v))|^2 (r_1^2 + r_2^2).
\end{aligned}$$

Also, let us note that during the course of the last computation we have proved that

$$\Delta_\rho(0, \phi, 0) = b(\phi)(r_1 f(\varphi_\phi(v)) + r_2 f^{\perp}(\varphi_\phi(v))). \tag{5.69}$$

By the assumptions for this section, we have $r_1^2 + r_2^2 \neq 0$, and therefore $\Delta_\rho(0, \phi, 0) \neq 0$. Thus, by an application of the implicit function theorem, there is a function $(\phi, \epsilon) \mapsto h(\phi, \epsilon)$ such that $h(\phi, 0) \equiv 0$ and

$$\langle \Delta(h(\phi, \epsilon), \phi, \epsilon),\ r_1 f(\varphi_\phi(v)) + r_2 f^{\perp}(\varphi_\phi(v)) \rangle \equiv 0.$$

Recall that $\Gamma \times \mathbb{T}$ is normally nondegenerate if

$$\text{Kernel}\, \delta_\zeta(\varphi_\phi(v), 0) = [f(\phi_\phi(v))].$$

Since this kernel is a one-dimensional subspace of the two-dimensional tangent space of the Poincaré section, the range of the infinitesimal displacement must also be one-dimensional, and therefore $r_1^2 + r_2^2 \neq 0$. Of course, this inequality also holds if either Γ is hyperbolic or the differential of the return time is nonzero.

The reduced displacement function is just the projection of the displacement onto a complement for the range of the infinitesimal displacement. For definiteness, let us consider the reduced displacement function $(\phi, \epsilon) \mapsto \widetilde{\Delta}(\phi, \epsilon)$ given by

$$\widetilde{\Delta}(\phi, \epsilon) = \langle \Delta(h(\phi, \epsilon), \phi, \epsilon),\ -r_2 f(\varphi_\phi(v)) + r_1 f^{\perp}(\varphi_\phi(v)) \rangle.$$

Since $\Delta(h(\phi, 0), \phi, 0) \equiv 0$, we have $\widetilde{\Delta}(\phi, 0) \equiv 0$ and

$$\widetilde{\Delta}(\phi, \epsilon) = \epsilon(\widetilde{\Delta}_\epsilon(\phi, 0) + O(\epsilon)).$$

If the *bifurcation function* $B : \mathbb{R} \to \mathbb{R}$ is defined by $B(\phi) := \widetilde{\Delta}_\epsilon(\phi, 0)$, then we have, as usual, the following proposition: *The simple zeros of B are the (ultra)subharmonic continuation points.* This ends the reduction step.

The identification of the bifurcation function B is accomplished with the aid of a simple computation. Indeed, using the definition of B, we have that

$$
\begin{aligned}
B(\phi) = \langle \Delta_\rho(h(\phi, 0), \phi, 0) h_\epsilon(\phi, 0) + \Delta_\epsilon(h(\phi, 0), \phi, 0), \\
- r_2 f(\varphi_\phi(v)) + r_1 f^\perp(\varphi_\phi(v)) \rangle.
\end{aligned}
\tag{5.70}
$$

To simplify this expression, apply identity (5.69) to obtain the representation

$$
B(\phi) = \langle \Delta_\epsilon(h(\phi, 0), \phi, 0), \ -r_2 f(\varphi_\phi(v)) + r_1 f^\perp(\varphi_\phi(v)) \rangle.
$$

Also, let us note that, as in the last section, $B(\phi) = \mathcal{Q}(\phi) P_\epsilon(\varphi_\phi(v), 0)$.

Using the equations (5.65) and (5.37), substitute the solution (5.33) of the nonhomogeneous variational equation for Δ_ϵ to obtain the formula

$$
\begin{aligned}
B(\phi) = \langle m\eta'(0) f(\varphi_\phi(v)) + \big(\mathcal{N}(\phi) + a(2\pi n/\omega)\mathcal{M}(\phi)\big) f(\varphi_\phi(v)) \\
+ b(2\pi n/\omega)\mathcal{M}(\phi) f^\perp(\varphi_\phi(v)), -r_2 f(\varphi_\phi(v)) + r_1 f^\perp(\varphi_\phi(v)) \rangle
\end{aligned}
$$

where, by Proposition 5.36,

$$
r_1 = a(2\pi n/\omega, \varphi_\phi(v)), \qquad r_2 = b(2\pi n/\omega, \varphi_\phi(v)) - 1.
$$

Hence, the bifurcation function is given by

$$
\begin{aligned}
B(\phi) = \big((1 - b(2\pi n/\omega, \varphi_\phi(v)))(mk + \mathcal{N}(\phi)) \\
+ a(2\pi n/\omega, \varphi_\phi(v))\mathcal{M}(\phi)\big) |f(\varphi_\phi(v))|^2.
\end{aligned}
$$

Define the *subharmonic bifurcation function* by

$$
\mathcal{C}(\phi) := (1 - b(2\pi n/\omega, \varphi_\phi(v)))(mk + \mathcal{N}(\phi)) + a(2\pi n/\omega, \varphi_\phi(v))\mathcal{M}(\phi)
\tag{5.71}
$$

where

$$
b(t, \varphi_\phi(v)) = \frac{|f(v)|^2}{|f(\varphi_{t+\phi}(v))|^2} e^{\int_0^t \operatorname{div} f(\varphi_{s+\phi}(v))\, ds},
$$

$$
a(t, \varphi_\phi(v)) = \int_0^t \big(2\kappa(s, \varphi_\phi(v)) |f(\varphi_{s+\phi}(v))|
$$
$$
- \operatorname{curl} f(\varphi_{s+\phi}(v))\big) b(s, \varphi_\phi(v))\, ds,
$$

$$
b(2\pi n/\omega, \varphi_\phi(v)) = b^n(2\pi/\omega, \varphi_\phi(v)) = (e^{\int_\Gamma \operatorname{div} f})^n,
$$

$$
a(2\pi n/\omega, \varphi_\phi(v)) = \Big(\sum_{j=0}^{n-1} b^j(2\pi/\omega, \varphi_\phi(v)) \Big)
$$
$$
\times \int_0^{2\pi/\omega} \big(2\kappa(t, \varphi_\phi(v)) |f(\varphi_{t+\phi}(v))|
$$
$$
- \operatorname{curl} f(\varphi_{t+\phi}(v))\big) b(t, \varphi_\phi(v))\, dt;
$$

and

$$\mathcal{M}(\phi) = \int_0^{2\pi n/\omega} \frac{1}{b(t,\phi)|f(\varphi_{t+\phi}(v))|^2} f(\varphi_{t+\phi}(v)) \wedge g(\varphi_{t+\phi}(v), t, 0)\, dt,$$

$$\mathcal{N}(\phi) = \int_0^{2\pi n/\omega} \frac{1}{|f(\varphi_{t+\phi}(v))|^2} \langle g(\varphi_{t+\phi}(v), t, 0), f(\varphi_{t+\phi}(v)) \rangle\, dt$$
$$- \int_0^{2\pi n/\omega} \frac{a(t,\phi)}{b(t,\phi)|f(\varphi_{t+\phi}(v))|^2} f(\varphi_{t+\phi}(v)) \wedge g(\varphi_{t+\phi}(v), t, 0)\, dt.$$

Remark 1. The function $\phi \mapsto b(2\pi n/\omega, \varphi_\phi(v))$ is constant, but the function $\phi \mapsto a(2\pi n/\omega, \varphi_\phi(v))$ may not be constant.

Theorem 5.43. *If Γ is an $(m:n)$ resonant unperturbed periodic solution of the periodically perturbed oscillator (5.68) such that $\Gamma \times \mathbb{T}$ is a normally nondegenerate unperturbed invariant torus for the system (5.39), then the simple zeros of the subharmonic bifurcation function $\phi \mapsto \mathcal{C}(\phi)$ are (ultra)subharmonic continuation points.*

By inspection of the formula for the subharmonic bifurcation function $\phi \mapsto \mathcal{C}(\phi)$, let us note that this function is periodic with period $2\pi/\omega$, the period of the resonant limit cycle Γ. This simple observation leads to an important application of Theorem 5.43, at least in the case where Γ is hyperbolic. In fact, the theorem provides a partial answer to the following question: What are the regions in the (η, ϵ) parameter space corresponding to the existence of $(m:n)$ (ultra)subharmonics of the system (5.68)? For this application it is traditional to view these regions in frequency-amplitude coordinates instead of period-amplitude coordinates. Thus, let us define $\Omega = 2\pi/\eta$. The subset $\mathcal{D}_{m/n}$ of all (Ω, ϵ) in the parameter space such that the corresponding system (5.68) has an $(m:n)$ (ultra)subharmonic is called the $(m:n)$ *entrainment domain*. In effect, relative to the geometric interpretation provided by the system (5.39), if (Ω, ϵ) is in $\mathcal{D}_{m/n}$, then there is a solution of the corresponding system on the perturbed invariant torus that wraps n times in the direction of ϕ, the parameter on Γ, and m times in the direction of the "time" given by τ in system (5.39).

Theorem 5.43 only applies for ϵ sufficiently small. Thus, it cannot provide an answer to the general question posed above. Nonetheless, this theorem does give valuable insight into the geometry of entrainment domains near $\epsilon = 0$. To see why this is so, note first that the frequency, in terms of equation (5.37), is given by

$$\Omega = \frac{m}{n}\omega - \frac{km^2}{2\pi n^2}\omega^2\epsilon + O(\epsilon^2), \tag{5.72}$$

and the (ultra)subharmonics correspond to the simple zeros of the subharmonic bifurcation function $\mathcal{C}$. Thus, using the definition of $\mathcal{C}$ we expect (ultra)subharmonics to exist whenever the detuning parameter k satisfies the equation

$$(1 - b(2\pi n/\omega, \varphi_\phi(v)))mk = -C(\phi) \tag{5.73}$$

for some $\phi \in \mathbb{R}$, where the new function C is defined by

$$C(\phi) = (1 - b(2\pi n/\omega, \varphi_\phi(v)))\mathcal{N}(\phi) + a(2\pi n/\omega, \varphi_\phi(v))\mathcal{M}(\phi).$$

The existence of solutions for equation (5.73) depends on the maximum and minimum values of the function C on the interval $0 \le \phi < 2\pi/\omega$. Let us denote these values by $C_{\min}$ and $C_{\max}$. Also, let us assume that the unperturbed resonant torus is attracting, that is, $b(2\pi n/\omega, \varphi_\phi(v)) < 1$. Under these assumptions, we have the following result: If

$$C_{\min} < (b-1)mk < C_{\max},$$

then $(m : n)$ (ultra)subharmonics exist *for sufficiently small* $|\epsilon|$. In other words, from equation (5.72), the lines in the frequency-amplitude space given by

$$L_1 := \{(\Omega, \epsilon) : \Omega = \frac{m}{n}\omega + \frac{C_{\min}}{(1-b)}\frac{m\omega^2}{2\pi n^2}\epsilon\},$$

$$L_2 := \{(\Omega, \epsilon) : \Omega = \frac{m}{n}\omega + \frac{C_{\max}}{(1-b)}\frac{m\omega^2}{2\pi n^2}\epsilon\} \tag{5.74}$$

are the tangent lines to the $(m : n)$ entrainment domain at $\epsilon = 0$. The shape of an entrainment domain (see, for example, Figure 5.6) suggested to Vladimir Arnold the shape of a tongue. Thus, entrainment domains are often referred to as *Arnold tongues*. They are also called *resonance zones* or *resonance horns*.

If $C(\phi) \equiv 0$, then the tangents L_1 and L_2 computed above coincide. In this case the tongue has vertical tangents provided that it extends all the way to the ϵ-axis.

If C has a simple zero, then the corresponding tongue is "open." Also, in the (Ω, ϵ)-coordinates the left boundary of the tongue corresponds to the ϕ coordinate on Γ giving the minimum value of C, while the right boundary corresponds to the maximum value of C. Thus, we see how the phase of the entrained solution shifts as the detuning parameter is changed so that $(1 - b(2\pi n/\omega))mk$ passes from the minimum to the maximum value of C. Finally, if a boundary is crossed as the detuning k is varied, say the boundary corresponding to the minimum of C, then it is clear that for k sufficiently small there are no (ultra)subharmonics, for k at the minimum of C there is a bifurcation point, and as k increases from this value, two branches of subharmonics bifurcate. This scenario is very common (generic, in fact) and is called a *saddle-node bifurcation*. It will be studied in detail in Chapter 8.

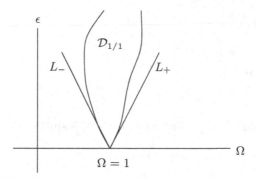

Figure 5.6: A schematic depiction of the $(1:1)$ entrainment domain $\mathcal{D}_{1/1}$ and its tangent lines $L_\pm := \{(\Omega, \epsilon) : \epsilon = \pm 2\,(\Omega - 1)\}$ at $(\Omega, \epsilon) = (1, 0)$ for the system (5.75).

Let us consider the family of differential equations

$$\dot{x} = -y + x(1 - x^2 - y^2),$$
$$\dot{y} = x + y(1 - x^2 - y^2) + \epsilon \cos(\Omega t) \qquad (5.75)$$

to illustrate some typical computations that are used to approximate the boundaries of entrainment domains (see [37]).

The unperturbed member of the family (5.75) has the unit circle as an attracting hyperbolic limit cycle with the corresponding solution starting at $(x, y) = (\cos\theta, \sin\theta)$ given by

$$x(t) = \cos(t + \theta), \qquad y(t) = \sin(t + \theta).$$

If $\Omega(\epsilon) := m/n + \Omega_1 \epsilon$, then the period of the forcing function is

$$\frac{2\pi}{\Omega(\epsilon)} = \frac{2\pi n}{m} - 2\pi \left(\frac{n}{m}\right)^2 \Omega_1 \epsilon + O(\epsilon^2).$$

Also, for this system $a \equiv 0$, and therefore

$$\mathcal{C}(\theta) = (1 - b(2\pi n))mk + (1 - b(2\pi n))\mathcal{N}(\theta, 2\pi n)$$
$$= \left(1 - e^{-4\pi n}\right) mk + \left(1 - e^{-4\pi n}\right) \int_0^{2\pi n} \cos(t + \theta) \cos(\Omega(0)t)\, dt.$$

Moreover, we have that

$$C(\theta) = \left(1 - e^{-4\pi n}\right) \int_0^{2\pi n} \cos(t + \theta) \cos(\frac{m}{n}t)\, dt$$

$$= \left(1 - e^{-4\pi n}\right) \times$$

$$\begin{cases} \pi n \cos \theta, & \frac{m}{n} = 1, \\ \frac{n^2}{m^2 - n^2} \left(\sin\theta + \frac{m+n}{2n}\sin(2\pi m - \theta) + \frac{m-n}{2n}\sin(2\pi m + \theta)\right), & \frac{m}{n} \neq 1 \end{cases}$$

$$= \begin{cases} \left(1 - e^{-4\pi n}\right) \pi n \cos\theta, & m = n, \\ 0, & m \neq n. \end{cases}$$

Thus, for $m = n$; that is, for the $(1:1)$ resonance, the tangents of the entrainment domain at the resonant point $(\Omega, \epsilon) = (1, 0)$ are

$$\epsilon = \pm 2 \left(\Omega - 1\right),$$

whereas, for the case $m \neq n$, the tangents have infinite slope.

The phase shift mentioned above is also easy to see in this example. The phase angle is θ. Also, if we use the equality $m = n$ and divide by a common factor, then the equation

$$k + \pi \cos\theta = 0$$

has the same roots as the zeros of the subharmonic bifurcation function. In particular, the detuning parameter k simply serves to translate the graph of the function $\theta \mapsto \pi \cos\theta$ in the vertical direction. Thus, at the left boundary of the tongue, $k = \pi$ and the phase of the entrained solution will be near $\theta = \pi$, whereas at the right hand boundary we have $k = -\pi$ and the phase will be near $\theta = 0$.

Exercise 5.44. Suppose that Γ is a hyperbolic limit cycle of the planar system $\dot{u} = f(u)$. Show that the linearized flow on the limit cycle attracts or repels normal vectors on the limit cycle at an exponential rate. Hint: The limit cycle has a characteristic multiplier that is not unity. Alternatively use the function b defined in Diliberto's theorem. Also, show that normal vectors on the invariant torus $\Gamma \times \mathbb{T}$ for the system

$$\dot{u} = f(u), \qquad \dot{\psi} = 1$$

on the phase cylinder where ψ is an angular variable are attracted exponentially, whereas tangent vectors have contraction rate zero.

Exercise 5.45. The theory of this chapter does not apply directly to determine the subharmonic solutions of the system

$$\dot{x} = y - x(1 - x^2 - y^2)^2 - \epsilon \cos t,$$
$$\dot{y} = -x - y(1 - x^2 - y^2)^2 + \epsilon \sin t.$$

Why? Develop an extension of the continuation theory to cover this case and use your extension to determine the subharmonics (see [35]).

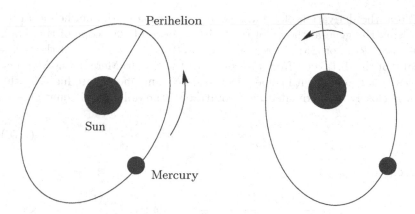

Figure 5.7: The point of closest approach to the Sun moves after each orbit in the direction of the revolution of Mercury. (The orbit of Mercury is nearly circular. The figure is not drawn to scale.)

5.3.6 Lindstedt Series and the Perihelion of Mercury

We have discussed in detail how to prove the existence of periodic solutions of nonlinear differential equations by continuation. In this section we will consider a procedure invented by Anders Lindstedt in 1882 that can be used to find useful series approximations for these periodic solutions. Lindstedt's method will be applied to the problem of the precession of the perihelion of Mercury—the most famous verification of the general theory of relativity— and in the next section it will used to determine the widths of entrainment domains for a forced van der Pol oscillator. The limitations of Lindstedt's method will also be briefly discussed.

Let us begin with the problem of the perihelion of Mercury. If a Cartesian coordinate system is fixed at the Sun, then the osculating ellipse traced out by the motion of Mercury is observed to *precess*. This means that the perihelion of Mercury—the point of closest approach to the Sun— changes after each revolution, moving in the direction of the motion of the planet (see Figure 5.7). In fact, the point of perihelion is observed to advance by approximately 43 seconds of arc per century. No satisfactory explanation of this phenomenon was known until after the introduction of the general theory of relativity by Albert Einstein. In particular, in 1915 Einstein found that his theory indeed predicts a precession of 43 seconds of arc per century—a stunning confirmation of his theory (see [214, Part I]). Shortly thereafter, Karl Schwarzschild (1916) found a solution of the gravitational field equations for a circularly symmetric body—the Sun— and he gave a rigorous derivation of the same relativistic correction to the Newtonian solution for the orbit of Mercury (see, for example, [231, p. 247] for more history).

While the derivation of Schwarzschild's solution of the perihelion problem from the general theory of relativity is beyond the scope of this book (see [138], [214], or [231] for readable accounts), it turns out that the reciprocal ρ of the distance r from the center of the Sun to Mercury, as Mercury moves on a geodesic with respect to the space-time metric produced by the Sun, is closely approximated by a solution of the differential equation

$$\frac{d^2\rho}{d\phi^2} = \gamma - \rho + \delta\rho^2 \tag{5.76}$$

where

$$\gamma := \frac{G_0 M(1 - \frac{G_0(m+M)}{ac^2})}{G_0(m+M)a(1-e^2)}, \qquad \delta := \frac{3G_0 M}{c^2}, \tag{5.77}$$

M is the mass of the Sun, m is the mass of Mercury, G_0 is the Newtonian gravitational constant, a is the semi-major axis of the elliptical orbit of Mercury, e is the eccentricity of the ellipse, and c is the speed of light (see [138] and [231]). We will predict the precession of the perihelion of Mercury from the differential equation (5.76).

Exercise 5.46. The constant γ in display (5.77) is usually given (see [231]) in the form

$$\gamma := \frac{c^2 b^2}{G_0 M}$$

where $b := r^2\, d\theta/ds$ and s is the proper time along the orbit of Mercury. Show that this formula agrees with the definition of γ in display (5.77). Hint: In relativity,

$$\frac{ds}{dt} = c\sqrt{1 - \frac{v^2}{c^2}};$$

that is, the proper time is the arc length along the orbit with respect to the space-time metric. Use the formula for the angular momentum given in Exercise 3.21. Approximate the velocity v along the orbit by $2\pi a/T$, where T is the orbital period, and use Kepler's third law, as in Exercise 3.21.

In view of the results in Section 3.2.2, especially the harmonic oscillator model (3.29) for Kepler motion, the system (5.76) with $\alpha = 0$ is exactly the same as the model predicted from Newton's theory. In fact, as we have seen, this model predicts a fixed elliptical orbit for Mercury. We will see that the perturbed orbit precesses.

The sizes of the parameters in equation (5.76) depend on the choice of the units of measurement. Thus, it is not meaningful to say that α is a small parameter. This basic problem is ubiquitous in applied mathematics. While some authors do not worry about the units, there is only one correct way to proceed: *rescale the variables so that the new system is dimensionless.*

For equation (5.76), if we define a new dependent variable $\eta := \rho/\gamma$, then the differential equation is recast in the form

$$\frac{d^2\eta}{d\phi^2} + \eta = 1 + \epsilon\eta^2 \tag{5.78}$$

where the ratio $\epsilon := \delta\gamma$ is dimensionless. To approximate ϵ, note that

$$\epsilon = \frac{3GM/c^2}{a(1-e^2)} \frac{GM}{G(m+M)} \left(1 - \frac{G(m+M)}{ac^2}\right),$$

m is much smaller than M, and $GM/(ac^2)$ is small, to obtain

$$\epsilon \approx \frac{3GM}{c^2 a(1-e^2)}, \tag{5.79}$$

and use the physical constants

$$\begin{aligned} G_0 &= 6.668 \times 10^{-11} \tfrac{\text{m}^3}{\text{kg·sec}^2}, & a &= (387)(149,598,845)\text{m}, \\ c &= 3 \times 10^8 \tfrac{\text{m}}{\text{sec}}, & m_{\text{Sun}} &= (332700)(5.977) \times 10^{24}\text{kg}, \\ e &= 0.206, \end{aligned}$$

$$\tag{5.80}$$

reported in [94] to compute the approximation

$$\epsilon \approx 7.973 \times 10^{-8}. \tag{5.81}$$

The differential equation (5.78) has two rest points in the phase plane: a center near the point with coordinates $(1,0)$, and a saddle near $(1/\epsilon, 0)$. Moreover, the orbit corresponding to the perturbed motion of Mercury corresponds to one of the periodic orbits surrounding the center (see Exercise 5.47).

Exercise 5.47. Show that the phase portrait of the system (5.78) has exactly two rest points: a saddle and a sink; approximate the positions of these rest points with power series in ϵ; and show that the orbit of Mercury corresponds to a periodic orbit. Note that it is *not* enough for this physical problem to prove the result for "sufficiently small epsilon." Rather, the value $\epsilon = \gamma\delta$ must be used! Hint: Initial conditions for the orbit of Mercury can be approximated from the physical data. The level sets of the energy corresponding to the differential equation (5.78) are invariant manifolds in the phase plane. In fact, one of them forms the boundary of the period annulus.

How can we find a useful approximation of the perturbed periodic orbit corresponding to the motion of Mercury? To answer this question, let us view ϵ as a parameter and observe that the differential equation (5.78) is analytic. Thus, the periodic solution $\phi \mapsto \eta(\phi, \epsilon)$ that we wish to approximate is given by an analytic function η *of two variables*. Also, this solution is an analytic function of the initial conditions. Thus, the perturbed solution can be expanded as a convergent power series in ϵ; at least this is true if ϵ is sufficiently small. We will come back to this problem in a moment. For now, let us assume that there is a series expansion of the form

$$\eta(\phi, \epsilon) = \eta_0(\phi) + \eta_1(\phi)\epsilon + \eta_2(\phi)\epsilon^2 + O(\epsilon^3). \qquad (5.82)$$

A natural idea is to substitute the series (5.82) into the differential equation (5.78), and then try to solve for the unknown Taylor coefficients by equating like powers of ϵ. In fact, if this is done, then (using dots to denote derivatives with respect to ϕ) the order zero equation is

$$\ddot{\eta}_0 + \eta_0 = 1. \qquad (5.83)$$

Note that we have some freedom in the choice of initial data for the solution of the differential equation (5.83). For example, if we consider the system in the phase plane, then there is an interval on the η-axis that lies to the right of the unperturbed rest point at $(1, 0)$ and contains one of the intersection points of our perturbed periodic orbit with the η-axis. In fact, this interval can be chosen to be a Poincaré section. Thus, we can suppose that the desired periodic orbit corresponding to the solution $\phi \mapsto \eta(\phi, \epsilon)$ starts at $\phi = 0$ on this section at a point with coordinate $1 + b$ for some $b = b(\epsilon) > 0$. In other words, for sufficiently small $\epsilon > 0$, we have the initial conditions $\eta(0, \epsilon) = 1 + b$ and $\dot{\eta}(0, \epsilon) = 0$. In particular, $\eta_0(0) = 1 + b$, $\dot{\eta}_0(0) = 0$, and the corresponding solution or the order zero differential equation is

$$\eta_0(\phi) = 1 + b\cos\phi.$$

Note that truncation at this order predicts elliptical motion for Mercury. In fact, the zero order approximation is just the solution of the harmonic oscillator model (3.29) of Kepler motion.

By using a trigonometric identity and some algebraic manipulation, the first order term in the series expansion of η is seen to be the solution of the initial value problem

$$\ddot{\eta}_1 + \eta_1 = \left(\frac{1}{\beta^2} + \frac{b^2}{2}\right) + \frac{2b}{\beta^2}\cos\phi + \frac{b^2}{2}\cos 2\phi,$$
$$\eta_1(0) = 0, \qquad \dot{\eta}_1(0) = 0, \qquad (5.84)$$

and, by an application of the variation of constants formula, the solution of this initial value problem has the form

$$\eta_1(\phi) = c_1 + c_2\phi\sin\phi + c_3\cos 2\phi$$

where c_1, c_2, and c_3 are nonzero constants.

We now have a problem: The first order approximation

$$\eta(\phi) \approx \eta_0(\phi) + \epsilon\eta_1(\phi)$$

is not periodic. Indeed, because one Fourier mode of the forcing function in the differential equation (5.84) is in resonance with the natural frequency of the harmonic oscillator, the expression for $\eta_1(\phi)$ contains the *secular* term $c_2\phi\sin\phi$. Indeed, the function $\phi \mapsto c_2\phi\sin\phi$ is unbounded as $\phi \to \infty$.

The word "secular" means an event that occurs once in a century. The inference is clear: Even if its coefficient is small, a secular term will eventually have arbitrarily large values. In particular, if there is a secular term in an approximation with a finite number of terms, then the approximation will not be periodic unless there is a fortuitous cancellation.

We started with a periodic function $\phi \mapsto \eta(\phi, \epsilon)$, but the first order term in its series expansion in powers of the perturbation parameter ϵ is not periodic. How can this be?

As an example to illustrate the reason for the appearance of secular terms, let us consider the harmonic oscillator with small detuning given by

$$\ddot{u} + (1 + \epsilon)^2 u = 0$$

with the initial conditions $u(0) = b$ and $\dot{u}(0) = 0$. For this example, we have that

$$u(t, \epsilon) = b\cos((1 + \epsilon)t) = b\cos t - (bt\sin t)\epsilon - \frac{1}{2}(bt^2\cos t)\epsilon^2 + O(\epsilon^3).$$

Hence, even though the series represents a periodic function, every finite order approximation obtained by truncation of the series is unbounded. Clearly, these finite order approximations are not useful over long time intervals. Also, note that the terms in this series expansion have the "wrong" period. Whereas the solution is periodic with period $2\pi/(1+\epsilon)$, the trigonometric terms on the right hand side all have period 2π.

Lindstedt observed that secular terms appear in the series for a perturbed periodic solution because the parameter-dependent frequency of the perturbed periodic orbit is not taken into account. He showed that the secular terms can be eliminated if the solution *and its frequency* are simultaneously expanded in powers of the perturbation parameter.

As an illustration of Lindstedt's method, let us consider a perturbed linear system of the form

$$\ddot{u} + \lambda^2 u = \epsilon f(u, \dot{u}, \epsilon) \tag{5.85}$$

that has a family of periodic solutions $t \mapsto u(t, \epsilon)$ with the initial conditions $u(0, \epsilon) = b$ and $\dot{u}(0, \epsilon) = 0$. In other words, the corresponding periodic orbits in the phase plane all pass through the point with coordinates $(b, 0)$. Also, let us define the function ω given by $\epsilon \mapsto \omega(\epsilon)$ such that the frequency of the periodic solution $t \mapsto u(t, \epsilon)$ is $\omega(\epsilon)$.

Lindstedt introduces a new independent variable

$$\tau = \omega(\epsilon)t$$

so that the desired periodic solution $t \mapsto u(t, \epsilon)$ is given by

$$u(t, \epsilon) = v(\omega(\epsilon)t, \epsilon)$$

where $\tau \mapsto v(\tau, \epsilon)$ is the 2π-periodic solution of the initial value problem

$$\omega^2(\epsilon)v'' + \lambda^2 v = \epsilon f(v, \omega(\epsilon)v', \epsilon), \qquad v(0, \epsilon) = b, \quad v'(0, \epsilon) = 0 \qquad (5.86)$$

and v' denotes the derivative of v with respect to τ.

Lindstedt's computational method is the following: Write the 2π-periodic function $\tau \mapsto v(\tau, \epsilon)$ and the frequency $\epsilon \mapsto \omega(\epsilon)$ as series

$$v(\tau, \epsilon) = v_0(\tau) + v_1(\tau)\epsilon + v_2(\tau)\epsilon^2 + \cdots,$$
$$\omega(\epsilon) = \lambda + \omega_1 \epsilon + \omega_2 \epsilon^2 + \cdots,$$

substitute these series into the differential equation (5.86), and then compute the unknown coefficients recursively by equating the terms with like powers of ϵ. Alternatively, the differential equations for the Taylor coefficients of v can be computed directly from the differential equation (5.86) as variational equations.

To determine the order zero coefficient, set $\epsilon = 0$ in equation (5.86) to see that v_0 is the solution of the initial value problem

$$\lambda^2(w'' + w) = 0, \qquad w(0) = b, \quad w'(0) = 0,$$

and therefore

$$v_0(\tau) = b\cos\tau. \qquad (5.87)$$

Next, let us note that $v_1(\tau) = v_\epsilon(\tau, 0)$. Hence, by differentiating both sides of equation (5.86) with respect to ϵ and evaluating at $\epsilon = 0$, the function v_1 is seen to be the solution of the initial value problem

$$\lambda^2(w'' + w) = f(b\cos\tau, -\lambda b\sin\tau, 0) + 2\lambda\omega_1 b\cos\tau,$$
$$w(0) = 0, \qquad w'(0) = 0. \qquad (5.88)$$

Because the function $\tau \mapsto v(\tau, \epsilon)$ is 2π-periodic independent of ϵ, so is the function $\tau \mapsto v_\epsilon(\tau, 0)$, and therefore the point $(b, 0)$ is a continuation point of periodic solutions in the phase plane for the (usual) first order system corresponding to the differential equation in display (5.88). By rescaling and then applying Theorem 5.1 to this first order system, it follows that

$$\int_0^{2\pi} (f(b\cos\tau, -\lambda b\sin\tau, 0) + 2\lambda\omega_1 b\cos\tau) \sin\tau \, d\tau = 0.$$

Hence, the Fourier series for the function $\tau \mapsto f(b\cos\tau, -\lambda b\sin\tau, 0)$, which has the form

$$A_0 + A_1\cos\tau + B_1\sin\tau + \sum_{n=2}^{\infty}(A_n\cos n\tau + B_n\sin n\tau),$$

must be such that $B_1 = 0$. If we impose this condition and also choose $\omega_1 = A_1/(2\lambda b)$, then the forcing function on the right hand side of the linear system (5.88) has no resonant term. Thus, the corresponding solution v_1 contains no secular terms, and it is periodic with period 2π.

Using the second order variational equation, Theorem 5.1, and an appropriate choice of ω_2, all secular terms can be eliminated in the corresponding linear system, and the function v_2 is therefore periodic with period 2π. In fact, this procedure can be repeated to determine all of the coefficients of the Taylor series in ϵ for the perturbed frequency $\omega(\epsilon)$ and the solution v. Moreover, it follows from our assumptions that the resulting series converge.

The original periodic solution is represented by a series of form

$$u(t, \epsilon) = v(\omega(\epsilon)t, \epsilon) = b\cos(\omega(\epsilon)t) + v_1(\omega(\epsilon)t)\epsilon + O(\epsilon^2) \tag{5.89}$$

where v_1 is determined above, and the frequency of the original periodic solution is given by

$$\omega(\epsilon) = \lambda + \frac{A_1}{2\lambda b}\epsilon + O(\epsilon).$$

Let us note that because the series coefficients of the series (5.89) depend on ϵ, the Lindstedt series expansion for u is *not* a Taylor series.

If the Lindstedt procedure is carried out to some *finite* order—the only possibility in most applied problems—then, to obtain an approximation to the desired periodic solution, we must substitute a truncation of the series for the frequency ω into a truncation of the Lindstedt series for the periodic solution. This leads to the question "How well does the truncated Lindstedt series approximate the original periodic solution?" The answer for the case considered here is that the difference between the nth order truncation and the solution is $O(\epsilon^{n+1})$ on a time interval of length C/ϵ for some constant $C > 0$. See [170] for a careful treatment of order estimates of this type.

For one-dimensional oscillators, the error estimate just mentioned for Lindstedt series can be obtained from the associated Taylor series for the same solution. The analysis for multidimensional differential equations is more complicated. For example, for Hamiltonian perturbations of multidimensional Hamiltonian systems, the Lindstedt series generally diverge! This famous result of Poincaré is very important in the history of mathematics. The divergence of these series suggests that the underlying dynamics must be very complex. In fact, this observation led Poincaré to several major results, for example, the discovery of chaotic dynamics in Hamiltonian dynamical systems (see [12], [21], [73], [145], [146], [160], [166], [167], and [214]). On the other hand, Lindstedt series are useful for approximating the periodic solutions that are obtained as continuations of periodic orbits of the type considered in this chapter. In fact, it is no accident that Theorem 5.1 is used to obtain the Lindstedt series for the example analyzed above. The bifurcation functions (called the *determining equations* in the context of Lindstedt series) can be used to obtain Lindstedt approximations for the continued periodic solutions in each case that we have discussed (see Exercise 5.49).

Let us return to the perihelion of Mercury.

To apply Lindstedt series to obtain an approximation for the precession of perihelion, introduce new variables

$$v := \eta - 1, \qquad \tau = \omega(\epsilon)\phi$$

into equation (5.78) so that

$$\omega^2(\epsilon)v'' + v = \epsilon(1 + v)^2$$

(where v' denotes $dv/d\tau$), and use equations (5.87) and (5.88) to show that $v_0(\tau) = b\cos\tau$ and v_1 is the solution of the initial value problem

$$w'' + w = \left(1 + \frac{b^2}{2}\right) + 2b(1 + \omega_1)\cos\tau + \frac{b^2}{2}\cos 2\tau$$

with initial conditions $w(0) = w'(0) = 0$. Thus, following Lindstedt's procedure, if $\omega_1 := -1$, then the secular terms are eliminated. In fact, in the original variables we have

$$\rho(\phi) = \gamma + b\gamma \cos\left((1 - \epsilon)\phi\right) + O(\epsilon). \tag{5.90}$$

Moreover, the lowest-order truncation of the Lindstedt series (5.90) that includes the relativistic correction yields the approximation

$$\rho(\phi) \approx \gamma + b\gamma \cos\left((1 - \epsilon)\phi\right). \tag{5.91}$$

In view of equation (5.91), the distance $r = 1/\rho$ of Mercury to the center of the Sun is approximated by

$$r \approx \frac{1}{\gamma\left(1 + b\cos\left((1 - \epsilon)\phi\right)\right)}. \tag{5.92}$$

Also, the perihelion for this elliptical orbit occurs when the argument of the cosine is a multiple of 2π. Thus, if the orbit starts at perihelion at $\phi = 0$, then after one revolution it returns to perihelion when $(1 - \epsilon)\phi = 2\pi$, that is, when ϕ has advanced by approximately

$$2\pi\epsilon = \frac{6\pi G_0 M}{a(1 - e^2)}$$

radians from the unperturbed value $\phi = 2\pi$.

Using the expression for Kepler's third law in Exercise 3.21 and the physical constants (5.80), the orbital period of Mercury is seen to be

$$T \approx 7.596 \times 10^6 \text{sec}.$$

In other words, Mercury orbits the Sun approximately 414.9 times in a century. Using the estimate for ϵ in display (5.81), the orbital advance of the perihelion per century is thus found to be 2.08×10^{-4} radians, or approximately 43 seconds of arc per century. (Can you imagine how Einstein must have felt when he computed this number?)

Exercise 5.48. For the perturbed harmonic oscillator $\ddot{u} + u = \epsilon u$, the natural frequency is "corrected" at first order in the perturbation parameter by $\omega(\epsilon) = 1 - \epsilon$. What is the first order correction if the perturbation is ϵu^2 or ϵu^3? What about ϵu^n.

Exercise 5.49. Discuss the application of Lindstedt's method to forced oscillators. For example, find the first order approximation for the solution(s) of the forced oscillator

$$\ddot{u} + u = \epsilon(\alpha \cos(\omega t) + b u^3).$$

Hint: Recall the theory in Section 5.3.2 for the continuation of periodic solutions in an isochronous period annulus. In particular, recall that we expect to find periodic solutions when the parameter ω is near a resonance, say $\omega(\epsilon) = 1 + \omega_1 \epsilon$. In this case, assume the value of the detuning ω_1 is known, and look for solutions (harmonics) with frequency ω. This search can be conducted within the geometry of the stroboscopic Poincaré map. Unlike the case of an autonomous perturbation; here the frequency is known, but the initial position of the solution in the Poincaré section is not known. Rather, the initial position, the continuation curve, is a function of ϵ. This suggests the introduction of a new time variable $\tau = \omega(\epsilon)t$ so that we can look for periodic solutions with period 2π of the scaled differential equation

$$\omega^2(\epsilon)v'' + v = \epsilon(\alpha \cos(\tau) - \beta u^3).$$

To apply the Lindstedt method, we must expand $v(t, \epsilon)$ as a power series in ϵ as before, but, because the initial position of the periodic orbit is not known, we must also expand the initial values $v(0, \epsilon)$ and $v'(0, \epsilon)$. The coefficients for these series expansions of the initial data and the function v are to be determined by equating coefficients. If

$$v(0, \epsilon) = \zeta_{10} + \zeta_{11}\epsilon + O(\epsilon^2), \qquad v'(0, \epsilon) = \zeta_{20} + \zeta_{21}\epsilon + O(\epsilon^2),$$

then the 2π-periodic zero order approximation is

$$v_0(\tau) = \zeta_{10} \cos \tau + \zeta_{20} \sin \tau.$$

The values of ζ_{10} and ζ_{20} are determined at the next order. Compute the first order approximation, consider the condition required to make the approximation

2π-periodic, and compare your result with the bifurcation equations obtained at the end of Section 5.3.2. Also, consider the form of the Lindstedt series in the original time variable.

Exercise 5.50. Compute to at the least second order in the small parameter the approximate period of the perturbed periodic orbit for the van der Pol oscillator (5.3) (see [6] and [69]).

5.3.7 Entrainment Domains for van der Pol's Oscillator

Consider the forced van der Pol oscillator in the form

$$\ddot{x} + \delta(x^2 - 1)\dot{x} + x = \epsilon \cos \Omega t. \tag{5.93}$$

We will use the formulas of Section 5.3.5 together with Lindstedt approximations to estimate—because the unperturbed system is not explicitly integrable—the widths of the entrainment domains for system (5.93).

For small δ, the second order Lindstedt approximation for the solution corresponding to the unperturbed limit cycle Γ is given by [226]

$$x(t) = 2\cos s + \left(\frac{3}{4}\sin s - \frac{1}{4}\sin 3s\right)\delta$$

$$+ \left(-\frac{1}{8}\cos s + \frac{3}{16}\cos 3s - \frac{5}{96}\cos 5s\right)\delta^2 + O(\delta^3) \tag{5.94}$$

where $s = (1 - \delta^2/16 + O(\delta^4))t$, and the approximate period of the limit cycle is $\tau := 2\pi(1 + \delta^2/16) + O(\delta^4)$. Moreover, these approximations are valid; that is, the difference between the approximation and the exact solution is bounded by a constant times δ^3 on the time scale of one period of the limit cycle Γ.

Recall the function C given in equation (5.73) and the formulas (5.74) used to determine the width of the entrainment domains. To use these formulas, let us approximate the extrema of the function C. This is accomplished by using the Lindstedt series (5.94) to approximate the phase plane parameterization of Γ given by

$$\theta \mapsto (x(t + \theta), \dot{x}(t + \theta)).$$

If the resulting formulas are inserted into C and the terms of like order are collected, then we obtain an approximation of the form

$$C(\theta) \approx c_1(\theta)\delta + c_2(\theta)\delta^2.$$

This approximation vanishes unless $m = n$ or $m = 3n$, a manifestation of the resonances that appear in the approximation of the limit cycle as well as the order of the approximation. At these resonances we have that

$$b(n\tau) = 1 - 2n\pi\delta + 2n^2\pi^2\delta^2 + O(\delta^3).$$

Also, for $m = n$ the function C is given by

$$C(\theta) = -(n^2\pi^2\cos\theta)\delta + \frac{1}{8}n^2\pi^2(\sin 3\theta - 3\sin 5\theta$$
$$+ \sin\theta + 8n\pi\cos\theta + 4\sin\theta\cos 2\theta + 6\sin\theta\cos 4\theta)\delta^2 + O(\delta^3),$$

while for $m = 3n$ it is given by

$$C(\theta) = -\frac{1}{8}\left(n^2\pi^2\sin 3\theta\right)\delta^2.$$

In order to approximate the extrema of C in case $m = n$, note that the extrema of the function $\theta \mapsto C(\theta)/\delta$ at $\delta = 0$ occur at $\theta = 0$ and $\theta = \pi$. The perturbed extrema are then approximated using the series expansion of the left hand side of the equation $C'(\theta) = 0$. In fact, for $m = n$ we have

$$C_{\min} = -n^2\pi^2\delta + n^3\pi^3\delta^2 + O(\delta^3), \qquad C_{\max} = n^2\pi^2\delta - n^3\pi^3\delta^2 + O(\delta^3),$$

while for $m = 3n$ we have

$$C_{\min} = -\frac{1}{8}n^2\pi^2\delta^2 + O(\delta^3), \qquad C_{\max} = \frac{1}{8}n^2\pi^2\delta^2 + O(\delta^3).$$

By inserting these expressions into the formulas (5.74) for the tangent lines of the entrainment domains at $\epsilon = 0$ we obtain for $m = n$ the $O(\delta^4)$ approximation

$$\epsilon = \pm\left(4 + \frac{1}{2}\delta^2\right)\left(\Omega - \left(1 - \frac{1}{16}\delta^2\right)\right),$$

while for $m = 3n$ we obtain the $O(\delta^3)$ approximation

$$\epsilon = \pm\left(\frac{32}{3}\delta^{-1} - \frac{32n\pi}{3} + \frac{4}{3}\delta - \frac{4n\pi}{3}\delta^2\right)\left(\Omega - 3\left(1 - \frac{1}{16}\delta^2\right)\right).$$

Of course, the accuracy of these computations can be improved and higher order resonances can be studied by starting with higher order Lindstedt approximations (see [6] and [69]). Also, the presence of the term containing δ^{-1} in the slope of the tangent line for the $(3 : 1)$ resonance indicates that the entrainment domain has nearly vertical tangents for small δ, and therefore this entrainment domain is very thin near the Ω-axis.

Exercise 5.51. Numerical values can be obtained from the approximation formulas in this section. For example, if $\delta = 0.1$ and $(m : n) = (1 : 1)$, then the tangents obtained from the Lindstedt series are approximately

$$\epsilon = \pm 4.005(\Omega - 0.999).$$

Find the entrainment domain for this case using a numerical simulation of the van der Pol oscillator, approximate the tangents to the entrainment domains using the results of your simulation, and compare the results with the approximations given by Lindstedt series (see [37]). Hint: Find the frequency ω of the unperturbed van der Pol limit cycle using a numerical simulation. Set up a grid of (Ω, ϵ) values for Ω near ω and ϵ near zero. Then, for each choice of these parameter values set initial conditions near the intersection of the unperturbed limit cycle with the x-axis, iterate the Poincaré map several times and test to see if the iterates converge to a fixed point. If they do, assume that entrainment has occurred and color the corresponding grid point. If no entrainment occurs, then leave the corresponding grid point uncolored. The entrainment domain will emerge from the display of the colored grid points.

5.3.8   Periodic Orbits of Multidimensional Systems with First Integrals

Consider a k-dimensional family of differential equations

$$\dot{x} = f(x, \lambda), \quad x \in \mathbb{R}^n, \quad \lambda \in \mathbb{R}^k \tag{5.95}$$

with the family of solutions $t \mapsto \phi(t, x, \lambda)$ such that $\phi(0, x, \lambda) = x$. We will discuss the continuation theory for periodic orbits of the family member with $\lambda = 0$.

The continuation theory for a periodic orbit, whose Floquet multiplier one has geometric multiplicity one, is a straightforward generalization of the theory for planar systems that has been discussed in detail in this chapter (see Exercises 5.11 and 5.20). In this section, we will discuss the continuation of periodic orbits in case the family (5.95) has m first integrals $H_1, H_2, \ldots, H_m$; that is, for each $j \in \{1, 2, \ldots, m\}$, we have $H_j : \mathbb{R}^n \times \mathbb{R}^k \to \mathbb{R}$ and $d/dt H_j(\phi(t, x, \lambda), \lambda) \equiv 0$. The theory uses a familiar idea—apply the implicit function theorem—but it is slightly more complicated because the existence of m independent first integrals increases the geometric multiplicity of the Floquet multiplier one to at least $m + 1$. In this case, a periodic orbit is usually contained in an m-parameter family of periodic orbits and it is this m-parameter family that persists.

The set of first integrals $\{H_1, H_2, \ldots, H_m\}$ is called *independent* on $U \times V \subseteq \mathbb{R}^n \times \mathbb{R}^k$ if $\{\operatorname{grad} H_1(x, \lambda), \operatorname{grad} H_1(x, \lambda), \ldots, \operatorname{grad} H_m(x, \lambda)\}$ is linearly independent whenever $(x, \lambda) \in U \times V$.

Theorem 5.52. *If the member of the family (5.95) corresponding to $\lambda = 0$ has a periodic orbit Γ_0 with period T_0 and the family has m independent first integrals on $U \times \{\lambda : |\lambda| < \lambda_0\}$ where $U \subseteq \mathbb{R}^n$ is an open neighborhood of Γ_0 and $\lambda_0 > 0$, then the geometric multiplicity of the Floquet multiplier one is at least $m + 1$. If, in addition, the geometric multiplicity of the Floquet multiplier one of Γ_0 is $m + 1$, then Γ_0 is contained in an m-parameter family $\Upsilon(\mu)$ (that is, μ is contained in an open subset of $\mathbb{R}^m$ and the image of Υ is an $m + 1$-dimensional manifold) of periodic orbits of $\dot{x} = f(x, 0)$ and this family persists. Its continuation $\Gamma(\lambda, \mu)$ is unique and defined for $|\lambda| < a$ and $|\mu| < a$ for some $a > 0$. Moreover the family of periodic* *orbits* $\mu \to \Gamma(\lambda, \mu)$, *which may be chosen such that* $\Gamma(0,0) = \Gamma_0$, *is such that each periodic orbit* $\Gamma(\lambda, \mu)$ *is contained in an* $(n - m)$-*dimensional invariant manifold* $M(\lambda) = \{x \in \mathbb{R}^n : H_1(x, \lambda) = a_1(\lambda, \mu), H_2(x, \lambda) = a_2(\lambda, \mu), \dots, H_m(x, \lambda) = a_m(\lambda, \mu)\}$, *where for each* $i \in \{1, 2, \dots, m\}$, $a_i :$ $\{\lambda : |\lambda| < a\} \times \{\mu : |\mu| < a\} \to \mathbb{R}$, *the corresponding period function* $T(\lambda, \mu)$ *is such that* $T(0,0) = T_0$, *and every function in sight is as smooth as the family (5.95).*

Proof. We will outline a proof.

Let the first integrals be $H_1, H_2, \dots, H_m$ and suppose that Γ_0 contains the point $x_0 \in \mathbb{R}^n$. By the hypothesis, the set

$$\{\operatorname{grad} H_1(x_0, 0), \operatorname{grad} H_2(x_0, 0), \dots, \operatorname{grad} H_m(x_0, 0)\}$$

is linearly independent. Also, by the definition of a first integral, the nonzero vector $f(x_0, 0)$ is perpendicular (with respect to the usual metric in $\mathbb{R}^n$) to the span of these gradients. Choose vectors $v_1, v_2, \dots, v_{n-m-1}$ in $\mathbb{R}^n$ so that the gradients together with these vectors span the hyperplane at x_0 that is perpendicular to $f(x_0, 0)$. Let Σ be a subset of this hyperplane Σ_0 that contains x_0 and is a Poincaré section for the family (5.95) for all $|\lambda| < \lambda_1$, where $\lambda_1 > 0$, and let $P : \Sigma \times \{\lambda : |\lambda| < \lambda_1\} \to \Sigma_0$ denote the associated family of Poincaré maps.

By the definition of a first integral, it follows that

$$H_i(P(\xi, \lambda), \lambda) = H_i(\xi, \lambda) \tag{5.96}$$

for all each $i \in \{1, 2, \dots, m\}$, $\xi \in \Sigma$ and $|\lambda| < \lambda_1$. Also, by the definition and a differentiation, we have that, for each $i \in \{1, 2, \dots, m\}$,

$$\operatorname{grad} H_i(x_0, 0)\phi_\xi(T(0,0), x_0, 0) = \operatorname{grad} H_i(x_0, 0),$$

that is, the number one is an eigenvalue of the transpose of the linear map $\phi_\xi(T(0,0), x_0, 0)$. Since the transpose has the same eigenvalues as the map, one is also an eigenvalue (hence, a Floquet multiplier) of $\phi_\xi(T(0,0), x_0, 0)$. Since the gradients are linearly independent, its geometric multiplicity is at least $m+1$. (Note: The relationship between the eigenvalues and eigenvectors of a linear transformation and its transpose is made clear by examining the transformation to Jordan canonical form; indeed, $BAB^{-1} = J$ if and only if $(B^{-1})^* A^* B^* = J^*$, where the upper star denotes the transpose.)

Define $\Lambda : \mathbb{R}^{n-1} \times \mathbb{R}^k \times \mathbb{R}^m \to \mathbb{R}^{n-1}$ by

$$\Lambda(\xi, \lambda, \mu) = (H_1(\xi, \lambda) - H_1(x_0, 0) + \mu_1, \dots, H_m(\xi, \lambda) - H_m(x_0, 0) + \mu_m,$$
$$(P(\xi, \lambda) - \xi) \cdot v_1, \dots, (P(\xi, \lambda) - \xi) \cdot v_{n-m-1}).$$

Note that $\Lambda(x_0, 0, 0) = 0$ and, by the multiplicity hypothesis, $\Lambda_\xi(x_0, 0, 0) :$ $\mathbb{R}^{n-1} \to \mathbb{R}^{n-1}$ is invertible. By the implicit function theorem, there is a positive number b and a unique function

$$\beta : \{\lambda : |\lambda| < b\} \times \{\mu : |\mu| < b\} \to \Sigma$$

such that $\Lambda(\beta(\lambda,\mu),\lambda,\mu) \equiv 0$.

Using equation (5.96) and the implicit solution, it follows that

$$H_i(P(\beta(\lambda,\mu),\lambda),\lambda) = H_i(\beta(\lambda,\mu),\lambda)$$

for $i \in \{1,2,\dots,m\}$ and

$$(P(\beta(\lambda,\mu),\lambda) - \beta(\lambda,\mu)) \cdot v_i = 0$$

for $i \in \{1,2,\dots,n-m-1\}$. Using the independence of the first integrals and the inverse function theorem applied to the map

$$\xi \mapsto (H_1(\xi,\lambda),\dots,H_m(\xi,\lambda),\xi \cdot v_1,\dots,\xi \cdot v_{m-n-1}),$$

it follows that $P(\beta(\lambda,\mu),\lambda) = \beta(\lambda,\mu)$ whenever $|\lambda|$ and $|\mu|$ are smaller than some positive number a.

In particular, for each fixed λ, there is an m-dimensional family of periodic solutions.   □

In addition to the existence of independent first integrals, the key assumption of Theorem 5.52 is on the multiplicity of the Floquet multiplier one. In particular, the result may not hold if this multiplicity is too large.

For a planar Hamiltonian system, the Hamiltonian is an independent first integral at a periodic orbit and the multiplicity of the Floquet multiplier is exactly two. Thus, by Theorem 5.52, a periodic orbit of a planar Hamiltonian system cannot be isolated; it must be contained in a one-parameter family of periodic orbits. In the case of more than one degree-of-freedom (corresponding to a four or higher even-dimensional system), it is necessary to check that the multiplicity of the Floquet multiplier is $m + 1$. In fact, for Hamiltonian systems, this multiplicity is often exceeded. The reason is simple to understand. But, some additional theory of Hamiltonian systems is required to prove a theorem that states the exact multiplicity.

Suppose that the functions $H_1, H_2, \dots, H_m$ are independent first integrals of the Hamiltonian system X_1 with Hamiltonian H_1. We have proved that if X_1 has a periodic orbit, then one is a Floquet multiplier with geometric multiplicity at least $m + 1$. Consider the Hamiltonian systems $X_2, \dots, X_m$ corresponding to the remaining integrals. Also, let ϕ_i^t denote the flow of X_i. It might happen that, for some j, the flow of X_j commutes with the flow of X_1; that is, $\phi_1^t \circ \phi_i^s = \phi_i^s \circ \phi_1^t$. In this case, at the point x_0 on a periodic orbit of X_1 with period T and by differentiation with respect to s at $x = 0$, we have the identity $D\phi_1^T(x_0)X_j(x_0) = X_j(x_0)$. In other words, $X_j(x_0)$ is an eigenvector corresponding to the Floquet multiplier one. It turns out that if the flow of $X_j(x_0)$ commutes with the flow of each of the vector fields $X_1, X_2, \dots, X_m$, then $X_j(x_0)$ is not in the span of the eigenvectors corresponding to the mere existence of first integrals. Thus, the multiplicity of the Floquet multiplier is at least $m+2$; and, in this case, Theorem 5.52 does not apply.

We will need a few new ideas to see why $X_j(x_0)$ is independent of the other eigenvectors with eigenvalue one.

For the Hamiltonian case, the dimension n must be even, say $n = 2N$. Let angle brackets denote the usual inner product on $\mathbb{R}^N$ and, for two vectors $v = (v_1, v_2)$ and $w = (w_1, w_2)$ in $\mathbb{R}^n = \mathbb{R}^N \times \mathbb{R}^N$, define the skew-symmetric bilinear form $\omega : \mathbb{R}^n \times \mathbb{R}^n \to \mathbb{R}$ by

$$\omega(v, w) = \langle v_1, w_2 \rangle - \langle v_2, w_1 \rangle.$$

Also, note that ω is nondegenerate; that is, if $\omega(v, w) = 0$ for all w, then $v = 0$. In general, a nondegenerate skew-symmetric bilinear form on a vector space is called a symplectic form, an object that plays a fundamental role in Hamiltonian mechanics. Indeed, in our case, we have the equation

$$\operatorname{grad} H_1(v) = \omega(X_1, v)$$

for all $v \in \mathbb{R}^n$. In general, suppose that we have a symplectic form assigned to each tangent space of an even-dimensional manifold M so that the assignment varies smoothly over the manifold and a function $H : M \to \mathbb{R}$. The Hamiltonian vector field with Hamiltonian H is defined to be the unique vector field X_H on M such that

$$dH(Y)_p = \omega(X_H, Y)_p$$

for all vector fields Y on the manifold and all $p \in M$. Thus, in the general theory of Hamiltonian systems, the symplectic form comes first; it is used to define the Hamiltonian structure. In our simple case, the manifold is $\mathbb{R}^n$ and the (usual) symplectic form is constant over the manifold. It produces the Hamiltonian vector field $(H_y, -H_x)$.

Returning to our Floquet multipliers, we will show that the eigenvector X_j is not in the span of the eigenvectors corresponding to the Floquet multipliers obtained by the existence of first integrals. Similar to the construction in the proof of Theorem 5.52 where the vectors $v_1, v_2, \ldots, v_{n-m-1}$ are defined, let us choose coordinates, on the Poincaré section Σ in a neighborhood of the point $x_0 \in \Sigma$ that lies on a periodic orbit of X_1, using the first integrals as the first m coordinates. Here, we take the coordinate map to be

$$x \mapsto (H_1(x), H_2(x), \ldots, H_m(x), x \cdot v_1, x \cdot v_2, \ldots, x \cdot v_{n-m-1}).$$

Since, for the Poincaré map P, we have the equation $H_i(P(\xi)) = H_i(\xi)$, the first m coordinates of $P(\xi)$ are simply $\xi_1, \xi_2, \ldots, \xi_m$. Hence, the derivative of P has the block form

$$\begin{pmatrix} I & 0 \\ A & B \end{pmatrix}$$

where I is the $m \times m$-identity and $B : \mathbb{R}^{n-m-1} \to \mathbb{R}^{n-m-1}$. The m Floquet multipliers obtained by the existence of the first integrals correspond to the diagonal elements of the $m \times m$-identity. Note that an eigenvector corresponding to one of these eigenvalues must have a nonzero element among its first m components. In fact, if the eigenvector has block form (u, v), then it must satisfy the equation $Au + Bv = v$. If $u = 0$, then $Bv = v$ and the eigenvalue belongs to B instead.

We say that X_i and X_j (and H_1 and H_j) are in involution if $\omega(X_i, X_j) = 0$. We need the following proposition.

Proposition 5.53. *The Hamiltonian vector fields X_i and X_j are in involution if and only if their flows commute.*

Assuming this result, we will show that if the Hamiltonian vector field X_1 has independent first integrals $H_1, H_2, \ldots, H_m$ and H_j is in involution with each of these first integrals, then the Floquet multiplier one has multiplicity at least $m + 2$.

Since X_j is in involution with the Hamiltonian vector field X_i, for each $i \in \{1, 2, \ldots, m\}$, we have that

$$\operatorname{grad} H_j \cdot X_i = \omega(X_j, X_i) = 0.$$

In other words, X_i does not have a component in the direction of the first m coordinates. This proves the claim.

It remains to prove Proposition 5.53. By employing more machinery (the calculus of differential forms), this fact has an elegant proof (see, for example, [2]), which holds for Hamiltonians on manifolds. We will simply give the ingredients of a proof for the manifold $\mathbb{R}^n$ with the usual constant symplectic form.

For simplicity of notation, let's consider two Hamiltonians H and K and their corresponding Hamiltonian vector fields $X = (H_y, -H_x)$ and $Y = (K_y, -K_x)$. These vector fields are in involution when

$$\langle H_x, K_y \rangle - \langle H_y, K_x \rangle = 0.$$

We define the Lie bracket $[X, Y]$ of X and Y by $[X, Y] = L_X Y$ (see Definition 4.12), and we claim that $[X, Y] = XY - YX$, where XY is the vector field obtained by taking the directional derivative of each component of Y in the direction X, etc. In our notation, the x-component of $[X, Y]$ is

$$\langle H_y, K_{yx} \rangle - \langle H_x, K_{yy} \rangle - (\langle K_y, H_{yx} \rangle - \langle K_x, H_{yy} \rangle).$$

By differentiating $\omega(X, Y)$ with respect to y, we find that this expression vanishes. The second component is the same as the derivative of the involution relation with respect to x. The next lemma finishes the proof of Proposition 5.53.

Lemma 5.54. *(1) On $\mathbb{R}^n$, $[X, Y] = XY - YX$. (2) The flows of two vector fields X and Y commute if and only if their Lie bracket vanishes.*

Proof. Let ϕ_t denote the flow of the vector field X and ψ_t denote the flow of Y.

By the definition of the Lie derivative

$$L_X Y(x) = \frac{d}{dt} D\phi_{-t}(\phi_t(x)) Y(\phi_t(x))\big|_{t=0}.$$

Let the vector field Z be defined by

$$Z(\phi_t(x)) = D\phi_{-t}(\phi_t(x)) Y(\phi_t(x)) \tag{5.97}$$

and note that $D\phi_t(x) Z(\phi_t(x)) = Y(\phi_t(x))$. Also, from the definition of a flow, note the identity

$$\frac{d}{dt} D\phi_t(x) = DX(\phi_t(x)) D\phi_t(x).$$

Using these formulas and differentiating with respect to t in equation (5.97), we have

$$D\phi_t(x) \frac{d}{dt} Z(\phi_t(x)) = DY(\phi_t(x)) X(\phi_t(x)) - DX(\phi_t(x)) Y(\phi_t(x));$$

and, after evaluation at $t = 0$, the identity

$$L_X Y = DY(x) X(x) - DX(x) Y(x).$$

In other words, $[X, Y] = XY - YX$.

Suppose that the flows commute; that is, $\phi_t \circ \psi_s = \psi_s \circ \phi_t$. Equivalently, we have the identity $\psi_s = \phi_{-t} \circ \psi_s \circ \phi_t$. By differentiating with respect to s and evaluating at $s = 0$, it follows that

$$D\phi_{-t}(\phi_t(x)) Y(\phi_t(x)) = Y(x).$$

Hence $[X, Y] = L_X Y = 0$.

Suppose that $[X, Y] = 0$. Note that, in general, if $d/dt h(\phi_t(x))|_{t=0} = 0$ for all x, then $t \mapsto h(\phi_t(x))$ is a constant function; indeed,

$$0 = \frac{d}{dt} h(\phi_t(\phi_s(x)))|_{t=0} = \frac{d}{dt} h(\phi_{s+t})|_{t=0} = \frac{d}{ds} h(\phi_s).$$

By applying this fact in the definition of the Lie derivative and using the hypothesis, we have that $t \mapsto D\phi_{-t}(\phi_t(x)) Y(\phi_t(x))$ is constant. Hence,

$$D\phi_{-t}(\phi_t(x)) Y(\phi_t(x)) = Y(x).$$

Let $\gamma_s := \phi_{-t} \circ \psi_s \circ \phi_t$ and note that

$$\begin{aligned}
\frac{d}{ds} \gamma_s(x) &= D\phi_{-t}(\psi_s \circ \phi_t(x)) Y(\psi_s \circ \phi_t(x)) \\
&= D\phi_{-t}(\phi_t \circ \gamma_s(x))) Y(\phi_t \circ \gamma_s(x)) \\
&= Y(\gamma_s(x)).
\end{aligned}$$

By the uniqueness of solutions of initial value problems, we must have that $\gamma_s = \psi_s$.

$\square$

Exercise 5.55. Suppose that A and B are matrices. Use Lemma 5.54 to prove that $e^{tA}e^{tB} = e^{t(A+B)}$ if and only if $AB = BA$.

Exercise 5.56. Prove that if $\dot{x} = f(x)$, for $x \in \mathbb{R}^n$, has n independent first integrals, then $f = 0$.

Exercise 5.57. Show that periodic orbits of the uncoupled harmonic oscillators model

$$H(q_1, q_2, p_1, p_2) = \frac{1}{2}(p_1^2 + p_2^2) + \frac{1}{2}(q_1^2 + q_2^2)$$

occur in four-parameter families. This suggests the existence of an independent integral in involution with the Hamiltonian. Find this integral.

Exercise 5.58. Prove that the system

$$\dot{x} = -y + x(x^2 + y^2 - 1), \qquad \dot{y} = x + y(x^2 + y^2 - 1)$$

does not have a (nonconstant) first integral.

Exercise 5.59. (a) Prove that the system

$$\dot{x} = -y + xy, \qquad \dot{y} = x + y^2$$

has a (nonconstant) first integral. (b) Prove that the system has a center at the origin.

Exercise 5.60. [Completely Integrable Hamiltonians] A Hamiltonian system with n degrees-of-freedom is called completely integrable if it has n independent first integrals. (a) Prove that the Hamiltonian system with Hamiltonian $H = p_1^2/2 + 1 - \cos q_2 + (p_2^2 + q_2^2)/2$ is completely integrable. (b) Prove that the system

$$\dot{q} = p,$$
$$\dot{p} = -\sin q + \frac{x^4}{4}\sin q,$$
$$\dot{x} = y,$$
$$\dot{y} = -x - x^3 \cos q \qquad (5.98)$$

has a (nonconstant) first integral. (d) Prove that system (5.98) is not completely integrable. Hint: Use theorem 5.52 (cf. [103, Section 4.8]).

5.4   Forced Oscillators

In this section we will apply our continuation theory to the oscillator

$$\ddot{x} + \epsilon h(x, \dot{x})\dot{x} + f(x) = \epsilon g(t) \qquad (5.99)$$

where the function $t \mapsto g(t)$, the external force, has period $2\pi/\Omega$. As usual, we will consider the differential equation (5.99) as the first order system

$$\dot{x} = y, \qquad \dot{y} = -f(x) + \epsilon(g(t) - h(x, y)y), \qquad (5.100)$$

and we will assume that the unperturbed system

$$\dot{x} = y, \qquad \dot{y} = -f(x) \qquad (5.101)$$

has a period annulus $\mathcal{A}$ containing a resonant periodic solution $\Gamma_{m/n}$ whose period $2\pi/\omega$ is in $(m : n)$ resonance with the period of g. Also, we will assume that the period function on $\mathcal{A}$ has a nonzero derivative at $\Gamma_{m/n}$.

Under the assumptions stated above, we have proved that the simple zeros of the function $\phi \mapsto M(\phi)$ given by

$$M(\phi) = \int_0^{n2\pi/\omega} y(t + \phi, \xi)\big(g(t) - h(x(t + \phi, \xi), y(t + \phi, \xi))y(t + \phi), \xi)\big) \, dt$$

$$(5.102)$$

are the continuation points for $(m : n)$ (ultra)subharmonics. Here ϕ may be viewed as a coordinate on $\Gamma_{m/n}$ and ξ is a point on $\Gamma_{m/n}$ that defines an origin for the coordinate ϕ. For simplicity, we choose ξ to lie on the x-axis.

Note that the integrand of the integral used to define M is periodic with period $n2\pi/\omega$. If we suppress the variable ξ and change the variable of integration to $s = t + \phi$, then

$$M(\phi) = \int_\phi^{\phi + n2\pi/\omega} \big(y(s)g(s - \phi) - h(x(s), y(s))y^2(s)\big) \, ds.$$

The function

$$\theta \mapsto \int_\theta^{\theta + n2\pi/\omega} \big(y(s)g(s - \phi) - h(x(s), y(s))y^2(s)\big) \, ds$$

is constant for each fixed value of ϕ. Thus, we can represent the bifurcation function in the following convenient form:

$$M(\phi) = \int_0^{n2\pi/\omega} \big(y(s)g(s - \phi) - h(x(s), y(s))y^2(s)\big) \, ds = I_1(\phi) + I_2,$$

where

$$I_1(\phi) := \int_0^{n2\pi/\omega} y(s)g(s - \phi) \, ds, \qquad I_2 := \int_0^{n2\pi/\omega} h(x(s), y(s))y^2(s) \, ds.$$

The function $t \mapsto (x(-t), -y(-t))$ is a solution of the unperturbed differential equation with initial value at the point $(\xi, 0)$. Thus, by the uniqueness of solutions, x is an even function and y is an odd function of time.

Because $s \mapsto y(s)$ is an odd function, we have the Fourier series

$$y(s) = \sum_{k=1}^{\infty} y_k \sin k\omega s, \qquad g(s) = g_0 + \sum_{k=1}^{\infty} g_k^c \cos k\Omega s + \sum_{k=1}^{\infty} g_k^s \sin k\Omega s$$

where all coefficients are *real*. With these representations, it is easy to compute

$$I_1(\phi) = \sum_{k=1}^{\infty} \sum_{\ell=1}^{\infty} y_k g_\ell^s \int_0^{n2\pi/\omega} \sin(k\omega s) \sin(\ell\Omega(s - \phi)) \, ds$$

$$+ \sum_{k=1}^{\infty} \sum_{\ell=1}^{\infty} y_k g_\ell^c \int_0^{n2\pi/\omega} \sin(k\omega s) \cos(\ell\Omega(s - \phi)) \, ds.$$

Moreover, taking into account the resonance relation $\Omega = m\omega/n$ and applying the change of variables $\theta = \omega s/n$, we have

$$I_1(\phi) = \frac{n}{\omega} \sum_{k=1}^{\infty} \sum_{\ell=1}^{\infty} y_k g_\ell^s \int_0^{2\pi} \sin(nk\theta) \sin(m\ell\theta - \ell\Omega\phi)$$

$$+ \frac{n}{\omega} \sum_{k=1}^{\infty} \sum_{\ell=1}^{\infty} y_k g_\ell^c \int_0^{2\pi} \sin(nk\theta) \cos(m\ell\theta - \ell\Omega\phi).$$

The integrals in the last formula vanish unless $k = mj$ and $\ell = nj$ for some integer $j > 0$. Thus, we obtain a simplification that yields the formula

$$I_1(\phi) = \frac{n\pi}{\omega} \left(\sum_{j=1}^{\infty} y_{mj} g_{nj}^s \cos(nj\Omega\phi) + \sum_{j=1}^{\infty} y_{mj} g_{nj}^c \sin(nj\Omega\phi) \right)$$

$$= \frac{n\pi}{\omega} \left(\sum_{j=1}^{\infty} y_{mj} g_{nj}^s \cos(mj\omega\phi) + \sum_{j=1}^{\infty} y_{mj} g_{nj}^c \sin(mj\omega\phi) \right).$$

In particular, $\phi \mapsto I_1(\phi)$ is a $2\pi/(m\omega)$-periodic function.

To simplify I_2, let us note that the corresponding integrand is $2\pi/\omega$-periodic, and therefore

$$I_2 = n \int_0^{2\pi/\omega} h(x(s), y(s)) y^2(s) \, ds.$$

We are interested in the simple zeros of M on the interval $0 \leq \phi < 2\pi/\omega$. Let us note that the graph of $I_1(\phi)$ over this interval repeats m times since $\phi \mapsto I_1(\phi)$ is $2\pi/(m\omega)$-periodic. The constant I_2 simply translates the graph of $I_1(\phi)$ to the graph of M. Also, if $I_1(\phi) - I_2$ has k zeros on $0 \leq \phi < 2\pi/(m\omega)$, then M has mk zeros.

Generally, a periodic function has an even number of zeros over one period. Hence, generally, there is a nonnegative integer N such that $k =$

$2N$, and M has an even number of zeros. Thus, we expect $2mN$ (ultra)subharmonics to continue from a given resonant periodic solution. It is important to note that this number will be large if m is large. In this regard, let us note that if the period of the unperturbed orbit $\Gamma_{m/n}$ is large, then there are resonances with m large.

In order for I_1 to be nonzero, the forcing must contain some Fourier modes that are the same as the modes present in the derivative y of the unperturbed solution corresponding to the periodic orbit $\Gamma_{m/n}$. It is not clear how to determine which modes are present in this unperturbed solution without solving the differential equation. But, because y is an odd function, we might expect all odd modes to be present.

Under the assumption that I_1 is not zero, M has zeros provided that I_2 is not too large in absolute value. In effect, I_2 serves to translate the graph of the periodic function I_1 in a vertical direction. This suggests that if the damping is too large, then there will be no periodic solutions that continue from the resonant unperturbed periodic orbits. In fact, there is a delicate relationship between the amplitude of I_1 and the magnitude of I_2 that is required to determine the global dynamics. The precise relationship that is required must be obtained from each choice of the model equation.

Example 5.61. Consider the damped periodically forced oscillator

$$\ddot{x} + \epsilon\alpha x + f(x) = \epsilon\beta \cos\Omega t.$$

Whereas $g_1^c = \beta$, all other Fourier modes vanish. Thus, on a resonant unperturbed orbit, if we use the notation of this section, we must have $n = 1$ and $j = 1$. In fact, we have

$$I_1(\phi) = \frac{\pi}{\omega} y_m\beta \sin(m\omega\phi), \quad I_2 = \alpha \int_0^{2\pi/\omega} y^2(s)\, ds,$$

$$M(\phi) = \frac{\pi}{\omega} y_m\beta \sin(m\omega\phi) - \alpha|y|_2^2$$

where the norm is the L^2-norm. Note that M has simple zeros if and only if

$$0 < \left(\frac{\pi y_m\beta}{\omega|y|_2^2\alpha}\right)^{-1} < 1.$$

In particular, if $y_m \neq 0$ and if the ratio α/β is sufficiently small, then there are $2m$ zeros.

To determine the number and positions of the continuable periodic orbits, we must determine the resonant periodic orbits in the period annuli of the unperturbed system (5.101); that is, we must determine the behavior of the period function associated with the given period annulus. While the period function must have a nonzero derivative at a resonant unperturbed periodic orbit to apply our first-order continuation theory, it is sometimes possible to determine if a resonant periodic orbit is continuable even in the case where the period function vanishes at this orbit (see [35] and [189]). But, the more basic problem of finding the critical points of a period function is nontrivial even for system (5.101) (see the survey [196], and also [41] and [42]).

Note that system (5.101) has all its rest points on the x-axis. If these rest points are all nondegenerate (their linearizations have nonzero eigenvalues), then the rest points will be either hyperbolic saddle points or centers. To prove this fact, recall that system (5.101) has a first integral. Indeed, if we view the differential equation $\ddot{x} + f(x) = 0$ as a model equation for a nonlinear spring, then we know that its total energy

$$H(x, y) = \frac{1}{2}y^2 + F(x)$$

where

$$F(x) := \int_0^x f(s)\, ds$$

is a first integral. Here the choice $F(0) = 0$ is arbitrary; the addition of a constant to H just redefines the "potential energy." Also, note that H is constant on the trajectories of the differential equation (5.101).

Without loss of generality, suppose that system (5.101) has a rest point at the origin. By our choice of the energy, $H(0, 0) = 0$. Also, since $f(0) = 0$, we also have that $H_x(0, 0) = 0$. By the assumption that the rest point is nondegenerate, we have $H_{xx}(0, 0) = f'(0) \neq 0$ and so

$$H(x, y) = \frac{1}{2}y^2 + \frac{f'(0)}{2}x^2 + O(x^3).$$

More generally, suppose $H : \mathbb{R}^n \to \mathbb{R}$. We say that H has a *singularity* at $0 \in \mathbb{R}^n$ if $H(0) = 0$ and $\operatorname{grad} H(0) = 0$. The singularity is called *nondegenerate* if $\det(\operatorname{Hess} H(0)) \neq 0$ where $\operatorname{Hess} H$ is the $n \times n$ matrix with components

$$\frac{\partial^2 H}{\partial x_i \partial x_j}(x_1, \ldots, x_n), \qquad i = 1, \ldots, n, \ j = 1, \ldots, n.$$

Theorem 5.62 (Morse's Lemma). *If $H : \mathbb{R}^n \to \mathbb{R}$, given by*

$$(x_1, \ldots, x_n) \mapsto H(x_1, \ldots, x_n),$$

has a nondegenerate singularity at the origin, then there is a smooth function $h : \mathbb{R}^n \mapsto \mathbb{R}^n$ such that $h(0) = 0$, $\det Dh(0) \neq 0$, and

$$H(h(x_1, \ldots, x_n)) = \sum_{i,j=1}^n \frac{\partial^2 H}{\partial x_i \partial x_j}(0) x^i x^j.$$

Informally, Morse's lemma states that there is a nonlinear change of coordinates defined in some neighborhood of the origin such that the function H in the new coordinate system is given by the quadratic form determined by the Hessian of H at the origin. See [158] for an elementary proof of Morse's lemma. The best proof uses more sophisticated machinery (see [2]).

Exercise 5.63. Prove Morse's lemma for $n = 1$. Hint: Reduce to the case where the function H is given by $H(x) = x^2 f(x)$ and $f(0) > 0$. Define $g(x) = x\sqrt{f(x)}$ and prove that there is a function h such that $g(h(x)) = x$. The desired change of coordinates is given by $x = h(z)$.

Exercise 5.64. Suppose that $H : \mathbb{R}^n \to \mathbb{R}$ has a nondegenerate singularity at the origin. Show that $\operatorname{Hess} H(0)$ has n real eigenvalues

$$\lambda_1^2, \dots, \lambda_k^2, -\lambda_{k+1}^2, \dots, -\lambda_n^2$$

where $\lambda_i \neq 0$ for $i = 1, \dots, n$. The number $n - k$ is called the *index* of the singularity. Prove the following corollary of the Morse lemma: There is a change of coordinates $h : \mathbb{R}^n \to \mathbb{R}^n$ such that

$$H(h(x_1, \dots, x_n)) = \sum_{i=1}^{k} \lambda_i^2 x_i^2 - \sum_{i=k+1}^{n} \lambda_i^2 x_i^2.$$

For system (5.101), it follows from the Morse lemma that there is a new coordinate system near each rest point such that the orbits of the system (5.101) all lie on level curves of the conic $y^2 + f'(0)x^2$. There are only two cases: If $f'(0) > 0$, then the origin is a center, and if $f'(0) < 0$, then the origin is a hyperbolic saddle.

Each center is surrounded by a period annulus $\mathcal{A}$. Let us suppose that there are rest points on the boundary of $\mathcal{A}$. In this case, there are either one or two hyperbolic saddle points on the boundary; the remainder of the boundary is composed of the stable and unstable manifolds of these saddle points. Because there are rest points on the boundary of $\mathcal{A}$, the corresponding period function grows without bound as its argument approaches the boundary of $\mathcal{A}$. In particular, the period annulus contains an infinite number of resonant periodic orbits, and among these there are orbits with arbitrarily large periods. Also, the period function approaches $2\pi/\sqrt{f'(0)}$, the period of the linearization of the system at the origin, as its argument approaches the origin. Thus, there is at least one unperturbed periodic solution with each preassigned period in the interval $(2\pi/\sqrt{f'(0)}, \infty)$. Let us also note that if the period function is not an increasing function, then there may be more than one unperturbed orbit in $\mathcal{A}$ with the same period. Also, if there is a rest point on the outer boundary of the period annulus, then the frequency of the resonant periodic orbits approaches zero as the resonant orbits approach the boundary.

Since the rational numbers are a dense subset of $\mathbb{R}$ and since the resonance relation has the form $n2\pi/\omega = m2\pi/\Omega$, there are infinitely many resonant periodic solutions in each subannulus containing two periodic orbits with different periods. In particular, a period annulus whose boundary contains rest points has a subannulus with this property. Thus, it should be clear that if the unperturbed system (5.101) has a period annulus containing periodic orbits with different periods, if the derivative of the period function does not vanish on each resonant orbit in the annulus, and if the damping is sufficiently small, then there will be a large number of perturbed (ultra)subharmonics.

Are there infinitely many (ultra)subharmonics? Our analysis does not answer this question. To see why, recall our main result: If the function M in display (5.102) has simple zeros along an $(m : n)$ resonant unperturbed periodic orbit, then for *sufficiently small* ϵ there are $2mN$ perturbed (ultra)subharmonics where N is some positive integer. But, if we consider an infinite number of resonant orbits, for example a sequence of periodic orbits that approach the boundary of our period annulus, then it might happen that the infinite number of requirements for ϵ to be sufficiently small cannot be satisfied simultaneously without taking $\epsilon = 0$. For this reason, we cannot conclude that there is an $\epsilon > 0$ such that the corresponding perturbed system has infinitely many periodic solutions even if (ultra)subharmonics continue from all resonant unperturbed periodic orbits. Thus, we are left with evidence that oscillators with an infinite number of (ultra)subharmonics exist, but we have no proof.

If an oscillator has an infinite number of hyperbolic periodic orbits of saddle type all contained in some compact subset of its extended phase space, then we might expect the dynamical behavior of the oscillator in a neighborhood of this set to be very complex: orbits in the neighborhood might tend to follow a stable manifold, pass by a saddle point, follow the motion on its unstable manifold, pass near another stable manifold, and then repeat the process. Whatever the exact nature of such a flow, it should be clear that we cannot hope to understand the dynamics of oscillators without considering this possible behavior. It turns out that by using some new ideas introduced in Chapter 6 we will be able to show that some periodically perturbed oscillators do indeed have an infinite number of (ultra)subharmonics and that their flows are "chaotic".

6
Homoclinic Orbits, Melnikov's Method, and Chaos

In the last chapter, we discussed the near resonance continuation theory for periodic orbits of periodically perturbed oscillators. For the case where the unperturbed oscillator has a regular period annulus, we found that there is generally an infinite number of resonances at which a first order perturbation theory can be used to prove the existence of perturbed periodic orbits. But, as mentioned previously, we cannot conclude from the results of our analysis that the perturbed oscillator has infinitely many periodic orbits. To do so would seem to require a condition that might be impossible to satisfy. Indeed, the nonzero amplitude of the perturbation would have to be made sufficiently small for each of an infinite sequence of continuations corresponding to an infinite sequence of resonant unperturbed periodic orbits that approaches the boundary of a period annulus. The subject of this chapter is a perturbation theory that is valid at the *boundary* of the period annulus. When the theory is applied, the amplitude of the perturbation is required to be sufficiently small only once.

Generally, the boundary of a period annulus for an unperturbed oscillator consists of one or more saddle points connected by homoclinic or heteroclinic orbits. Let us define a saddle connection to be an orbit whose α- and ω-limit sets are hyperbolic saddle points. A saddle connection is called a *homoclinic orbit* if its α- and ω-limit sets coincide. On the other hand, the saddle connection is called a *heteroclinic orbit* if its α-limit set is disjoint from its ω-limit set.

If the saddle points on the boundary of our period annulus are hyperbolic, then they persist along with their stable and unstable manifolds. For simplicity, let us consider the case where there is just one hyperbolic

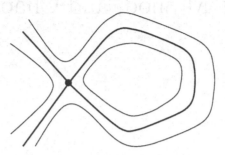

Figure 6.1: A homoclinic loop.

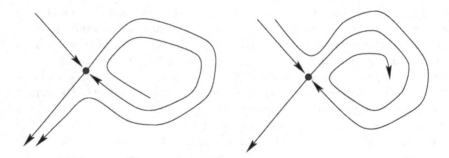

Figure 6.2: Possible phase portraits of a planar system after perturbation of a system with a homoclinic orbit.

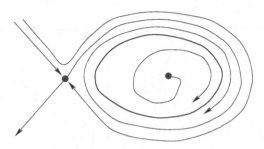

Figure 6.3: A homoclinic loop bifurcation: A periodic orbit appears after a perturbation that breaks a homoclinic loop.

saddle point p on the boundary of our period annulus such that this rest point is "connected to itself" by a homoclinic orbit as in Figure 6.1. If the perturbation is autonomous, then the portions of the perturbed stable and unstable manifolds at p that form the homoclinic loop either coincide or separate into one of the two configurations depicted in Figure 6.2. For a periodic nonautonomous perturbation, we will consider the corresponding (stroboscopic) Poincaré map. The saddle point p is a fixed (or periodic) point for the unperturbed Poincaré map and the homoclinic orbit lies on the invariant stable and unstable manifolds of p. After perturbation, the perturbed stable and unstable manifolds can coincide, split, or cross. The main problem addressed in this chapter is the determination of the relative positions of the perturbed invariant manifolds for both the autonomous and nonautonomous cases.

For autonomous perturbations, the splitting of saddle connections is important because it is related to the existence of limit cycles. For example, suppose that the perturbed configuration of stable and unstable manifolds is as depicted in the right hand panel of Figure 6.2. If the perturbation of the rest point at the inner boundary of the unperturbed period annulus is a source and the perturbation is sufficiently small, then no additional rest points appear; and, by the Poincaré–Bendixson theorem, there must be at least one periodic orbit "inside" the original homoclinic loop (see Figure 6.3).

For the case of a periodic perturbation, the most interesting case occurs when the perturbed stable and unstable manifolds of the Poincaré map cross. For the case of a homoclinic loop, a point of intersection is called a *transverse homoclinic point* for the Poincaré map if the stable and unstable manifolds meet transversally, that is, the sum of their tangent spaces at the crossing point is equal to the tangent space of the two-dimensional Poincaré section at this point. (There is an analogous concept for transverse heteroclinic points.)

If there is a transverse homoclinic point, then, by a remarkable theorem called the Smale–Birkhoff theorem, there is a nearby "chaotic invariant

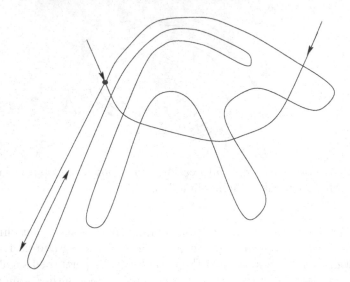

Figure 6.4: Part of a homoclinic tangle for the stable and unstable manifolds of a saddle fixed point of a Poincaré map.

set." A weak version of this theorem states that if there is a transverse homoclinic point, then the perturbed Poincaré map has infinitely many unstable periodic points in a small neighborhood of the unperturbed homoclinic loop. But even more is true: There is a compact invariant set that contains these periodic points and also infinitely many nonperiodic solutions that "wander as randomly as a sequence of coin tosses" in the vicinity of the boundary of the original period annulus [203]. Moreover, the trajectories of solutions starting in this invariant set are "sensitively dependent" on their initial conditions; that is, no matter how close we take their initial conditions, the corresponding points on two different trajectories will be at least half of the diameter of the invariant set apart at some finite future time. The existence of such an invariant set is what we mean when we say the system is *chaotic* (see, for example, the mathematical references [103], [166], [194], [203], [205], and [233], as well as the general references [14], [73], [105], and [145]).

Although the proof of the existence of "chaotic dynamics" in the presence of a transverse homoclinic point requires several new ideas that we will not discuss here, it is very easy to see why the existence of a transverse homoclinic point must lead to complicated dynamics. Note first that the forward iterates of a transverse homoclinic point, themselves all transverse homoclinic points, must approach the corresponding saddle point along its stable manifold. Because these points also lie on the unstable manifold of *the same saddle point*, the unstable manifold must stretch and fold as shown schematically in Figure 6.4. This *homoclinic tangle* is responsible for the existence of a chaotic invariant set.

The chaotic invariant sets in the homoclinic tangle are similar to hyperbolic saddle points in the sense that these chaotic invariant sets have both stable and unstable manifolds. Thus, roughly speaking, many solutions of the corresponding differential equation (which have their initial points near one of these chaotic invariant sets) will approach the chaotic invariant set along the direction of the stable manifold on some long time-scale, but eventually these trajectories leave the vicinity of the chaotic invariant set along the direction of the stable manifold. Such an orbit will exhibit *transient chaos*. This is what usually happens if the differential equation is not conservative. On the other hand, for Hamiltonian systems where the dimension of the phase space is not more than four (for a mechanical system this means that there are not more than two degrees-of-freedom), these "transient orbits" are often constrained to some neighborhood of the original homoclinic loop. In this case, they continually revisit the chaotic invariant sets obtained from the transverse homoclinic points and they exhibit chaotic effects for all time. Finally, there are dissipative systems that contain "chaotic attractors," compact chaotic invariant sets that attract all nearby orbits. These chaotic sets are not necessarily associated with transverse homoclinic points. Chaotic attractors are poorly understood. For example, it is generally very difficult to prove the existence of a chaotic attractor for a system of differential equations. On the other hand, it is not at all difficult to "see" a chaotic attractor using numerical simulations (see Exercise 6.1).

We will show how to detect the splitting of saddle connections by defining a function that determines the separation of the perturbed invariant manifolds as a function of the bifurcation parameters. It turns out that the appropriate function is the limit of the subharmonic Melnikov function as the base points on the periodic orbits approach the boundary. For example, for the case of the forced oscillator (5.100) where we have defined the subharmonic Melnikov function M in display (5.102), the function we seek is the limit of M as ξ approaches a point on the boundary that is not a rest point. The limit function, again denoted by M, is again called the *Melnikov function*. In fact, the Melnikov function for the differential equation (5.100) is given by

$$M(\phi) = \int_{-\infty}^{\infty} y(t + \phi, \xi)\big(g(t) - h(x(t + \phi, \xi), y(t + \phi, \xi))y(t + \phi, \xi)\big)\, dt$$

or, after the obvious change of variables, by

$$M(\phi) = \int_{-\infty}^{\infty} y(s, \xi)\big(g(s - \phi) - h(x(s, \xi), y(s, \xi))y(s, \xi)\big)\, ds$$

where ξ is a base point on the boundary and the coordinate ϕ specifies position on the boundary.

For an autonomous perturbation, the Melnikov function does not depend on the initial point for the unperturbed solution on the boundary. In this case the sign of the Melnikov function determines the direction in which the invariant manifolds split. For a time periodic perturbation, the Melnikov function does depend on the initial point on the boundary, and its simple zeros correspond to positions where the perturbed stable and unstable manifolds intersect transversally for sufficiently small $\epsilon \neq 0$.

The derivation and analysis of the Melnikov function for autonomous perturbations is of course a special case of its derivation for nonautonomous perturbations. Since the analysis for autonomous perturbations is conceptually simpler, we will give a detailed discussion of this case first.

Exercise 6.1. Write a report on numerical simulations of the Lorenz system

$$\dot{x} = \sigma(y - x), \quad \dot{y} = \rho x - y - xz, \quad \dot{z} = -\beta z + xy$$

(see the original paper of Edward N. Lorenz [147] or any book on dynamical systems theory). Start by setting the parameter values $\beta = \frac{8}{3}$, $\rho = 28$, and $\sigma = 10$. Choose an initial condition in the first quadrant, for instance near the unstable manifold of the saddle point at the origin, integrate forward in time, and display the resulting approximate orbit using three-dimensional graphics. The "Lorenz butterfly attractor" will appear. Also graph one of the observables, say $t \mapsto y(t)$, and compare the time series you obtain with the graph of a periodic function. Choose a second initial condition near the initial condition you started with and plot together the two simulated graphs of $t \mapsto y(t)$. Note that these graphs will stay close together for a while (as they must due to the smoothness of solutions with respect to initial conditions), but eventually they will diverge. In fact, after a transient, the evolution of the two solutions will appear to be completely unrelated. For this reason, it is impossible to predict the position of the state vector from the initial conditions over long time-periods with an accuracy that is small compared with the diameter of the attractor; the evolution of the system is extremely sensitive to changes in the initial conditions. This is the signature of a chaotic flow.

6.1   Autonomous Perturbations: Separatrix Splitting

Consider the planar system

$$\dot{u} = f(u, \lambda), \qquad u \in \mathbb{R}^2, \quad \lambda \in \mathbb{R}^n \tag{6.1}$$

and let $\xi_0 \in \mathbb{R}^2$ be a regular point for the unperturbed system

$$\dot{u} = f(u, 0). \tag{6.2}$$

As usual, let $t \mapsto u(t, \xi, \lambda)$ denote the solution of the differential equation (6.1) such that $u(0, \xi, \lambda) = \xi$, define $f^{\perp}(u) = Rf(u, 0)$ where

$$R := \begin{pmatrix} 0 & -1 \\ 1 & 0 \end{pmatrix},$$

and let $t \mapsto \Psi(t, \xi)$ denote the flow of the orthogonal system $\dot{u} = f^{\perp}(u)$. Here, of course, $t \mapsto \Psi(t, \xi_0)$ is transverse to $t \mapsto u(t, \xi_0, 0)$ at ξ_0.

Define

$$\Sigma := \{\Psi(t, \xi_0) : t \in \mathbb{R}\}, \tag{6.3}$$

and suppose that we have devised some construction that produces two families of solutions of the differential equation (6.1), each parametrized by λ, whose members are all transverse to Σ such that at $\lambda = 0$ the corresponding solutions coincide with the unperturbed solution $t \mapsto u(t, \xi_0, 0)$. Our objective is to obtain some information about the rate of separation of the solutions belonging to the two parametrized families of solutions. In fact, we will obtain a general conclusion about this separation rate following the presentation given by Stephen Schecter [202]. This result will then be used to address the problem of breaking saddle connections.

Suppose that our construction produces two smooth functions $\rho^i : \mathbb{R}^n \to \mathbb{R}$, $i = 1, 2$, given by $\lambda \mapsto \rho^i(\lambda)$ such that $\rho^i(0) = 0$ where $\rho^i(\lambda)$ gives the point of intersection of the ith family with Σ. We desire information about the separation of the two solution families of the differential equation (6.1) given by

$$\gamma^i(t, \lambda) := u(t, \Psi(\rho^i(\lambda), \xi_0), \lambda), \qquad i = 1, 2. \tag{6.4}$$

Let us view these families as "variations" of the unperturbed solution; that is, γ^i is a family of solutions containing the unperturbed solution at $\lambda = 0$

$$\gamma^i(t, 0) = u(t, \xi_0, 0), \qquad i = 1, 2.$$

Also, γ^i has initial value on the transverse section Σ. In fact,

$$\gamma^i(0, \lambda) = \Psi(\rho^i(\lambda), \xi_0).$$

The separation of the variations from the unperturbed solution is defined by the function $\lambda \mapsto \rho^1(\lambda) - \rho^2(\lambda)$; it measures a signed distance between the points where our variations cross Σ. Of course, in a perturbation problem, it is unlikely that we will be given the functions ρ^1 and ρ^2 explicitly. At best, we will be able to infer their existence (probably by an application of the implicit function theorem). Nonetheless, for small λ, a good approximation of the separation is given by the *separation function* sep : $\mathbb{R}^n \to \mathbb{R}$ defined by

$$\begin{aligned} \mathrm{sep}(\lambda) &:= \langle \Psi(\rho^1(\lambda), \xi_0) - \Psi(\rho^2(\lambda), \xi_0), f^{\perp}(\xi_0) \rangle \\ &= f(\xi_0, 0) \wedge (\Psi(\rho^1(\lambda), \xi_0) - \Psi(\rho^2(\lambda), \xi_0)). \end{aligned}$$

Let us note that, sep$(0) = 0$. Also, sep$(\lambda) = 0$ if and only if the solutions $\gamma^1(t, \lambda)$ and $\gamma^2(t, \lambda)$ are identical. This last fact follows because a solution of an initial value problem is unique.

As usual, we can determine the local nature of $\mathcal{S} := \{\lambda : \text{sep}(\lambda) = 0\}$ provided that there is at least one $j = 1, \ldots, n$ such that $\text{sep}_{\lambda_j}(0) \neq 0$. In fact, if this condition holds, then (by the implicit function theorem) $\mathcal{S}$ is a surface of dimension $n - 1$ passing through $0 \in \mathbb{R}^n$ whose normal vector at this point is just grad(sep)(0).

What have we done so far? In analogy with our continuation theory for periodic solutions, we have defined a function akin to the displacement function and we have reduced the study of its zero set to an application of the implicit function theorem. Let us make this reduction useful by identifying the partial derivatives of the separation function.

To identify the partial derivatives of the separation function *using the original differential equation* (6.1), we expect to solve a variational equation. But to obtain a nontrivial variational equation we must have some time dependence in the separation function. This requirement motivates the definition of the *time-dependent separation function* $S : \mathbb{R} \times \mathbb{R}^n \to \mathbb{R}$ given by

$$S(t, \lambda) := \langle \gamma^1(t, \lambda) - \gamma^2(t, \lambda), f^\perp(\varphi_t(\xi_0)) \rangle$$
$$= f(\varphi_t(\xi_0), 0) \wedge (\gamma^1(t, \lambda) - \gamma^2(t, \lambda))$$

where φ_t is the flow of the system (6.2). Since $S(0, \lambda) = \text{sep}(\lambda)$, the main idea—originally due to Melnikov—is to compute the desired partial derivatives of the separation function sep from the corresponding partial derivatives of the time-dependent separation function S.

Let us define two auxiliary functions

$$S^i(t, \lambda) := \langle \gamma^i(t, \lambda), f^\perp(\varphi_t(\xi_0)) \rangle$$
$$= f(\varphi_t(\xi_0), 0) \wedge \gamma^i(t, \lambda)$$

for $i = 1, 2$, and note that $S(t, \lambda) = S^1(t, \lambda) - S^2(t, \lambda)$. To compute the required partial derivatives, start with the formula

$$S^i_{\lambda_j}(t, 0) = f(\varphi_t(\xi_0), 0) \wedge \gamma^i_{\lambda_j}(t, 0), \tag{6.5}$$

and use the solution $t \mapsto \gamma^i(t, \lambda)$ of the differential equation (6.1) to obtain the variational equation

$$\dot{\gamma}^i_{\lambda_i}(t, 0) = f_u(\varphi_t(\xi_0), 0)\gamma^i_{\lambda_j}(t, 0) + f_{\lambda_j}(\varphi_t(\xi_0), 0). \tag{6.6}$$

Next, define $A(t) := f_u(\varphi_t(\xi_0), 0)$ and use equation (6.5) to obtain the differential equation

$$\dot{S}^i_{\lambda_j}(t, 0) = f_u(\varphi_t(\xi_0), 0)f(\varphi_t(\xi_0), 0) \wedge \gamma^i_{\lambda_j}(t, 0) + f(\varphi_t(\xi_0), 0) \wedge \dot{\gamma}^i_{\lambda_j}(t, 0)$$
$$= A(t)f(\varphi_t(\xi_0), 0) \wedge \gamma^i_{\lambda_j}(t, 0) + f(\varphi_t(\xi_0), 0) \wedge A(t)\gamma^i_{\lambda_j}(t, 0)$$
$$+ f(\varphi_t(\xi_0), 0) \wedge f_{\lambda_j}(\varphi_t(\xi_0), 0). \tag{6.7}$$

Formula (6.7) can be simplified by an application of the following easily proved proposition from vector analysis: If A is a 2×2 matrix and $v, w \in \mathbb{R}^2$, then

$$Av \wedge w + v \wedge Aw = (\operatorname{tr} A)v \wedge w.$$

In fact, with the aid of this proposition, we have

$$\dot{S}^i_{\lambda_j}(t,0) = \operatorname{div} f(\varphi_t(\xi_0),0)f(\varphi_t(\xi_0),0) \wedge \gamma^i_{\lambda_j}(t,0)$$
$$+ f(\varphi_t(\xi_0),0) \wedge f_{\lambda_j}(\varphi_t(\xi_0),0)$$
$$= \operatorname{div} f(\varphi_t(\xi_0),0)S^i_{\lambda_j}(t,0) + f(\varphi_t(\xi_0),0) \wedge f_{\lambda_j}(\varphi_t(\xi_0),0). \quad (6.8)$$

The differential equation (6.8) is a linear variational equation for the function $t \mapsto S^i_{\lambda_j}(t,0)$. To solve it, let us assume that we know the behavior of $\gamma^1(t,0)$ as $t \to -\infty$ and the behavior of $\gamma^2(t,0)$ as $t \mapsto \infty$. If we define

$$K(t) := e^{-\int_0^t \operatorname{div} f(\varphi_t(\xi_0),0)\, ds}$$

and integrate both sides of the differential equation

$$\frac{d}{dt}\left(K(t)S^i_{\lambda_j}(t,0)\right) = K(t)f(\varphi_t(\xi_0)) \wedge f_{\lambda_j}(\varphi_t(\xi_0),0),$$

then we obtain the identities

$$S^1_{\lambda_j}(0,0) = K(t)S^1_{\lambda_j}(t,0) + \int_t^0 K(s)f(\varphi_s(\xi_0)) \wedge f_{\lambda_j}(\varphi_s(\xi_0),0)\, ds,$$

$$-S^2_{\lambda_j}(0,0) = -K(t)S^2_{\lambda_j}(t,0) + \int_0^t K(s)f(\varphi_s(\xi_0)) \wedge f_{\lambda_j}(\varphi_s(\xi_0),0)\, ds.$$

Note that the right hand side of each of these identities is *constant* with respect to t. Using this fact, the desired partial derivative is given by

$$\operatorname{sep}_{\lambda_j}(0) = \lim_{t\to-\infty} \left[K(t)f(\varphi_t(\xi_0)) \wedge \gamma^1_{\lambda_j}(t,0)\right.$$
$$\left. + \int_t^0 K(s)f(\varphi_s(\xi_0)) \wedge f_{\lambda_j}(\varphi_s(\xi_0),0)\, ds\right]$$
$$+ \lim_{t\to\infty} \left[-K(t)f(\varphi_t(\xi_0)) \wedge \gamma^2_{\lambda_j}(t,0)\right.$$
$$\left. + \int_0^t K(s)f(\varphi_s(\xi_0)) \wedge f_{\lambda_j}(\varphi_s(\xi_0),0)\, ds\right]. \quad (6.9)$$

We reiterate that the indicated limits exist because the quantities in square brackets are constants. Of course, the summands of each expression in square brackets are not necessarily constants.

The representation (6.9) of the partial derivatives of the separation function is useful because it is general. We will return to the main topic of this section and apply this result to the perturbation of saddle connections.

Suppose that ξ_0 denotes a point on a saddle connection for system (6.2) connecting the hyperbolic saddle points p_0 and q_0 (maybe $p_0 = q_0$); that is,

$$\lim_{t \to -\infty} \varphi_t(\xi_0) = p_0, \qquad \lim_{t \to \infty} \varphi_t(\xi_0) = q_0.$$

Also, let Σ denote the section at ξ_0 defined in display (6.3). By the implicit function theorem, if λ is sufficiently small, then there are perturbed hyperbolic saddle points

$$p_\lambda = p_0 + O(\lambda), \qquad q_\lambda = q_0 + O(\lambda)$$

for the system (6.1). Define $t \mapsto \gamma^1(t, \lambda)$ to be the solution of the system (6.1) with initial condition on Σ (as in equation (6.4)) that lies on the unstable manifold of p_λ, and let $t \mapsto \gamma^2(t, \lambda)$ denote the corresponding solution on the stable manifold of the hyperbolic saddle point q_λ. By Theorem 4.1, the stable and unstable manifolds γ^i, $i = 1, 2$, intersect the fixed curve Σ. To see this, add the equation $\dot{\lambda} = 0$ to the system (6.1) and use the smoothness of the center stable manifold corresponding to each rest point of the augmented system corresponding to the saddle points p_0 and q_0.

We will outline a proof of the following proposition:

Proposition 6.2. *If system (6.1) with $\lambda = 0$ has a saddle connection and if γ^1 and γ^2, as in display (6.4), are defined to be solutions on the unstable and stable manifolds of the perturbed saddle points, then*

$$\lim_{t \to -\infty} K(t) f(\varphi_t(\xi_0)) \wedge \gamma^1_{\lambda_j}(t, 0) = 0, \tag{6.10}$$

$$\lim_{t \to \infty} K(t) f(\varphi_t(\xi_0)) \wedge \gamma^2_{\lambda_j}(t, 0) = 0. \tag{6.11}$$

Moreover,

$$\mathrm{sep}_{\lambda_j}(0) = \int_{-\infty}^{\infty} e^{-\int_0^t \mathrm{div}\, f(\varphi_s(\xi_0),0)\, ds} f(\varphi_t(\xi_0),0) \wedge f_{\lambda_j}(\varphi_t(\xi_0),0)\, dt. \tag{6.12}$$

The important formula (6.12) for the partial derivatives of the separation function with respect to the parameters was probably known to Poincaré. It was also discovered independently by several different authors (see, for example, [156], [193], and [208]). In spite of this history, the integral is now most often called the Melnikov integral.

Since $\mathrm{sep}(0) = 0$, the Taylor series of the separation function at $\lambda = 0$ is

$$\mathrm{sep}(\lambda) = \sum_{j=1}^{n} \lambda_j \, \mathrm{sep}_{\lambda_j}(0) + O(|\lambda|^2). \tag{6.13}$$

In particular, if $n = 1$ and $\epsilon := \lambda_1$, then

$$\mathrm{sep}(\epsilon) = \epsilon(\mathrm{sep}_\epsilon(0) + O(\epsilon)). \tag{6.14}$$

Therefore, if $\text{sep}_\epsilon(0) \neq 0$ and if $|\epsilon|$ is sufficiently small, then formula (6.12) can be used to determine the sign of $\text{sep}(\epsilon)$, and therefore the splitting direction of the perturbed stable and unstable manifolds relative to the direction determined by $f^\perp(\xi_0)$.

An outline for a proof of the limit (6.11) will be given; a proof for the limit (6.10) can be constructed similarly.

View the vector field f as a mapping $f : \mathbb{R}^2 \times \mathbb{R}^n \to \mathbb{R}^2$. Since $f(q_0, 0) = 0$ and since $f_u(q_0, 0) : \mathbb{R}^2 \to \mathbb{R}^2$ is a nonsingular linear transformation (it has no eigenvalue on the imaginary axis in the complex plane by the hyperbolicity of q_0), the implicit function theorem implies there is a map $q : \mathbb{R}^n \to \mathbb{R}^2$ defined near $\lambda = 0$ such that $q(0) = q_0$ and $f(q(\lambda), \lambda) \equiv 0$. By the continuous dependence of the eigenvalues of a matrix on its coefficients, we have that $q(\lambda)$ is a hyperbolic saddle point for $|\lambda|$ sufficiently small.

As mentioned above, the stable manifold of $q(\lambda)$ varies smoothly with λ by Theorem 4.1. In particular, the function $(t, \lambda) \mapsto \gamma^2(t, \lambda)$ depends smoothly on t and λ, and $\lim_{t\to\infty} \gamma^2(t, \lambda) = q(\lambda)$. The matrix $f_u(q_0, 0)$ has two real eigenvalues $-\mu_1 < 0 < \mu_2$. Moreover, as $t \to \infty$ the curve $t \mapsto \gamma^2(t, 0)$ approaches the saddle point q_0 tangent to the eigenspace corresponding to the eigenvalue $-\mu$.

By an affine change of coordinates, if necessary, we may as well assume that q_0 is located at the origin and the unperturbed differential equation $\dot{u} = f(u, 0)$ has the form

$$\dot{x} = -\mu x + f_1(x, y), \qquad \dot{y} = \nu y + f_2(x, y) \tag{6.15}$$

where both μ and ν are positive constants, and the functions f_1 and f_2 together with their first order partial derivatives vanish at the origin. In these coordinates, the stable manifold of the hyperbolic saddle point at the origin is given by the graph of a smooth function $h : U \to \mathbb{R}$ where $U \subset \mathbb{R}$ is an open interval containing the origin on the x-axis. Moreover, because the stable manifold is tangent to the x-axis at the origin, we have $h(0) = 0$ and $h'(0) = 0$.

The estimate that we will use to compute the limit (6.11) is the content of the following proposition.

Proposition 6.3. *If $|x_0|$ is sufficiently small, then there is a positive constant c such that the solution of the system (6.15) starting at $(x_0, h(x_0))$ satisfies the estimate*

$$|x(t)| + |y(t)| \leq ce^{-\mu t}, \qquad t \geq 0.$$

The next lemma (compare Theorem 4.1) will be used to prove Proposition 6.3.

Lemma 6.4. *If $t \mapsto x(t)$ is the solution of the initial value problem*

$$\dot{x} = -\mu x + g(x), \qquad x(0) = x_0,$$

where $\mu > 0$ and $g : \mathbb{R} \to \mathbb{R}$ is a smooth function such that $g(0) = 0$ and $g'(0) = 0$, then there are constants $\epsilon > 0$ and $c > 0$ such that $|x(t)| \leq ce^{-\mu t}$ for $t \geq 0$ whenever $|x_0| < \epsilon$.

Proof. The function defined by

$$G(x) := \begin{cases} x^{-2}g(x), & x \neq 0 \\ 0, & x = 0 \end{cases}$$

is continuous at $x = 0$. Thus, there is some constant C such that $|G(x)| \leq C$ for sufficiently small $|x|$.

For $x \neq 0$, we have

$$-\frac{\dot{x}}{x^2} - \frac{\mu}{x} = -G(x).$$

If $y := 1/x$, then $\dot{y} - \mu y = -G(x(t))$ and

$$e^{-\mu t}y(t) = y(0) - \int_0^t e^{-\mu s}G(x(s))\, ds.$$

Thus, we have the estimate

$$|e^{-\mu t}y(t)| \geq |y(0)| - \int_0^t e^{-\mu s}|G(x(s)|\, ds.$$

For sufficiently small $|x_0|$,

$$|y(0)| \leq |e^{-\mu t}y(t)| + C \int_0^t e^{-\mu s}\, ds. \tag{6.16}$$

To prove this last inequality, use the following simple proposition: if $|x_0|$ is sufficiently small, then $|x(t)| < |x_0|$ for $t \geq 0$. It is proved by observing that the point $x = 0$ is an attracting rest point for our one-dimensional differential equation. A weaker assumption would also be sufficient. For example, it suffices to assume that for $|x_0|$ sufficiently small, the interval $(-|x_0|, |x_0|)$ is positively invariant. This follows immediately by considering the direction of the vector field corresponding to our differential equation at the end points of the appropriately chosen interval.

After an elementary integration, inequality (6.16) states that

$$|y(0)| \leq |e^{-\mu t}y(t)| + \frac{C}{\mu}(1 - e^{-\mu t}).$$

Moreover, because $t \geq 0$, it follows that

$$|y(0)| \leq |e^{-\mu t}y(t)| + \frac{C}{\mu},$$

and therefore

$$\frac{1}{|x_0|} - \frac{C}{\mu} \leq \frac{1}{e^{\mu t}|x(t)|}.$$

If $|x_0| > 0$ is sufficiently small, then

$$\frac{1}{|x_0|} - \frac{C}{\mu} > \frac{1}{c} > 0$$

for some $c > 0$. Thus, we have that

$$|x(t)| \leq ce^{-\mu t}.$$

$\square$

Exercise 6.5. Under the same hypotheses as in Lemma 6.4, prove that

$$\lim_{t \to \infty} e^{\mu t} x(t)$$

exists and is not equal to zero (see [202]).

Let us prove Proposition 6.3.

Proof. Consider the change of coordinates for the system (6.15) given by

$$p = x, \qquad q = y - h(x).$$

In these coordinates, the saddle point stays fixed at the origin, but the stable manifold is transformed to the p-axis. The restriction of the transformed differential equation to the p-axis is given by

$$\dot{p} = -\mu p + f_1(p, h(p)). \tag{6.17}$$

If $g(p) := f_1(p, h(p))$, then all the hypotheses of the lemma are satisfied, and we conclude that there is some $|p_0| \neq 0$ such that solutions of the differential equation (6.17) satisfy $|p(t)| \leq ce^{-\mu t}$ for some $c > 0$ and all $t \geq 0$ whenever $|p(0)| < |p_0|$. In the original coordinates, the corresponding solution on the stable manifold is given by $x(t) = p(t)$, $y(t) = h(x(t))$. Since $y = h(x)$ is tangent to the x-axis at $x = 0$, there is a constant $c_1 > 0$ such that $|h(x)| < c_1 x^2$ for $|x|$ sufficiently small. Thus, if the initial value of the solution of the differential equation (6.15) on the stable manifold is sufficiently close to the origin, then there is a positive constant c such that

$$|x(t)| + |y(t)| = |x(t)| + |h(x(t))| \leq |x(t)|(1 + c_1|x(t)|) \leq ce^{-\mu t}. \quad \square$$

To conclude our discussion of the limit (6.11), we must analyze the asymptotic behavior of the functions $K(t)$, $f(\varphi_t(\xi_0), 0)$, and $\gamma^2_{\lambda_j}(t, 0)$. Let us note first that since $f(u, 0)$ is Lipschitz, we have

$$\|f(u, 0)\| = \|f(u, 0) - f(0, 0)\| \leq L\|u\|$$

for some constant $L > 0$. By the proposition,

$$\|f(\varphi_t(\xi_0), 0)\| \leq Lce^{-\mu t}.$$

Likewise, using the smoothness of $u \mapsto f(u,0)$, we have

$$\operatorname{div} f(u,0) = \operatorname{tr} f_u(0,0) + R(u)$$

where, for sufficiently small $||u||$, there is a constant $c_2 > 0$ such that the remainder R satisfies $||R(u)|| \le c_2||u||$. Thus

$$
\begin{aligned}
K(t) &= e^{-\int_0^t \nu - \mu \, du} e^{-\int_0^t R(u(s)) \, ds} \\
&\le e^{(\mu-\nu)t} e^{c_2 \int_0^t ce^{-\mu s} \, ds} \\
&\le c_3 e^{(\mu-\nu)t}
\end{aligned}
$$

for some $c_3 > 0$. It follows that

$$\lim_{t \to \infty} K(t) f(\varphi_t(\xi_0), 0) = 0.$$

To complete the argument, we will show that $|\gamma^2_{\lambda_j}(t,0)|$ is bounded. For this, we use the smoothness of the stable manifold with respect to the parameter λ. There is no loss of generality if we assume that the hyperbolic saddle $q(\lambda)$ remains at the origin with its stable manifold tangent to the x-axis as in system (6.15). Indeed, this geometry can be achieved by a parameter-dependent affine change of coordinates. More precisely, there is a smooth function $(x, \lambda) \mapsto h(x, \lambda)$ defined near $(x, y) = (0,0)$ such that the stable manifold is the graph of the function $x \mapsto h(x, \lambda)$. Of course, we also have that $h(0, \lambda) \equiv 0$ and $h_x(0, \lambda) \equiv 0$. Using this representation of the stable manifold,

$$\gamma^2(t, \lambda) = (x(t, \lambda), h(x(t, \lambda), \lambda))$$

where $t \mapsto x(t, \lambda)$ is a solution of a differential equation

$$\dot{x} = -\mu x + g(x, \lambda)$$

similar to differential equation (6.17). After differentiation, we find that

$$\gamma^2_{\lambda_j}(t, 0) = (x_{\lambda_j}(t, 0), h_x(x(t, 0), 0) x_{\lambda_j}(t, 0) + h_{\lambda_j}(x(t, 0), 0)).$$

By the smoothness of the function h, both $h_x(x, 0)$ and $h_{\lambda_j}(x, 0)$ are bounded for x in some fixed but sufficiently small interval containing $x = 0$. Thus, the boundedness of the function $t \mapsto \gamma^2_{\lambda_j}(t, 0)$ will be proved once we show that $t \mapsto x_{\lambda_j}(t, 0)$ is bounded as $t \to \infty$. To obtain this bound, note that the function $t \mapsto x_{\lambda_j}(t, 0)$ is a solution of the variational equation

$$\dot{w} = -\mu w + g_x(x(t, 0), 0)w + g_{\lambda_j}(x(t, 0), 0). \tag{6.18}$$

Because $g_x(0, 0) = 0$ and the function g is smooth, we have the estimate

$$|g_x(x, 0)| \le c_1 |x|$$

for some $c_1 > 0$. In addition, since $g(0, \lambda) \equiv 0$, it follows that $g_{\lambda_j}(0,0) = 0$. Also, by the smoothness of g, the partial derivative g_{λ_j} is locally Lipschitz. In fact, there is some $c_2 > 0$ such that

$$|g_{\lambda_j}(x,0)| = |g_{\lambda_j}(x,0) - g_{\lambda_j}(0,0)| \le c_2|x|.$$

With the obvious choice of notation, the differential equation (6.18) has the form

$$\dot{w} = (-\mu + \alpha(t))w + \beta(t) \tag{6.19}$$

and the solution

$$w(t) = e^{-\mu t}e^{\int_0^t \alpha(s)\, ds}\left(w(0) + \int_0^t e^{\mu s}e^{-\int_0^s \alpha(\tau)\, d\tau}\beta(s)\, ds\right).$$

By Proposition 6.3, there is a constant $c_3 > 0$ such that

$$|\alpha(t)| \le c_3 e^{-\mu t}, \qquad |\beta(t)| \le c_3 e^{-\mu t},$$

for $t \ge 0$. Also, let us note that

$$\int_0^t |\alpha(s)|\, ds \le \frac{c_3}{\mu}(1 - e^{-\mu t}) < \frac{c_3}{\mu}.$$

Thus, we obtain the following growth estimate for the solution of the differential equation (6.19):

$$|w(t)| \le e^{-\mu t}e^{c_3/\mu}|w(0)| + e^{-\mu t}e^{c_3/\mu}c_3 e^{c_3/\mu}t.$$

In particular, $|w(t)|$ is bounded for $t \ge 0$. This completes the proof.

As an application of our result on the splitting of separatrices, let us consider the damped van der Pol oscillator

$$\ddot{x} + \epsilon(x^2 - 1)\dot{x} + x - c^2 x^3 = 0$$

where $c > 0$ and ϵ is a small parameter. If, as usual, we define $\dot{x} = y$, then the energy for the unperturbed system is given by

$$H(x, y) = \frac{1}{2}y^2 + \frac{1}{2}x^2 - \frac{1}{4}c^2 x^4.$$

The unperturbed Hamiltonian system

$$\dot{x} = y, \qquad \dot{y} = -x + c^2 x^3$$

has a pair of hyperbolic saddle points at $(x, y) = (\pm 1/c, 0)$ and a center at the origin surrounded by a regular period annulus. The boundary of the period annulus is a pair of heteroclinic orbits of the unperturbed system that both lie on the curve with energy $1/(4c^2)$.

The Melnikov integral has the form

$$M = \int_{-\infty}^{\infty} y^2 (1 - x^2) \, dt.$$

Using the equality $\dot{x}/y = 1$ and the energy relation, let us note that the time parameter on the heteroclinic orbits is given by

$$t = \int_0^x \left(\frac{1}{2c^2} - \sigma^2 + \frac{c^2}{2} \sigma^4 \right)^{-1/2} d\sigma.$$

After integration, this fact yields the solution

$$x(t) = \frac{1}{c} \tanh(t/\sqrt{2}), \qquad y(t) = \frac{1}{c\sqrt{2}} \operatorname{sech}^2(t/\sqrt{2}),$$

and the formula

$$M = \frac{1}{2c^2} \int_{-\infty}^{\infty} \operatorname{sech}^4(t/\sqrt{2})(1 - \frac{1}{c^2} \tanh^2(t/\sqrt{2})) \, dt.$$

This elementary integral can be evaluated using the substitution $u = \tanh(t/\sqrt{2})$ to obtain the value

$$M = \frac{2\sqrt{2}}{15c^2} \left(5 - \frac{1}{c^2} \right).$$

If, for example, $c^2 < \frac{1}{5}$, then $M < 0$ and both heteroclinic orbits break. If in addition $\epsilon > 0$ is sufficiently small, then the system will have a limit cycle surrounding the origin. (Why?)

Exercise 6.6. Discuss the splitting of saddle connections for the damped Duffing equation

$$\ddot{x} + \epsilon x - x + c^2 x^3 = 0.$$

Does the perturbed system have limit cycles?

Exercise 6.7. A heteroclinic orbit of a planar Hamiltonian system does not persist under a general (autonomous) Hamiltonian perturbation. Prove that a homoclinic loop of a planar Hamiltonian system persists under (autonomous) Hamiltonian perturbation. Determine the fate of the heteroclinic orbits in the phase plane for the mathematical pendulum when it is perturbed in the family

$$\dot{\theta} = v, \qquad \dot{v} = -\sin\theta + \epsilon$$

as ϵ varies in the closed unit interval. Repeat the exercise for the perturbed pendulum system viewed as a family on the phase cylinder?

6.2 Periodic Perturbations: Transverse Homoclinic Points

In this section we will consider periodic perturbations of a planar Hamiltonian oscillator

$$\dot{x} = H_y(x, y), \qquad \dot{y} = -H_x(x, y) \qquad (6.20)$$

whose phase portrait has a homoclinic loop as depicted in Figure 6.1. Our main objective is to prove that if the Melnikov function defined on the homoclinic loop has simple zeros, then the periodically perturbed oscillator has transverse homoclinic points.

There are at least two reasons for the unnecessary restriction to unperturbed *Hamiltonian* systems. First, because Hamiltonian vector fields are divergence free, the Liouville factor

$$e^{-\int_0^t \operatorname{div} f(\varphi_t(\xi_0),0)\, ds}$$

is constant. Therefore, the expression for the Melnikov integral is simplified (see, for example, formula (6.9)). The second reason is the recognition that for the most important applications of the theory, the unperturbed systems are Hamiltonian.

To avoid writing the components of system (6.20), let us define the vector $\nu = (x, y)$ and, with a slight abuse of notation, the function

$$f(\nu) := (H_y(\nu), -H_x(\nu))$$

so that differential equation (6.20) has vector form

$$\dot{\nu} = f(\nu).$$

Also, let us suppose that $g : \mathbb{R}^2 \times \mathbb{R} \times \mathbb{R} \to \mathbb{R}^2$ is a function given by $(\nu, t, \epsilon) \mapsto g(\nu, t, \epsilon)$ that is $2\pi/\Omega$-periodic in t. The corresponding periodically perturbed oscillator is given in vector form by

$$\dot{\nu} = f(\nu) + \epsilon g(\nu, t, \epsilon),$$

and in component form by

$$\dot{x} = H_y(x, y) + \epsilon g_1(x, y, t, \epsilon),$$
$$\dot{y} = -H_x(x, y) + \epsilon g_2(x, y, t, \epsilon). \qquad (6.21)$$

Let us denote the flow of the unperturbed Hamiltonian system (6.20) by φ_t, the homoclinic loop at the hyperbolic saddle point ν_0 for the unperturbed system (6.20) by Γ, and the solution of the perturbed system (6.21) by $t \mapsto V(t, \nu, \epsilon)$ where $\nu \in \mathbb{R}^2$ and $V(0, \nu, \epsilon) \equiv \nu$. Also, as usual, let us define the (stroboscopic) parametrized Poincaré map $P : \mathbb{R}^2 \times \mathbb{R} \to \mathbb{R}^2$ by

$$P(\nu, \epsilon) := V(2\pi/\Omega, \nu, \epsilon).$$

Finally, the Melnikov function $\mathcal{M} : \Gamma \to \mathbb{R}$ for the perturbed oscillator (6.21) is defined by

$$\mathcal{M}(\zeta) := \int_{-\infty}^{\infty} f(\varphi_t(\zeta)) \wedge g(\varphi_t(\zeta), t, 0) \, dt \qquad (6.22)$$

where, of course, $f \wedge g := f_1 g_2 - g_1 f_2$.

The main result of this section on the existence of transverse homoclinic points is stated in the following theorem.

Theorem 6.8. *If $|\epsilon|$ is sufficiently small, then the parametrized Poincaré map for the system (6.21) has a hyperbolic saddle fixed point $\nu(\epsilon)$ such that $\nu(\epsilon) = \nu_0 + O(\epsilon)$. If ζ_0 is a simple zero of the Melnikov function $\mathcal{M}$ defined on Γ and $|\epsilon| \neq 0$ is sufficiently small, then the corresponding Poincaré map (at this value of ϵ) has a transverse homoclinic point relative to the stable and unstable manifolds of the hyperbolic fixed point $\nu(\epsilon)$. If, on the other hand, $\mathcal{M}$ has no zeros and $|\epsilon| \neq 0$ is sufficiently small, then the stable and unstable manifolds of $\nu(\epsilon)$ do not intersect.*

For the applications of Theorem 6.8, it is often convenient to work with a local coordinate on the homoclinic loop Γ. In fact, if we choose some point z on Γ, then the homoclinic orbit is parametrized by the corresponding solution of the differential equation, for example, by $\ell \mapsto \varphi_{-\ell}(z)$. Thus, the function $M : \mathbb{R} \to \mathbb{R}$ defined by

$$M(\ell) := \mathcal{M}(\varphi_{-\ell}(z)) = \int_{-\infty}^{\infty} f(\varphi_{t-\ell}(z)) \wedge g(\varphi_{t-\ell}(z), t, 0) \, dt \qquad (6.23)$$

is a local representation of the Melnikov function. Moreover, by the change of variables $\sigma := t - \ell$, we also have the useful identity

$$M(\ell) = \int_{-\infty}^{\infty} f(\varphi_\sigma(z)) \wedge g(\varphi_\sigma(z), \sigma + \ell, 0) \, d\sigma. \qquad (6.24)$$

As an important example, let us consider the first order system equivalent to the periodically forced pendulum

$$\ddot{\theta} + \lambda \sin \theta = \epsilon a \sin \Omega t$$

on the phase cylinder, that is, the system

$$\dot{\theta} = v,$$
$$\dot{v} = -\lambda \sin \theta + \epsilon a \sin \Omega t \qquad (6.25)$$

where θ is an angular coordinate modulo 2π. The unperturbed phase cylinder system has a hyperbolic saddle point with coordinates $(\theta, v) = (\pi, 0)$ and two corresponding homoclinic loops. Moreover, the unperturbed system is Hamiltonian with respect to the total energy

$$H(\theta, v) := \frac{1}{2} v^2 - \lambda \cos \theta,$$

and both homoclinic loops lie on the energy surface in the phase cylinder corresponding to the graph of the energy relation

$$v^2 = 2\lambda(1 + \cos\theta).$$

Using the equality $(1/v)d\theta/dt = 1$, the energy relation, and the identity $1 + \cos\theta = 2\cos^2(\theta/2)$, we have the unperturbed system restricted to the upper homoclinic orbit given by the scalar differential equation

$$2\sqrt{\lambda} = \frac{1}{\cos(\theta/2)}\frac{d\theta}{dt}.$$

If we impose the initial condition $\theta(0) = 0$, then the initial value problem has the elementary implicit solution

$$\frac{1}{2}\ln\left(\frac{1 + \sin(\theta/2)}{1 - \sin(\theta/2)}\right) = \sqrt{\lambda}\,t,$$

or equivalently the solution

$$t \mapsto \theta(t) = 2\arcsin(\tanh(\sqrt{\lambda}t)).$$

The corresponding solution of the pendulum equation

$$\theta = 2\arcsin(\tanh(\sqrt{\lambda}t)) = 2\arctan(\sinh(\sqrt{\lambda}t)),$$
$$v = 2\sqrt{\lambda}\operatorname{sech}(\sqrt{\lambda}t) \tag{6.26}$$

with the initial condition $(\theta, v) = (0, 2\sqrt{\lambda})$ is easily determined by substitution of $\theta(t)$ into the energy relation or by differentiation of the function $t \mapsto \theta(t)$ with respect to t.

In view of the solution (6.26) on the upper homoclinic loop, the Melnikov function (6.24) for the periodically forced pendulum is given by

$$M(\ell) := 2a\sqrt{\lambda}\int_{-\infty}^{\infty}\operatorname{sech}(\sqrt{\lambda}\sigma)\sin(\Omega(\sigma + \ell))\,d\sigma.$$

By using the trigonometric identity for the sine of the sum of two angles and by observing that the function $\sigma \mapsto \operatorname{sech}(\sqrt{\lambda}\sigma)\sin(\Omega\sigma)$ is odd, the formula for M can be simplified to the identity

$$M(\ell) = 2a\sqrt{\lambda}\sin(\Omega\ell)\int_{-\infty}^{\infty}\operatorname{sech}(\sqrt{\lambda}\sigma)\cos(\Omega\sigma)\,d\sigma$$

where the value of the improper integral is given by

$$\int_{-\infty}^{\infty}\operatorname{sech}(\sqrt{\lambda}\sigma)\cos(\Omega\sigma)\,d\sigma = \frac{\pi}{\sqrt{\lambda}}\operatorname{sech}\left(\frac{\pi\Omega}{2\sqrt{\lambda}}\right). \tag{6.27}$$

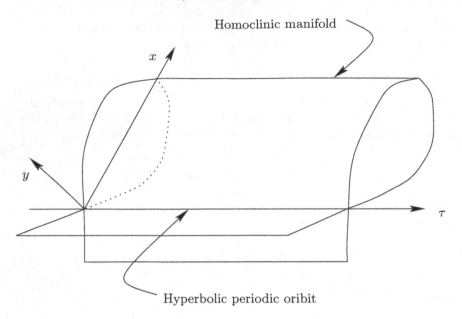

Figure 6.5: Phase portrait for the system (6.30) on the phase cylinder. The homoclinic manifold is the cylinder over the homoclinic loop of the corresponding planar Hamiltonian system.

The function M has infinitely many simple zeros given by

$$\left\{\ell = \frac{m\pi}{\Omega} : m \in \mathbb{Z}\right\}.$$

Thus, by Theorem 6.8, the Poincaré map for the system (6.25) has transverse homoclinic points. (Treat yourself to an aesthetic experience. Find a few quiet hours, sit alone, avoid all computer algebra systems, review the elements of complex analysis, and then use the residue calculus to compute the value of the improper integral (6.27). Pure as light, let Cauchy's theorem, a crown jewel of 19th century mathematics, shine within.)

In preparation for the proof of Theorem 6.8, let us recast the differential equation (6.21) as the first order system on the phase cylinder $\mathbb{R}^2 \times \mathbb{T}$ given by

$$\begin{aligned}
\dot{x} &= H_y(x, y) + \epsilon G_1(x, y, \tau, \epsilon), \\
\dot{y} &= -H_x(x, y) + \epsilon G_2(x, y, \tau, \epsilon), \\
\dot{\tau} &= \Omega
\end{aligned} \tag{6.28}$$

where τ is an angular variable modulo 2π and

$$G_i(x, y, \tau, \epsilon) := g_i(x, y, \tau/\Omega, \epsilon)$$

for $i = 1, 2$. Also, let us note that the corresponding vector form of system (6.28) is

$$
\begin{aligned}
\dot{V} &= f(V) + \epsilon G(V, \tau, \epsilon), \\
\dot{\tau} &= \Omega.
\end{aligned}
\tag{6.29}
$$

The unperturbed system

$$
\begin{aligned}
\dot{x} &= H_y(x, y), \\
\dot{y} &= -H_x(x, y), \\
\dot{\tau} &= \Omega
\end{aligned}
\tag{6.30}
$$

has a two-dimensional *homoclinic manifold* S corresponding to the homoclinic loop of the corresponding planar Hamiltonian system as sketched in Figure 6.5. Note that the original hyperbolic saddle point of the planar Hamiltonian system corresponds to a hyperbolic periodic orbit γ of system (6.30) that has two-dimensional stable and unstable manifolds, denoted $W^s(\gamma)$ and $W^u(\gamma)$, respectively. Moreover, the homoclinic manifold is contained in $W^s(\gamma) \cup W^u(\gamma)$.

To obtain a coordinate system on the homoclinic manifold, let us recall that the local coordinate on the homoclinic loop is given by the function $\ell \mapsto \varphi_{-\ell}(z)$ where z is fixed in Γ. The manifold S is parametrized in the same manner. In fact, if $p \in S$, then there is a unique point $(\ell, \tau) \in \mathbb{R} \times \mathbb{T}$ such that

$$
p = (\varphi_{-\ell}(z), \tau).
$$

In other words, the map

$$
(\ell, \tau) \mapsto (\varphi_{-\ell}(z), \tau)
$$

is a global chart whose image covers the manifold S.

We are interested in the fate of the homoclinic manifold for $\epsilon \neq 0$. The first observation is that the periodic orbit γ is continuable for sufficiently small $|\epsilon|$ and its continuation is a hyperbolic periodic orbit $\gamma(\epsilon)$ with a two-dimensional stable manifold $W^s(\gamma(\epsilon))$ and a two-dimensional unstable manifold $W^u(\gamma(\epsilon))$. The persistence of γ, and hence the first statement of Theorem 6.8, follows easily from the results of Chapter 5. The existence of the perturbed stable and unstable manifolds follows from results similar to those in Chapter 4. In fact, the existence of the perturbed invariant manifolds can be proved from the existence of invariant manifolds for the hyperbolic fixed point of the perturbed Poincaré map. The Hartman–Grobman theorem for diffeomorphisms in Chapter 4 can be used to obtain the existence of *continuous* invariant manifolds at the hyperbolic fixed point of the Poincaré map corresponding to the hyperbolic saddle point ν_0. The proof of

the smoothness of these invariant sets is analogous to the proof of smoothness given in Chapter 4 for the invariant stable and unstable manifolds at
a hyperbolic rest point of a differential equation.

We will prove a version of Theorem 6.8 that takes into account the geometry of the homoclinic manifold. The formulation of this result requires
an extension of the Melnikov function (6.23) to a function, also denoted by
the symbol M, that is defined on the homoclinic manifold S by

$$M(\ell, \tau) := \int_{-\infty}^{\infty} f(\varphi_{t-\ell}(z)) \wedge G(\varphi_{t-\ell}(z), \Omega t + \tau, 0) \, dt. \tag{6.31}$$

The statement in Theorem 6.8 concerning the existence of a transverse
homoclinic point is an easy consequence of the following result.

Theorem 6.9. *If there is a point in S with coordinates (ℓ, τ) such that*

$$M(\ell, \tau) = 0, \qquad M_\ell(\ell, \tau) \neq 0,$$

*and if $|\epsilon| \neq 0$ is sufficiently small, then the stable manifold $W^s(\gamma(\epsilon))$ and
the unstable manifold $W^u(\gamma(\epsilon))$ intersect transversally at a point in the
phase cylinder $O(\epsilon)$ close to the point $(\varphi_{-\ell}(z), \tau)$.*

A point p of transversal intersection of the stable and unstable manifolds
of the hyperbolic periodic orbit γ in the phase cylinder corresponds to a
point of transversal intersection of the stable and unstable manifolds of
the corresponding hyperbolic fixed point of the perturbed Poincaré map.
In fact, the corresponding point of transversal intersection on the Poincaré
section may be taken to be the first intersection of the orbit through p with
the Poincaré section.

The proof of Theorem 6.9 will require some additional notation and two
lemmas.

Let us measure the splitting of the stable and unstable manifolds relative
to the unperturbed homoclinic manifold S. To be precise, note first that
there is a natural choice for a normal vector at each point $p := (\varphi_{-\ell}(z), \tau) \in
S$, namely the vector

$$N(\ell, \tau) = (\varphi_{-\ell}(z), \tau, \eta(\ell), 0)$$

with base point $(\varphi_{-\ell}(z), \tau)$ and the first component of its principal part
given by

$$\eta(\ell) := DH(\varphi_{-\ell}(z)) = (H_x(\varphi_{-\ell}(z)), H_y(\varphi_{-\ell}(z))).$$

Of course, the tangent space to S at the point p is generated by the two
vectors

$$(H_y(\varphi_{-\ell}(z)), -H_x(\varphi_{-\ell}(z)), 0), \qquad (0, 0, 1)$$

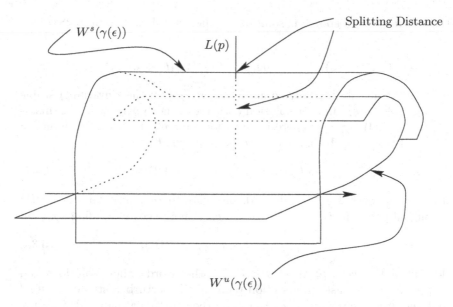

Figure 6.6: Perturbed stable and unstable manifolds. The splitting distance is computed with respect to the lines in the direction of the normals to the homoclinic manifold.

where the base point is suppressed and the last component is in $\mathbb{R}$, the tangent space to the circle $\mathbb{T}$ at the point with angle τ. Note that both of these basis vectors are orthogonal to the vector $N(\ell, \tau)$ with respect to the usual inner product of $\mathbb{R}^3$.

The unperturbed stable and unstable manifolds are transverse to the line $L(p)$ through the point p on $\mathcal{S}$ with direction vector $N(\ell, \tau)$. Thus, for a small perturbation, the perturbed stable and unstable manifolds must also intersect $L(p)$ transversally (see Figure 6.6). The idea is to use the distance between the intersection points of the perturbed invariant manifolds and the line $L(p)$ as a measure of the distance between the perturbed manifolds at the point $p \in \mathcal{S}$. But there is a problem: The perturbed invariant manifolds might intersect the line more than once, perhaps even an infinite number of times. Thus, it is not clear which intersection points to choose in order to measure the distance at p between the perturbed stable and unstable manifolds.

Suppose that $p^s(\epsilon)$ is a point on $L(p) \cap W^s(\gamma(\epsilon))$, and $p^u(\epsilon)$ is a point on $L(p) \cap W^u(\gamma(\epsilon))$. Also, recall that the point p depends on the coordinates ℓ and τ. If, in components relative to the phase cylinder,

$$p^s(\epsilon) = (z^s(\ell, \tau, \epsilon), \tau), \qquad p^u(\epsilon) = (z^u(\ell, \tau, \epsilon), \tau),$$

then there are corresponding solutions of the perturbed system (6.28) given
by

$$t \mapsto (V^s(t, z^s(\ell, \tau, \epsilon), \epsilon), \; \tau + \Omega t), \qquad t \mapsto (V^u(t, z^u(\ell, \tau, \epsilon), \epsilon), \; \tau + \Omega t).$$

Of course, the solution corresponding to $p^s(\epsilon)$ is in the (invariant) stable
manifold $W^s(\gamma(\epsilon))$ and the solution corresponding to $p^u(\epsilon)$ is in the unstable manifold $W^u(\gamma(\epsilon))$. There is one choice for $p^s(\epsilon)$ among all points in
$L(p) \cap W^s(\gamma(\epsilon))$ such that the corresponding solution

$$t \mapsto (V^s(t, z^s(\ell, \tau, \epsilon), \epsilon), \tau + \Omega t), \tag{6.32}$$

does not intersect $L(p)$ for all $t > 0$. Likewise, there is one choice for $p^u(\epsilon)$
among all points in $L(p) \cap W^u(\gamma(\epsilon))$ such that the corresponding solution

$$t \mapsto (V^u(t, z^u(\ell, \tau, \epsilon), \epsilon), \; \tau + \Omega t), \tag{6.33}$$

does not intersect $L(p)$ for all $t < 0$. In other words, these solutions are,
respectively, the "last" intersection point of the perturbed stable manifold
and the "first" intersection of the perturbed unstable manifold with the
line $L(p)$. While it is intuitively clear that these special intersections points
exist, this fact can be proved (see, for example, [233, p. 495]). At any rate,
let us use these special intersection points to measure the distance between
the perturbed stable and unstable manifolds.

Lemma 6.10. *If $p \in S$ and $|\epsilon|$ is sufficiently small, then the first com-
ponents of the solutions (6.32) and (6.33) corresponding to the last intersection point $p^s(\epsilon)$ and the first intersection point $p^u(\epsilon)$ on $L(p)$ have the
following representations:*

$$V^s(t, z^s(\ell, \tau, \epsilon), \epsilon) = \varphi_{t-\ell}(z) + \epsilon r^s(t) + O(\epsilon^2), \quad t \geq 0,$$
$$V^u(t, z^u(\ell, \tau, \epsilon), \epsilon) = \varphi_{t-\ell}(z) + \epsilon r^u(t) + O(\epsilon^2), \quad t \leq 0 \tag{6.34}$$

*where the functions $r^s : (0, \infty) \to \mathbb{R}^2$ and $r^u : (-\infty, 0) \to \mathbb{R}^2$ given by
$r^s(t) = V_\epsilon^s(t, z^s(\ell, \tau, 0), 0)$ and $r^u(t) = V_\epsilon^u(t, z^u(\ell, \tau, 0), 0)$ are bounded on
the indicated infinite time intervals.*

Proof. We will prove the result for the solutions on the stable manifold;
the result for the unstable manifold is similar. Also, we will suppress the
variables ℓ and τ by using the notation

$$V^s(t, \epsilon) := V^s(t, z^s(\ell, \tau, \epsilon), \epsilon), \qquad V^u(t, \epsilon) := V^u(t, z^u(\ell, \tau, \epsilon), \epsilon).$$

The basic estimate required to prove the lemma is obtained with an
application of Gronwall's inequality (Lemma 2.1). Fix ℓ and τ. Also, recall
that $t \mapsto V^s(t, \epsilon)$ is a solution of the differential equation

$$\dot{V} = F(V, t, \epsilon) := f(V) + \epsilon G(V, t, \epsilon),$$

and $t \mapsto \varphi_{t-\ell}(z)$ is a solution of the differential equation $\dot{V} = F(V, t, 0)$. By integration, we have that

$$V^s(t, \epsilon) - z^s(\ell, \tau, \epsilon) = \int_0^t F(V^s(\sigma, \epsilon), \sigma, \epsilon) \, d\sigma,$$

$$\varphi_{t-\ell}(z) - \varphi_{-\ell}(z) = \int_0^t F(\varphi_{\sigma-\ell}(z), \sigma, \epsilon) \, d\sigma. \tag{6.35}$$

Both solutions belong to the projection to the V-plane of a stable manifold of a periodic orbit in the phase cylinder. Thus, both solutions for $t \geq 0$ lie in a compact subset K of the plane. By the smoothness of the function F, there is a Lipschitz constant $C_1 > 0$ such that

$$|F(V_1, t, \epsilon) - F(V_1, t, 0)| \leq C(|V_1 - V_2| + |\epsilon|)$$

for V_i, $i = 1, 2$ in K and $|\epsilon|$ sufficiently small. Also, by the smoothness of the stable manifold with respect to ϵ, if $|\epsilon|$ is sufficiently small, then there is a constant $C_2 > 0$ such that

$$|z^s(\ell, \tau, \epsilon) - \varphi_{-\ell}(z)| \leq C_2 \epsilon. \tag{6.36}$$

If we subtract the equations in display (6.35) and use the inequalities just mentioned, then we obtain the estimate

$$|V^s(t, \epsilon) - \varphi_{t-\ell}(z)| \leq \epsilon C_2 + \epsilon C_1 t + C_1 \int_0^t |V^s(\sigma, \epsilon) - \varphi_{\sigma-\ell}(z)| \, d\sigma.$$

Hence, by an application of Gronwall's inequality,

$$|V^s(t, \epsilon) - \varphi_{t-\ell}(z)| \leq \epsilon(C_2 + C_1 t)e^{C_1 t}. \tag{6.37}$$

Recall that $\nu(\epsilon)$ denotes the perturbed hyperbolic saddle point and ν_0 the hyperbolic saddle point for the planar Hamiltonian system. By a simple application of the implicit function theorem, it follows that

$$\nu(\epsilon) = \nu_0 + O(\epsilon).$$

Since the solutions in the inequality (6.37) belong to the respective stable manifolds of $\nu(\epsilon)$ and ν_0, there is some constant $C_3 > 0$ and some $T > 0$ such that if $t > T$, then

$$|V^s(t, \epsilon) - \varphi_{t-\ell}(z)| \leq \epsilon C_3. \tag{6.38}$$

Therefore, if $|\epsilon|$ is sufficiently small, then, by the Gronwall estimate (6.37) for $0 \leq t \leq T$ and the estimate (6.38) for $t > T$, there is a constant $C > 0$ such that

$$|V^s(t, \epsilon) - \varphi_{t-\ell}(z)| \leq \epsilon C \tag{6.39}$$

for all $t \geq 0$.

Because the solution V is a smooth function of the parameter ϵ, there is a smooth remainder R such that

$$V^s(t, \epsilon) = \varphi_{t-\ell}(z) + \epsilon r^s(t) + \epsilon^2 R(t, \epsilon).$$

Thus, using the inequality (6.39), we have that

$$\epsilon |r^s(t) + \epsilon R(t, \epsilon)| = |V^s(t, \epsilon) - \varphi_{t-\ell}(z)| \leq \epsilon C.$$

Finally, let us divide this estimate by ϵ and then set $\epsilon = 0$ to obtain the desired result: $|r^s(t)| \leq C$ for $t \geq 0$.   □

Let us define the distance between the perturbed stable and unstable manifolds at $p = (\varphi_\ell(z), \tau)$ to be

$$
\begin{aligned}
\operatorname{sep}(\ell, \tau, \epsilon) &:= \frac{\langle p_\epsilon^u - p_\epsilon^s, N(\ell, \tau) \rangle}{|N(\ell, \tau)|} \\
&= \frac{\langle z^u(\ell, \tau, \epsilon) - z^s(\ell, \tau, \epsilon), \eta(\ell) \rangle}{|\eta(\ell)|} \\
&= \frac{DH(\varphi_{-\ell}(z))(z^u(\ell, \tau, \epsilon) - z^s(\ell, \tau, \epsilon))}{|\eta(\ell)|} \quad (6.40)
\end{aligned}
$$

Because $\operatorname{sep}(\ell, \tau, 0) \equiv 0$, we have the representation

$$
\begin{aligned}
\operatorname{sep}(\ell, \tau, \epsilon) &= \operatorname{sep}(\ell, \tau, 0) + \epsilon \operatorname{sep}_\epsilon(\ell, \tau, 0) + O(\epsilon^2) \\
&= \epsilon(\operatorname{sep}_\epsilon(\ell, \tau, 0) + O(\epsilon)). \quad (6.41)
\end{aligned}
$$

Also, by differentiation with respect to ϵ in equation (6.40), it follows that the leading order coefficient of the separation function is given by

$$\operatorname{sep}_\epsilon(\ell, \tau, 0) = \frac{\bar{M}(\ell, \tau)}{|DH(\varphi_{-\ell}(z))|} \quad (6.42)$$

where

$$\bar{M}(\ell, \tau) := DH(\varphi_{-\ell}(z))(V_\epsilon^u(0, z^u(\ell, \tau, 0), 0) - V_\epsilon^s(0, z^s(\ell, \tau, 0), 0)). \quad (6.43)$$

In particular, up to a normalization, $\bar{M}(\ell, \tau)$ is the leading coefficient in the expansion (6.41).

Lemma 6.11. *The function $\bar{M}$ defined in display (6.43) is equal to the Melnikov function defined in display (6.31); that is, if a point on the homoclinic manifold is given in coordinates by (ℓ, τ), then*

$$\bar{M}(\ell, \tau) = M(\ell, \tau).$$

Proof. (The proof of this lemma is similar to the proof of Proposition 6.2.)
Define the time-dependent Melnikov function

$$m(t, \ell, \tau) := DH(\varphi_{t-\ell}(z))(V_\epsilon^u(t, 0) - V_\epsilon^s(t, 0))$$

where ℓ and τ are suppressed as in the proof of Lemma 6.10, and note that
$m(0, \ell, \tau) = \bar{M}(\ell, \tau)$. Also, define two more auxiliary functions m^s and m^u
by

$$m^s(t, \ell, \tau) = DH(\varphi_{t-\ell}(z))V_\epsilon^s(t, 0), \quad m^u(t, \ell, \tau) = DH(\varphi_{t-\ell}(z))V_\epsilon^u(t, 0)$$

so that $m(t, \ell, \tau) = m^u(t, \ell, \tau) - m^s(t, \ell, \tau)$. If m^* denotes either m^u or m^s,
and likewise V^* denotes V^s or V^u, then

$$\begin{aligned}\dot{m}^*(t, \ell, \tau) = D^2H(\varphi_{t-\ell}(z))[f(\varphi_{t-\ell}(z)), V_\epsilon^*(t, 0)] \\ + DH(\varphi_{t-\ell}(z))\dot{V}_\epsilon^*(t, 0).\end{aligned} \tag{6.44}$$

Let us also recall that $t \mapsto V^*(t, \epsilon)$ is defined to be a solution of the system (6.21); that is,

$$\dot{V}^* = f(V^*) + \epsilon G(V^*, \Omega t + \tau, \epsilon). \tag{6.45}$$

By differentiation of equation (6.45) with respect to ϵ at $\epsilon = 0$ we obtain
the second variational equation

$$\dot{V}_\epsilon^* = Df(\varphi_{t-\ell}(z))V_\epsilon^* + G(\varphi_{t-\ell}(z), \Omega t + \tau, 0). \tag{6.46}$$

Let us substitute the expression for $\dot{V}_\epsilon^*$ given by equation (6.46) into the
differential equation (6.44) and rearrange the terms to obtain

$$\dot{m}^*(t, \ell, \tau) = DH(\varphi_{t-\ell}(z))G(\varphi_{t-\ell}(z), \Omega t + \tau, 0) + B(t)V_\epsilon^*(t, 0) \tag{6.47}$$

where $B(t)$ is the linear transformation of $\mathbb{R}^2$ given by

$$\begin{aligned}B(t)V := D^2H(\varphi_{t-\ell}(z))[f(\varphi_{t-\ell}(z)), V] \\ + DH(\varphi_{t-\ell}(z))Df(\varphi_{t-\ell}(z))V.\end{aligned} \tag{6.48}$$

Also, by differentiating both sides of the identity

$$DH(\xi)f(\xi) \equiv 0$$

with respect to $\xi \in \mathbb{R}^2$, let us observe that

$$D^2H(\xi)f(\xi) + DH(\xi)Df(\xi) \equiv 0.$$

Thus, it follows that $B(t) \equiv 0$, and the differential equation (6.47) for m^*
reduces to

$$\dot{m}^*(t, \ell, \tau) = DH(\varphi_{t-\ell}(z))G(\varphi_{t-\ell}(z), \Omega t + \tau, 0). \tag{6.49}$$

By integration of equation (6.49) separately for m^s and m^u, the following formulas are obtained:

$$m^s(t, \ell, \tau) - m^s(0, \ell, \tau) = \int_0^t DH(\varphi_{\sigma - \ell}(z)) G(\varphi_{\sigma - \ell}(z), \Omega\sigma + \tau, 0) \, d\sigma,$$

$$m^u(0, \ell, \tau) - m^u(t, \ell, \tau) = \int_{-t}^0 DH(\varphi_{\sigma - \ell}(z)) G(\varphi_{\sigma - \ell}(z), \Omega\sigma + \tau, 0) \, d\sigma.$$

$$(6.50)$$

In view of Lemma 6.34, the function $t \mapsto V_\epsilon^s(t, 0)$ is bounded. Also, because DH vanishes at the hyperbolic saddle point ν_0, we have that

$$\lim_{t \to \infty} DH(\varphi_{t - \ell}(z)) = 0,$$

and therefore

$$\lim_{t \to \infty} m^s(t, \ell, \tau) = 0.$$

It follows that the improper integral on the right hand side of the first equation in display (6.50) converges and

$$-m^s(0, \ell, \tau) = \int_0^\infty DH(\varphi_{t - \ell}(z)) G(\varphi_{t - \ell}(z), \Omega t + \tau, 0) \, dt.$$

Similarly, we have that

$$m^u(0, \ell, \tau) = \int_{-\infty}^0 DH(\varphi_{t - \ell}(z)) G(\varphi_{t - \ell}(z), \Omega t + \tau, 0) \, dt.$$

To complete the proof, simply note the equality

$$m^u(0, \ell, \tau) - m^s(0, \ell, \tau) = \bar{M}(\ell, \tau)$$

and that the sum of the integral representations of the quantities $m^u(0, \ell, \tau)$ and $-m^s(0, \ell, \tau)$ is just the Melnikov integral.   □

As a consequence of Lemma 6.11 and the representation of the separation function (6.41), we have now proved that

$$\text{sep}(\ell, \tau, \epsilon) = \epsilon\left(\frac{M(\ell, \tau)}{|DH(\varphi_{-\ell}(z))|} + O(\epsilon)\right).$$

$$(6.51)$$

In other words, *the Melnikov function (properly normalized) is the leading order term in the series expansion of the separation function in powers of the perturbation parameter.* This is the key result of Melnikov theory.

Let us now prove Theorem 6.9.

Proof. For notational convenience, let us define

$$S(\ell, \tau, \epsilon) := \frac{M(\ell, \tau)}{|\eta(\ell)|} + O(\epsilon).$$

where $\eta(\ell) = DH(\varphi_{-\ell}(z))$ so that formula (6.51) is recast in the form

$$\mathrm{sep}(\ell, \tau, \epsilon) = \epsilon S(\ell, \tau, \epsilon).$$

If $M(\ell_0, \tau_0) = 0$ and $M_\ell(\ell_0, \tau_0) \neq 0$, then $(\ell_0, \tau_0, 0)$ is a zero of S such that $S_\ell(\ell_0, \tau_0, 0) \neq 0$. Therefore, by the implicit function theorem, there is a real-valued function h defined on some product neighborhood of $\epsilon = 0$ and $\tau = \tau_0$ such that $h(0, \tau_0) = \ell_0$ and $S(h(\epsilon, \tau), \tau, \epsilon) \equiv 0$. Or, in other words, using the definition of the separation function, our result implies that if $|\epsilon|$ is sufficiently small, then the stable and unstable manifolds intersect at the points given by

$$(V^s(0, z^s(h(\epsilon, \tau), \epsilon), \epsilon), \tau) \equiv (V^u(0, z^u(h(\epsilon, \tau), \epsilon), \epsilon), \tau), \qquad (6.52)$$

or equivalently at the points

$$(z^s(h(\epsilon, \tau), \epsilon), \tau) \equiv (z^u(h(\epsilon, \tau), \epsilon), \tau). \qquad (6.53)$$

To complete the proof we will show that if $|\epsilon| \neq 0$ is sufficiently small, then the stable and unstable manifolds intersect transversally at the point given in display (6.53).

Let us note that the curves in the phase cylinder given by

$$\ell \mapsto (z^s(\ell, \tau_0, \epsilon), \tau_0), \qquad \tau \mapsto (z^s(\ell_0, \tau, \epsilon), \tau)$$

both lie in $W^s(\gamma(\epsilon))$. Therefore, the vectors

$$(z^s_\ell(\ell_0, \tau_0, \epsilon), 0), \qquad (z^s_\tau(\ell_0, \tau_0, \epsilon), 1)$$

span the tangent space to $W^s(\gamma(\epsilon))$ at the intersection point with coordinates (ℓ_0, τ_0). Indeed, since $S_\ell(\ell_0, \tau_0, 0) \neq 0$, it follows from the definition of the separation function and the continuity with respect to ϵ that if $|\epsilon| \neq 0$ is sufficiently small, then $z^s_\ell(\ell_0, \tau_0, \epsilon) \neq 0$. Thus, the first tangent vector is nonzero. Because the second component of the second tangent vector is nonzero, the two vectors are linearly independent. Similarly, the vectors

$$(z^u_\ell(\ell_0, \tau_0, \epsilon), 0), \qquad (z^u_\tau(\ell_0, \tau_0, \epsilon), 1)$$

span the tangent space to the unstable manifold at the intersection point.

The stable and unstable manifolds meet transversally provided that three of the four tangent vectors given above span $\mathbb{R}^3$. To determine a linearly independent subset of these tangent vectors, we will use the definition of the Melnikov function and Lemma 6.34.

First, in view of the equalities

$$M(\ell_0, \tau_0) = 0, \qquad z^u(\ell_0, \tau_0, \epsilon) = z^s(\ell_0, \tau_0, \epsilon),$$

and the definition of the Melnikov function, let us note that

$$\frac{\partial}{\partial \ell}\left(\frac{DH(\varphi_{-\ell}(z))(z^u(\ell, \tau_0, \epsilon) - z^s(\ell, \tau_0, \epsilon))}{|\eta(\ell)|}\right)\bigg|_{\ell=\ell_0} = \epsilon\left(\frac{M_\ell(\ell_0, \tau_0)}{|\eta(\ell_0)|} + O(\epsilon)\right)$$

and

$$\frac{\partial}{\partial \ell}\left(\frac{DH(\varphi_{-\ell}(z))(z^u(\ell, \tau_0, \epsilon) - z^s(\ell, \tau_0, \epsilon))}{|\eta(\ell)|}\right)\bigg|_{\ell=\ell_0} =$$
$$\left(\frac{DH(\varphi_{-\ell_0}(z))(z^u_\ell(\ell_0, \tau_0, \epsilon) - z^s_\ell(\ell_0, \tau_0, \epsilon))}{|\eta(\ell_0)|}\right).$$

By combining the results of these computations, we have that

$$DH(\varphi_{-\ell_0}(z))(z^u_\ell(\ell_0, \tau_0, \epsilon) - z^s_\ell(\ell_0, \tau_0, \epsilon)) = \epsilon(M_\ell(\ell_0, \tau_0) + O(\epsilon)). \quad (6.54)$$

Set $t = 0$ and $\tau = \tau_0$ and differentiate both sides of both equations in display (6.34) with respect to ℓ at $\ell = \ell_0$ to obtain the representations

$$z^s_\ell(\ell_0, \tau_0, \epsilon) = -f(\varphi_{-\ell}(z)) + \epsilon z^s_{\epsilon\ell}(\ell_0, \tau_0, 0) + O(\epsilon^2),$$
$$z^u_\ell(\ell_0, \tau_0, \epsilon) = -f(\varphi_{-\ell}(z)) + \epsilon z^u_{\epsilon\ell}(\ell_0, \tau_0, 0) + O(\epsilon^2).$$

Thus, by substitution into the equation (6.54), let us note that

$$\epsilon\big(DH(\varphi_{-\ell_0}(z))(z^u_{\epsilon\ell}(\ell_0, \tau_0, 0) - z^s_{\epsilon\ell}(\ell_0, \tau_0, 0)) + O(\epsilon)\big)$$
$$= \epsilon(M_\ell(\ell_0, \tau_0) + O(\epsilon)). \quad (6.55)$$

Also, since the determinant of a matrix is a multilinear form with respect to the columns of the matrix, it follows by an easy computation using the definition of the Hamiltonian vector field f that

$$\det\left(z^u_\ell(\ell_0, \tau_0, \epsilon), z^s_\ell(\ell_0, \tau_0, \epsilon)\right) = \epsilon\big(\det\left(-f, z^s_{\epsilon\ell}\right) + \det\left(z^u_{\epsilon\ell}, -f\right) + O(\epsilon)\big)$$
$$= \epsilon(DH(\varphi_{-\ell_0}(z)(z^u_{\epsilon\ell}(\ell_0, \tau_0, 0)$$
$$- z^s_{\epsilon\ell}(\ell_0, \tau_0, 0)) + O(\epsilon)). \quad (6.56)$$

In view of the equations (6.55) and (6.56) we have that

$$\det\left(z^u_\ell(\ell_0, \tau_0, \epsilon), z^s_\ell(\ell_0, \tau_0, \epsilon)\right) = M_\ell(\ell_0, \tau_0) + O(\epsilon).$$

Therefore, the determinant is not zero, and the vectors

$$z^u_\ell(\ell_0, \tau_0, \epsilon), \qquad z^s_\ell(\ell_0, \tau_0, \epsilon)$$

are linearly independent. Hence, due to the independence of these vectors, the tangent vectors

$$(z^u_\ell(\ell_0, \tau_0, \epsilon), 0), \quad (z^s_\ell(\ell_0, \tau_0, \epsilon), 0), \quad (z^s_\tau(\ell_0, \tau_0, \epsilon), 1)$$

are linearly independent, as required. As a result, if $|\epsilon| \neq 0$ is sufficiently small, then the perturbed stable and unstable manifolds meet transversally at the base point of these tangent vectors.   □

Exercise 6.12. Discuss the existence of transverse homoclinic points for the periodically perturbed Duffing oscillator

$$\ddot{x} - x + x^3 = \epsilon \sin(\Omega t).$$

Exercise 6.13. Discuss the existence of transverse homoclinic points for the periodically perturbed damped pendulum

$$\ddot{\theta} + \omega^2 \sin \theta = \epsilon g(\theta, t)$$

where

$$g(\theta, t) := -\lambda \dot{\theta} + \sin(\Omega t).$$

How does the existence of transverse homoclinic points depend on the parameters? What happens if the sinusoidal time periodic external force is replaced by a smooth periodic function $p(t)$. What happens if the viscous damping term is replaced by $-\lambda \dot{\theta}^2$?

Exercise 6.14. Discuss the existence of transverse homoclinic points for the parametrically excited pendulum

$$\ddot{\theta} + (\omega^2 + \epsilon \cos(\Omega t)) \sin \theta = 0.$$

Exercise 6.15. Discuss the existence of transverse homoclinic points for the pendulum with "feedback control"

$$\ddot{\theta} + \sin \theta + \alpha \theta - \beta = \epsilon(-\lambda \dot{\theta} + \gamma \cos(\Omega t)$$

(see [236]). The "Melnikov analysis" of this system seems to require numerical approximations of the Melnikov integral. Compute an approximation of the Melnikov integral and find parameter values where your computations suggest the existence of simple zeros. Plot some orbits of the stroboscopic Poincaré map to obtain an approximation of its phase portrait. Also find parameter values where a numerical experiment suggests the corresponding dynamical system has sensitive dependence on initial conditions.

Exercise 6.16. Using formula (6.31), prove that $M_\ell(\ell, \tau) \neq 0$ if and only if $M_\tau(\ell, \tau) \neq 0$. Note that a corollary of this result is the conclusion of Theorem 6.9 under the hypothesis that $M(\ell, \tau) = 0$ and $M_\tau(\ell, \tau) \neq 0$.

6.3 Origins of ODE: Fluid Dynamics

The description of the motion of fluids is a central topic in practical scientific research with a vast literature in physics, engineering, mathematics, and computation. The basic model is a system of partial differential equations of evolution type. Thus, as might be expected, many specializations of this model lead to ordinary differential equations. In fact, some of the most interesting and most important problems in ordinary differential equations have their origin in fluid dynamics.

The purpose of this section is to briefly discuss the Euler and Navier–Stokes model equations; to derive a system of ordinary differential equations, called the ABC system, that has been used to describe the steady state motion of an ideal fluid in a certain ideal situation; and to discuss the dynamics of the ABC system as an application of our analysis of perturbed oscillators.

Caution: Treat this section as "a finger pointing at the moon."

6.3.1   The Equations of Fluid Motion

Let us consider a fluid with constant density ρ confined to some region $\mathcal{R}$ in space, and let us assume that the motion of the fluid is given by the time-dependent velocity field $u : \mathcal{R} \times \mathbb{R} \to \mathbb{R}^3$ with $(\xi, t) \mapsto u(\xi, t)$. The position of a particle of the moving fluid is given by a smooth curve $t \mapsto \gamma(t)$ in $\mathcal{R}$. Thus, the momentum of this fluid particle is $\rho u(\gamma(t), t)$, and, according to Newton's law, the motion of the particle is given by the differential equation

$$\rho \frac{d}{dt}(u(\gamma(t), t)) = F$$

where F denotes the sum of the forces. Although a fluid is always subjected to the force of gravity and perhaps to other external body forces, let us ignore these forces and consider only the constitutive force laws that model the internal shear forces that are essential to our understanding of the physical nature of fluids, just as Hooke's law is the essential constitutive force law for springs.

Internal fluid forces can be derived from more basic physical laws (see, for example, [56] and [139]); however, let us simply note that the basic force law is

$$F = \mu \Delta u - \operatorname{grad} P$$

where μ is a constant related to the viscosity of the fluid, P is called the fluid pressure, and the Laplacian operates on the velocity field componentwise. Of course, the gradient and the Laplacian derivatives are with respect to the space variables only. Let us also note that the viscosity term is a function of the fluid velocity, but the pressure is a second unknown dependent variable in the system. Thus, we will have to have two equations in the two unknown functions u and P.

Using Newton's law and the constitutive force law, the equation of motion for a fluid is

$$\rho\left(\frac{\partial u}{\partial t}(\xi, t) + Du(\xi, t)u(\xi, t)\right) = \mu \Delta u(\xi, t) - \operatorname{grad} P(\xi, t)$$

where D denotes differentiation with respect to the space variables. In fluid mechanics, if x, y, z are the Cartesian coordinates in $\mathbb{R}^3$ and e_x, e_y, e_z are the usual unit direction vectors (here the subscripts denote coordinate directions, not partial derivatives), then the gradient operator

$$\nabla := \frac{\partial}{\partial x} e_x + \frac{\partial}{\partial y} e_y + \frac{\partial}{\partial z} e_z$$

is introduced and the advection term $(Du)u$ is rewritten, using the usual inner product, in the form $\langle u, \nabla \rangle u$, or more commonly as $u \cdot \nabla u$. Here, ∇ acts componentwise on the vector field u.

The fluid density must satisfy the continuity equation (3.92)

$$\frac{\partial \rho}{\partial t} + \operatorname{div}(\rho u) = 0.$$

Thus, under our assumption that the density is constant (homogeneous fluid), we must have that u is divergence free. This is equivalent to the assumption that the fluid is incompressible.

Because our fluid is confined to a region of space, some boundary conditions must be imposed. In fact, physical experiments show that the correct boundary condition is $u \equiv 0$ on the boundary ∂R of the region R. To demonstrate this fact yourself, consider cleaning a metal plate by using a hose to spray it with water; for example, try cleaning a dirty automobile. As the pressure of the water increases, the size of the particles of dirt that can be removed decreases. But, it is very difficult to remove all the dirt by spraying alone. This can be checked by polishing with a clean cloth. In fact, the velocity of the spray decreases rapidly in the boundary layer near the plate. Dirt particles with sufficiently small diameter are not subjected to flow velocities that are high enough to dislodge them.

By introducing units of length, time, and velocity (that is, $x \mapsto x/L$, $t \mapsto t/T$ and $u \mapsto u/U$) and introducing the kinematic viscosity $\nu := \mu/\rho$, the system of equations for the velocity field and the pressure can be rescaled to the dimensionless form of the Navier–Stokes equations for an incompressible fluid in $\mathcal{R}$ given by

$$\frac{\partial u}{\partial t} + u \cdot \nabla u = \frac{1}{\operatorname{Re}} \Delta - \operatorname{grad} p,$$
$$\operatorname{div} u = 0,$$
$$u = 0 \quad \text{in } \partial \mathcal{R} \tag{6.57}$$

where $\operatorname{Re} := LU/\nu$ is the (dimensionless) Reynolds number. The existence of this scaling is important: If two flows have the same Reynold's number, then the flows have the same dynamics. For example, flow around a *scaled model* of an airplane in a wind tunnel might be tested at the same Reynold's number expected for the airplane under certain flight conditions. Perhaps the same Reynold's number can be obtained by increasing the velocity in the wind tunnel to compensate for the smaller length scale of the model. In principle, the behavior of the model is then exactly the same as the real aircraft.

Euler's equations for fluid motion can be viewed as an idealization of the Navier–Stokes equations for a fluid with zero viscosity. These equations have the form

$$\frac{\partial u}{\partial t} + u \cdot \nabla u = -\operatorname{grad} p,$$

$$\operatorname{div} u = 0,$$

$$\langle u, \eta \rangle = 0 \quad \text{in } \partial \mathcal{R} \tag{6.58}$$

where η is the outward unit normal vector field on $\partial \mathcal{R}$. Note that the "no slip" boundary condition for the Navier–Stokes equations is replaced by the condition that there is no fluid passing through the boundary. The reason for the physically unrealistic Euler boundary conditions is to ensure that Euler's partial differential equations are "well posed", that is, they have unique solutions depending continuously on initial conditions.

A naive expectation is that the limit of a family of solutions of the Navier–Stokes equations as the Reynold's number increases without bound is a solution of Euler's equations. After all, the term $\Delta u/\mathrm{Re}$ would seem to go to zero as $\mathrm{Re} \to \infty$. Note, however, the possibility that the second derivatives of the velocity field are unbounded in the limit. For this and other reasons, the limiting behavior of the Navier–Stokes equations for large values of the Reynold's number is not yet completely understood. Thus, the dynamical behavior of the family as the Reynold's number grows without bound is a fruitful area of research.

Flow in A Pipe

As an example of the solution of a fluid flow problem, let us consider perhaps the most basic example of the subject: flow in a round pipe.

If we choose cylindrical coordinates r, θ, z with the z-axis being the axis of symmetry of a round pipe with radius a, then it seems natural to expect that there are *some* flow regimes for which the velocity field has its only nonzero component in the axial direction of the pipe; that is, the velocity field has the form

$$u(r, \theta, z, t) = (0, 0, u_z(r, \theta, z, t)) \tag{6.59}$$

where the components of this vector field are taken with respect to the basis vector fields e_r, e_θ, e_z that are defined in terms of the usual basis of

Euclidean space by

$$e_r := (\cos\theta, \sin\theta, 0), \quad e_\theta := (-\sin\theta, \cos\theta, 0), \quad e_z := (0, 0, 1).$$

Let us express the Euler and the Navier–Stokes equations in cylindrical coordinates. Recall that if f is a function and $F = F_r e_r + F_\theta e_\theta + F_z e_z$ is a vector field on Euclidean space, then in cylindrical coordinates,

$$\nabla f = \frac{\partial f}{\partial r} e_r + \frac{1}{r} \frac{\partial f}{\partial \theta} e_\theta + \frac{\partial f}{\partial z} e_z,$$

$$\operatorname{div} f = \frac{1}{r} \frac{\partial}{\partial r}(r F_r) + \frac{1}{r} \frac{\partial F_\theta}{\partial \theta} + \frac{\partial F_z}{\partial z},$$

$$\Delta f = \frac{1}{r} \frac{\partial}{\partial r}\left(r \frac{\partial f}{\partial r}\right) + \frac{1}{r^2} \frac{\partial^2 f}{\partial \theta^2} + \frac{\partial^2 f}{\partial z^2}. \tag{6.60}$$

To obtain the Navier–Stokes equations in cylindrical coordinates, consider the unknown velocity field $u = u_r e_r + u_\theta e_\theta + u_z e_z$. Write this vector field in the usual Cartesian components by using the definitions of the direction fields given above, insert the result into the Navier–Stokes equations, and then compute the space derivatives using the operators given in display (6.60). If we multiply the first two of the resulting component equations—the equations in the directions e_x and e_y—by the matrix

$$\begin{pmatrix} \cos\theta & \sin\theta \\ -\sin\theta & \cos\theta \end{pmatrix},$$

then we obtain the equivalent system

$$\frac{\partial u_r}{\partial t} + (u \cdot \nabla) u_r - \frac{1}{r} u_\theta^2 = \frac{1}{\operatorname{Re}}\left(\Delta u_r - \frac{1}{r^2}\left(u_r + 2\frac{\partial u_\theta}{\partial \theta}\right)\right) - \frac{\partial p}{\partial r},$$

$$\frac{\partial u_\theta}{\partial t} + (u \cdot \nabla) u_\theta + \frac{1}{r} u_r u_\theta = \frac{1}{\operatorname{Re}}\left(\Delta u_\theta - \frac{1}{r^2}\left(u_\theta - 2\frac{\partial u_r}{\partial \theta}\right)\right) - \frac{1}{r} \frac{\partial p}{\partial \theta},$$

$$\frac{\partial u_z}{\partial t} + (u \cdot \nabla) u_z = \frac{1}{\operatorname{Re}} \Delta u_z - \frac{\partial p}{\partial z},$$

$$\operatorname{div} u = 0. \tag{6.61}$$

The Euler equations in cylindrical coordinates for the fluid motion in the pipe are obtained from system (6.61) by deleting the terms that are divided by the Reynold's number. If the velocity field u has the form given in equation (6.59), then u automatically satisfies the Euler boundary condition at the wall of the pipe. Thus, the Euler equations for this velocity field u and scaled pressure p reduce to the system

$$\frac{\partial p}{\partial r} = 0, \quad \frac{\partial p}{\partial \theta} = 0, \quad \frac{\partial u_z}{\partial t} + u_z \frac{\partial u_z}{\partial z} = -\frac{\partial p}{\partial z} = 0, \quad \frac{\partial u_z}{\partial z} = 0.$$

It follows that p must be a function of z only, and

$$\frac{\partial u_z}{\partial t} = -\frac{\partial p}{\partial z}. \tag{6.62}$$

If we now differentiate equation (6.62) with respect to z, then we see immediately that $\partial^2 p/\partial z^2 = 0$. Therefore, $p = p_0 + p_1 z$ for some constants p_0 and p_1, and we must also have that $u_z = -p_1 t + u_0$ for some constant u_0. Let us note that if we were to impose the no slip boundary conditions (which are not correct for Euler flow), then the only possible solution is $u_z = 0$ and $p = p_0$. In particular, we cannot impose a nonzero initial fluid velocity.

There are two cases for Euler flow (with the no penetration boundary condition): If $p_1 = 0$, then the pressure is constant in the pipe and the velocity field is constant. This is called *plug flow*. If $p_1 \neq 0$, then both the pressure and the velocity become unbounded as time passes to infinity. Both cases are not physically realistic. For example, the flow in the first case does not satisfy the experimentally observed fact that the velocity of the flow is larger in the center of the pipe than at its wall. Nonetheless, because of its mathematical simplicity, plug flow is often used as a model. For example, plug flow is often used to model flow in tubular reactors studied in chemical engineering.

What about Navier–Stokes flow?

If we consider the same pipe, the same coordinate system, and the same hypothesis about the direction of the velocity field, then the Navier–Stokes equations reduce to

$$\frac{\partial p}{\partial r} = 0, \quad \frac{\partial p}{\partial \theta} = 0, \quad \frac{\partial u_z}{\partial t} + u_z \frac{\partial u_z}{\partial z} = \frac{1}{\mathrm{Re}} \Delta u_z - \frac{\partial p}{\partial z} = 0, \quad \frac{\partial u_z}{\partial z} = 0,$$

with the no slip boundary condition at the wall of the pipe given by

$$u_z(a, \theta, z, t) \equiv 0.$$

This system of equations is already difficult to solve! Nevertheless, we can obtain a solution if we make two additional assumptions: The velocity field is in steady state and it is symmetric with respect to rotations about the central axis of the pipe. With these assumptions, if we take into account the equation $\partial u_z/\partial z = 0$, then it suffices to solve the single equation

$$\frac{1}{\mathrm{Re}} \left(\frac{1}{r} \frac{\partial}{\partial r} \left(r \frac{\partial u_z}{\partial r} \right) \right) = p_z.$$

Because $p_r = 0$ and $p_\theta = 0$, we have that p_z depends only on z while the left hand side of the last equation depends only on r. Thus, the functions on both sides of the equation must have the same constant value, say c. If this is the case, then $p = cz + p_0$.

The remaining ordinary differential equation

$$ru_z''(r) + u_z'(r) = (c\mathrm{Re})r$$

with the initial condition $u_z(a) = 0$ has the *continuous* solution

$$u_z(r) = \frac{1}{4}c\mathrm{Re}\,(r^2 - a^2).$$

Thus, we have derived the result that the steady state velocity field u predicted by the Navier–Stokes equations is parabolic with respect to the radial coordinate. This flow field is physically realistic, at least if the Reynold's number is sufficiently small; it is called *Poiseuille flow*.

Exercise 6.17. Consider Poiseuille flow in a section of length L of an infinite round pipe with radius a. If the pressure is p_{in} at the inlet of the section and the flow speed at the center of the pipe is v_{in}, then determine the pressure at the outlet. What happens in the limit as the Reynold's number grows without bound? Compare with the prediction of plug flow.

Using the vector identity

$$\frac{1}{2}\,\mathrm{grad}(u \cdot u) = u \times \mathrm{curl}\,u + u \cdot \nabla u$$

where $\cdot$ denotes the usual inner product on Euclidean space, let us rewrite Euler's equation in the form

$$u_t - u \times \mathrm{curl}\,u = \mathrm{grad}(-\frac{1}{2}(u \cdot u) - p).$$

With the definition $\alpha := -\frac{1}{2}|u|^2 - p$, we obtain Bernoulli's form of Euler's equations

$$u_t = u \times \mathrm{curl}\,u + \mathrm{grad}\,\alpha,$$
$$\mathrm{div}\,u = 0,$$
$$u \cdot \eta = 0 \quad \text{in } \partial\mathcal{R}. \tag{6.63}$$

Potential Flow

Let us consider an important specialization of Bernoulli's form of Euler's equations: potential flow in two space dimensions. The idea is the following. Assume that the velocity field u is the gradient of a potential f so that $u = \mathrm{grad}\,f$. Substitution into system (6.63), using the identity $\mathrm{curl}(\mathrm{grad}\,u) = 0$ and some rearrangement, gives the equations of motion

$$\mathrm{grad}(\frac{\partial f}{\partial t} + \frac{1}{2}|\,\mathrm{grad}\,f|^2 - p) = 0, \qquad \Delta f = 0. \tag{6.64}$$

As a result, we see immediately that the quantity

$$\frac{\partial f}{\partial t} + \frac{1}{2}|\operatorname{grad} f|^2 - p$$

is constant with respect to the space variables. In particular, if u is a steady state velocity field, then there is a constant c such that

$$p = c - \frac{1}{2}|u|^2; \tag{6.65}$$

that is, the pressure is a constant minus half the square of the velocity. This is *Bernoulli's law*.

In view of the second equation of system (6.64), the potential f is a harmonic function. Therefore, by considering a hypothetical flow on a two-dimensional plane with Cartesian coordinates (x, y) and velocity field $u = (\dot{x}, \dot{y})$, the potential f is *locally* the real part of a holomorphic function, say $h = f + i\psi$. Moreover, the pair f, ψ satisfies the Cauchy–Riemann equations

$$\frac{\partial f}{\partial x} = \frac{\partial \psi}{\partial y}, \qquad \frac{\partial f}{\partial y} = -\frac{\partial \psi}{\partial x}.$$

Thus, the assumption that $u = \operatorname{grad} f$ implies the fluid motions are solutions of an ordinary differential equation that can be viewed in two different ways: as the gradient system

$$\dot{x} = \frac{\partial f}{\partial x}, \qquad \dot{y} = \frac{\partial f}{\partial y};$$

or the Hamiltonian system

$$\dot{x} = \frac{\partial \psi}{\partial y}, \qquad \dot{y} = -\frac{\partial \psi}{\partial x}. \tag{6.66}$$

The function ψ, a Hamiltonian function for system (6.66), is called the *stream function*. The orbits of system (6.66), called *stream lines,* all lie on level sets of ψ. Let us also note that because the stream lines are orbits of a gradient system, there are no periodic fluid motions *in a region where the function h is defined.*

It should be clear that function theory can be used to study planar potential flow. For example, if ψ is a harmonic function defined in a simply connected region of the complex plane such that the boundary of the region is a level set of ψ, then ψ is the imaginary part of a holomorphic function defined in the region, and therefore ψ is the stream function of a steady state flow. This fact can be used to find steady state solutions of Euler's equations in many regions of the complex plane.

As an example, let us start with plug flow in a pipe with radius a and notice that every planar slice containing the axis of the pipe is invariant under the flow. In fact, if we view the strip

$$S := \{(x, y) : 0 < y < 2a\}$$

as such a slice where we have taken x as the axial direction, then the plug flow solution of Euler's equations in $\mathcal{S}$ is given by the velocity field $u = (0, c)$ and the pressure $p = p_0$ where c and p_0 are constants. This is a potential flow, with potential $f(x, y) = cx$, stream function $\psi(x, y) = cy$, and *complex potential* $h(x, y) = cz = c(x + iy)$.

Suppose that Q is an invertible holomorphic function defined on $\mathcal{S}$ and that $\mathcal{R}$ is the image of $\mathcal{S}$ under Q, then $w \mapsto h(Q^{-1}(w))$ for $w \in \mathcal{R}$ is a holomorphic function on $\mathcal{R}$. Moreover, by writing $h = f + i\psi$, it is easy to see that $w \mapsto \psi(Q^{-1}(w))$ is a stream function for a steady state potential flow in $\mathcal{R}$. In particular, stream lines of ψ map to stream lines of $w \mapsto \psi(Q^{-1}(w))$.

For example, let us note that $w := Q(z) = \sqrt{z}$ has a holomorphic branch defined on the strip $\mathcal{S}$ such that this holomorphic function maps $\mathcal{S}$ into the region in the first quadrant of the complex plane bounded above by the parabola $\{(\sigma, \tau) : \sigma\tau = a\}$. In fact, $Q^{-1}(w) = w^2$ so that

$$x = \sigma^2 - \tau^2, \qquad y = 2\sigma\tau.$$

The new "flow at a corner" has the complex potential $h(Q^{-1}(w)) = cw^2 = c(\sigma^2 - \tau^2 + 2i\sigma\tau)$. Thus, the velocity field is

$$u = (2c\sigma, -2c\tau).$$

The corresponding pressure is found from Bernoulli's equation (6.65). In fact, there is a constant p_1 such that

$$p = p_1 - 2c^2(\sigma^2 + \tau^2). \tag{6.67}$$

The stream lines for the flow at a corner are all parabolas.

The flow near a wall is essentially plug flow. In fact, if we consider, for example, the flow field on a vertical line orthogonal to the σ-axis, say the line with equation $\sigma = \sigma_0$, then the velocity field near the wall, where $\tau \approx 0$, is closely approximated by the constant vector field $(2c\sigma_0, 0)$. In other words, the velocity profile is nearly linear.

Exercise 6.18. Consider the plug flow vector field $u = (c, 0)$ defined in a horizontal strip in the upper half plane of width $2a$. Find the push forward of u into the first quadrant with respect to the map $Q(z) = \sqrt{z}$ with inverse $Q^{-1}(w) = w^2$. Is this vector field a solution of Euler's equations at the corner? Explain.

A Boundary Layer Problem

We have just seen that planar steady state Euler flow has stream lines that are (locally) orbits of both a Hamiltonian differential equation and a gradient differential equation. Moreover, in our example of flow at a corner, the velocity profile near the walls is linear. What about planar steady state Navier–Stokes flow?

Let us again consider the physical problem of flow at a corner (see [155, p. 222]). By physical reasoning, we might expect that the most prominent difference between Euler flow and Navier–Stokes flow at a corner is produced near the walls at the corner. The stream lines of the Euler flow are bent near the corner, but the velocity of the flow field does not approach zero at the walls—the fluid in the Euler model moves as if it had zero viscosity. For the Navier–Stokes flow, where the viscosity of the fluid is taken into account, the fluid velocity vanishes at the walls. On the other hand, the Navier–Stokes flow far away from the corner would be expected to be essentially the same as the Euler flow.

In our model, the fluid velocity field is assumed to be divergence free. Because we are working in *two* space dimensions, this assumption implies that there is a stream function; that is, the velocity field is Hamiltonian. In fact, if the planar coordinates at the corner are renamed to x, y and the velocity field u has components v, w so that the associated differential equation for the fluid motion is

$$\dot{x} = v(x,y), \qquad \dot{y} = w(x,y),$$

then the orbits of this system correspond to solutions of the exact first order differential equation $dy/dx = w/v$. Recall that the differential equation is exact if the corresponding differential one-form $w\,dx - v\,dy$ is closed; that is, if $\partial w/\partial y + \partial v/\partial x = 0$. Thus, there is a (locally defined) function $\psi(x, y)$ such that $\partial\psi/\partial x = -w$ and $\partial\psi/\partial y = v$; that is, ψ is a stream function for the flow. This result is proved in elementary courses in differential equations; it is also a special case of Poincaré's lemma: *If $n > 0$, then a closed form on a simply connected region of $\mathbb{R}^n$ is exact.*

Using the stream function for the Euler flow at the corner given by $(x, y) \mapsto 2cxy$ and some physical reasoning, we might guess that the stream function for the corresponding Navier–Stokes flow is given by $\psi(x,y) = xg(y)$ for some function g to be determined. Of course, we are free to assume our favorite form for this stream function. The problem is to show that there is a corresponding solution of the Navier–Stokes equations and to use this solution to predict the velocity profile for the flow near the corner.

For the stream function $\psi(x, y) = xg(y)$, the velocity field is

$$(v(x,y), w(x,y)) = (xg'(y), -g(y)). \tag{6.68}$$

Because the formula for the pressure for the Euler flow is given by equation (6.67) and the unknown function g depends only on the second space variable, let us postulate that the pressure for the Navier–Stokes flow is given by

$$p(x, y) = p_0 - 2c^2(x^2 + G(y)) \tag{6.69}$$

where p_0 is a constant, and G is a function to be determined.

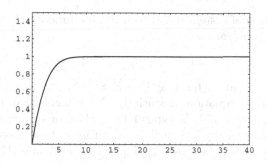

Figure 6.7: Plot of $f'(t/\epsilon)$ versus t for the solution of the Falkner-Skan boundary value problem (6.72) with $\epsilon = 1/10$.

The steady state Navier–Stokes equations are

$$v\frac{\partial v}{\partial x} + w\frac{\partial v}{\partial y} = \frac{1}{\text{Re}}\Delta v - \frac{\partial p}{\partial x},$$

$$v\frac{\partial w}{\partial x} + w\frac{\partial w}{\partial y} = \frac{1}{\text{Re}}\Delta w - \frac{\partial p}{\partial y},$$

$$\frac{\partial v}{\partial x} + \frac{\partial w}{\partial y} = 0 \tag{6.70}$$

with the boundary condition that the velocity field $(v(x,y), w(x,y))$ vanishes at the wall.

If the velocity field (6.68) and the pressure (6.69) are inserted into the Navier–Stokes equations (6.70), the system reduces to the equations

$$\frac{1}{\text{Re}}g''' + gg'' - (g')^2 + 4c^2 = 0, \qquad G' = \frac{1}{2c^2}(gg' + \frac{1}{\text{Re}}g'') \tag{6.71}$$

with the boundary conditions

$$g(0) = 0, \qquad g'(0) = 0.$$

We also have made the assumption that the velocity field (6.68) is the same as the Euler velocity field $(2cx, -2cy)$ far away from the wall. Ideally, we must have $2cx \approx xg'(y)$ for large y, that is,

$$\lim_{y\to\infty} g'(y) = 2c.$$

We will be able to solve for the pressure and thus construct the desired solution of the system (6.70) provided that there is a solution of the first equation of system (6.71) with the specified initial and boundary conditions. Let us rescale with $g := 2cf$ and define $\epsilon = 1/(c\text{Re})$ to reduce our quest for a solution of system (6.70) to finding a function f that solves the boundary value problem

$$\epsilon f''' + f f'' - (f')^2 + 1 = 0, \qquad f(0) = f'(0) = 0, \quad f'(\infty) = 1 \qquad (6.72)$$

for $\epsilon > 0$ a small parameter (see Exercises 1.11, 1.146, and 6.19). The ordinary differential equation, essentially the Falkner–Skan equation (see [62], [83], [116], and [137]), is typical of a class of equations that arise in "boundary layer theory," the origin of an important class of "singular perturbation problems," (see [132], [134], [170], [175], and [179]).

Exercise 6.19. A proof of the existence of a solution of the boundary value problem (6.72) is not trivial. This exercise outlines the main ingredients for a geometric proof. Recall Exercise 1.146 and recast the boundary value problem as follows: For $\epsilon > 0$ sufficiently small, determine the existence of a solution of the system

$$\dot{x} = y, \quad \dot{y} = z, \quad \epsilon \dot{z} = y^2 - xz - 1 \qquad (6.73)$$

such that $x(0) = 0$, $y(0) = 0$, and $\lim_{\tau \to \infty} y(\tau) = 1$. Equivalently, we may consider the fast-time system

$$\dot{x} = \epsilon y, \quad \dot{y} = \epsilon z, \quad \dot{z} = y^2 - xz - 1 \qquad (6.74)$$

and again seek a solution with $x(0) = 0$, $y(0) = 0$, and $\lim_{t \to \infty} y(t) = 1$. (a) For system (6.74) with $\epsilon = 0$, show that the plane $\{(x, y, z) : y = 1\}$ is invariant and the open ray $\{(x, y, z) : x > 0, y = 1, z = 0\}$ is invariant with basin of attraction $\{(x, y, z) : x > 0, y = 1\}$. (b) For $\epsilon > 0$ and small, the qualitative structure of part (a) persists. Show that the line $\{(x, y, z) : y = 1, z = 0\}$ is invariant for system (6.74) for all ϵ. (c) By part (a), the half-plane $\{(x, y, z) : x > 0, y = 1\}$ is the stable manifold of the open ray $\{(x, y, z) : x > 0, y = 1, z = 0\}$. This qualitative structure will persist for sufficiently small ϵ. Justify this statement. (d) This is the hard part. If $\epsilon > 0$, then the two-dimensional stable manifold for the open ray $\{(x, y, z) : x > 0, y = 1, z = 0\}$ intersects the plane $\{(x, y, z) : y = 0\}$ and the curve of intersection meets the line $\{(x, y, z) : x = 0, y = 0\}$ exactly once. Show that if this statement is true, then the original boundary value problem (6.73) has a unique solution. (e) Reproduce Figure (6.7). Hint: Set $x(0) = 0$ and $y(0) = 0$ in system (6.74) and use Newton's method to find a zero of the function $\zeta \to y(40, \zeta) - 1$, where $t \mapsto y(t, \zeta)$ is the second component of the solution of the system with the initial value $y(0, \zeta) = 0$. This numerical method is called *shooting*. Note: There is nothing special about the number 40; it is just a choice for a large experimental value of the fast time. (f) Support the claim in part (d) by numerical experiments. In particular, graph an approximation of the intersection curve of the stable manifold and the plane.

6.3.2 ABC Flows

The dynamics of a fluid that is predicted by Euler's equations (6.58) depend on the region that confines the flow and on the initial velocity field. In this section, we will study the fluid dynamics of an ideal family of steady state solutions that are periodic in the entire space relative to all three directions.

Let us seek a steady state solution u, a rest point of the infinite dimensional flow given by Euler's equations in Bernoulli's form (6.63), that is periodic in each space variable with period 2π. If there is such a steady state, then it exists on all of $\mathbb{R}^3$ so no additional boundary condition is necessary. In effect, the usual boundary condition for Euler's equations is replaced by the periodicity requirements. For this reason our requirements are called *periodic boundary conditions*. Also, if we like, we can view the solution as a vector field on the (compact) three-dimensional torus $\mathbb{T}^3$ defined by considering each of the Cartesian coordinates of $\mathbb{R}^3$ modulo 2π. We will consider the special class of steady states given as solutions of the system

$$u \times \operatorname{curl} u = 0, \qquad \operatorname{div} u = 0. \tag{6.75}$$

System (6.75) has many solutions, but certainly the most famous are the velocity fields of the form

$$u = (A \sin z + C \cos y, B \sin x + A \cos z, C \sin y + B \cos x)$$

where A, B, and C are constants. These vector fields generate the ABC flows (see [12], [16], [38], [44], [95], and [92]). The corresponding system of ordinary differential equations

$$\begin{aligned}
\dot{x} &= A \sin z + C \cos y, \\
\dot{y} &= B \sin x + A \cos z, \\
\dot{z} &= C \sin y + B \cos x
\end{aligned} \tag{6.76}$$

is a useful test example for the behavior of steady state Euler flow.

By rescaling the system and the time parameter, and by reordering the variables if necessary, all the interesting cases for different parameter values can be reduced to the consideration of parameters satisfying the inequalities $A = 1 \geq B \geq C \geq 0$. To obtain a perturbation problem, let us consider the system with $A = 1 > B = \beta > C = \epsilon$ where ϵ is a small parameter. Also, to simplify some formulas to follow, let us introduce a translation of the first variable $x \mapsto x + \pi/2$. The ABC system that we will study then has the form

$$\begin{aligned}
\dot{x} &= \sin z + \epsilon \cos y, \\
\dot{y} &= \beta \cos x + \cos z, \\
\dot{z} &= -\beta \sin x + \epsilon \sin y
\end{aligned} \tag{6.77}$$

where $0 < \beta < 1$, and $\epsilon \geq 0$ is a small parameter.

Note that the subsystem

$$\dot{x} = \sin z, \qquad \dot{z} = -\beta \sin x \tag{6.78}$$

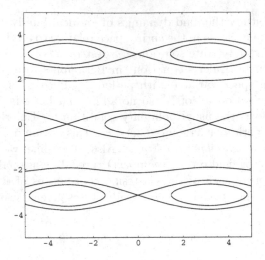

Figure 6.8: Computer generated phase portrait for system (6.78) with $\beta = 0.16$.

of system (6.77) is Hamiltonian with respect to the Hamiltonian function

$$\mathcal{H}(x, z) := \beta \cos x + \cos z. \tag{6.79}$$

Of course, the function $\mathcal{H}$ is constant on orbits of system (6.78).

A typical phase portrait for system (6.78) is depicted in Figure 6.8. Independent of the choice of β, there is a rest point at the origin surrounded by a period annulus $\mathcal{A}$ whose outer boundary consists of two hyperbolic saddle points with coordinates $(\pm \pi, 0)$ together with the heteroclinic orbits connecting these saddles (see Exercise 6.20). If we view the system on $\mathbb{T}^3$, then these saddle points coincide, and the boundary of the period annulus is just one saddle and a homoclinic orbit.

Exercise 6.20. Prove the statements made in this section about the phase portrait of system (6.78).

Each orbit Γ_h in $\mathcal{A}$ corresponds to a level set of $\mathcal{H}$ given by

$$\Gamma_h := \{(x, z) : \mathcal{H}(x, z) = h\}$$

for some h in the range $1 - \beta < h < 1 + \beta$. The boundary of the period annulus corresponds to the level set with $h = 1 - \beta$.

On each orbit in the closure of the period annulus $\mathcal{A}$ for the unperturbed system (6.78) we have that

$$\dot{y} = \beta \cos x + \cos z = h$$

for some "energy" $h > 0$. It follows that $\dot{y}$ is positive everywhere in an open neighborhood of the closure of $\mathcal{A}$. Let us therefore view y as a time-like variable for the perturbed system and consider the associated system

$$x' = \frac{\sin z}{\beta \cos x + \cos z} + \epsilon \frac{\cos y}{\beta \cos x + \cos z},$$

$$z' = \frac{-\beta \sin x}{\beta \cos x + \cos z} + \epsilon \frac{\sin y}{\beta \cos x + \cos z} \qquad (6.80)$$

where $'$ denotes differentiation with respect to y. Of course, if we find a solution

$$y \mapsto (x(y, \epsilon), z(y, \epsilon)) \qquad (6.81)$$

of system (6.80), then there are corresponding solutions

$$t \mapsto (x(y(t), \epsilon), y(t), z(y(t), \epsilon)) \qquad (6.82)$$

of system (6.77) obtained by solving the equation

$$\dot{y} = \beta \cos x(y, \epsilon) + \cos z(y, \epsilon). \qquad (6.83)$$

Let us notice that system (6.80) with $\epsilon = 0$ is the same as system (6.78) up to a reparametrization of the independent variable. Moreover, the unperturbed system (6.80) is a Hamiltonian system with respect to the Hamiltonian function (6.79). Finally, we have the following useful proposition: If $y \mapsto (x(y), z(y))$ is the solution of the unperturbed system (6.80) with the initial condition $x(0) = 0$, $z(0) = z_0$, then

$$-x(-y) = x(y), \qquad z(-y) = z(y), \qquad (6.84)$$

that is, x is odd and z is even. To see this, consider the new functions u and v defined by

$$(u(y), v(y)) = (-x(-y), z(-y))$$

and verify that the function $y \mapsto ((u(y), v(y))$ is a solution of the unperturbed system (6.80) with the initial condition $u(0) = 0$, $v(0) = z_0$.

6.3.3 Chaotic ABC Flows

The unperturbed system (6.80) has heteroclinic cycles. For example, a cycle is formed by the hyperbolic saddle points at $(x, z) = (\pm\pi, 0)$ and their connecting orbits. (Note that this cycle is also the boundary of a period annulus.) Or, if we view the system on the phase cylinder obtained by considering the variable x as an angle, then this cycle has only one saddle point and it is connected by two distinct homoclinic orbits. In this section we will see that for all but one value of the parameter β in the interval $0 < \beta < 1$, the Melnikov function along these heteroclinic orbits has simple zeros. Thus, for sufficiently small $\epsilon > 0$, system (6.80), and of course the corresponding original ABC system, has a chaotic invariant set. This result serves as an interesting application of our perturbation theory. It suggests that "real" fluids have chaotic motions.

Let us recall that the unperturbed heteroclinic orbits lie on the set

$$\{(x, z) : \cos z + \beta \cos x = 1 - \beta\}, \tag{6.85}$$

and let us consider the unperturbed solution $y \mapsto (x(y), z(y))$ starting at the point $(0, \arccos(1 - 2\beta))$. The Melnikov function is given (up to a nonzero scalar multiple) by

$$\mathcal{M}(\phi) = \frac{1}{(1 - \beta)^2} \int_{-\infty}^{\infty} (\sin(z(y + \phi)) \sin y + \beta \sin(x(y + \phi)) \cos y) \, dy. \tag{6.86}$$

This integral is easily transformed to the more useful representation

$$\mathcal{M}(\phi) = \frac{\sin \phi}{(1 - \beta)^2} \int_{-\infty}^{\infty} (\beta \sin x(s) \sin s - \sin z(s) \cos s) \, ds \tag{6.87}$$

by first changing the independent variable in the integral (6.86) to $s := y + \theta$ and then by using the sum formulas for sine and cosine together with the facts that the function $y \mapsto \sin x(y)$ is odd and $y \mapsto \sin z(y)$ is even. If, in addition, we apply integration by parts to obtain the formula

$$\int_{-\infty}^{\infty} \sin z(s) \cos s \, ds = \frac{\beta}{1 - \beta} \int_{-\infty}^{\infty} \cos z(s) \sin x(s) \sin s \, ds,$$

and substitute for $\cos z(s)$ from the energy relation in display (6.85), then we have the identity

$$\int_{-\infty}^{\infty} \sin z(s) \cos s \, ds = \int_{-\infty}^{\infty} \beta \sin x(s) \sin s \, ds$$
$$- \frac{\beta^2}{1 - \beta} \int_{-\infty}^{\infty} \sin x(s) \cos x(s) \sin s \, ds.$$

Finally, by substitution of this identity into equation (6.87), we obtain the following representation for the Melnikov function

$$\mathcal{M}(\phi) = \frac{\beta^2 \sin \phi}{(1 - \beta)^3} \int_{-\infty}^{\infty} \sin x(s) \cos x(s) \sin s \, ds. \tag{6.88}$$

Of course it is now obvious that the Melnikov function will have infinitely many simple zeros along the heteroclinic orbit provided that the integral

$$I_s := \int_{-\infty}^{\infty} \sin x(s) \cos x(s) \sin s \, ds.$$

does not vanish.

To determine if I_s is not zero, let us consider a method to evaluate this improper integral. The first step is to find explicit formulas for the unperturbed solution. Note that $z(y) > 0$ along the heteroclinic orbit. Integrate the unperturbed differential equation

$$\frac{x'(y)}{\sin z(y)} = \frac{1}{1-\beta}$$

on the interval $(0, y)$, and use the energy relation to obtain the equation

$$\frac{y}{1-\beta} = \int_0^{x(y)} \frac{1}{\sqrt{1 - ((1-\beta) - \beta \cos s)^2}} \, ds$$

$$= \frac{1}{\beta} \int_0^{x(y)} \frac{1}{\sqrt{(1 + \cos s)((2 - \beta(1 + \cos s))}} \, ds.$$

The form of the last integrand suggests the substitution $u = 1 + \cos s$, which transforms the integral so that the last equality becomes

$$\frac{y\sqrt{\beta}}{1-\beta} = -\int_2^{1+\cos x(\tau)} \frac{1}{u\sqrt{(2-u)(2-\beta u)}} \, du.$$

Using the indefinite integral

$$\int \frac{1}{u\sqrt{(2-u)(2-\beta u)}} \, du = -\frac{1}{2} \ln \left(\frac{4\sqrt{(2-u)(2-\beta u)} - 2(\beta+1)u + 8}{u} \right)$$

and a simple algebraic computation, we have the equality

$$\cos x(y) = -\frac{(\beta-1)e^{4cy} + 2(3-\beta)e^{2cy} + \beta - 1}{(\beta-1)e^{4cy} - 2(\beta+1)e^{2cy} + \beta - 1}$$

where

$$c := \frac{\sqrt{\beta}}{1-\beta}.$$

Also, by the trigonometric identity $\sin^2 x + \cos^2 x = 1$, we have that

$$\sin x(y) = -4\sqrt{1-\beta}\frac{e^{cy}(e^{cy} - 1)}{(\beta-1)e^{4cy} - 2(\beta+1)e^{2cy} + \beta - 1}.$$

Define

$$F(w) := -4(1-\beta)^{-3/2} \frac{w(w^2-1)((1-\beta)w^4 + 2(\beta-3)w^2 + 1 - \beta)}{(w^4 + 2\frac{1+\beta}{1-\beta}w^2 + 1)^2}$$

and note that

$$I_s = \int_{-\infty}^{\infty} F(e^{cy}) \sin y \, dy$$

$$= \int_{-\infty}^{\infty} F(e^{\zeta\sqrt{\beta}}) \sin((1-\beta)\zeta) \, d\zeta.$$

Also, note that the poles of the integrand of I_s correspond to the zeros of the denominator of F. To determine these zeros let us write the denominator in the factored form

$$w^4 + 2\frac{1+\beta}{1-\beta}w^2 + 1 = (w^2 - u_1)(w^2 - u_2)$$

where

$$u_1 := \frac{\sqrt{\beta}-1}{\sqrt{\beta}+1}, \qquad u_2 := \frac{\sqrt{\beta}+1}{\sqrt{\beta}-1}.$$

The poles corresponding to $e^{2\zeta\sqrt{\beta}} = u_1$ are

$$\zeta = \frac{1}{2\sqrt{\beta}}\left(\ln(-u_1) + \pi i + 2k\pi i\right), \qquad k \in \mathbb{Z},$$

where $\mathbb{Z}$ denotes the set of integers, and the poles corresponding to u_2 are

$$\zeta = \frac{1}{2\sqrt{\beta}}\left(\ln(-u_2) - \pi i + 2k\pi i\right), \qquad k \in \mathbb{Z}.$$

The locations of the poles suggest integration around the rectangle Γ in the complex plane whose vertices are T, $T + i\pi/\sqrt{\beta}$, $-T + i\pi/\sqrt{\beta}$, and $-T$. In fact, for sufficiently large $T > 0$, Γ encloses exactly two poles of the integrand, namely,

$$\zeta_1 := \frac{1}{2\sqrt{\beta}}\left(\ln(-u_1) + \pi i\right), \qquad \zeta_2 := \frac{1}{2\sqrt{\beta}}\left(\ln(-u_2) + \pi i\right).$$

The function F defined above is odd. It also has the following property: If $w \neq 0$, then $F(1/w) = -F(w)$. Using these facts, the identity $\sin\zeta = (e^{i\zeta} - e^{-i\zeta})/(2i)$, and a calculation, our integral can be recast in the form

$$I_s = -i\int_{-\infty}^{\infty} F(e^{\zeta\sqrt{\beta}}) e^{i(1-\beta)\zeta} \, d\zeta.$$

For notational convenience, define

$$\mathcal{K} := \mathcal{K}(\beta) = \frac{(1-\beta)\pi}{2\sqrt{\beta}},$$

and also consider the contour integral

$$\int_\Gamma F\left(e^{\zeta\sqrt{\beta}}\right)e^{i(1-\beta)\zeta}\,d\zeta.$$

The corresponding path integral along the upper edge of Γ is just $e^{-2\mathcal{K}}$ multiplied by the path integral along the lower edge. Also, by using the usual estimates for the absolute value of an integral, it is easy to see that the path integrals along the vertical edges of Γ approach zero as T increases without bound. Thus, the real improper integral I_s is given by

$$I_s = -i\left(1 + e^{-2\mathcal{K}}\right)^{-1}\int_\Gamma F\left(e^{\zeta\sqrt{\beta}}\right)e^{i(1-\beta)\zeta}\,d\zeta$$

$$= 2\pi\left(1 + e^{-2\mathcal{K}}\right)^{-1}\left(\mathrm{Res}(\zeta_1) + \mathrm{Res}(\zeta_2)\right) \tag{6.89}$$

where the residues are computed relative to the function G given by

$$G(\zeta) := F\left(e^{\zeta\sqrt{\beta}}\right)e^{i(1-\beta)\zeta}.$$

Define

$$F_1(w) := (w^2 - u_1)^2 F(w), \qquad F_2(w) := (w^2 - u_2)^2 F(w)$$

and compute the Laurent series of G at ζ_1 and ζ_2 to obtain the following residues:

$$\mathrm{Res}(\zeta_1) = \frac{e^{i(1-\beta)\zeta_1}}{4\beta u_1^2}\left(\sqrt{\beta}e^{\zeta_1\sqrt{\beta}}F_1'\left(e^{\zeta_1\sqrt{\beta}}\right) - \left(2\sqrt{\beta} - i(1-\beta)\right)F_1\left(e^{\zeta_1\sqrt{\beta}}\right)\right),$$

$$\mathrm{Res}(\zeta_2) = \frac{e^{i(1-\beta)\zeta_2}}{4\beta u_2^2}\left(\sqrt{\beta}e^{\zeta_2\sqrt{\beta}}F_2'\left(e^{\zeta_2\sqrt{\beta}}\right) - \left(2\sqrt{\beta} - i(1-\beta)\right)F_2\left(e^{\zeta_2\sqrt{\beta}}\right)\right).$$

To simplify the sum of the residues, let us define

$$\mathcal{A} := \cos\left(\frac{1-\beta}{2\sqrt{\beta}}\ln(-u_2)\right), \qquad \mathcal{B} := \sin\left(\frac{1-\beta}{2\sqrt{\beta}}\ln(-u_2)\right)$$

so that

$$e^{i(1-\beta)\zeta_1} = e^{-\mathcal{K}}(\mathcal{A} - \mathcal{B}i), \qquad e^{i(1-\beta)\zeta_2} = e^{-\mathcal{K}}(\mathcal{A} + \mathcal{B}i),$$

and let us note that since $u_1 = 1/u_2$, we have

$$e^{\zeta_1\sqrt{\beta}}e^{\zeta_2\sqrt{\beta}} = -1.$$

Also, note that the function F_1 is odd, F_1' is even, and verify the following identities:

$$F_1(1/w) = -\frac{1}{w^4 u_2^2}F_2(w),$$

$$F_1'(1/w) = \frac{1}{w^2 u_2^2}F_2'(w) - \frac{4}{w^3 u_2^2}F_2(w).$$

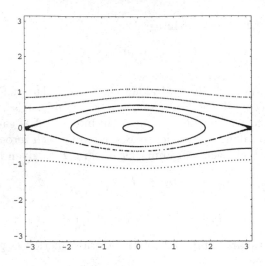

Figure 6.9: Some orbits of the stroboscopic Poincaré map for system (6.80) with $\epsilon = 0.01$ and $\beta = 0.1$.

Finally, for notational convenience, define $\mathcal{L} := \sqrt{-u_2}$.

Using the notation and the identities just mentioned, the residues are given by

$$\text{Res}(\zeta_1) = \frac{e^{-\kappa}}{4\beta u_2^2}\big((\mathcal{A} - \mathcal{B}i)(-i\mathcal{L}\sqrt{\beta}\,F_2'(i\mathcal{L}) + \big(2\sqrt{\beta} + i(1 - \beta)\big)F_2(i\mathcal{L})\big),$$

$$\text{Res}(\zeta_2) = \frac{e^{-\kappa}}{4\beta u_2^2}\big((\mathcal{A} + \mathcal{B}i)(i\mathcal{L}\sqrt{\beta}\,F_2'(i\mathcal{L}) - \big(2\sqrt{\beta} - i(1 - \beta)\big)F_2(i\mathcal{L})\big).$$

Thus, in view of formula (6.89), we have

$$I_s = \frac{\pi e^{-\kappa}}{\beta u_2^2(1 + e^{-2\kappa})}\big(\mathcal{B}\sqrt{\beta}\,\big(- 2iF_2(i\mathcal{L}) - \mathcal{L}F_2'(i\mathcal{L})\big) + \mathcal{A}(1 - \beta)iF_2(i\mathcal{L})\big).$$

$$(6.90)$$

The quantities

$$F_2'(i\mathcal{L}) = \frac{4(1 + \sqrt{\beta})^2(\beta + 2\sqrt{\beta}) - 1)}{\sqrt{1 - \beta}\,(1 - \sqrt{\beta})\beta^{3/2}},$$

$$-iF_2(i\mathcal{L}) = 4\Big(\frac{1 + \sqrt{\beta}}{1 - \sqrt{\beta}}\Big)^{1/2}\frac{(1 + \sqrt{\beta})^2}{\sqrt{1 - \beta}\,(1 - \sqrt{\beta})\beta}$$

are real and $-iF_2(i\mathcal{L})$ is nonzero for $0 < \beta < 1$. Also, if the identity

$$2 + \mathcal{L}\frac{F_2'(i\mathcal{L})}{iF_2(i\mathcal{L})} = \frac{1 - \beta}{\sqrt{\beta}}$$

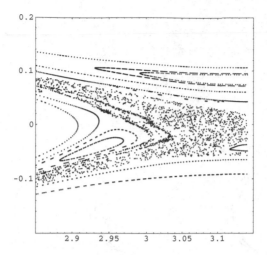

Figure 6.10: Blowup of Figure (6.9) near the unperturbed hyperbolic saddle point at $(x, z) = (\pi, 0)$. Several orbits are depicted.

is inserted into equation (6.90), then

$$I_s = \frac{\pi(1 - \beta)e^{-\mathcal{K}}}{\beta u_2^2(1 + e^{-2\mathcal{K}})}(-iF_2(i\mathcal{L}))(\mathcal{B} - \mathcal{A}).$$

Remark 2. The computation of the Melnikov function for the ABC flow given here follows the analysis in [38] where there are a few computational errors that are repaired in the analysis of this section. In particular, the final value of I_s reported in [38] is not correct.

Clearly, $I_s = 0$ if and only if $\mathcal{A} = \mathcal{B}$, or equivalently if

$$\tan\Big(\frac{1 - \beta}{2\sqrt{\beta}} \ln\Big(\frac{1 + \sqrt{\beta}}{1 - \sqrt{\beta}}\Big)\Big) = 1.$$

The last equation has exactly one root for β in the open unit interval: $\beta \approx 0.3$. Thus, except at this one parameter value, our computation proves that the perturbed stable and unstable manifolds intersect transversally, and, as a result, the corresponding perturbed flow is chaotic in a zone near these manifolds.

The results of a simple numerical experiment with the dynamics of system (6.80) are depicted in Figure 6.9 and Figure 6.10. These figures each depict several orbits of the stroboscopic Poincaré map—the independent variable y is viewed as an angular variable modulo 2π. Figure 6.10 is a blowup of a portion of Figure 6.9 near the vicinity of the unperturbed saddle point at $(x, z) = (\pi, 0)$. The results of this experiment suggest some of the fine structure in the stochastic layer that forms after breaking the heteroclinic orbits of the unperturbed Poincaré map. As predicted, the orbit structure appears to be very complex (see [16]).

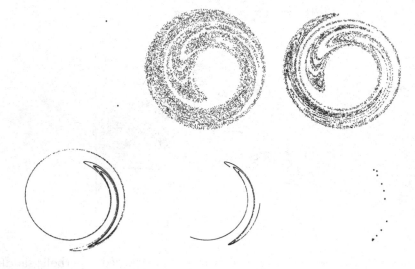

Figure 6.11: The figure depicts the attractor for the stroboscopic Poincaré map for the system $\ddot{\theta} + \mu\dot{\theta} + \sin\theta = -1/10 + 2\cos(2t)\sin\theta$ where, from left to right, $\mu = 0.03,\ 0.0301,\ 0.1,\ 0.5,\ 0.56,$ and 0.65.

Exercise 6.21. Reproduce Figures 6.9 and 6.10. The value $\epsilon = 0.01$ was used to obtain an easily reproducible picture. Note that our theory only predicts the existence of chaotic invariant sets for *sufficiently small* ϵ. Probably $\epsilon = 0.01$ is too big. Perform a series of numerical experiments to illustrate how the stochastic layer changes as ϵ changes for both smaller and larger values of ϵ.

Exercise 6.22. Discuss the statement: "The ABC system is conservative." Note that system (6.80) is a perturbed Hamiltonian system with no damping. The nature of chaotic invariant sets for dissipative systems can be quite different from the chaotic invariant sets for Hamiltonian systems. In particular, dissipative systems can have chaotic attractors. Roughly speaking, a chaotic attractor S is a compact invariant set with a dense orbit such that S contains the ω-limit set of every orbit in an open neighborhood of S. Although it is very difficult to prove the existence of chaotic attractors, numerical evidence for their existence is abundant. Consider the stroboscopic Poincaré map on the phase cylinder for the parametrically excited pendulum with damping and torque given by

$$\ddot{\theta} + \mu\dot{\theta} + \sin\theta = -\tau + a\cos(2t)\sin\theta.$$

Let the usual coordinates on the phase cylinder be (v, θ) where $v := \dot{\theta}$. It is convenient to render the graphics in a new coordinate system on the cylinder that flattens a portion of the cylinder into an annulus on the plane. For example, in Figure 6.11 iterates of the Poincaré map are plotted in the (x, y)-plane with

$$x = (2(4-v))^{1/2}\cos\theta, \qquad y = (2(4-v))^{1/2}\sin\theta$$

for the system

$$\ddot{\theta} + \mu\dot{\theta} + \sin\theta = -\frac{1}{10} + 2\cos(2t)\sin\theta$$

for six different values of μ. In each case a *single* orbit is depicted. The same picture is obtained independent of the initial value for the iterations as long as the first few iterations are not plotted. Thus, it appears that each depicted orbit is near an attractor. Reproduce Figure 6.11. Also, explore other regions of the parameter space of the oscillator by performing numerical experiments. To learn more about chaotic attractors, see, for example, [103], [194], and [233].

Periodic Orbits of ABC Flows

In the last section we proved that the ABC system has chaotic invariant sets for some choices of the parameters. If such a set exists as a consequence of the transversal intersection of stable and unstable manifolds near an unperturbed heteroclinic cycle, then it follows from a general theory (one which we have not presented here) that the chaotic invariant set contains infinitely many periodic orbits. But, this result does not tell us if any of the unperturbed periodic orbits in the various resonant tori are continuable. Although the rigorous determination of the continuable unperturbed periodic orbits seems to be a difficult problem which is not yet completely solved, we will use this problem as a vehicle to introduce some new techniques.

Before we begin the continuation analysis, let us note that if we find a continuable subharmonic of the unperturbed system (6.80), then there is a corresponding family of *periodic* solutions of system (6.77). To see this, let us suppose that the family of solutions (6.81) is a continuation of an unperturbed periodic orbit with period $2\pi m$ for some positive integer m, and let us consider the solutions of the equation (6.83) with the initial condition $y(0) = 0$. Because the family (6.81) at $\epsilon = 0$ is a periodic orbit of the unperturbed system (6.80), there is a number h such that

$$\beta\cos x(y,0) + \cos z(y,0) = h.$$

Thus, $t \mapsto ht$ is a solution of equation (6.83). Since this solution is complete, if ϵ is sufficiently small, then the solution $t \mapsto y(t,\epsilon)$ of system (6.83) such that $y(0,\epsilon) = 0$ exists at least on the interval

$$0 \leq t \leq \frac{2\pi m}{h}.$$

Moreover, there is a positive function $\epsilon \mapsto \eta(\epsilon)$ such that $\eta(0) = 2\pi m/h$ and

$$y(\eta(\epsilon),\epsilon) = 2\pi m.$$

The corresponding vector function

$$t \mapsto (x(y(t,\epsilon),\epsilon), y(t,\epsilon), z(y(t,\epsilon),\epsilon)) \tag{6.91}$$

is a solution of system (6.77). Moreover, we have, for example, the equations

$$x(y(\eta(\epsilon), \epsilon), \epsilon) = x(2\pi n, \epsilon) = x(0, \epsilon) = x(y(0, \epsilon), \epsilon);$$

that is, the function $s \mapsto x(y(s, \epsilon), \epsilon)$ is periodic with period $\eta(\epsilon)$. Of course, the same is true for the function $s \mapsto x(z(s, \epsilon), \epsilon)$, and it follows that, for each fixed small ϵ, the function (6.91) is a periodic solution of the ABC system.

As we have seen, the unperturbed system (6.80) has a period annulus $\mathcal{A}$ surrounding the origin whose boundary contains hyperbolic saddle points. These saddle points are fixed points of the stroboscopic Poincaré map that persist under perturbation (see Exercise 6.23). In the perturbed system their continuations are unstable periodic orbits, as are the corresponding periodic orbits of the ABC system. This fact is important for proving the hydrodynamic instability of the ABC systems (see [92]).

Exercise 6.23. Prove that the hyperbolic saddle points in the phase plane for the unperturbed system (6.80) viewed as periodic orbits in the corresponding phase cylinder persist as hyperbolic periodic orbits under perturbation in system (6.80) and that these perturbed periodic orbits are hyperbolic saddle type periodic orbits for the corresponding ABC system.

As the periodic orbits in $\mathcal{A}$ approach the outer boundary of this period annulus, the corresponding periods increase without bound. Therefore, the period annulus $\mathcal{A}$ is certainly not isochronous. We might expect this period annulus to be regular. But, because this is not always the case, the perturbation analysis for this problem is complicated. Note, however, that the continuation theory will apply if we can find a resonant unperturbed periodic orbit Γ for the system (6.80) such that the derivative of the associated period function does not vanish at Γ and simultaneously the associated subharmonic Melnikov function (5.67) has simple zeros.

As discussed in the last section, periodic orbits in $\mathcal{A}$ are in one-to-one correspondence with their energy h, with $1 - \beta < h < 1 + \beta$. Also, let us consider the corresponding period function $h \mapsto T(h)$ where $T(h)$ is the minimum period of the periodic orbit denoted $\Gamma(h)$. Because the perturbation terms are periodic with period 2π, the periodic orbit $\Gamma(h)$ is in $(m : n)$ resonance provided that

$$2\pi m = nT(h).$$

Let us fix h and assume for the moment that $T'(h) \neq 0$ so that the required regularity assumption is satisfied, and let us consider the solution $y \mapsto (x(y), z(y))$ of the unperturbed system with orbit $\Gamma(h)$ and initial condition $(x(0), y(0)) = (0, \arccos(h - \beta))$. Because the divergence of the unperturbed system vanishes, simple zeros of the function

$$M(\phi) = \frac{1}{h^2} \int_0^{2\pi m} (\sin(z(y + \phi)) \sin y + \beta \sin(x(y + \phi)) \cos y) \, dy$$

correspond to continuation points. By an argument similar to the one used to derive equation (6.88), it is easy to show that

$$M(\phi) = \frac{\beta^2}{h^3} \sin \phi \int_0^{2\pi m} \sin x(s) \cos x(s) \sin s \, ds.$$

Thus, using our continuation analysis, in particular Theorem 5.38, we have proved the following proposition.

Proposition 6.24. *Suppose that* $\Gamma(h)$ *is an* $(m : n)$ *resonant unperturbed periodic orbit of system (6.80) with energy* h *in a period annulus with period function* T *and* $s \mapsto (x(s), y(s))$ *is an unperturbed solution with orbit* $\Gamma(h)$. *If* $T'(h) \neq 0$ *and*

$$I(h) := \int_0^{2\pi m} \sin(2x(s)) \sin s \, ds \neq 0, \tag{6.92}$$

then there are $2m$ *continuation points on* $\Gamma(h)$.

To apply Proposition 6.24 to prove that there are in fact $2m$ continuation points on the orbit $\Gamma(h)$ we must show that $I(h) \neq 0$ and $T'(h) \neq 0$. But, even if we cannot do this rigorously, our analysis is still valuable. For example, if we fix β, then we can use numerical approximations to graph the functions I and T as an indication of the validity of the requirements. This is probably a more reliable method than a direct search for the periodic solutions by numerical integration of the perturbed differential equations.

There is no simple argument to show that $I(h) \neq 0$. In fact, for most resonances, $I(h)$ vanishes and our first order method fails. A precise statement is the content of the next proposition.

Proposition 6.25. *If* $n \neq 1$, *then* $I(h) = 0$.

Proof. To prove the proposition, use the periodicity of the integrand to recast the integral as

$$I(h) = \int_{-\pi m}^{\pi m} \sin(2x(s)) \sin s \, ds.$$

Then, by the change of variables $s = m\sigma$ and the resonance relation we have that

$$I(h) = m \int_{-\pi}^{\pi} \sin \left(2x \left(\frac{nT(h)}{2\pi} \sigma \right) \right) \sin m\sigma \, ds.$$

The function

$$t \mapsto \sin\left(2x\left(\frac{T(h)}{2\pi}t\right)\right)$$

is odd and 2π-periodic. Thus, it can be represented by a (convergent) Fourier sine series, say

$$\sum_{\nu=1}^{\infty} b_\nu(h) \sin \nu t.$$

If this series is evaluated at $t = n\sigma$ and inserted into the integral, all but one of the summands vanish. The exceptional term is

$$b_\nu(h) \int_{-\pi}^{\pi} \sin n\nu\sigma \sin m\sigma \, d\sigma$$

with $n\nu = m$. Since m and n are relatively prime, this term can only be nonzero if $n = 1$ and $\nu = m$, as required. Moreover, $I(h) \neq 0$ if and only if the Fourier coefficient $b_\nu(h) \neq 0$.   □

Exercise 6.26. Prove: If $t \mapsto y(t)$ is an odd periodic function with period $2\pi/\omega$ and $2\pi n/\omega = 2\pi m/\Omega$ for relatively prime integers m and n with $n > 1$, then

$$\int_0^{2\pi n/\omega} y(t) \sin \Omega t \, dt = 0.$$

Although an antiderivative for the integrand of $I(h)$ at an $(m : 1)$ resonance cannot be expressed in elementary functions, this integral can be evaluated using Jacobi elliptic functions. We will indicate the procedure for doing this below. Unfortunately, the resulting value seems to be too complex to yield a simple statement of precisely which of the $(m : 1)$ resonances are excited at first order. Therefore we will not give the full derivation here. Instead, we will use this problem to introduce the Jacobi elliptic functions and the Picard–Fuchs equation for the period function. For a partial result on the existence of continuable periodic orbits see [38] and Exercise 6.27. In fact, most of the $(m : 1)$ resonances are excited.

Exercise 6.27. This exercise is a research project. For which $(m : 1)$ resonances of system (6.80) is the integral (6.92) not zero?

Let us now glimpse into the wonderful world of elliptic integrals, a gem of 19th century mathematics that remains a very useful tool in both modern pure and applied mathematics (see [30] and [234]). Perhaps the best way to approach the subject of special functions is to view it in analogy with trigonometry. The trigonometric functions are so familiar that we tend not to notice how they are used. Often, we operate with these functions simply by using their properties—periodicity and trigonometric identities. We do not consider their values, except at a few special values of their arguments. The complete elliptic integrals and the Jacobi elliptic functions that we will mention below can be treated in the same way. Of course, it is clear why the trigonometric functions show up so often: Circles appear everywhere in mathematics! The reason why elliptic functions show up so often is deeper; perhaps after more familiarity with the subject this reason will become apparent.

What are elliptic functions? For $0 \leq \phi \leq \pi/2$ and $0 \leq k \leq 1$, define

$$u := u(\phi, k) = \int_0^\phi \frac{1}{\sqrt{1 - k^2 \sin^2 \theta}} \, d\theta.$$

The Jacobi elliptic functions are functions of two variables defined as follows:

$$\mathrm{sn}(u, k) := \sin \phi, \quad \mathrm{cn}(u, k) := \cos \phi, \quad \mathrm{dn}(u, k) := \sqrt{1 - k^2 \, \mathrm{sn}^2(u, k)}$$

where the argument k it is called the *elliptic modulus*. The complete elliptic integrals of the first and second kinds are defined, respectively, by

$$K(k) := \int_0^{\pi/2} \frac{1}{\sqrt{1 - k^2 \sin^2 \theta}} \, d\theta, \qquad E(k) := \int_0^{\pi/2} \sqrt{1 - k^2 \sin^2 \theta} \, d\theta.$$

The domain of the Jacobi elliptic functions can be extended to the entire complex plane where each of these functions is "doubly periodic"; for example, sn has the periods $4K(k)$ and $2iK(\sqrt{1 - k^2})$, and cn has the periods $4K(k)$ and $2K(k) + 2iK(\sqrt{1 - k^2})$. In fact, more generally, a doubly periodic meromorphic function for which the ratio of its periods is not real is called an *elliptic function*. By the definitions of the Jacobi elliptic functions, we have the identities

$$\mathrm{sn}^2(u, k) + \mathrm{cn}^2(u, k) = 1, \qquad \mathrm{dn}^2(u, k) + k^2 \, \mathrm{sn}^2(u, k) = 1.$$

These are just two simple examples of the many relations and identities that are known.

Exercise 6.28. Consider the solution $t \mapsto (x(t), y(t))$ of the system of differential equations $\dot{x} = -y$, $\dot{y} = x$ with the initial condition $x(0) = 1$ and $y(0) = 0$, and *define* the sine and cosine functions by

$$(x(t), y(t)) = (\cos t, \sin t).$$

Prove the basic trigonometric identities and periodicity properties of the sine and cosine using this definition. Also, prove that

$$\theta = \int_0^{\sin\theta} \frac{1}{\sqrt{1-s^2}}\,ds.$$

Suppose that $0 < k < 1$ and consider the solution of the system of differential equations

$$\dot{x} = yz, \qquad \dot{y} = -xz, \qquad \dot{z} = -k^2 xy$$

with initial condition $(x(0), y(0), z(0)) = (0, 1, 1)$. Show that this solution is given by

$$(x(t), y(t), z(t)) = (\operatorname{sn}(t, k), \operatorname{cn}(t, k), \operatorname{dn}(t, k)).$$

If this solution is taken as the definition of the Jacobi elliptic functions, then it is possible to derive many of the most important properties of these functions without too much difficulty (see [22, p. 137]).

Exercise 6.29. Consider the pendulum model given by $\ddot{\theta} + \lambda \sin\theta = 0$, define the phase plane in the usual manner by defining a new variable $v := \dot{\theta}$, and note that there is a center at the origin of the phase plane. The period function for the corresponding period annulus is not constant. Fill in the details of the following derivation of a formula for this period function.

If the periodic orbit meets the θ-axis at $\theta = \theta_0$, then the energy surface corresponding to the periodic orbit is the graph of the relation

$$v^2 = 2\lambda(\cos\theta - \cos\theta_0).$$

Note that $d\theta/dt = v$ and consider the symmetries of the periodic orbit to deduce that the period T of the orbit is given by

$$T = \frac{4}{\sqrt{2\lambda}} \int_0^{\theta_0} \frac{1}{\sqrt{\cos\theta - \cos\theta_0}}\,d\theta.$$

Use the identity $\cos\theta = 1 - 2\sin^2(\theta/2)$ to rewrite both of the terms $\cos\theta$ and $\cos\theta_0$ in the integrand, and then change variables in the integral using

$$\sin\phi = \frac{\sin(\theta/2)}{\sin(\theta_0/2)}$$

to obtain the formula

$$T = \frac{4}{\sqrt{\lambda}} \int_0^{\pi/2} \frac{1}{\sqrt{1 - k^2 \sin^2\phi}}\,d\phi = \frac{4}{\sqrt{\lambda}} K(k), \qquad k = \sin(\theta_0/2).$$

Show that the limit of the period function as the periodic orbits approach the origin is $T(0) := 2\pi/\sqrt{\lambda}$ and that the period function grows without bound as the periodic orbits approach the outer boundary of the period annulus. Suppose that the bob of a physical pendulum is pulled out $15°$, $30°$, or $90°$ from the downward vertical position and released from rest. Approximate the periods of the corresponding periodic motions using a numerical integration or a careful analysis of the series expansion of K in powers of k. What percent error is made if these periods are approximated by $T(0)$? (Galileo is said to have deduced that the period of the librational motion of a pendulum does not depend on its amplitude. He made this deduction while sitting in a cathedral and observing a chandelier swinging in the breeze blowing through an open window. Discuss his theory in light of your approximations.)

How do the elliptic functions arise for the ABC flows? To answer this question, let us consider the solution $y \mapsto (x(y), z(y))$ of the unperturbed system (6.80) defined above with the initial condition

$$x(0) = 0, \qquad z(0) = \arccos((h - \beta))$$

and note that the corresponding orbit $\Gamma(h)$ meets the positive x-axis at the point with coordinates

$$(\arccos((h - 1)/\beta), 0).$$

The first equation of the unperturbed system (6.80) can be rewritten in the form

$$\frac{1}{\sin z(y)} x'(y) = \frac{1}{h}.$$

If we restrict attention for the moment to the portion of $\Gamma(h)$ in the first quadrant with $y > 0$, then after integration, we have the identity

$$\frac{y}{h} = \int_0^y \frac{x'(\tau)}{\sin z(\tau)} \, d\tau.$$

If we apply the change of variables $s = x(\tau)$ followed by $t = -\cos s$ and rearrange the integrand, then we have the identity

$$\frac{\beta y}{h} = \int_c^{-\cos x(y)} \frac{1}{\sqrt{a - t} \sqrt{b - t} \sqrt{t - c} \sqrt{t - d}} \, dt$$

where

$$a := 1, \qquad b := \frac{1 - h}{\beta}, \qquad c := -1, \qquad d := -\frac{1 + h}{\beta}$$

and $a > b \geq -\cos x(y) > c > d$. This integral can be evaluated using the Jacobi elliptic functions (see [30, p. 112]) to obtain

$$\frac{\beta y}{h} = \sqrt{\beta} \, \mathrm{sn}^{-1}(\sin \phi, k)$$

where

$$k^2 = \frac{(1 + \beta)^2 - h^2}{4\beta}, \qquad \sin \phi = \left(\frac{2\beta(1 - \cos x(y))}{(1 - h + \beta)(1 + h - \beta \cos x(y))} \right)^{1/2}.$$

It follows that

$$\cos x(y) = \frac{1 - \mathcal{A}^2 \operatorname{sn}^2(\beta y/h, k)}{1 - \mathcal{B}^2 \operatorname{sn}^2(\beta y/h, k)}$$

with

$$\mathcal{A}^2 := \frac{(1 - h + \beta)(1 + h)}{2\beta}, \qquad \mathcal{B}^2 := \frac{1 - h + \beta}{2},$$

and, using the trigonometric identity $\sin^2 \theta + \cos^2 \theta = 1$, we also have

$$\sin x(y) = \sqrt{2} \sqrt{\mathcal{A}^2 - \mathcal{B}^2} \, \frac{\operatorname{sn}(\beta y/h, k) \operatorname{dn}(\beta y/h, k)}{1 - \mathcal{B}^2 \operatorname{sn}^2(\beta y/h, k)}.$$

Moreover, it is easy to see that the solution formulas for $\sin x(y)$ and $\cos x(y)$ are valid for all y.

The function sn has real period $4K$; and therefore, the period of $\Gamma(h)$ is given by

$$T = \frac{4h}{\sqrt{\beta}} K(k(h)) = 8\sqrt{C^2 - k^2} \, K(k)$$

where

$$C^2 = \frac{(1 + \beta)^2}{4\beta}.$$

Because $dh/dk < 0$, the critical points of T are in one-to-one correspondence with the critical points of the period function viewed as a function of the elliptic modulus k.

There is a beautiful approach to the study of the monotonicity properties of T that is based on the following observation: The derivatives of the complete elliptic integrals E and K can be expressed as linear combinations (with function coefficients) of the same complete elliptic integrals. In fact, we have

$$E'(k) = \frac{E(k) - K(k)}{k}, \qquad K'(k) = \frac{E(k) - (1 - k^2)K(k)}{k(1 - k^2)}.$$

Of course, this means that K'' and E'' can also be expressed in the same manner. As a result, the *three* expressions for $T(k)$, $T'(k)$, and $T''(k)$ are all linear combinations of the *two* functions $E(k)$ and $K(k)$. Thus, T, T', and T'' must be linearly dependent; that is, T satisfies a second order differential equation. In fact, T satisfies the Picard–Fuchs equation

$$\frac{C^2 - k^2}{k(1 - k^2)} T'' + \frac{k^4 + (1 - 3C^2)k^2 + C^2}{k^2(1 - k^2)^2} T'$$

$$+ \frac{(1 - 2C^2)k^2 + C^2(2 - C^2)}{k(C^2 - k^2)(1 - k^2)^2} T = 0.$$

The function T is positive, $0 < k < 1$, and $1 < C^2 < \infty$. By the Picard–Fuchs equation, if $T'(k) = 0$, then the sign of $T''(k)$ is the same as the sign of the expression

$$\mathcal{C} := C^2(C^2 - 2) + (2C^2 - 1)k^2.$$

We also have the Taylor series expansion

$$T(k) = 4\pi C + \frac{\pi(C^2 - 2)}{C}k^2 + O(k^4).$$

These facts are the key ingredients required to prove the following two propositions: 1) *If $C^2 > 2$, then T has no critical points.* 2) *T has at most two critical points.* The proofs are left as exercises.

Exercise 6.30. (a) Prove: If f and g are two functions such that f', f'', g', and g'' are all linear combinations (with function coefficients) of f and g, then every linear combination T of f and g is a solution of a homogeneous second order ODE. (b) Find a second order ODE satisfied by the function $x \mapsto a\sin x + b\cos x$ where a and b are constants. Prove that this function does not have a positive relative minimum. (c) Repeat part (b) for the function $x \mapsto x\sin x + x\cos x$. (d) Repeat part (b) for the function $x \mapsto J_\nu(x) + J'_\nu(x)$ where J_ν is the Bessel function of the first kind of order ν and a is a constant. Hint: Recall Bessel's equation $x^2 J''_\nu + x J'_\nu + (x^2 - \nu^2)J_\nu = 0$. (e) Formulate a general theorem that uses properties of the coefficients of the second order ODE satisfied by T and the asymptotics of T at the origin to imply that T is a monotone function. Apply your result to prove that the function $T : (0,1) \to \mathbb{R}$ given by $T(k) = 2E(k) - (2 - k^2)K(k)$ is negative and monotone decreasing (see [42, page 290]).

7
Averaging

This chapter is an introduction to the method of averaging—a far-reaching and rich mathematical subject that has many important applications. Our approach to the subject is through perturbation theory; for example, we will discuss the existence of periodic orbits for periodically forced oscillators. In addition, some ideas will be introduced that have implications beyond the scope of this book.

We have already discussed in Section 3.2 applications of the method of averaging to various perturbations of a Keplerian binary. While an understanding of these applications is not required as background for the mathematical theory in this chapter, a review of the Keplerian perturbation problem in celestial mechanics is highly recommended as a wonderful way to gain an appreciation for the subject at hand. For further study there are many excellent mathematical treatments of the theory and applications of the method of averaging (see, for example, [12], [14], [106], [146], [170], [201] and [103], [134], [145], [175], [233]).

7.1 The Averaging Principle

Let us consider a family of differential equations given by

$$\dot{u} = f(u) + \epsilon g(u, t, \epsilon), \qquad u \in \mathbb{R}^n \tag{7.1}$$

where the perturbation term is periodic in time with period $\eta > 0$. Also, let us suppose that the unperturbed system

$$\dot{u} = f(u), \qquad u \in \mathbb{R}^n \tag{7.2}$$

is completely integrable (see Exercise 5.60). In this case, a theorem of Joseph Liouville states that the phase space for the unperturbed system is foliated by invariant tori (see [12]). We will not prove this result; instead, we will simply assume that a region of the phase space of our unperturbed system is foliated by invariant tori. In the planar case, this is exactly the assumption that the unperturbed system has a period annulus.

The method of averaging is applied after our system is transformed to the standard form

$$\dot{I} = \epsilon F(I, \theta), \qquad \dot{\theta} = \omega(I) + \epsilon G(I, \theta).$$

The new coordinates (I, θ), called action-angle variables, are always defined under our integrability assumption on the unperturbed system. To illustrate this result, we will construct the action-angle variables for the harmonic oscillator and outline the construction for a general planar Hamiltonian system; the Delaunay elements, defined for the Kepler problem in Section 3.2, provide a more substantial example.

The harmonic oscillator is the Hamiltonian system

$$\dot{q} = p, \qquad \dot{p} = -\omega^2 q, \tag{7.3}$$

with Hamiltonian $H : \mathbb{R} \times \mathbb{R} \to \mathbb{R}$, which represents the total energy, given by

$$H(q, p) := \frac{1}{2}p^2 + \frac{1}{2}\omega^2 q^2.$$

For this system, the entire punctured plane is a period annulus, the periodic orbits correspond to regular level sets of H, and each level set is an ellipse. Similarly, if a general one-degree-of-freedom Hamiltonian system

$$\dot{q} = \frac{\partial H}{\partial p}, \qquad \dot{p} = -\frac{\partial H}{\partial q} \tag{7.4}$$

with Hamiltonian $H : \mathbb{R} \times \mathbb{R} \to \mathbb{R}$ has a period annulus $\mathcal{A}$, then each periodic orbit in $\mathcal{A}$ is a subset of a regular level set of H.

In case $\mathcal{A}$ is a period annulus for the Hamiltonian system (7.4), let $\mathcal{O}(q_0, p_0)$ denote the periodic orbit that passes through the point $(q_0, p_0) \in \mathcal{A}$, and note that $\mathcal{O}(q_0, p_0)$ is a subset of the regular energy surface

$$\{(q, p) \in \mathbb{R}^2 : H(q, p) = H(q_0, p_0)\}.$$

The function $I : \mathcal{A} \to \mathbb{R}$ defined by

$$I(q_0, p_0) := \frac{1}{2\pi} \int_{\mathcal{O}(q_0, p_0)} p \, dq \tag{7.5}$$

is called the *action variable* for the Hamiltonian system on the period annulus. Its value at (q_0, p_0) is the normalized area of the region in the phase space enclosed by the periodic orbit $\mathcal{O}(q_0, p_0)$. (The action variable should not be confused with the action integral of a mechanical system, which is the integral of its Lagrangian over a motion.)

For the harmonic oscillator, the action variable at $(q_0, p_0) \neq (0,0)$ is $1/(2\pi)$ multiplied by the area enclosed by the ellipse $\mathcal{O}(q_0, p_0)$ with equation

$$\frac{1}{2}p^2 + \frac{1}{2}\omega^2 q^2 = \frac{1}{2}a^2$$

where $a := (p_0^2 + \omega^2 q_0^2)^{1/2}$. The area of the ellipse (π times the product of the lengths of its semimajor and semiminor axes) is a^2/ω. Therefore, the action variable for the harmonic oscillator is proportional to its Hamiltonian; in fact,

$$I(q, p) = \frac{a^2}{2\omega} = \frac{1}{\omega}H(q, p).$$

Since the Hamiltonian is constant on orbits, the action variable is a first integral, that is, $\dot{I} = 0$.

For our planar Hamiltonian system (7.4), let Σ be a Poincaré section in the period annulus $\mathcal{A}$, and let $T : \mathcal{A} \to \mathbb{R}$ denote the period function; it assigns to $(q, p) \in \mathcal{A}$ the period of the periodic orbit $\mathcal{O}(q, p)$. Also, let $t \mapsto (Q(t), P(t))$ denote the solution corresponding to $\mathcal{O}(q, p)$ that has its initial value on Σ. Since, by the definition of the action variable,

$$I(q, p) = \frac{1}{2\pi} \int_0^{T(q,p)} P(s)\dot{Q}(s)\, ds,$$

the function $t \mapsto I(q(t), p(t))$ is constant for every solution $t \mapsto (q(t), p(t))$ on $\mathcal{O}(q, p)$; that is, $\dot{I} = 0$.

To define the angle variable, let us first define the time map $\tau : \mathcal{A} \to \mathbb{R}$; it assigns to each point $(q, p) \in \mathcal{A}$ the minimum positive nonnegative time required to reach (q, p) along the solution of the system that starts at the intersection point of the orbit $\mathcal{O}(q, p)$ and the section Σ. With this notation, the angular variable θ is defined by

$$\theta(q, p) := \frac{2\pi}{T(q, p)}\tau(q, p). \tag{7.6}$$

For the harmonic oscillator, every periodic orbit has the same period $2\pi/\omega$. Moreover, if we take the section Σ to be the positive q-axis, then

$$q = \frac{a}{\omega}\cos(\omega\tau(q, p)), \quad p = -a\sin(\omega\tau(q, p)).$$

Since $\dot{q} = p$, we have that

$$-a\sin(\omega\tau(q, p))\frac{d}{dt}(\tau(q, p)) = -a\sin(\omega\tau(q, p)),$$

and therefore, $\dot{\tau} = 1$. Hence, the angular variable satisfies the differential equation $\dot{\theta} = \omega$.

In the general case, the frequency of the periodic orbit may be a nonconstant function of the action variable, and the differential equation for the angular variable has the form $\dot{\theta} = \omega(I)$. To prove this fact, note first that if $t \mapsto (q(t), p(t))$ is a solution, then $t \mapsto T(q(t), p(t))$ is a constant function whose value is the period of the corresponding periodic orbit. For notational convenience, let $t \mapsto (Q(t), P(t))$ denote the solution corresponding to the same periodic orbit that has its initial value on the section Σ. Using the definition of τ, we have the identities

$$Q(\tau(q(t), p(t))) = q(t), \quad P(\tau(q(t), p(t))) = p(t).$$

In particular, $t \mapsto (Q(\tau(q(t), p(t))), P(\tau(q(t), p(t))))$ is a solution of the original Hamiltonian system. More generally, suppose that the function $t \mapsto u(t)$ is a complete solution of the differential equation $\dot{u} = f(u)$ and $t \mapsto u(\gamma(t))$ is also a solution for some function $\gamma : \mathbb{R} \to \mathbb{R}$. Clearly $t \mapsto u(t + \gamma(0))$ is a solution of the differential equation with the same initial condition as $t \mapsto u(\gamma(t))$. Hence, $u(\gamma(t)) = u(t + \gamma(0))$ for all $t \in \mathbb{R}$; and, by differentiating both sides of this identity with respect to t, we have that

$$f(u(\gamma(t)))\dot{\gamma}(t) = f(u(t + \gamma(0))).$$

Since $u(\gamma(t)) = u(t + \gamma(0))$ and f does not vanish along the orbit, we have the equation $\dot{\gamma} = 1$. By applying this result to the Hamiltonian system, it follows that $\dot{\tau} = 1$, and therefore, $\dot{\theta} = 2\pi/T(q, p)$. Clearly, T is constant on each orbit $\mathcal{O}(q, p)$. Since the action variable increases as the areas bounded by the periodic orbits in the period annulus increase, the frequency $\omega(q, p) := 2\pi/T(q, p)$ can be viewed as a function of the action variable and $\dot{\theta} = \omega(I)$, as required.

In Section 7.3, we will prove that the function

$$(q, p) \mapsto (I(q, p), \theta(q, p))$$

defines a polar coordinate chart on an annular subset of $\mathcal{A}$. It follows that the change to action-angle variables is nonsingular, and the Hamiltonian system in action-angle variables, that is, the system

$$\dot{I} = 0, \qquad \dot{\theta} = \omega(I),$$

can be viewed as a system of differential equations on the phase cylinder, which is the product of a line with coordinate I and a one-dimensional torus with coordinate θ (see Section 1.8.5).

More generally, a multidimensional integrable system has an invariant manifold that is topologically the cross product of a Cartesian space $\mathbb{R}^M$ and a torus $\mathbb{T}^N$. In this case, action-angle variables I and θ can be defined in $\mathbb{R}^M \times \mathbb{T}^N$ such that the integrable system is given by

$$\dot{I} = 0, \qquad \dot{\theta} = \omega(I) \qquad\qquad (7.7)$$

where $I \in \mathbb{R}^M$ and $\theta \in \mathbb{T}^N$ are vector variables. This is the standard form for an integrable system, the starting point for classical perturbation theory.

The method of averaging is a powerful tool that is used to obtain and analyze approximate solutions for perturbations of integrable systems, that is, for systems of differential equations of the form

$$\dot{I} = \epsilon F(I, \theta), \qquad \dot{\theta} = \omega(I) + \epsilon G(I, \theta) \tag{7.8}$$

where θ is a vector of angular variables defined modulo 2π, the functions $\theta \mapsto F(I, \theta)$ and $\theta \mapsto G(I, \theta)$ are 2π-periodic, and $|\epsilon|$ is considered to be small. Poincaré called the analysis of system (7.8) "the fundamental problem of dynamical systems."

In physical applications, mathematical models are rarely formulated directly in action-angle variables. As illustrated by the analysis of the perturbed Kepler problem in Section 3.2, the transformation to action-angle variables can be a formidable task. On the other hand, the benefits of working with a system in the standard form (7.8) often justify the effort to obtain the coordinate transformation.

An immediate benefit derived from the transformation to action-angle variables is the simplicity of the geometry of the unperturbed dynamics of system (7.7). In fact, the solution with the initial condition $(I, \theta) = (I_0, \theta_0)$ is given by $I(t) \equiv I_0$ and $\theta(t) = \omega(I_0)t + \theta_0$. Note that the action variables specify a torus in the phase space, and the angle variables evolve linearly on this torus.

Definition 7.1. Suppose that I_0 is in $\mathbb{R}^M$. The N-dimensional invariant torus

$$\{(I, \theta) \in \mathbb{R}^M \times \mathbb{T}^N : I = I_0\}$$

for the system (7.7) is *resonant* if there is a nonzero integer vector K of length N such that $\langle K, \omega(I_0) \rangle = 0$ where $\langle \ \rangle$ denotes the usual inner product. In this case we also say that the frequencies, the components of the vector $\omega(I_0)$, are in resonance.

If an invariant torus for the system (7.7) is not resonant, then every orbit on the torus is dense. In case $N = 2$, every orbit on a resonant torus is periodic. Matters are not quite so simple for $N > 2$ where the existence of a resonance relation does not necessarily mean that all orbits on the corresponding invariant torus are periodic. This is just one indication that the dynamics of systems with more than two frequencies is in general quite different from the dynamics of systems with one frequency. But, in all cases, the existence of resonant tori plays a central role in the analysis of the perturbed dynamical system.

Some aspects of the near resonance behavior of the planar case of system (7.1) are discussed in detail in Chapter 5, especially the continuation theory for resonant unperturbed periodic solutions. As we have seen, this special case, and the general multidimensional time-periodic system (7.1) can be viewed as systems on a phase cylinder by the introduction of a new angular variable so that the extended system is given by

$$\dot{u} = f(u) + \epsilon g(u, \tau, \epsilon), \qquad \dot{\tau} = 1. \tag{7.9}$$

If $u \in \mathbb{R}^2$ and the system $\dot{u} = f(u)$ has a period annulus $\mathcal{A}$, then it is integrable. In this case, a subset of the three-dimensional phase space for the extended system (7.9) at $\epsilon = 0$ is filled with invariant two-dimensional tori corresponding to the periodic orbits in $\mathcal{A}$. Thus, there is one action variable, which has a constant value on each periodic orbit, and two angle variables. One of the angular variables is τ; the other is the angle variable defined for the action-angle variables of the unperturbed planar system.

The basic idea that leads to the development of the method of averaging arises from an inspection of system (7.8). In particular, since $|\epsilon|$ is assumed to be small and the time derivatives of the action variables are all proportional to ϵ, the action variables will remain close to their constant unperturbed values (while undergoing small-amplitude high-frequency oscillations due to the relatively fast angular velocities) on a time interval whose length is inversely proportional to $|\epsilon|$. Thus, on this time interval, we would expect to obtain a close approximation to the slow evolution of the action variables (with their small-amplitude high-frequency oscillations removed) by averaging the (slow) action variables over the (fast) angle variables.

In most applications, we are interested in the evolution of the action variables, not the angle variables. For example, let us note that the semimajor axis of the orbit of a planet about a star is given by an action variable. While a determination of the planet's exact position in the sky requires the specification of an action and an angle variable, a prediction of the long-term evolution of the relative distance between the planet and the star is probably more interesting (and more realistic) than a prediction of the planet's exact position relative to the star at some point in the distant future.

The Averaging Principle. *Let* $t \mapsto (I(t), \theta(t))$ *be a solution of system (7.8), and let* $t \mapsto J(t)$ *denote the solution of the corresponding initial value problem*

$$\dot{J} = \epsilon \bar{F}(J), \qquad J(0) = I(0) \tag{7.10}$$

(called the averaged system) where $\bar{F}$ *is the function defined by*

$$\bar{F}(J) := \frac{1}{(2\pi)^N} \int_{\mathbb{T}^N} F(J, \theta) \, d\theta.$$

If $\epsilon > 0$ *is small, then there are constants* $C > 0$ *and* $\tau > 0$ *such that* $|I(t) - J(t)| < \epsilon C$ *as long as* $0 \leq \epsilon t < \tau$; *that is, the solution of the averaged system is a useful approximation of the evolution of the action variables of system (7.8).*

The averaging principle has a long history, which is deeply rooted in perturbation problems that arise in celestial mechanics (see, for example, [201]); but, the averaging principle is not a theorem. Nevertheless, in a physical application, it might be reasonable to replace a mathematical model, which is given in the form of the differential equation (7.8), with the corresponding averaged system (7.10), to use the averaged system to make a prediction, and to then test the prediction against the results of a physical experiment.

The next theorem (the *averaging theorem*) validates the averaging principle under the hypothesis that there is exactly one angle variable. In this case, there is a 2π-periodic change of variables for system (7.8) such that the transformed differential equation decouples, and the first-order truncation with respect to ϵ of its action variables is exactly the averaged system (7.10). The existence of such "averaging transformations" is the essential ingredient used in the proof of the averaging theorem and the central issue of the subject.

Theorem 7.2 (Averaging Theorem). *Suppose that system (7.8) is defined on $U \times \mathbb{T}$ where U is an open subset of $\mathbb{R}^M$.*

(i) *If there is some number λ such that $\omega(I) > \lambda > 0$ for all $I \in U$, then there is a bounded open ball $\mathcal{B}$ contained in U, a number $\epsilon_1 > 0$, and a smooth function $k : \mathcal{B} \times \mathbb{T} \to \mathcal{B}$ such that for each $I \in \mathcal{B}$ the function $\theta \mapsto k(I, \theta)$ is 2π-periodic, the function $I \to I + \epsilon k(I, \theta)$ is invertible on $\mathcal{B}$ for $0 \leq \epsilon < \epsilon_1$, and the change of coordinates given by*

$$L = I + \epsilon k(I, \theta) \tag{7.11}$$

transforms the system (7.8) to the form

$$\dot{L} = \epsilon \bar{F}(L) + \epsilon^2 F_1(L, \theta, \epsilon), \quad \dot{\theta} = \omega(L) + \epsilon G_1(L, \theta, \epsilon) \tag{7.12}$$

where

$$\bar{F}(L) = \frac{1}{2\pi} \int_0^{2\pi} F(L, \theta) \, d\theta$$

and both of the functions F_1 and G_1 are 2π-periodic with respect to their second arguments.

(ii) *If in addition $T > 0$, $\mathcal{B}_0$ is an open ball whose closure is contained in the interior of $\mathcal{B}$, and if for each $I_0 \in \mathcal{B}$ the number $\tau(I_0)$ denotes the largest number less than or equal to T such that the solution of the averaged system (7.10) with initial condition $J(0) = I_0$ is in the closure of $\mathcal{B}_0$ for $0 \leq t \leq \tau(I_0)$, then there are positive numbers*

$\epsilon_2 \leq \epsilon_1$ *and* C *such that for each* $I_0 \in \mathcal{B}_0$ *and for* $0 \leq \epsilon < \epsilon_2$
all solutions $t \mapsto (I(t), \theta(t))$ *of the system (7.8) with initial value*
$I(0) = I_0$ *are approximated by the solution* $t \mapsto J(t)$ *of the averaged*
system (7.10) with $J(0) = I_0$ *as follows:*

$$|I(t) - J(t)| < C\epsilon$$

on the time interval given by $0 \leq \epsilon t < \tau(I_0)$.

Proof. To prove statement (i), define a new function $\widetilde{F}$ on $U \times \mathbb{T}$ given by

$$\widetilde{F}(L, \theta) := F(L, \theta) - \bar{F}(L),$$

and let k denote the solution of the differential equation

$$\frac{\partial k}{\partial \theta}(L, \theta) = -\frac{1}{\omega(L)} \widetilde{F}(L, \theta) \tag{7.13}$$

with the initial condition $k(L, 0) = 0$; that is, k is given by

$$k(L, \theta) = -\frac{1}{\omega(L)} \int_0^\theta \widetilde{F}(L, s) \, ds.$$

The function k is defined on $U \times \mathbb{T}$, and the function $\theta \mapsto k(L, \theta)$ is
2π-periodic. To prove the periodicity, fix L, define the new function

$$\widehat{k}(\theta) = k(L, \theta + 2\pi) - k(L, \theta),$$

and note that $\widehat{k}(0) = 0$ and $\widehat{k}'(\theta) \equiv 0$. (Our definition of k is given using
the assumption that there is only one angle. The existence of a correspond-
ing 2π-periodic function is problematic when there are several angles (see
Exercise 7.4).)

We will show that the relation $L = I + \epsilon k(I, \theta)$ defines a coordinate
transformation by proving that the corresponding function $(I, \theta) \mapsto (L, \theta)$
is invertible. To this end, consider the smooth function $K : U \times U \times \mathbb{T} \times \mathbb{R} \to \mathbb{R}^M$ given by

$$(I, L, \theta, \epsilon) \mapsto I + \epsilon k(I, \theta) - L.$$

For each point $\xi = (L, \theta)$ in $c\ell(\mathcal{B}) \times \mathbb{T}$ (where $c\ell$ denotes the closure of the
set), we have that $K(L, L, \theta, 0) = 0$ and the partial derivative $K_I(L, L, \theta, 0)$
is the identity transformation of $\mathbb{R}^M$. Therefore, by the implicit function
theorem, there is a product neighborhood $\Gamma_\xi \times \gamma_\xi$ contained in $(U \times \mathbb{T}) \times \mathbb{R}$
and containing the point $(\xi, 0)$, and a smooth function $H^\xi : \Gamma_\xi \times \gamma_\xi \to U$
such that

$$H^\xi(L, \theta, \epsilon) + \epsilon k(H^\xi(L, \theta, \epsilon), \theta) = L$$

for all $((L, \theta), \epsilon) \in \Gamma_\xi \times \gamma_\xi$. In other words, the function $L \mapsto H^\xi(L, \theta, \epsilon)$ is a local inverse for the function $I \mapsto I + \epsilon k(I, \theta)$. Moreover, if $(I, (L, \theta), \epsilon) \in U \times \Gamma_\xi \times \gamma_\xi$ is such that $K(I, L, \theta, \epsilon) = 0$, then $I = H^\xi(L, \theta, \epsilon)$.

Since $c\ell(\mathcal{B}) \times \mathbb{T}$ is compact, there is a finite collection of the neighborhoods $\Gamma_\xi \times \gamma_\xi$ that cover $\mathcal{B} \times \mathbb{T}$. Let Γ denote the union of the corresponding Γ_ξ, and let γ denote the intersection of the corresponding intervals on the real line. We have $\mathcal{B} \subset \Gamma$, and there is some ϵ_0 such that γ contains the closed interval $[0, \epsilon_0]$.

The function k has a global Lipschitz constant $\mathrm{Lip}(k)$ on the compact set $c\ell(\mathcal{B}) \times \mathbb{T}$. Let us define $\epsilon_1 > 0$ such that

$$\epsilon_1 < \min \left\{ \frac{1}{\mathrm{Lip}(k)}, \epsilon_0 \right\}.$$

If $\theta \in \mathbb{T}$ and $0 \le \epsilon \le \epsilon_1$, then the map $I \mapsto I + \epsilon k(I, \theta)$ is injective. In fact, if

$$I_1 + \epsilon k(I_1, \theta) = I_2 + \epsilon k(I_2, \theta),$$

then

$$|I_1 - I_2| = |\epsilon| \, \mathrm{Lip}(k) |I_1 - I_2| < |I_1 - I_2|,$$

and therefore $I_1 = I_2$. It follows that there is a function $H : \Gamma \times \mathbb{T} \times [0, \epsilon_1] \to \mathcal{B}$ such that H is the "global" inverse; that is,

$$H(L, \theta, \epsilon) + \epsilon k(H(L, \theta, \epsilon), \theta) = L.$$

By the uniqueness of the smooth local inverses H^ξ, the function H must agree with each function H^ξ on the intersection of their domains. Thus, H is smooth and we have defined a coordinate transformation $L := I + \epsilon k(I, \theta)$ on $\mathcal{B} \times \mathbb{T} \times [0, \epsilon_1]$. Moreover, by expanding H in a Taylor series at $\epsilon = 0$ and with the first order remainder given by $\mathcal{H}$, we see that $I = L + \epsilon \mathcal{H}(I, \theta, \epsilon)$. It is also easy to check that $\theta \mapsto \mathcal{H}(I, \theta, \epsilon)$ is a 2π-periodic function.

Using the coordinate transformation, we have that

$$\dot{L} = \dot{I} + \epsilon \frac{\partial k}{\partial I} \dot{I} + \epsilon \frac{\partial k}{\partial \theta} \dot{\theta}$$

$$= \epsilon F(I, \theta) + \epsilon \frac{\partial k}{\partial I}(I, \theta)(\epsilon F(I, \theta)) + \epsilon \frac{\partial k}{\partial \theta}(I, \theta)(\omega(I) + \epsilon G(I, \theta))$$

$$= \epsilon \left(F(I, \theta) + \frac{\partial k}{\partial \theta}(I, \theta) \omega(I) \right) + \epsilon^2 \alpha(I, \theta)$$

where

$$\alpha(I, \theta) := \frac{\partial k}{\partial I}(I, \theta) F(I, \theta) + \frac{\partial k}{\partial \theta}(I, \theta) G(I, \theta).$$

Using the inverse transformation and Taylor's theorem, there is a function F_1 such that

$$\dot{L} = \epsilon(F(L,\theta) + \frac{\partial k}{\partial \theta}(L,\theta)\omega(L)) + \epsilon^2 F_1(L,\theta,\epsilon). \qquad (7.14)$$

After the formula for the partial derivative of k (equation (7.13)) is inserted into the equation (7.14), the new differential equation is given by

$$\dot{L} = \epsilon(\widetilde{F}(L,\theta) + \bar{F}(L) - \widetilde{F}(L,\theta)) + \epsilon^2\beta(L,\theta,\epsilon).$$

Thus, the coordinate transformation (7.11) applied to the system (7.8) yields a new system of the form

$$\dot{L} = \epsilon\bar{F}(L) + \epsilon^2 F_1(L,\theta,\epsilon),$$
$$\dot{\theta} = \omega(L) + \epsilon G_1(L,\theta,\epsilon). \qquad (7.15)$$

This completes the proof of statement (i).

To prove the asymptotic estimate in statement (ii), consider the differential equation for $L - J$ obtained by subtracting the averaged system from the first differential equation of the system (7.15) and then integrate to obtain

$$L(t) - J(t) = L(0) - J(0) + \epsilon \int_0^t \bar{F}(L(s)) - \bar{F}(J(s))\, ds$$

$$+ \epsilon^2 \int_0^t F_1(L(s),\theta(s),\epsilon)\, ds.$$

If $\mathrm{Lip}(\bar{F}) > 0$ is a Lipschitz constant for $\bar{F}$, and if B is an upper bound for the function

$$(L,\theta,\epsilon) \mapsto |F_1(L,\theta,\epsilon)|$$

on the compact space $c\ell(\mathcal{B}) \times \mathbb{T} \times [0,\epsilon_1]$, then we have the estimate

$$|L(t) - J(t)| \leq |L(0) - J(0)| + \epsilon\,\mathrm{Lip}(\bar{F}) \int_0^t |L(s) - J(s)|\, ds + \epsilon^2 B t \qquad (7.16)$$

as long as $L(t)$ remains in $c\ell(\mathcal{B})$.

An application of the specific Gronwall lemma from Exercise 2.3 to the inequality (7.16) yields the following estimate

$$|L(t) - J(t)| \leq \left(|L(0) - J(0)| + \epsilon\frac{B}{\mathrm{Lip}(\bar{F})}\right)e^{\epsilon\,\mathrm{Lip}(\bar{F})t}.$$

We also have that

$$L(0) = I(0) + \epsilon k(I(0),\theta(0)), \qquad I(0) = J(0).$$

Thus, if $0 \le \epsilon t \le \tau(I(0))$ (where τ is defined in the statement of the theorem), then there is a constant C_0 such that

$$|L(t) - J(t)| \le C_0 \epsilon \qquad (7.17)$$

as long as $L(t)$ remains in $c\ell(\mathcal{B})$.

Note that $L(t)$ is in $c\ell(\mathcal{B})$ whenever $|L(t) - J(t)|$ is less than the minimum distance between the boundaries of $\mathcal{B}_0$ and $\mathcal{B}$. If $0 < \epsilon_2 < \epsilon_1$ is chosen so that C_0/ϵ_2 is less than this distance, then the estimate (7.17) ensures that $L(t)$ is in $c\ell(\mathcal{B})$ on the time interval $0 \le \epsilon t \le \tau(I(0))$.

Finally, let C_1 be an upper bound for the function

$$(L, \theta) \mapsto |k(L, \theta)|$$

and note that for t in the range specified above we have the inequality

$$|I(t) - J(t)| \le |I(t) - L(t)| + |L(t) - J(t)| \le \epsilon C_1 + \epsilon C_0.$$

Therefore, with $C := C_0 + C_1$, we have the required asymptotic estimate.

$$\square$$

Exercise 7.3. (a) Write van der Pol's equation

$$\ddot{x} + \epsilon(x^2 - 1)\dot{x} + \omega^2 x = 0.$$

as a first order system and transform to action angle variables. Hint: Use $\dot{x} = \omega y$. (b) Compute the averaged equation and show that it has a hyperbolic rest point. (c) Compare the results of part (b) and Section 5.1.

Exercise 7.4. Consider the PDE

$$\omega_1 k_\theta(\theta, \phi) + \omega_2 k_\phi(\theta, \phi) = g(\theta, \phi)$$

where g is 2π-periodic with respect to the angular variables θ and ϕ. This differential equation can be viewed as a PDE on the two-dimensional torus $\mathbb{T}^2$. Does the PDE have a 2π-periodic solution? More precisely, find sufficient conditions on g, ω_1, and ω_2 such that periodic solutions exist. Hint: Look for a solution k represented as a Fourier series. Note that expressions of the form $m\omega_1 + n\omega_2$, where m and n are integers, will appear in certain denominators. If there are resonances (that is, integers m and n such that $m\omega_1 + n\omega_2 = 0$) or an infinite sequence of "small denominators," then the Fourier series for k, which is determined by equating coefficients after substitution of the Fourier series in the PDE, will not converge for most choices of g. On the other hand, convergence is possible if the frequencies satisfy an appropriate Diophantine condition $|m\omega_1 + n\omega_2| \ge C(|m| + |n|)^{-\alpha}$, where $C > 0$ and $\alpha > 0$ are constants and the inequality holds for every nonzero integer vector (m, n). There is a vast literature on the problem of small divisors (see, for example, [146]).

7.2 Averaging at Resonance

In this section we will demonstrate a remarkable fact: Some of the most important features of the dynamical behavior near a resonance of a generic multidimensional oscillator are determined by an associated one-degree-of-freedom oscillator that resembles a perturbed pendulum with torque. We will also give some examples to show that the averaging principle is not always applicable in multifrequency systems.

Let us consider the system (7.1) with $u \in \mathbb{R}^{M+N}$ where the period of the perturbation is $\eta = 2\pi/\Omega$ and where the unperturbed system has an invariant set that is foliated by N-dimensional invariant tori. In this case, if action-angle variables $(I, \varphi) \in \mathbb{R}^M \times \mathbb{T}^N$ are introduced, then the differential equation is expressed in the form

$$\dot{I} = \epsilon F(I, \varphi, t) + O(\epsilon^2),$$
$$\dot{\varphi} = \omega(I) + \epsilon G(I, \varphi, t) + O(\epsilon^2). \tag{7.18}$$

Moreover, it is 2π-periodic in each component of the N-dimensional vector of angles and $2\pi/\Omega$-periodic in time. By introducing an additional angular variable τ, system (7.18) is equivalent to the autonomous system with M action variables and $N + 1$ angle variables given by

$$\dot{I} = \epsilon F(I, \varphi, \tau/\Omega) + O(\epsilon^2),$$
$$\dot{\varphi} = \omega(I) + \epsilon G(I, \varphi, \tau/\Omega) + O(\epsilon^2),$$
$$\dot{\tau} = \Omega. \tag{7.19}$$

Let us suppose that there is a resonance relation

$$\langle K, \omega(I) \rangle = n\Omega \tag{7.20}$$

where K is an integer vector of length M, and n is an integer such that the components of K and the integer n have no common factors. The corresponding set

$$\mathcal{R}_{K,n} := \{(I, \varphi, \tau) : \langle K, \omega(I) \rangle = n\Omega\},$$

which is generally a hypersurface in the phase space, is called a *resonance manifold*. Our goal is to describe the perturbed dynamics of the system (7.19) near this resonance manifold. To do this, we will use yet another set of new coordinates that will be introduced informally and abstractly. In practice, as we will demonstrate later, the appropriate new coordinates are chosen using the ideas of the abstract construction; but, their precise definition depends on special features of the system being studied.

The set

$$\mathcal{A}_{K,n} := \{I : \langle K, \omega(I) \rangle = n\Omega\},$$

is the intersection of the resonance manifold with the "action space." This set is generally a manifold in $\mathbb{R}^M$, sometimes also called a resonance manifold. To distinguish $\mathcal{A}_{K,n}$ from $\mathcal{R}_{K,n}$, let us call $\mathcal{A}_{K,n}$ the *resonance layer* associated with the resonance relation (7.20).

A point in the action space is determined by its distance from the resonance layer and by its projection to the resonance layer. In particular, there are local coordinates defined in a neighborhood of the resonance layer, or at least near a portion of this manifold, given by

$$ r = \langle K, \omega(I) \rangle - n\Omega, \qquad z = A(I) $$

where r is a measure of the distance of the point with action variable I to the resonance layer and the $(M-1)$-dimensional vector z is the vector coordinate of the projection, denoted by the smooth map A, of the point I to the resonance layer $\mathcal{A}_{K,n}$.

For $\epsilon \neq 0$, let us define the new stretched distance-coordinate $\rho = r/\sqrt{\epsilon}$ and new angular variables

$$ \psi = \langle K, \varphi \rangle - n\tau, \qquad \chi = B(\varphi, \tau) $$

where the vector function $B : \mathbb{T}^{N+1} \to \mathbb{T}^{N+1}$ is chosen so that the transformation to the new angles is invertible. Of course, B must also be 2π-periodic in each component of φ and in τ.

In the coordinates ρ, z, ψ, and χ, system (7.19) has the form

$$
\begin{aligned}
\dot{\rho} &= \sqrt{\epsilon} \langle K, D\omega(I) F(I, \phi, \tau/\Omega) \rangle + O(\epsilon), \\
\dot{z} &= O(\epsilon), \\
\dot{\psi} &= \sqrt{\epsilon}\rho + O(\epsilon^{3/2}), \\
\dot{\chi} &= O(1)
\end{aligned}
\tag{7.21}
$$

where I, ϕ, and τ are viewed as functions of ρ, z, ψ, and χ.

In system (7.21), ρ and ψ are slow variables, the $M-1$ variables represented by the vector z are "super slow," and χ is an N-dimensional vector of fast variables. In keeping with the averaging principle, we will average over the fast (angular) variables, although we have provided no theoretical justification for doing so unless $N = 1$. The system of differential equations obtained by averaging over the fast variables in system (7.21) is called the *partially averaged system at the resonance*.

To leading order in $\mu := \sqrt{\epsilon}$, the partially averaged system at the resonance is

$$
\begin{aligned}
\dot{\rho} &= \mu \langle K, D\omega(I) F^*(I, \psi) \rangle, \\
\dot{z} &= 0, \\
\dot{\psi} &= \mu\rho
\end{aligned}
\tag{7.22}
$$

where F^* is obtained by averaging the function

$$ \chi \mapsto F(I, \phi(\psi, \chi), \tau(\psi, \chi)/\Omega). $$

Here we have used the names of the original variables for the corresponding averaged variables even though this is a dangerous practice. The solutions of the partially averaged equations are not the same as the solutions of the original system.

The function F^* is periodic in its second argument with period some integer multiple of 2π. In particular, using Fourier series, there is a constant vector $c(I)$ and a vector-valued periodic function $\psi \mapsto h(I, \psi)$ with zero average such that

$$F^*(I, \psi) = c(I) + h(I, \psi).$$

Also, we can easily obtain the expansion, in powers of μ, of the function $D\omega$ expressed in the new coordinates. In fact, because

$$I(r, z) = I(\mu\rho, z) = I(0, z) + O(\mu),$$

it follows that $D\omega(I) = D\omega(I(0, z)) + O(\mu)$.

Under the generic assumption

$$\langle K, D\omega(I(0, z))F^*(I(0, z), \psi) \rangle \neq 0$$

(that is, the vector field corresponding to the averaged system is transverse to the resonance manifold); and, in view of the form of system (7.22) and the above definitions, there are real-valued functions $z \mapsto p(z)$ and $(z, \psi) \mapsto q(z, \psi)$ such that the function $(z, \phi) \mapsto p(z) + q(z, \phi)$ is not identically zero and the first order approximation to the partially averaged system has the form

$$\begin{aligned}
\dot{\rho} &= \mu(p(z) + q(z, \psi)), \\
\dot{z} &= 0, \\
\dot{\psi} &= \mu\rho.
\end{aligned} \tag{7.23}$$

Finally, by defining a *slow time* variable $s = \mu t$ and taking into account that z is a constant of the motion, we will view system (7.23) as the following parametrized family of differential equations with parameter z:

$$\frac{d\rho}{ds} = p(z) + q(z, \psi), \qquad \frac{d\psi}{ds} = \rho; \tag{7.24}$$

or, equivalently,

$$\frac{d^2\psi}{ds^2} - q(z, \psi) = p(z) \tag{7.25}$$

where the function $\psi \mapsto q(z, \psi)$ is periodic with average zero. For instance, q might be given by $q(z, \psi) = -\lambda \sin \psi$ for some $\lambda > 0$.

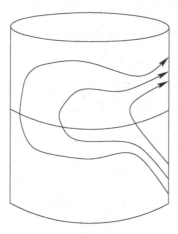

Figure 7.1: Phase portrait of pendulum with "large" constant torque. All orbits pass through the resonant value of the action variable.

In accordance with the usual physical interpretation of the differential equation (7.25), we have just obtained a wonderful result: *Near a resonance, every oscillator behaves like a pendulum influenced by a constant torque.*

The precise nature of the dynamical behavior near the resonance depends on the functions p and q in the differential equation (7.25) and the perturbation terms that appear in the higher order approximations of the partially averaged system. In particular, let us note that the coefficients of the pendulum equation are functions of the super slow variables. Thus, they vary slowly with the slow time. Although a rigorous description of the motion predicted by the partially averaged system is highly nontrivial and not yet completely understood, our result certainly provides a fundamental insight into the near resonance dynamics of oscillators. Also, this result provides a very good reason to study the dynamics of perturbed pendulum models.

Consider the simple pendulum with constant torque (equation (7.25) with $p(z) = c$ and $q(z, \psi) := -\lambda \sin \psi$) given by

$$\dot\rho = c - \lambda \sin \psi, \qquad \dot\psi = \rho \tag{7.26}$$

where $\lambda > 0$ and $c \geq 0$. The phase space of this system is the cylinder $(\rho, \psi) \in \mathbb{R} \times \mathbb{T}$. Also, let us note that the circle given by the equation $\rho = 0$ would correspond to the resonance manifold in our original oscillator.

If $c/\lambda > 1$, then $\dot\rho > 0$, and it is clear that all trajectories *pass through* the resonance manifold as depicted in Figure 7.1. If, on the other hand, $c/\lambda < 1$, then there are two rest points on the resonance manifold, a saddle and a sink, and the phase portrait will be as depicted in Figure 7.2. In particular, some orbits still pass through the resonance manifold, but now

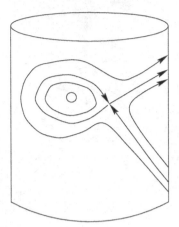

Figure 7.2: Phase portrait for pendulum with "small" constant torque. The region bounded by the homoclinic orbit corresponds to the trajectories that are captured into resonance. The corresponding action variable oscillates around its resonant value.

the periodic orbits surrounded by the homoclinic loop are *captured into the resonance*. These orbits correspond to orbits for which an action variable librates near its resonant value on a long time-scale. In the pendulum model, the libration goes on for ever. On the other hand, if a pendulum system is obtained by partial averaging at a resonance, then its coefficients are expected to vary slowly with time. In particular, the ratio c/λ will change over time and perhaps reach a value that exceeds one. In this case, the corresponding action variable can drift away from its resonance value.

If the averaging procedure is carried to the next higher order in μ, then a typical perturbation that might appear in the pendulum model is a small viscous friction. For example, the perturbed system might be

$$\dot{\rho} = c - \mu\rho - \lambda\sin\psi, \qquad \dot{\psi} = \rho. \qquad (7.27)$$

The phase portrait of this system on the phase cylinder for the case $c/\lambda < 1$ is depicted in Figure 7.3. Note that there is a "thin" set of trajectories, some with their initial point far from the resonance manifold, that are eventually captured into the resonance. Again, because the coefficients of system (7.27) will generally vary slowly with time, it is easy to imagine the following scenario will occur for the original system: An action variable of our multidimensional oscillator evolves toward a resonance, it is captured into the resonance and begins to librate about its resonant value. After perhaps a long sojourn near the resonance, the action variable slowly drifts away from its resonant value. Meanwhile, the same action variable for a solution with a slightly different initial condition evolves toward the reso-

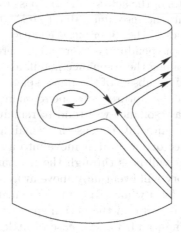

Figure 7.3: Phase portrait for pendulum with "small" constant torque and "small" viscous friction. A thin strip of trajectories are captured into the resonance. The corresponding action variable with initial condition in the strip moves toward its resonant value and then begins to oscillate around its resonant value.

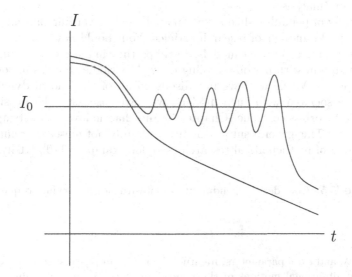

Figure 7.4: Two schematic time signals of an action variable I are depicted for orbits with slightly different initial conditions. One time trace passes through the resonant value $I = I_0$; the other is captured into the resonance on a long time-scale before it leaves the vicinity of its resonant value.

nant value, but the values of the action variable pass through the resonance without oscillating about the resonant value (see Figure 7.4).

Although the dynamics of the differential equation (7.25) are similar to the dynamics of the simple pendulum, there are generally several alternating saddles and centers along the resonance manifold. After a perturbation, there can be several "thin" subsets of the phase space corresponding to trajectories that are eventually captured into the resonance. Again trajectories that are captured into a resonance will tend to remain near the resonance manifold on a long time interval. But as the remaining super slow action variables drift, the trajectory will often move into a region near the resonance manifold where it will pass through the resonance. After it reaches this region, the trajectory will eventually move away from the influence of the resonance—at least for a while. To complicate matters further, the set of resonance manifolds is dense in the action space (the rational numbers are dense in the real line); and, for the case of at least three angle variables, resonance manifolds corresponding to different integer vectors can intersect. Thus, there is a complex web, called the *Arnold web*, of resonance manifolds that each influence the perturbed motion of nearby orbits. Although the precise dynamics in the phase space and the corresponding fluctuations of the action variables is usually very difficult to analyze, the resonance capture mechanism, which is partly responsible for the complexity of the motions in phase space for dissipative systems, is made reasonably clear by our analysis.

The study of pendulum-like equations with slowly varying parameters is the subject of hundreds of research articles. You should now see why there is so much interest in such models. Perhaps the simplest case to analyze is the pendulum with periodic forcing or with periodic changes in some of its parameters. While we have not discussed all of the known dynamical behavior associated with such models, we have discussed the possibility that periodic orbits continue (Chapter 5) and chaotic invariant sets appear (Chapter 6). This general subject area is certainly not closed; it remains a fruitful area of mathematical research (see, for example, [110], [164]).

Exercise 7.5. Consider the pendulum model with slowly varying torque given by

$$\dot{\rho} = a\sin(\sqrt{\epsilon}t) - \lambda\sin\psi, \qquad \dot{\psi} = \rho$$

where a, λ, and ϵ are parameters. Identify the region in phase space corresponding to the librational motions of the pendulum at the parameter value $\epsilon = 0$. Determine (by numerical integration if necessary) the behavior in forward and backward time of the corresponding solutions for the system with $\epsilon > 0$ that have initial conditions in the librational region.

Exercise 7.6. Consider the phase modulated pendulum (remember that our pendulum model (7.27) is only a special case of the type of equation that is obtained by partial averaging) given by

$$\ddot{\psi} + \sin(\psi + a\sin(\epsilon t)) = 0.$$

What can you say about the dynamics?

Exercise 7.7. Show that resonance manifolds do not intersect in systems with just two angle variables, but that they can intersect if there are three or more angles.

Because a trajectory can be captured into resonance, the averaging principle is not generally valid for systems with more than one angular variable. To see why, note that a solution of the averaged system might pass through a resonance while the corresponding solution of the original system is captured into the resonance. If this occurs, then the norm of the difference of the evolving action variables and the corresponding averaged variables, given by $|I(t) - J(t)|$, may grow to a value that is $O(1)$ in the perturbation parameter as time evolves and the solution $t \mapsto I(t)$ is trapped in the resonance. In particular, this scenario would violate the expected estimate; that is, $|I(t) - J(t)| < C_1\epsilon$.

A complete analysis for the dynamics of multifrequency systems is not known. Thus, this is an area of much current research (see, for example, [12], [14], [145], and [201]). One of the most important issues is to determine the "diffusion rates" for the action variables to leave the vicinity of a resonance and to arrive at a second resonance. The long term stability of the models of the motion of many-body systems, for example our solar system, is essentially bound up with this question. This is currently one of the great unsolved problems in mathematics.

A concrete counterexample to the validity of averaging for the case of two or more angles is provided by the system

$$\dot{I}_1 = \epsilon,$$
$$\dot{I}_2 = \epsilon\cos(\theta_2 - \theta_1),$$
$$\dot{\theta}_1 = I_1,$$
$$\dot{\theta}_2 = I_2, \tag{7.28}$$

introduced in [201] (see Exercise 7.8).

Exercise 7.8. Find the averaged system for the oscillator (7.28) and the general analytical solution of the averaged system. Show that a solution of the original system is given by

$$I_1(t) = \epsilon t + I_0,$$
$$I_2(t) = I_1(t),$$
$$\theta_1(t) = \epsilon\frac{1}{2}t^2 + I_0 t + \theta_1(0),$$
$$\theta_2(t) = \theta_1(t).$$

For these solutions, show that the estimate expected from the averaging theorem (Theorem 7.2) is not valid.

Let us note that system (7.28) has a resonance manifold given by the resonance relation $I_2 - I_1 = 0$. As prescribed above in our partial averaging procedure, consider new coordinates defined by

$$\sqrt{\epsilon}\,\rho = I_2 - I_1, \quad z = I_2, \quad \psi = \theta_2 - \theta_1, \quad \chi = \theta_2,$$

and note that system (7.28), when expressed in these coordinates, is given by

$$
\begin{aligned}
\dot{\rho} &= \sqrt{\epsilon}\,(\cos\psi - 1), \\
\dot{z} &= \epsilon \cos\psi, \\
\dot{\psi} &= \sqrt{\epsilon}\,\rho, \\
\dot{\chi} &= z.
\end{aligned}
\tag{7.29}
$$

Averaging over the fast angle χ in system (7.29) produces the partially averaged system

$$
\begin{aligned}
\dot{\bar{\rho}} &= \sqrt{\epsilon}\,(\cos\bar{\psi} - 1), \\
\dot{\bar{z}} &= 0, \\
\dot{\bar{\psi}} &= \sqrt{\epsilon}\,\bar{\rho}.
\end{aligned}
\tag{7.30}
$$

For each fixed $\bar{z}$, there is an orbit $\mathcal{O}_1$ whose ω-limit set is the rest point $(\bar{\rho}, \bar{z}, \bar{\psi}) = (0, \bar{z}, 0)$ and a second orbit $\mathcal{O}_2$ with this rest point as its α-limit set. (Prove this!) The trajectories corresponding to the orbit $\mathcal{O}_1$ are all captured into the resonance relative to the first order approximation of the partially averaged system, and the rest point is captured for all time. But the action variable corresponding to $\bar{z}$, a super slow variable, drifts from its initial value so that the trajectories corresponding to the orbit $\mathcal{O}_1$ eventually pass through the resonance. This example thus provides a clear illustration of the mechanism that destroys the possibility that the averaged system—not the partially averaged system—gives a good approximation to the full system over a long time-scale. In effect, the averaged system for this example does not "feel the influence of the resonance."

Exercise 7.9. The partially averaged system (7.30) is obtained by averaging over just one angle. Thus, the averaging theorem ensures that under appropriate restrictions the partially averaged system is a good approximation to the original system. Formulate appropriate restrictions on the domain of definition of the partially averaged system and determine an appropriate time scale for the validity of averaging. Give a direct proof that your formulation is valid.

Exercise 7.10. In applied mathematics, often only the lowest order resonances are considered; they seem to have the most influence on the dynamics. As an example to illustrate why this observation might be justified, consider the near resonance dynamics of system (7.28) at a "high" order resonance given by the resonance relation $mI_1 = nI_2$ where m and n are relatively prime, and $m \neq n$. Show that there are integers k and ℓ such that the matrix

$$R := \begin{pmatrix} m & -n \\ k & \ell \end{pmatrix}$$

is unimodular. Next, define new angular coordinates by

$$\begin{pmatrix} \phi_1 \\ \phi_2 \end{pmatrix} = R \begin{pmatrix} \theta_1 \\ \theta_2 \end{pmatrix}.$$

Also, define new action variables by

$$\sqrt{\epsilon}\,\rho = mI_1 - nI_2, \qquad z = kI_1 + \ell I_2.$$

Change to the new coordinates, find the partially averaged system, and show that, in this approximation, all orbits pass through the resonance. Does this mean that the averaging principle is valid for orbits starting near the higher order resonances in this example?

Let us consider the system (7.1) with $u \in \mathbb{R}^2$; that is, a planar periodically perturbed oscillator. Furthermore, let us assume that action-angle variables have been introduced as in system (7.18). In this case, the resonance layer given by the resonance relation $m\omega(I) = n\Omega$ is (generically) a point $I = I_0$ in the one-dimensional action space.

To determine the partially averaged system at the resonance layer given by $I = I_0$, let ρ denote the scaled distance to the resonance layer; that is,

$$\sqrt{\epsilon}\,\rho = I - I_0.$$

Also, let $\tau = \Omega t$, and introduce a new angular variable by

$$\psi = m\phi - n\tau.$$

Then, to first order in the perturbation parameter $\sqrt{\epsilon}$, the differential equation in these new coordinates is given by the system

$$\dot{\rho} = \sqrt{\epsilon}\,F(I_0, \psi/m + n\tau/m, \tau/\Omega) + O(\epsilon),$$
$$\dot{\psi} = \sqrt{\epsilon}\,m\omega'(I_0)\rho + O(\epsilon),$$
$$\dot{\tau} = \Omega$$

where ρ and ψ are slow variables, and τ (corresponding to the time variable in our nonautonomous perturbation) is a fast angular variable.

By the averaging theorem, there is a change of coordinates such that the transformed system, to leading order in $\mu = \sqrt{\epsilon}$, is given by

$$\dot{J} = \mu \bar{F}(\theta), \qquad \dot{\theta} = \mu m \omega'(I_0) J \qquad (7.31)$$

where

$$\bar{F}(\theta) := \frac{1}{2\pi m} \int_0^{2\pi m} F(I_0, \theta/m + n\tau/m, \tau/\Omega) \, d\tau.$$

Under the assumption that $\omega'(I_0) \neq 0$—in other words, under the assumption that the unperturbed resonant periodic orbit corresponding to the action variable $I = I_0$ is normally nondegenerate—the averaged system for $\epsilon > 0$ has a nondegenerate rest point at (J_0, θ_0) if and only if $J_0 = 0$ and the function $\bar{F}$ has θ_0 as a simple zero.

Note that the solution of the system

$$\dot{J} = \mu \bar{F}(\theta), \qquad \dot{\theta} = \mu m \omega'(I_0) J, \qquad \dot{\tau} = \Omega$$

starting at the point $(J, \theta, \tau) = (0, \theta_0, 0)$ is a periodic orbit, and in addition if the rest point is hyperbolic, then this periodic orbit is hyperbolic. We would like to conclude that there is a corresponding periodic orbit for the original oscillator. This fact is implied by the following more general theorem.

Theorem 7.11. *Consider the system*

$$\dot{I} = \epsilon F(I, \theta) + \epsilon^2 F_2(I, \theta, \epsilon),$$
$$\dot{\theta} = \omega(I) + \epsilon G(I, \theta, \epsilon) \qquad (7.32)$$

where $I \in \mathbb{R}^M$ and $\theta \in \mathbb{T}$, where F, F_2, and G are 2π-periodic functions of θ, and where there is some number c such that $\omega(I) > c > 0$. If the averaged system has a nondegenerate rest point and ϵ is sufficiently small, then system (7.32) has a periodic orbit. If in addition $\epsilon > 0$ and the rest point is hyperbolic, then the periodic orbit has the same stability type as the hyperbolic rest point; that is, the dimensions of the corresponding stable and unstable manifolds are the same.

Proof. The averaged differential equation is given by $\dot{J} = \epsilon \bar{F}(J)$ where $\bar{F}$ is the average of the function $\theta \mapsto F(I, \theta)$. Let us suppose that J_0 is a nondegenerate rest point of the averaged system; that is, $\bar{F}(J_0) = 0$ and the derivative $D\bar{F}(J_0)$ is an invertible transformation.

By the averaging theorem, if ϵ is sufficiently small, then there is a 2π-periodic change of coordinates of the form $J = I + \epsilon L(I, \theta)$, defined in an open set containing $\{J_0\} \times \mathbb{T}$, such that system (7.32) in these new coordinates is given by

$$\dot{J} = \epsilon \bar{F}(J) + O(\epsilon^2),$$
$$\dot{\theta} = \omega(J) + O(\epsilon). \qquad (7.33)$$

Let $t \mapsto (J(t,\xi,\epsilon), \theta(t,\xi,\epsilon))$ denote the solution of the system (7.33) such that $J(0,\xi,\epsilon) = \xi$ and $\theta(0,\xi,\epsilon) = 0$. By an application of the implicit function theorem, there is a smooth function $(\xi,\epsilon) \mapsto T(\xi,\epsilon)$ that is defined in a neighborhood of $(J,\theta) = (J_0, 0)$ such that $T(J_0, 0) = 2\pi/\omega(J_0)$ and $\theta(T(\xi,\epsilon), \xi, \epsilon) \equiv 2\pi$. Moreover, let us define a (parametrized) Poincaré map, with the same domain as the transit time map T, by

$$P(\xi,\epsilon) = J(T(\xi,\epsilon), \xi, \epsilon).$$

By expanding the function $\epsilon \mapsto P(\xi,\epsilon)$ into a Taylor series at $\epsilon = 0$, we obtain

$$P(\xi,\epsilon) = J(T(\xi,0),\xi,0) + \epsilon(\dot{J}(T(\xi,0),\xi,0)T_\epsilon(\xi,0) \\ + J_\epsilon(T(\xi,0),\xi,0)) + O(\epsilon^2).$$

Note that $J(T(\xi,0),\xi,0) \equiv \xi$ and $\dot{J}(T(\xi,0),\xi,0) = 0$. Moreover, the function $t \mapsto J_\epsilon(t,\xi,0)$ is the solution of the variational initial value problem given by

$$\dot{W} = \bar{F}(J(t,\xi,0)), \qquad W(0) = 0.$$

Using the identities $J(t,\xi,0) \equiv \xi$ and $T(\xi,0) \equiv 2\pi/\omega(\xi)$, it follows that $J_\epsilon(t,\xi,0) = t\bar{F}(\xi)$ and

$$P(\xi,\epsilon) = \xi + \epsilon \frac{2\pi}{\omega(\xi)} \bar{F}(\xi) + O(\epsilon^2). \tag{7.34}$$

Consider the displacement function $\delta(\xi,\epsilon) := P(\xi,\epsilon) - \xi$ and note that its zeros correspond to the fixed points of the Poincaré map. Also, the zeros of the displacement function are the same as the zeros of the reduced displacement function defined by

$$\Delta(\xi,\epsilon) := \frac{2\pi}{\omega(\xi)} \bar{F}(\xi) + O(\epsilon).$$

By easy computations, it follows that $\Delta(J_0, 0) = 0$ and

$$\Delta_\xi(J_0, 0) = \frac{2\pi}{\omega(J_0)} D\bar{F}(J_0).$$

Hence, by an application of the implicit function theorem, there is a function $\epsilon \mapsto \beta(\epsilon)$ defined on some interval containing $\epsilon = 0$ such that $\beta(0) = J_0$ and such that for each ϵ in the domain of β, the vector $\beta(\epsilon) \in \mathbb{R}^M$ is a fixed point of the Poincaré map $\xi \mapsto P(\xi,\epsilon)$. In particular, $(J,\theta) = (\beta(\epsilon), 0)$ is the initial condition for a periodic orbit of the system (7.33). Since the original system (7.32) is obtained from system (7.33) by an (appropriately periodic) change of coordinates, there are corresponding periodic orbits in the original system.

Finally, to determine the stability type of the periodic orbit, we must compute the derivative of the Poincaré map with respect to the space variable. Using the series expansion (7.34), if the derivative with respect to ξ is evaluated at the initial point $\xi = \beta(\epsilon)$ of the perturbed periodic orbit and the result is expanded in a Taylor series at $\epsilon = 0$, the following formula is obtained:

$$P_\xi(\beta(\epsilon), \epsilon) = I + \epsilon \frac{2\pi}{\omega(J_0)} D\bar{F}(J_0) + O(\epsilon), \tag{7.35}$$

where, in deference to tradition, I in this formula is the identity map of $\mathbb{R}^M$, not the variable I in the original differential equation.

Abstractly, the matrix equation (7.35) has the form

$$P - I = \epsilon(A + \mathcal{R}(\epsilon))$$

where A is infinitesimally hyperbolic with, say, N eigenvalues with positive real parts and $M - N$ eigenvalues with negative real parts. If ϵ is sufficiently small, then the matrix $A + \mathcal{R}(\epsilon)$ has the same number of such eigenvalues. If in addition $\epsilon > 0$, then the matrix $\epsilon(A + \mathcal{R}(\epsilon))$ has the same number of such eigenvalues that are all as close to the origin in the complex plane as desired. Since there are only a finite number of eigenvalues and the eigenvalues of P are exactly the eigenvalues of the matrix $\epsilon(A + \mathcal{R}(\epsilon))$ shifted one unit to the right in the complex plane, it follows that, for sufficiently small positive ϵ, the matrix P has N eigenvalues outside the unit circle and $M - N$ eigenvalues inside the unit circle, as required. The proof that this structure is preserved by the inverse of the averaging transformation and is therefore inherited by the original system is left to the reader.   □

The partially averaged system (7.31) obtained above is given more precisely by the system

$$\dot{J} = \mu\bar{F}(\theta) + O(\mu^2), \qquad \dot{\theta} = \mu m \omega'(I_0)J + O(\mu^2) \tag{7.36}$$

where the presence of perturbation terms is indicated by the order symbol. Let us assume that $\omega'(I_0) > 0$ and consider some of the possible phase portraits of this system. The phase portrait (of the phase *plane*) of the first order approximation of system (7.36) in case $\bar{F} = 0$ is depicted in Figure 7.5. The J-axis, the intersection of the resonance manifold with the (J, θ)-plane, consists entirely of rest points. A higher order analysis is required to determine the dynamics of the perturbed system. The phase portrait for the first order approximation in case $\bar{F}$ has fixed sign (taken here to be positive) is shown in Figure 7.6. In this case all orbits pass through the resonance. A typical phase portrait for the case where $\bar{F}$ has simple zeros is depicted in Figure 7.7. There are several regions corresponding to librational motions where orbits are permanently captured into resonance. Finally, in Figure 7.8, two possible phase portraits of the stroboscopic Poincaré map

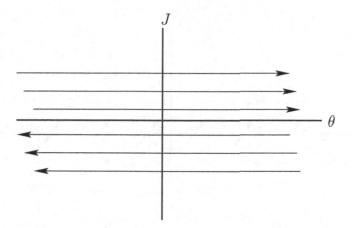

Figure 7.5: Phase portrait of the first order approximation of the partially averaged system (7.36) in case $\bar{F} = 0$.

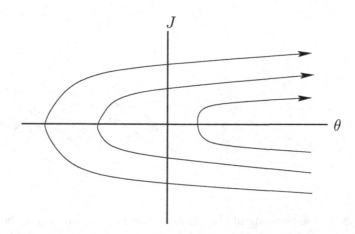

Figure 7.6: Phase portrait of the first order approximation of the partially averaged system (7.36) in case $\bar{F}$ is a positive function.

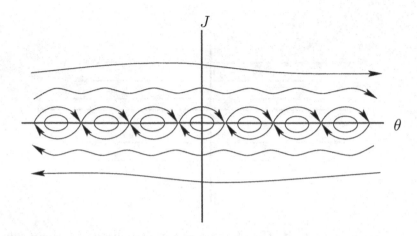

Figure 7.7: Phase portrait of the first order approximation of the partially averaged system (7.36) in case $\bar{F}$ has simple zeros.

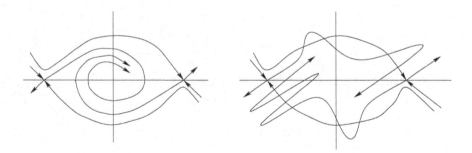

Figure 7.8: Phase portrait of the stroboscopic Poincaré map for the perturbed system (7.36). The left panel depicts entrainment, the right panel depicts chaos.

of the perturbed system are illustrated. Whereas the left panel corresponds to resonance capture—in the context of a periodically perturbed oscillator this would also be called entrainment—the right hand panel corresponds to *transient chaos;* that is, the chaotic invariant set is of saddle type so that nearby orbits approach the chaotic set along a stable manifold, they "feel" the chaos on some finite time scale, and they eventually drift away along an unstable manifold.

Exercise 7.12. In Theorem 7.11, suppose that the rest point is nondegenerate but not hyperbolic. What can be said about the stability type of the corresponding periodic orbit?

Exercise 7.13. Compare and contrast the continuation theory for periodic orbits of planar periodically perturbed oscillators given in Chapter 5 and the theory presented in this chapter.

Exercise 7.14. Consider the following modification of an example introduced in [100] and [103], namely, the system

$$\dot{x} = \quad y(1 - x^2 - y^2) + \epsilon[\delta x - x(x^2 + y^2) + \gamma x \cos(\Omega t)],$$
$$\dot{y} = -x(1 - x^2 - y^2) + \epsilon[\delta y - y(x^2 + y^2)] \qquad (7.37)$$

where δ, γ, and Ω are positive constants and ϵ is a small parameter.

Here the action-angle variables are trigonometric. Show that (I, θ) defined by the transformation

$$x = \sqrt{2I} \sin\theta, \qquad y = \sqrt{2I} \cos\theta$$

are action-angle variables for the system (7.37). The square root is employed to make the transformation have Jacobian equal to one. This is important in Hamiltonian mechanics where it is desirable to have coordinate transformations that respect the Hamiltonian structure—such transformations are called symplectic or canonical. At any rate, to find continuable periodic orbits, consider the $(m : n) = (2 : 1)$ resonance. Partially average the system at this resonance and use Theorem 7.11 to conclude that the original system has periodic orbits for small $\epsilon > 0$.

There are some interesting dynamics going on in this example. Try some numerical experiments to approximate the phase portrait of the stroboscopic Poincaré map. What is the main feature of the dynamics? Can you see the subharmonic solutions near the $(2 : 1)$ resonance? In addition to the references given above, look at [39].

Exercise 7.15. If a *linear* oscillator is periodically perturbed, then its response is periodic with the same frequency as the perturbation. On the other hand, the amplitude of the response depends on the frequency. In particular, the amplitude is large if the input frequency is (nearly) resonant with the natural frequency of the oscillator. A lot of important scientific work and a lot of engineering has been accomplished under the impression that the above statements are true when the first sentence begins with the phrase "If an oscillator is periodically forced ... ." By reading to this point in this book you are in a strong position to challenge

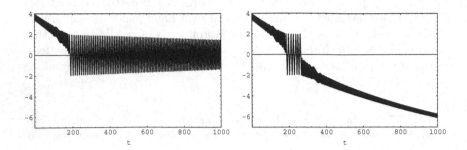

Figure 7.9: The figure depicts the response signal for $v := \dot\theta$ versus time for the system $\dot\theta = v$, $\dot v = -\sin\theta - \epsilon(m_1 + m_2 v - B\cos(\Omega(t - t_0))\sin\theta)$ with $t_0 = 0$, $\Omega = 2$, $m_1 = 10$, $m_2 = 1$, $B = 32$, and $\epsilon = .001$. The left panel depicts an orbit that is captured into resonance; the initial condition is $(\theta, \dot\theta) = (0, 3.940252)$. The right panel depicts the corresponding signal for the orbit with initial condition $(\theta, \dot\theta) = (0, 3.940253)$.

these statements when the word "linear" is left out. Prove that the statements are true for linear oscillators and give examples to show that nonlinear oscillators do not always behave so simply. Suppose that a nonlinear oscillator, say $\dot x = f(x)$, is periodically perturbed with a periodic perturbation of frequency Ω and the function $t \mapsto x_i(t)$ is observed where x_i is one of the component functions of a solution $t \mapsto (x_1(t), \dots, x_n(t))$. Will the signal $t \mapsto x_i(t)$ retain some "trace" of the periodic input? For example, consider the power spectrum of this function, that is, the square of the absolute value of its Fourier transform. Will the frequency Ω have a large amplitude in the power spectrum? Try some numerical experiments. The previous question does not have a simple answer. But questions of this type arise all the time in physics and engineering where we are confronted with multivariable systems that are often far too complex to be analyzed with analytic methods. Discuss the reasons why the study of simple models might be valuable for understanding complex systems.

Exercise 7.16. Consider the system

$$\dot\theta = v, \qquad \dot v = -\sin\theta - \epsilon(m_1 + m_2 v - B\cos(\Omega(t - t_0))\sin\theta),$$

a parametrically excited pendulum with damping and torque. Reproduce the Figure 7.9 as an illustration of passage through resonance. Determine an approximate neighborhood of the point $(\theta, \dot\theta) = (0, 3.940252)$ corresponding to the initial conditions for orbits that are captured into resonance. Can you automate a criterion for "capture into resonance"? Explore other regions of the parameter space by using numerical experiments.

7.3   Action-Angle Variables

To use the theory presented so far in this chapter we must be able to express our oscillator in action-angle variables. In practice, the construction of action-angle variables is a formidable task—recall the construction of the Delaunay variables in Chapter 3. For linear oscillators the appropriate coordinate change can be constructed using polar coordinates, while the construction of action-angle variables for the pendulum requires the use of Jacobi elliptic functions. A general construction of action-angle variables for planar oscillators is presented in this section. The construction uses some of the ideas discussed in Chapter 5.

Let us consider a differential equation of the form

$$\dot{u} = f(u) + \epsilon g(u, t) \tag{7.38}$$

where the unperturbed system

$$\dot{u} = f(u) \tag{7.39}$$

has a period annulus $\mathcal{A}$. We will construct action-angle variables near a periodic orbit Γ contained in $\mathcal{A}$. The differential equation (7.39), expressed in the new coordinates that we denote by I and ϑ, has the form

$$\dot{I} = 0, \qquad \dot{\vartheta} = \omega(I).$$

Interpreted geometrically, these new coordinates are related to polar coordinates in that I is a radial variable and ϑ is an angular variable. In fact, whereas I is constant on each periodic solution, ϑ changes linearly on each periodic solution. In case the system (7.39) is Hamiltonian, the new coordinates reduce to the usual action-angle variables on $\mathcal{A}$.

With reference to system (7.39), define the orthogonal system

$$\dot{u} = f^{\perp}(u), \qquad u \in X \tag{7.40}$$

where, in oriented local coordinates, $f^{\perp}(u) := Jf(u)$ with

$$J = \begin{pmatrix} 0 & -1 \\ 1 & 0 \end{pmatrix}.$$

We mention that J rotates vectors in the plane through a positive angle of $\pi/2$ radians. The same symbol J is often used in this context with the opposite sign.

Let φ_t denote the flow of the differential equation (7.39) and let ψ_t denote the flow of the differential equation (7.40). Also, for vectors ξ_1 and ξ_2 in $\mathbb{R}^2$, define $\xi_1 \wedge \xi_2 := \langle \xi_1, J\xi_2 \rangle$, where the brackets denote the usual inner product in $\mathbb{R}^2$.

A periodic orbit Γ of (7.39) has an orientation determined by its time parameterization. To specify an orientation, we define $\varepsilon = \varepsilon(f) = 1$ in case for each $\zeta \in \Gamma$ the vector $f^{\perp}(\zeta)$ is the outer normal at ζ. If $f^{\perp}(\zeta)$ is the inner normal, then $\varepsilon := -1$. Also, the orientation of the period annulus $\mathcal{A}$ is defined to be the orientation inherited from its constituent periodic solutions.

Choose a point $\zeta \in \mathcal{A}$ and note that there is an open interval $U \subset \mathbb{R}$ containing the origin such that the image of the map $\rho \mapsto \psi_{\rho}(\zeta)$ for $\rho \in U$ is a section Σ_{ζ} transverse to the orbits of system (7.39) in $\mathcal{A}$. Also, define $\Upsilon : U \times \mathbb{R} \to \mathcal{A}$ by

$$\Upsilon(\rho, \phi) = \varphi_{\phi}(\psi_{\rho}(\zeta)). \tag{7.41}$$

Clearly, Υ is smooth. In fact, Υ is a covering map, that is, a periodic coordinate system on $\mathcal{A}$. We will see below that Υ defines "flow box" coordinates: coordinates that straighten out the flow in a neighborhood of the periodic orbit containing the point ζ.

To construct the action-angle variables, let us begin by considering the derivative of the map Υ defined in display (7.41). Diliberto's theorem (Theorem 5.5) states that if

$$b(t, \zeta) := \frac{\|f(\zeta)\|^2}{\|f(\varphi_t(\zeta))\|^2} e^{\int_0^t \operatorname{div} f(\varphi_s(v)) \, ds},$$

$$a(t, \zeta) := \int_0^t \left(2\kappa(s, \zeta) \|f(\varphi_s(\zeta))\| - \operatorname{curl} f(\varphi_s(\zeta)) \right) b(s, \zeta) \, ds, \tag{7.42}$$

where κ denotes the signed scalar curvature along the curve $t \mapsto \varphi_t(\zeta)$, $\zeta \in \mathcal{A}$, then

$$D\Upsilon(\rho, \phi) \frac{\partial}{\partial \phi} = f(\Upsilon(\rho, \phi)),$$

$$D\Upsilon(\rho, \phi) \frac{\partial}{\partial \rho} = b(\phi, \psi_{\rho}(v)) f^{\perp}(\Upsilon(\rho, \phi)) + a(\phi, \psi_{\rho}(v)) f(\Upsilon(\rho, \phi)).$$

In other words, the matrix representation of the derivative $D\Upsilon(\rho, \phi)$ relative to the ordered bases $\{\partial/\partial \rho, \partial/\partial \phi\}$ and $\{f^{\perp}(\Upsilon(\rho, \phi)), f(\Upsilon(\rho, \phi))\}$ is given by

$$D\Upsilon(\rho, \phi) = \begin{pmatrix} b(\phi, \psi_{\rho}(v)) & 0 \\ a(\phi, \psi_{\rho}(v)) & 1 \end{pmatrix}.$$

Since b does not vanish for $\zeta \in \mathcal{A}$, it follows that Υ is a local diffeomorphism and in fact Υ is a covering map onto its image.

To express the original system (7.38) in (ρ, ϕ)-coordinates, note first that there are smooth functions $(u, t) \mapsto p(u, t)$ and $(u, t) \mapsto q(u, t)$ such that

$$g(u, t) = p(u, t) f^{\perp}(u) + q(u, t) f(u) \tag{7.43}$$

for all $(u, t) \in \mathcal{A} \times \mathbb{R}$. Thus, to change system (7.38) to the new coordinates, we simply solve for

$$j(u, t) \frac{\partial}{\partial \rho} + k(u, t) \frac{\partial}{\partial \phi}$$

in the matrix equation

$$\begin{pmatrix} b & 0 \\ a & 1 \end{pmatrix} \begin{pmatrix} j \\ k \end{pmatrix} = \begin{pmatrix} \epsilon p \\ 1 + \epsilon q \end{pmatrix}$$

to obtain

$$\begin{pmatrix} j \\ k \end{pmatrix} = \begin{pmatrix} \epsilon \frac{1}{b} p \\ 1 + \epsilon(q - \frac{a}{b} p) \end{pmatrix}.$$

It follows that system (7.39) in the new coordinates is given by

$$\dot{\rho} = \epsilon \frac{1}{b(\phi, \psi_\rho(v))} p(\Upsilon(\rho, \phi), t),$$

$$\dot{\phi} = 1 + \epsilon \Big(q(\Upsilon(\rho, \phi), t) - \frac{a(\phi, \psi_\rho(v))}{b(\phi, \psi_\rho(v))} p(\Upsilon(\rho, \phi), t) \Big). \tag{7.44}$$

To compress notation, let us write (7.44) in the form

$$\dot{\rho} = \epsilon Q(\rho, \phi, t), \qquad \dot{\phi} = 1 + \epsilon R(\rho, \phi, t). \tag{7.45}$$

Define a second change of coordinates by

$$\rho = \beta(I), \qquad \phi = \alpha(I)\vartheta \tag{7.46}$$

where $I \mapsto \alpha(I)$ and $I \mapsto \beta(I)$ are smooth functions to be specified below. Here, since the coordinate transformation must be invertible, we need only assume that $\alpha(I)\beta'(I) \neq 0$. In the (I, ϑ)-coordinates, system (7.44) has the form

$$\dot{I} = \epsilon \frac{1}{\beta'(I)} Q(\beta(I), \alpha(I)\vartheta, t),$$

$$\dot{\vartheta} = \frac{\dot{\phi} - \vartheta \alpha'(I)\dot{I}}{\alpha(I)} \tag{7.47}$$

$$= \frac{1}{\alpha(I)} + \epsilon \Big(\frac{1}{\alpha(I)} R(\beta(I), \alpha(I)\vartheta, t) - \vartheta \frac{\alpha'(I)}{\alpha(I)\beta'(I)} Q(\beta(I), \alpha(I)\vartheta, t) \Big).$$

To specify the functions α and β we require two auxiliary functions—the period function and the area function. To define the period function, recall that the image of the map $\rho \mapsto \psi_\rho(\zeta)$ for $\rho \in U$ is a section for the unperturbed flow on the period annulus $\mathcal{A}$. The period function on $\mathcal{A}$ relative to this section is the map $\widetilde{T} : U \to \mathbb{R}$ that assigns to each $\rho \in U$ the minimum period of the solution of system (7.39) that passes through the point $\phi_\rho(\zeta) \in \mathcal{A}$. In the "standard" case, $\mathcal{A}$ is an annulus whose inner boundary is a rest point. In this case, we define the area function $\zeta \mapsto A(\zeta)$; it assigns to each $\zeta \in \mathcal{A}$ the area enclosed by the unperturbed solution through ζ.

The function β is defined to be the solution of the initial value problem

$$\frac{d\rho}{dI} = \varepsilon \frac{2\pi}{\widetilde{T}(\rho)} \frac{1}{||f(\psi_\rho(\zeta))||^2}, \qquad \rho(I_0) = 0 \tag{7.48}$$

where in the standard case $I_0 = A(\zeta)/(2\pi)$, and in the case where $\mathcal{A}$ has a nontrivial inner boundary $I_0 = 0$. The choice of initial condition for the standard case agrees with tradition; a different choice of initial condition simply results in a constant translation of the "action" variable. The function α is defined by

$$\alpha(I) := -\varepsilon \frac{\widetilde{T}(\beta(I))}{2\pi} \tag{7.49}$$

where $\varepsilon = \pm 1$ according to the orientation of the period annulus $\mathcal{A}$.

Using the definition $T(I) := \widetilde{T}(\beta(I))$, the system (7.47) has the form

$$\dot{I} = \varepsilon\epsilon \frac{T(I)}{2\pi} ||f(\psi_\rho(\zeta))||^2 Q(\beta(I), \alpha(I)\vartheta, t),$$

$$\dot{\vartheta} = -\varepsilon \frac{2\pi}{T(I)} - \varepsilon\epsilon \left(\frac{2\pi}{T(I)} R(\beta(I), \alpha(I)\vartheta, t) \right.$$

$$\left. + \vartheta \frac{T'(I)}{2\pi} ||f(\psi_\rho(\zeta))||^2 Q(\beta(I), \alpha(I)\vartheta, t) \right). \tag{7.50}$$

From equation (7.43), we have the identities

$$p = \frac{1}{||f||^2} \langle g, f^\perp \rangle = \frac{1}{||f||^2} f \wedge g, \qquad q = \frac{1}{||f||^2} \langle f, g \rangle.$$

In view of system (7.44), the system (7.50) can be rewritten in the form

$$\dot{I} = \varepsilon\epsilon \frac{T(I)}{2\pi} \mathcal{E}(I, \vartheta) f(\Upsilon(\beta(I), \alpha(I)\vartheta)) \wedge g(\Upsilon(\beta(I), \alpha(I)\vartheta), t),$$

$$\dot{\vartheta} = -\varepsilon \frac{2\pi}{T(I)}$$

$$- \varepsilon\epsilon \left[\frac{2\pi}{T(I)} ||f(\Upsilon(\beta(I), \alpha(I)\vartheta))||^{-2} \langle f, g \rangle + \left(\vartheta \frac{T'(I)}{2\pi} ||f(\psi_{\beta(I)}(\zeta))||^2 \right. \right.$$

$$\left. \left. - \frac{2\pi}{T(I)} a(\alpha(I)\vartheta, \psi_{\beta(I)}(\zeta)) \right) ||f(\phi_{\beta(I)}(\zeta))||^{-2} \mathcal{E}(I, \vartheta) f \wedge g \right] \tag{7.51}$$

where

$$\mathcal{E}(I, \vartheta) := e^{-\int_0^{\alpha(I)\vartheta} \operatorname{div} f(\Upsilon(\beta(I), \alpha(I)s))\, ds}.$$

Again, for notational convenience, let us write the first order system (7.51) in the compact form

$$\dot{I} = \epsilon F(I, \vartheta, t), \qquad \dot{\vartheta} = \omega(I) + \epsilon G(I, \vartheta). \tag{7.52}$$

Note that both F and G are 2π-periodic in ϑ and $2\pi/\Omega$-periodic in t. Thus, we have transformed the original perturbed system to action-angle coordinates.

To prove that the action-angle coordinate transformation

$$u = \Upsilon(\beta(I), \alpha(I)\vartheta) \tag{7.53}$$

is canonical in case the unperturbed system is Hamiltonian, it suffices to show the transformation is area preserving, that is, the Jacobian of the transformation is unity. In fact, the Jacobian is

$$\det\left[\begin{pmatrix} -f_2(u) & f_1(u) \\ f_1(u) & f_2(u) \end{pmatrix} \begin{pmatrix} a(\phi, \psi_\rho(\zeta)) & 1 \end{pmatrix} \begin{pmatrix} \beta'(I) & 0 \\ \alpha'(I)\vartheta & \alpha(I) \end{pmatrix}\right]$$

$$= \frac{\|f(u)\|^2}{\|f(\psi_\rho(\zeta))\|^2} b(\phi, \psi_\rho(\zeta)).$$

But, if f is a Hamiltonian vector field, then $\operatorname{div} f = 0$, and

$$b(\phi, \psi_\rho(\zeta)) = \frac{\|f(\psi_\rho(\zeta))\|^2}{\|f(u)\|^2},$$

as required. Moreover, in case f is the Hamiltonian vector field for the Hamiltonian H, we have $f(u) = -J \operatorname{grad} H(u)$. Recall that $\rho = \beta(I)$ and define $h := H(\psi_\rho(\zeta))$. Then,

$$\frac{dI}{dh} = \varepsilon\frac{\widetilde{T}(\rho(h))}{2\pi}.$$

Thus, the derivative of the action variable with respect to energy is the normalized energy-period function, as it should be.

8

Local Bifurcation

Consider the family of differential equations

$$\dot{u} = f(u, \lambda), \quad u \in \mathbb{R}^n, \quad \lambda \in \mathbb{R}. \tag{8.1}$$

In case $f(u_0, \lambda_0) = 0$, the differential equation with parameter value $\lambda = \lambda_0$ has a rest point at u_0 and the linearized system at this point is given by

$$\dot{W} = f_u(u_0, \lambda_0)W. \tag{8.2}$$

If the eigenvalues of the linear transformation $f_u(u_0, \lambda_0) : \mathbb{R}^n \mapsto \mathbb{R}^n$ are all nonzero, then the transformation is invertible, and by an application of the implicit function theorem there is a curve $\lambda \mapsto \beta(\lambda)$ in $\mathbb{R}^n$ such that $\beta(\lambda_0) = u_0$ and $f(\beta(\lambda), \lambda) \equiv 0$. In other words, for each λ in the domain of β the point $\beta(\lambda) \in \mathbb{R}^n$ corresponds to a rest point for the member of the family (8.1) at the parameter value λ.

Recall that if all eigenvalues of the linear transformation $f_u(u_0, \lambda_0)$ have nonzero real parts, then the transformation is called *infinitesimally hyperbolic* and the rest point u_0 is called *hyperbolic*. Also, in this case, since the eigenvalues of $Df(u, \lambda)$ depend continuously on u and the parameter λ, if $|\lambda - \lambda_0|$ is sufficiently small, then the rest point $u = \beta(\lambda)$ of the differential equation (8.1) at the parameter value λ has the same stability type as the rest point $u_0 = \beta(\lambda_0)$. In particular, if the rest point u_0 is hyperbolic, then for sufficiently small λ the perturbed rest point $\beta(\lambda)$ is also hyperbolic.

If $f_u(u_0, \lambda_0)$ is not infinitesimally hyperbolic, then there is at least one eigenvalue with zero real part. It turns out that the topology of the local phase portrait of the corresponding differential equation (8.1) at this rest point may change under perturbation (that is, the local phase portrait at u_0 may change for λ close but not equal to λ); if it does, we will say that a *bifurcation* occurs. For example, the phase portrait for a nearby differential equation may have no rest points or several rest points in the vicinity of the original rest point. In this chapter, we will consider such bifurcations in case the linear transformation $f_u(u_0, \lambda_0)$ has a simple zero eigenvalue; that is, a zero eigenvalue with algebraic (and geometric) multiplicity one, or a pair of pure imaginary complex conjugate eigenvalues each with algebraic multiplicity one, and we will describe some of the "generic" bifurcations that occur under these conditions.

Although only the loss of stability at a rest point of a differential equation will be discussed, the basic results presented here can be modified to cover the case of the loss of stability of a fixed point of a map; and in turn the modified theory can be applied to the Poincaré map to obtain a bifurcation theory for periodic orbits. The extension of bifurcation theory from rest points to periodic orbits is only the beginning of a vast subject that has been developed far beyond the scope of this book. For example, the loss of stability of a general invariant manifold can be considered. On the other hand, bifurcation theory is by no means complete: Many interesting problems are unresolved. (See the books [8] and [58] for detailed and wide ranging results on bifurcations of planar vector fields, and [11], [57], [97], [98], [103], [151], [208], [232], and [233] for more general bifurcation theory.)

Exercise 8.1. Prove that the eigenvalues of an $n \times n$ matrix depend continuously on the components of the matrix.

8.1   One-Dimensional State Space

In this section, some of the general concepts of bifurcation theory will be illustrated in their simplest form by an analysis of the most important bifurcation associated with rest points of scalar differential equations, namely, the saddle-node bifurcation. In addition, we will discuss how bifurcation problems arise in applied mathematics.

8.1.1   The Saddle-Node Bifurcation

Consider the family of differential equations

$$\dot{u} = \lambda - u^2, \quad u \in \mathbb{R}, \quad \lambda \in \mathbb{R} \tag{8.3}$$

and note that if $f(u, \lambda) := \lambda - u^2$, then

$$f(0,0) = 0, \quad f_u(0,0) = 0, \quad f_{uu}(0,0) = -2, \quad f_\lambda(0,0) = 1.$$

Also, the rest points for members of this family are given by $\lambda = u^2$. Thus, if $\lambda < 0$, then there are no rest points; if $\lambda = 0$, then there is one rest point called a *saddle-node* (the system matrix for the linearization has a simple zero eigenvalue); and if $\lambda > 0$, then there are two rest points given by $u = \pm\sqrt{\lambda}$, one stable and the other unstable. This family provides an example of a *saddle-node bifurcation* (see Figure 1.6 for the bifurcation diagram).

Roughly speaking, the one-parameter family (8.1) has a saddle-node bifurcation at (u_0, λ_0) if its bifurcation diagram near (u_0, λ_0), with the point (u_0, λ_0) translated to the origin, is similar to Figure 1.6. The next proposition states the precise conditions that the bifurcation diagram must satisfy and sufficient conditions for a saddle-node bifurcation to occur at $(u, \lambda) = (0,0)$ in case system (8.1) is a scalar differential equation; a formal definition and a more general theorem on saddle-node bifurcation (Theorem 8.12) will be formulated and proved in Section 8.2.

Proposition 8.2. *Suppose that the differential equation (8.1) is given by a smooth (parameter-dependent) vector field $(u, \lambda) \mapsto f(u, \lambda)$. If*

$$f(0,0) = 0, \quad f_u(0,0) = 0, \quad f_{uu}(0,0) \neq 0, \quad f_\lambda(0,0) \neq 0,$$

then there is a saddle-node bifurcation at $(u, \lambda) = (0,0)$. In particular, there is a number $p_0 > 0$ and a unique smooth curve β in $\mathbb{R} \times \mathbb{R}$ given by $p \mapsto (p, \gamma(p))$ for $|p| < p_0$ such that each point in the range of β corresponds to a rest point, and the range of β is quadratically tangent to $\mathbb{R} \times \{0\}$; that is,

$$f(p, \gamma(p)) \equiv 0, \qquad \gamma(0) = \gamma'(0) = 0, \quad \gamma''(0) \neq 0.$$

Moreover, the stability type of the rest points corresponding to β changes at $p = 0$; that is, $p \mapsto f_u(p, \gamma(p))$ changes sign at $p = 0$. Also, $\gamma''(0) = -f_{uu}(0,0)/f_\lambda(0,0)$.

Proof. Because $f_\lambda(0,0) \neq 0$, we can apply the implicit function theorem to obtain the existence of a curve $p \mapsto \gamma(p)$ such that $\gamma(0) = 0$ and $f(p, \gamma(p)) \equiv 0$ for $|p| < p_0$ where p_0 is some positive real number. Since the derivative of the function $p \mapsto f(p, \gamma(p))$ is zero, we have the identity

$$f_u(p, \gamma(p)) + f_\lambda(p, \gamma(p))\gamma'(p) = 0.$$

In particular,

$$f_u(0,0) + f_\lambda(0,0)\gamma'(0) = 0,$$

and, in view of the hypotheses, $\gamma'(0) = 0$. Since the second derivative of the function $p \mapsto f(p, \gamma(p))$ is also zero, we have the equation

$$f_{uu}(0,0) + f_\lambda(0,0)\gamma''(0) = 0.$$

By rearrangement of this equation and by the hypotheses of the proposition, it follows that

$$\gamma''(0) = -\frac{f_{uu}(0,0)}{f_\lambda(0,0)} \neq 0.$$

Finally, because the derivative of the map $p \mapsto f_u(p, \gamma(p))$ at $p = 0$ is the nonzero number $f_{uu}(0,0)$, this map indeed changes sign at $p = 0$. $\square$

8.1.2 A Normal Form

If f satisfies all the hypotheses of Proposition 8.2, then by an application of the preparation theorem (Theorem 5.15) this function can be factored in the form

$$f(u, \lambda) = (a_0(u) + \lambda)U(u, \lambda)$$

where $a_0(0) = 0$ and $U(0,0) \neq 0$. Thus, the flow of the differential equation

$$\dot{u} = f(u, \lambda) \tag{8.4}$$

is topologically equivalent to the flow of the differential equation $\dot{u} = a_0(u) + \lambda$ on some open neighborhood of the origin by the identity homeomorphism. Or, if you like, the two differential equations are equivalent by a rescaling of time (see Proposition 1.30). Moreover, taking into account our hypotheses $f_u(0,0) = 0$ and $f_{uu}(0,0) \neq 0$, we have that $a_0'(0) = 0$ and $a_0''(0) \neq 0$. As a result, the function a is given by

$$a_0(u) = \frac{1}{2}a_0''(0)u^2 + O(u^3).$$

By the Morse lemma (see Exercise 5.63), there is a change of coordinates $u = \mu(y)$ with $\mu(0) = 0$ that transforms the differential equation (8.4) into the form

$$\dot{y} = \frac{1}{\mu'(y)}(\lambda \pm y^2)$$

where, of course, $\mu'(y) \neq 0$ because the change of coordinates is invertible. By a final rescaling of time and, if necessary, a change in the sign of λ, we obtain the equivalent differential equation

$$\dot{y} = \lambda - y^2. \tag{8.5}$$

The family (8.5) is a *normal form* for the saddle-node bifurcation: Every one-parameter family of scalar differential equations that satisfies the hypotheses of Proposition 8.2 at a point in the product of the phase space and the parameter space can be (locally) transformed to this normal form by a (nonlinear) change of coordinates and a rescaling of time. In this context, the differential equation (8.5) is also called a *versal deformation* or a *universal unfolding* of the saddle-node.

The reader may suspect that the use of such terms as "versal deformation" and "universal unfolding" is indicative of a rich and mature underlying theory. This is indeed the case. Moreover, there are a number of excellent books on this subject. For example, the book of Vladimir Arnold [11] has a masterful exposition of the "big ideas" of bifurcation theory while the books of Martin Golubitsky and David G. Schaeffer [97] and Golubitsky, Ian Stewart, and Schaeffer [98] contain a more comprehensive study of the subject (see also [57] and [58]).

In the next two sections we will explore some of the philosophy of bifurcation theory and discuss how bifurcation problems arise in applied mathematics.

8.1.3 Bifurcation in Applied Mathematics

Is bifurcation theory important in applied mathematics? To discuss this question, let us suppose that we have a model of a physical system given by a family of differential equations that depends on some parameters. We will consider the process that might be used to identify these parameters and the value of the resulting model for making physical predictions.

In a typical scenario, a model has "system parameters" and "control parameters." System parameters specify the measurements of intrinsic physical properties, whereas control parameters correspond to adjustments that can be made while maintaining the integrity of the physical system. By changing the control parameters in the mathematical model, we can make predictions so as to avoid expensive physical experiments. Also, we can explore the phenomena that occur over the range of the control parameters.

Ideally, system parameters are identified by comparing predictions of the model with experimental data. But, for a realistic model with several system parameters, the parameter identification will almost always require a complicated analysis. In fact, parameter identification is itself a fascinating and important problem in applied mathematics that is not completely solved. Let us simply note that the parameter identification process will not be exact. Indeed, if an approximation algorithm is combined with experimental data, then some uncertainty is inevitable.

Suppose the model system of differential equations that we obtain from our parameter identification process contains a degenerate rest point for some choices of the control parameters. Have we just been unlucky? Can we adjust the parameters to avoid the degeneracy? What does the appearance of a degenerate rest point tell us about our original model?

Let us first consider the case where there are no control parameters. If, for example, our original model is given by the differential equation (8.5) and our parameter identification process results in specifying the system parameter value $\lambda = 0$ so that the corresponding differential equation has a degenerate rest point, then it would seem that we have been very unlucky. Indeed, predictions from the model with $\lambda = 0$ would seem to be quite unreliable. By an arbitrarily small change in the estimated value of the system parameter, we can construct a model differential equation with two hyperbolic rest points or no rest points at all. The choice $\lambda = 0$ for the system parameter produces a model that is not structurally stable. On the other hand, by arbitrarily small changes of the system parameter, we can produce two structurally stable models with completely different qualitative behavior (see Exercise 8.3).

Clearly, it is important to know if the choice of system parameters produces a structurally unstable model or a model that is "close" to one that is structurally unstable; if this is the case, then it is important to analyze the qualitative behavior of the models that are produced by small changes in the system parameters. Whereas in the scalar model (8.5) the analysis is transparent, it is not at all obvious how we might detect such structural instabilities in a multiparameter or multidimensional model. On the other hand, because system parameters are viewed as fixed once they are identified, we can theoretically avoid the structural instabilities by simply reassigning the system parameters.

For the record, two vector fields defined on the same state space are called *topologically equivalent* if there is a homeomorphism of the state space that maps all orbits of the first vector field onto orbits of the second vector field and preserves the direction of time along all the orbits (the time parameterization of the orbits is ignored). Of course, if two vector fields are topologically equivalent, then their phase portraits are qualitatively the same. A vector field (and the corresponding differential equation) is called *structurally stable* if there is an open set of vector fields in the C^1 topology that contains the given vector field, and all vector fields in this open set are topologically equivalent to the given vector field. The idea is that the topological type of a structurally stable vector field is not destroyed by a small smooth perturbation (recall Exercise 1.116).

While it might seem reasonable to suspect that most models are structural stable (for instance, we might expect that the set of structurally stable vector fields is open and dense in the C^1 topology), this is not the case. On the other hand, there is a rich mathematical theory of structural stability. In particular, deep theorems in this subject state necessary and sufficient conditions for a vector field to be structurally stable. An introduction to these results is given in the book of Stephen Smale [204] and the references therein. From the perspective of applied mathematics, the definition of structural instability is perhaps too restrictive. A system is deemed unstable if its topological type is destroyed by an *arbitrary* C^1 perturbation. But in mathematical modeling the differential equations that arise are not arbitrary. Rather, they are derived from physical laws. Thus, the structural stability of a model *with respect to its parameters*—the subject matter of bifurcation theory—is often a more important consideration than the C^1 structural stability of the model.

Let us now consider a model system that does contain control parameters. For example, let us suppose that the original system is given by the differential equation

$$\dot{u} = \lambda - au^2$$

where a is a system parameter and λ is a control parameter. If our parameter identification algorithm produces a nonzero value for the system parameter a, then our model is a one-parameter family of differential equations that has a saddle-node at the control parameter value $\lambda = 0$. Moreover, if $\lambda = 0$ is in the range of the control parameter, then this instability is *unavoidable* for all nearby choices of the system parameter. This observation suggests the reason why bifurcation theory is important in the analysis of models given by *families* of differential equations: While a nondegenerate member of a family may be obtained by a small change of its parameter, all sufficiently small perturbations of the *family* may contain members with a degeneracy. We will discuss this essential fact in more detail in the next section.

Exercise 8.3. Consider the set $\mathcal{S}$ of all smooth functions defined on $\mathbb{R}$ endowed with the $C^1([0,1])$ topology; that is, the distance between f and g in $\mathcal{S}$ is

$$\|f - g\| = \|f - g\|_0 + \|f' - g'\|_0$$

where the indicated C^0-norm is just the usual supremum norm over the unit interval. Also, let S denote the subset of $\mathcal{S}$ consisting of the functions $f \in \mathcal{S}$ that satisfy the following properties: (i) $f(0) \neq 0$ and $f(1) \neq 0$. (ii) If a is in the open interval $(0,1)$ and $f(a) = 0$, then $f'(a) \neq 0$. Prove that each element in S is structurally stable relative to $\mathcal{S}$. Also, prove that S is an open and dense subset of $\mathcal{S}$.

8.1.4 Families, Transversality, and Jets

A structurally unstable system might occur at some parameter value in a family of differential equations. This possibility leads to the question "Is such a degeneracy avoidable for some family obtained by an arbitrarily small perturbation of the given family?" We might also ask if a system in a structurally stable family can contain a nonhyperbolic rest point.

One way to gain some insight into the questions that we have just asked, is to construct a geometric interpretation of the space of vector fields as in Figure 8.1. Indeed, let us consider the space of all smooth vector fields and the subset of all vector fields that have a nonhyperbolic rest point. Suppose that vector fields are represented heuristically by points in usual Euclidean three-dimensional space and degenerate vector fields are represented by the points on a hypersurface $\mathcal{D}$. Since the complement of the set $\mathcal{D}$ is dense, if f is a point in $\mathcal{D}$, then there are points in the complement of $\mathcal{D}$ that

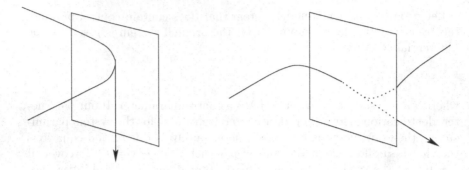

Figure 8.1: Two families of vector fields, represented as curves, meet the set of structurally unstable vector fields represented by hyperplanes. The family in the left hand illustration is tangent to the hyperplane. A small perturbation produces a family consisting entirely of structurally stable vector fields. In contrast, all sufficiently small perturbations of the family depicted as the curve (which might be tangent to the hyperplane) in the right hand illustration have structurally unstable members.

are arbitrarily close to f. By analogy, if our geometric interpretation is faithful, then there is an arbitrarily small C^1 perturbation of our vector field f that is nondegenerate; that is, the corresponding system has only hyperbolic rest points. This is indeed the case if we restrict our vector fields to compact domains.

Next, consider a one-parameter family of vector fields as a curve in the space of all smooth vector fields, and suppose that this curve meets the hypersurface $\mathcal{D}$ that represents the degenerate vector fields. If the curve meets the surface so that its tangent vector at the intersection point is not tangent to the surface—we call this a transversal intersection—then every sufficiently small deformation of the curve will have a nonempty transversal intersection with $\mathcal{D}$. In other words, the degeneracy (depicted by the intersection of the curve with the hypersurface) cannot be removed by perturbation of the curve. By analogy, if our original family of vector fields meets a "surface" corresponding to a degenerate set in the space of all vector fields (which is infinite dimensional) and if the intersection of the curve with this degenerate surface is transversal, then the degeneracy cannot be removed by a small deformation of the family. This is one of the main reasons why bifurcation theory is important in applied mathematics when we are studying a model that is given by a *family* of differential equations.

The geometric picture we have discussed gives the correct impression for structural instabilities due to the nonhyperbolicity of rest points, the subject of this chapter. Indeed, we will show how to make a precise interpretation of this geometry for scalar vector fields. There is, however, an important warning: Our picture is misleading for some more complicated structural instabilities, a topic that is beyond the scope of this book (see, for example, [192] and [204]).

Let us identify the set of all scalar vector fields with the space of smooth functions $C^\infty(\mathbb{R}, \mathbb{R})$. In view of Proposition 8.2, only a finite set of the partial derivatives of a scalar family is required to determine the presence of a saddle-node bifurcation. In fact, this observation is the starting point for the construction of a finite dimensional space, called the space of k-jets, that corresponds to the ambient space in our geometric picture.

Although the "correct" definition of the space of k-jets requires the introduction of vector bundles (see, for example, [3]), we will enjoy a brief glimpse of this theory by considering the special case of the construction for the space $C^\infty(\mathbb{R}, \mathbb{R})$ where everything is so simple that the general definition of a vector bundle can be avoided.

Consider the space $\mathbb{R} \times C^\infty(\mathbb{R}, \mathbb{R})$ and let k denote a nonnegative integer. We will say that two elements (x, f) and (y, g) in the product space are equivalent if

$$(x, f(x), f'(x), f''(x), \dots, f^{(k)}(x)) = (y, g(y), g'(y), g''(y), \dots, g^{(k)}(y))$$

where the equality is in the vector space $\mathbb{R}^{k+2}$. The set of all equivalence classes is denoted $J^k(\mathbb{R}, \mathbb{R})$ and called the *space of k-jets*.

Let us denote the equivalence class determined by (x, f) with the symbol $[x, f]$ and define the natural projection π^k of $J^k(\mathbb{R}, \mathbb{R})$ into $\mathbb{R}$ by $\pi^k([x, f]) = x$. The *k-jet extension* of $f \in C^\infty(\mathbb{R}, \mathbb{R})$ is the map $j^k(f) : \mathbb{R} \to J^k(\mathbb{R}, \mathbb{R})$ defined by

$$j^k(f)(u) = [u, f].$$

Because $\pi^k(j^k(f)(u)) \equiv u$, the k-jet extension is called a section of the fiber bundle with total space $J^k(\mathbb{R}, \mathbb{R})$, base $\mathbb{R}$, and projection π^k. The fiber over the base point $x \in \mathbb{R}$ is the set $\{[x, f] : f \in C^\infty(\mathbb{R}, \mathbb{R})\}$. Also, let us define $\mathcal{Z}^k$ to be the image of the zero section of $J^k(\mathbb{R}, \mathbb{R})$; that is, $\mathcal{Z}^k$ is the image of the map $\zeta : \mathbb{R} \to J^k(\mathbb{R}, \mathbb{R})$ given by $\zeta(u) = [u, 0]$.

The k-jet bundle can be "realized" by a choice of local coordinates. In fact, the usual choice for the local coordinates is determined by the map $\Phi^k : J^k(\mathbb{R}, \mathbb{R}) \to \mathbb{R} \times \mathbb{R}^{k+1}$ defined by

$$\Phi^k([u, f]) = (u, f(u), f'(u), \dots, f^k(u)).$$

It is easy to check that Φ^k is well-defined and that we have the commutative diagram

$$
\begin{array}{ccc}
J^k(\mathbb{R}, \mathbb{R}) & \xrightarrow{\Phi^k} & \mathbb{R} \times \mathbb{R}^{k+1} \\
\downarrow{\scriptstyle \pi^k} & & \downarrow{\scriptstyle \pi_1} \\
\mathbb{R} & \xrightarrow{\text{identity}} & \mathbb{R}
\end{array}
$$

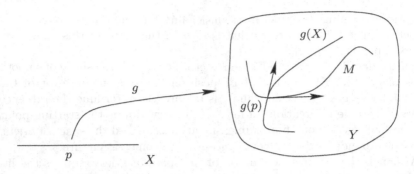

Figure 8.2: The sum of the tangent space to $g(X)$ at $g(p)$ and the tangent space to M at $g(p)$ is the tangent space to Y at $g(p)$. In this case, the map $g : X \to Y$ is transverse to the submanifold $M \subset Y$.

where π_1 is the projection onto the first factor of $\mathbb{R} \times \mathbb{R}^{k+1}$. Thus, $J^k(\mathbb{R}, \mathbb{R})$ is identified with $\mathbb{R} \times \mathbb{R}^{k+1}$ as a smooth manifold. Also, the set $\mathcal{Z}$ is given in the local coordinates by $Z := \mathbb{R} \times \{0\}$. The jet space is the desired finite dimensional space that incorporates all the data needed to consider bifurcations that depend only on a finite number of partial derivatives of a family of scalar vector fields.

We will need the concept of transversality of a map and a submanifold (see Figure 8.2).

Definition 8.4. Suppose that $g : X \to Y$ is a smooth map and M denotes a submanifold of the manifold Y. We say that the map g is *transverse* to M at a point $p \in X$ if either $g(p) \notin M$, or $g(p) \in M$ and the sum of the tangent space of M at $g(p)$ and the range of the derivative $Dg(p)$ (both viewed as linear subspaces of the tangent space of Y at p) is equal to the entire tangent space of Y at $g(p)$. The function g is said to be transverse to the manifold M if it is transverse to M at every point of X.

The next theorem—stated here with some informality—is a far-reaching generalization of the implicit function theorem.

Theorem 8.5 (Thom's Transversality Theorem). *The set S of maps whose k-jet extensions are transverse to a submanifold M of the space of k-jets is a dense subset of the space of all sufficiently smooth maps; moreover, S is a countable intersection of open dense sets. In addition, if M is closed in the space of k-jets, then S is open.*

To make Thom's theorem precise, we would have to define topologies on our function spaces. The usual C^r topology is induced by the norm defined as the sum of the suprema of the absolute values of the partial derivatives of a function up to order r. But, this topology is not defined on the space $C^r(\mathbb{R}, \mathbb{R})$ because some of the functions in this space are unbounded or have an unbounded partial derivative. To get around this problem, we can restrict attention to functions defined on a compact domain in $\mathbb{R}$, or we can use one of the two useful topologies on $C^r(\mathbb{R}, \mathbb{R})$ called the *weak* and the *strong topology*. Roughly speaking, if f is a function, $\alpha > 0$, and K is a compact subset of $\mathbb{R}$, then a basic open set in the weak topology, also called the compact open topology, is defined to be the set of functions g such that the distance between f and g in the C^r-norm, relative to the compact set K, is less than the positive number α. The strong topology is similar, but it includes the neighborhoods defined by requiring that functions be close on (infinite) families of compact subsets of their domains. The strong topology is important because some of its open neighborhoods control the size of the function and its partial derivatives "at infinity." These topologies are the same if the functions in $C^r(\mathbb{R}, \mathbb{R})$ are all restricted to a compact set. In this case, the corresponding function space is the usual Banach space of C^r functions defined on the compact set. The important observation for our discussion is that Thom's theorem is valid for both the weak and strong topologies. (See the book of Morris Hirsch [120] for a precise definition of these topologies and a proof of Thom's theorem.)

A set is called *residual* if it is the (countable) intersection of open and dense subsets. By Baire's theorem, every residual set in a complete metric space is dense (see [199]). Also, a property that holds on a residual set is called *generic*. It turns out that even though the weak and strong topologies on $C^\infty(\mathbb{R}, \mathbb{R})$ are not metrizable, the set $C^\infty(\mathbb{R}, \mathbb{R})$ is a Baire space with respect to these topologies; that is, with respect to these topologies, a countable intersection of open and dense sets is dense. Using these notions, Thom's transversality theorem can be restated as follows: *The property of transversal intersection is generic.*

As a simple example of an application of Thom's theorem, let us consider the transversality of the 0-jet extensions of functions in $C^\infty(\mathbb{R}, \mathbb{R})$ with the image of the zero section. Note that by the definition of transversality the 0-jet extension of $f \in C^\infty(\mathbb{R}, \mathbb{R})$ is transversal to the image of the zero section $\mathcal{Z}^0$ at $u \in \mathbb{R}$ if either $j^0(f)(u) \neq [u, 0]$, or $j^0(f)(u) = [u, 0]$ and the image of the derivative of the 0-jet extension $j^0(f)$ at u plus the tangent space to $\mathcal{Z}^0$ at $[u, 0]$ is the tangent space to $J^0(\mathbb{R}, \mathbb{R})$ at $[u, 0]$. We will determine this transversality condition more explicitly and use Thom's theorem to state a fact about the genericity of vector fields with hyperbolic rest points.

The differentiability of the map $j^0(f)$ and the properties of its derivative are local properties that can be determined in the local coordinate representation of the jet bundle. In fact, with respect to the local coordinates mentioned above, the local representative of the map $j^0(f)$ is $u \mapsto \Phi^0(j^0(f)(u))$. In other words, the local representation of $j^0(f)$ is the map $F : \mathbb{R} \to \mathbb{R} \times \mathbb{R}$ defined by $u \to (u, f(u))$; and, in these coordinates, the range of the derivative of F is spanned by the vector $(1, f'(u))$.

The local representation of $\mathcal{Z}^0$ is given by the linear manifold $Z^0 := \{(x, y) \in \mathbb{R} \times \mathbb{R} : y = 0\}$. Hence, the tangent space of Z^0 at each of its points can be identified with Z^0. Moreover, let us note that Z^0, viewed as a subspace of $\mathbb{R} \times \mathbb{R}$, is spanned by the vector $(1, 0)$.

The 0-jet extension of the function f is transverse to the zero section at the point $u \in \mathbb{R}$ if and only if $f'(u) \neq 0$; it is transverse to the zero section if it is transverse at every $u \in \mathbb{R}$. In other words, the 0-jet extension of f is transverse to the zero section if and only if all zeros of f are nondegenerate; or equivalently if and only if all rest points of the corresponding differential equation $\dot{u} = f(u)$ are hyperbolic.

By Thom's theorem, if f is in $C^\infty(\mathbb{R}, \mathbb{R})$, then there is an arbitrarily small perturbation of f such that the corresponding differential equation has only hyperbolic rest points. Moreover, the set of all scalar differential equations with hyperbolic rest points is open.

The proof of Thom's theorem is not trivial. But, for the simple case that we are considering, part of Thom's result is a corollary of the implicit function theorem. In fact, we will show that if f has finitely many nondegenerate zeros, then every sufficiently small perturbation of f has the same property.

Consider the Banach space $C^1(\mathbb{R}, \mathbb{R})$ consisting of all elements of $C^1(\mathbb{R}, \mathbb{R})$ that are bounded in the C^1-norm. Suppose that $f \in C^1(\mathbb{R}, \mathbb{R})$ has only nondegenerate zeros and consider the map $\rho : \mathbb{R} \times C^1(\mathbb{R}, \mathbb{R}) \to \mathbb{R}$ given by $(u, f) \mapsto f(u)$. This map is smooth (see Exercise 8.8). Moreover, if $\rho(u_0, f_0) = 0$, then we have $\rho_u(u_0, f_0) = f'(u_0) \neq 0$. Thus, there is a map $f \mapsto \beta(f)$ defined on a neighborhood U of f_0 in $C^1(\mathbb{R}, \mathbb{R})$ with image in an open subset $V \subset \mathbb{R}$ such that $\beta(f_0) = u_0$ and $f(\beta(f)) \equiv 0$. Moreover, if $(u, f) \in V \times U$ and $f(u) = 0$, then $u = \beta(f)$. In other words, every function in the neighborhood U has exactly one zero in V. Also, there are open subsets $U_0 \subseteq U$ and $V_0 \subseteq V$ such that for each function f in U_0 we have $f'(u) \neq 0$ whenever $u \in V_0$. Hence, every function in U_0 has a unique nondegenerate zero in V_0. If, in addition, the function f has only finitely many zeros, then we can be sure that every perturbation of f has only nondegenerate zeros.

Exercise 8.6. Consider the set of differential equations of the form $\dot{u} = f(u)$, where $u \in \mathbb{R}^n$, that have a finite number of rest points, and show that the subset of these systems with hyperbolic rest points is open and dense in the C^1 topology.

We have used jet spaces to analyze the perturbations of scalar differential equations that have only hyperbolic rest points. We will discuss differential equations with saddle-nodes and show that the conditions required for a saddle-node are the same as the conditions for a certain jet extension map to be transversal to the zero section of a jet bundle.

Consider the 1-jet extensions of smooth scalar maps and the image of the zero section $Z^1 \subset J^1(\mathbb{R}, \mathbb{R})$. If $j^1(f)(u) \in Z^1$, then f has a saddle-node at u; that is, $f(u) = 0$ and $f'(u) = 0$. Thus, to study the saddle-node bifurcation, we must consider *families* of maps in $C^\infty(\mathbb{R}, \mathbb{R})$. In fact, we will identify these families as elements of the space $C^\infty(\mathbb{R} \times \mathbb{R}, \mathbb{R})$ where a typical element f is given by a function of two variables $(u, \lambda) \mapsto f(u, \lambda)$.

Let us define a new jet bundle with total space $J^{(1,0)}(\mathbb{R} \times \mathbb{R}, \mathbb{R})$ consisting of all equivalence classes of triples $(u, \lambda, f) \in \mathbb{R} \times \mathbb{R} \times C^\infty(\mathbb{R} \times \mathbb{R}, \mathbb{R})$ where the triples (v, δ, f) and (w, ν, g) are equivalent if

$$v = w, \quad \delta = \nu, \quad f(v, \delta) = g(w, \nu), \quad f_u(v, \delta) = g_u(w, \nu),$$

and the bundle projection is given by $[u, \lambda, f(u, \lambda), f_u(u, \lambda)] \mapsto (u, \lambda)$. Our somewhat nonstandard jet space $J^{(1,0)}(\mathbb{R} \times \mathbb{R}, \mathbb{R})$ may be viewed as a space of families of sections of the 1-jet bundle of functions in $C^\infty(\mathbb{R}, \mathbb{R})$.

The $(1, 0)$-jet extension of $f \in C^\infty(\mathbb{R} \times \mathbb{R}, \mathbb{R})$ is the map

$$j^{(1,0)}(f) : \mathbb{R} \times \mathbb{R} \to J^{(1,0)}(\mathbb{R} \times \mathbb{R}, \mathbb{R})$$

given by $j^{(1,0)}(f)(u, \lambda) = [u, \lambda, f(u, \lambda), f_u(u, \lambda)]$, and the image of the zero section $\mathcal{Z}^{(1,0)}$ is the set of all equivalence classes of triples of the form $(u, \lambda, 0)$.

Let us note that the local representative of the $(1, 0)$-jet extension is given by

$$(u, \lambda) \mapsto (u, \lambda, f(u, \lambda), f_u(u, \lambda)).$$

Note also that the $(1, 0)$-jet extension is transverse to the zero section $\mathcal{Z}^{(1,0)}$ at a point (u, λ) where $f(u, \lambda) = 0$ and $f_u(u, \lambda) = 0$ if the following obtains: The vector space sum of

(*i*) the range of the derivative of the local representative of the $(1, 0)$-jet extension; and

(*ii*) the tangent space of the local representation of $\mathcal{Z}^{(1,0)}$ at $(u, \lambda, 0, 0)$

is equal to the entire space $\mathbb{R}^4$. In other words, these vector subspaces are

(*i*) the span of the vectors

$$(1, 0, f_u(u, \lambda), f_{uu}(u, \lambda)) \quad \text{and} \quad (0, 1, f_\lambda(u, \lambda), f_{\lambda u}(u, \lambda)); \quad \text{and}$$

(*ii*) the span of the vectors $(1, 0, 0, 0)$ and $(0, 1, 0, 0)$.

This transversality condition is met provided that

$$f_\lambda(u, \lambda) \neq 0 \quad \text{and} \quad f_{uu}(u, \lambda) \neq 0,$$

exactly the conditions for a nondegenerate saddle-node bifurcation!

Just as for the case of nondegenerate zeros, the subset of all families of smooth maps that have a saddle-node bifurcation is dense, and this set can be identified as the countable intersection of open and dense subsets of the space $C^\infty(\mathbb{R} \times \mathbb{R}, \mathbb{R})$. Moreover, by using the implicit function theorem, it is easy to prove that if a family has a saddle-node bifurcation at some point, then a small perturbation of this family also has a saddle-node bifurcation at a nearby point. Thus, by translating our criteria for saddle node bifurcations into the geometric language of transversality, we have developed an approach to showing that saddle-node bifurcations can be unavoidable in all sufficiently small perturbations of some one-parameter *families* of maps; and, as a result, we have a positive answer to the question "Is bifurcation theory important?"

In the remainder of this chapter we will not pursue the ideas that we have discussed in this section. Rather, we will only consider sufficient conditions to obtain nondegenerate bifurcation in one-parameter families. Transversality theory can be applied in each case that we will consider to show that, in an appropriate sense, the bifurcations are generic.

We have discussed the unavoidability of the saddle-node bifurcation in one-parameter families of maps. This leads to the question "Are saddle-nodes unavoidable in two-parameter families of maps?" The answer is "yes." In fact, nothing new happens for the saddle-node bifurcation relative to multiparameter families of maps. The reason is that the set corresponding to the saddle-node has codimension one in an appropriate function space. On the other hand, bifurcation theory in families with two or more parameters is generally much more difficult than the theory for one-parameter families because global features of the dynamics must be taken into account (see, for example, [57], [58], [103], and [233]).

Exercise 8.7. Formulate and prove a theorem based on the implicit function theorem that can be used to show that a small perturbation of a family of maps with a saddle-node bifurcation has a nearby saddle-node bifurcation.

Exercise 8.8. Prove: The map $\mathbb{R} \times C^1([a, b], \mathbb{R}) \mapsto \mathbb{R}$ given by $(u, f) \mapsto f(u)$ is smooth.

Exercise 8.9. Prove: There is a saddle-node bifurcation for some values of the parameter λ in the family

$$\dot{u} = \cos \lambda - u \sin u.$$

Exercise 8.10. Draw the bifurcation diagram for the scalar family of differential equations

$$\dot{x} = \lambda x - x^2.$$

The bifurcation at $\lambda = 0$ is called *transcritical*. Prove a proposition similar to Proposition 8.2 for the existence of a transcritical bifurcation.

Exercise 8.11. (a) Draw the bifurcation diagram for the scalar family of differential equations

$$\dot{x} = \lambda x - x^3.$$

The bifurcation at $\lambda = 0$ is called the *supercritical pitchfork*. (b) Prove a proposition similar to Proposition 8.2 for the existence of a supercritical pitchfork bifurcation. (c) Compare and contrast the supercritical pitchfork bifurcation with the *subcritical pitchfork bifurcation* whose normal form is

$$\dot{x} = \lambda x + x^3.$$

8.2 Saddle-Node Bifurcation by Lyapunov–Schmidt Reduction

We will consider the saddle-node bifurcation for the n-dimensional family of differential equations (8.1) given by

$$\dot{u} = f(u, \lambda), \quad u \in \mathbb{R}^n, \ \lambda \in \mathbb{R}.$$

It should be clear from the previous section that sufficient conditions for the saddle-node bifurcation to occur do not mention the solutions of the differential equations in this family; rather, the analysis so far requires only knowledge of the function $f : \mathbb{R}^n \times \mathbb{R} \to \mathbb{R}^n$ that defines the family of vector fields associated with our family of differential equations. In view of this fact, we say that $u_0 \in \mathbb{R}^n$ is a *saddle-node* for $f : \mathbb{R}^n \times \mathbb{R} \to \mathbb{R}^n$ at λ_0 if $f(u_0, \lambda_0) = 0$, the linear transformation $f_u(u_0, \lambda_0) : \mathbb{R}^n \to \mathbb{R}^n$ has zero as an eigenvalue with algebraic multiplicity one, and all other eigenvalues have nonzero real parts. Also, a *saddle-node bifurcation* is said to occur at a saddle-node $u = u_0$ for the parameter value $\lambda = \lambda_0$ if the following conditions are met:

SNB1 *There is a number $p_0 > 0$ and a smooth curve $p \mapsto \beta(p)$ in $\mathbb{R}^n \times \mathbb{R}$ such that $\beta(0) = (u_0, \lambda_0)$ and $f(\beta(p)) \equiv 0$ for $|p| < p_0$.*

SNB2 *The curve β has a quadratic tangency with $\mathbb{R}^n \times \{\lambda_0\}$ at (u_0, λ_0). More precisely, if the components of β are defined by*

$$\beta(p) = (\beta_1(p), \beta_2(p)),$$

then $\beta_2(0) = \lambda_0$, $\beta_2'(0) = 0$, and $\beta_2''(0) \neq 0$.

SNB3 *If $p \neq 0$, then the matrix $f_u(\beta(p))$ is infinitesimally hyperbolic. Also, exactly one eigenvalue of the matrix crosses the imaginary axis with nonzero speed at the parameter value $p = 0$.*

The next theorem, called the *saddle-node bifurcation theorem*, gives sufficient generic conditions for a saddle-node bifurcation to occur.

Theorem 8.12. *Suppose that $f : \mathbb{R}^n \times \mathbb{R} \mapsto \mathbb{R}^n$ is a smooth function, $u = u_0$ is a saddle-node for f at $\lambda = \lambda_0$, and the kernel of the linear transformation $f_u(u_0, \lambda_0) : \mathbb{R}^n \to \mathbb{R}^n$ is spanned by the nonzero vector $k \in \mathbb{R}^n$. If $f_\lambda(u_0, \lambda_0) \in \mathbb{R}^n$ and $f_{uu}(u_0, \lambda_0)(k, k) \in \mathbb{R}^n$ are both nonzero and both not in the range of $f_u(u_0, \lambda_0)$, then there is a saddle-node bifurcation at $(u, \lambda) = (u_0, \lambda_0)$ (that is SNB1, SNB2, and SNB3 are met). Moreover, among all C^∞ one-parameter families that have a saddle-node, those that undergo a saddle-node bifurcation form an open and dense subset.*

The second derivatives that appear in the statement of Theorem 8.12 are easily understood from the correct point of view. Indeed, suppose that $g : \mathbb{R}^n \to \mathbb{R}^n$ is a smooth function given by $u \mapsto g(u)$ and recall that its (first) derivative Dg is a map from $\mathbb{R}^n$ into the linear transformations of $\mathbb{R}^n$; that is, $Dg : \mathbb{R}^n \to L(\mathbb{R}^n, \mathbb{R}^n)$. If $u, v, w \in \mathbb{R}^n$, then the derivative of g at u in the direction w is denoted by $Dg(u)w$. Consider the map $u \mapsto Dg(u)w$. If $g \in C^2$, then its derivative at $u \in \mathbb{R}^n$ in the direction v is defined by

$$\frac{d}{dt}Dg(u + tv)w\Big|_{t=0} = (D^2g(u)w)v = D^2g(u)(w, v).$$

Hence, if $g \in C^2$, then to compute the second derivative D^2g, it suffices to compute the first derivative of the map $u \mapsto Dg(u)w$. Also, note that the function $(w, v) \mapsto D^2g(u)(w, v)$ is bilinear and symmetric. The linearity follows from the linearity of the first derivatives; the symmetry is a restatement of the equality of mixed partial derivatives.

Exercise 8.13. Suppose that $f : \mathbb{R} \times \mathbb{R} \times \mathbb{R} \to \mathbb{R} \times \mathbb{R}$ is given by

$$f(x, y, \lambda) = (\lambda - x^2 + xy, -2y + x^2 + y^2)$$

and $u := (x, y)$. Show that f satisfies all the hypotheses of Theorem 8.12 at $(u, \lambda) = (0, 0)$. Draw the phase portrait of $\dot{u} = f(u, \lambda)$ near $(u, \lambda) = (0, 0)$ for $\lambda < 0$, $\lambda = 0$, and $\lambda > 0$.

Exercise 8.14. Prove that if $g \in C^2$, then $D^2g(u)(v, w) = D^2g(u)(w, v)$.

We now turn to the proof of Theorem 8.12.

Proof. Assume, with no loss of generality, that $u = 0$ is a saddle-node for f at $\lambda = 0$. Also, assume that zero is an eigenvalue of the linearization $f_u(0, 0) : \mathbb{R}^n \to \mathbb{R}^n$ with algebraic multiplicity one, and the kernel $\mathcal{K}$ of this linear transformation is one-dimensional, say $\mathcal{K} = [k]$.

Using the Lyapunov–Schmidt reduction and linear algebra, let us choose an $(n-1)$-dimensional complement $\mathcal{K}^\perp$ to $\mathcal{K}$ in $\mathbb{R}^n$ whose basis is

$$k_2^\perp, \ldots, k_n^\perp.$$

Corresponding to these choices, there is a coordinate transformation $\Psi : \mathbb{R} \times \mathbb{R}^{n-1} \to \mathbb{R}^n$ given by

$$(p, q) \mapsto pk + \sum_{i=2}^{n} q_i k_i^\perp$$

where, in the usual coordinates of $\mathbb{R}^{n-1}$, the point q is given by $q = (q_2, \ldots, q_n)$. Likewise, the range $\mathcal{R}$ of $f_u(0, 0)$ is $(n-1)$-dimensional with a one-dimensional complement $\mathcal{R}^\perp$. Let $\Pi : \mathbb{R}^n \to \mathcal{R}$ and $\Pi^\perp : \mathbb{R}^n \to \mathcal{R}^\perp$ be corresponding complementary linear projections.

With the notation defined above, consider the map $\varrho : \mathbb{R} \times \mathbb{R}^{n-1} \times \mathbb{R} \to \mathcal{R}$ given by $(p, q, \lambda) \mapsto \Pi f(\Psi(p, q), \lambda)$. Since $f(0,0) = 0$, we have that $\varrho(0,0,0) = 0$. From equation (5.60) of the abstract formulation of the Lyapunov–Schmidt reduction, we see that $\varrho_q(0,0,0)$ is invertible as a linear transformation $\mathbb{R}^{n-1} \to \mathbb{R}^{n-1}$. Thus, there is a function $h : \mathbb{R} \times \mathbb{R} \to \mathbb{R}^{n-1}$ given by $(p, \lambda) \mapsto h(p, \lambda)$ with $h(0,0) = 0$ such that for (p, λ) in a sufficiently small neighborhood of the origin in $\mathbb{R}^{n-1} \times \mathbb{R}$ we have

$$\Pi f(\Psi(p, h(p, \lambda), \lambda)) \equiv 0. \tag{8.6}$$

It is instructive to check the invertibility of the derivative directly. In fact, we have

$$\varrho_q(0,0,0) = \Pi f_u(0,0)\Psi_q(0,0).$$

But $\Psi_q(0,0) : \mathbb{R}^{n-1} \to \mathbb{R}^n$ is given by

$$\Psi_q(0,0)q = \sum_{i=2}^{n} q_i k_i^{\perp}.$$

Hence, the range of Ψ_q is the complement of the Kernel $f_u(0,0)$ previously chosen. Also Ψ_q is an isomorphism onto its range. On the complement of its kernel, $f_u(0,0)$ is an isomorphism onto its range and Π is the identity on this range. In other words, $\varrho_q(0,0,0)$ is an isomorphism.

Viewed geometrically, the function h defines a two-dimensional surface in $\mathbb{R} \times \mathbb{R}^{n-1} \times \mathbb{R}$ given by $\{(p, h(p, \lambda), \lambda) : (p, \lambda) \in \mathbb{R}^n \times \mathbb{R}\}$ which lies in the zero set of ϱ. In addition, the (Lyapunov–Schmidt) reduced function is $\tau : \mathbb{R} \times \mathbb{R} \to \mathcal{R}^{\perp}$ defined by

$$(p, \lambda) \mapsto \Pi^{\perp} f(\Psi(p, h(p, \lambda)), \lambda).$$

Of course, if (p, λ) is a zero of τ, then $f(\Psi(p, h(p, \lambda)), \lambda) = 0$.

We have $\tau(0,0) = 0$. If $\tau_\lambda(0,0) \neq 0$, then by the implicit function theorem there is a unique curve $p \mapsto \gamma(p)$ in $\mathbb{R}$ such that $\gamma(0) = 0$ and $\tau(p, \gamma(p)) \equiv 0$. Moreover, in this case, it follows that

$$f(\Psi(p, h(p, \gamma(p))), \gamma(p)) \equiv 0.$$

In other words, the image of the function β defined by

$$p \mapsto (\Psi(p, h(p, \gamma(p))), \gamma(p))$$

is a curve in the zero set of $f(u, \lambda)$ that passes through the point $(u, \lambda) = (0,0)$.

To show SNB1, we will prove the inequality $\tau_\lambda(0,0) \neq 0$. Let us note first that

$$\tau_\lambda(0,0) = \Pi^{\perp}(f_u(0,0)\Psi_q(0,0)h_\lambda(0,0) + f_\lambda(0,0)).$$

Since $\Pi^{\perp}$ projects to the complement of the range of $f_u(0,0)$, the last formula reduces to

$$\tau_\lambda(0,0) = \Pi^{\perp} f_\lambda(0,0).$$

By hypothesis, $f_\lambda(0,0)$ is a nonzero vector that is not in $\mathcal{R}$; therefore, $\tau_\lambda(0,0) \neq 0$, as required.

To prove SNB2, we will show that $\gamma'(0) = 0$ and $\gamma''(0) \neq 0$. Note first that the derivative of the identity $\tau(p, \gamma(p)) \equiv 0$ with respect to p is given by

$$\tau_p(p, \gamma(p)) + \tau_\lambda(p, \gamma(p))\gamma'(p) \equiv 0. \tag{8.7}$$

Moreover, if we set $p = 0$ and use the equality $\gamma(0) = 0$, then

$$\tau_p(0,0) + \tau_\lambda(0,0)\gamma'(0) = 0.$$

Next, recall that $\tau_\lambda(0,0) \neq 0$. Also, use the definition of τ to compute

$$\tau_p(p, \lambda) = \Pi^{\perp} f_u(\Psi(p, h(p, \lambda)), \lambda)\big(\Psi_p(p, h(p, \lambda)) + \Psi_q(p, h(p, \lambda))h_p(p, \lambda)\big), \tag{8.8}$$

and, in particular,

$$\tau_p(0,0) = \Pi^{\perp} f_u(0,0)\big(\Psi_p(0,0) + \Psi_q(0,0)h_p(0,0)\big).$$

Because $\Pi^{\perp}$ projects to the complement of the range of $f_u(0,0)$, it follows that $\tau_p(0,0) = 0$, and therefore $\gamma'(0) = 0$. Using this fact and equation (8.7), we obtain the equality

$$\tau_{pp}(0,0) + \tau_\lambda(0,0)\gamma''(0) = 0.$$

Thus, it follows that

$$\gamma''(0) = -\frac{\tau_{pp}(0,0)}{\tau_\lambda(0,0)}.$$

To prove the inequality $\tau_{pp}(0,0) \neq 0$, first use equation (8.8) and recall that $\Pi^{\perp}$ projects to the complement of the range of $f_u(0,0)$ to obtain the equality

$$\tau_{pp}(0,0) = \Pi^{\perp} f_{uu}(0,0)\big(\Psi_p(0,0) + \Psi_q(0,0)h_p(0,0)\big)^2$$

where "the square" is shorthand for the argument of the bilinear form $f_{uu}(0,0)$ on $\mathbb{R}^n$.

Next, differentiate both sides of the identity (8.6) with respect to p at $p = 0$ to obtain the equation

$$\Pi f_u(0,0)\big(\Psi_p(0,0) + \Psi_q(0,0)h_p(0,0)\big) = 0. \tag{8.9}$$

Because Π projects to the range of $f_u(0,0)$, equation (8.9) is equivalent to the equation

$$f_u(0,0)(\Psi_p(0,0) + \Psi_q(0,0)h_p(0,0)) = 0,$$

and therefore the vector

$$\Psi_p(0,0) + \Psi_q(0,0)h_p(0,0)$$

is in the kernel $\mathcal{K}$ of $f_u(0,0)$. But by the definition of Ψ we have $\Psi_p(0,0) = k \in \mathcal{K}$ and $\Psi_q(0,0)h_p(0,0) \in \mathcal{K}^\perp$. Thus, $h_p(0,0) = 0$, and it follows that $\tau_{pp}(0,0) \neq 0$ if and only if

$$f_{uu}(0,0)(k,k) \neq 0, \qquad f_{uu}(0,0)(k,k) \notin \mathcal{R}. \tag{8.10}$$

This completes the proof of SNB2.

To prove SNB3, and thus complete the proof of the theorem, let us consider the curve β of rest points given by $p \mapsto (\Psi(p, h(p, \gamma(p)), \gamma(p))$. We must show that the matrix $f_u(\beta(p))$ is invertible for small nonzero $p \in \mathbb{R}$ and a single eigenvalue of $f_u(\beta(p))$ passes through zero with nonzero speed at $p = 0$. In other words, the rest points on the curve β are hyperbolic for $p \neq 0$, and there is a generic change of stability at $p = 0$. Of course, the first condition follows from the second.

To analyze the second condition, consider the eigenvalues of $f_u(\beta(p))$. By the hypothesis of the theorem, there is exactly one zero eigenvalue at $p = 0$. Thus, there is a curve $p \mapsto \sigma(p)$ in the complex plane such that $\sigma(0) = 0$ and such that $\sigma(p)$ is an eigenvalue of $f_u(\beta(p))$. Also, there is a corresponding eigenvector $V(p)$ such that

$$f_u(\beta(p))V(p) = \sigma(p)V(p), \tag{8.11}$$
$$V(0) = k.$$

By differentiating both sides of the identity (8.11) with respect to p at $p = 0$ and simplifying the result, we obtain the equation

$$f_{uu}(0,0)(k,k) + f_u(0,0)V'(0) = \sigma'(0)k$$

and its projection

$$\Pi^\perp f_{uu}(k,k) = \sigma'(0)\Pi^\perp k.$$

In view of the inequality (8.10), we have that $\Pi^\perp f_{uu}(0,0)(k,k) \neq 0$, and therefore $\sigma'(0)$ is a nonzero real number. $\square$

Exercise 8.15. Prove: With the notation as in the proof of Theorem 8.12, if $\Pi^\perp k = 0$ and $n \geq 2$, then zero is an eigenvalue of $f_u(0,0)$ with multiplicity at least two.

Exercise 8.16. Suppose that $A : \mathbb{R}^n \to \mathbb{R}^n$ is a linear transformation with exactly one zero eigenvalue. Show that there is a nonzero "left eigenvector" $w \in \mathbb{R}^n$ such that $w^{\mathrm{T}} A = 0$. Also, show that v is in the range of A if and only if $\langle v, w \rangle = 0$. Discuss how this exercise gives a method to verify the hypotheses of Theorem 8.12.

Exercise 8.17. Verify the existence of a saddle-node bifurcation for the function $f : \mathbb{R}^2 \times \mathbb{R} \to \mathbb{R}^2$ given by

$$f(x, y, \lambda) = (\lambda - x^2, -y).$$

Exercise 8.18. Determine the bifurcation diagram for the phase portrait of the differential equation

$$x\ddot{x} + a\dot{x}^2 = b$$

where a and b are parameters.

Exercise 8.19. [Hamiltonian saddle-node] Suppose that

$$\dot{u} = f(u, \lambda), \qquad u \in R^2 \tag{8.12}$$

is a planar *Hamiltonian* family with parameter $\lambda \in \mathbb{R}$. Prove that if $f(u_0, \lambda_0) = 0$ and the corresponding linearization at u_0 has a zero eigenvalue, then this eigenvalue has algebraic multiplicity two. In particular, a planar Hamiltonian system cannot have a saddle-node. Define (u_0, λ_0) to be a *Hamiltonian saddle-node* at λ_0 if $f(u_0, \lambda_0) = 0$ and $f_u(u_0, \lambda_0)$ has a zero eigenvalue with geometric multiplicity one. A *Hamiltonian saddle-node bifurcation* occurs if the following conditions hold:

- There exist $s_0 > 0$ and a smooth curve γ in $R^2 \times R$ such that $\gamma(0) = (u_0, \lambda_0)$ and $f(\gamma(s)) \equiv 0$ for $|s| < s_0$.
- The curve of critical points γ is quadratically tangent to $R^2 \times \{\lambda_0\}$ at (u_0, λ_0).
- The Lyapunov stability type of the rest points on the curve γ changes at $s = 0$.

Prove the following proposition formulated by Jason Bender [24]: *Suppose that the origin in $\mathbb{R}^2 \times \mathbb{R}$ is a Hamiltonian saddle-node for (8.12) and $k \in R^2$ is a nonzero vector that spans the one-dimensional kernel of the linear transformation $f_u(0, 0)$. If the two vectors $f_\lambda(0, 0) \in R^2$ and $f_{uu}(0, 0)(k, k) \in R^2$ are nonzero and not in the range of $f_u(0, 0)$, then a Hamiltonian saddle-node bifurcation occurs at the origin.*

Reformulate the hypotheses of the proposition in terms of the Hamiltonian for the family so that there is no mention of the vector k. Also, discuss the Hamiltonian saddle-node bifurcation for the following model of a pendulum with feedback control

$$\dot{x} = y, \qquad \dot{y} = -\sin x - \alpha x + \beta$$

(see [236]). Generalize the proposition to Hamiltonian systems on $\mathbb{R}^{2n}$. (See [160] for the corresponding result for Poincaré maps at periodic orbits of Hamiltonian systems.)

8.3   Poincaré–Andronov–Hopf Bifurcation

Consider the family of differential equations

$$\dot{u} = F(u, \lambda), \qquad u \in \mathbb{R}^N, \quad \lambda \in \mathbb{R}^M \tag{8.13}$$

where λ is a vector of parameters.

Definition 8.20. An ordered pair $(u_0, \lambda_0) \in \mathbb{R}^N \times \mathbb{R}^M$ consisting of a parameter value λ_0 and a rest point u_0 for the corresponding member of the family (8.13) is called a *Hopf point* if there is a curve C in $\mathbb{R}^N \times \mathbb{R}^M$, called an *associated curve*, that is given by $\epsilon \mapsto (C_1(\epsilon), C_2(\epsilon))$ and satisfies the following properties:

(i) $C(0) = (u_0, \lambda_0)$ and $F(C_1(\epsilon), C_2(\epsilon)) \equiv 0$.

(ii) The linear transformation given by the derivative $F_u(C_1(\epsilon), C_2(\epsilon))$: $\mathbb{R}^N \to \mathbb{R}^N$ has a pair of nonzero complex conjugate eigenvalues $\alpha(\epsilon) \pm \beta(\epsilon) i$, each with algebraic (and geometric) multiplicity one. Also, $\alpha(0) = 0$, $\alpha'(0) \neq 0$, and $\beta(0) \neq 0$.

(iii) Except for the eigenvalues $\pm \beta(0) i$, all other eigenvalues of $F_u(u_0, \lambda_0)$ have nonzero real parts.

Our definition says that a one-parameter family of differential equations has a Hopf point if a single pair of complex conjugate eigenvalues, associated with the linearizations of a corresponding family of rest points, crosses the imaginary axis in the complex plane with nonzero speed at the parameter value of the bifurcation point, whereas all other eigenvalues have nonzero real parts. We will show that if some additional generic assumptions are met, then there are members of the family (8.13) that have a limit cycle near the Hopf point.

Let us show first that it suffices to consider the bifurcation for a planar family of differential equations associated with the family (8.13).

Because the linear transformation given by the derivative $F_u(u_0, \lambda_0)$ at the Hopf point (u_0, λ_0) has exactly two eigenvalues on the imaginary axis, the results in Chapter 4, especially equation (4.24), can be used to show that there is a center manifold reduction for the family (8.13) that produces a family of planar differential equations

$$\dot{u} = f(u, \lambda), \qquad u \in \mathbb{R}^2, \quad \lambda \in \mathbb{R}^M, \tag{8.14}$$

with a corresponding Hopf point. Moreover, there is a product neighborhood $U \times V \subset \mathbb{R}^N \times \mathbb{R}^M$ of the Hopf point (u_0, λ_0) such that if $\lambda \in V$ and the corresponding member of the family (8.13) has a bounded orbit in U, then this same orbit is an invariant set for the corresponding member of the planar family (8.14). Thus, it suffices to consider the bifurcation of limit cycles from the Hopf point of this associated planar family.

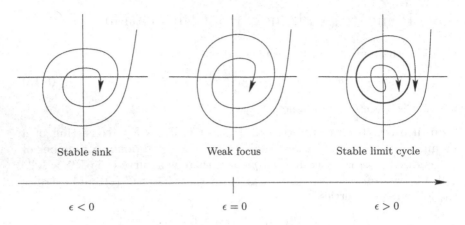

Stable sink   Weak focus   Stable limit cycle

$\epsilon < 0$ $\epsilon = 0$ $\epsilon > 0$

Figure 8.3: Supercritical Hopf bifurcation: A limit cycle emerges from a weak focus as the bifurcation parameter is increased.

There are important technical considerations related to the smoothness and uniqueness of the planar family obtained by a center manifold reduction at a Hopf point. For example, let us note that by the results in Chapter 4 if the family (8.13) is C^1, then the augmented family, obtained by adding a new equation corresponding to the parameters, has a local C^1 center manifold. But this result is not strong enough for the proof of the Hopf bifurcation theorem given below. In fact, we will require the reduced planar system (8.14) to be C^4. Fortunately, the required smoothness can be proved. In fact, using the fiber contraction principle as in Chapter 4, together with an induction argument, it is possible to prove that if $0 < r < \infty$ and the family (8.13) is C^r, then the reduced planar system at the Hopf point is also C^r in a neighborhood of the Hopf point. Let us also note that whereas local center manifolds are not necessarily unique, it turns out that all rest points, periodic orbits, homoclinic orbits, et cetera, that are sufficiently close to the original rest point, are on every center manifold. Thus, the bifurcation phenomena that are determined by reduction to a center manifold do not depend on the choice of the local center manifold (see, for example, [58]).

Let us say that a set $S \subset \mathbb{R}^N$ has *radii* (r_1, r_2) *relative to a point* p if $r_1 \geq 0$ is the radius of the smallest $\mathbb{R}^N$-ball centered at p that contains S and the distance from S to p is $r_2 \geq 0$.

Definition 8.21. The planar family (8.14) has a *supercritical Hopf bifurcation* at a Hopf point with associated curve $\epsilon \mapsto (c_1(\epsilon), c_2(\epsilon))$ if there are three positive numbers ϵ_0, K_1, and K_2 such that for each ϵ in the open interval $(0, \epsilon_0)$ the differential equation $\dot{u} = f(u, c_2(\epsilon))$ has a hyperbolic limit cycle with radii

$$(K_1\sqrt{\epsilon} + O(\epsilon),\ K_2\sqrt{\epsilon} + O(\epsilon))$$

relative to the rest point $u = c_1(\epsilon)$. The bifurcation is called *subcritical* if there is a similar limit cycle for the systems with parameter values in the range $-\epsilon_0 < \epsilon < 0$. Also, we say that the family (8.13) has a supercritical (respectively, subcritical) Hopf bifurcation at a Hopf point if the corresponding (center manifold) reduced system (8.14) has a supercritical (respectively, subcritical) Hopf bifurcation.

To avoid mentioning several similar cases as we proceed, let us consider only Hopf points such that the parametrized eigenvalues $\alpha \pm \beta\, i$ satisfy the additional assumptions

$$\alpha'(0) > 0, \qquad \beta(0) > 0. \tag{8.15}$$

In particular, we will restrict attention to the supercritical Hopf bifurcation as depicted in Figure 8.3.

Under our standing hypothesis (8.15), a rest point on the associated curve $\epsilon \mapsto c(\epsilon)$ of the Hopf point is a stable hyperbolic focus for the corresponding system (8.14) for $\epsilon < 0$ and an unstable hyperbolic focus for $\epsilon > 0$. We will introduce an additional hypothesis that implies "weak attraction" toward the rest point u_0 at the parameter value λ_0. In this case, there is a stable limit cycle that "bifurcates from this rest point" as ϵ increases through $\epsilon = 0$. This change in the qualitative behavior of the system as the parameter changes is the bifurcation that we wish to describe, namely, the supercritical Hopf bifurcation.

Before defining the notion of weak attraction, we will simplify the family (8.14) by a local change of coordinates and a reduction to one-parameter. Note that, after the translation $v = u - c_1(\epsilon)$, the differential equation (8.14) becomes

$$\dot{v} = f(v + c_1(\epsilon), \lambda)$$

with $f(0 + c_1(\epsilon), c_2(\epsilon)) \equiv 0$. In particular, in the new coordinates, the associated rest points remain at the origin for all values of the parameter ϵ. Thus, it suffices to consider the family (8.14) to be of the form

$$\dot{u} = f(u, \lambda), \qquad u \in \mathbb{R}^2, \quad \lambda \in \mathbb{R}, \tag{8.16}$$

only now with a Hopf point at $(u, \lambda) = (0, 0) \in \mathbb{R}^2 \times \mathbb{R}$ and with the associated curve c given by $\lambda \mapsto (0, \lambda)$.

Proposition 8.22. *If $(u, \lambda) = (0, 0) \in \mathbb{R}^2 \times \mathbb{R}$ is a Hopf point for the family (8.16) with associated curve $\lambda \mapsto (0, \lambda)$ and eigenvalues $\alpha(\lambda) \pm \beta(\lambda)\, i$, then there is a smooth parameter-dependent linear change of coordinates of the form $u = L(\lambda)z$ that transforms the system matrix $A(\lambda) := f_u(0, \lambda)$ of the linearization at the origin along the associated curve into the Jordan normal form*

$$\begin{pmatrix} \alpha(\lambda) & -\beta(\lambda) \\ \beta(\lambda) & \alpha(\lambda) \end{pmatrix}.$$

Proof. Suppose that $w(\lambda) = u_1(\lambda) + u_2(\lambda) i$ is a (nonzero) eigenvector for the eigenvalue $\alpha(\lambda) + \beta(\lambda) i$. We will show that there is an eigenvector of the form

$$\begin{pmatrix} 1 \\ 0 \end{pmatrix} - \begin{pmatrix} v_1(\lambda) \\ v_2(\lambda) \end{pmatrix} i.$$

To prove this fact, it suffices to find a family of complex numbers $c(\lambda) + d(\lambda) i$ such that

$$(c + d\,i)(u_1 + u_2\,i) = \begin{pmatrix} 1 \\ 0 \end{pmatrix} - \begin{pmatrix} v_1 \\ v_2 \end{pmatrix} i$$

for a family of numbers $v_1, v_2 \in \mathbb{R}$ where the minus sign is inserted to determine a convenient orientation. Equivalently, it suffices to solve the equation

$$cu_1 - du_2 = \begin{pmatrix} 1 \\ 0 \end{pmatrix},$$

which is expressed in matrix form as follows:

$$(u_1, -u_2) \begin{pmatrix} c \\ d \end{pmatrix} = \begin{pmatrix} 1 \\ 0 \end{pmatrix}.$$

Since the eigenvectors w and $\bar{w}$ corresponding to the distinct eigenvalues $\alpha \pm \beta i$ are linearly independent and

$$(u_1, -u_2) \begin{pmatrix} 1 & 1 \\ -i & i \end{pmatrix} = (w, \bar{w}),$$

it follows that $\det [u_1, -u_2] \neq 0$, and therefore we can solve (uniquely) for the vector (c, d).

Using this fact, we have the eigenvalue equation

$$A\left(\begin{pmatrix} 1 \\ 0 \end{pmatrix} - i \begin{pmatrix} v_1 \\ v_2 \end{pmatrix} \right) = (\alpha + i\beta)\left(\begin{pmatrix} 1 \\ 0 \end{pmatrix} - i \begin{pmatrix} v_1 \\ v_2 \end{pmatrix} \right),$$

as well as its real and imaginary parts

$$A \begin{pmatrix} 1 \\ 0 \end{pmatrix} = \alpha \begin{pmatrix} 1 \\ 0 \end{pmatrix} + \beta \begin{pmatrix} v_1 \\ v_2 \end{pmatrix}, \qquad A \begin{pmatrix} v_1 \\ v_2 \end{pmatrix} = -\beta \begin{pmatrix} 1 \\ 0 \end{pmatrix} + \alpha \begin{pmatrix} v_1 \\ v_2 \end{pmatrix}. \quad (8.17)$$

Hence, if

$$L := \begin{pmatrix} 1 & v_1 \\ 0 & v_2 \end{pmatrix},$$

then

$$AL = L \begin{pmatrix} \alpha & -\beta \\ \beta & \alpha \end{pmatrix}.$$

Again, since the vectors u_1 and u_2 are linearly independent, so are the following nonzero scalar multiples of these vectors

$$\begin{pmatrix} 1 \\ 0 \end{pmatrix}, \quad \begin{pmatrix} v_1 \\ v_2 \end{pmatrix}.$$

Thus, we have proved that the matrix L is invertible. Moreover, we can solve explicitly for v_1 and v_2. Indeed, using the equations (8.17), we have

$$(A - \alpha I) \begin{pmatrix} 1 \\ 0 \end{pmatrix} = \beta \begin{pmatrix} v_1 \\ v_2 \end{pmatrix}.$$

If we now set

$$A = \begin{pmatrix} a_{11} & a_{12} \\ a_{21} & a_{22} \end{pmatrix},$$

then

$$v_1 = \frac{a_{11} - \alpha}{\beta}, \quad v_2 = \frac{a_{21}}{\beta}.$$

Here $\beta := \beta(\lambda)$ is not zero at $\lambda = 0$, so the functions $\lambda \mapsto v_1(\lambda)$ and $\lambda \mapsto v_2(\lambda)$ are smooth. Finally, the change of coordinates $v = L(\lambda)z$ transforms the family of differential equations (8.16) to $\dot{z} = L^{-1}(\lambda)f(L(\lambda)z, \lambda)$, and the linearization of the transformed equation at $z = 0$ is given by

$$\begin{pmatrix} \alpha(\lambda) & -\beta(\lambda) \\ \beta(\lambda) & \alpha(\lambda) \end{pmatrix}.$$

The matrix function $\lambda \mapsto L^{-1}(\lambda)$ is also smooth at the origin. It is given by

$$L^{-1} = \frac{1}{v_2} \begin{pmatrix} v_2 & -v_1 \\ 0 & 1 \end{pmatrix}$$

where $1/v_2(\lambda) = \beta(\lambda)/a_{21}(\lambda)$. But, if $a_{21}(\lambda) = 0$, then the linearization has real eigenvalues, in contradiction to our hypotheses.   □

By Proposition 8.22, there is no loss of generality if we assume that the differential equation (8.16) has the form

$$\begin{aligned} \dot{x} &= \alpha(\lambda)x - \beta(\lambda)y + g(x, y, \lambda), \\ \dot{y} &= \beta(\lambda)x + \alpha(\lambda)y + h(x, y, \lambda) \end{aligned} \tag{8.18}$$

where the functions g and h together with their first partial derivatives with respect to the space variables vanish at the origin; the real functions $\lambda \mapsto \alpha(\lambda)$ and $\lambda \mapsto \beta(\lambda)$ are such that $\alpha(0) = 0$ (the real part of the linearization must vanish at $\lambda = 0$) and, by our standing assumption, $\alpha'(0) > 0$ (the derivative of the real part does not vanish at $\lambda = 0$); and, by the assumption that $\beta(0) > 0$, the eigenvalues $\alpha(\lambda) \pm i\beta(\lambda)$ are nonzero complex conjugates for $|\lambda|$ sufficiently close to zero. Moreover, there is no loss of generality if we assume that $\beta(0) = 1$. Indeed, this normalization can be achieved by a reparametrization of time in the family (8.18).

We will seek a periodic orbit of the family (8.18) near the origin of the coordinate system by applying the implicit function theorem to find a zero of the associated displacement function that is defined along the x-axis. For this application of the implicit function theorem we have to check that the displacement function has a smooth extension to the origin. While it is clear that the displacement has a continuous extension to the origin—define its value at the rest point to be zero—it is not clear that the extended displacement function is smooth. Indeed, the proof that the return map exists near a point p on a Poincaré section is based on the implicit function theorem and requires that the vector field be transverse to the section at p. But this condition is not satisfied at the origin for members of the family (8.18) because the vector field vanishes at this rest point.

Let us show that the displacement function for the system (8.18) is indeed smooth by using the blowup construction discussed in Section 1.8.5. The idea is that we can bypass the issue of the smoothness of the displacement at the origin for the family (8.18) by blowing up at the rest point. In fact, by changing the family (8.18) to polar coordinates we obtain the family

$$\dot{r} = \alpha(\lambda)r + p(r,\theta,\lambda), \qquad \dot{\theta} = \beta(\lambda) + q(r,\theta,\lambda) \qquad (8.19)$$

where

$$p(r,\theta,\lambda) := g(r\cos\theta, r\sin\theta, \lambda)\cos\theta + h(r\cos\theta, r\sin\theta, \lambda)\sin\theta,$$

$$q(r,\theta,\lambda) := \frac{1}{r}\big(h(r\cos\theta, r\sin\theta, \lambda)\cos\theta - g(r\cos\theta, r\sin\theta, \lambda)\sin\theta\big).$$

Since $(x,y) \mapsto g(x,y,\lambda)$ and $(x,y) \mapsto h(x,y,\lambda)$ and their first partial derivatives vanish at the origin, the function q in system (8.19) has a removable singularity at $r = 0$. Thus, the system is smooth. Moreover, by the change to polar coordinates, the rest point at the origin in the plane has been blown up to the circle $\{0\} \times \mathbb{T}$ on the phase cylinder $\mathbb{R} \times \mathbb{T}$. In our case, where $\beta(\lambda) \neq 0$, the rest point at the origin (for each choice of that parameter λ) corresponds to the periodic orbit on the cylinder given by the solution $r(t) \equiv 0$ and $\theta(t) = \beta(\lambda)t$. A Poincaré section on the cylinder for these periodic orbits, for example the line $\theta = 0$, has a smooth (parametrized) return map that is equivalent to the corresponding return map on the x-axis for the family (8.18). Thus, if we blow down—that is, project back to the plane—then the image of our transversal section is a smooth section for the flow with a smooth return map and a smooth return time map. In particular, both maps are smooth at the origin. In other words, the displacement function on the x-axis of the plane is conjugate (by the change of coordinates) to the smooth displacement function defined on the line $\theta = 0$ in the cylinder.

We will take advantage of the geometry on the phase cylinder: There our bifurcation problem concerns the bifurcation of periodic orbits *from a periodic orbit* rather than the bifurcation of periodic orbits from a rest point. Indeed, Hopf bifurcation on the phase cylinder is analogous to bifurcation from a multiple limit cycle as in our previous discussion following the Weierstrass preparation theorem (Theorem 5.15) on page 384.

For the generic case, we will soon see that the limit cycle, given on the cylinder by the set $\{(r, \theta) : r = 0\}$ for the family (8.19) at $\lambda = 0$, has multiplicity three. But, unlike the general theory for bifurcation from a multiple limit cycle with multiplicity three, the Hopf bifurcation has an essential new feature revealed by the geometry of the blowup: The bifurcation is symmetric with respect to the set $\{(r, \theta) : r = 0\}$. More precisely, each member of the family (8.19) is invariant under the change of coordinates given by

$$R = -r, \qquad \Theta = \theta - \pi. \tag{8.20}$$

While this symmetry has many effects, it should at least be clear that if a member of the family (8.19) has a periodic orbit that does not coincide with the set $\{(r, \theta) : r = 0\}$, then the system has two periodic orbits: one in the upper half cylinder, and one in the lower half cylinder. Also, if the set $\{(r, \theta) : r = 0\}$ is a limit cycle, then it cannot be semistable, that is, attracting on one side and repelling on the other (see Exercise 8.23). The geometry is similar to the geometry of the pitchfork bifurcation (see Exercise 8.11 and Section 8.4).

The general theory of bifurcations with symmetry is an important topic that is covered in detail in the excellent books [97] and [98].

Exercise 8.23. Prove: If the set $\Gamma := \{(r, \theta) : r = 0\}$ on the cylinder is a limit cycle for the member of the family (8.19) at $\lambda = 0$, then this limit cycle is not semistable. State conditions that imply Γ is a limit cycle and conditions that imply it is a hyperbolic limit cycle.

By our hypotheses, if $|r|$ is sufficiently small, then the line $\{(r, \theta) : \theta = 0\}$ is a transversal section for the flow of system (8.19) on the phase cylinder. Moreover, as we have mentioned above, there is a smooth displacement function defined on this section. In fact, let $t \mapsto (r(t, \xi, \lambda), \theta(t, \xi, \lambda))$ denote the solution of the differential equation (8.19) with the initial condition

$$r(0, \xi, \lambda) = \xi, \qquad \theta(0, \xi, \lambda) = 0,$$

and note that (because $\beta(0) = 1$)

$$\theta(2\pi, 0, 0) = 2\pi, \qquad \dot{\theta}(2\pi, 0, 0) = \beta(0) \neq 0.$$

By an application of the implicit function theorem, there is a product neighborhood $U_0 \times V_0$ of the origin in $\mathbb{R} \times \mathbb{R}$, and a function $T : U_0 \times V_0 \to \mathbb{R}$ such that $T(0,0) = 2\pi$ and $\theta(T(\xi, \lambda), \xi, \lambda) \equiv 2\pi$. Thus, the desired displacement function $\delta : U_0 \times V_0 \to \mathbb{R}$ is defined by

$$\delta(\xi, \lambda) := r(T(\xi, \lambda), \xi, \lambda) - \xi. \qquad (8.21)$$

The displacement function (8.21) is complicated by the presence of the implicitly defined return-time function T, a difficulty that can be avoided by yet another change of coordinates. Indeed, since $T(0,0) = 2\pi$ and $\dot\theta(t, 0, 0) = \beta(0) \neq 0$, it follows from the continuity of the functions T and θ and the implicit function theorem that there is a product neighborhood $U \times V$ of the origin with $U \times V \subseteq U_0 \times V_0$ such that for each $(\xi, \lambda) \in U \times V$ the function $t \mapsto \theta(t, \xi, \lambda)$ is invertible on some bounded time interval containing $T(\xi, \lambda)$ (see Exercise 8.28). Moreover, if the inverse function is denoted by $s \mapsto \theta^{-1}(s, \xi, \lambda)$, then the function $\rho : \mathbb{R} \times U \times V \to \mathbb{R}$ defined by

$$\rho(s, \xi, \lambda) = r(\theta^{-1}(s, \xi, \lambda), \xi, \lambda)$$

is a solution of the initial value problem

$$\frac{d\rho}{ds} = \frac{\alpha(\lambda)\rho + p(\rho, s, \lambda)}{\beta(\lambda) + q(\rho, s, \lambda)}, \qquad \rho(0, \xi, \lambda) = \xi$$

and

$$\rho(2\pi, \xi, \lambda) = r(T(\xi, \lambda), \xi, \lambda).$$

If we rename the variables ρ and s to *new variables* r and θ, then the displacement function $\delta : U \times V \to \mathbb{R}$ as defined in equation (8.21) with respect to the original variable r is also given by the formula

$$\delta(\xi, \lambda) = r(2\pi, \xi, \lambda) - \xi \qquad (8.22)$$

where $\theta \mapsto r(\theta, \xi, \lambda)$ is the solution of the initial value problem

$$\frac{dr}{d\theta} = \frac{\alpha(\lambda)r + p(r, \theta, \lambda)}{\beta(\lambda) + q(r, \theta, \lambda)}, \qquad r(0, \xi, \lambda) = \xi. \qquad (8.23)$$

In particular, with respect to the differential equation (8.23), the "return time" does not depend on the position ξ along the Poincaré section or the value of the parameter λ; rather, it has the constant value 2π.

Definition 8.24. Suppose that $(u, \lambda) = (0, 0) \in \mathbb{R}^2 \times \mathbb{R}$ is a Hopf point for the family (8.16). The corresponding rest point $u = 0$ is called a *weak attractor* (respectively, a *weak repeller*) if the associated displacement function (8.22) is such that $\delta_{\xi\xi\xi}(0,0) < 0$ (respectively, $\delta_{\xi\xi\xi}(0,0) > 0$). In addition, the Hopf point $(u, \lambda) = (0, 0)$ is said to have *multiplicity one* if $\delta_{\xi\xi\xi}(0,0) \neq 0$.

Theorem 8.25 (Hopf Bifurcation Theorem). *If the family of differential equations (8.16) has a Hopf point at $(u, \lambda) = (0,0) \in \mathbb{R}^2 \times \mathbb{R}$ and the corresponding rest point at the origin is a weak attractor (respectively, a weak repeller), then there is a supercritical (respectively, subcritical) Hopf bifurcation at this Hopf point.*

Proof. Let us assume that the family (8.16) is C^4. By Proposition 8.22, there is a smooth change of coordinates that transforms the family (8.16) into the family (8.18). Moreover, because $\beta(0) \neq 0$, the function

$$S(r, \theta, \lambda) := \frac{\alpha(\lambda)r + p(r, \theta, \lambda)}{\beta(\lambda) + q(r, \theta, \lambda)},$$

and therefore the family of differential equations

$$\frac{dr}{d\theta} = S(r, \theta, \lambda), \tag{8.24}$$

is as smooth as the original differential equation (8.16); that is, it is at least in class C^4.

The associated displacement function δ defined in equation (8.22) is given by the C^4 function

$$\delta(\xi, \lambda) := r(2\pi, \xi, \lambda) - \xi \tag{8.25}$$

where $\theta \mapsto r(\theta, \xi, \lambda)$ is the solution of the differential equation (8.24) with initial condition $r(0, \xi, \lambda) = \xi$. Moreover, each function $\xi \mapsto \delta(\xi, \lambda)$ is defined in a neighborhood of $\xi = 0$ in $\mathbb{R}$.

Since $\delta(0, \lambda) \equiv 0$, the displacement function is represented as a series,

$$\delta(\xi, \lambda) = \delta_1(\lambda)\xi + \delta_2(\lambda)\xi^2 + \delta_3(\lambda)\xi^3 + O(\xi^4),$$

whose first-order coefficient is given by

$$\delta_1(\lambda) = \delta_\xi(0, \lambda) = r_\xi(2\pi, 0, \lambda) - 1$$

where $\theta \mapsto r_\xi(\theta, 0, \lambda)$ is the solution of the variational initial value problem

$$\frac{dr_\xi}{d\theta} = S_r(0, \theta, \lambda)r_\xi = \frac{\alpha(\lambda)}{\beta(\lambda)}r_\xi, \qquad r_\xi(0, 0, \lambda) = 1.$$

Hence, by solving the scalar first order linear differential equation, we have that

$$\delta_1(\lambda) = r_\xi(2\pi, 0, \lambda) - 1 = e^{2\pi\alpha(\lambda)/\beta(\lambda)} - 1.$$

Moreover, since $\alpha(0) = 0$, it follows that

$$\delta(\xi, 0) = \xi^2\big(\delta_2(0) + \delta_3(0)\xi + O(\xi^2)\big).$$

Note that if $\delta_2(0) \neq 0$, then $\delta(\xi, 0)$ has constant sign for sufficiently small $|\xi| \neq 0$, and therefore the trajectories of the corresponding system (8.19) at $\lambda = 0$ do not spiral around the origin of its phase plane (draw a picture); equivalently, the periodic orbit $\{(r, \theta) : r = 0\}$ on the phase cylinder is a semistable limit cycle. But using the assumptions that $\alpha(0) = 0$ and $\beta(0) \neq 0$ and Exercise 8.23, this qualitative behavior cannot occur. In particular, the existence of a semistable limit cycle on the phase cylinder violates the symmetry (8.20), which carries over to the differential equation (8.23). In fact, if $\theta \to r(\theta, \xi, \lambda)$ is a solution of equation (8.23), then so is the function $\theta \to -r(\theta + \pi, \xi, \lambda)$. For all of these equivalent reasons, we have that $\delta_2(0) = 0$.

Consider the function $\Delta : \mathbb{R} \times \mathbb{R} \to \mathbb{R}$ defined on the domain of the displacement function by

$$\Delta(\xi, \lambda) = \delta_1(\lambda) + \delta_2(\lambda)\xi + \delta_3(\lambda)\xi^2 + O(\xi^3),$$

and note that

$$\Delta(0, 0) = e^{2\pi\alpha(0)/\beta(0)} - 1 = 0,$$
$$\Delta_\xi(0, 0) = \delta_2(0) = 0,$$
$$\Delta_{\xi\xi}(0, 0) = 2\delta_3(0) = \delta_{\xi\xi\xi}(0, 0)/3 \neq 0,$$
$$\Delta_\lambda(0, 0) = 2\pi\alpha'(0)/\beta(0) > 0.$$

By Proposition 8.2, the function Δ has a saddle-node bifurcation at $\xi = 0$ for the parameter value $\lambda = 0$. In particular, there is a curve $\xi \mapsto (\xi, \gamma(\xi))$ in $\mathbb{R} \times \mathbb{R}$ with $\gamma(0) = 0$, $\gamma'(0) = 0$, and $\gamma''(0) \neq 0$ such that $\Delta(\xi, \gamma(\xi)) \equiv 0$. As a result, we have that

$$\delta(\xi, \gamma(\xi)) = \xi\Delta(\xi, \gamma(\xi)) \equiv 0,$$

and therefore if $\lambda = \gamma(\xi)$, then there is a periodic solution of the corresponding member of the family (8.18) that meets the Poincaré section at the point with coordinate ξ.

For the remainder of the proof, let us assume that $\delta_{\xi\xi\xi}(0, 0) < 0$; the case where $\delta_{\xi\xi\xi}(0, 0) > 0$ is similar.

By Proposition 8.2, we have the inequality

$$\gamma''(0) = -\frac{\Delta_{\xi\xi}(0, 0)}{\Delta_\lambda(0, 0)} = -\frac{\beta(0)}{6\pi\alpha'(0)}\delta_{\xi\xi\xi}(0, 0) > 0,$$

and therefore the coefficient of the leading-order term of the series

$$\lambda = \gamma(\xi) = \frac{\gamma''(0)}{2}\xi^2 + O(\xi^3)$$

does not vanish. Hence, the position coordinate $\xi > 0$ corresponding to a periodic solution is represented as follows by a power series in $\sqrt{\lambda}$:

$$\xi = \left(-\lambda\frac{12\pi\alpha'(0)}{\beta(0)\delta_{\xi\xi\xi}(0, 0)}\right)^{1/2} + O(\lambda). \tag{8.26}$$

Thus, the distance from the periodic orbit to the origin is of the form $K_2\sqrt{k}+O(k)$ for some constant K_2. Using the construction in the discussion following display (8.21), it is easy to see that the function $S = S(r, \theta, \lambda)$ can be restricted to a compact subset of its domain with no loss of generality for our bifurcation analysis. By continuity, the magnitude of the partial derivative S_r is bounded by some constant $K > 0$ on such a compact set. Note that $S(0, \theta, \lambda) \equiv 0$ and apply the mean value theorem to S to obtain the inequality

$$|r(\theta, \xi, \lambda)| = |\xi| + \int_0^\theta |S(r, \phi, \lambda)| \, d\phi \le |\xi| + K \int_0^\theta |r| \, d\phi.$$

By an application of Gronwall's inequality, we have that

$$|r(\theta, \xi, \lambda)| \le \xi e^{2\pi K}$$

on our periodic solution. Hence, the periodic solution lies in a ball whose radius is $K_1\sqrt{k} + O(k)$ for some constant K_1, as required.

The proof will be completed by showing that the periodic solution corresponding to ξ given by the equation (8.26) is a stable hyperbolic limit cycle.

Consider the Poincaré map defined by

$$P(\xi, \lambda) := \delta(\xi, \lambda) + \xi = \xi(\Delta(\xi, \lambda) + 1)$$

and note that

$$P_\xi(\xi, \lambda) = \xi \Delta_\xi(\xi, \lambda) + \Delta(\xi, \lambda) + 1.$$

At the periodic solution we have $\lambda = \gamma(\xi)$, and therefore

$$P_\xi(\xi, \gamma(\xi)) = \xi \Delta_\xi(\xi, \gamma(\xi)) + 1.$$

Moreover, because $\Delta(\xi, \gamma(\xi)) \equiv 0$, we have the identity

$$\Delta_\xi(\xi, \gamma(\xi)) = -\Delta_\lambda(\xi, \gamma(\xi))\gamma'(\xi).$$

Using the relations $\Delta_\lambda(0, 0) > 0$, $\gamma'(0) = 0$, and $\gamma''(0) > 0$, it follows that if $\xi > 0$ is sufficiently small, then $\gamma'(\xi) > 0$ and $-\Delta_\lambda(\xi, \gamma(\xi)) < 0$; hence, $\Delta_\xi(\xi, \gamma(\xi)) < 0$ and $0 < P_\xi(\xi, \gamma(\xi)) < 1$. In other words, the periodic solution is a hyperbolic stable limit cycle.   □

While the presentation given in this section discusses the most important ideas needed to understand the Hopf bifurcation, there are a few unresolved issues. Note first that sufficient conditions for the Hopf bifurcation are given only for a two-dimensional system obtained by restriction to a center manifold, not for the original system of differential equations. In particular, the definition of a weak attractor is only given for two-dimensional systems. Also, we have not discussed an efficient method to determine the sign of the third space-derivative of the displacement function, an essential step for practical applications of the Hopf bifurcation theorem. For a resolution of the first issue see [151]; the second issue is addressed in the next section.

Exercise 8.26. Consider the two systems

$$\dot{r} = \lambda r \pm r^3, \qquad \dot{\theta} = 1 + ar^2,$$

where (r, θ) are polar coordinates. Show that the $+$ sign system has a supercritical Hopf bifurcation and the $-$ sign system has a subcritical Hopf bifurcation. The given systems are normal forms for the Hopf bifurcation.

Exercise 8.27. Show that the system

$$\dot{x} = \lambda x - y + xy^2, \qquad \dot{y} = x + \lambda y + y^3$$

has a subcritical Hopf bifurcation. Hint: Change to polar coordinates and compute (explicitly) the Poincaré map defined on the positive x-axis. Recall that Bernoulli's equation $\dot{z} = a(t)z + b(t)z^{n+1}$ is transformed to a linear equation by the change of variables $w = z^{-n}$.

Exercise 8.28. Suppose that $K \subseteq \mathbb{R}$ and $W \subseteq \mathbb{R}^k$ are open sets, $g : K \times W \to \mathbb{R}$ is a smooth function, and $T > 0$. If $[0, T] \subset K$, $0 \in W$, and $g_t(t, 0) \neq 0$ for all $t \in K$, then there are open product neighborhoods $I \times U \subseteq K \times W$ and $J \times V \subseteq \mathbb{R} \times \mathbb{R}^k$ with $[0, T] \subset I$ and $0 \in U$ and a smooth function $h : J \times V \to \mathbb{R}$ such that $h(g(t, u), u) = t$ whenever $(t, u) \in I \times U$. Hint: Consider the function $G : K \times W \times \mathbb{R} \to \mathbb{R}$ given by $G(t, w, s) = g(t, w) - s$ and note that $G(t, w, g(t, w)) \equiv 0$ and $G_t(t, w, g(t, w)) \neq 0$. Apply the implicit function theorem to obtain a function h such that $G(h(s, w), w, s) \equiv 0$ and $h(g(t, w), w) = t$. The implicit function is only locally defined but it is unique. Use the uniqueness to show that h is defined globally on an appropriate product neighborhood.

8.3.1   Multiple Hopf Bifurcation

The hypothesis in the Hopf bifurcation theorem, which states that a Hopf point has multiplicity one, raises at least two important questions: How can we check the sign of the third space-derivative $\delta_{\xi\xi\xi}(0, 0)$ of the displacement function? What happens if $\delta_{\xi\xi\xi}(0, 0) = 0$? The answers to these questions will be discussed in this section.

For the second question, let us note that (in the proof of the Hopf bifurcation theorem) the condition $\delta_{\xi\xi\xi}(0, 0) \neq 0$ ensures that the series representation of the displacement function has a nonzero coefficient at the lowest possible order. If this condition is not satisfied because $\delta_{\xi\xi\xi}(0, 0) = 0$, then the Hopf point is called *multiple* and the corresponding Hopf bifurcation is called a *multiple Hopf bifurcation*.

Let us consider the multiple Hopf bifurcation for the case of a planar vector field that depends on a vector of parameters. More precisely, we will consider the parameter λ in $\mathbb{R}^M$ and a corresponding family of differential equations

$$\dot{u} = f(u, \lambda), \qquad u \in \mathbb{R}^2 \tag{8.27}$$

with the following additional properties: the function f is real analytic; at the parameter value $\lambda = \lambda^*$, the origin $u = 0$ is a rest point for the differential equation $\dot{u} = f(u, \lambda^*)$; and the eigenvalues of the linear transformation $f_u(0, \lambda^*)$ are nonzero pure imaginary numbers. Under these assumptions, the displacement function δ is represented by a convergent power series of the form

$$\delta(\xi, \lambda) = \sum_{j=1}^{\infty} \delta_j(\lambda) \xi^j. \tag{8.28}$$

Definition 8.29. The rest point at $u = 0$, for the member of the family (8.27) at the parameter value $\lambda = \lambda^*$, is called a *weak focus of order k* if k is a positive integer such that

$$\delta_1(\lambda^*) = \cdots = \delta_{2k}(\lambda^*) = 0, \qquad \delta_{2k+1}(\lambda^*) \neq 0.$$

It is not difficult to show—a special case is proved in the course of the proof of the Hopf bifurcation theorem—that if $\delta_1(\lambda^*) = \cdots = \delta_{2k-1}(\lambda^*) = 0$, then $\delta_{2k}(\lambda^*) = 0$. In fact, this is another manifestation of the symmetry given in display (8.20).

The next theorem is a corollary of the Weierstrass preparation theorem (Theorem 5.15).

Proposition 8.30. *If the family (8.27) has a weak focus of order k at $u = 0$ for the parameter value $\lambda = \lambda^*$, then at most k limit cycles appear in a corresponding multiple Hopf bifurcation. More precisely, there is some $\epsilon > 0$ and some $\nu > 0$ such that $\dot{u} = f(u, \lambda)$ has at most k limit cycles in the open set $\{u \in \mathbb{R}^2 : |u| < \nu\}$ whenever $|\lambda - \lambda^*| < \epsilon$.*

While Proposition 8.30 states that at most k limit cycles appear in a multiple Hopf bifurcation at a weak focus of order k, additional information about the set of coefficients $\{\delta_{2j+1}(\lambda) : j = 0, \ldots, k\}$ is required to determine precisely how many limit cycles appear. For example, to obtain the maximum number k of limit cycles, it suffices to have these coefficients be independent in the following sense: There is some $\delta > 0$ such that for each $j \leq k$ and each $\epsilon > 0$, if $|\lambda_0 - \lambda^*| < \delta$ and

$$\delta_1(\lambda_0) = \delta_2(\lambda_0) = \cdots = \delta_{2j-1}(\lambda_0) = 0, \qquad \delta_{2j+1}(\lambda_0) \neq 0,$$

then there is a point λ_1 such that $|\lambda_1 - \lambda_0| < \epsilon$ and

$$\delta_1(\lambda_1) = \cdots = \delta_{2j-3}(\lambda_1) = 0, \qquad \delta_{2j-1}(\lambda_1)\delta_{2j+1}(\lambda_1) < 0.$$

The idea is that successive odd order coefficients can be obtained with opposite signs by making small changes in the parameter vector. The reason why this condition is important will be made clear later.

Before we discuss the multiple Hopf bifurcation in more detail, let us turn to the computation of the coefficients of the displacement function. To include the general case where the vector field depends on a vector of parameters, and in particular to include multiparameter bifurcations, we will consider an analytic family of differential equations of the form

$$\dot{x} = \epsilon x - y + p(x, y), \qquad \dot{y} = x + \epsilon y + q(x, y) \qquad (8.29)$$

where p and q together with their first order partial derivatives vanish at the origin. Moreover, the coefficients of the Taylor series representations of p and q at the origin are considered as parameters along with the parameter ϵ that is the real part of the eigenvalues of the linearization at the origin. We will show how to compute the Taylor coefficients of the Taylor series (8.28) corresponding to the displacement function for the system at $\epsilon = 0$. As a convenient notation, which is consistent with the notation used in the Hopf bifurcation theorem, let us consider the displacement function at the origin for the family (8.29) to be the function given by $(\xi, \epsilon) \mapsto \delta(\xi, \epsilon)$ where the additional parameters are suppressed.

Set $\epsilon = 0$ in system (8.29), change to polar coordinates, and consider the initial value problem (similar to the initial value problem (8.23))

$$\frac{dr}{d\theta} = \frac{r^2 A(r, \theta)}{1 + rB(r, \theta)}, \qquad r(0, \xi) = \xi \qquad (8.30)$$

where ξ is viewed as the coordinate on the Poincaré section corresponding to a segment of the line $\{(r, \theta) : \theta = 0\}$. The solution r is analytic and it is represented by a series of the form

$$r(\theta, \xi) = \sum_{j=1}^{\infty} r_j(\theta)\xi^j. \qquad (8.31)$$

As we proceed, note that only the first few terms of this series are required to determine $\delta_{\xi\xi\xi}(0, 0)$. Therefore, the analyticity of the family (8.30) is *not* necessary to verify the nondegeneracy condition for the Hopf bifurcation.

In view of the initial condition for the solution of the differential equation (8.31), it follows that $r_1(\theta) \equiv 1$ and $r_j(0) = 0$ for all $j \geq 2$. Hence, if the series (8.31) is inserted into the differential equation (8.30) and like powers of ξ are collected, then the sequence $\{r_j(\theta)\}_{j=2}^{\infty}$ of coefficients can be found recursively.

Since $r_1(\theta) \equiv 1$, the displacement function has the representation

$$\delta(\xi, 0) = r(2\pi, \xi) - \xi = r_2(2\pi)\xi^2 + r_3(2\pi)\xi^3 + O(\xi^4).$$

Also, since $\delta_2(0) = 0$, it follows that $r_2(2\pi) = 0$, and therefore

$$\delta(\xi, 0) = r_3(2\pi)\xi^3 + O(\xi^4). \tag{8.32}$$

Thus, we have proved that

$$\delta_{\xi\xi\xi}(0, 0) = 3! r_3(2\pi). \tag{8.33}$$

Since the coefficient $r_3(2\pi)$ can be computed using power series, formula (8.33) gives a method to compute the derivative $\delta_{\xi\xi\xi}(0, 0)$ (see Exercise 8.32 and the formula (8.43)).

Exercise 8.31. In the context of Definition 8.29, show that if $\delta_1(\lambda^*) = \cdots = \delta_{2k-1}(\lambda^*) = 0$, then $\delta_{2k}(\lambda^*) = 0$.

Exercise 8.32. Find an expression for $r_3(2\pi)$ in terms of the Taylor coefficients of the functions p and q in equation (8.29). Hint: Only the coefficients of order two and three are required.

The method proposed above for determining the Taylor coefficients of the displacement function has the advantage of conceptual simplicity; its disadvantage is the requirement that a differential equation be solved to complete each step of the algorithm. We will describe a more computationally efficient procedure—introduced by Lyapunov—that is purely algebraic. The idea of the procedure is to recursively construct *polynomial* Lyapunov functions for the system (8.29) that can be used to determine the stability of the rest point at the origin.

To implement Lyapunov's procedure, let

$$p(x, y) = \sum_{j=2}^{\infty} p_j(x, y), \qquad q(x, y) = \sum_{j=2}^{\infty} q_j(x, y)$$

where p_j and q_j are homogeneous polynomials of degree j for each $j = 2, \ldots, \infty$; let V denote the proposed Lyapunov function represented *formally* as the series

$$V(x, y) = \frac{1}{2}(x^2 + y^2) + \sum_{j=3}^{\infty} V_j(x, y) \tag{8.34}$$

where each V_j is a homogeneous polynomial of degree j; and let

$$X(x, y) := (-y + p(x, y))\frac{\partial}{\partial x} + (x + q(x, y))\frac{\partial}{\partial y}$$

denote the vector field associated with the system (8.29).

Exercise 8.33. Suppose that V, in display (8.34), is an analytic function. Show that there is a neighborhood U of the origin such that $V(x,y) > 0$ for $(x,y) \in U \setminus (0,0)$.

To determine the stability of the rest point at the origin for the system corresponding to the vector field X, let us begin by defining the *Lie derivative* of V in the direction of the vector field X by

$$(\mathcal{L}_X V)(x,y) = \frac{d}{dt}V(\varphi_t(x,y))\Big|_{t=0} = \mathrm{grad}V(x,y) \cdot X(x,y)$$

where φ_t denotes the flow of X. Also, let us recall the discussion of Lyapunov's direct method in Section 1.7. In particular, using the language of Lie derivatives, recall that if $V(x,y) > 0$ and $\mathcal{L}_X V(x,y) \leq 0$ on some punctured neighborhood of the origin, then V is called a *Lyapunov function* for system (8.29) at $(x,y) = (0,0)$, and we have the following theorem:

Theorem 8.34. *If V is a Lyapunov function at $(x,y) = (0,0)$ for the system (8.29) at $\epsilon = 0$ and $\mathcal{L}_X V(x,y) < 0$ for each point (x,y) in some punctured neighborhood of the origin, then the rest point at the origin is asymptotically stable.*

Lyapunov's idea for applying Theorem 8.34 to the system (8.29) at $\epsilon = 0$ is to construct the required function V recursively. We will explain this construction and also show that it produces the coefficients of the Taylor series of the displacement function.

Define $\mathcal{H}_n$ to be the vector space of all homogeneous polynomials of degree n in the variables x and y. Also, consider the vector field on $\mathbb{R}^2$ given by $R(x,y) = (x,y,-y,x)$, and observe that if V is a function defined on $\mathbb{R}^2$, then the Lie derivative $\mathcal{L}_R V$ can be viewed as the action of the *linear* differential operator $\mathcal{L}_R$, defined by

$$\mathcal{L}_R := -y\frac{\partial}{\partial x} + x\frac{\partial}{\partial y},$$

on V. In particular, $\mathcal{L}_R$ acts on the vector space $\mathcal{H}_n$ as follows:

$$(\mathcal{L}_R V)(x,y) = -yV_x(x,y) + xV_y(x,y)$$

(see Exercise 8.35).

Exercise 8.35. Prove that $\mathcal{H}_n$ is a finite dimensional vector space, compute its dimension, and show that the operator $\mathcal{L}_R$ is a linear transformation of this vector space.

Using the definition of the Lie derivative, we have that

$$\mathcal{L}_X V(x,y) = \Big(x + \sum_{j=3}^{\infty} V_{jx}(x,y)\Big)\Big(-y + \sum_{j=2}^{\infty} p_j(x,y)\Big)$$
$$+ \Big(y + \sum_{j=3}^{\infty} V_{jy}(x,y)\Big)\Big(x + \sum_{j=2}^{\infty} q_j(x,y)\Big)$$

where the subscripts x and y denote partial derivatives. Moreover, if we collect terms on the right hand side of this identity according to their degrees, then

$$\mathcal{L}_X V(x,y) = x p_2(x,y) + y q_2(x,y) + (\mathcal{L}_R V_3)(x,y) + O((x^2 + y^2)^2)$$

where $x p_2(x,y) + y q_2(x,y) \in \mathcal{H}_3$.

Proposition 8.36. *If n is an odd integer, then $\mathcal{L}_R : \mathcal{H}_n \to \mathcal{H}_n$ is a linear isomorphism.*

Assuming for the moment the validity of Proposition 8.36, it follows that there is some $V_3 \in \mathcal{H}_3$ such that

$$(\mathcal{L}_R V_3)(x,y) = -x p_2(x,y) - y q_2(x,y).$$

Hence, with this choice of V_3, the terms of order three in the expression for $\mathcal{L}_X V$ vanish, and this expression has the form

$$\mathcal{L}_X V(x,y) = x p_3(x,y) + y q_3(x,y) + V_{3x}(x,y)p_2(x,y) + V_{3y}(x,y)q_2(x,y)$$
$$+ (\mathcal{L}_R V_4)(x,y) + O((x^2 + y^2)^{5/2}).$$

Proposition 8.37. *If n is an even integer, say $n = 2k$, then the linear transformation $\mathcal{L}_R : \mathcal{H}_n \to \mathcal{H}_n$ has a one-dimensional kernel generated by $(x^2 + y^2)^k \in \mathcal{H}_n$. Also, the homogeneous polynomial $(x^2 + y^2)^k$ generates a one-dimensional complement to the range of $\mathcal{L}_R$.*

Assuming the validity of Proposition 8.37, there is a homogeneous polynomial $V_4 \in \mathcal{H}_4$ such that

$$\mathcal{L}_X V(x,y) = L_4(x^2 + y^2)^2 + O((x^2 + y^2)^{5/2}) \qquad (8.35)$$

where L_4 is a constant with respect to the variables x and y.

Equation (8.35) is useful. Indeed, if $L_4 \neq 0$, then the function

$$V(x,y) = \frac{1}{2}(x^2 + y^2) + V_3(x,y) + V_4(x,y) \qquad (8.36)$$

determines the stability of the rest point at the origin. More precisely, if $L_4 < 0$, then V is a Lyapunov function in some sufficiently small neighborhood of the origin and the rest point is stable. If $L_4 > 0$, then the rest point is unstable (to see this fact just reverse the direction of time).

Remark 3. These stability results do not require the vector field X to be analytic. Also, the formal computations with the series V are justified because the Lyapunov function (8.36) that is a requisite for applying Theorem 8.34 turns out to be a polynomial.

It should be clear that if $L_4 = 0$, then by the same procedure used to obtain L_4 we can produce a new V such that the leading term of the expression for $\mathcal{L}_X V$ is $L_6(x^2 + y^2)^3$, and so on. Moreover, we have a useful stability theorem.

Theorem 8.38. *If $L_{2n} = 0$, $n = 2, \ldots, N$, but $L_{2N+2} \neq 0$, then the stability of the rest point at the origin is determined: If $L_{2N+2} < 0$, then the rest point is stable. If $L_{2N+2} > 0$, then the rest point is unstable.*

The constant L_{2k} is called the kth *Lyapunov quantity.* By Theorem 8.38 and the algorithm for computing these Lyapunov quantities, we have a method for constructing Lyapunov functions at linear centers of planar systems. If after a finite number of steps a nonzero Lyapunov quantity is obtained, then we can produce a polynomial Lyapunov function and use it to determine the stability of the rest point.

What happens if all Lyapunov quantities vanish? This question is answered by the Lyapunov center theorem [144]:

Theorem 8.39 (Lyapunov Center Theorem). *If the vector field X is analytic and $L_{2n} = 0$ for each integer $n \geq 2$, then the origin is a center. Moreover, the formal series for V is convergent in a neighborhood of the origin and it represents a function whose level sets are orbits of the differential equation corresponding to X.*

Exercise 8.40. Write a program using an algebraic processor that upon input of system (8.29) and an integer N outputs L_{2n}, $n = 2, \ldots, N$. Use your program to compute L_4 for the system (8.29) in case the coefficients of p and q are regarded as parameters. Hint: Look ahead to page 590.

We will prove Propositions 8.36 and 8.37 on page 584. But, before we do so, let us establish the relationship between the Taylor coefficients of the displacement function and the Lyapunov quantities.

Proposition 8.41. *Suppose that $\xi \mapsto \delta(\xi, 0)$ is the displacement function for the system (8.29) at $\epsilon = 0$, and L_{2n} for $n \geq 2$, are the corresponding Lyapunov quantities. If k is a positive integer and $L_{2j} = 0$ for the integers $j = 1, \ldots, k - 1$, then*

$$\frac{\partial^{2k-1}\delta}{\partial \xi^{2k-1}}(0,0) = (2k-1)!2\pi L_{2k}.$$

In particular, $\delta_{\xi\xi\xi}(0,0) = 3!2\pi L_4$.

Proof. We will prove only the last statement of the theorem; the general proof is left as an exercise.

By equation (8.33), we have that $\delta_{\xi\xi\xi}(0,0) = 3!r_3(2\pi)$. Thus, it suffices to show that $r_3(2\pi) = 2\pi L_4$.

In polar coordinates, the polynomial

$$V(x,y) = \frac{1}{2}(x^2 + y^2) + V_3(x,y) + V_4(x,y)$$

in display (8.36) is given by

$$V := V(r\cos\theta, r\sin\theta) = \frac{1}{2}r^2 + r^3 V_3(\cos\theta, \sin\theta) + r^4 V_4(\cos\theta, \sin\theta).$$

Define $\rho = \sqrt{2V}$ and let $r := r(\theta, \xi)$ denote the (positive) solution of the initial value problem (8.30). If we also define $v_j(\theta) := 2V_j(\cos\theta, \sin\theta)$ for $j = 3, 4$, then we have

$$\rho = (r^2(1 + v_3(\theta)r + v_4(\theta)r^2))^{1/2}$$
$$= r(1 + \frac{v_3(\theta)}{2}r + \phi(\theta)r^2 + O(r^3)) \tag{8.37}$$

where $\phi(\theta) = v_4(\theta)/2 - (v_3(\theta))^2/8$. Moreover, if $r \geq 0$ is sufficiently small, then ρ is represented as indicated in display (8.37).

Define

$$\Delta(\xi) := \rho(2\pi, \xi) - \rho(0, \xi),$$

and use the initial condition $r(0, \xi) = \xi$ together with equation (8.37), to express Δ in the form

$$\Delta(\xi) = r(2\pi, \xi)(1 + \frac{v_3(2\pi)}{2}r(2\pi, \xi) + \phi(2\pi)r^2(2\pi, \xi))$$
$$- \xi(1 + \frac{v_3(0)}{2}\xi + \phi(0)\xi^2) + O(\xi^4) + O(r^4(2\pi, \xi)).$$

Also, since $v_j(0) = v_j(2\pi)$, we have the equation

$$\Delta(\xi) = r(2\pi, \xi)(1 + \frac{v_3(0)}{2}r(2\pi, \xi) + \phi(0)r^2(2\pi, \xi))$$
$$- \xi(1 + \frac{v_3(0)}{2}\xi + \phi(0)\xi^2) + O(\xi^4) + O(r^4(2\pi, \xi)).$$

Using formula (8.31), namely,

$$r(2\pi, \xi) = \xi + r_3(2\pi)\xi^3 + O(\xi^4),$$

it is easy to show that

$$\Delta(\xi) = r_3(2\pi)\xi^3 + O(\xi^4). \tag{8.38}$$

Also, using a direct computation, we have

$$
\begin{aligned}
\Delta(\xi) &= \rho(2\pi, \xi) - \rho(0, \xi) \\
&= \int_0^{2\pi} \frac{d\rho}{d\theta}(\theta, \xi)\, d\theta = \int_0^{2\pi} \frac{1}{\rho} \frac{dV}{d\theta}\, d\theta = \int_0^{2\pi} \frac{1}{\rho} \frac{dV}{dt} \frac{dt}{d\theta}\, d\theta \\
&= \int_0^{2\pi} \frac{1}{r(1 + \frac{1}{2} v_3(\theta) r + O(r^2))} (L_4 r^4 + O(r^6)) \frac{1}{1 + rB(r, \theta)}\, d\theta \\
&= \int_0^{2\pi} L_4 r^3 + O(r^4)\, d\theta \\
&= \int_0^{2\pi} L_4 (\xi + r_3(\theta)\xi^3 + O(\xi^4))^3 + O(\xi^4)\, d\theta \\
&= 2\pi L_4 \xi^3 + O(\xi^4).
\end{aligned}
$$

By equating the last expression for Δ to the expression in display (8.38), it follows that $r_3(2\pi) = 2\pi L_4$, as required. $\square$

Exercise 8.42. Our definition of the Lyapunov quantities depends on the basis for the complement of the range of $\mathcal{L}_R : \mathcal{H}_n \to \mathcal{H}_n$ for each even integer n. If a different basis is used, "Lyapunov quantities" can be defined in a similar manner. Describe how these quantities are related to the original Lyapunov quantities.

Exercise 8.43. Describe all Hopf bifurcations for the following equations:

1. $\dot{x} = y$, $\dot{y} = -x + \epsilon y - ax^2 y$ where $a \in \mathbb{R}$.

2. $\dot{x} = \epsilon x - y + p(x, y)$, $\dot{y} = x + \epsilon y + q(x, y)$ where p and q are homogeneous quadratic polynomials.

3. $\ddot{x} + \epsilon(x^2 - 1)\dot{x} + x = 0$ where $\epsilon \in \mathbb{R}$.

4. $\dot{x} = (x - \beta y)x + \epsilon y$, $\dot{y} = (x^2 - y)y$ where $\beta, \epsilon \in \mathbb{R}$.

Propositions 8.36 and 8.37 can be proved in a variety of ways. The proof given here uses some of the elementary ideas of Lie's theory of symmetry groups for differential equations (see [178]). The reader is encouraged to construct a purely algebraic proof.

Proof. Recall that the operator $\mathcal{L}_R : \mathcal{H}_n \to \mathcal{H}_n$ defines Lie differentiation in the direction of the vector field given by $R(x, y) = (x, y, -y, x)$. Geometrically, the vector field R represents the infinitesimal (positive) rotation of the plane centered at the origin. Its flow is the linear (positive) rotation given by

$$
\varphi_t(x, y) = e^{tA} \begin{pmatrix} x \\ y \end{pmatrix} = \begin{pmatrix} \cos t & -\sin t \\ \sin t & \cos t \end{pmatrix} \begin{pmatrix} x \\ y \end{pmatrix} \tag{8.39}
$$

where

$$A := \begin{pmatrix} 0 & -1 \\ 1 & 0 \end{pmatrix}.$$

If $f : \mathbb{R}^2 \to \mathbb{R}$ denotes a smooth function and $z := (x, y)$, then we have

$$\mathcal{L}_R f(z) = \frac{d}{dt}(f(\varphi_t(z)))\Big|_{t=0}.$$

A fundamental proposition in Lie's theory is the following statement: *If h is an* infinitesimally invariant *function with respect to a vector field X (that is, $\mathcal{L}_X h = 0$), then h is constant along integral curves of X.* This simple result depends on the group property of the flow of X.

To prove Lie's proposition in our special case, first define $h(t, z) = f(\varphi_t(z))$ and compute

$$\frac{d}{dt} h(t, z)\Big|_{t=s} = \lim_{\tau \to 0} \frac{1}{\tau}[h(s + \tau, z) - h(s, z)]$$

$$= \lim_{\tau \to 0} \frac{1}{\tau}[f(\varphi_\tau(\varphi_s(z))) - f(\varphi_s(z))]$$

$$= \mathcal{L}_R f(\varphi_s(z)). \tag{8.40}$$

If $s \mapsto \mathcal{L}_R f(\varphi_s(z))$ vanishes identically (that is, f is infinitesimally invariant), then the function $t \mapsto h(t, z)$ is a constant, and therefore $f(\varphi_t(z)) = f(z)$ for each $t \in \mathbb{R}$. For our special case where $\mathcal{L}_R : \mathcal{H}_n \to \mathcal{H}_n$ and $H \in \mathcal{H}_n$ we have the following corollary: *The homogeneous polynomial H is in the kernel of $\mathcal{L}_R$ if and only if H is rotationally invariant.*

If n is odd and $H \in \mathcal{H}_n$ is in the kernel of $\mathcal{L}_R$, then

$$H(x, y) = H(\varphi_\pi(x, y)) = H(-x, -y) = (-1)^n H(x, y) = -H(x, y),$$

and therefore $H = 0$. In other words, since $\mathcal{H}_n$ is finite dimensional, the linear operator $\mathcal{L}_R$ is invertible.

If n is even and $H \in \mathcal{H}_n$ is rotationally invariant, then

$$H(\cos\theta, \sin\theta) = H(1, 0), \qquad 0 \le \theta < 2\pi.$$

Moreover, since $H \in \mathcal{H}_n$ is homogeneous, we also have that

$$H(r\cos\theta, r\sin\theta) = r^n H(1, 0);$$

in other words, $H(x, y) = H(1, 0)(x^2 + y^2)^{n/2}$. Thus, the kernel of $\mathcal{L}_R$ is one-dimensional and it is generated by the homogeneous polynomial $(x^2 + y^2)^{n/2}$.

We will show that the polynomial $(x^2 + y^2)^{n/2}$ generates a complement to the range of $\mathcal{L}_R$. Because the kernel of $\mathcal{L}_R$ is one-dimensional, its range has codimension one. Thus, it suffices to show that the nonzero vector $(x^2 + y^2)^{n/2}$ is not in the range.

If there is some $H \in \mathcal{H}_n$ such that $\mathcal{L}_R H(x, y) = (x^2 + y^2)^{n/2}$, then choose $z = (x, y) \neq 0$ and note that $\mathcal{L}_R H(\varphi_t(z)) = ||z||^n \neq 0$. By the formula (8.40), the function $t \mapsto H(\varphi_t(z))$ is the solution of the initial value problem

$$\dot{u} = ||z||^n, \qquad u(0) = H(z),$$

and therefore $H(\varphi_t(z)) = ||z||^n t + H(z)$. Since $t \mapsto H(\varphi_t(z))$ is 2π-periodic, it follows that $||z|| = 0$, in contradiction. $\square$

We have just developed all the ingredients needed to detect (multiple) Hopf bifurcation from a weak focus of finite order for a family of the form

$$\dot{x} = \lambda_1 x - y + p(x, y, \lambda), \qquad \dot{y} = x + \lambda_1 y + q(x, y, \lambda) \qquad (8.41)$$

where $\lambda = (\lambda_1, \dots, \lambda_N)$ is a vector-valued parameter. For simplicity, let us assume that the coefficients of the Taylor series at the origin for the functions $(x, y) \mapsto p(x, y, \lambda)$ and $(x, y) \mapsto q(x, y, \lambda)$ are polynomials in the components of λ. Also, let us recall that the Lyapunov quantities are computed at the parameter values where $\lambda_1 = 0$. Thus, as a convenient notation, let $\Lambda = (0, \lambda_2, \dots, \lambda_N)$ be the vector variable for points in this hypersurface of the parameter space so that the Lyapunov quantities are functions of the variables $\lambda_2, \dots, \lambda_N$. Moreover, if k is a positive integer and for some fixed Λ^* in the hypersurface we have $L_{2j}(\Lambda^*) = 0$ for $j = 2, \dots, k-1$, and $L_{2k}(\Lambda^*) \neq 0$, then by Proposition 8.30 at most $k-1$ limit cycles appear near the origin of the phase plane for the members of the family corresponding to parameter values λ with $|\lambda - \Lambda^*|$ sufficiently small.

If $L_{2k}(\Lambda^*) = 0$ for each integer $k \geq 2$, then the theory discussed so far does not apply because the bifurcation point does not have finite multiplicity. To include this case, the rest point at the origin is called an *infinite order weak focus* if the Taylor coefficients of the displacement function at the origin are such that $\delta_j(\Lambda^*) = 0$ for all integers $j \geq 1$. We will briefly discuss some of the beautiful ideas that can be used to analyze the bifurcations for this case (see for example [23], [42], [197], and [240]).

The starting point for the general theory is the observation encoded in the following proposition.

Proposition 8.44. *The Lyapunov quantities for an analytic family are polynomials in the Taylor coefficients of the corresponding analytic family of vector fields.*

Hence, each Lyapunov quantity can be computed, in principle, in a finite number of steps. The proof of the proposition starts on page 589.

As we have seen, the Lyapunov quantities are closely related to the Taylor coefficients of the displacement function. Thus, there is good reason to believe that the problem of the appearance of limit cycles at an infinite order weak focus can be approached by working with polynomials.

Let us recall that the displacement function for the family (8.41) has the form

$$\delta(\xi, \lambda) = \delta_1(\lambda)\xi + \sum_{j=2}^{\infty} \delta_j(\lambda)\xi^j$$

where $\delta_1(\lambda) = e^{2\pi\lambda_1} - 1$. Moreover, if the rest point at the origin is an infinite order weak focus at $\lambda = \Lambda^*$, then $\delta_j(\Lambda^*) = 0$ for each integer $j \geq 1$.

Two analytic functions are said to define the same *germ*, at a point in the intersection of their domains if they agree on an open set containing this point; or equivalently if they have the same Taylor series at this point. The set of all germs of analytic functions of the variables $\lambda_1, \dots, \lambda_N$ at the point Λ^*, that is, convergent power series in powers of

$$\lambda_1, \lambda_2 - \lambda_2^*, \dots, \lambda_N - \lambda_N^*,$$

has (by the Hilbert basis theorem) the structure of a Noetherian ring (see, for example, [29]). Therefore, the chain of its ideals

$$(\delta_1) \subseteq (\delta_1, \delta_2) \subseteq (\delta_1, \delta_2, \delta_3) \subseteq \cdots,$$

must stabilize. More precisely, there is an ideal

$$(\delta_1, \delta_2, \delta_3, \dots, \delta_K)$$

that contains all subsequent ideals in the chain; in other words, there is an ideal generated by a *finite* initial segment of Taylor coefficients of the displacement function that contains *all* of the Taylor coefficients. Hence, for each positive integer J, there is a set of analytic functions $\{\mu_{Jk}(\lambda) : k = 1, \dots, K\}$ such that

$$\delta_J(\lambda) = \sum_{k=1}^{K} \mu_{Jk}(\lambda)\delta_k(\lambda). \qquad (8.42)$$

By using the representation (8.42) and a *formal* calculation, it is easy to obtain the following series expansion for the displacement function:

$$\delta(\xi, \lambda) = \delta_1(\lambda)\xi(1 + \sum_{j=K+1}^{\infty} \mu_{j1}(\lambda)\xi^{j-1})$$

$$+ \delta_2(\lambda)\xi^2(1 + \sum_{j=K+1}^{\infty} \mu_{j2}(\lambda)\xi^{j-2})$$

$$+ \cdots + \delta_K(\lambda)\xi^K(1 + \sum_{j=K+1}^{\infty} \mu_{jK}(\lambda)\xi^{j-K}).$$

While it is certainly not obvious that this formal rearrangement of the Taylor series of the displacement function is convergent, this result has been proved (see, for example, [42] and the references therein). By an inspection of this series, it is reasonable to expect and not too difficult to prove that if $|\xi|$ and $|\lambda - \lambda^*|$ are sufficiently small, then the appearance of limit cycles is determined by an analysis of the zero set of the function

$$B(\xi, \lambda) := \delta_1(\lambda)\xi + \delta_2(\lambda)\xi^2 + \cdots + \delta_K(\lambda)\xi^K.$$

In particular, because B is a polynomial of degree K in the variable ξ, the displacement function δ cannot have more than K "local" zeros.

It turns out that, by the symmetry of the problem, only the odd order Taylor coefficients of the displacement function are important. In fact, there is some positive integer k such that the initial segment of Taylor coefficients given by $(\delta_1, \delta_3, \delta_5, \ldots, \delta_{2k+1})$ generates the ideal of all Taylor coefficients. In this case, the multiple bifurcation point is said to have *order* k. Of course, the reason for this definition is that at most k local limit cycles can appear after perturbation of a bifurcation point of order k. Indeed, let us note that the origin $\xi = 0$ accounts for one zero of the displacement function and each limit cycle accounts for two zeros because such a limit cycle must cross both the positive and the negative ξ-axis. Since the displacement function has at most $2k + 1$ zeros, there are at most k local limit cycles.

As mentioned previously, additional conditions must be satisfied to determine the exact number of limit cycles. For example, let us suppose that the function

$$B_1(\lambda, \xi) := \delta_1(\lambda) + \delta_2(\lambda)\xi + \cdots + \delta_{2k+1}(\lambda)\xi^{2k}$$

is such that $\delta_{2k+1}(\Lambda_1) > 0$ and $\delta_j(\Lambda_1) = 0$ for $j = 1, \ldots, 2k$. For example, this situation might arise if we found that $L_{2j}(\Lambda_1)$ vanishes for $j = 1, \ldots k$ and then noticed that the value of the polynomial L_{2k+2} at Λ_1 is positive. At any rate, if there is a parameter value Λ_2 so that

$$\delta_{2k+1}(\Lambda_2) > 0, \qquad \delta_{2k}(\Lambda_2) < 0, \qquad \delta_j(\Lambda_2) = 0$$

for $j = 1, \ldots, 2k - 1$, and $|\Lambda_2 - \Lambda_1|$ is sufficiently small, then the function $\xi \mapsto B_1(\xi, \Lambda_2)$ will have two zeros near $\xi = 0$, one positive zero and one zero at the origin. By continuity, if $|\Lambda_3 - \Lambda_2|$ is sufficiently small, then the corresponding function at the parameter value Λ_3 also has continuations of these zeros. Moreover, if there is a choice of Λ_3 in the required open subset of the parameter space such that $\delta_{2k-1}(\Lambda_3) > 0$, then $B_1(\xi, \Lambda_3)$ has three zeros, and so on. Well, almost. We have used Λ_j for $j = 1, 2, 3, \ldots$ to indicate that $\lambda_1 = 0$. But, at the last step, where $\delta_1(\lambda) = e^{2\pi\lambda_1} - 1$ is adjusted, we can take a nonzero value of λ_1.

To implement the theory just outlined, we must compute some finite set of Taylor coefficients, say $\{\delta_j : j = 1, \ldots, 2k + 1\}$, and then *prove* that the ideal generated by these Taylor coefficients contains all subsequent Taylor coefficients. This is a difficult problem that has only been solved in a few special cases. The most famous result of this type was proved by Nikolai N. Bautin [23] for quadratic systems—that is, for

$$\dot{x} = \epsilon x - y + p(x, y), \qquad \dot{y} = x + \epsilon y + q(x, y)$$

where p and q are homogeneous quadratic polynomials and where λ is the vector consisting of ϵ and the coefficients of p and q. In this case, Bautin showed that the ideal of all Taylor coefficients is generated by $(\delta_1, \delta_3, \delta_5, \delta_7)$. Thus, at most three limit cycles can bifurcate from the origin. Moreover, it is possible to construct an example where three limit cycles do appear (see [23] and [240]).

From the above remarks, it should be clear that it is not easy to count the exact number of limit cycles of a polynomial system. Indeed, this is the content of Hilbert's 16th problem: *Is there a bound for the number of limit cycles of a polynomial system in terms of the degrees of the polynomials that define the system?* This problem is not solved, even for quadratic systems. The best result obtained so far is the following deep theorem of Yulij Il'yashenko [126] and Jean Ecalle [79].

Theorem 8.45. *A polynomial system has at most a finite number of limit cycles.*

(See the book of Il'yashenko [126] and the review [36] for a mathematical history of the work on Hilbert's problem, and see [187] for a complete bibliography of quadratic systems theory.)

The remainder of this section is devoted to the promised proof that the Lyapunov quantities for an analytic system are polynomials in the Taylor coefficients of the vector field corresponding to the system and to a description of an algorithm that can be used to compute the Lyapunov quantities.

Consider the vector field

$$X(x, y) := (-y\frac{\partial}{\partial x} + x\frac{\partial}{\partial y}) + \left(\sum_{j=2}^{\infty} p_j(x, y)\frac{\partial}{\partial x} + \sum_{j=2}^{\infty} q_j(x, y)\frac{\partial}{\partial y} \right)$$

where $p_j, q_j \in \mathcal{H}_j$ for each integer $j \geq 1$. Also, let

$$V(x, y) := \frac{1}{2}(x^2 + y^2) + \sum_{j=3}^{\infty} V_j(x, y)$$

where $V_j \in \mathcal{H}_j$ for each $j \geq 1$. The Lie derivative of V in the direction X is given by

$$
\mathcal{L}_X V = \sum_{j=3}^{\infty} \mathcal{L}_R V_j + x \sum_{j=2}^{\infty} p_j + y \sum_{j=2}^{\infty} q_j + \sum_{j=2}^{\infty} p_j \sum_{j=3}^{\infty} V_{jx} + \sum_{j=2}^{\infty} q_j \sum_{j=3}^{\infty} V_{jy}
$$

$$
= \sum_{j=3}^{\infty} \mathcal{L}_R V_j + \sum_{j=2}^{\infty} (x p_j + y q_j) + \sum_{j=2}^{\infty} \left(\sum_{i=0}^{j-2} p_{j-i} V_{(i+3)x} + q_{j-i} V_{(i+3)y} \right)
$$

$$
= \mathcal{L}_R V_3 + x p_2 + y q_2 + \sum_{j=4}^{\infty} \Big(\mathcal{L}_R V_j + x p_{j-1} + y q_{j-1}
$$

$$
+ \sum_{i=0}^{j-4} p_{j-i-2} V_{(i+3)x} + q_{j-i-2} V_{(i+3)y} \Big).
$$

For each even integer $j \geq 2$, let $\Pi_j : \mathcal{H}_j \to \mathcal{H}_j$ denote the linear projection whose kernel is the range of the operator $\mathcal{L}_R$ and whose range is our one-dimensional complement to the range of the operator $\mathcal{L}_R$; that is, the subspace of $\mathcal{H}_{2j}$ generated by the vector $(x^2 + y^2)^j$. Also, for each integer $j \geq 4$, define $H_j \in \mathcal{H}_j$ by

$$
H_j := x p_{j-1} + y q_{j-1} + \sum_{i=0}^{j-4} \left(p_{j-i-2} V_{(i+3)x} + q_{j-i-2} V_{(i+3)y} \right)
$$

so that

$$
\mathcal{L}_X V = \mathcal{L}_R V_3 + x p_2 + y q_2 + \sum_{j=4}^{\infty} \left(\mathcal{L}_R V_j + H_j \right).
$$

The following algorithm can be used to compute the Lyapunov quantities:

> Input $(k, p_2, \dots, p_{2k-1}, q_2, \dots, q_{2k-1})$
> $\quad V_3 := -\mathcal{L}_R^{-1}(x p_2 + y q_2)$
> For j from 4 to $2k$ do
> $\quad$ If j is odd, then $V_j := -\mathcal{L}_R^{-1}(x p_{j-1} + y q_{j-1} + H_j)$
> $\quad$ If j is even, then
> $\qquad L_j := \Pi_j(H_j)/(x^2 + y^2)^{j/2}$
> $\qquad V_j := -\mathcal{L}_R^{-1}\big(H_j - L_j(x^2 + y^2)^{j/2}\big)$
> End for loop;
> Output $(L_4, L_6, \dots, L_{2k})$.

To implement the algorithm, it is perhaps best to first choose a basis for each vector space $\mathcal{H}_j$ and then to represent the linear transformations Π_j and $\mathcal{L}_R$ in this basis (see Exercise 8.47).

Remark 4. The value of L_4 in case

$$p_j(x,y) = \sum_{i=0}^{j} a_{j-i,i} x^{j-i} y^i, \qquad q_j(x,y) = \sum_{i=0}^{j} b_{j-i,i} x^{j-i} y^i$$

is given by

$$L_4 = \frac{1}{8}(a_{20}a_{11} + b_{21} + 3a_{30} - b_{02}b_{11}$$
$$+ 3b_{03} + 2b_{02}a_{02} - 2a_{20}b_{20} - b_{20}b_{11} + a_{12} + a_{02}a_{11}). \quad (8.43)$$

The sign of this quantity is the same as the sign of the third Taylor coefficient of the displacement function. Thus, the sign of L_4 can be used to determine the stability of a weak focus as required in the Hopf bifurcation theorem.

Finally, we will show that if k is a positive integer, then the Lyapunov quantity L_{2k} is a polynomial in the Taylor coefficients of p and q at the origin. To prove this fact, note that

$$\mathcal{L}_R V_{2k} + H_{2k} - L_{2k}(x^2 + y^2)^k = 0.$$

Moreover, the linear flow of the vector field R is given by

$$\varphi_t(x,y) = e^{tA}\begin{pmatrix} x \\ y \end{pmatrix} \text{ where } A = \begin{pmatrix} 0 & -1 \\ 1 & 0 \end{pmatrix},$$

and, by Exercise 8.46, the projection Π_{2k} is represented by

$$\Pi_{2k}H(x,y) = \frac{1}{2\pi}\int_0^{2\pi} H(\varphi_t(x,y))\, dt. \quad (8.44)$$

Since the rotationally invariant elements of $\mathcal{H}_{2k}$ are in the complement of the range of $\mathcal{L}_R$, the composition $\Pi_{2k}\mathcal{L}_R$ is equal to the zero operator, and therefore

$$L_{2k}(x^2 + y^2)^k = \Pi_{2k}H_{2k}(x,y).$$

In particular, the desired Lyapunov quantity is given by the integral

$$L_{2k} = \frac{1}{2\pi}\int_0^{2\pi} H_k(\cos t, \sin t)\, dt.$$

Hence, by inspection of the algorithm for computing the Lyapunov quantities, it is clear that L_{2k} is a polynomial in the coefficients of the polynomials

$$p_2, \dots, p_{2k-1}, q_2, \dots, q_{2k-1}.$$

Exercise 8.46. Demonstrate that the representation (8.44) is valid by showing: a) Π_{2k} is linear, b) $\Pi_{2k}H$ is rotationally invariant, and c) $\Pi_{2k}(x^2 + y^2)^k = (x^2 + y^2)^k$.

Exercise 8.47. Prove the following statements. The set $B := \{x^{n-i}y^i \mid i = 0, \dots, n\}$ is a basis for $\mathcal{H}_n$ and $\mathcal{L}_R$ has the following $(n+1) \times (n+1)$ matrix representation with respect to the given (ordered) basis:

$$\mathcal{L}_R = \begin{pmatrix} 0 & 1 & 0 & 0 & 0 & \cdots \\ -n & 0 & 2 & 0 & 0 & \cdots \\ 0 & 1-n & 0 & 3 & 0 & \cdots \\ 0 & 0 & 2-n & 0 & 4 & \cdots \\ \cdot & \cdot & \cdot & \cdot & \cdot & \cdots \end{pmatrix}.$$

The kernel of $\mathcal{L}_R$ on $\mathcal{H}_{2k}$, for $k \geq 2$, is generated by the vector

$$K = (B_{k,0}, 0, B_{k,1}, 0, B_{k,2}, \dots, 0, B_{k,k})$$

where the numbers

$$B_{k,j} = \frac{k!}{j!(k-j)!}$$

are the binomial coefficients, and

$$\{(a_1, \dots, a_{2k}, 0) : (a_1, \dots, a_{2k}) \in \mathbb{R}^{2k}\}$$

is a vector space complement of the kernel. The operator $\mathcal{L}_R$ on $\mathcal{H}_{2k}$ is represented by the matrix $(\ell_1, \dots, \ell_{2k+1})$ partitioned by the indicated columns. The matrix representation for $\mathcal{L}_R$, restricted to this complement of the kernel, is given by $(\ell_1, \dots, \ell_{2k}, 0)$. Consider $V, H \in \mathcal{H}_{2k}$ and the associated matrix equation

$$(\ell_1, \dots, \ell_n, K)V = H,$$

where the matrix is partitioned by columns and H is represented in the basis B. The matrix is invertible. If the solution V is given by the vector $(a_1, \dots, a_{2k}, L)$, then H is given by $H = \sum_{j=1}^{2k} a_j \ell_j + LK$ where L is the corresponding Lyapunov quantity. The projection Π_{2k} is given by $\Pi_{2k}(H) := LK$.

Exercise 8.48. Determine the stability of the rest point at the origin for the system

$$\dot{x} = -y - x^2 + xy, \qquad \dot{y} = x + 2xy.$$

Exercise 8.49. Discuss the Hopf bifurcation for the following systems:

1. $\dot{x} = \epsilon x - y - x^2 + xy, \qquad \dot{y} = x + \epsilon y + 2xy.$
2. $\dot{x} = \epsilon x - y - x^2 + \epsilon xy, \qquad \dot{y} = x + \epsilon y + 2y^2.$
3. $\dot{x} = x(x - \beta y) + \epsilon y, \qquad \dot{y} = y(x^2 - y).$

Exercise 8.50. (a) Show that the system

$$\dot{u} = v, \qquad \dot{v} = -u - \frac{1}{\sqrt{5}}(2u^2 + 3uv - 2v^2)$$

has a center at the origin. Hint: Use Exercise 1.134 and separate variables to find a first integral. (b) Show by computation that the Lyapunov quantities L_2, L_4, and L_6 all vanish.

Exercise 8.51. Consider the quadratic system in Bautin normal form:

$$\dot{x} = \lambda_1 x - y - \lambda_3 x^2 + (2\lambda_2 + \lambda_5)xy + \lambda_6 y^2,$$
$$\dot{y} = x + \lambda_1 y + \lambda_2 x^2 + (2\lambda_3 + \lambda_4)xy - \lambda_2 y^2.$$

Find the corresponding Lyapunov quantities L_2, L_4, and L_6. Construct a curve in the parameter space with a supercritical Hopf bifurcation. Construct a quadratic system with two limit cycles surrounding the origin and a quadratic system with three limit cycles surrounding the origin. (If you need help, see [43] or [183, p. 449].)

Exercise 8.52. The family

$$\ddot{\theta} + \sin\theta - \Omega\cos\theta\sin\theta = I\dot{\theta},$$

where Ω and I are real parameters, is a simple model in dimensionless form of a whirling pendulum with a feedback control. Discuss the existence of a Hopf bifurcation for the rest point at the origin in the phase plane at the control coefficient value $I = 0$. How does the existence of a Hopf bifurcation depend on the rotation speed Ω? Draw the bifurcation diagram.

Exercise 8.53. Consider the following model for the dimensionless concentrations x and y of certain reacting chemicals

$$\dot{x} = a - x - \frac{4xy}{1+x^2}, \qquad \dot{y} = bx\left(1 - \frac{y}{1+x^2}\right),$$

and the curve C in the first quadrant of the (a,b)-parameter space given by $b = 3a/5 - 25/a$. Prove that a supercritical Hopf bifurcation occurs when a curve in the parameter space crosses C from above. This exercise is taken from [218, p. 256] where the derivation of the model and typical phase portraits are described.

Exercise 8.54. [Normal Forms] The computation of Lyapunov quantities is a special case of a procedure for simplifying vector fields near a rest point. To describe the procedure, called reduction to normal form, suppose that $\dot{x} = f(x)$ is a smooth system on $\mathbb{R}^n$ with a rest point at the origin. Let $A := Df(0)$ and expand f in its Taylor series to order k at the origin to obtain the representation

$$f(x) = Ax + \sum_{j=2}^{k} f_j(x) + O(|x|^{k+1})$$

where f_j is the jth-order term of the expansion, whose components are homogeneous polynomials of degree j. To simplify the differential equation, we first simplify the linear terms to Jordan normal form via a linear transformation $x = By$. (a) Show that the transformed system is given by

$$\dot{y} = B^{-1}ABy + \sum_{j=2}^{k} B^{-1}f_j(By) + O(|y|^{k+1}),$$

which is equivalent to the differential equation

$$\dot{x} = Jx + \sum_{j=2}^{k} f_j^1(x) + O(|x|^{k+1})$$

where $J := B^{-1}AB$ and $f_j^1 := B^{-1} \circ f_j \circ B$.

(b) To simplify the quadratic terms, consider a transformation of the form $x = y + h(y)$ where h has quadratic homogeneous polynomial components. Show that this transformation is invertible in a neighborhood of the origin.

(c) Show that the change of variables in part (b) transforms the differential equation to the form

$$\dot{y} = (I + Dh(y))^{-1}(J(y + h(y)) + \sum_{j=2}^{k} f_j^1(y + h(y)) + O(|y|^{k+1}),$$

which gives (to second order)

$$\dot{y} = Jy + (Jh(y) - Dh(y)Jy) + f_2^1(y) + O(|y|^3).$$

(d) Note that in part (c) the previously simplified first order term is not changed by the coordinate transformation. Also, define H_j to be the transformations of $\mathbb{R}^n$ whose components are homogeneous polynomials of degree j. Consider the operator L_J (the Lie derivative in the direction J) where J is viewed as a vector field. Prove that $L_J : H_2 \to H_2$, L_J is linear, and $(L_J h)(y) = Dh(y)Jy - Jh(y)$.

(e) Define $\mathcal{R}$ to be the range of L_J and choose a complement $\mathcal{K}$ of the linear subspace $\mathcal{R}$ in H_2. The element $f_2^1 \in H_2$ can be represented in the form $f_2^1 = R_2^1 + K_2^1$ where $R_2^1 \in \mathcal{R}$ and $K_2^1 \in \mathcal{K}$. Solve the equation $L_J h = R_2^1$ for h and show that this choice leads to the transformed differential equation

$$\dot{y} = Jy + f_2^2(y) + \sum_{j=3}^{k} f_j^2(y) + O(|y|^{k+1})$$

where $f_2^2 \in \mathcal{K}$. At this point the original differential equation is in normal form to order two.

(f) Show that the transformation to normal form can be continued to order k. Hint: At each order j there is a transformation of the from $x = y + h(y)$ where $h \in H_j$ such that after this transformation the terms of order less than j remain unchanged and the new differential equation is in normal form to order j; that is, the jth order term is in a space complementary to the range of L_J in H_j.

(g) Suppose that $n = 2$ and

$$A = \begin{pmatrix} 0 & -1 \\ 1 & 0 \end{pmatrix}.$$

determine a normal form for an arbitrary planar system of the form $\dot{x} = Ax + F(x)$ where $F(0) = DF(0) = 0$ and F is smooth.

(h) Show that with appropriate choices of complementary subspaces the normal form for part (g) is given by

$$\dot{r} = \sum_{j=1}^{k} a_{2j+1} r^{2j+1}, \qquad \dot{\theta} = 1 + \sum_{j=1}^{k} b_{2j} r^{2j}.$$

(i) Repeat part (h) for the matrix

$$A = \begin{pmatrix} \alpha & -\beta \\ \beta & \alpha \end{pmatrix}.$$

to obtain the normal form for the Hopf bifurcation. With appropriate choices, its polar coordinate representation is

$$\dot{r} = \alpha r + \sum_{j=1}^{k} a_{2j+1} r^{2j+1}, \qquad \dot{\theta} = \beta + \sum_{j=1}^{k} b_{2j} r^{2j}$$

(cf. Exercise 8.26). Note: There is an extensive literature on normal forms (see, for example, [58], [103], [171], [201], and [233]).

8.4   Dynamic Bifurcation

In this section we will introduce dynamic bifurcation theory and discuss an important delay phenomenon associated with slow passage through a bifurcation point. Previously we considered qualitative changes in the phase portraits for members of a family of differential equations such as

$$\dot{u} = f(u, \lambda), \qquad u \in \mathbb{R}^n, \quad \lambda \in \mathbb{R}$$

for different values of the parameter λ. This is called static bifurcation theory; the parameter is changed in our analysis, but it *does not change with time*. While static bifurcation theory might seem to give correct results when the parameter is varied slowly with time—maybe the parameter is changed by moving a control dial, new phenomena may occur that are not explained by static bifurcation theory. We will analyze some of these phenomena associated with bifurcations in systems such as

$$\dot{\lambda} = \epsilon g(u, \lambda, \epsilon), \qquad \dot{u} = f(u, \lambda, \epsilon)$$

where the dynamic parameter λ is viewed as a dependent variable and ϵ is a (small) parameter. This subject is called dynamic bifurcation theory.

Let us begin by formulating and proving a proposition about certain maps associated with autonomous planar systems. Consider a differential equation $\dot{u} = X(u)$ on $\mathbb{R}^2$ with flow φ_t and two curves Σ and Σ' in $\mathbb{R}^2$ that are both transverse to the vector field X (that is, X is never tangent to either curve). The differential equation induces a *section map* $P : \Sigma \to \Sigma'$ if there is a smooth function $T : \Sigma \to \mathbb{R}$ such that the function P defined on Σ by $P(\sigma) = \varphi_{T(\sigma)}(\sigma)$ has values in Σ'.

Proposition 8.55. *Suppose that*

$$\dot{x} = f(x, y), \qquad \dot{y} = yg(x, y)$$

is a C^1 differential equation on $\mathbb{R}^2$; $a, b \in \mathbb{R}$ with $a < b$; $\Sigma_b := \{(x, y) \in \mathbb{R}^2 : x = b\}$; and, for each $\delta > 0$, $\Sigma_a^{\delta} := \{(x, y) \in \mathbb{R}^2 : x = a \text{ and } |y| < \delta\}$. If $f(x, 0) > 0$ for all x in the closed interval $[a, b]$ and $\delta > 0$ is sufficiently *small, then the differential equation induces a section map P from Σ_a^δ to Σ_b such that, in the usual local coordinate on Σ_a^δ given by $(a, \eta) \mapsto \eta$,*

$$P'(0) = e^{\int_a^b \frac{g(x,0)}{f(x,0)}\, dx}.$$

Proof. Let $t \mapsto (x(t, \eta), y(t, \eta))$ denote the family of solutions of the differential equation with the initial conditions

$$x(0, \eta) \equiv a, \qquad y(0, \eta) \equiv \eta.$$

Since the x-axis is invariant, $y(t, 0) \equiv 0$. Also, since $f(x, 0) > 0$ on the interval $[a, b]$, there is a number $T_0 > 0$ such that $x(T_0, 0) = b$ and $\dot{x}(T_0, 0) > 0$. Hence, by the implicit function theorem, there is some number $\delta > 0$ and a smooth function $T : [-\delta, \delta] \to \mathbb{R}$ such that $T(0) = T_0$ and $x(T(\eta), \eta) \equiv b$ for all $\eta \in [-\delta, \delta]$. It follows that the desired section map (in local coordinates) exists and is given by

$$P(\eta) = y(T(\eta), \eta).$$

Moreover, by using the identity $\dot{y}(t, 0) \equiv 0$, the derivative of the section map at $\eta = 0$ is given by

$$P'(0) = \dot{y}(T(0), 0)T'(0) + y_\eta(T(0), 0) = y_\eta(T(0), 0).$$

Note that

$$x_\eta(0, \eta) \equiv 0, \qquad y_\eta(0, \eta) \equiv 1$$

and $t \mapsto (x_\eta(t, \eta), y_\eta(t, \eta))$ is the solution of the variational initial value problem

$$\dot{u} = f_x(x(t, \eta), y(t, \eta))u + f_y(x(t, \eta), y(t, \eta))v,$$
$$\dot{v} = y(t, \eta)g_x(x(t, \eta), y(t, \eta))u$$
$$+ [y(t, \eta)g_y(x(t, \eta), y(t, \eta)) + g(x(t, \eta), y(t, \eta))]v,$$
$$u(0) = 0,$$
$$v(0) = 1.$$

After evaluation at $\eta = 0$, the second equation decouples from the system, and it follows immediately that

$$P'(0) = v(T(0)) = e^{\int_0^{T(0)} g(x(s,0),0)\, ds}$$

where $\dot{x}(t, 0) = f(x(t, 0), 0)$, $x(0, 0) = 0$, and $x(T(0), 0) = b$. By introducing the new variable $\xi = x(s, 0)$, we have $d\xi = f(\xi, 0)\, ds$ and

$$P'(0) = e^{\int_a^b \frac{g(\xi,0)}{f(\xi,0)}\, d\xi},$$

as required. $\qquad\square$

Let us now consider real numbers $a < b$ and a family of differential equations on $\mathbb{R}^2$ given by

$$\dot{x} = \epsilon F(x, y, \epsilon), \qquad \dot{y} = yG(x, y, \epsilon) \qquad (8.45)$$

where ϵ is the parameter and $F(x, 0, 0) > 0$ whenever $x \in [a, b]$. If $|\epsilon|$ is sufficiently small, then $F(x, 0, \epsilon) > 0$ whenever $x \in [a, b]$. Hence, using the notation of Proposition 8.55, if $\epsilon > 0$ is sufficiently small, then

$$P'(0, a, b, \epsilon) = e^{\int_a^b \frac{G(\xi, 0, \epsilon)}{\epsilon F(\xi, 0, \epsilon)} \, d\xi}.$$

Consider the orbit of system (8.45) starting at the point (a, η), for $|\eta|$ small, and note that if $P'(0, a, b, \epsilon) < 1$, then this orbit meets Σ_b at a point (b, σ) where $|\sigma| < |\eta|$; that is, the orbit is attracted toward the x-axis. On the other hand, if $P'(0, a, b, \epsilon) > 1$, then the orbit is repelled from the x-axis. The critical value $P'(0, a, b, \epsilon) = 1$ occurs if and only if

$$\int_a^b \frac{G(\xi, 0, \epsilon)}{F(\xi, 0, \epsilon)} \, d\xi = 0. \qquad (8.46)$$

For system (8.45) with $\epsilon = 0$, the x-axis is an invariant set consisting entirely of rest points. Moreover, the linearized system at the rest point $(x, 0)$ has the system matrix

$$\begin{pmatrix} 0 & 0 \\ 0 & G(x, 0, 0) \end{pmatrix}.$$

Note that if $G(x, 0, 0) < 0$, then the one-dimensional stable manifold of the rest point is normal to the x-axis and solutions starting on this manifold are attracted to the rest point exponentially fast. On the other hand, if $G(x, 0, 0) > 0$, then the rest point has a one-dimensional unstable manifold and solutions are repelled. The rest point $(x, 0)$ is called a *turning point* if $G(x, 0, 0) = 0$.

Let us assume that $a < c < b$ and $(c, 0)$ is a turning point such that $G(x, 0, 0) < 0$ for $x < c$ and $G(x, 0, 0) > 0$ for $x > c$. Using what we have proved about the derivative of the section map, it is clear that if $\epsilon > 0$ is sufficiently small, then a solution starting near the x-axis at a point (x, y) with $x < c$ will be attracted (in a very short time) to the vicinity of the x-axis, drift slowly along the x-axis at least until its first component is larger than c, and eventually it will be repelled (in a very short time) from the vicinity of the x-axis. The next (somewhat imprecise) theorem—we have not defined the concept "leaves the vicinity"—identifies the point on the x-axis where the solution leaves.

Theorem 8.56. *For system (8.45), suppose that $a < c$ and $(c, 0)$ is a turning point such that $G(x, 0, 0) < 0$ for $x < c$ and $G(x, 0, 0) > 0$ for $x > c$, and $\eta \neq 0$ is such that the family of solutions $t \mapsto (x(t, a, \epsilon), y(t, a, \epsilon))$ with*

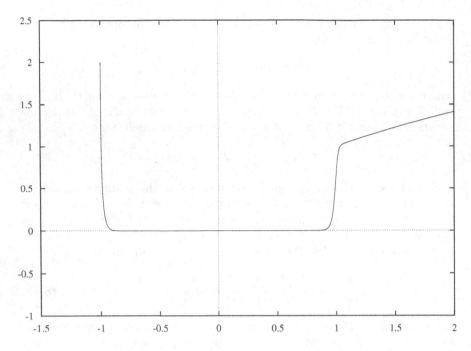

Figure 8.4: The solution in phase space (y versus μ) for system (8.47) with $\epsilon = 0.02$ and initial conditions $\mu = -1$ and $y = 2$.

$x(0, a, \epsilon) = a$ and $y(0, a, \epsilon) = \eta$ is defined for sufficiently small $|\epsilon|$. Then, as ϵ approaches zero from above, the point on the x-axis where the solution leaves the vicinity of the x-axis approaches $(b, 0)$, where b is the smallest number larger than c that satisfies the equation (8.46).

Proof. The solution will be repelled from the x-axis near $(b, 0)$ whenever the derivative of the section map $P'(0, a, b, \epsilon)$ is positive. The sign of this derivative is determined by the integral equation (8.46). If b satisfies the equation, then the orbit is repelled from the vicinity of the x-axis after t increases past the time T when $x(T) = b$. But, the derivative at such a point can be made arbitrarily large for sufficiently small $\epsilon > 0$.   □

To apply Theorem 8.56, let us first consider the static supercritical pitchfork bifurcation whose normal form is given by the family

$$\dot{y} = y(\mu - y^2).$$

The point $y = 0$ is a rest point for every member of the family. If the parameter μ is negative, then this rest point is stable; if μ is positive, it is unstable. Moreover, a pair of stable rest points $y = \pm\sqrt{\mu}$ exist for $\mu > 0$.

Imagine that the parameter μ is slowly changed from a negative value to a positive value. The steady state of the system would be expected to change from the stable steady state $y = 0$, for negative values of the parameter μ, to one of the new stable steady states $y = \pm\sqrt{\mu}$ as soon as μ is positive. To test the validity of this scenario, suppose that $\epsilon > 0$ is small and the state of the system is governed by the differential equation

$$\dot{\mu} = \epsilon, \qquad \dot{y} = y(\mu - y^2) \tag{8.47}$$

so that the "parameter" μ changes slowly with time. By an application of formula (8.46), the solution starting at the point $(-\mu_0, y_0)$ is attracted to the y-axis. It drifts along this invariant set—that is, it stays near the "steady state" $y = 0$—until $\mu \approx \mu_0$. Then, it leaves the vicinity of this invariant set and approaches one of the new stable "steady states." Of course, system (8.47) has no rest points. But, the μ-axis is an invariant set for the dynamic bifurcation system; it corresponds to the rest points at $y = 0$ for the "static system" ($\epsilon = 0$). Also, there are orbits of the dynamic bifurcation system that lie near the parabola given by $\mu = y^2$.

In contrast to the static bifurcation scenario, the dynamic bifurcation is delayed; a solution $t \mapsto (\mu(t), y(t))$, which starts near the μ-axis at time $t = 0$ and whose first component eventually increases through the critical value $\mu = 0$, does not immediately approach a stable steady state (that is, one of the invariant sets near the graphs of the functions $\mu \mapsto \pm\sqrt{\mu}$). Rather, the solution remains near the "unstable" μ-axis until $\mu(t)$ exceeds $-\mu(0)$, a value that can be much larger than the static bifurcation value $\mu = 0$ (see Figure 8.4).

The existence of a subcritical pitchfork bifurcation in a physical system often signals the possibility of a dramatic change in the dynamics. To see why, consider first the corresponding normal form dynamic bifurcation system

$$\dot{\mu} = \epsilon, \qquad \dot{y} = y(\mu + y^2). \tag{8.48}$$

A solution of system (8.48) starting near the negative μ-axis at $(-\mu_0, y_0)$ is attracted toward the invariant μ-axis, and it remains nearby until $\mu \approx \mu_0$. But, in contrast to the supercritical pitchfork bifurcation, when $\mu > \mu_0$ the magnitude of this solution grows rapidly without being bounded by the ghosts of the corresponding static bifurcation's stable steady states. Thus, such a solution would be expected to move far away from equilibrium toward some distant attractor.

Two simple examples of slow passage through a bifurcation, where the bifurcation parameter is a linear function of time, have just been discussed. The main feature of these examples is the existence of two time-scales: fast attraction and slow passage. Indeed, recognition of the existence of different time scales in a dynamical problem is often the key to understanding seemingly exotic phenomena. As a final example of this type, we will see that slow passage through a subcritical pitchfork bifurcation coupled with an appropriate *nonlinear* change in the bifurcation parameter can be viewed

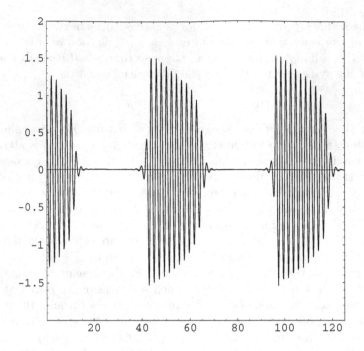

Figure 8.5: A plot of x versus t for system (8.49) with parameter values $a = 0.7$, $\eta = 0.1$, and $\omega = 3$ and initial conditions $(x, y, u) = (1, 0, -0.5)$.

as an explanation for "bursting," a dynamical behavior observed, for example, in the electrical behavior of neurons (see [131]).

Let us consider the system

$$\dot{z} = (u + i\omega)z + 2z|z|^2 - z|z|^4,$$
$$\dot{u} = \eta(a - |z|^2) \tag{8.49}$$

where $z = x + iy$ is (for notational convenience) a complex state variable and a, ω, and η are real parameters. The behavior of the state variable x versus t for a typical solution of system (8.49) with the parameter η small and $0 < a < 1$ is depicted in Figure 8.5. Note that the state of the system seems to alternate between periods of quiescence followed by bursts of oscillations. To reveal the underlying mechanism that produces this behavior, let us change to polar coordinates $z = re^{i\theta}$ and note that the angular variable decouples in the equivalent system

$$\dot{u} = \eta(a - r^2),$$
$$\dot{r} = ur + 2r^3 - r^5,$$
$$\dot{\theta} = \omega. \tag{8.50}$$

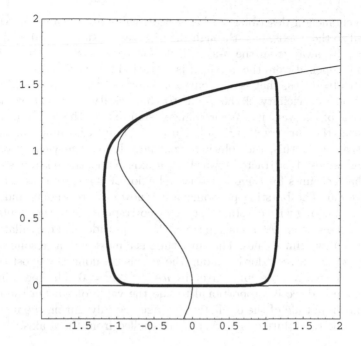

Figure 8.6: Plot (r versus u) of the limit cycle for system (8.50) with parameters $a = 0.7$ and $\eta = 0.1$ together with the line $r = 0$ and the curve $u + 2r^2 - r^4 = 0$.

Hence, we can determine the dynamical behavior of system (8.49) by analyzing the phase portrait of the two-dimensional subsystem for the amplitude r and the "bifurcation variable" u.

System (8.50) resembles our normal form model (8.48) for the subcritical pitchfork bifurcation. At least we can recognize the subcritical pitchfork bifurcation for the system

$$\dot{u} = \eta a, \qquad \dot{r} = r(u + 2r^2).$$

Here, η plays the role of ϵ and u plays the role of μ. By an analysis of the corresponding static subcritical pitchfork bifurcation for system (8.50), that is, the bifurcations of rest points for the system

$$\dot{r} = ur + 2r^3 - r^5 \tag{8.51}$$

where u is the bifurcation parameter, we see that for $u > 0$ there are two stable steady states which are not associated with the pitchfork bifurcation; they exist because of the additional term r^5 in the equation for the rest points of the differential equation (8.51). With the addition of the dynamic bifurcation parameter, given by the first equation of system (8.50), the delay phenomenon described in this section will occur. Solutions will be attracted to the u-axis, pass through the bifurcation value $u = 0$, and then move rapidly away from the u-axis. But, for system (8.50), after such a solution is repelled from the u-axis, it is attracted to the ghost of the static stable steady states, that is, the curve $u + 2r^2 - r^4 = 0$. In addition, in transit from the vicinity of the u-axis to the vicinity of this curve, the u component of the velocity vector changes sign. Thus, the solution moves slowly near the curve $u + 2r^2 - r^4 = 0$ at least until its first component is negative. After this, the solution eventually leaves the vicinity of this curve and is rapidly attracted toward the u-axis. Thus, there is a hysteresis effect that accounts for the oscillatory behavior of the system as depicted in Figure 8.6. The bursting phenomena are now easy to understand: The angle θ is changing with constant frequency corresponding to the oscillatory nature of the system. At the same time, the amplitude of the oscillation is changing on two time-scales. The amplitude is almost zero as a solution on the limit cycle moves "slowly" along the u-axis or along the ghost of the static stable steady state curve given by $u + 2r^2 - r^4 = 0$. The speed in this slow regime is directly proportional to the the value of ηa. On the other hand, the amplitude of the oscillation changes rapidly during the intervals of time while the solution is attracted or repelled from the u-axis.

Exercise 8.57. Predict the behavior of the solution of the system

$$\dot{\mu} = \epsilon, \qquad \dot{y} = y(\sin(\mu) + y^2 - y^4)$$

with the initial condition $(\mu, y) = (-1, 1/2)$ under the assumption that $\epsilon > 0$ is small.

Exercise 8.58. Find the general solution of the differential equation

$$\dot{y} = y(\epsilon t - y^2)$$

corresponding to system (8.47). Can you explain the behavior of the solution depicted in Figure 8.4? Hint: The differential equation is a form of Bernoulli's equation.

Exercise 8.59. Study systems (8.49) and (8.50) analytically and numerically for various values of the parameters. How does the character of the bursting change as a is increased from $a = 0$ to $a = 1$ (statically) for fixed values of η and ω? (The answer is in [131].) Prove that the planar system, consisting of the first two equations of system (8.50), has a limit cycle.

Exercise 8.60. Show that slow passage through a subcritical pitchfork bifurcation for system (8.50) corresponds to slow passage through a Hopf bifurcation for system (8.49).

References

[1] Abraham, R. and J. E. Marsden (1978). *Foundations of Mechanics*, 2nd ed. Reading: The Benjamin/Cummings Pub. Co.

[2] Abraham, R., J. E. Marsden, and T. Ratiu (1988). *Manifolds, Tensor Analysis, and Applications*, 2nd ed. New York: Springer-Verlag.

[3] Abraham, R. and J. Robbin (1967). *Transversal Mappings and Flows.* New York: W. A. Benjamin, Inc.

[4] Acheson, D. (1997). *From calculus to chaos : an introduction to dynamics.* Oxford : New York : Oxford University Press.

[5] Ahlbrandt, C. D. and J. Ridenhour (2003). Floquet theory for time scales and Putzer representations of matrix logarithms. *J. Diff. Eqs. Appl.* (1)**9**, 77–92.

[6] Andersen, C. and J. Geer (1982). Power series expansions for frequency and period of the limit cycle of the van der Pol equation. *SIAM J. Appl. Math.* **42**, 678–693.

[7] Andronov, A. A., E. A. Leontovich, I. I. Gordon, and A. G. Maier (1973). *Qualitative Theory of Second-Order Dynamic Systems.* New York: John Wiley & Sons.

[8] Andronov, A. A., E. A. Leontovich, I. I. Gordon, and A. G. Maier (1973). *Theory of Bifurcations of Dynamic Systems on a Plane.* New York: John Wiley & Sons.

[9] Andronov, A. A., A. A. Vitt, and S. E. Khaiken (1966). *Theory of Oscillators*. Oxford: Pergamon Press.

[10] Arnold, V. I. (1973). *Ordinary Differential Equations*. Cambridge: M. I. T. Press.

[11] Arnold, V. I. (1982). *Geometric Methods in the Theory of Ordinary Differential Equations*. New York: Springer-Verlag.

[12] Arnold, V. I. (1978). *Mathematical Methods of Celestial Mechanics*. New York: Springer-Verlag.

[13] Arnold, V. I., ed. (1988). *Dynamical Systems I*. New York: Springer-Verlag.

[14] Arnold, V. I., ed. (1988). *Dynamical Systems III*. New York: Springer-Verlag.

[15] Arnold, V. I., ed. (1990). *Huygens and Barrow, Newton and Hooke*, E. J. F. Primrose, trans. Basel: Birkhäuser-Verlag.

[16] Arnold, V. I. and B. A. Khesin (1998). *Topological Methods in Hydrodynamics*. New York: Springer-Verlag.

[17] Aronson D. G. and H. F. Weinberger (1978). Multidimensional nonlinear diffusion arising in population genetics. *Adv. Math.* **30**, 33–76.

[18] Arscott F. M. (1964). *Periodic Differential Equations*. New York: MacMillan.

[19] Artés, J. C., F. Dumortier, and J. Llibre (2005). *Qualitative Theory of Planar Differential Systems*. Book Manuscript.

[20] Ashbaugh, M., C. Chicone, and R. Cushman (1991). The twisting tennis racket. *J. of Dyn. and Diff. Eqs.* **3**, 67–85.

[21] Barrow-Green, J. (1997). *Poincaré and the Three Body Problem*. Providence: Amer. Math. Soc.

[22] Bates, L. M. and R. H. Cushman (1997). *Global Aspects of Classical Integrable Systems*. Boston: Birkhäuser-Verlag.

[23] Bautin, N. N. (1954). On the number of limit cycles which appear with the variation of coefficients from an equilibrium position of focus or center type. *Amer. Math. Soc. Transl.* **100**, 1–19.

[24] Bender, J. (1997). Chaotic dynamics in a planar Hamiltonian system. *Master's Project*. Columbia: University of Missouri.

[25] Benguria, R. D. and M. C. Depassier (1996). Variational characterization of the speed of propagation of fronts for the nonlinear diffusion equation. *Comm. Math. Phy.* **175**, 221–227.

[26] Bisplinghoff, R. L. and H. Ashley (1975). *Principles of Aeroelasticity.* New York: Dover Publications, Inc.

[27] Bliss, G. A. (1925). *Calculus of Variations.* Carus Math. Monographs, Mathematical Association of America.

[28] Bogoliubov, N. N. and Y. A. Mitropolsky (1961). *Asymptotic Methods in the Theory of Nonlinear Oscillations.* Delhi: Hindustan Pub. Corp.

[29] Brieskorn, E. and H. Knörrer (1986). *Plane Algebraic Curves.* Basel: Birkhäuser-Verlag.

[30] Byrd, P. F. and M. D. Friedman (1971). *Handbook of Elliptic Integrals for Engineers and Scientists,* 2nd ed. Berlin: Springer-Verlag.

[31] Carathéodory, C. (1954). *Theory of Functions of a Complex Variable,* **1**, pp. 229–236. New York: Chelsea.

[32] Chen, M., X.-Y. Chen, and J. K. Hale (1992). Structural stability for time-periodic one-dimensional parabolic equations. *J. Diff. Eqs.* **96**(2), 355–418.

[33] Chicone, C. (1986). Limit cycles of a class of polynomial vector fields in the plane. *J. Diff. Eqs.* **63**(1), 68–87.

[34] Chicone, C. (1987). The monotonicity of the period function for planar Hamiltonian vector fields. *J. Diff. Eqs.* **69**, 310–321.

[35] Chicone, C. (1992). Bifurcation of nonlinear oscillations and frequency entrainment near resonance. *SIAM J. Math. Anal.* **23**(6), 1577–1608.

[36] Chicone, C. (1993). Review of *Finiteness Theorems for Limit Cycles,* by Yu. Il'yashenko. *Bull. Amer. Math. Soc.* **28**(1), 123–130.

[37] Chicone, C. (1994). Lyapunov–Schmidt reduction and Melnikov integrals for bifurcation of periodic solutions in coupled oscillators. *J. Diff. Eqs.* **112**, 407–447.

[38] Chicone, C. (1995). A geometric approach to regular perturbation theory with an application to hydrodynamics. *Trans. Amer. Math. Soc.* **374**(12), 4559–4598.

[39] Chicone, C. (1997). Invariant tori for periodically perturbed oscillators. *Publ. Mat.* **41**, 57–83.

[40] Chicone, C., J. Critser and J. Benson (2005). Exact solutions of the Jacobs two parameter flux model and cryobiological applications. *J. Cryobiology*, **50**(3), 308–316.

[41] Chicone, C. and F. Dumortier (1993). Finiteness for critical points of the period function for analytic vector fields on the plane. *Nonlinear Analy.* **20**(4), 315-335.

[42] Chicone, C. and M. Jacobs (1989). Bifurcation of critical periods for plane vector fields. *Trans. Amer. Math. Soc.* **312**, 433–486.

[43] Chicone, C. and M. Jacobs (1991). Bifurcation of limit cycles from quadratic isochrones. *J. Diff. Eqs.* **91**, 268–327.

[44] Chicone, C. and Yu. Latushkin (1997). The geodesic flow generates a fast dynamo: an elementary proof. *Proc. Amer. Math. Soc.* **125**(11), 3391–3396.

[45] Chicone, C. and Yu. Latushkin (1999). *Evolution Semigroups in Dynamical Systems and Differential Equations*. Providence: American Mathematical Society.

[46] Chicone, C. and W. Liu (1999). On the continuation of an invariant torus in a family with rapid oscillation. *SIAM J. Math. Analy.* **31**, 386–415.

[47] Chicone, C. and W. Liu (2004). Asymptotic phase revisited. *J. Diff. Eqs.* **204**(1), 227–246.

[48] Chicone, C., B. Mashhoon, and D. G. Retzloff (1996). Gravitational ionization: periodic orbits of binary systems perturbed by gravitational radiation. *Ann. Inst. H. Poincaré.* **64**(1), 87–125.

[49] Chicone, C., B. Mashhoon, and D. G. Retzloff (1996). On the ionization of a Keplerian binary system by periodic gravitational radiation. *J. Math Physics.* **37**, 3997–1416.

[50] Chicone, C., B. Mashhoon, and D. G. Retzloff (1997). Addendum: On the ionization of a Keplerian binary system by periodic gravitational radiation. *J. Math Physics.* **38**(1), 554.

[51] Chicone, C., B. Mashhoon, and D. G. Retzloff (1997). Gravitational ionization: a chaotic net in the Kepler system. *Class. Quantum Grav.* **14**, 699–723.

[52] Chicone, C., B. Mashhoon, and D. G. Retzloff (1999). Chaos in the Kepler system. *Class. Quantum Grav.* **16**, 507–527.

[53] Chicone, C. and J. Sotomayor (1986). On a class of complete polynomial vector fields in the plane. *J. Diff. Eqs.* **61**(3), 398–418.

[54] Chicone, C. and R. Swanson (2000). Linearization via the Lie derivative. *Electron. J. Diff. Eqns.*, Monograph 02 (http://ejde.math.swt.edu).

[55] Chicone, C. and J. Tian (1982). On general properties of quadratic systems. *Proc. Amer. Math. Soc.* **89**(3), 167–178.

[56] Chorin, A. J. and J. E. Marsden (1990). *A Mathematical Introduction to Fluid Mechanics,* 2nd ed. New York: Springer-Verlag.

[57] Chow, S. N. and J. K. Hale (1982). *Methods of Bifurcation Theory.* New York: Springer-Verlag.

[58] Chow, S. N., C. Li, and D. Wang (1994). *Normal Forms and Bifurcation of Planar Vector Fields.* Cambridge: Cambridge University Press.

[59] Coddington, E. A. and N. Levinson (1955). *Theory of Ordinary Differential Equations.* New York: McGraw–Hill Book Co.

[60] Colwell, P. (1992). Bessel functions and Kepler's equation. *Amer. Math. Monthly.* **99**, 45–48.

[61] Colwell, P. (1993). *Solving Kepler's Equation over Three Centuries.* Richmond: Willmann-Bell, Inc.

[62] Coppel, W. A. (1960). On a differential equation of boundary-layer theory. *Phil. Trans. Roy. Soc. London, Ser. A.* **253**, 101–36.

[63] Corbera, M. and J. Llibre (2002). On symmetric periodic orbits of the elliptic Sitnikov problem via the analytic continuation method. Celestial Mechanics (Evanston 1999), 91–97, *Contemporary Mathematics,* Vol. 292. Providence: American Mathematical Society.

[64] Cushman, R. (1983). Reduction, Brouwer's Hamiltonian, and the critical inclination, *Celest. Mech.* **31**, 401–429. Corrections: (1984). *Celest. Mech.* **33**, 297.

[65] Cushman, R. (1984). Normal form for Hamiltonian vectorfields with periodic flow. In *Differential Geometric Methods in Mathematical Physics.* S. Sternberg Ed. Reidel, Dordrecht: Reidel, 125–145.

[66] Cushman, R. and J. C. van der Meer (1987). Orbiting dust under radiation pressure. In *Proceedings of the* XV*th International Conference on Differential Geometric Methods in Theoretical Physics.* H. Doebner and J. Henning Eds. Singapore: World Scientific, 403-414.

[67] Cushman, R. (1988). An analysis of the critical inclination problem using singularity theory. *Celest. Mech.* **42**, 39–51.

[68] Cushman, R. (1992). A survey of normalization techniques applied to perturbed Keplerian systems. *Dynamics reported*. New series **1**, 54–112.

[69] Dadfar, M., J. Geer, and C. Andersen (1984). Perturbation analysis of the limit cycle of the free van der Pol equation. *SIAM J. Appl. Math.* **44**, 881–895.

[70] Davenport, J. H., Y. Siret, and E. Tournier (1993). *Computer Algebra: Systems and Algorithms for Algebraic Computation,* 2nd ed. London: Academic Press, Ltd.

[71] Davis, H. T. (1960). *Introduction to Nonlinear Differential and Integral Equations.* New York: Dover Pub. Inc.

[72] Devaney, R.L. (1986). *An Introduction to Chaotic Dynamical Systems.* Benjamin/Cummings: Menlo Park, CA.

[73] Diacu, F. and P. Holmes (1996). *Celestial Encounters : The Origins of Chaos and Stability.* Princeton: Princeton University Press.

[74] Diliberto, S. P. (1950). On systems of ordinary differential equations. In *Contributions to the Theory of Nonlinear Oscillations,* Annals of Mathematics Studies, **20**, pp. 1–38. Princeton: Princeton University Press.

[75] Dumortier, F. (1977). Singularities of vector fields on the plane. *J. Diff. Eqs.* **23**(1), 53–106.

[76] Dumortier, F. (1978). *Singularities of Vector Fields.* Rio de Janeiro: IMPA Monograph, No. 32.

[77] Dumortier, F. (2005). Asymptotic phase and invariant foliations near periodic orbits. *Proc. Amer. Math. Soc.,* In press.

[78] Dunford, N. and J. T. Schwartz (1958). *Linear Operators.* New York: Interscience Publications, Inc.

[79] Écalle, J. (1992). *Introduction aux fonctions analysables et preuve constructive de la conjecture du Dulac.* Actualities Math. Paris: Hermann.

[80] Elbialy, M. S. (2001). Local contractions of Banach spaces and spectral gap conditions *J. Funct. Anal* **182**(1), 108–150.

[81] Evans, L. C. (1998). *Partial Differential Equations.* Providence: American Mathematical Society.

[82] Ewing, G. M. (1969). *Calculus of Variations with Applications.* New York: W. W. Norton & Company, Inc.

[83] Falkner, V. M. and S. W. Skan (1931). Solution of boundary layer equations. *Philos. Mag.* **7**(12), 865–896.

[84] Farkas, M. (1994). *Periodic Motions.* New York: Springer-Verlag.

[85] Fenichel, N. (1971). Persistence and smoothness of invariant manifolds for flows. *Indiana U. Math. J.* **21**, 193–226.

[86] Fermi, E., S. Ulam, and J. Pasta (1974). Studies of nonlinear problems I. In *Nonlinear Wave Motion*, Lect. Appl. Math., *Amer. Math. Soc.* **15**, 143–155.

[87] Feynman, R. P., R. B. Leighton, and M. Sands (1964). *The Feynman Lectures on Physics I–III.* Reading: Addison–Wesley Pub. Co.

[88] Fife, P. C. (1979). *Mathematical Aspects of Reacting and Diffusing Systems.* Lect. Notes Biomath., **28**. New York: Springer-Verlag.

[89] Flanders, H. (1963). *Differential Forms with Applications to the Physical Sciences.* New York: Dover Publications, Inc.

[90] Fletcher, C. A. J. (1982). Burgers' equation: a model for all reasons. In *Numerical Solutions of Partial Differential Equations.* J. Noye, ed. Amsterdam: North–Holland Pub. Co.

[91] Frantz, M. (2006). Some graphical solutions of the Kepler problem, Amer. Math. Monthly., **113**(1), 47–56.

[92] Friedlander, S. A. Gilbert, and M. Vishik (1993). Hydrodynamic instability and certain ABC flows. *Geophys. Astrophys. Fluid Dyn.* **73**(1-4), 97–107.

[93] Friedrich, H. and D. Wintgen, (1989). The hydrogen atom in a uniform magnetic field—an example of chaos. *Phys. Reports.* **183**(1), 37–79.

[94] Geyling, F. T. and H. R. Westerman (1971). *Introduction to Orbital Mechanics.* Reading: Addison–Wesley Pub. Co.

[95] Ghil, M. and S. Childress (1987). *Topics in Geophysical Fluid Dynamics : Atmospheric Dynamics, Dynamo Theory, and Climate Dynamics.* New York: Springer-Verlag.

[96] Gleick, J. (1987). *Chaos: Making a New Science.* New York: Viking.

[97] Golubitsky, M. and D. G. Schaeffer (1985). *Singularities and Groups in Bifurcation Theory*, 1. New York: Springer-Verlag.

[98] Golubitsky, M., I. Stewart, and D.G. Schaeffer (1988). *Singularities and Groups in Bifurcation Theory*, 2. New York: Springer-Verlag.

[99] Gonzales, E. A. (1969). Generic properties of polynomial vector fields at infinity. *Trans. Amer. Math. Soc.* **143**, 201–222.

[100] Greenspan, B and P. Holmes (1984). Repeated resonance and homoclinic bifurcation in a periodically forced family of oscillators. *SIAM J. Math. Analy.* **15**, 69–97.

[101] Grimshaw, R. (1990). *Nonlinear Ordinary Differential Equations.* Oxford: Blackwell Scientific Pub.

[102] Grobman, D. (1959). Homeomorphisms of systems of differential equations (Russian). *Dokl. Akad. Nauk.* **128**, 880–881.

[103] Guckenheimer, J. and P. Holmes (1986). *Nonlinear Oscillations, Dynamical Systems, and Bifurcations of Vector Fields,* 2nd ed. New York: Springer-Verlag.

[104] Gutiérrez, C. (1995). A solution to the bidimensional global asymptotic stability conjecture. *Ann. Inst. H. Poincaré Anal. Non Linéaire.* **12**(6), 627–671.

[105] Gutzwiller, M. C. (1990). *Chaos in Classical and Quantum Mechanics.* New York: Springer-Verlag.

[106] Hale, J. K. (1980). *Ordinary Differential Equations*, 2nd ed. Malabar: R. E. Krieger Pub. Co.

[107] Hale, J. K., L.T. Magalhães, and W. M. Oliva (1984). *An Introduction to Infinite Dimensional Dynamical Systems—Geometric Theory.* New York: Springer-Verlag.

[108] Hale, J. K. (1988). *Asymptotic Behavior of Dissipative Systems.* Providence: American Mathematical Society.

[109] Hall, R. (1984). Resonance zones in two-parameter families of circle homeomorphisms. *SIAM J. Math. Analy.* **15**, 1075–1081.

[110] Haller, G. (1999). *Chaos Near Resonance.* New York: Springer-Verlag.

[111] Hahn, W. (1967). *Stability of Motion.* A. P. Baartz, trans. New York: Springer-Verlag.

[112] Harris, W. A., J. P. Fillmore, and D. R. Smith (2001). Matrix exponentials—another approach. *SIAM Rev.* **43**(4), 694–706.

[113] Hartman, P. (1960). A lemma in the theory of structural stability of differential equations. *Proc. Amer. Math. Soc.* **11**, 610–620.

[114] Hartman, P. (1960). On local homeomorphisms of Euclidean space. *Bol. Soc. Mat. Mexicana* **5** (2), 220–241.

[115] Hartman, P. (1964). *Ordinary Differential Equations.* New York: John Wiley & Sons, Inc.

[116] Hastings, S. P. and W. C. Troy (1988). Oscillating solutions of the Falkner–Skan equation for positive β. *J. Diff. Eqs.* **71**, 123–144.

[117] Hayashi, C. (1964). *Nonlinear Oscillations in Physical Systems.* New York: McGraw–Hill Book Co.

[118] Henry, D. (1981). *Geometric Theory of Semilinear Parabolic Equations.* New York: Springer-Verlag.

[119] Howe, R. (1983). Very basic Lie theory. *Amer. Math. Monthly.* **90**, 600–623; Correction (1984). **91**, 247.

[120] Hirsch, M. (1976). *Differential Topology.* New York: Springer-Verlag.

[121] Hirsch, M. and C. Pugh (1970). Stable manifolds and hyperbolic sets. In *Global Analysis XIV, Amer. Math. Soc.* 133–164.

[122] Hirsch, M., C. Pugh, and M. Shub (1977). *Invariant Manifolds,* Lecture Notes in Math., **583**. New York: Springer-Verlag.

[123] Hirsch, M. and S. Smale (1974). *Differential Equations, Dynamical Systems, and Linear Algebra.* New York: Academic Press.

[124] Holmes, P. and J. Marsden (1981). A partial differential equation with infinitely many periodic orbits and chaotic oscillations of a forced beam. *Arch. Rat. Mech. Analy.* **76**, 135–165.

[125] Hubbard, J. and B. Burke (2002). *Vector calculus, linear algebra, and differential forms. A unified approach,* 2nd ed. Upper Saddle River, NJ: Prentice Hall, Inc.

[126] Il'yashenko, Yu. S. (1991). *Finiteness Theorems for Limit Cycles,* Transl. Math. Mono., **94**. Providence: American Mathematical Society.

[127] Ince, I. L. (1956) *Ordinary Differential Equations.* New York: Dover.

[128] Iooss, G. and M. Adelmeyer (1992). *Topics in Bifurcation Theory and Applications.* River Edge: World Scientific Pub. Company, Inc.

[129] Isaacson, E and H. Keller (1966). *Analysis of Numerical Methods.* New York: John Wiley & Sons, Inc.

[130] Iserles, A. (2002). Expansions that grow on trees. *Not. Amer. Math. Soc.* **49**(4), 430–440.

[131] Izhikevich, E. M. (2001). Synchronization of elliptic bursters. *SIAM Rev.* **43**(2), 315–344.

[132] Jones, C. K. R. T. (1995). Geometric singular perturbation theory. *Dynamical systems (Montecatini Terme, 1994).* Lecture Notes in Math. **1609**. Berlin: Springer-Verlag, 44–118.

[133] Katok, A. B and B. Hasselblatt (1995). *Introduction to the Modern Theory of Dynamical Systems.* Cambridge: Cambridge University Press.

[134] Kevorkian, J. and J. D. Cole (1981). *Perturbation Methods in Applied Mathematics.* New York: Springer-Verlag.

[135] Kovalevsky, J. (1967). *Introduction to Celestial Mechanics,* Astrophysics and Space Science Library, **7**. New York: Springer-Verlag.

[136] Kolmogorov, A. N., I. G. Petrovskii, and N. S. Piskunov (1991). A study of the diffusion equation with increase in the amount of substance, and its applications to a biological problem. In *Selected Works of A. N. Kolmogorov,* V. M. Tikhomirov, ed. Dordrecht: Kluwer Academic Pub. Group.

[137] Kundu, P. K. and I. M. Cohen (2004). *Fluid Mechanics,* 3rd ed., Amsterdam: Elsevier Academic Press.

[138] Landau, L. D. and E. M. Lifshitz (1951). *Classical Theory of Fields.* Reading: Addison–Wesley Press.

[139] Landau L. D. and E. M. Lifshitz (1987). *Fluid Mechanics,* 2nd ed., J. B. Sykes and W. H. Reid, trans. New York: Pergamon Press.

[140] Lang, S. (1962). *Introduction to Differentiable Manifolds.* New York: Interscience Pub.

[141] Lefschetz, S. (1977). *Differential Equations: Geometric Theory,* 2nd ed. New York: Dover Publications, Inc.

[142] Levi, M. (1988). Stability of the inverted pendulum—a topological explanation. *SIAM Rev.* **30**, 639–644.

[143] Levi, M., C. Hoppensteadt, and M. Miranker (1978). Dynamics of the Josephson junction. *Quart. Appl. Math.* **36**, 167–198.

[144] Liapounoff, M. A. (1947). *Problème général de la stabilité du movement.* Princeton: Princeton University Press.

[145] Lichtenberg, A. J. and M. A. Lieberman (1982). *Regular and Stochastic Motion.* New York: Springer-Verlag.

[146] Lochak, P. and C. Meunier (1988). *Multiphase Averaging for Classical Systems,* H. S. Dumas, trans. New York: Springer-Verlag.

[147] Lorenz, E. N. (1963). Deterministic non-periodic flow. *J. Atmos. Sci.* **20**, 131–141.

[148] Loud, W. S. (1964). Behavior of the periods of solutions of certain plane autonomous systems near centers. *Contributions Diff. Eqs.* **3**, 21–36.

[149] Magnus, W. and S. Winkler (1979). *Hill's Equation.* New York: Dover Publications, Inc.

[150] Massey, W. S. (1967). *Algebraic Topology: An Introduction.* New York: Springer-Verlag.

[151] Marsden, J. E. and M. McCracken (1976). *The Hopf Bifurcation and Its Applications.* New York: Springer-Verlag.

[152] McCuskey, S. W. (1963). *Introduction to Celestial Mechanics.* Reading: Addison–Wesley Pub. Co.

[153] McGehee, R. P. and B. B. Peckham (1994). *Resonance surfaces for forced oscillators.* Experiment. Math. (3)**3**, 221–244.

[154] McGehee, R. P. and B. B. Peckham (1996). *Arnold flames and resonance surface folds.* Internat. J. Bifur. Chaos Appl. Sci. Engrg. (2)**6** (1996), 315–336.

[155] McLachlan, N. W. (1958). *Ordinary Non-linear Differential Equations in Engineering and Physical Sciences,* 2nd ed. New York: Oxford University Press.

[156] Melnikov, V. K. (1963). On the stability of the center for time periodic perturbations. *Trans. Moscow Math. Soc.* **12**, 1–57.

[157] Miller, R. K and A. N. Michel (1982). *Ordinary Differential Equations.* New York: Academic Press.

[158] Milnor, J. (1963). *Morse Theory.* Princeton: Princeton University Press.

[159] Milnor, J. (1978). Analytic proofs of the "hairy ball theorem" and the Brouwer fixed-point theorem. *Amer. Math. Monthly* **85**(7), 521–524.

[160] Meyer, K. R. and G. R. Hall (1992). *Introduction to Hamiltonian Dynamical Systems and the N-Body Problem.* New York: Springer-Verlag.

[161] Meyer, K. R. and D. S. Schmidt (1977). Entrainment domains. *Funkcialaj Edvacioj.* **20**, 171–192.

[162] Minorsky, N. (1962). *Nonlinear Oscillations.* Princeton: D. Van Nostrand Company, Inc.

[163] Mironenko, V. I. (1986). *Reflection function and periodic solutions of differential equations.* (Russian) Minsk: Universitet-skoe.

[164] Morozov, A. D. (1998). *Quasi-Conservative Systems: Cycles, Resonances and Chaos.* Series A, **30**, Singapore: World Scientific.

[165] Moser, J. (1969). On a theorem of Anosov. *J. Diff. Eqns.* **5**, 411–440.

[166] Moser, J. (1973). *Stable and Random Motions in Dynamical Systems,* Annals of Math. Studies, **77**. Princeton: Princeton University Press.

[167] Moser, J. (1978/79). Is the solar system stable? *Math. Intelligencer* **1**(2), 65–71.

[168] Munkres, J. R. (1975). *Topology: A First Course.* Englewood Cliffs: Prentice–Hall, Inc.

[169] Munkres, J. R. (1984). *Elements of Algebraic Topology.* Reading: Addison–Wesley Pub. Co.

[170] Murdock, J. A. (1991). *Perturbations: Theory and Methods.* New York: Wiley–Interscience Pub.

[171] Murdock, J. A. (2003). *Normal Forms and Unfoldings for Local Dynamical Systems.* New York: Springer-Verlag.

[172] Murdock, J. A. and C. Robinson (1980). Qualitative dynamics from asymptotic expansions: Local theory. *J. Diff. Eqns.* **36**, 425–441.

[173] Murray, J. D. (1980). *Mathematical Biology.* New York: Springer-Verlag.

[174] Mustafa, O. G. and Y. V. Rogovchenko (2005). Estimates for domains of local invertibility of diffeomorphisms. *Proc. Amer. Math. Soc.*, In Press.

[175] Nayfeh, A. H. (1973). *Perturbation Methods.* New York: John Wiley & Sons.

[176] Nemytskii, V. V. and V. V. Stepanov (1989). *Qualitative Theory of Differential Equations.* New York: Dover Publications, Inc.

[177] Oden, J. T. (1979). *Applied Functional Analysis: A First Course for Students of Mechanics and Engineering Science.* Englewood Cliffs: Prentice–Hall, Inc.

[178] Olver, P. J. (1986). *Applications of Lie Groups to Differential Equations*. New York: Springer-Verlag.

[179] O'Malley, R. E. (1991). *Singular Perturbation Methods for Ordinary Differential Equations*. New York: Springer-Verlag.

[180] Palais, R. P. (1997). The symmetries of solitons. *Bull. Amer. Math. Soc.* **34**, 339–403.

[181] Palis, J. and W. deMelo (1982). *Geometric Theory of Dynamical Systems: An Introduction*. New York: Springer-Verlag.

[182] Pao, C. V. (1992). *Nonlinear Parabolic and Elliptic Equations*. New York: Plenum Press.

[183] Perko, L. (1996). *Differential Equations and Dynamical Systems*, 2nd ed. New York: Springer-Verlag.

[184] Petty, C. M. and J. V. Breakwell (1960). Satellite orbits about a planet with rotational symmetry. *J. Franklin Inst.* **270**(4), 259–282.

[185] Poincaré, H. (1881). Mémoire sur les courbes définies par une équation différentielle. *J. Math. Pures et Appl.* **7**(3), 375–422; (1882). **8**, 251–296; (1885). **1**(4), 167–244; (1886). **2**, 151–217; all reprinted (1928). *Oeuvre,* Tome I. Paris: Gauthier–Villar.

[186] Pugh, C. (1969). On a theorem of P. Hartman. *Amer. J. Math.* **91**, 363–367.

[187] Reyn, J. W. (1994). *A Bibliography of the Qualitative Theory of Quadratic Systems of Differential Equations in the Plane*, 3rd ed., Tech. Report 94-02. Delft: TU Delft.

[188] Rayleigh, Lord (1883). On maintained vibrations. *Phil. Mag.* **15**, 229.

[189] Rhouma, M. B. H. and C. Chicone (2000). On the continuation of periodic orbits. *Meth. Appl. Analy.* **7**, 85–104 .

[190] Robbin, J. W. (1968). On the existence theorem for differential equations. *Proc. Amer. Math. Soc.* **19**, 1005–1006.

[191] Robbin, J. W. (1971). Stable manifolds of semi-hyperbolic fixed points. *Illinois J. Math.* **15**(4), 595–609.

[192] Robbin, J. W. (1972). Topological conjugacy and structural stability for discrete dynamical systems. *Bull. Amer. Math. Soc.* **78**(6), 923–952.

[193] Robbin, J. W. (1981). Algebraic Kupka–Smale theory. *Lecture Notes in Math.* **898**, 286–301.

[194] Robinson, C. (1995). *Dynamical Systems: Stability, Symbolic Dynamics, and Chaos.* Boca Raton: CRC Press.

[195] Roberts, A. J. (1997). Low-dimensional modelling of dynamical systems. http://xxx.lanl.gov/abs/chao-dyn/9705010.

[196] Rothe, F. (1993). Remarks on periods of planar Hamiltonian systems. *SIAM J. Math. Anal.* **24**(1), 129–154.

[197] Roussarie, R. (1998). *Bifurcation of Planar Vector Fields and Hilbert's Sixteenth Problem.* Basel: Birkhaäuser-Verlag.

[198] Rousseau, C. and B. Toni (1997). Local bifurcations of critical periods in the reduced Kukles system. *Can. J. Math.* **49**(2), 338–358.

[199] Rudin, W. (1966). *Real and Complex Analysis.* New York: McGraw–Hill Book Co.

[200] Sánchez, D (2002). *Ordinary Differential Equations: A Brief Eclectic Tour.* Washington DC: The Mathematical Association of America.

[201] Sanders, J. A. and F. Verhulst (1985). *Averaging Methods in Nonlinear Dynamical Systems.* New York: Springer-Verlag.

[202] Schecter, S. (1990). Simultaneous equilibrium and heteroclinic bifurcation of planar vector fields via the Melnikov integral. *Nonlinearity.* **3**, 79–99.

[203] Smale, S. (1965). Diffeomorphisms with many periodic points. In *Differential and Combinatorial Topology,* S. S. Cairns, ed. Princeton: Princeton Univ. Press 63–80.

[204] Smale, S. (1980). *The Mathematics of Time: Essays on Dynamical Systems, Economic Processes and Related Topics.* New York: Springer-Verlag.

[205] Smale, S. (1998). Finding a horseshoe on the beaches of Rio. *Math. Intelligencer.* **20**, 39–62.

[206] Smoller, J. (1980). *Shock Waves and Reaction-Diffusion Equations.* New York: Springer-Verlag.

[207] Sontag, E. D. (1990). *Mathematical Control Theory.* New York: Springer-Verlag.

[208] Sotomayor, J. (1974). Generic one-parameter families of vector fields on two-dimensional manifolds. *Publ. Math. IHES* **43**, 5–46.

[209] Sotomayor, J. (1979). *Lições de equações diferenciais ordinárias.* Rio de Janerio: IMPA.

[210] Sotomayor, J. (1990). Inversion of smooth mappings. *Z. Angew. Math. Phys.* **41**, 306–310.

[211] Sotomayor, J. (1996). Private communication. Lleida Spain.

[212] Spanier, E. H. (1981). *Algebraic Topology,* corrected reprint. New York: Springer-Verlag.

[213] Spivak, M. (1965). *Calculus on Manifolds.* New York: W. A. Benjamin, Inc.

[214] Sternberg, S. (1969). *Celestial Mechanics,* Parts I and II. New York: W. A. Benjamin, Inc.

[215] Stoker, J. J. (1950). *Nonlinear Vibrations.* New York: John Wiley & Sons.

[216] Strang, G. and G. J. Fix (1973). *An Analysis of the Finite Element Method.* Englewood Cliffs: Prentice–Hall, Inc.

[217] Strauss, W. A. (1992). *Partial Differential Equations: An Introduction.* New York: John Wiley & Sons.

[218] Strogatz, S. (1994). *Nonlinear Dynamics and Chaos.* Reading: Addison–Wesley Pub. Co.

[219] Takens, F. (1974). Singularities of vector fields. *Publ. Math. IHES* **43**, 47–100.

[220] Temam, R. (1988). *Infinite-dimensional Dynamical Systems in Mechanics and Physics.* New York: Springer-Verlag.

[221] Trotter, H. F. (1959). On the product of semi-groups of operators. *Proc. Amer. Math. Soc.* **10**, 545–551.

[222] van der Pol, B. (1926). On relaxation oscillations. *Phil. Mag.* **2**, 978–992.

[223] van der Pol, B. and J. van der Mark (1927). Frequency demultiplication. *Nature.* **120**, 363–364.

[224] van der Pol, B. and J. van der Mark (1929). The heartbeat considered as a relaxation oscillation and an electrical model of the heart. *Arch. Néerl. de Physiol. de L'homme et des Animaux.* **14**, 418–443.

[225] Varadarajan, V. S. (1974). *Lie Groups, Lie Algebras and their Representations,* New York: Springer-Verlag.

[226] Verhulst, F. (1989). *Nonlinear Differential Equations and Dynamical Systems,* New York: Springer-Verlag.

[227] Vinograd, R. E. (1957). The inadequacy of the method of characteristic exponents for the study of nonlinear equations. *Mat. Sbornik.*
41(83), 431–438.

[228] Watson, G. N. (1966). *A Treatise on the Theory of Bessel Functions,*
2nd ed. Cambridge: Cambridge University Press.

[229] Weinberger, H. (1999). An example of blowup produced by equal
diffusions. *J. Diff. Eqns.* (1)**154**, 225–237.

[230] Weissert, T. (1997). *The Genesis of Simulation in Dynamics: Pur-
suing the Fermi–Ulam–Pasta Problem.* New York: Springer-Verlag.

[231] Weyl, H. (1952). *Space–Time–Matter,* H. L. Brose, trans. New York:
Dover Publications, Inc.

[232] Wiggins, S. W. (1988). *Global Bifurcations and Chaos.* New York:
Springer-Verlag.

[233] Wiggins, S. W. (1990). *Introduction to Applied Nonlinear Dynamical
Systems and Chaos.* New York: Springer-Verlag.

[234] Whittaker, E. T. and G. N. Watson (1927). *A Course of Modern
Analysis,* 1996 reprint. Cambridge: Cambridge University Press.

[235] Winkel, R. (2000). A transfer principle in the real plane from nonsingular algebraic curves to polynomial vector fields. *Geom. Dedicata.*
79, 101–108.

[236] Yagasaki, K. (1994). Chaos in a pendulum with feedback control.
Nonlinear Dyn. **6**, 125–142.

[237] Yakubovich, V. A. and V. M. Starzhinskii (1975). *Linear Differential
Equations with Periodic Coefficients,* D. Louvish, trans. New York:
John Wiley & Sons.

[238] Yorke, J. A. (1969). Periods of periodic solutions and the Lipschitz
constant. *Proc. Amer. Math. Soc.* **22**, 509–512.

[239] Zeidler, E. (1991). *Applied Functional Analysis: Applications to
Mathematical Physics.* New York: Springer-Verlag.

[240] Żołądek, H. (1994). Quadratic systems with center and their perturbations. *J. Diff. Eqs.* **109**(2), 223–273.

Index